Lecture Notes in Computational Science and Engineering 76

Editors:

For further volumes:
http://www.springer.com/series/3527

Jan S. Hesthaven • Einar M. Rønquist
Editors

Spectral and High Order Methods for Partial Differential Equations

Selected papers from the ICOSAHOM '09 conference, June 22-26, Trondheim, Norway

Springer

Editors
Jan S. Hesthaven
Brown University
Division of Applied Mathematics
182 George Street
Providence, RI 02912
USA
Jan.Hesthaven@Brown.edu

Einar M. Rønquist
Norwegian University of Science
and Technology
Department of Mathematical Sciences
7491 Trondheim
Norway
ronquist@math.ntnu.no

ISBN 978-3-642-15336-5 e-ISBN 978-3-642-15337-2
DOI 10.1007/978-3-642-15337-2
Springer Heidelberg Dordrecht London New York

Mathematics Subject Classification (2010): 65M70, 65N35, 65N30, 74S25, 76M10, 76M22, 78M10, 78M22

Cover illustration: The nudg++ team - Tim Warburton (Rice University), Nigel Nunn, Nico Gödel (Helmut-Schmidt-University, University of the Federal Armed Forces Hamburg)

Cover design: deblik, Berlin

Printed on acid-free paper

Springer is part of Springer Science+Business Media (www.springer.com)

Foreword

This volume presents selected papers from the eigth ICOSAHOM (International Conference On Spectral and High Order Methods) conference which was held at the Norwegian University of Science and Technology (NTNU), Trondheim, Norway, during the week June 22–26, 2009. These papers were refereed by members of the scientific committee of ICOSAHOM as well as by other leading scientists.

The first ICOSAHOM conference was held in Como, Italy, in 1989. At that point the importance of high order methods was recognized and it was deciced to organize a series of conferences to be held every 3 years (Montpelier, France, 1992; Houston, TX, USA, 1995; Tel Aviv, Israel, 1998; Uppsala, Sweden, 2001; Providence, RI, USA, 2004; and Beijing, China, 2007).

At the conference in Beijing in 2007, it was decided to organize the conferences every other year. This decision was made partly due to the growing importance and the growing activity in this field. From the interest seen at ICOSAHOM 2009, this seems to have been an appropriate decision. The number of registered participants was over 200, while the total number of talks was 215, comprising nine invited talks, 153 talks in 19 different topic-specific minisymposia, and 53 talks in various contributed sessions. The ICOSAHOM conferences remain the main meeting place for researchers with interest in the theoretical, applied and computational aspects of high order methods for the numerical solution of partial differential equations.

The content of the proceedings is organized as follows. First, contributions from the invited speakers are included, listed in alphabetical order according to the invited speaker. Next, contributions from the speakers at all the minisymposia are included, listed in alphabetical order according to the first author of each paper. Finally, contributions from the speakers at the various contributed sessions are included, also listed in alphabetical order according to the the first author.

As part of the conference, a special minisymposium was organized in memory of David Gottlieb who passed away in December 2008, and who left an indelible mark on the field of applied mathematics in general and spectral methods in particular.

The success of the meeting was ensured through the generous financial support by the Research Council of Norway, the National Science Foundation (NSF), the Norwegian University of Science and Technology (NTNU – through the Faculty of Information Technology, Mathematics and Electrical Engineering and through the Program for Computational Science and Visualization), and Simula Research Laboratory (through Center for Biomedical Computing).

Finally, the conference could not have happened without the invaluable support and assistance of conference coordinator Anne Kajander. Special thanks also go to Tormod Bjøntegaard for all his contributions. The assistance from the members of the Numerical Analysis Group and from the graduate students at the Department of Mathematical Sciences at NTNU is also gratefully acknowledged.

Jan S. Hesthaven
Einar M. Rønquist

Contents

hp-FEM for the Contact Problem with Tresca Friction in Linear Elasticity: The Primal Formulation

P. Dörsek and J.M. Melenk

Abstract We present an a priori analysis of the *hp*-version of the finite element method for the primal formulation of frictional contact in linear elasticity. We introduce a new limiting case estimate for the interpolation error at Gauss and Gauss-Lobatto quadrature points. An *hp*-adaptive strategy is presented; numerical results show that this strategy can lead to exponential convergence.

1 Introduction

We study the *hp*-version of the finite element method (*hp*-FEM) APPLIED to a contact problem with Tresca friction in two-dimensional linear elasticity. In contrast to the more realistic Coulomb friction model, Tresca friction leads to a convex minimisation problem, which is simpler from a mathematical point of view. Nevertheless, the efficient numerical treatment of Tresca friction problems is important since solvers for such problems are building blocks for solvers for Coulomb friction problems (see [17, Sect. 2.5.4]).

The mathematical formulation of the frictional contact problem as a minimisation problem is provided in [10] and can be shown to be equivalent to a variational inequality of the second kind. First order h-version approximations have been available since the 1980s, see [12, 13], where the approximations can actually be chosen to be conforming and the nondifferentiable functional can be evaluated exactly. When moving to higher order discretisations, it is highly impractical to retain these properties. For the closely related variational inequalities of the first kind stemming from non-frictional obstacle and contact problems, Maischak and Stephan analysed *hp*-boundary element methods in [21, 22], and obtained convergence rates under certain regularity assumptions on the exact solution; they also presented an adaptive strategy based on a multilevel estimator. Results for the frictional contact problem in

J.M. Melenk (✉) and P. Dörsek
Vienna University of Technology, Wiedner Hauptstraße 8-10, 1040 Vienna
e-mail: melenk@tuwien.ac.at

J.S. Hesthaven and E.M. Rønquist (eds.), *Spectral and High Order Methods for Partial Differential Equations*, Lecture Notes in Computational Science and Engineering 76,
DOI 10.1007/978-3-642-15337-2_1,

the hp-boundary element method were next provided in [5]; however, the variational crimes associated with approximating the nondifferentiable friction functional j, which is clearly necessary in a high order context, were not addressed. In [15], this discretisation error was analysed.

In the present article, we focus on two issues: Firstly, we provide an a priori analysis for the errors arising from a discretisation of the non-differentiable friction functional j. We proceed in a different way than it was done in [15] and base our analysis on a new limiting case interpolation error estimate for functions in the Besov space $\mathrm{B}^{1/2}_{2,1}(a,b)$. Secondly, we show numerically for a two-dimensional model problem from [16] that hp-adaptivity can yield exponential convergence.

2 Problem Formulation

Let $\Omega \subseteq \mathbb{R}^2$ be a polygonal domain. We decompose its boundary Γ with normal vector $\boldsymbol{\nu}$ into three relatively open, disjoint parts Γ_D, Γ_N and Γ_C. On Γ_D with $|\Gamma_\mathrm{D}| > 0$ we prescribe homogeneous Dirichlet conditions, on Γ_N Neumann conditions with given traction $\mathbf{t}$, and on Γ_C contact conditions with Tresca friction, where the friction coefficient g is assumed to be constant. The volume forces are denoted by $\mathbf{F}$. Furthermore, we assume that contact holds on the entirety of Γ_C. For simplicity of exposition, we will assume that Γ_C is a single edge of Ω.

We denote by $\mathrm{H}^s(\Omega)$ the usual Sobolev spaces on Ω, and similarly on the boundary parts, with norms defined through the Slobodeckij seminorms (see [26]). The dual space of $\mathrm{H}^s(\Gamma_\mathrm{C})$ is denoted by $(\mathrm{H}^s(\Gamma_\mathrm{C}))'$. The Besov spaces $\mathrm{B}^s_{2,q}(\Omega)$, $s \in (k, k+1)$, $k \in \mathbb{N}_0$, $q \in [1,\infty]$, are defined as the interpolation spaces $(\mathrm{H}^k(\Omega), \mathrm{H}^{k+1}(\Omega))_{s-k,q}$ (note that the J- and the K-method of interpolation generate the same spaces with equivalent norms, see e.g. [27, Lemma 24.3]).

We employ standard notation of linear elasticity: $\boldsymbol{\varepsilon}_{ij}(\mathbf{v}) := \frac{1}{2}\left(\frac{\partial v_i}{\partial x_j} + \frac{\partial v_j}{\partial x_i}\right)$ denotes the small strain tensor and $\boldsymbol{\sigma}(\mathbf{v}) := \mathsf{C}\boldsymbol{\varepsilon}(\mathbf{v})$ the stress tensor. Here, C is the Hooke tensor, which is assumed to be uniformly positive definite. A vector field $\boldsymbol{\mu}$ on Γ_C can be decomposed in its normal component $\mu_n := \boldsymbol{\mu} \cdot \boldsymbol{\nu}$ and its tangential component $\boldsymbol{\mu}_t := \boldsymbol{\mu} - (\boldsymbol{\mu} \cdot \boldsymbol{\nu})\boldsymbol{\nu}$. With the trace operator $\gamma_{0,\Gamma_\mathrm{D}}\colon (\mathrm{H}^1(\Omega))^2 \to (\mathrm{H}^{1/2}(\Gamma_\mathrm{D}))^2$, we set

$$V := \left\{\mathbf{v} \in \left(\mathrm{H}^1(\Omega)\right)^2 : \gamma_{0,\Gamma_\mathrm{D}}(\mathbf{v}) = 0\right\}. \tag{1}$$

Next, we define the bilinear form $a\colon V \times V \to \mathbb{R}$, the linear form $L : V \to \mathbb{R}$ and the convex, nondifferentiable functional $j : V \to \mathbb{R}$ by

$$a(\mathbf{v},\mathbf{w}) := \int_\Omega \boldsymbol{\sigma}(\mathbf{v}) : \boldsymbol{\varepsilon}(\mathbf{w})\mathrm{d}\mathbf{x}, \tag{2}$$

$$L(\mathbf{v}) := \int_\Omega \mathbf{F}\cdot\mathbf{v}\mathrm{d}\mathbf{x} + \int_{\Gamma_\mathrm{N}} \mathbf{t}\cdot\mathbf{v}\mathrm{d}s_\mathbf{x}, \qquad j(\mathbf{v}) := \int_{\Gamma_\mathrm{C}} g|\mathbf{v}_t|\mathrm{d}s_\mathbf{x}. \tag{3}$$

The primal version of the continuous linearly elastic contact problem with Tresca friction then reads:

$$\text{Find the minimiser } \mathbf{u} \in V \text{ of } J(\mathbf{v}) := \frac{1}{2}a(\mathbf{v},\mathbf{v}) - L(\mathbf{v}) + j(\mathbf{v}). \qquad (4)$$

As is well-known, this minimiser can also be characterised by (see [10])

$$a(\mathbf{u},\mathbf{v}-\mathbf{u}) + j(\mathbf{v}) - j(\mathbf{u}) \geq L(\mathbf{v}-\mathbf{u}) \qquad \forall \mathbf{v} \in V. \qquad (5)$$

The unique solvability of (4) follows by standard arguments since the Hooke tensor C is uniformly positive definite and $|\Gamma_{\mathrm{D}}| > 0$, see [16, 17, 19].

Choosing a discrete finite-dimensional subspace $V_N \subseteq V$ and a discretisation $j_N : V_N \to \mathbb{R}$ of j, we obtain the discrete primal formulation:

$$\text{Find the minimiser } \mathbf{u}_N \in V_N \text{ of } J_N(\mathbf{v}) := \frac{1}{2}a(\mathbf{v},\mathbf{v}) - L(\mathbf{v}) + j_N(\mathbf{v}). \qquad (6)$$

Let $\mathcal{T}_N$ be a shape regular triangulation of Ω consisting of affine triangles or quadrilateral elements $K \in \mathcal{T}_N$ with diameter $h_{N,K}$; no hanging nodes are admitted for simplicity. Assume that the boundary parts Γ_{C}, Γ_{D} and Γ_{N} are resolved by the mesh. For each $K \in \mathcal{T}_N$, let $p_{N,K} \in \mathbb{N}$ be a polynomial degree. We assume that neighboring elements have comparable polynomial degrees, i.e. there exists $C_{\mathrm{poly}} > 0$ independent of $\mathcal{T}_N$ such that

$$C_{\mathrm{poly}}^{-1}\, p_{N,K} \leq p_{N,K'} \leq C_{\mathrm{poly}}\, p_{N,K} \qquad \forall K, K' \in \mathcal{T}_N \text{ with } \overline{K} \cap \overline{K'} \neq \emptyset. \qquad (7)$$

Letting F_K be the affine (if K is a triangle) or bilinear (if K is a quadrilateral) element map, we set

$$V_N := \{\mathbf{v}_N \in V : \mathbf{v}_N|_K \circ F_K \in \Pi^{p_{N,K}}(K) \text{ for all } K \in \mathcal{T}_N\}, \qquad (8)$$

where $\Pi^p(K)$ is the space of polynomials of (total) degree p if K is a triangle and $\Pi^p(K)$ is the tensor product space $\mathcal{Q}^p$ of polynomials of degree p in each variable (see [26, p. 178, (4.4.30)]) if K is a quadrilateral.

We denote the set of edges on the contact boundary by $\mathcal{E}_{\mathrm{C},N}$, that is,

$$\mathcal{E}_{\mathrm{C},N} := \{E : E \subset \Gamma_{\mathrm{C}} \text{ is an edge of } \mathcal{T}_N\}. \qquad (9)$$

We see that for every $E \in \mathcal{E}_{\mathrm{C},N}$, there exists a unique $K_E \in \mathcal{T}_N$ such that E is an edge of K_E.

3 A Priori Error Estimates

We obtain the discretisations j_N of the functionals j by a quadrature formula: For each edge $E \in \mathcal{E}_{C,N}$, let $\widetilde{G}_{E,q}$ be the points of either the Gauss or Gauss-Lobatto quadrature on E with $q+1$ points, together with the corresponding weights $\omega_{E,q,\mathbf{x}}$ for $\mathbf{x} \in E$, which are obtained by applying an affine transformation from the reference edge $\hat{E} = (-1,+1)$ to E. Then, for $\mathbf{v}_N \in V_N$, and choosing a vector $(q_{N,E})_{E \in \mathcal{E}_{C,N}} \subset \mathbb{N}_0$, we define

$$j_N(\mathbf{v}_N) := \sum_{E \in \mathcal{E}_{C,N}} j_{N,E}(\mathbf{v}_N), \qquad \text{where} \tag{10}$$

$$j_{N,E}(\mathbf{v}_N) := \sum_{\mathbf{x} \in \widetilde{G}_{E,q_{N,E}}} g|\mathbf{v}_{N,t}(\mathbf{x})|\omega_{E,q_{N,E},\mathbf{x}}. \tag{11}$$

Note, in particular, that j_N is well-defined, as $\mathbf{v}_N$ is continuous on $\overline{\Omega}$, and thus also on $\overline{\Gamma_C}$. We shall assume that there exists a constant $C_{\text{quad}} > 0$ independent of N and E such that

$$C_{\text{quad}}^{-1} p_{N,K_E} \le q_{N,E} \le C_{\text{quad}}\, p_{N,K_E}. \tag{12}$$

The main result of this section is:

Theorem 3.1. *Let $\mathbf{u} \in \mathrm{H}^{3/2}(\Omega)$ be the solution of* (4) *and $\mathbf{u}_N \in V_N$ be the solution of* (6) *where j_N is chosen as in* (10), (11). *Assuming* (12), *we have*

$$\|\mathbf{u} - \mathbf{u}_N\|_{\mathrm{H}^1(\Omega)} \le C_{\mathbf{u}} \max_{K \in \mathcal{T}_N} h_{N,K}^{1/4} p_{N,K}^{-1/4} (1 + \sqrt[4]{\ln p_{N,K}}), \tag{13}$$

where $C_{\mathbf{u}}$ depends on $\mathbf{u}$, the shape-regularity of $\mathcal{T}_N$, and C_{poly}, C_{quad}.

The proof of Theorem 3.1 is given in Sect. 3.3.

3.1 An Interpolation Error Estimate for $\mathrm{B}^{1/2}_{2,1}$*-Functions*

In [2], error estimates for the one-dimensional Gauss-Lobatto i_N and Gauss interpolation operators j_N are proved, namely, for $u \in \mathrm{H}^{1/2+\varepsilon}(\hat{E})$,

$$\|u - i_N u\|_{\mathrm{L}^2(\hat{E})} + \|u - j_N u\|_{\mathrm{L}^2(\hat{E})} \le C_\varepsilon N^{-1/2-\varepsilon} |u|_{\mathrm{H}^{1/2+\varepsilon}(\hat{E})}, \tag{14}$$

where $\hat{E} := (-1,+1)$ is the reference element and $\varepsilon > 0$ arbitrary. As functions in $\mathrm{H}^{1/2}(\hat{E})$ are not necessarily continuous, the choice $\varepsilon = 0$ is not admissible. Thus, we consider the Besov space $\mathrm{B}^{1/2}_{2,1}(\hat{E}) = \left(\mathrm{L}^2(\hat{E}), \mathrm{H}^1(\hat{E})\right)_{1/2,1}$, which is defined as the J-method interpolation space of $\mathrm{L}^2(\hat{E})$ and $\mathrm{H}^1(\hat{E})$ with parameters $\theta = 1/2$ and $q = 1$, and consists of continuous functions.

The main result is:

Theorem 3.2. *There exists $C > 0$ independent of $N \in \mathbb{N}$ such that*

$$\|u - i_N u\|_{L^2(\hat{E})} + \|u - j_N u\|_{L^2(\hat{E})} \le C N^{-1/2} \|u\|_{B^{1/2}_{2,1}(\hat{E})} \quad \forall u \in B^{1/2}_{2,1}(\hat{E}).$$

We shall only provide proofs for the case of Gauss-Lobatto interpolation; for Gauss interpolation, one proceeds analogously.

The following result is a multiplicative variant of [1, Lemme III.1.4] obtained by applying the Gagliardo-Nirenberg-Sobolev inequality instead of the Sobolev imbedding theorem:

Lemma 3.3. *There exists $C > 0$ such that for all bounded intervals $(a, b) \subset \mathbb{R}$ and all $\psi \in H^1(a, b)$*

$$\|\psi\|^2_{L^\infty(a,b)} \le C \left(\frac{1}{b-a} \|\psi\|^2_{L^2(a,b)} + \|\psi\|_{L^2(a,b)} \|\psi'\|_{L^2(a,b)} \right). \tag{15}$$

Let $\eta_{N,i} = \cos(\xi_{N,i})$ and $\rho_{N,i}$, $i = 0, \ldots, N$, be the nodes and weights of the Gauss-Lobatto quadrature with $N + 1$ points. With the Lagrange interpolation polynomials $L_{N,j}(t) := \prod_{k \ne j} \frac{t - \eta_{N,k}}{\eta_{N,j} - \eta_{N,k}}$, $j = 0, \ldots, N$, we define the Gauss-Lobatto interpolation operator $i_N \colon C([-1, +1]) \to \mathcal{P}^N$ by

$$i_N u := \sum_{j=0}^{N} u(\eta_{N,j}) L_{N,j}. \tag{16}$$

By applying the sharper estimate given in Lemma 3.3 in the proof of [1, Théorème III.1.15], we obtain the following multiplicative result:

Proposition 3.4. *There exists $C > 0$ such that for all $N \in \mathbb{N}$ and all $u \in H^1(\hat{E})$ we have the bound*

$$\begin{aligned} \|i_N u\|^2_{L^2(\hat{E})} \le C \big(& N^{-2} \left(|u(-1)|^2 + |u(1)|^2 \right) + \|u\|^2_{L^2(\hat{E})} \\ & + N^{-1} \|u\|_{L^2(\hat{E})} \|u' \sqrt{1 - x^2}\|_{L^2(\hat{E})} \big). \end{aligned} \tag{17}$$

Remark 3.5. Proposition 3.4 is a special case of the following, more general result. Let $H^{k,\alpha}(\hat{E})$ be the space of all functions with

$$\|v\|^2_{H^{k,\alpha}(\hat{E})} := \sum_{\ell=0}^{k} \int_{-1}^{+1} |u^{(\ell)}(x)|^2 (1 - x^2)^{\alpha + \ell} dx < \infty, \tag{18}$$

and set $L^{2,\alpha}(\hat{E}) := H^{0,\alpha}(\hat{E})$. These spaces were also considered in [14, Sect. 3]. One can show for the Gauss-Jacobi-Lobatto interpolant i_N^α with $\alpha > -1$ (see [8, Appendix])

$$\|i_N^\alpha u\|^2_{L^{2,\alpha}(\hat{E})} \lesssim \|u\|^2_{L^{2,\alpha}(\hat{E})} + N^{-1}\|u\|_{L^{2,\alpha}(\hat{E})}\|u'\|_{L^{2,\alpha+1}(\hat{E})} + N^{-2-2\alpha}\left(u(-1)^2 + u(+1)^2\right) \tag{19}$$

for all $u \in H^{1,\alpha}(-1,+1) \cap C([-1,+1])$. For the special case $\alpha = -1/2$, i.e. the Chebyshev-Lobatto interpolation operator, one can show additionally (see [8, Appendix] for details)

$$\|i_N^{-1/2} u\|^2_{L^{2,-1/2}(\hat{E})} \lesssim \|u\|^2_{L^{2,-1/2}(\hat{E})} + N^{-1}\|u\|_{L^{2,-1/2}(\hat{E})}\|u\|_{H^{1,-1/2}(\hat{E})} \tag{20}$$

and that $i_N^{-1/2}$ is stable on $H^{1,-1/2}(-1,+1)$ as well as on the interpolation space $\left(L^{2,-1/2}(\hat{E}), H^{1,-1/2}(\hat{E})\right)_{1/2,1}$. ∎

Combining Lemma 3.3 with Proposition 3.4 yields:

Corollary 3.6. *There exists $C > 0$ such that for all $N \in \mathbb{N}$ and all $u \in H^1(\hat{E})$ there holds*

$$\|i_N u\|_{L^2(\hat{E})} \le C\left(\|u\|_{L^2(\hat{E})} + N^{-1/2}\|u\|^{1/2}_{L^2(\hat{E})}\|u\|^{1/2}_{H^1(\hat{E})}\right). \tag{21}$$

A key step towards the proof of Theorem 3.2 is the following result:

Theorem 3.7. *Let $T_N : C([-1,+1]) \to \mathcal{P}^N$, $N \in \mathbb{N}$, be continuous linear operators satisfying for a $C > 0$ independent of $N \in \mathbb{N}$*

$$T_N p = p \quad \text{for } p \in \mathcal{P}^N \quad \text{and} \tag{22}$$

$$\|T_N u\|_{L^2(\hat{E})} \le C\left(\|u\|_{L^2(\hat{E})} + N^{-1/2}\|u\|^{1/2}_{L^2(\hat{E})}\|u\|^{1/2}_{H^1(\hat{E})}\right) \quad \forall u \in H^1(\hat{E}). \tag{23}$$

Then there exists a constant $C > 0$ such that for all $N \in \mathbb{N}$

$$\|u - T_N u\|_{L^2(\hat{E})} \le C N^{-1/2}\|u\|_{B^{1/2}_{2,1}(\hat{E})} \quad \text{for all } u \in B^{1/2}_{2,1}(\hat{E}). \tag{24}$$

Note that $T_N : B^{1/2}_{2,1}(\hat{E}) \to \mathcal{P}^N$ is well-defined and continuous as we have the continuous injection $B^{1/2}_{2,1}(\hat{E}) \hookrightarrow C([-1,+1])$ (see [27]).

Proof. We shall first prove the multiplicative error estimate

$$\|u - T_N u\|_{L^2(\hat{E})} \lesssim N^{-1/2}\|u\|^{1/2}_{L^2(\hat{E})}\|u\|^{1/2}_{H^1(\hat{E})}. \tag{25}$$

To that end, we start by observing that [23, Prop. A.2] provides sequence of operators $\pi_N : L^2(\hat{E}) \to \mathcal{P}^N$ with

$$\|\pi_N u\|_{L^2(\hat{E})} \lesssim \|u\|_{L^2(\hat{E})} \qquad \text{for all } u \in L^2(\hat{E}), \tag{26}$$
$$\|u - \pi_N u\|_{L^2(\hat{E})} \lesssim N^{-1}\|u\|_{H^1(\hat{E})} \qquad \text{for all } u \in H^1(\hat{E}), \tag{27}$$
$$\text{and} \quad \|\pi_N u\|_{H^1(\hat{E})} \lesssim \|u\|_{H^1(\hat{E})} \qquad \text{for all } u \in H^1(\hat{E}). \tag{28}$$

As $T_N \circ \pi_N = \pi_N$, we see by (23) that

$$\begin{aligned}
\|u - T_N u\|_{L^2(\hat{E})} &\le \|u - \pi_N u\|_{L^2(\hat{E})} + \|T_N(u - \pi_N u)\|_{L^2(\hat{E})} \\
&\lesssim \|u - \pi_N u\|_{L^2(\hat{E})} + N^{-1/2}\|u - \pi_N u\|_{L^2(\hat{E})}^{1/2}\|u - \pi_N u\|_{H^1(\hat{E})}^{1/2} \\
&\lesssim N^{-1/2}\|u\|_{L^2(\hat{E})}^{1/2}\|u\|_{H^1(\hat{E})}^{1/2}.
\end{aligned} \tag{29}$$

A careful analysis of the proof of [27, Theorem 25.3] shows that this yields

$$\|u - T_N u\|_{L^2(\hat{E})} \lesssim N^{-1/2}\|u\|_{B^{1/2}_{2,1}(\hat{E})}, \tag{30}$$

that is, the claimed estimate. □

Theorem 3.2 follows by combining Corollary 3.6 and Theorem 3.7.

3.2 A Polynomial Inverse Estimate

We need the following inverse estimate:

Lemma 3.8 (Generalised $B^{1/2}_{2,1}$-$H^{1/2}$ *p*-version inverse inequality). *There exists a constant $C > 0$ such that for all polynomials $q \in \mathcal{P}^p$ and all $\kappa \in \mathbb{R}$,*

$$\big\| |q| - \kappa \big\|_{B^{1/2}_{2,1}(\hat{E})} \le C(1 + \sqrt{\ln p})\left(|q|_{H^{1/2}(\hat{E})} + |\kappa - \bar{q}|\right), \tag{31}$$

where $\bar{q} := \frac{1}{2}\int_{-1}^{+1} |q(x)| dx$ is the integral mean of $|q|$.
The particular choices $\kappa = \bar{q}$ and $\kappa = 0$ lead to

$$\big\| |q| - \bar{q} \big\|_{B^{1/2}_{2,1}(\hat{E})} \le C(1 + \sqrt{\ln p})\left(|q|_{H^{1/2}(\hat{E})}\right),$$
$$\big\| |q| \big\|_{B^{1/2}_{2,1}(\hat{E})} \le C(1+\sqrt{\ln p})\left(|q|_{H^{1/2}(\hat{E})} + |\bar{q}|\right) \le C(1 + \sqrt{\ln p})\|q\|_{H^{1/2}(\hat{E})}.$$

Proof. We use the K-method of interpolation (see [26, 27]). Let

$$K(t, u) := \inf_{v \in H^1(\hat{E})} \left[\|u - v\|^2_{L^2(\hat{E})} + t^2\|v\|^2_{H^1(\hat{E})}\right]^{1/2}. \tag{32}$$

By [6, p. 193, (7.4)], we see that for arbitrary $\varepsilon \in (0, 1]$,

$$\begin{aligned}\big\||q|-\kappa\big\|_{B^{1/2}_{2,1}(\hat{E})} &\sim \int_0^1 t^{-1/2}K(t,|q|-\kappa)\frac{dt}{t}\\ &= \int_0^{\varepsilon} t^{-1/2}K(t,|q|-\kappa)\frac{dt}{t} + \int_{\varepsilon}^1 t^{-1/2}K(t,|q-\kappa|)\frac{dt}{t}. \end{aligned} \tag{33}$$

For $0 < s \leq 1$ we recall the equivalence of norms

$$\|w\|^2_{H^s(\hat{E})} \sim |w|^2_{H^s(\hat{E})} + \left(\frac{1}{2}\int_{-1}^{+1} w(x)dx\right)^2, \tag{34}$$

which is proved by a standard compactness argument as for the Deny-Lions Lemma (see also [27, Lemma 11.1]). Hence, for $0 < s \leq 1$, we have

$$\|w - \frac{1}{2}\int_{-1}^{+1} w(x)dx\|^2_{H^s(\hat{E})} \sim |w|^2_{H^s(\hat{E})} \quad \text{for all } w \in H^s(\hat{E}). \tag{35}$$

To treat the first integral in (33) we choose $v = |q| - \kappa$ in (32) so that

$$\begin{aligned}\int_0^{\varepsilon} t^{-1/2}K(t,|q|-\kappa)\frac{dt}{t} &\leq \int_0^{\varepsilon} t^{1/2}\big\||q|-\kappa\big\|_{H^1(\hat{E})}\frac{dt}{t}\\ &= 2\sqrt{\varepsilon}\big\||q|-\kappa\big\|_{H^1(\hat{E})} \leq 2\sqrt{\varepsilon}\left(\big\||q|-\bar{q}\big\|_{H^1(\hat{E})} + \sqrt{2}|\kappa-\bar{q}|\right)\\ &\lesssim 2\sqrt{\varepsilon}\left(\big||q|\big|_{H^1(\hat{E})} + \sqrt{2}|\kappa-\bar{q}|\right). \end{aligned} \tag{36}$$

Note that $\big||q|\big|_{H^1(\hat{E})} = |q|_{H^1(\hat{E})}$. The inverse inequality in [3, p. 100, Theorem III.4.2] together with (35) implies

$$\sqrt{\varepsilon}|q|_{H^1(\hat{E})} \leq \sqrt{\varepsilon}\|q-\bar{q}\|_{H^1(\hat{E})} \lesssim \sqrt{\varepsilon}p\|q-\bar{q}\|_{H^{1/2}(\hat{E})} \lesssim \sqrt{\varepsilon}p|q|_{H^{1/2}(\hat{E})}. \tag{37}$$

For the second integral in (33), the Cauchy-Schwarz inequality for the measure $\frac{dt}{t}$ applied to the functions $t \mapsto 1$ and $t \mapsto t^{-1/2}K(t,|q|-\kappa)$ and the characterization of $\big\||q|-\kappa\big\|_{H^{1/2}(\hat{E})}$ by the K-method of interpolation yield

$$\begin{aligned}\int_{\varepsilon}^1 t^{-1/2}K(t,|q|-\kappa)\frac{dt}{t} &\leq \sqrt{\int_{\varepsilon}^1\frac{dt}{t}}\sqrt{\int_{\varepsilon}^1\left(t^{-1/2}K(t,|q|-\kappa)\right)^2\frac{dt}{t}}\\ &\lesssim \sqrt{\ln\frac{1}{\varepsilon}}\big\||q|-\kappa\big\|_{H^{1/2}(\hat{E})} \lesssim \sqrt{\ln\frac{1}{\varepsilon}}\left(\big\||q|-\bar{q}\big\|_{H^{1/2}(\hat{E})} + |\kappa-\bar{q}|\right)\\ &\lesssim \sqrt{\ln\frac{1}{\varepsilon}}\left(\big||q|\big|_{H^{1/2}(\hat{E})} + |\kappa-\bar{q}|\right), \end{aligned} \tag{38}$$

where the last step again follows from (35). Additionally, by the definition of the $H^{1/2}$-seminorm, we see easily that $\big||q|\big|_{H^{1/2}(\hat{E})} \leq |q|_{H^{1/2}(\hat{E})}$, which yields

$$\int_{\varepsilon}^{1} t^{-1/2} K(t, |q| - \kappa) \frac{dt}{t} \lesssim \sqrt{\ln \frac{1}{\varepsilon}} \left(|q|_{H^{1/2}(\hat{E})} + |\kappa - \bar{q}| \right), \tag{39}$$

We set $\varepsilon := \frac{1}{p^2}$ and obtain by inserting (37), (39) in (33)

$$\begin{aligned} \big\| |q| - \kappa \big\|_{B^{1/2}_{2,1}(\hat{E})} &\lesssim (1 + \sqrt{\ln p}) |q|_{H^{1/2}(\hat{E})} + (p^{-1} + \sqrt{\ln p}) |\kappa - \bar{q}| \\ &\le (1 + \sqrt{\ln p}) \left(|q|_{H^{1/2}(\hat{E})} + |\kappa - \bar{q}| \right). \end{aligned} \tag{40}$$

□

3.3 Convergence Rates: Proof of Theorem 3.1

We now prove a convergence rate result for the primal formulation of the friction problem. We follow in style the article [4]. A similar estimate was derived in [15, Lemma 4.1] using different techniques.

In the following, $\mathbf{u}$ and $\mathbf{u}_N$ denote the solutions of (4) and (6).

Proposition 3.9. *Define* $S_{\mathbf{u}}(\mathbf{v}) := a(\mathbf{u}, \mathbf{v}) - L(\mathbf{v})$. *Then, for all* $\mathbf{v}_N \in V_N$,

$$\begin{aligned} a(\mathbf{u} - \mathbf{u}_N, \mathbf{u} - \mathbf{u}_N) &\le a(\mathbf{u} - \mathbf{u}_N, \mathbf{u} - \mathbf{v}_N) + S_{\mathbf{u}}(\mathbf{v}_N - \mathbf{u}) \\ &\quad + j_N(\mathbf{v}_N) - j(\mathbf{v}_N) + j(\mathbf{u}_N) - j_N(\mathbf{u}_N) + j(\mathbf{v}_N - \mathbf{u}). \end{aligned} \tag{41}$$

Proof. It follows from (6) that

$$\begin{aligned} &a(\mathbf{u} - \mathbf{u}_N, \mathbf{u} - \mathbf{u}_N) = \\ &\quad a(\mathbf{u} - \mathbf{u}_N, \mathbf{u} - \mathbf{v}_N) + a(\mathbf{u}, \mathbf{v}_N - \mathbf{u}_N) - a(\mathbf{u}_N, \mathbf{v}_N - \mathbf{u}_N) \\ &\quad \le a(\mathbf{u} - \mathbf{u}_N, \mathbf{u} - \mathbf{v}_N) + a(\mathbf{u}, \mathbf{v}_N - \mathbf{u}_N) - L(\mathbf{v}_N - \mathbf{u}_N) + j_N(\mathbf{v}_N) - j_N(\mathbf{u}_N) \\ &\quad \le a(\mathbf{u} - \mathbf{u}_N, \mathbf{u} - \mathbf{v}_N) + S_{\mathbf{u}}(\mathbf{v}_N - \mathbf{u}_N) + j_N(\mathbf{v}_N) - j_N(\mathbf{u}_N). \end{aligned}$$

Since for all $\mathbf{v} \in V$ we have

$$\begin{aligned} S_{\mathbf{u}}(\mathbf{v}_N - \mathbf{u}_N) &= S_{\mathbf{u}}(\mathbf{v}_N - \mathbf{u}) + S_{\mathbf{u}}(\mathbf{u} - \mathbf{v}) + S_{\mathbf{u}}(\mathbf{v} - \mathbf{u}_N) \\ &\le S_{\mathbf{u}}(\mathbf{v}_N - \mathbf{u}) + j(\mathbf{v}) - j(\mathbf{u}) + S_{\mathbf{u}}(\mathbf{v} - \mathbf{u}_N), \end{aligned} \tag{42}$$

we obtain

$$\begin{aligned} a(\mathbf{u} - \mathbf{u}_N, \mathbf{u} - \mathbf{u}_N) &\le a(\mathbf{u} - \mathbf{u}_N, \mathbf{u} - \mathbf{v}_N) + S_{\mathbf{u}}(\mathbf{v}_N - \mathbf{u}) + S_{\mathbf{u}}(\mathbf{v} - \mathbf{u}_N) \\ &\quad + j_N(\mathbf{v}_N) - j(\mathbf{u}) + j(\mathbf{v}) - j_N(\mathbf{u}_N). \end{aligned} \tag{43}$$

Choose now $\mathbf{v} = \mathbf{u}_N$ and note that $j(\mathbf{v}_N) \le j(\mathbf{v}_N - \mathbf{u}) + j(\mathbf{u})$, and thus $-j(\mathbf{u}) \le -j(\mathbf{v}_N) + j(\mathbf{u} - \mathbf{v}_N)$. The claim then follows. □

Proposition 3.9 shows that the main task is to estimate the error introduced by approximating j by j_N on V_N. This will be done now.

Theorem 3.10. *For every $E \in \mathcal{E}_{C,N}$ let $K_E \in \mathcal{T}_N$ be such that E is an edge of K_E. For $\mathbf{w}_N \in V_N$, set*

$$j_E(\mathbf{w}_N) := \int_E g|\mathbf{w}_N| \mathrm{d}s_{\mathbf{x}}. \tag{44}$$

Then, we have the estimates

$$|j_E(\mathbf{w}_N) - j_{N,E}(\mathbf{w}_N)| \leq C\left(h_{N,K_E}(1 + \sqrt{\ln p_{N,K_E}})q_{N,E}^{-1/2}|\mathbf{w}_N|_{\mathrm{H}^1(K_E)}\right) \tag{45}$$

$$\leq C\left(h_{N,K_E}(1 + \sqrt{\ln p_{N,K_E}})p_{N,K_E}^{-1/2}|\mathbf{w}_N|_{\mathrm{H}^1(K_E)}\right). \tag{46}$$

Here, $C > 0$ depends only on the shape regularity of $\mathcal{T}_N$ and C_{quad}.

Proof. It is clear that

$$j_{N,E}(\mathbf{w}_N) = g \int_E i_{E,q_{N,E}}|\mathbf{w}_N| \mathrm{d}s_{\mathbf{x}}, \tag{47}$$

where $i_{E,q_{N,E}}$ denotes the local interpolation operator on E at the $q_{N,E} + 1$ Gauss-Lobatto points. Therefore, we have to estimate

$$\begin{aligned} |j_E(\mathbf{w}_N) - j_{N,E}(\mathbf{w}_N)| &= |g| \, \| \, |\mathbf{w}_N| - i_{E,q_{N,E}}|\mathbf{w}_N| \, \|_{L^1(E)} \\ &\leq C h_E^{1/2} \| \, |\mathbf{w}_N| - i_{E,q_{N,E}}|\mathbf{w}_N| \, \|_{L^2(E)}. \end{aligned} \tag{48}$$

Theorem 3.2 provides a constant $C > 0$ such that

$$\big\| |\mathbf{w}| - i_q|\mathbf{w}| \big\|_{\mathrm{L}^2(\hat{E})} \leq C q^{-1/2} \|\mathbf{w}\|_{\mathrm{B}^{1/2}_{2,1}(\hat{E})} \qquad \forall q \in \mathbb{N}. \tag{49}$$

Apply now a scaling argument: Let $\Phi_E : \hat{E} \to E$ be an invertible, affine mapping. As $i_{E,q_{N,E}}$ reproduces constant functions, we have for any $\kappa \in \mathbb{R}$,

$$\begin{aligned} \big\| |\mathbf{w}_N| - i_{E,q_{N,E}}|\mathbf{w}_N| \big\|_{\mathrm{L}^2(E)} &= \|(|\mathbf{w}_N| - \kappa) - i_{E,q_{N,E}}(|\mathbf{w}_N| - \kappa)\|_{\mathrm{L}^2(E)} \\ &= \frac{h_{N,E}^{1/2}}{2} \big\|(|\mathbf{w}_N \circ \Phi_E| - \kappa) - i_{q_{N,E}}(|\mathbf{w}_N \circ \Phi_E| - \kappa)\big\|_{\mathrm{L}^2(\hat{E})} \\ &\leq C h_{N,E}^{1/2} q_{N,E}^{-1/2} \big\| |\mathbf{w}_N \circ \Phi_E| - \kappa \big\|_{\mathrm{B}^{1/2}_{2,1}(\hat{E})}. \end{aligned} \tag{50}$$

With the choice $\kappa := \frac{1}{2}\int_{\hat{E}} |\mathbf{w}_N \circ \Phi_E| \mathrm{d}x$, Lemma 3.8 gives

$$\big\| |\mathbf{w}_N \circ \Phi_E| - \kappa \big\|_{\mathrm{B}^{1/2}_{2,1}(\hat{E})} \lesssim \left(1 + \sqrt{\ln p_{N,E}}\right) |\mathbf{w}_N \circ \Phi_E|_{\mathrm{H}^{1/2}(\hat{E})}. \tag{51}$$

Thus, inserting (51) in (50) and scaling produces

$$\| |\mathbf{w}_N| - i_{E,q_{N,E}} |\mathbf{w}_N| \|_{L^2(E)} \lesssim h_{N,E}^{1/2} (1 + \sqrt{\ln p_{N,E}}) q_{N,E}^{-1/2} |\mathbf{w}_N|_{H^{1/2}(E)}.$$

Finally, inserting this in (48) we get with the trace theorem

$$\begin{aligned} |j_E(\mathbf{w}_N) - j_{N,E}(\mathbf{w}_N)| &\lesssim h_{N,E} (1 + \sqrt{\ln p_{N,E}}) q_{N,E}^{-1/2} |\mathbf{w}_N|_{H^{1/2}(E)} \\ &\lesssim h_{N,K_E} (1 + \sqrt{\ln p_{N,K_E}}) q_{N,E}^{-1/2} |\mathbf{w}_N|_{H^1(K_E)}. \end{aligned}$$

This proves (45). The bound (46) follows from (46) and (12). □

Let h_N, p_N and q_N be the local mesh width, polynomial degree and quadrature order, respectively, and introduce the local approximation quantification

$$\omega_N := h_N^{1/2} p_N^{-1/2} (1 + \sqrt{\ln p_N}). \tag{52}$$

Corollary 3.11. *Set $\mathcal{S}_N := \bigcup_{E \in \mathcal{E}_{C,N}} K_E$. Let ω_N be given by (52). Then there exists $C > 0$ independent of N such that for every $\mathbf{w}_N \in V_N$*

$$|j(\mathbf{w}_N) - j_N(\mathbf{w}_N)| \le C \|\omega_N \nabla \mathbf{w}_N\|_{L^2(\mathcal{S}_N)} \le C \|\omega_N \nabla \mathbf{w}_N\|_{L^2(\Omega)}. \tag{53}$$

Proof. Applying Theorem 3.10 to $\mathbf{w}_N$ and summing over $E \in \mathcal{E}_{C,N}$, we obtain by the discrete Cauchy-Schwarz inequality and the trace theorem that

$$\begin{aligned} |j_N(\mathbf{w}_N) - j(\mathbf{w}_N)| &\le \sum_{E \in \mathcal{E}_{C,N}} |j_E(\mathbf{w}_N) - j_{N,E}(\mathbf{w}_N)| \\ &\lesssim \sum_{E \in \mathcal{E}_{C,N}} h_{N,E}^{1/2} h_{N,E}^{1/2} (1 + \sqrt{\ln p_{N,K_E}}) q_{N,E}^{-1/2} |\mathbf{w}_N|_{H^1(K_E)} \\ &\le \left(\sum_{E \in \mathcal{E}_{C,N}} h_{N,E} \right)^{1/2} \left(\sum_{E \in \mathcal{E}_{C,N}} h_{N,E} (1 + \ln p_{N,K_E}) q_{N,K_E}^{-1} |\mathbf{w}_N|_{H^1(K_E)}^2 \right)^{1/2} \\ &= |\Gamma_C|^{1/2} \left(\sum_{E \in \mathcal{E}_{C,N}} h_{N,E} (1 + \ln p_{N,K_E}) q_{N,K_E}^{-1} |\mathbf{w}_N|_{H^1(K_E)}^2 \right)^{1/2} \\ &\lesssim \|h_N^{1/2} (1 + \sqrt{\ln p_N}) q_N^{-1/2} \nabla \mathbf{w}_N\|_{L^2(\mathcal{S}_N)} \end{aligned}$$

□

Theorem 3.12. *There exists $C > 0$ independent of N such that the following is true. Set $\mathcal{S}_N := \bigcup_{E \in \mathcal{E}_{C,N}} K_E$, and let $\mathbf{u}_N$ and $\mathbf{u}$ be the solutions of* (6) *and* (4), *respectively. Let ω_N be given by (52). Then:*

$$\|\mathbf{u}-\mathbf{u}_N\|_{\mathrm{H}^1(\Omega)} \le C \inf_{\mathbf{v}_N\in V_N}\Big(\|\omega_N\nabla\mathbf{u}_N\|_{\mathrm{L}^2(\mathcal{S}_N)} + \|\omega_N\nabla\mathbf{v}_N\|_{\mathrm{L}^2(\mathcal{S}_N)} \\ + \|\mathbf{u}-\mathbf{v}_N\|_{\mathrm{H}^1(\Omega)} + \|\mathbf{u}-\mathbf{v}_N\|^2_{\mathrm{H}^1(\Omega)} + |S_{\mathbf{u}}(\mathbf{u}-\mathbf{v}_N)|\Big)^{1/2}.$$

Before proving Theorem 3.12, we remark that we can trivially estimate

$$\|\omega_N\nabla\mathbf{u}_N\|_{\mathrm{L}^2(\mathcal{S}_N)} + \|\omega_N\nabla\mathbf{v}_N\|_{\mathrm{L}^2(\mathcal{S}_N)} \le \|\omega_N\nabla\mathbf{u}_N\|_{\mathrm{L}^2(\Omega)} + \|\omega_N\nabla\mathbf{v}_N\|_{\mathrm{L}^2(\Omega)}. \tag{54}$$

This estimate is rather generous: for quasi-uniform meshes the strip $\mathcal{S}_N$ has area $O(h)$, and hence the left-hand side of (54) can be expected to be significantly smaller than the right-hand of (54).

Proof. We employ Proposition 3.9. By the V-boundedness and V-ellipticity of a and the V-boundedness of j, we see that

$$\|\mathbf{u}-\mathbf{u}_N\|^2_{\mathrm{H}^1(\Omega)} \lesssim \|\mathbf{u}-\mathbf{u}_N\|_{\mathrm{H}^1(\Omega)}\|\mathbf{u}-\mathbf{v}_N\|_{\mathrm{H}^1(\Omega)} + \|\mathbf{v}_N-\mathbf{u}\|_{\mathrm{H}^1(\Omega)} \\ + j_N(\mathbf{v}_N) - j(\mathbf{v}_N) + j(\mathbf{u}_N) - j_N(\mathbf{u}_N) + S_{\mathbf{u}}(\mathbf{v}_N-\mathbf{u}),$$

from which it follows by applying the ε-trick that

$$\|\mathbf{u}-\mathbf{u}_N\|^2_{\mathrm{H}^1(\Omega)} \lesssim \|\mathbf{u}-\mathbf{v}_N\|^2_{\mathrm{H}^1(\Omega)} + \|\mathbf{u}-\mathbf{v}_N\|_{\mathrm{H}^1(\Omega)} \\ + |j_N(\mathbf{v}_N) - j(\mathbf{v}_N)| + |j_N(\mathbf{u}_N) - j(\mathbf{u}_N)| + S_{\mathbf{u}}(\mathbf{v}_N-\mathbf{u}).$$

Applying Corollary 3.11 to $\mathbf{u}_N$ and $\mathbf{v}_N$, the result now follows by the local equivalence of p_N and q_N. □

Clearly, choosing $\mathbf{v}_N \in V_N$ to be the best approximation of $\mathbf{u}$ with respect to the H^1-norm proves that $\|h_N^{1/2}(1+\sqrt{\ln p_N})p_N^{-1/2}\mathbf{v}_N\|_{\mathrm{H}^1(\Omega)}$ stays bounded and converges with a rate of $h_N^{1/2}(1+\sqrt{\ln p_N})p_N^{-1/2}$, and $\|\mathbf{u}-\mathbf{v}_N\|_{\mathrm{H}^1(\Omega)} \to 0$ if $h_N/p_N \to 0$. It still remains to show that $\|\mathbf{u}_N\|_{\mathrm{H}^1(\Omega)}$ stays bounded.

Lemma 3.13. *The norms in* $\mathrm{H}^1(\Omega)$ *of the solutions* $\mathbf{u}_N$ *of* (6) *stay bounded for* $N \to \infty$.

Proof. Choose $\mathbf{v}_N = 0$. Then, as $j_N(\mathbf{w}_N) \ge 0$ for all $\mathbf{w}_N \in V_N$,

$$a(\mathbf{u}_N,\mathbf{u}_N) \le L(\mathbf{u}_N) - j_N(\mathbf{u}_N) \le L(\mathbf{u}_N). \tag{55}$$

The result now follows by the coercitivity of a and the boundedness of L. □

Remark 3.14. Lemma 3.13 shows convergence of the primal method if $\bigcup_N V_N$ is dense in V. This can also be shown similarly as in [15] using Glowinski's theorem. ■

Finally, Theorem 3.1 now easily follows from Theorem 3.12, the trivial bound (54), and hp-approximation results given, for example, in [23].

4 Numerical Experiments

4.1 A Posteriori Error Estimation

One way to realise numerically the minimisation problem (6) is by dualisation. Specifically, we assume that the quadrature points $\tilde{G}_{E,q_{N,E}}$ are the Gauss points and that

$$q_{N,E} \geq p_{N,K_E} - 1 \qquad \forall E \in \mathcal{E}_{C,N}. \tag{56}$$

We introduce the bilinear forms b and b_N by

$$b(\mathbf{u},\lambda) := g\int_{\Gamma_C} \mathbf{u}_t \lambda \, ds_{\mathbf{x}}, \quad b_N(\mathbf{u},\lambda) := g \sum_{E\in\mathcal{E}_{C,N}} \sum_{\mathbf{x}\in\tilde{G}_{E,q_{N,E}}} \omega_{E,q_{N,E},\mathbf{x}} \mathbf{u}_t(\mathbf{x})\lambda(\mathbf{x}),$$

$$W_N := \{\lambda \in L^2(\Gamma_C) : \lambda|_E \in \mathcal{P}^{q_{N,E}} \quad \forall E \in \mathcal{E}_{C,N}\},$$
$$\Lambda_N := \{\lambda \in W_N : |\lambda(\mathbf{x})| \leq 1 \quad \forall \mathbf{x} \in \tilde{G}_{E,q_{N,E}} \quad \forall E \in \mathcal{E}_{C,N}\},$$

where, in the present 2D setting, we view the tangential component $\mathbf{u}_t$ of $\mathbf{u}$ as a scalar function in the definition of b and b_N. It is easy to see that $j_N(\mathbf{u}) = \sup_{\lambda\in\Lambda_N} b_N(\mathbf{u},\lambda)$. Hence, the minimisation problem (6) can be reformulated as a saddle point problem of finding $(\mathbf{u}_N, \lambda_N) \in V_N \times \Lambda_N$ such that

$$a(\mathbf{u}_N, \mathbf{v}) \qquad + b_N(\mathbf{v}, \lambda_N) = L(\mathbf{v}) \qquad \forall v \in V_N, \tag{57a}$$
$$b_N(\mathbf{u}_N, \mu - \lambda_N) \qquad \leq 0 \qquad \forall \mu \in \Lambda_N. \tag{57b}$$

(57) has solutions; the component $\mathbf{u}_N$ is the unique solution of (6), which justifies our using the same symbol. Any Lagrange multiplier λ_N can be used for a posteriori error estimation. Indeed, exploiting the fact that $b(\mathbf{v},\lambda) = b_N(\mathbf{v},\lambda)$ for all $\mathbf{v} \in V_N$ and $\lambda \in W_N$, one can proceed as in [7, Sec. 4] to show the following result (see [8, Appendix] for the details):

Theorem 4.1. *Assume (56). Let* $\mathbf{u}$, $\mathbf{u}_N$ *solve (4), (6), and let* λ_N *be a Langrange multiplier satisfying (57). Then* $\|\mathbf{u} - \mathbf{u}_N\|^2_{H^1(\Omega)} \leq C\eta_N^2$, *where the* error indicator

$$\eta_N^2 := \sum_{K\in\mathcal{T}_N} \eta_{N,K}^2 \tag{58}$$

is defined in terms of element error indicators

$$\eta_{N,K}^2 := h_{N,K}^2 p_{N,K}^{-2} \|\mathbf{r}_K\|^2_{L^2(K)} + h_{N,K} p_{N,K}^{-1} \sum_{E\subseteq\partial K} \|\mathbf{R}_E\|^2_{L^2(E)} \tag{59}$$
$$+ j_{\partial K\cap\Gamma_C}(\mathbf{u}_N) - b_{\partial K\cap\Gamma_C}(\mathbf{u}_N, \tilde{\lambda}_N) + g^2\|\lambda_N - \tilde{\lambda}_N\|^2_{(H^{1/2}(\partial K\cap\Gamma_C))'};$$

here, the element residuals $\mathbf{r}_K$ *and the edge jumps* $\mathbf{R}_E$ *are given by*

$$\mathbf{r}_K := -\operatorname{div}\sigma(\mathbf{u}_N) - \mathbf{F}, \qquad \mathbf{R}_E := \begin{cases} \frac{1}{2}\left[\sigma(\mathbf{u}_N)\cdot\boldsymbol{\nu}\right]_E & \text{if } E \subset \Omega, \\ (\sigma(\mathbf{u}_N)\cdot\boldsymbol{\nu})_t + g\lambda_N & \text{if } E \subset \Gamma_{\mathrm{C}} \\ \sigma(\mathbf{u}_N)\cdot\boldsymbol{\nu} - \mathbf{t} & \text{if } E \subset \Gamma_{\mathrm{N}}, \\ 0 & \text{if } E \subset \Gamma_{\mathrm{D}}. \end{cases}$$

Finally, $\tilde{\lambda}_N$ is the $L^2(\Gamma_{\mathrm{C}})$-projection of λ_N onto Λ_N.

Remark 4.2. In our numerical experiments, we estimate the error indicator η_N further by replacing the $(H^{1/2})'$-norm by the L^2-norm and estimating rather generously the contributions of $j_{\partial K\cap\Gamma_{\mathrm{C}}}(\mathbf{u}_N) - b_{\partial K\cap\Gamma_{\mathrm{C}}}(\mathbf{u}_N, \tilde{\lambda}_N)$ for those edges $E \subset \Gamma_{\mathrm{C}}$ where $\lambda_N|_E \neq \tilde{\lambda}_N|_E$. See [7, Remark 4.3] for details. ■

4.2 Numerical Examples

We consider the two-dimensional numerical problem of [16, Example 6.12]. Let $\Omega = (0,4)\times(0,4)$, assume homogeneous Dirichlet conditions on $\Gamma_{\mathrm{D}} := \{4\}\times(0,4)$, frictional contact on $\Gamma_{\mathrm{C}} := (0,4)\times\{0\}$, and Neumann conditions on $\Gamma_{\mathrm{N}} := (\{0\}\times(0,4)) \cup ((0,4)\times\{4\})$, where $\mathbf{t}(0,s) = (150(5-s), -75)\,\mathrm{daN\,mm^{-2}}$ for $s \in (0,4)$ and $\mathbf{t} = 0$ on $(0,4)\times\{4\}$. The elasticity parameters are chosen to be $E = 1{,}500\,\mathrm{daN\,mm^{-2}}$ and $\nu = 0.4$, the friction coefficient is $g = 450\,\mathrm{daN\,mm^{-2}}$. We assume plane stress conditions.

We perform six numerical experiments: h-uniform and h-adaptive methods with polynomial degrees 2 and 3; a p-uniform method starting with polynomial degree 2; and an hp-adaptive method starting with polynomial degree 3. The initial meshes are uniform and consist of 16 squares.

Quadrilateral meshes with hanging nodes are used. We require the "one hanging node rule" and that all irregular nodes be one-irregular. Differing polynomial degrees on neighbouring elements are resolved by using the minimum rule on the edge. For the discretisation of j, we choose Gaussian quadrature and $q_{N,E} = p_{N,K_E} - 1$ for $E \in \mathcal{E}_{\mathrm{C},N}$, i.e. we use p_{N,K_E} quadrature points in (11). As described in Sect. 4.1, the minimisation problem (6) is recast in primal-dual form and the resulting first kind variational inequality is solved with the MPRGP algorithm (see [9]). As a by-product, we obtain a Lagrange multiplier $\lambda_N \in W_N$, which is used to define the error indicators of (58). These are plotted in Fig. 1. All calculations were done using `maiprogs`, [20]. Static condensation of the internal degrees of freedom was realized with `pardiso`, [18, 24, 25].

In the hp-adaptive scheme, each adaptive step refines those 20% of the elements that have the largest error indicators (59). The decision of whether to perform an h-refinement or a p-enrichment is based on [7, Algorithm 5.1] with $\delta = 1$. The essential idea of that algorithm is similar to Strategy II of [11]: A p-enrichment for an element K can only be done if two conditions are met: (1) the coefficients of the Legendre expansion of the displacement field decay sufficiently rapidly and (2), if

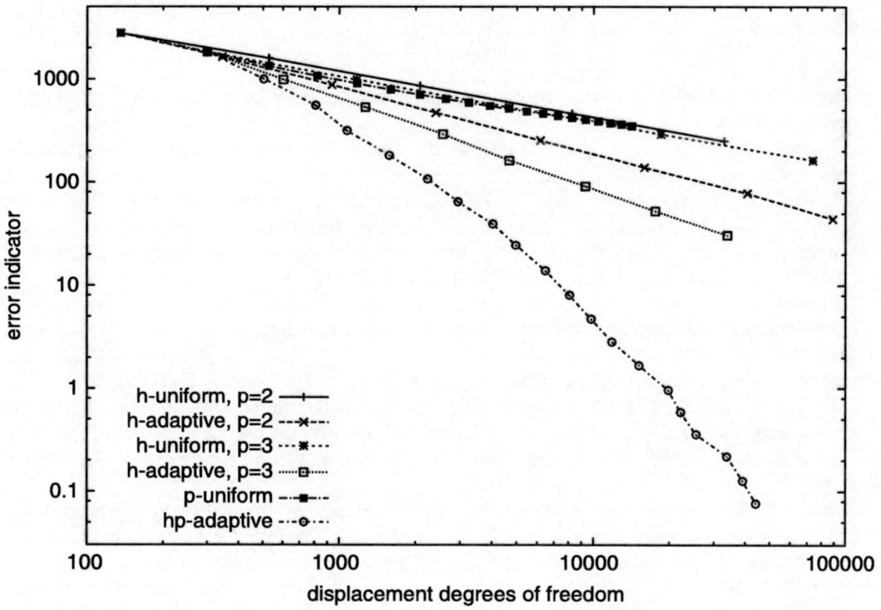

Fig. 1 Error indicator η_N vs. problem size

K has an edge E on the contact boundary Γ_C, then λ_N satisfies $\|\lambda_N\|_{L^\infty(E)} \leq 1$. This last condition $\|\lambda_N\|_{L^\infty(E)} \leq 1$ is strictly enforced by ensuring that an upper bound for $\|\lambda_N\|_{L^\infty(E)}$ is bounded by 1. This upper bound is obtained by expanding the polynomial $\lambda_N|_E$ into a Legendre series, computing the extrema of the leading quadratic part explicitly and estimating the remainder with the triangle inequality; we refer to [7] for details, where a similar strategy is employed in the context of a primal-dual formulation.

Figure 1 shows the error indicators for the two uniform and adaptive h-methods, the uniform p-method and the hp-adaptive method. Assuming that the error behaves like $\|\mathbf{u} - \mathbf{u}_N\|_{H^1(\Omega)} = CN^{-\alpha}$ in the uniform h- and p-versions and the adaptive h-versions, we obtain by a least squares fit rates of about $\alpha = 0.44$ for the h-uniform and $\alpha = 0.33$ for the p-uniform methods and about $\alpha = 0.64$ and $\alpha = 0.87$ for the adaptive schemes with polynomial degrees 2 and 3, respectively. For the hp-adaptivity, we obtain $\gamma = 0.35$, assuming an error behaviour of the form $\|\mathbf{u} - \mathbf{u}_N\|_{H^1(\Omega)} = C\exp(-\gamma N^{1/3})$.

Acknowledgements The first author gratefully acknowledges partial support by the FWF grant W8.

References

1. Bernardi, C., Maday, Y.: Approximations spectrales de problèmes aux limites elliptiques. Springer, Paris (1992)
2. Bernardi, C., Maday, Y.: Polynomial interpolation results in Sobolev spaces. J. Comput. Appl. Math. **43**(1–2), 53–80 (1992)
3. Bernardi, C., Dauge, M., Maday, Y.: Polynomials in the Sobolev World. Preprints of the Laboratories J.-L. Lions (2007). http://hal.archives-ouvertes.fr/hal-00153795/en/
4. Brezzi, F., Hager, W.W., Raviart, P.A.: Error estimates for the finite element solution of variational inequalities. Numer. Math. **28**(4), 431–443 (1977)
5. Chernov, A., Maischak, M., Stephan, E.P.: *hp*-mortar boundary element method for two-body contact problems with friction. Math. Methods Appl. Sci. **31**(17), 2029–2054 (2008). 10.1002/mma.1005
6. DeVore, R.A., Lorentz, G.G.: Constructive approximation. Springer, Berlin (1993)
7. Dörsek, P., Melenk, J.M.: Adaptive *hp*-FEM for the contact problem with Tresca friction in linear elasticity: The primal-dual formulation and a posteriori error estimation. Tech. Rep. 37/2009, Institute for Analysis and Scientific Computing, Vienna University of Technology (2009)
8. Dörsek, P., Melenk, J.M.: Adaptive *hp*-FEM for the contact problem with Tresca friction in linear elasticity: The primal formulation. Tech. Rep. 36/2009, Institute for Analysis and Scientific Computing, Vienna University of Technology (2009). http://www.asc.tuwien.ac.at
9. Dostál, Z., Schöberl, J.: Minimizing quadratic functions subject to bound constraints with the rate of convergence and finite termination. Comput. Optim. Appl. **30**(1), 23–43 (2005)
10. Duvaut, G., Lions, J.L.: Inequalities in mechanics and physics. Springer, Berlin (1976)
11. Eibner, T., Melenk, J.M.: An adaptive strategy for *hp*-FEM based on testing for analyticity. Comput. Mech. **39**(5), 575–595 (2007)
12. Glowinski, R.: Numerical methods for nonlinear variational problems. Springer, New York (1984)
13. Glowinski, R., Lions, J.L., Trémolières, R.: Numerical analysis of variational inequalities. vol. 8. North-Holland, Amsterdam (1981)
14. Guo, B., Heuer, N.: The optimal rate of convergence of the *p*-version of the boundary element method in two dimensions. Numer. Math. **98**(3), 499–538 (2004)
15. Gwinner, J.: On the p-version approximation in the boundary element method for a variational inequality of the second kind modelling unilateral contact and given friction. Appl. Numer. Math. (2008). 10.1016/j.apnum.2008.12.027
16. Han, W.: A posteriori error analysis via duality theory. Springer, New York (2005)
17. Hlaváček, I., Haslinger, J., Nečas, J., Lovíšek, J.: Solution of variational inequalities in mechanics. Springer, New York (1988)
18. Karypis, G., Kumar, V.: A fast and high quality multilevel scheme for partitioning irregular graphs. SIAM J. Sci. Comput. **20**(1), 359–392 (electronic) (1998)
19. Kikuchi, N., Oden, J.T.: Contact problems in elasticity: a study of variational inequalities and finite element methods. Society for Industrial and Applied Mathematics (SIAM), Philadelphia, PA (1988)
20. Maischak, M.: Manual of the software package maiprogs. Tech. Rep. 48, Institut für Angewandte Mathematik, Universität Hannover (2001)
21. Maischak, M., Stephan, E.P.: Adaptive *hp*-versions of BEM for Signorini problems. Appl. Numer. Math. **54**(3–4), 425–449 (2005)
22. Maischak, M., Stephan, E.P.: Adaptive *hp*-versions of boundary element methods for elastic contact problems. Comput. Mech. **39**(5), 597–607 (2007)
23. Melenk, J.M.: *hp*-interpolation of nonsmooth functions and an application to *hp*-a posteriori error estimation. SIAM J. Numer. Anal. **43**(1), 127–155 (electronic) (2005)
24. Schenk, O., Gärtner, K.: Solving unsymmetric sparse systems of linear equations with PARDISO. In: Computational science – ICCS 2002, Part II (Amsterdam), *Lecture Notes in Comput. Sci.*, vol. 2330, pp. 355–363. Springer, Berlin (2002)

25. Schenk, O., Gärtner, K.: On fast factorization pivoting methods for sparse symmetric indefinite systems. Electron. Trans. Numer. Anal. **23**, 158–179 (electronic) (2006)
26. Schwab, C.: *p*- and *hp*-finite element methods. Numerical Mathematics and Scientific Computation. The Clarendon Press Oxford University Press, New York (1998)
27. Tartar, L.: An introduction to Sobolev spaces and interpolation spaces, *Lecture Notes of the Unione Matematica Italiana*, vol. 3. Springer, Berlin (2007)

On Multivariate Chebyshev Polynomials and Spectral Approximations on Triangles

Brett N. Ryland and Hans Z. Munthe-Kaas

Abstract In this paper we describe the use of multivariate Chebyshev polynomials in computing spectral derivations and Clenshaw–Curtis type quadratures. The multivariate Chebyshev polynomials give a spectrally accurate approximation of smooth multivariate functions. In particular we investigate polynomials derived from the A_2 root system. We provide analytic formulas for the gradient and integral of A_2 bivariate Chebyshev polynomials. This yields triangular based Clenshaw–Curtis quadrature and spectral derivation algorithms with $\mathcal{O}(N \log N)$ computational complexity. Through linear and nonlinear mappings, these methods can be applied to arbitrary triangles and non-linearly transformed triangles. A MATLAB toolbox and a C++ library have also been developed for these methods.

1 Introduction

Classical Chebyshev polynomials are among the most important building blocks in approximation theory. We recall some of their beautiful and immensely useful properties:

- Polynomial interpolation in Chebyshev zeros and Chebyshev extremal points converges exponentially fast for analytic functions on $[-1, 1]$ and with super-algebraic speed for smooth functions.
- The Lebesgue constant for Chebyshev interpolation grows logarithmically in the number of interpolation points N.
- Chebyshev polynomials are orthogonal both with a continuous weighted inner product and also with discrete inner products based on Gauss-Chebyshev or Gauss-Chebyshev-Lobatto quadrature (nodes in Chebyshev zeros or Chebyshev extremal points).

Brett N. Ryland and Hans Z. Munthe-Kaas (✉)
Department of Mathematics, University of Bergen, Postbox 7803, 5020 Bergen, Norway
e-mail: nappers@gmail.com, hans.munthe-kaas@math.uib.no

J.S. Hesthaven and E.M. Rønquist (eds.), *Spectral and High Order Methods for Partial Differential Equations*, Lecture Notes in Computational Science and Engineering 76,
DOI 10.1007/978-3-642-15337-2_2,

- Chebyshev polynomial interpolation is equivalent to discrete Fourier cosine transform under a change of variables, thus all basic operations can be computed in $\mathcal{O}(N \log N)$ operations using FFT. This includes transforms between nodal values and expansion coefficients, mesh refinement and coarsening, products, integration and derivations.
- Integration and derivation of Chebyshev polynomial expansions can be done exactly. This leads to the highly accurate Clenshaw–Curtis quadrature and spectral Chebyshev derivation.

Chebyshev polynomials are also frequently used for multidimensional approximations. The standard approach is to construct multivariate polynomials as tensor products of univariate polynomials. However, this approach limits the application of multivariate Chebyshev approximations to rectangular and brick shaped domains, and domains that can be constructed from these by, e.g., spectral elements.

In this paper we will discuss families of multivariate Chebyshev polynomials obtained by an alternative construction, where tensor products of univariate polynomials appears as just one particular case. The construction is based on central ideas in group theory and representation theory. The construction was first done for particular cases by Koornwinder [12, 13] and later in full generality by Hoffman and Withers [10], see also [1, 5, 15]. Applications of these polynomials in numerical algorithms were discussed in [17]. For an introduction to the group theoretic background of this paper we refer to [6, 11].

This work is in particular motivated by the goal to construct spectral type discretisations on domains subdivided into triangles and simplices. Approximation theory on triangles have been discussed in various contexts, see [2–4, 7–9, 18, 20, 22]. Fast Fourier type transforms for the symmetric functions that appear in the context of multivariate Chebyshev polynomials are discussed in [16, 19].

The construction of multivariate Chebyshev polynomials starts by looking at periodic exponential functions through a kaleidoscope of mirrors acting on $\mathbb{R}^d$. More specifically we ask:

Which polytopes $S \subset \mathbb{R}^d$ generate a periodic tessellation of $\mathbb{R}^d$ under reflections about its faces?

Up to group isomorphisms there is just one such S for $d = 1$, four for $d = 2$ and seven for $d = 3$. For $d = 1$ we can take the domain $S = [0, \pi] \subset \mathbb{R}$, which generates a 2π periodic tessellation. For $d = 2$ the four possibilities are S being a rectangle or a triangle with 60°–60°–60°, 45°–45°–90° or 30°–60°–90° angles. In $d = 3$, the possible polytopes are prisms with base polygon being one of the four 2-d cases, as well as three particular tetrahedra. For each of these polytopes, there exists a family of multivariate Chebyshev polynomials, which are orthogonal on a domain which is the image of S under a certain change of variables. The classification of all these polytopes S, and thus also the classification of multivariate Chebyshev polynomials, is done in terms of Dynkin diagrams. This is a graph with d nodes, each node representing one mirror in $\mathbb{R}^d$. The nodes are not connected if the mirrors are orthogonal, connected with one edge if the mirrors meet at 60°, two edges if they meet at 45° and three edges for 30°. If two sets of nodes are totally disconnected, the mirrors form two subsets which are mutually orthogonal. In this case the corresponding polynomials become tensor products of each of the connected components. Thus, it is sufficient to classify only the *connected* Dynkin diagrams. The complete classification of connected Dynkin diagrams is shown.

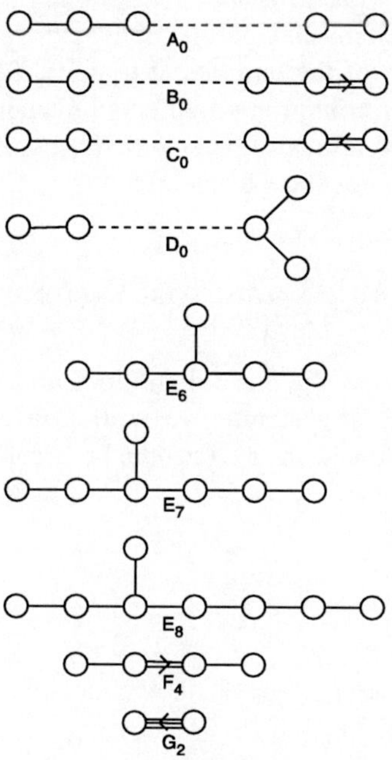

The classical univariate polynomials are represented by the diagram consisting of a single dot, A_1. Tensor product polynomials in d dimensions is represented as a diagram of d separate dots. The 2-d triangles $60°-60°-60°$, $45°-45°-90°$ and $30°-60°-90°$ are given by the diagrams A_2, B_2 and G_2.

Given such a domain S, we find a domain of periodicity P such that $S \subset P \subset \mathbb{R}^d$. The construction of multivariate Chebyshev polynomials goes as follows, exemplified by the classical case:

1. Take the Fourier basis functions on P. Classical: $S = [0, \pi]$, $P = [-\pi, \pi]$, Fourier basis $\exp(ik\theta)$.
2. Using reflection symmetries, fold the Fourier basis to symmetric functions on S. Classical: $\cos(k\theta) = \frac{1}{2}(\exp(ik\theta) + \exp(-ik\theta))$.
3. Use the symmetrised generators for the Fourier basis to change variables. This turns the symmetrised Fourier basis into Chebyshev polynomials on a transformed domain $\widetilde{S}$. Classical: $x = \frac{1}{2}(\exp(i\theta) + \exp(-i\theta)) = \cos(\theta)$, $\widetilde{S} = [-1, 1]$.

It should be remarked that these polynomials enjoy most of the beautiful properties of their univariate cousins. However, in the classical case the transformed domain $\widetilde{S}$ is still an interval, whereas in the general case S is typically a simplex or a product of simplices, e.g., a prism, while the transformed domain $\widetilde{S}$ is a more complicated domain which is usually non-convex and often has cusps in the corners. Coping with the shape of $\widetilde{S}$ is the main difficulty in the practical use of the multivariate Chebyshev polynomials.

This paper is organised as follows. In Sect. 2 we review basic properties of root systems and multivariate Chebyshev polynomials with special emphasis on the A_2 case related to symmetries of the equilateral triangle. Section 3 treats new spectrally accurate methods for computing gradients on triangles. Section 4 discusses Clenshaw–Curtis type quadratures for triangles. Lastly, in Sect. 5 we demonstrate the algorithms through numerical experiments.

2 Chebyshev Polynomials and Root Systems

In this section we will review the basic definitions and properties of multivariate Chebyshev polynomials. Root systems give explicit information about the reflections and translations discussed in the introduction, and are hence important for computational algorithms.

2.1 Root Systems

A root system Φ on a vector space V is a collection of vectors satisfying the following four conditions [11]:

i. Φ spans V
ii. If $\alpha \in \Phi$, then $c\alpha \in \Phi \iff c \in \{1, -1\}$
iii. If $\alpha \in \Phi$, then Φ is closed under the reflection $\sigma_\alpha = I - 2\frac{\alpha\alpha^T}{\alpha^T\alpha}$
iv. Integrality condition: $\alpha, \beta \in \Phi \implies < \beta, \alpha >= 2\frac{(\alpha,\beta)}{(\alpha,\alpha)} \in \mathbb{Z}$, where $(\alpha, \beta) = \alpha^T\beta$.

In one dimension, the only root system is given by $\Phi = \{\alpha, -\alpha\}$, where α is a non-zero vector. In two dimensions, there are four root systems, which are shown in Fig. 1. The first of these is a tensor product of the 1-dimensional root system, while the remainder are irreducible. A complete classification of irreducible root systems is given by the Dynkin diagrams A_n, B_n, C_n, D_n, E_6, E_7, E_8, F_4 and G_2.

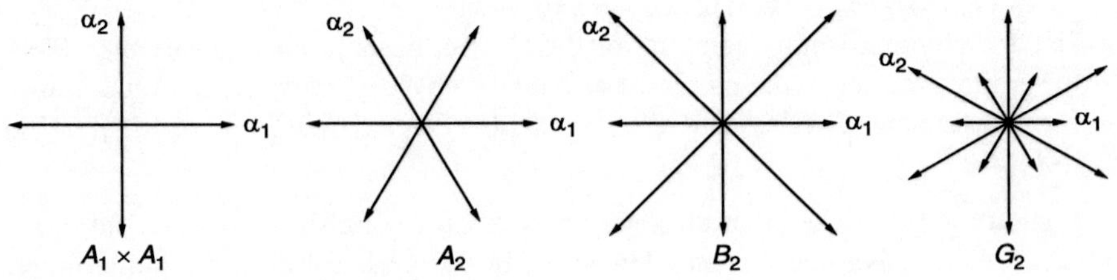

Fig. 1 Root systems in two dimensions

2.2 Multivariate Chebyshev Polynomials

Following [17], let Φ be a root system with a basis $\{\alpha_1, \dots, \alpha_d\}$ where the α_i are simple positive roots of Φ. Corresponding to this root system is the Weyl group $W = \{\sigma_\alpha | \alpha \in \Phi\}$, where σ_α is the reflection in the hyperplane orthogonal to α. Expressed in the basis $\{\alpha_j\}$, W is generated by the integer matrices

$$\widetilde{\sigma}_i = I - e_i e_i^T C, \qquad i = 1, \dots d,$$

where C is the Cartan matrix $C_{jk} = 2\frac{\alpha_j^T \alpha_k}{\alpha_j^T \alpha_j}$, I is the identity matrix, and $\{e_i\}$ is the standard basis on $\mathbb{R}^d$.

The *root lattice* is the set of all translations generated by the roots $\Lambda = \text{span}_{\mathbb{Z}}\{\alpha_1, \dots, \alpha_d\}$, and the *affine Weyl group* $\widetilde{W} = \Lambda \rtimes W$ is the group generated by both translations and reflections along the roots. The fundamental domain of $\widetilde{W}$ is a polytope $S \subset \mathbb{R}^d$ and the periodicity domain P is the parallelepiped spanned by the simple positive roots α_j.

In terms of the basis $\frac{1}{2\pi}\{\alpha_1/|\alpha_1|, \dots, \alpha_d/|\alpha_d|\}$, we identify P with the abelian group $G = (\mathbb{R}/2\pi\mathbb{Z})^d$, whose dual group is $\widehat{G} = \mathbb{Z}^d$. The Fourier basis for periodic functions is defined via the pairing

$$(k, \theta) := \mathrm{e}^{ik\cdot\theta},$$

where $k \in \widehat{G}$ and $\theta \in G$. Via the symmetries of the Weyl group, one can define the symmetrised pairing

$$(k, \theta)_s := \frac{1}{|W|}\sum_{g\in W}(k, g\theta) = \frac{1}{|W|}\sum_{g\in W}(g^T k, \theta),$$

where $|W|$ is the number of elements in the Weyl group W.

Note that in the same way that G can be recovered (up to periodicity) from the fundamental domain by application of elements from the Weyl group, one can define a dual fundamental domain (i.e., a fundamental domain in $\widehat{G}$) such that $\widehat{G}$ is recovered from the dual fundamental domain by application of the transpose of elements from the Weyl group.

We can now define the multivariate Chebyshev polynomials in the following way:

Definition 1. The multivariate Chebyshev polynomial of degree k is given by

$$T_k(z) := (k, \theta)_s, \tag{1}$$

where $z_j(\theta) = (e_j, \theta)_s$ for $j = 1, \dots, d$, and the e_j are the standard basis vectors in $\mathbb{R}^d$.

The multivariate Chebyshev polynomials are related to each other via the following relations

$$\begin{aligned} T_0 &= 1, \\ T_{e_j} &= z_j, \\ T_k &= T_{g^T k} \text{ for } g \in W, \\ T_{-k} &= \overline{T_k}, \\ T_k T_l &= \frac{1}{|W|} \sum_{g \in W} T_{k+g^T l} = \frac{1}{|W|} \sum_{g \in W} T_{l+g^T k}. \end{aligned} \tag{2}$$

These relations clearly show that the multivariate Chebyshev polynomials are indeed polynomials in the z_j.

A multivariate function may be expanded in an infinite weighted sum over multivariate Chebyshev polynomials,

$$f(z) = \sum_{k \in \widehat{G}} a(k) T_k(z),$$

where the coefficients $a(k)$ can be obtained via a Fourier transform, i.e.,

$$\begin{aligned} \sum_{k \in \widehat{G}} a(k) T_k(z) &= \frac{1}{|W|} \sum_{k \in \widehat{G}} a(k) \sum_{g \in W} (g^T k, \theta) \\ &= \frac{1}{|W|} \sum_{k \in \widehat{G}} \sum_{g \in W} a(g^T k)(g^T k, \theta) \\ &= \sum_{k \in \widehat{G}} a(k)(k, \theta) \end{aligned}$$

since $a(k) = a(g^T k)$ for all $g \in W$. Thus $a(k) = \widehat{f_s}(\theta)$, where $f_s(\theta)$ is the pull-back of $f(z)$ to the periodicity domain P. Note that $f_s(\theta)$ is a symmetric function, $f_s(g\theta) = f_s(\theta)$ for all $g \in W$.

To do numerical computations, we discretise P with a regular lattice and sample $f(z)$. It is essential that this lattice respects all the symmetries of $\widetilde{W}$. There are several ways to accomplish this. One possibility is to downscale the root lattice Λ by a factor N so that P contains an $N \times N$ grid. This grid is invariant under the action of $\widetilde{W}$. Thus we obtain a finite polynomial approximation

$$f(z) \approx P_N(z) = \sum_{k \in \widehat{G}_N} a_N(k) T_k(z), \tag{3}$$

where $\widehat{G}_N = (\mathbb{Z}/N\mathbb{Z})^d$ is the d-dimensional N-periodic integer lattice. If the approximating polynomial, $P_N(z)$, is evaluated at the set of points $\{z^* = z(\theta_j)\}$, where $j \in \widehat{G}_N$ and $\theta_j = 2\pi j/N \in \mathbb{R}^d$, then the expansion of $P_N(z)$ is

$$P_N(z(\theta_j)) = \sum_{k_d=0}^{N-1} \cdots \sum_{k_1=0}^{N-1} a_N(k) \exp(ik \cdot \theta_j),$$

which is simply an unnormalised d-dimensional inverse discrete Fourier transform. Thus, the $a_N(k)$ are given by

$$a_N(k) = \frac{1}{N^d} \sum_{j_d=0}^{N-1} \cdots \sum_{j_1=0}^{N-1} P_N(z(\theta_j)) \exp(ik \cdot \theta_j), \tag{4}$$

which can be computed quickly via the d-dimensional fast Fourier transform $\mathscr{F}_d$,

$$\{a_N(k)\} = \frac{1}{N^d} \mathscr{F}_d \{P_N(z^*)\}.$$

In the sequel, we omit the subscript N and write just $a(k) \equiv a_N(k)$.

2.3 *The A_2 Root System*

The Weyl group for the A_2 root system is

$$W = \left\{ \begin{bmatrix} 1 & 0 \\ 0 & 1 \end{bmatrix}, \begin{bmatrix} -1 & 1 \\ 0 & 1 \end{bmatrix}, \begin{bmatrix} 1 & 0 \\ 1 & -1 \end{bmatrix}, \begin{bmatrix} -1 & 1 \\ -1 & 0 \end{bmatrix}, \begin{bmatrix} 0 & -1 \\ 1 & -1 \end{bmatrix}, \begin{bmatrix} 0 & -1 \\ -1 & 0 \end{bmatrix} \right\}. \tag{5}$$

Here the fundamental domain is that of the $\widetilde{A}_2$ affine Weyl group, which is an equilateral triangle in the plane (see Fig. 2), and the dual fundamental domain is the set of points in Fourier space given by $\{k \in (\mathbb{Z}/N\mathbb{Z})^2 | k_2 \le k_1, 2k_1 + k_2 \le N\}$.

With this Weyl group, (2) gives rise to the following recursion formulas

$$T_{0,0} = 1, \quad T_{1,0} = z_1, \quad T_{0,1} = z_2 = T_{-1,0} = \overline{z_1},$$
$$T_{n,0} = 3z_1 T_{n-1,0} - 3z_2 T_{n-2,0} + T_{n-3,0},$$
$$T_{n,m} = \frac{1}{2}(3T_{n,0}\overline{T_{m,0}} - T_{n-m,0}).$$

However, it is more convenient, in practice, to work in the real-valued coordinates

$$x_1 = \frac{1}{2}(z_1 + z_2) = \frac{1}{3}(\cos(\theta_1) + \cos(\theta_2) + \cos(\theta_1 - \theta_2)),$$
$$x_2 = \frac{1}{2i}(z_1 - z_2) = \frac{1}{3}(\sin(\theta_1) - \sin(\theta_2) - \sin(\theta_1 - \theta_2)).$$

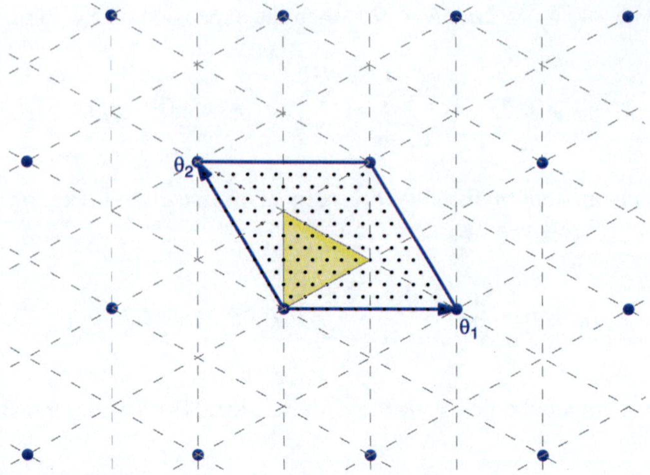

Fig. 2 The A_2 root system showing fundamental domain (*yellow triangle*), translation group (*blue hexagonal lattice*) and downscaled lattice (*small black dots*) for the downscaling factor $N = 12$. The *blue arrows* indicate the θ coordinates

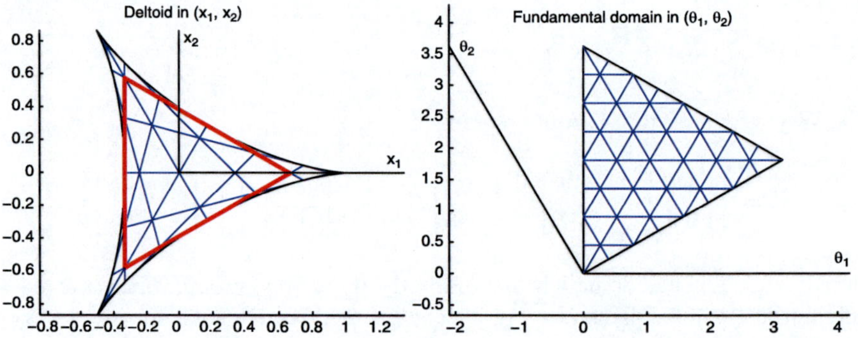

Fig. 3 $N = 12$ equally spaced tangent lines to the deltoid in x and θ coordinates. The points $\{x^*(\theta)\}$ lie at the intersection of these lines, cf. Fig. 2

Clearly, the multivariate Chebyshev polynomials are also polynomials in these coordinates.

While the multivariate Chebyshev polynomials are defined in a unit cell of the lattice, they exist in θ coordinates as multiple images (not necessarily whole) of the fundamental domain. For the A_2 root system, this is the equilateral triangle on the right of Fig. 3. However, in the x coordinates, this triangle is mapped to a deltoid as shown on the left of Fig. 3.

Lemma 1. *The Lebesgue constant $\lambda(N)$ for the points x^* grows as $\mathcal{O}((\log(N))^2)$.*

In higher dimensions the Lebesgue constant grows as $\mathcal{O}((\log(N))^d)$. The proof of this result relies on properties of the multidimensional Dirichlet kernel, see also [14].

3 Computing Gradients

Let us consider the gradient of the approximating function $P_N(z)$,

$$\nabla_z f(z) \approx \nabla_z P_N(z) = \sum_{k \in \widehat{G}_N} a(k) \nabla_z T_k(z). \tag{6}$$

For the univariate Chebyshev polynomials, the derivative of the approximating polynomial can be written exactly as a weighted sum over Chebyshev polynomials via the relation,

$$\partial_z P_N(z) = \sum_{k=0}^{N} a(k) \partial_z T_k(z) = \sum_{k=0}^{N-1} b(k) T_k(z), \tag{7}$$

where the $b(k)$ are calculated recursively by

$$b(k-1) = b(k+1) + 2ka(k) \qquad \text{for } k = N-1, \dots, 1, \tag{8}$$

with $b(N+1) = b(N) = 0$. This recursion formula can be obtained by substituting the relation

$$2T_k(z) = \frac{\partial_z T_{k+1}(z)}{k+1} - \frac{\partial_z T_{k-1}(z)}{k-1},$$

into (7) and matching terms.

In the general multivariate case the coefficients of the gradient are vectors $b(k) \in \mathbb{C}^d$. It can be shown that these satisfy the recursion

$$|W| k a(k) = \sum_{l=1}^{d} \sum_{\gamma \in W} b(k - \gamma^T e_l)_l (\gamma^T e_l). \tag{9}$$

The $b(m)$ in (9) can be obtained by setting $b(M) = 0$ and $a(M) = 0$ for M outside the dual fundamental domain and iteratively determining the $b(m)$ as m approaches the origin.

3.1 Gradients in the A_2 Root System

As mentioned earlier, we prefer to work in the real-valued (x_1, x_2) coordinates when dealing with the A_2 root system. The effect of this is that the gradient of the approximating polynomial becomes

$$\nabla_x P_N(x) = J_x(z)^{-T} J_z(\theta)^{-T} \sum_{k \in \widehat{G}_N} a(k) \nabla_\theta T_k(\theta) = J_x(z)^{-T} \sum_{m \in \widehat{G}_N} b(m) T_m(\theta), \tag{10}$$

where $J_z(\theta)$ is the Jacobian of the transformation between the θ and z domains, and

$$J_x(z)^{-T} = \begin{bmatrix} 1 & 1 \\ \mathrm{i} & -\mathrm{i} \end{bmatrix}.$$

The recursion formula (9) can then be used to determine the $b(m)$ from the $a(k)$ via the following 3 steps:

1. Sort the k in the dual fundamental domain by $k_1 + k_2 = c$ for $c = 0, 1, \ldots$, and with increasing k_2 for each c (see Fig. 4).
2. Assume that $b(k) = 0$ for k not in the dual fundamental domain and apply the stencils in Fig. 5 to each of the k in the fundamental domain in the reverse order to the sorting in step 1.
   - If $k_2 > 0$, use stencil (a) to obtain $b(k_1, k_2)_1$:

     $$b(k_1, k_2)_1 = 3(k_1+1)a(k_1+1, k_2) + b(k_1 + 2, k_2 - 1)_1 + b(k_1 + 2, k_2)_2 \\ -b(k_1, k_2 + 1)_2$$

     otherwise, use stencil (b):

     $$b(k_1, 0)_1 = 3(k_1 + 1)a(k_1 + 1, 0) + b(k_1 + 1, 1)_1 + b(k_1 + 2, 0)_2 - b(k_1, 1)_2.$$

   - If $k_1 > 0$, use stencil (c) to obtain $b(k_1, k_2)_2$:

     $$b(k_1, k_2)_2 = 3(k_2 + 1)a(k_1, k_2 + 1) + b(k_1 - 1, k_2+2)_2 + b(k_1, k_2+2)_1 \\ -b(k_1 + 1, k_2)_1$$

     otherwise use stencil (d):

     $$b(0, 0)_2 = 3a(0, 1) + b(1, 1)_2 + b(0, 2)_1 - b(1, 0)_1.$$

   - Apply the conjugate symmetry:

     $$b(k_2, k_1)_1 = \overline{b(k_1, k_2)_2}, \\ b(k_2, k_1)_2 = \overline{b(k_1, k_2)_1}.$$

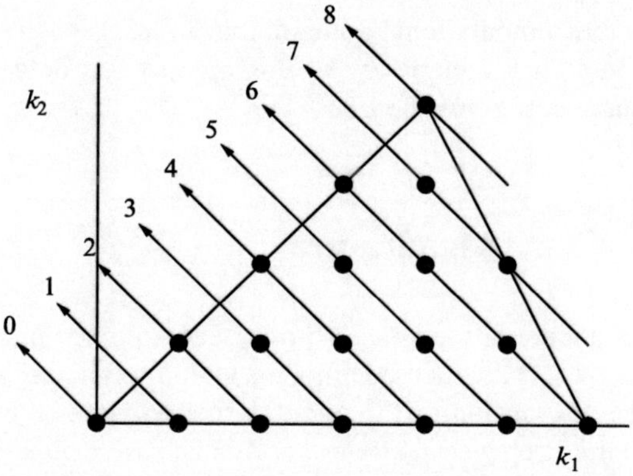

Fig. 4 Sorted dual fundamental domain for the A_2 root system with $N = 12$

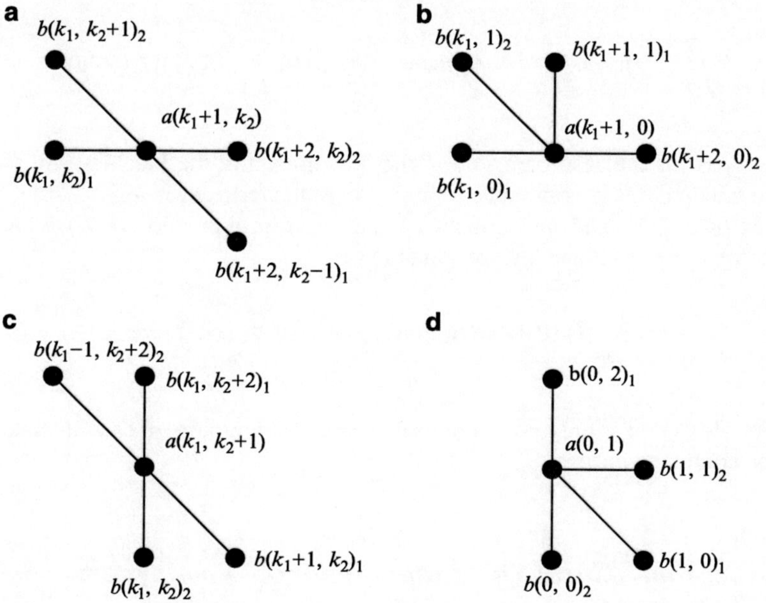

Fig. 5 Stencils for use in step 2

3. Apply the symmetries from the symmetry group W to obtain the rest of the coefficients:

$$b(g^T k) = b(k) \quad \text{for } g \in W.$$

By using this recursion formula, one can calculate the gradient of a function in $\mathcal{O}(N^2 + \alpha(N^2 \log N^2))$ time, where α indicates the amount of the time spent performing the fast Fourier transform and its inverse. Note that there are $\mathcal{O}(N^2)$

sample points in the computational domain, thus in terms of the number of sample points, the computational complexity of this approach is only marginally greater than linear, the main cost being the FFT.

4 Clenshaw–Curtis Quadrature

Clenshaw–Curtis quadrature is a well-known technique for univariate integration. For a function on $[-1, 1]$ the method amounts to approximating a given function by a finite Chebyshev expansion and integrating this polynomial exactly. For smooth functions this method behaves nearly as good as Gaussian quadrature [21].

As with the univariate Chebyshev polynomials, the integral of a multivariate function may be evaluated rapidly with a multivariate Clenshaw–Curtis quadrature technique.

$$\int_{\Omega_z} f(z)\mathrm{d}z \approx \int_{\Omega_z} P_N(z)\mathrm{d}z = \sum_{k\in\widehat{G}_N} a(k) \int_{\Omega_\theta} T_k(\theta)|J_z(\theta)|\mathrm{d}\theta, \tag{11}$$

where $|J_z(\theta)|$ is the absolute value of the determinant of the Jacobian and Ω_z and Ω_θ are the fundamental domains in z and θ coordinates respectively.

Since both $T_k(\theta)$ and the Jacobian $J_z(\theta)$ have terms of the form (k, θ) as building blocks, we can expand the integrals in (11) as

$$\int_{\Omega_\theta} T_k(\theta)|J_z(\theta)|\mathrm{d}\theta = \sum_{\kappa\in\widehat{G}_N} b(\kappa) \int_{\Omega_\theta} (\kappa, \theta)\mathrm{d}\theta, \tag{12}$$

for some $b(\kappa)$ that is only non-zero near $\kappa = k$. Furthermore, these integrals need only be computed once.

4.1 Clenshaw–Curtis Quadrature in the A_2 Root System

For the A_2 root system, the determinant of the Jacobian $J_x(\theta)$ is

$$\begin{aligned}\Gamma &= \frac{1}{9}(\sin(\theta_1+\theta_2) + \sin(\theta_1 - 2\theta_2) - \sin(2\theta_1 - \theta_2)) \\ &= -\frac{\mathrm{i}}{2}\cdot\frac{1}{9}\Big(\Big(\begin{bmatrix}1\\1\end{bmatrix},\theta\Big) - \Big(\begin{bmatrix}-1\\-1\end{bmatrix},\theta\Big) + \Big(\begin{bmatrix}1\\-2\end{bmatrix},\theta\Big) \\ &\quad - \Big(\begin{bmatrix}-1\\2\end{bmatrix},\theta\Big) + \Big(\begin{bmatrix}-2\\1\end{bmatrix},\theta\Big) - \Big(\begin{bmatrix}2\\-1\end{bmatrix},\theta\Big)\Big),\end{aligned}$$

which is zero on the boundary of the fundamental domain and negative within it. Furthermore, the orientation of the deltoid in the x coordinates is opposite to that of the fundamental domain in θ coordinates. Thus, (12) becomes

$$\int_{\Omega_\theta} T_k(\theta)|J_x(\theta)|\mathrm{d}\theta = \frac{\mathrm{i}}{2}\cdot\frac{1}{9}\cdot\frac{1}{|W|}\sum_{\kappa\in S}\delta_\kappa\int_{\Omega_\theta}(\kappa,\theta)\mathrm{d}\theta \tag{13}$$

for the set S consisting of $g^T k + l$, where $g \in W$,

$$l \in \left\{\begin{bmatrix}1\\1\end{bmatrix},\begin{bmatrix}-1\\2\end{bmatrix},\begin{bmatrix}2\\-1\end{bmatrix},\begin{bmatrix}-2\\1\end{bmatrix},\begin{bmatrix}1\\-2\end{bmatrix},\begin{bmatrix}-1\\-1\end{bmatrix}\right\},$$

and

$$\delta_\kappa = \begin{cases} 1 & \text{if } l \in \left\{\begin{bmatrix}1\\1\end{bmatrix},\begin{bmatrix}1\\-2\end{bmatrix},\begin{bmatrix}-2\\1\end{bmatrix}\right\},\\ -1 & \text{if } l \in \left\{\begin{bmatrix}-1\\-1\end{bmatrix},\begin{bmatrix}-1\\2\end{bmatrix},\begin{bmatrix}2\\-1\end{bmatrix}\right\}.\end{cases}$$

The fundamental domain in θ coordinates is the region bounded by

$$\theta_1 + \theta_2 = 2\pi,$$
$$\theta_1 = 2\theta_2,$$
$$\theta_2 = 2\theta_1.$$

Therefore, the integral $\int_{\Omega_\theta}(\kappa,\theta)\mathrm{d}\theta$ becomes

$$\int_{\Omega_\theta}(\kappa,\theta)\mathrm{d}\theta = \int_0^{\frac{2\pi}{3}}\int_{\frac{1}{2}\theta_2}^{2\theta_2}(\kappa,\theta)\mathrm{d}\theta_1\mathrm{d}\theta_2 + \int_{\frac{2\pi}{3}}^{\frac{4\pi}{3}}\int_{\frac{1}{2}\theta_2}^{2\pi-\theta_2}(\kappa,\theta)\mathrm{d}\theta_1\mathrm{d}\theta_2,$$

which evaluates to

$$\int_{\Omega_\theta}(\kappa,\theta)\mathrm{d}\theta = \begin{cases} \frac{2\pi^2}{3} & \text{if } \kappa_1=\kappa_2=0,\\ \frac{4\pi\mathrm{i}}{3\kappa_1} & \text{if } \kappa_1\neq 0, \kappa_2+\frac{1}{2}\kappa_1=0,\\ -\frac{2\pi\mathrm{i}}{3\kappa_1} & \text{if } \kappa_1\neq 0, \kappa_2+2\kappa_1=0,\\ -\frac{2\pi\mathrm{i}}{3\kappa_1} & \text{if } \kappa_1\neq 0, \kappa_2-\kappa_1=0,\\ \frac{3}{2\kappa_2^2}(-(\kappa_2,\frac{4\pi}{3})+2(\kappa_2,\frac{2\pi}{3})-1) & \text{if } \kappa_1=0, \kappa_2\neq 0,\\ -\frac{1}{\kappa_1}(\frac{1}{\kappa_2+2\kappa_1}((\kappa_2+2\kappa_1,\frac{2\pi}{3})-1) & \\ \quad -\frac{1}{\kappa_2+\frac{1}{2}\kappa_1}((\kappa_2+\frac{1}{2}\kappa_1,\frac{4\pi}{3})-1) & \\ \quad +\frac{1}{\kappa_2-\kappa_1}(\kappa_2-\kappa_1,\frac{2\pi}{3})((\kappa_2-\kappa_1,\frac{2\pi}{3})-1)) & \text{otherwise.} \end{cases} \tag{14}$$

Note that due to the symmetries in $\widehat{G}_N$, one need only evaluate $\int_{\Omega_\theta} T_k(\theta)|J_x(\theta)|\mathrm{d}\theta$ for values of k in the dual fundamental domain. The full integral (11) can then be computed by summing over just the k lying in the dual fundamental domain and weighting each $a(k)$ by the size of the orbit of k under the group action of W. Curiously, due to the symmetries of $T_k(\theta)$ and $|J_x(\theta)|$, we have the following lemma.

Lemma 2. *The integral* $\int_{\Omega_\theta} T_k(\theta)|J_x(\theta)|\mathrm{d}\theta$ *is zero unless* $k_1 = k_2$ *or* $|k_1 - k_2| = 3$.

Proof. First, let us rewrite (13) as

$$\begin{aligned}\int_{\Omega_\theta} T_k(\theta)|J_x(\theta)|\mathrm{d}\theta &= \frac{\mathrm{i}}{18|W|}\sum_{\kappa\in S}\delta_\kappa \int_{\Omega_\theta}(\kappa,\theta)\mathrm{d}\theta \\ &= \frac{\mathrm{i}}{18|W|}\sum_{m}\delta_m \sum_{g\in W}\delta_g \int_{\Omega_\theta}(g^T m,\theta)\mathrm{d}\theta,\end{aligned}$$

where

$$m \in k + \eta^T l = \left\{\begin{bmatrix}k_1+1\\k_2+1\end{bmatrix},\begin{bmatrix}k_1-1\\k_2+2\end{bmatrix},\begin{bmatrix}k_1-2\\k_2+1\end{bmatrix},\begin{bmatrix}k_1-1\\k_2-1\end{bmatrix},\begin{bmatrix}k_1+1\\k_2-2\end{bmatrix},\begin{bmatrix}k_1+2\\k_2-1\end{bmatrix}\right\}$$

and $\delta_m \in \{1,-1,1,-1,1,-1\}$ respectively. The set $S = g^T m$ is then given by

$$S = \left\{\begin{bmatrix}m_1\\m_2\end{bmatrix},\begin{bmatrix}-m_1\\m_1+m_2\end{bmatrix},\begin{bmatrix}-m_1-m_2\\m_1\end{bmatrix},\begin{bmatrix}-m_2\\-m_1\end{bmatrix},\begin{bmatrix}m_2\\-m_1-m_2\end{bmatrix},\begin{bmatrix}m_1+m_2\\-m_2\end{bmatrix}\right\} \tag{15}$$

with $\delta_g \in \{1,-1,1,-1,1,-1\}$ respectively, such that $\delta_\kappa = \delta_m\delta_g$.

Note that if $k_1 - k_2 \in \{-3,0,3\}$, then for some $\kappa \in S$ we have that $\kappa_1 = \kappa_2$, $\kappa_1 + 2\kappa_2 = 0$ or $2\kappa_1 + \kappa_2 = 0$. In this case, the first four lines of (14) play a part in the sum and the integral can be non-zero. On the other hand, if $k_1 - k_2 \notin \{-3,0,3\}$, then none of the first four lines of (14) play a part in the sum.

Let us consider the reduced sum

$$\sum_{g\in W}\delta_g \int_{\Omega_\theta}\left(g^T m,\theta\right)\mathrm{d}\theta. \tag{16}$$

Now, if one of the $\kappa \in g^T m$ in this sum is of the form $\kappa_1 = 0$, $\kappa_2 \neq 0$, then it can be easily seen from (15) that S reduces to three distinct pairs, each of which have δ_g being of opposite sign. Thus, the terms in (16) coming from the fifth line of (14) identically cancel.

It just remains to consider (16) where all of the integrals $\int_{\Omega_\theta}(g^T m,\theta)\mathrm{d}\theta$ are evaluated by the last line of (14). Directly evaluating this sum, one finds that after some simple but tedious manipulation, (16) reduces to

$$\sum_{g\in W} \delta_g \int_{\Omega_\theta} \left(g^T m, \theta\right) \mathrm{d}\theta = \frac{4\left(m_1^2 + m_1 m_2 + m_2^2\right)^2 \left(\overline{\alpha} - \alpha^2\right)}{m_1 m_2 \left(m_2^2 - m_1^2\right)\left(m_1 + 2m_2\right)\left(m_2 + 2m_1\right)},$$

where $\alpha = \left(\kappa_2 - \kappa_1, \frac{2\pi}{3}\right) = \left(\kappa_2 + 2\kappa_1, \frac{2\pi}{3}\right) = \overline{\left(\kappa_2 + \frac{1}{2}\kappa_1, \frac{4\pi}{3}\right)}$ and $\overline{\alpha} - \alpha^2 = 0$ since $\kappa_2 - \kappa_1 \in \mathbb{Z}$ for all k.

Thus, the integral $\int_{\Omega_\theta} T_k(\theta)|J_x(\theta)|\mathrm{d}\theta$ over the deltoid is zero unless $k_1 = k_2$ or $|k_1 - k_2| = 3$. □

5 Triangles

The reader is no doubt aware that deltoids are not, in general, good shapes for decomposing surfaces into. Rather, it is much more desirable to decompose a surface into a number of triangles, which can then be mapped to a standard triangle either with linear or non-linear maps.

The difficulty now arises as to how to apply the multivariate Chebyshev polynomials (which naturally live on the deltoid) to such a triangle. We envisage two possible methods of doing this:

1. Stretch the deltoid to the triangle with corners at the corners of the deltoid.
2. Use the equilateral triangle that is inscribed in the deltoid.

Method (1) is appealing, since all of the data points in the deltoid lie within the triangle. Straightening maps are discussed in [17]. However, due to the shape of the deltoid, with its singularities at the corners, we are not able to straighten the deltoid to a triangle without compromising spectral convergence.

On the other hand, method (2) has the advantage that it does not require any further mappings (the gradient algorithm can be used directly and the Clenshaw–Curtis quadrature algorithm can be used with only a small modification to the integral of (k, θ) over the fundamental domain). However, this method makes use of the data points that lie outside the triangle to obtain the coefficients $a(k)$. This has implications for the use of these algorithms in spectral and spectral element methods on domains with boundaries, which will be discussed in a later paper.

5.1 *Clenshaw–Curtis Quadrature Over a Triangle*

Since we are restricting the domain of integration, Ω_x, to the equilateral triangle inscribed within the deltoid (red triangle in Fig. 3), the integrals $\int_{\Omega_\theta}(\kappa, \theta)\mathrm{d}\theta$ in Sect. 4 must be modified. They become

$$\int_{\Omega_\theta} (\kappa, \theta)\mathrm{d}\theta = \int_{\frac{\pi}{3}}^{\frac{2\pi}{3}} \int_{\frac{\pi}{3}}^{\theta_2 + \frac{\pi}{3}} (\kappa, \theta)\mathrm{d}\theta_1 \mathrm{d}\theta_2 + \int_{\frac{2\pi}{3}}^{\pi} \int_{\theta_2 - \frac{\pi}{3}}^{\pi} (\kappa, \theta)\mathrm{d}\theta_1 \mathrm{d}\theta_2,$$

where Ω_θ is the restriction of the fundamental domain to the equilateral triangle in θ coordinates.

As with the integral over the deltoid, this integral can be evaluated directly to give

$$\int_{\Omega_\theta} (\kappa,\theta)\mathrm{d}\theta = \begin{cases} \frac{\pi^2}{3} & \text{if } \kappa_1=\kappa_2=0, \\ \frac{\mathrm{i}\pi}{3\kappa_2}((\kappa_2,\frac{\pi}{3})-(\kappa_2,\pi))+\frac{1}{\kappa_2^2}(2(\kappa_2,\frac{2\pi}{3})-(\kappa_2,\pi)-(\kappa_2,\frac{\pi}{3})) & \text{if } \kappa_1=0, \kappa_2\neq 0, \\ \frac{\mathrm{i}\pi}{3\kappa_1}((\kappa_1,\frac{\pi}{3})-(\kappa_1,\pi))+\frac{1}{\kappa_1^2}(2(\kappa_1,\frac{2\pi}{3})-(\kappa_1,\pi)-(\kappa_1,\frac{\pi}{3})) & \text{if } \kappa_2=0, \kappa_1\neq 0, \\ \frac{2}{\kappa_1^2}(1-\cos(\kappa_1\frac{\pi}{3}))+\frac{2\pi}{3\kappa_1}\sin(\kappa_1\frac{\pi}{3}) & \text{if } \kappa_1+\kappa_2=0, \\ \frac{1}{\kappa_1\kappa_2}((\kappa_1,\frac{\pi}{3})((\kappa_2,\frac{2\pi}{3})-(\kappa_2,\frac{\pi}{3}))-(\kappa_1,\pi)((\kappa_2,\pi)-(\kappa_2,\frac{2\pi}{3}))) & \\ \quad +\frac{1}{\kappa_1(\kappa_2+\kappa_1)}((\kappa_1,\frac{2\pi}{3})((\kappa_2,\pi)+(\kappa_2,\frac{\pi}{3})) & \\ \quad -(\kappa_2,\frac{2\pi}{3})((\kappa_1,\pi)+(\kappa_1,\frac{\pi}{3}))) & \text{otherwise.} \end{cases} \tag{17}$$

Again, the symmetries of $T_k(\theta)$ and $|J_x(\theta)|$ restrict the values of k for which the integral $\int_{\Omega_\theta} T_k(\theta)|J_x(\theta)|\mathrm{d}\theta$ is non-zero. We obtain the following lemma.

Lemma 3. *The integral* $\int_{\Omega_\theta} T_k(\theta)|J_x(\theta)|\mathrm{d}\theta$ *is zero unless* $k_1 - k_2 = 0 \pmod 3$.

Proof. As with the proof of Lemma 2, we consider the reduced sum

$$\sum_{g\in W} \delta_g \int_{\Omega_\theta} (g^T m,\theta)\mathrm{d}\theta, \tag{18}$$

where m and δ_g are given in Lemma 2 and the integrals are evaluated by (17).

If $m_1 = 0, m_2 = 0$ or $m_1 + m_2 = 0$, then it can be easily seen from (15) that the $g^T m$ occur in pairs with δ_g being of opposite sign. Thus the contributions to (18) from the first four lines of (17) identically cancel with each other and we need only consider the contributions coming from the last line of (17).

Directly evaluating (18) where the integrals are evaluated by the last line of (17) gives

$$\sum_{g\in W} \delta_g \int_{\Omega_\theta} (g^T m,\theta)\mathrm{d}\theta = \frac{2\mathrm{i}(m_1\alpha + m_2\beta)}{m_1 m_2(m_1+m_2)},$$

where

$$\begin{aligned}\alpha = &-\sin(m_1\pi/3) + \sin((m_1-2m_2)\pi/3) + \sin((m_2+3m_1)\pi/3) \\ &+ \sin((m_1+3m_2)\pi/3) - \sin((m_1+m_2)\pi/3) + \sin((3m_1+2m_2)\pi/3)\end{aligned}$$

and

$$\begin{aligned}\beta = &-\sin(m_2\pi/3) + \sin((m_2-2m_1)\pi/3) + \sin((m_1+3m_2)\pi/3) \\ &+ \sin((m_2+3m_1)\pi/3) - \sin((m_2+m_1)\pi/3) + \sin((3m_2+2m_1)\pi/3).\end{aligned}$$

Now, let $a = m_2 - m_1$, such that α and β become

$$\alpha = -\sin(m_1\pi/3)\,(1 + 2\cos(2a\pi/3)) - \sin(2m_1\pi/3)\,(\cos(a\pi) + 2\cos(a\pi/3))$$

and

$$\beta = -\sin(m_2\pi/3)\,(1 + 2\cos(2a\pi/3)) - \sin(2m_2\pi/3)\,(\cos(a\pi) + 2\cos(a\pi/3))\,,$$

which are only non-zero if $a = 0 \pmod 3$.

Since $m_1 - m_2 = k_1 - k_2 \pmod 3$, we find that the integral $\int_{\Omega_\theta} T_k(\theta)|J_x(\theta)|\mathrm{d}\theta$ over the triangle is only non-zero if $k_1 - k_2 = 0 \pmod 3$. □

5.2 *Nonlinear Transformations*

Given a nonlinear mapping $\phi : \mathbb{R}^2 \mapsto \mathbb{R}^2$ such that $y = \phi(x)$, the Jacobian of this map can be calculated numerically at each point $y(x)$ as

$$J_y(x) = \begin{bmatrix} \frac{\partial y_1}{\partial x_1} & \frac{\partial y_1}{\partial x_2} \\ \frac{\partial y_2}{\partial x_1} & \frac{\partial y_2}{\partial x_2} \end{bmatrix}$$

using the gradient algorithm of Sect. 3.1.

The gradient of a function $f(y)$ on the nonlinear triangle can then be obtained by multiplying the gradient obtained from the gradient algorithm by the inverse transpose of the Jacobian at each point in the domain. i.e.,

$$\nabla_y f(y) = J_y(x)^{-T}\nabla_x f(x). \tag{19}$$

Integrals using Clenshaw–Curtis quadrature can also be performed on nonlinear triangles. This is done by transforming the integral back to the x domain and performing the integration there. For example,

$$\int_{\Omega_y} F(f(y), \nabla_y f(y))\mathrm{d}y = \int_{\Omega_x} |J_y(x)| F(f(\phi(x)), J_y(x)^{-T}\nabla_y f(\phi(x)))\mathrm{d}x \tag{20}$$

where Ω_y is the nonlinear triangle and Ω_x is the equilateral triangle (see Fig. 6). This integral can then be evaluated by Clenshaw–Curtis quadrature and numerical gradient computations.

The computational cost of computing the Jacobian of the map at each point in the domain is approximately twice the cost of computing a gradient. However, for a fixed mapping the Jacobian can be precomputed. The further computational costs associated with the above modifications to the gradient and Clenshaw–Curtis quadrature are also $\mathcal{O}(N^2)$ and only marginally increase the running time of the algorithms.

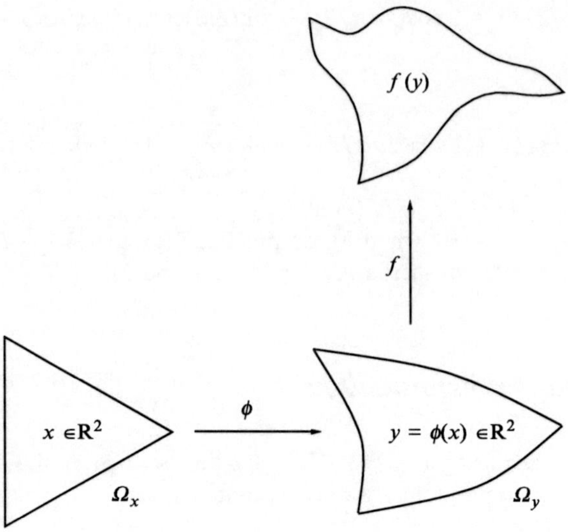

Fig. 6 A function $f(y)$ is defined on a nonlinear triangle Ω_y, which is mapped from the equilateral triangle Ω_x by the map ϕ

5.3 Linear Transformations

If the mapping ϕ happens to be a linear mapping, then the gradient and Clenshaw–Curtis quadrature techniques for the transformed triangles simplify as the Jacobian $J_y(x)$ is a constant matrix for such transformations.

Thus, when calculating the gradient, the inverse transpose of the Jacobian need only be calculated once, however, it must still be applied to each point in the domain. Furthermore, in Clenshaw–Curtis quadrature, the constant factor of $|J_y(x)|$ in the integral of $P_N(y)$ can be applied after the integral of $P_N(x)$ is calculated. That is, (20) becomes

$$\int_{\Omega_y} F(f(y), \nabla_y f(y))\mathrm{d}y = |J_y(x)| \int_{\Omega_x} F(f(\phi(x)), J_y(x)^{-T} \nabla_y f(\phi(x)))\mathrm{d}x \tag{21}$$

These simplifications provide a marginal improvement in the running times of the gradient and Clenshaw–Curtis algorithms.

6 Numerics

In this section, we show timing and convergence results for the gradient and Clenshaw–Curtis quadrature algorithms using the test function $\exp(\sin(y_1)\sin(y_2))$ on the triangle with corners $\{(0,0),(0,1),(1,0)\}$. We also show timing and

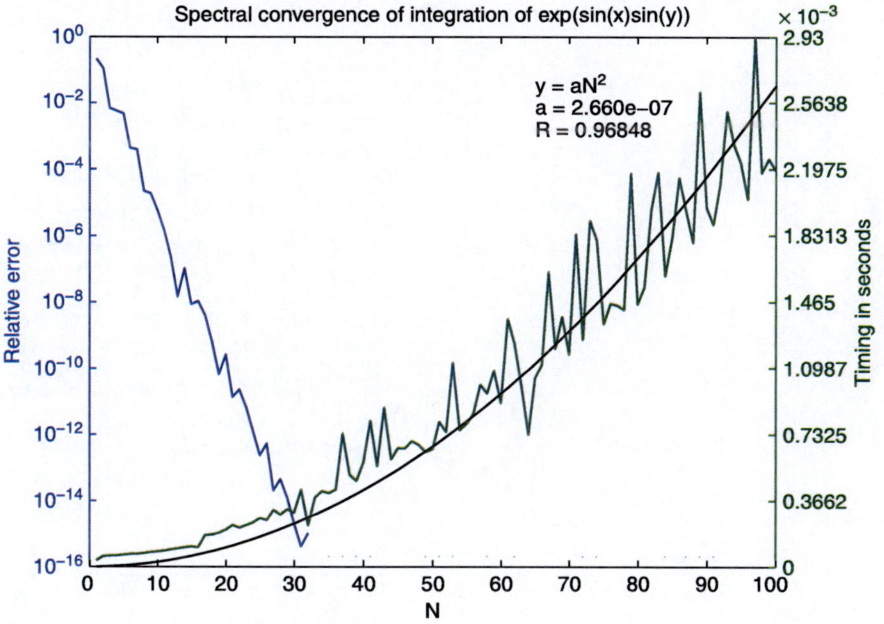

Fig. 7 Integration of $\exp(\sin(y_1)\sin(y_2))$ on the triangle

convergence results for the calculation of the surface area of a spherical triangle, which requires the use of both algorithms on a non-linearly transformed triangle. All computations are performed using a combination of MEX and MATLAB R2007b on a single 2.4 GHz processor.

Quadratic reference timing curves (fitted using least squares) have also been plotted to emphasise the efficiency of the algorithms. Again, we would like to emphasise that there are $\mathcal{O}(N^2)$ points within the fundamental domain, of which, just over half lie within the triangle, thus our methods achieve spectral accuracy with nearly linear computational complexity in the number of sample points.

We begin with Clenshaw–Curtis quadrature of our test function on the triangle. Figure 7 shows the spectral rate of convergence of the Clenshaw–Curtis quadrature algorithm for a sufficiently smooth function. Most of the variation in the timing curve comes from the two dimensional fast Fourier transform (which is performed by FFTW) and is reproducible (cf. Figs. 8 and 9).

If instead of the test function $\exp(\sin(y_1)\sin(y_2))$, one uses a monomial of degree p, one finds that the Clenshaw–Curtis quadrature algorithm becomes exact for N greater than some value (typically around $N/2$). This occurs because the non-zero Fourier coefficients of the approximation $P_N(x)$ for a monomial of degree p are limited to a hexagon of radius p centered at the origin of the dual fundamental domain. The required value of N to make the integration exact is then the smallest value of N such that the dual fundamental domain contains all of the k such that

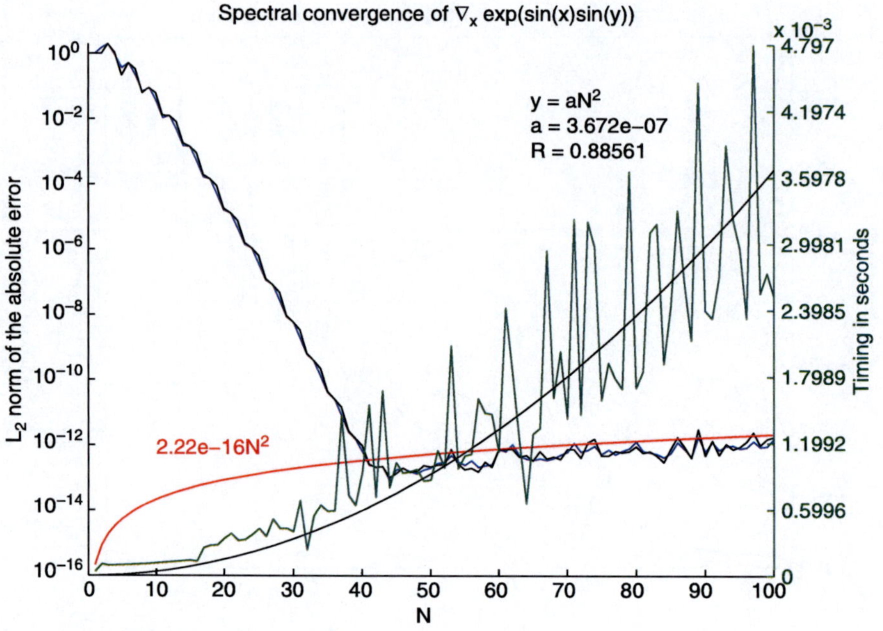

Fig. 8 Gradient of $\exp(\sin(y_1)\sin(y_2))$ on the triangle

both $a(k) \neq 0$ and $\int_{\Omega_\theta} T_k(\theta)|J_x(\theta)|\mathrm{d}\theta \neq 0$. A similar result holds true for the gradient algorithm.

We now proceed to the calculation of the gradient of our test function using the recursion formula of Sect. 3. Figure 8 shows that the gradient algorithm also has a spectral rate of convergence. The two curves on the left of this figure are the L_2 norms of the absolute error in the y_1 and y_2 components of the gradient. The numerical error in the gradient for large N is entirely due to accumulated round-off error and grows as $\mathcal{O}(\epsilon N^2)$, where ϵ is machine precision. This phenomenon is consistent with the general rule that numerical integration is stable while numerical differentiation is unstable.

Lastly, in Fig. 9, we show convergence and timing results for our gradient and Clenshaw–Curtis quadrature algorithms applied to a nonlinear triangle. Instead of showing these separately, we demonstrate their use by calculating the surface area of the spherical triangle on the unit sphere with corners $(\hat{x}_1, \hat{x}_2, \hat{x}_3)$, where $\hat{x}_i$ are normalised versions of the x_i:

$$x_1 = \begin{bmatrix} -\frac{1}{3} \\ -\frac{1}{\sqrt{3}} \\ 1 \end{bmatrix}, \quad x_2 = \begin{bmatrix} -\frac{1}{3} \\ \frac{1}{\sqrt{3}} \\ 1 \end{bmatrix}, \quad x_3 = \begin{bmatrix} \frac{2}{3} \\ 0 \\ 1 \end{bmatrix}.$$

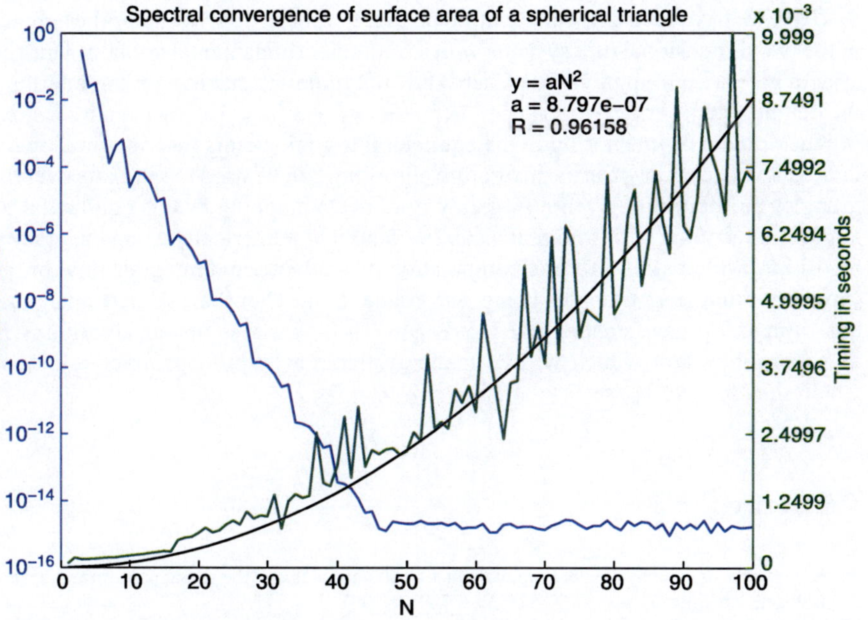

Fig. 9 Surface area of a spherical triangle with corners $(\hat{x}_1, \hat{x}_2, \hat{x}_3)$

The surface area of a function $f(y_1, y_2)$ is calculated by first computing the gradient $\nabla_y f(y)$ and then by integrating over the domain the function

$$S = \sqrt{1 + \left(\frac{\partial f}{\partial y_1}\right)^2 + \left(\frac{\partial f}{\partial y_2}\right)^2}.$$

From Fig. 9 one can clearly see that the gradient and Clenshaw–Curtis quadrature algorithms perform as well and almost as fast for nonlinear maps as they do for linear maps.

7 Summary

In this paper, we have constructed a family of multivariate Chebyshev polynomials based on a symmetric extension of the fundamental domain of the affine Weyl group associated with a root system. Based on these multivariate Chebyshev polynomials, we have developed algorithms to approximate the gradient and the integral of functions over the fundamental domain associated with a root system. These algorithms are spectrally accurate and extremely fast, with computational complexity dominated by FFTs.

Here, we have focussed our attention on the A_2 root system, which is the simplest of the two dimensional root systems with a triangular fundamental domain. This root system gives rise to multivariate Chebyshev polynomials that live on the interior of the deltoid $3(x_1^2 + x_2^2)^2 - 8x_1(x_1^2 - 3x_2^2) + 6(x_1^2 + x_2^2) = 1$. However, by restricting the domain of integration to the equilateral triangle that is inscribed within this deltoid, the Clenshaw–Curtis quadrature algorithm can be used to integrate over this triangle. Furthermore, given a (possibly nonlinear) mapping of this equilateral triangle, the Jacobian of this mapping can be computed numerically using the gradient algorithm, which allows for the computation of gradients and integrals on arbitrary (possibly nonlinear) triangles using our gradient and Clenshaw–Curtis quadrature algorithms. We have created MATLAB and C++ libraries of our algorithms for the A_2 root system, which will be made available at http://hans.munthe-kaas.no/Chebyshev for public use.

References

1. R. J. Beerends. Chebyshev polynomials in several variables and the radial part of the Laplace–Beltrami operator. *Trans. AMS*, 328(2):779–814, 1991
2. L. P. Bos. Bounding the Lebegue function for lagrange interpolation in a simplex. *J. Approx. Theory*, 38:43–59, 1983
3. Q. Chen and I. Babuska. Aprroximate optimal points for polynomial interpolation of real functions in an interval and a triangle. *Comp. Meth. App. Mech. Eng.*, 128(3-4):405–417, 1995
4. M. Dubiner. Spectral methods on triangles and other domains. *J. Sci. Comput.*, 6(4):345–390, 1991
5. R. Eier and R. Lidl. A class of orthogonal polynomials in k variables. *Math. Ann.*, 260:93–99, 1982
6. A. F. Fässler and E. Stiefel. *Group theoretical methods and their applications*. Birkhäuser, Boston, 1992
7. F. X. Giraldo and T. Warburton. A nodal triangle-based specral element method for the shallow water equations on the sphere. *J. Comp. Phys.*, 2005
8. W. Heinrichs. Improved Lebesgue constant on the triangle. *J. Comp. Phys.*, 207:625–638, 2005
9. J. S. Hesthaven. From electrostatics to almost optimal nodal sets for polynomial interpolation in a simplex. *SIAM J. Numer. Anal.*, 35(2):655–676, 1998
10. M. E. Hoffman and W. D. Withers. Generalized Chebyshev polynomials associated with affine Weyl groups. *Trans. AMS*, 308(1):91–104, 1988
11. J. E. Humphreys. *Introduction to Lie algebras and representation theory*. Springer, New York, 1970
12. T. Koornwinder. Orthogonal polynomials in two variables which are eigenfunctions of two algebraically independent partial differential operators I–IV. *Indag. Math.*, 36:48–66 and 357–381, 1974
13. T. Koornwinder. Two-variable analogues of the classical orthogonal polynomials. In R. Askey, editor, *Theory and application of special functions*. Academic Press, New York, 1975
14. H. Li, J. Sun, and Y. Xu. Discrete Fourier analysis, cubature and interpolation on a hexagon and a triangle. *SIAM J. Numer. Anal.*, 46:1653–1681, 2008
15. R. Lidl. Tchebyscheffpolynome in mehreren Variabelen. *J. Reine Angew. Math.*, 273:178–198, 1975
16. H. Munthe-Kaas. Symmetric FFTs; a general approach. In *Topics in linear algebra for vector- and parallel computers, PhD thesis*. NTNU, Trondheim, Norway, 1989. Available at: http://hans.munthe-kaas.no

17. H.Z. Munthe-Kaas. On group Fourier analysis and symmetry preserving discretizations of PDEs. *J. Phys. A Math. Gen.*, 39(19):5563–5584, 2006
18. R. Pasquetti and F. Rapetti. Spectral element methods on triangles and quadrilaterals: comparisons and applications. *J. Comp. Phys.*, 198(1):349–362, 2004
19. M. Puschel and M. Rötteler. Cooley–Tukey FFT like algorithm for the Discrete Triangle Transform. In *Proc. 11th IEEE DSP Workshop*, 2004
20. M. A. Taylor, B. A. Wingate, and R. E. Vincnent. An Algorithm for Computing Fekete point in the triangle. *SIAM J. Numer. Anal.*, 38(5):1707–1720, 2000
21. L.N. Trefethen. Is Gauss quadrature better than Clenshaw–Curtis? *SIAM Rev.*, 50(1):67, 2008
22. T. Warburton. An explicit construction of interpolation nodes on the simplex. *J. Eng. Math.*, 56(3):247–262, 2006

Stochastic Spectral Galerkin and Collocation Methods for PDEs with Random Coefficients: A Numerical Comparison

Joakim Bäck, Fabio Nobile, Lorenzo Tamellini, and Raul Tempone

Abstract Much attention has recently been devoted to the development of Stochastic Galerkin (SG) and Stochastic Collocation (SC) methods for uncertainty quantification. An open and relevant research topic is the comparison of these two methods. By introducing a suitable generalization of the classical sparse grid SC method, we are able to compare SG and SC on the same underlying multivariate polynomial space in terms of accuracy vs. computational work. The approximation spaces considered here include isotropic and anisotropic versions of Tensor Product (TP), Total Degree (TD), Hyperbolic Cross (HC) and Smolyak (SM) polynomials. Numerical results for linear elliptic SPDEs indicate a slight computational work advantage of isotropic SC over SG, with SC-SM and SG-TD being the best choices of approximation spaces for each method. Finally, numerical results corroborate the optimality of the theoretical estimate of anisotropy ratios introduced by the authors in a previous work for the construction of anisotropic approximation spaces.

Keywords Elliptic equations · Multivariate polynomial approximation · PDEs with random data · Smolyak approximation · Stochastic collocation methods · Stochastic Galerkin methods · Uncertainty quantification

1 Introduction

Nowadays, we observe a widespread need for including uncertainty in mathematical models and quantify its effect on given outputs of interest used in decision making. Such uncertainty may reflect, on one side, our ignorance or inability to properly

J. Bäck and R. Tempone
Applied Mathematics and Computational Science, KAUST, Saudi Arabia
e-mail: joakim.back.09@ucl.ac.uk, raul.tempone@kaust.edu.sa

F. Nobile (✉) and L. Tamellini
MOX, Department of Mathematics "F. Brioschi", Politecnico di Milano, Italy
e-mail: fabio.nobile@polimi.it, lorenzo.tamellini@mail.polimi.it

J.S. Hesthaven and E.M. Rønquist (eds.), *Spectral and High Order Methods for Partial Differential Equations*, Lecture Notes in Computational Science and Engineering 76,
DOI 10.1007/978-3-642-15337-2_3, © Springer-Verlag Berlin Heidelberg 2011

characterize all input parameters of the mathematical model; on the other side, it may describe intrinsic variability of the event we model. Probability theory offers a natural framework to describe uncertainty, where all uncertain inputs are treated as random variables or more generally as random fields.

Monte Carlo Sampling (MCS) is probably the most natural and widely used technique to forward propagate the input randomness onto the system response or specific quantities of interest. While being very flexible and easy to implement, MCS features a very slow convergence and does not exploit the possible regularity that the solution might have with respect to the input variables.

Much attention has been recently devoted towards alternative methods which exploit such regularity and achieve sometimes a better convergence rate. Stochastic Galerkin (SG) and Stochastic Collocation (SC) are examples of such methods for uncertainty quantification. An open and relevant research topic is the comparison of these two approaches. This work provides, on a couple of numerical examples, a fair comparison between the performances of SG and SC methods *with the same underlying approximation space.*

Traditionally, the SG method approximates the solution in a multivariate polynomial space of given total degree (see e.g. [11, 13, 27] and references therein), or in anisotropic tensor product polynomial spaces [2, 8, 14]. Other global polynomial spaces has been considered recently, see for instance [5, 24], as well as different approximation spaces such as piecewise polynomials [2, 12, 25].

On the other hand the SC method adopted so far for SPDEs follows the classical Smolyak construction, see e.g. [9, 16, 26] and the references therein. It is very relevant to this work the fact that the sparse collocation method considered in [16, 26] leads to an approximate solution in a polynomial space, which we call hereafter Smolyak space, that differs from the total degree polynomial space most commonly used in SG approximation.

In this work we will consider several choices of multivariate polynomial spaces, namely: tensor product (TP), total degree (TD), hyperbolic cross (HC) and Smolyak (SM) spaces. We consider on the one hand, SG approximations in either of these spaces. On the other hand, we propose a generalization of the classical sparse collocation method that allows us to achieve approximations in these same spaces. By following this path, we are able to compare the two alternative approaches (SG vs. SC) given the same underlying multivariate polynomial space.

Once both SG and SC are posed on the same approximation space the second ingredient in a fair comparison is the computational work associated to each of them for the same level of accuracy. Since SC entails the solution of a number of *uncoupled* deterministic problems, its corresponding computational work is directly proportional to the number of collocation points. On the other hand, SG entails the solution of a large system of *coupled* deterministic problems whose size corresponds to the number of stochastic degrees of freedom (sdof). This can be achieved by an iterative strategy, here chosen to be a Preconditioned Conjugate Gradient solver following [18]. Therefore, a natural approximation of its computational work is given by the product of the number of sdof times the number of iterations performed.

This work assesses, on a numerical example having eight input random variables, the performances of the SG and SC methods in terms of accuracy vs. (estimated) computational cost. The numerical study shows that the two approaches have comparable performances. Actually, SC seems to be more efficient for relative errors larger than 10^{-4}, whereas SG is better for smaller errors.

The second numerical example that we propose contains four input random variables that have largely different influence on the solution. It is thus suited for anisotropic approximations, where higher polynomials degrees are used to discretize the dependence on the random variables that have a greater influence on the solution. We introduce anisotropic versions of both the SG and SC methods and compare their performances for different choices of anisotropy ratios. The results show that theoretically derived anisotropy ratios following [15] have the best performance and that our formula for the optimal anisotropy ratios is sharp.

2 Problem Setting

Let D be a convex bounded polygonal domain in $\mathbb{R}^d$ and $(\Omega, \mathcal{F}, P)$ be a complete probability space. Here Ω is the set of outcomes, $\mathcal{F} \subset 2^{\Omega}$ is the σ-algebra of events and $P : \mathcal{F} \to [0, 1]$ is a probability measure. Consider the stochastic linear elliptic boundary value problem: find a random function, $u : \Omega \times \overline{D} \to \mathbb{R}$, such that P-almost everywhere in Ω, or in other words almost surely (a.s.), the following equation holds:

$$\begin{cases} -\operatorname{div}(a(\omega, \mathbf{x})\nabla u(\omega, \mathbf{x})) = f(\mathbf{x}) & \mathbf{x} \in D, \\ u(\omega, \mathbf{x}) = 0 & \mathbf{x} \in \partial D. \end{cases} \tag{1}$$

where the operators div and ∇ imply differentiation with respect to the physical coordinate only.

The theory presented in this work extends straightforwardly to the case of a random forcing term $f = f(\omega, \mathbf{x})$ as well as to a non homogeneous, possibly random, Dirichlet datum on the boundary. For easiness of presentation, we will consider the case where the randomness appears only in the diffusion coefficient, which is, however, the most difficult case, since the solution u depends nonlinearly on it, whereas it depends linearly on the forcing term and boundary data.

We will make the following assumptions on the random diffusion coefficient:

(A1) $a(\omega, \mathbf{x})$ *is strictly positive and bounded with probability 1, i.e., there exist* $a_{\min} > 0$ *and* $a_{\max} < \infty$ *such that*

$$P(a_{\min} \le a(\omega, \mathbf{x}) \le a_{\max}, \ \forall \mathbf{x} \in \overline{D}) = 1$$

(A2) $a(\omega, \mathbf{x})$ *has the form*

$$a(\omega,\mathbf{x}) = b_0(\mathbf{x}) + \sum_{n=1}^{N} y_n(\omega) b_n(\mathbf{x}) \tag{2}$$

where $\mathbf{y} = [y_1,\dots,y_N]^T : \Omega \to \mathbb{R}^N$, *is a vector of independent random variables.*

We denote by $\Gamma_n = y_n(\Omega)$ the image set of the random variable y_n, $\Gamma = \Gamma_1 \times \ldots \times \Gamma_N$, and we assume that the random vector $\mathbf{y}$ has a joint probability density function $\rho : \Gamma \to \mathbb{R}_+$ that factorizes as $\rho(\mathbf{y}) = \prod_{n=1}^N \rho_n(y_n)$, $\forall \mathbf{y} \in \Gamma$. Observe that for assumption (A1) to hold, the image set Γ has to be a bounded set in $\mathbb{R}^N$.

After assumption (A2), the solution u of (1) depends on the single realization $\omega \in \Omega$ only through the value taken by the random vector $\mathbf{y}$. We can therefore replace the probability space $(\Omega, \mathcal{F}, P)$ with $(\Gamma, B(\Gamma), \rho(\mathbf{y})d\mathbf{y})$, where $B(\Gamma)$ denotes the Borel σ-algebra on Γ and $\rho(\mathbf{y})d\mathbf{y}$ is the distribution measure of the vector $\mathbf{y}$.

Finally, we introduce the functional space $H^1(D)$ of square integrable functions in D with square integrable distributional derivatives; its subspace $H_0^1(D)$ of functions with zero trace on the boundary, and the space $L^2_\rho(\Gamma)$ of square integrable functions on Γ with respect to the measure $\rho(\mathbf{y})d\mathbf{y}$.

We are now in the position to write a weak formulation of problem (1):
find $u \in H_0^1(D) \otimes L^2_\rho(\Gamma)$ *such that* $\forall v \in H_0^1(D) \otimes L^2_\rho(\Gamma)$

$$\int_\Gamma \int_D \left(b_0(\mathbf{x}) + \sum_{n=1}^N y_n b_n(\mathbf{x}) \right) \nabla u(\mathbf{x},\mathbf{y}) \cdot \nabla v(\mathbf{x},\mathbf{y})\, \rho(\mathbf{y})\, d\mathbf{x}\, d\mathbf{y}$$
$$= \int_\Gamma \int_D f(\mathbf{x}) v(\mathbf{x},\mathbf{y})\, \rho(\mathbf{y})\, d\mathbf{x}\, d\mathbf{y}. \tag{3}$$

Under assumption (A1), a straightforward application of the Lax-Milgram lemma allows to prove that there exists a unique solution to problem (3) for any $f \in L^2(D)$. Moreover, the following estimate holds:

$$\|\nabla u\|_{L^2(D)\otimes L^2_\rho(\Gamma)} \le \frac{C_p}{a_{\min}} \|f\|_{L^2(D)}$$

where C_p is the Poincaré constant such that $\|u\|_{L^2(D)} \le C_p \|\nabla u\|_{L^2(D)}$ for any $u \in H_0^1(D)$.

It is well known (see e.g. [3, 14]) that the solution depends analytically on each parameter $y_n \in \Gamma_n$. In particular, denoting $\Gamma_n^* = \prod_{j \ne n} \Gamma_j$ and $\mathbf{y}_n^*$ an arbitrary element of Γ_n^*, there exists a constant M and regions $\Sigma_n \subset \mathbb{C}$ in the complex plane for $n = 1,\dots,N$, with $\Sigma_n \supset \Gamma_n$, in which the solution $u(\mathbf{x}, y_n, \mathbf{y}_n^*)$ admits an analytic continuation $u(\mathbf{x}, z, \mathbf{y}_n^*)$, $z \in \Sigma_n$. Moreover

$$\max_{z \in \Sigma_n} \max_{\mathbf{y}_n^* \in \Gamma_n^*} \|\nabla u(\cdot, z, \mathbf{y}_n^*)\|_{H^1(D)} \le M, \qquad \text{for } n = 1,\dots,N.$$

2.1 *Finite Element Approximation in the Physical Space*

Let $\mathcal{T}_h$ be a triangulation of the physical domain D and $V_h(D) \subset H_0^1(D)$ a finite element space of piecewise continuous polynomials on $\mathcal{T}_h$, with dimension $N_h = dim(V_h(D))$. We introduce the *semi-discrete* problem: *find* $u_h \in V_h(D) \otimes L^2_\rho(\Gamma)$ *such that* $\forall v_h \in V_h(D)$

$$\int_D \left(b_0(\mathbf{x}) + \sum_{n=1}^{N} y_n b_n(\mathbf{x}) \right) \nabla u_h(\mathbf{x}, \mathbf{y}) \cdot \nabla v_h(\mathbf{x})\, d\mathbf{x} = \int_D f(\mathbf{x}) v_h(\mathbf{x})\, d\mathbf{x}, \quad (4)$$

ρ-a.e. in Γ.

Problem (4) admits a unique solution for almost every $\mathbf{y} \in \Gamma$. Moreover, u_h satisfies the same analyticity result as the continuous solution u.

Let $\{\phi_i\}_{i=1}^{N_h}$ be a Lagrangian basis of $V_h(D)$ and consider the expansion of the semi-discrete solution as $u_h(\mathbf{x}, \mathbf{y}) = \sum_{i=1}^{N_h} u_i(\mathbf{y})\phi_i(\mathbf{x})$. Denoting by $\mathbf{U}(\mathbf{y}) = [u_1(\mathbf{y}), \ldots, u_{N_h}(\mathbf{y})]^T$ the vector of nodal values as functions of the random variables $\mathbf{y}$, problem (4) can be written in algebraic form as

$$\left(K_0 + \sum_{n=1}^{N} y_n K_n \right) \mathbf{U}(\mathbf{y}) = \mathbf{F}, \qquad \rho\text{-a.e. in } \Gamma \quad (5)$$

where $(K_n)_{ij} = \int_D b_n(\mathbf{x}) \nabla\phi_j(\mathbf{x}) \cdot \nabla\phi_i(\mathbf{x})$, for $n = 0, \ldots, N$, are deterministic stiffness matrices and $\mathbf{F}_i = \int_D f(\mathbf{x})\phi_i(\mathbf{x})$ is a deterministic right hand side.

In writing (5) we have heavily exploited the fact that the random diffusion coefficient is an affine function of the random variables y_n. This allows of an efficient evaluation of the stochastic stiffness matrix $A(\mathbf{y}) = K_0 + \sum_{n=1}^{N} y_n K_n$ in any point $\mathbf{y} \in \Gamma$ and greatly simplifies the implementation of the SG method that will be presented in the next section.

3 Polynomial Approximation in the Stochastic Dimension

We seek a further approximation of $u_h(\cdot, \mathbf{y})$ with respect to $\mathbf{y}$ by global polynomials, which is sound because of the analyticity of the semi-discrete solution with respect to the input random variables $\mathbf{y}$.

In this work we aim at comparing numerically several choices of multivariate polynomials spaces. We remark that the choice of the polynomial space is critical when the number of input random variables, N, is large, since the number of stochastic degrees of freedom might grow very fast with N, even exponentially, for instance when isotropic tensor product polynomials are used, cf. (6). This effect is known as the *curse of dimensionality*.

Let $w \in \mathbb{N}$ be an integer index denoting the level of approximation and $\mathbf{p} = (p_1, \dots, p_N)$ a multi-index. We introduce a sequence of increasing index sets $\Lambda(w)$ such that $\Lambda(0) = \{(0, \dots, 0)\}$ and $\Lambda(w) \subseteq \Lambda(w+1)$, for $w \geq 0$. Finally, we denote by $\mathbb{P}_{\Lambda(w)}(\Gamma)$ the multivariate polynomial space

$$\mathbb{P}_{\Lambda(w)}(\Gamma) = span\left\{\prod_{n=1}^{N} y_n^{p_n}, \text{ with } \mathbf{p} \in \Lambda(w)\right\}$$

and seek a *fully discrete* approximation $u_{hw} \in V_h(D) \otimes \mathbb{P}_{\Lambda(w)}(\Gamma)$.

In the following we consider four possible choices of index sets:

Tensor product polynomial space (TP)

$$\Lambda(w) = \{\mathbf{p} \in \mathbb{N}^N : \max_{n=1\dots,N} p_n \leq w\} \tag{6}$$

Total degree polynomial space (TD)

$$\Lambda(w) = \{\mathbf{p} \in \mathbb{N}^N : \sum_{n=1}^{N} p_n \leq w\} \tag{7}$$

Hyperbolic cross space (HC)

$$\Lambda(w) = \{\mathbf{p} \in \mathbb{N}^N : \prod_{n=1}^{N} (p_n + 1) \leq w + 1\} \tag{8}$$

Smolyak polynomial space (SM)

$$\Lambda(w) = \{\mathbf{p} \in \mathbb{N}^N : \sum_{n=1}^{n} f(p_n) \leq f(w)\}, \text{ with } f(p) = \begin{cases} 0, & p = 0 \\ 1, & p = 1 \\ \lceil \log_2(p) \rceil, & p \geq 2 \end{cases} \tag{9}$$

TP and TD spaces are the most common choices. The first suffers greatly from the curse of dimensionality and is impractical for a large dimension N. The second has a reduced curse of dimensionality and has been widely used in SG approximations (see e.g. [11, 13, 17, 23, 27]). HC spaces have been introduced in [1] in the context of approximation of periodic functions by trigonometric polynomials. Recently they have been used to solve elliptic PDEs in high dimension in [21]. Finally, the SM space is an unusual choice in the context of SG approximations. The reason for introducing it will be made clear later, as this space appears naturally when performing interpolation on a sparse grid following the Smolyak construction (see Sect. 3.2). Observe that the Smolyak space is similar to the hyperbolic cross space; indeed, the HC index set can be equivalently written as $\Lambda^{HC}(w) = \{\mathbf{p} \in \mathbb{N}^N : \sum_{n=1}^N \log_2(p_n+1) \leq \log_2(w+1)\}$. Other polynomial spaces have been introduced e.g. in [24].

It is also useful to introduce *anisotropic* versions of these spaces. Let $\boldsymbol{\alpha} = (\alpha_1, \ldots, \alpha_N) \in \mathbb{R}_+^N$ be a vector of positive weights, and $\alpha_{\min} = \min_n \boldsymbol{\alpha}$. The anisotropic version of the spaces previously defined reads:

Anisotropic tensor product polynomial space (ATP)

$$\Lambda(w) = \{\mathbf{p} \in \mathbb{N}^N : \max_{n=1\ldots,N} \alpha_n p_n \leq \alpha_{\min} w\} \tag{10}$$

Anisotropic total degree polynomial space (ATD)

$$\Lambda(w) = \{\mathbf{p} \in \mathbb{N}^N : \sum_{n=1}^N \alpha_n p_n \leq \alpha_{\min} w\} \tag{11}$$

Anisotropic hyperbolic cross space (AHC)

$$\Lambda(w) = \{\mathbf{p} \in \mathbb{N}^N : \prod_{n=1}^N (p_n+1)^{\frac{\alpha_n}{\alpha_{\min}}} \leq w+1\} \tag{12}$$

Anisotropic Smolyak polynomial space (ASM)

$$\Lambda(w) = \{\mathbf{p} \in \mathbb{N}^N : \sum_{n=1}^N \alpha_n f(p_n) \leq \alpha_{\min} f(w)\} \tag{13}$$

In all cases introduced except for the Smolyak space, the maximum polynomial degree used in each direction y_n does not exceed the index w and there is at least one direction (corresponding to the minimum weight $\alpha_{\min}$) for which the monomial y_n^w is in the polynomial space. For the Smolyak space this property holds only if $\log_2(w)$ is integer.

In the next sections we introduce and compare two possible ways of obtaining a *fully-discrete* approximation $u_{hw} \in V_h(D) \otimes \mathbb{P}_{\Lambda(w)}(\Gamma)$, namely Galerkin projection and collocation on a suitable sparse grid.

3.1 Stochastic Galerkin Approximation

The Stochastic Galerkin (SG) – Finite Element approximation consists in restricting the weak formulation (3) to the subspace $V_h(D) \otimes \mathbb{P}_{\Lambda(w)}(\Gamma)$ and reads: *find* $u_{hw}^{SG} \in V_h(D) \otimes \mathbb{P}_{\Lambda(w)}(\Gamma)$ *such that* $\forall v_{hw} \in V_h(D) \otimes \mathbb{P}_{\Lambda(w)}(\Gamma)$

$$\int_{\Gamma}\int_{D}\left(b_0(\mathbf{x})+\sum_{n=1}^{N} y_n b_n(\mathbf{x})\right)\nabla u_{hw}^{SG}(\mathbf{x},\mathbf{y})\cdot\nabla v_{hw}(\mathbf{x},\mathbf{y})\,\rho(\mathbf{y})\,d\mathbf{x}\,d\mathbf{y}$$

$$=\int_{\Gamma}\int_{D} f(\mathbf{x})v_{hw}(\mathbf{x},\mathbf{y})\,\rho(\mathbf{y})\,d\mathbf{x}\,d\mathbf{y}. \tag{14}$$

Let $\{\psi_{n,p}\}_{p=0}^{\infty}$ be the sequence of orthonormal polynomials in Γ_n with respect to the weight ρ_n, i.e. for any $n = 1,\dots,N$ and $p \geq 0$

$$\int_{\Gamma_n}\psi_{n,p}(t)v(t)\rho_n(t)\,dt = 0 \quad \forall v \in \mathbb{P}_{p-1}(\Gamma_n).$$

Given a multi-index $\mathbf{p} = (p_1,\dots,p_N)$, let $\psi_{\mathbf{p}}(\mathbf{y}) = \prod_{n=1}^{N}\psi_{n,p_n}(y_n)$ be the product of one dimensional orthonormal polynomials. Then a basis for the space $\mathbb{P}_{\Lambda(w)}(\Gamma)$ is given by $\{\psi_{\mathbf{p}}, \mathbf{p}\in\Lambda(w)\}$ and the SG solution can be expanded as

$$u_{hw}^{SG}(\mathbf{x},\mathbf{y}) = \sum_{\mathbf{p}\in\Lambda(w)} u_{\mathbf{p}}(\mathbf{x})\psi_{\mathbf{p}}(\mathbf{y}) = \sum_{\mathbf{p}\in\Lambda(w)}\sum_{i=1}^{N_h} u_{\mathbf{p},i}\phi_i(\mathbf{x})\psi_{\mathbf{p}}(\mathbf{y}). \tag{15}$$

Given this expansion and exploiting the orthonormality of the basis $\{\psi_{\mathbf{p}}(\mathbf{y})\}$, one can easily compute mean and variance of u_{hw}^{SG} as $\mathbb{E}\left[u_{hw}^{SG}\right](\mathbf{x}) = u_{\mathbf{0}}(\mathbf{x})$ and $\mathbb{V}\mathrm{ar}\left[u_{hw}^{SG}\right](\mathbf{x}) = \sum_{\mathbf{p}\in\Lambda(w)} u_{\mathbf{p}}^2(\mathbf{x}) - \mathbb{E}\left[u_{hw}^{SG}\right]^2(\mathbf{x})$.

Let $\mathbf{U}_{\mathbf{p}} = [u_{\mathbf{p},1},\dots,u_{\mathbf{p},N_h}]^T$ be the vector of nodal values of the finite element solution corresponding to the $\mathbf{p}$ multi-index. Then inserting expression (15) into (14) and recalling the definition of the deterministic stiffness matrices K_n, we obtain the *system of* $N_w = dim(\mathbb{P}_{\Lambda(w)}(\Gamma))$ *coupled finite element problems*

$$K_0\mathbf{U}_{\mathbf{p}} + \sum_{n=1}^{N}\sum_{\mathbf{q}\in\Lambda(w)} G_{\mathbf{p},\mathbf{q}}^{n} K_n \mathbf{U}_{\mathbf{q}} = \mathbf{F}\delta_{\mathbf{0}\mathbf{p}}, \qquad \forall \mathbf{p}\in\Lambda(w) \tag{16}$$

where $G_{\mathbf{p},\mathbf{q}}^{n} = \int_{\Gamma} y_n\psi_{\mathbf{p}}(\mathbf{y})\psi_{\mathbf{q}}(\mathbf{y})\rho(\mathbf{y})\,d\mathbf{y}$ and δ_{ij} is the usual Kroneker symbol. $G_{\mathbf{p},\mathbf{q}}^{n}$ can be explicitly calculated via the well known three terms relation for orthogonal polynomials, see e.g. [10, 20].

The resulting matrix of the algebraic system (16) is highly sparse, symmetric and positive definite. See e.g. [18] for sparsity plots. For its solution we consider a Preconditioned Conjugate Gradient (PCG) method with block diagonal preconditioner $P_{\mathbf{q},\mathbf{q}} = K_0 + \sum_{n=1}^{N} G_{\mathbf{q},\mathbf{q}}^{n} K^n$ as suggested in [18]. It follows easily from assumption (A1) that the condition number of the preconditioned matrix is independent of the discretization parameters both in the physical and stochastic spaces, see [7, 19] for a detailed analysis of the condition number of the SG matrix.

Each PCG iteration implies the solution of N_w deterministic problems with matrix $P_{\mathbf{q},\mathbf{q}}$. If the finite element discretization is relatively coarse and the dimension of the probability space is moderate, a Cholesky factorization of all matrices $P_{\mathbf{q},\mathbf{q}}$ could be computed once and for all. In general, this strategy could lead to excessive memory requirements and an iterative method should be preferred. Observe that in certain cases (e.g. for uniform random variables) all blocks are equal and this reduces considerably the computational burden.

Let us now denote by W_{FE} the cost for solving one deterministic problem and by N_{iter} the number of PCG iterations. In this work we focus on the computational cost for solving the linear system (16) and neglect the time for assembling the full stochastic matrix, which highly depends on how much the computer code has been optimized. Therefore, we can estimate the total cost W_{SGFE} for SG – finite element as

$$W_{SGFE} \approx N_w * W_{FE} * N_{iter}. \tag{17}$$

This estimate will be used to compare the SG method with the SC method in the numerical tests presented in Sect. 4.

3.2 Stochastic Collocation Approximation on Sparse Grids

The Stochastic Collocation (SC) – Finite Element method consists in collocating the semi-discrete problem (4) in a set of points $\{\boldsymbol{\theta}_j \in \Gamma, j = 1, \dots, M_w\}$, i.e., computing the solutions $u_h(\cdot, \boldsymbol{\theta}_j)$ and building a global polynomial approximation u_{hw}^{SC} (not necessarily interpolatory) upon those evaluations: $u_{hw}^{SC}(\mathbf{x}, \mathbf{y}) = \sum_{j=1}^{M_w} u_h(\mathbf{x}, \boldsymbol{\theta}_j)\tilde{\psi}_j(\mathbf{y})$ for suitable multivariate polynomials $\{\tilde{\psi}_j\}_{j=1}^{M_w}$.

We consider here a generalization of the classical Smolyak construction (see e.g. [4,22]) to build a multivariate polynomial approximation on a sparse grid. For each direction y_n we introduce a sequence of one dimensional polynomial interpolant operators of increasing order: $\mathcal{U}_n^{m(i)} : C^0(\Gamma_n) \to \mathbb{P}_{m(i)-1}(\Gamma_n)$. Here $i \geq 1$ denotes the level of approximation and $m(i)$ the number of collocation points used to build the interpolation at level i, with the requirement that $m(1) = 1$ and $m(i) < m(i+1)$ for $i \geq 1$. In addition, let $m(0) = 0$ and $\mathcal{U}_n^{m(0)} = 0$. In this work the collocation points $\{\theta_{n,j}^{(i)}, j = 1, \dots, m(i)\}$ for the one dimensional interpolation formula $\mathcal{U}_n^{m(i)}$ will be taken as the Gauss points with respect to the weight ρ_n, that is the zeros of the orthogonal polynomial $\psi_{n,m(i)}$. To simplify the presentation of the sparse grid approximation (18), we now introduce the difference operators

$$\Delta_n^{m(i)} = \mathcal{U}_n^{m(i)} - \mathcal{U}_n^{m(i-1)}.$$

Given an integer $w \geq 0$ and a multi-index $\mathbf{i} = (i_1, \dots, i_N) \in \mathbb{N}_+^N$, $\mathbf{i} \geq \mathbf{1}$, we introduce a function $g : \mathbb{N}_+^N \to \mathbb{N}$ strictly increasing in each argument and define a sparse grid approximation of u_h as

$$u_{hw}^{SC} = \mathcal{S}_w^{m,g}[u_h] = \sum_{\mathbf{i} \in \mathbb{N}_+^N : g(\mathbf{i}) \leq w} \bigotimes_{n=1}^{N} \Delta_n^{m(i_n)}(u_h). \tag{18}$$

The previous formula implies evaluation of the function u_h in a finite set of points $\mathcal{H}_w^{m,g} \subset \Gamma$ (*sparse grid*). From the construction (18) one can easily build the corresponding quadrature formula, and evaluate e.g. $\mathbb{E}\left[u_{hw}^{SC}\right](\mathbf{x}) = \sum_{j=1}^{M_\omega} \omega_j u_h(\mathbf{x}, \boldsymbol{\theta}_j)$ and $\mathbb{V}\mathrm{ar}\left[u_{hw}^{SC}\right] = \omega_j u_h^2(\mathbf{x}, \boldsymbol{\theta}_j) - \mathbb{E}\left[u_{hw}^{SC}\right]^2(\mathbf{x})$. To fully characterize the sparse approximation operator $\mathcal{S}_w^{m,g}$ one has to provide the two strictly increasing functions $m : \mathbb{N}_+ \to \mathbb{N}_+$ and $g : \mathbb{N}_+^N \to \mathbb{N}$. The first defines the relation between the level i and the number of points $m(i)$ in the corresponding one dimensional polynomial interpolation formula $\mathcal{U}^{m(i)}$, while the second characterizes the set of multi-indices used to construct the sparse approximation. Since m is not surjective in $\mathbb{N}^+$ (unless it is affine) we introduce a *left inverse* $m^{-1}(k) = \min\{i \in \mathbb{N}_+ : \ m(i) \geq k\}$. Observe that with this choice m^{-1} is a (non-strictly) increasing function satisfying $m^{-1}(m(i)) = i$, and $m(m^{-1}(k)) \geq k$.

Let $\mathbf{m}(\mathbf{i}) = (m(i_1), \ldots, m(i_N))$ and consider the polynomial order set

$$\Lambda^{m,g}(w) = \{\mathbf{p} \in \mathbb{N}^N, \ \ g(\mathbf{m}^{-1}(\mathbf{p}+\mathbf{1})) \leq w\}.$$

The following result characterizes the polynomial space underlying the sparse approximation $\mathcal{S}_w^{m,g}[u_h]$:

Proposition 1.

(a) For any $f \in C^0(\Gamma)$*, we have* $\mathcal{S}_w^{m,g}[f] \in \mathbb{P}_{\Lambda^{m,g}(w)}$.
(b) Moreover, $\mathcal{S}_w^{m,g}[v] = v, \ \ \forall v \in \mathbb{P}_{\Lambda^{m,g}(w)}$.

Proof. Let us denote by $\mathbb{P}_{\mathbf{m}(\mathbf{i})-1}$ the tensor product polynomial space

$$\mathbb{P}_{\mathbf{m}(\mathbf{i})-1} = span\left\{\prod_{n=1}^{N} y_n^{p_n}, \ \ p_n \leq m(i_n) - 1\right\}.$$

Clearly we have that $\bigotimes_{n=1}^{N} \Delta_n^{m(i_n)}(f) \in \mathbb{P}_{\mathbf{m}(\mathbf{i})-1}(\Gamma)$ and

$$\begin{aligned}
\mathcal{S}_w^{m,g}[f] \in span\Big\{ & \bigcup_{\mathbf{i}\in\mathbb{N}_+^N :\, g(\mathbf{i})\leq w} \mathbb{P}_{\mathbf{m}(\mathbf{i})-1}(\Gamma)\Big\} \\
\equiv span\Big\{ & \bigcup_{\mathbf{i}\in\mathbb{N}_+^N :\, g(\mathbf{i})\leq w} span\{\prod_{n=1}^{N} y_n^{p_n}, \ \ \mathbf{p} \leq \mathbf{m}(\mathbf{i}) - \mathbf{1}\}\Big\} \\
\equiv span\Big\{ & \bigcup_{\mathbf{i}\in\mathbb{N}_+^N :\, g(\mathbf{i})\leq w} span\{\prod_{n=1}^{N} y_n^{p_n}, \ \ \mathbf{m}^{-1}(\mathbf{p}+\mathbf{1}) \leq \mathbf{i}\}\Big\} \\
\equiv span\{ & \prod_{n=1}^{N} y_n^{p_n}, \ \ g(\mathbf{m}^{-1}(\mathbf{p}+\mathbf{1})) \leq w\} =: \mathbb{P}_{\Lambda^{m,g}(w)}(\Gamma).
\end{aligned}$$

This proves (a). Due to linearity in (18), to prove point (b) we only need to show that the approximation formula $\mathcal{S}_w^{m,g}$ is exact for all monomials $\prod_{n=1}^N y_n^{p_n}$ with $\mathbf{p} \in \Lambda^{m,g}(w)$. We have

$$\mathcal{S}_w^{m,g}\left[\prod_{n=1}^N y_n^{p_n}\right] = \sum_{\mathbf{i}\in\mathbb{N}_+^N:\, g(\mathbf{i})\le w} \bigotimes_{n=1}^N \Delta_n^{m(i_n)} \mathbf{y}^{\mathbf{p}}$$

$$= \sum_{\mathbf{i}\in\mathbb{N}_+^N:\, g(\mathbf{i})\le w} \prod_{n=1}^N \left((\mathcal{U}^{m(i_n)} - \mathcal{U}^{m(i_n-1)}) y_n^{p_n} \right).$$

Observe that $\mathcal{U}^{m(i_n)} y_n^{p_n}$ will be an exact interpolation whenever $m(i_n) \ge p_n+1$ and therefore the term $\prod_{n=1}^N (\mathcal{U}^{m(i_n)} - \mathcal{U}^{m(i_n-1)}) y_n^{p_n}$ will vanish if any of the $m(i_n-1) \ge p_n + 1$ or equivalently if there exists at least one n such that $i_n \ge m^{-1}(p_n+1)+1$. Let $\bar{i}_n = m^{-1}(p_n+1)$ for $n = 1, \ldots, N$. The multi-index $\bar{\mathbf{i}} = (\bar{i}_1, \ldots, \bar{i}_N)$ satisfies the constraint $g(\bar{\mathbf{i}}) \le p$.

Then, the previous formula reduces to

$$\mathcal{S}_w^{m,g}\left[\prod_{n=1}^N y_n^{p_n}\right] = \sum_{\mathbf{i}\le\bar{\mathbf{i}}} \prod_{n=1}^N \left((\mathcal{U}^{m(i_n)} - \mathcal{U}^{m(i_n-1)}) y_n^{p_n} \right)$$

$$= \prod_{n=1}^N \sum_{i_n=0}^{\bar{i}_n} \left((\mathcal{U}^{m(i_n)} - \mathcal{U}^{m(i_n-1)}) y_n^{p_n} \right) = \prod_{n=1}^N \mathcal{U}^{m(\bar{i}_n)} y_n^{p_n}.$$

The final result follows from the fact that $m(\bar{i}_n) = m(m^{-1}(p_n+1)) \ge p_n+1$ and therefore the interpolant $\mathcal{U}^{m(\bar{i}_n)}$ is exact for $y_n^{p_n}$. □

Remark 1. Observe that in the previous Lemma we have never used the assumption that the one dimensional interpolants are based on Gauss points. Hence, the previous result still holds for interpolants based on arbitrary (distinct) knots and for an arbitrary strictly increasing function $m(i)$.

We recall that the most typical choice of m and g is given by (see [4, 22])

$$m(i) = \begin{cases} 1, & \text{for } i = 1 \\ 2^{i-1}+1, & \text{for } i > 1 \end{cases} \quad \text{and } g(\mathbf{i}) = \sum_{n=1}^N (i_n - 1)$$

This choice of m, combined with the choice of Clenshaw–Curtis interpolation points (extrema of Chebyshev polynomials) leads to nested sequences of one dimensional interpolation formulas and a reduced sparse grid. In the same vein, it is possible to show that the underlying polynomial space associated to the operator $\mathcal{S}_w^{m,g}$ is the Smolyak space $\mathbb{P}_{\Lambda(w)}$ defined in (9).

On the other hand, if we choose $m(i) = i$, it is easy to find functions g for the construction of sparse collocation approximations in the polynomial spaces

Table 1 Sparse approximation formulas and corresponding underlying polynomial space

Approximation space	SC: m, g	SG: $\Lambda(w)$
Tensor product (TP)	$m(i) = i$ $g(\mathbf{i}) = \max_n(i_n - 1) \le w$	$\{\mathbf{p} \in \mathbb{N}^N : \max_n p_n \le w\}$
Total degree (TD)	$m(i) = i$ $g(\mathbf{i}) = \sum_n (i_n - 1) \le w$	$\{\mathbf{p} \in \mathbb{N}^N : \sum_n p_n \le w\}$
Hyperbolic cross (HC)	$m(i) = i$ $g(\mathbf{i}) = \prod_n (i_n) \le w + 1$	$\{\mathbf{p} \in \mathbb{N}^N : \prod_n (p_n + 1) \le w + 1\}$
Smolyak (SM)	$m(i) = \begin{cases} 2^{i-1} + 1, & i > 1 \\ 1, & i = 1 \end{cases}$ $g(\mathbf{i}) = \sum_n (i_n - 1) \le w$	$\{\mathbf{p} \in \mathbb{N}^N : \sum_n f(p_n) \le f(w)\}$ $f(p) = \begin{cases} 0, & p = 0 \\ 1, & p = 1 \\ \lceil \log_2(p) \rceil & p \ge 2 \end{cases}$

introduced in Sect. 3, namely tensor product (6), total degree (7) and hyperbolic cross (8) spaces. Table 1 summarizes several available. It is also straightforward to build the corresponding anisotropic sparse approximation formulas.

Let now $\mathcal{H}_w^{m,g}$ be the sparse grid associated to the formula $\mathcal{S}_w^{m,g}$ and $M_w = \#\mathcal{H}_w^{m,g}$ the number of distinct collocation points in $\mathcal{H}_w^{m,g}$. To form the sparse collocation solution $u_{h,w}$ we only have to solve M_w *independent* deterministic problems. Observe, however, that in general the number of points M_w is much larger than the dimension N_w of the corresponding polynomial space $\mathbb{P}_{\Lambda^{m,g}(w)}$. The computational cost of the SC – Finite Element method can therefore be estimated as

$$W_{SCFE} \approx M_w * W_{FE}, \tag{19}$$

to be compared with the cost of the SG – Finite Element method in the same polynomial space, given by (17).

4 Numerical Results

4.1 Test Case 1: Isotropic Problem

In this first test case we consider a thermal diffusion problem in the form of (1) defined in the unit square $[0, 1]^2$, with homogeneous Dirichlet boundary conditions and stochastic conductivity coefficient that depends on a finite, small, number of random variables. The coefficient is chosen in such a way that each random input has more or less the same influence on the solution (isotropic problem).

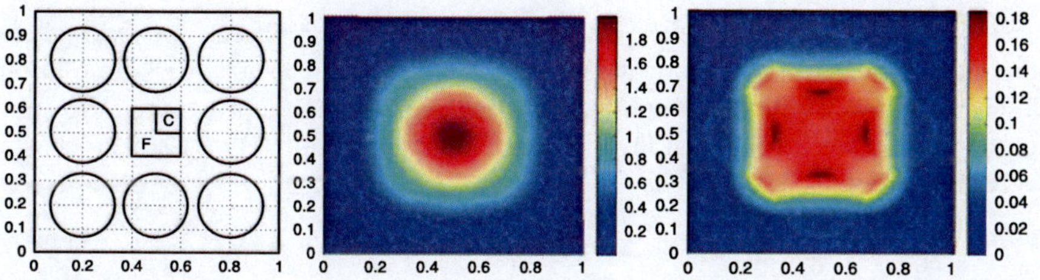

Fig. 1 *Left*: geometry for test case 1. *Middle*: expected value of the solution. *Right*: standard deviation of the solution

Figure 1 (left) shows the geometry of the test case. The forcing term is deterministic, $f(\mathbf{x}) = 100\chi_F(\mathbf{x})$, where $\chi_F(\mathbf{x})$ is the indicator function of F, a square subdomain with side length equal to 0.2, centered in the domain. The material features 8 circular inclusions with radius $r = 0.13$ and symmetrically distributed with respect to the center of the square, each with a uniformly distributed random conductivity. Let $\chi_n(\mathbf{x}), n = 1, \ldots, 8$ be the indicator function for each circle. The expression of the stochastic conductivity coefficient is then in the form of (2), with $b_n(\mathbf{x}) = \chi_n(\mathbf{x})$:

$$a(\omega, \mathbf{x}) = b_0(\mathbf{x}) + \sum_{n=1}^{8} y_n(\omega)\chi_n(\mathbf{x}), \quad \text{with } b_0 = 1 \text{ and } y_n(\omega) \sim \mathcal{U}(-0.99, -0.2).$$

As a consequence, the basis functions $\psi_{n,p}$ for SG methods will be Legendre polynomials orthonormal with respect to the uniform probability measure in $[-0.99, -0.2]$, and the collocation points for SC will be the corresponding Gauss points.

We will compare the accuracy of the Stochastic Galerkin (SG) and Stochastic Collocation (SC) methods by looking at statistical indicators of two quantities of interest:

- $\psi_1(u) = \int_F u(\mathbf{x})d\mathbf{x}$
- $\psi_2(u) = \int_C \partial_x u(\mathbf{x})d\mathbf{x}$.

The quantity $\psi_2(u)$ is defined only on C, the upper right part of F, since by symmetry its expected value on F is 0 whatever (isotropic) Galerkin or Collocation approximation is considered.

Let u_p be an approximate solution (computed either with SG or SC) and u_{ex} the exact solution. For both quantities ψ_1 and ψ_2 we will check the convergence of the following errors:

- Error in the mean: $\varepsilon_{\text{mean}}\left[\psi_j\right] = |\mathbb{E}\left[\psi_j(u_p)\right] - \mathbb{E}\left[\psi_j(u_{ex})\right]|$
- Error in the variance: $\varepsilon_{\text{var}}\left[\psi_j\right] = |\mathbb{V}\text{ar}\left[\psi_j(u_p)\right] - \mathbb{V}\text{ar}\left[\psi_j(u_{ex})\right]|$
- Error in L^2 norm: $\varepsilon_{\text{norm}}\left[\psi_j\right] = \sqrt{\mathbb{E}\left[(\psi_i(u_p) - \psi_i(u_{ex}))^2\right]}$.

Since we do not know the exact solution for this problem, we will check the convergence of the statistical indicators with respect to an overkill solution, which we consider close enough to the exact one. To this end we take the solution computed with SG-TD at level 9, which has approximately 24,000 stochastic degrees of freedom (sdof). The L^2 error will be calculated via a MCS approximation, i.e., $\varepsilon_{\text{norm}}\left[\psi_j\right] \simeq \frac{1}{M}\left(\sum_{l=1}^{M}\left[\psi_j(u_p(\mathbf{y}_l)) - \psi_j(u_{ex}(\mathbf{y}_l))\right]^2\right)^{1/2}$, where $\mathbf{y}_l$, $l = 1,\ldots,M$, are M randomly chosen points in Γ. To this end we have used $M = 1{,}000$ points.

We remark that here and in the following test all the computations are performed on the same physical mesh, which is supposed to be refined enough to solve adequately the elliptic problem for every value $\mathbf{y}$ of the random variables. Moreover notice that, as stated in Sect. 2.1, the FEM solution and the exact solution have the same regularity with respect to the stochastic variables. Therefore we expect the convergence in the stochastic dimension not to be affected by space discretization.

We have compared the performances of the SG and Collocation methods with the four choices of polynomial spaces presented in Table 1. In our convergence plots we have also added the performance of the classical MCS method.

Figure 2 shows the error $\varepsilon_{\text{mean}}\left[\psi_1\right]$ vs. the estimated computational cost (normalized to the cost W_{FE} of a deterministic solve) given by formula (17) for SG methods and (19) for SC methods. For the MCS method the cost is simply $M * W_{FE}$, where M is the number of samples used. The MCS has been repeated 20 times and only the average error over the 20 repetitions is shown.

As one can see, MCS has the worst performance, followed by tensor product polynomial spaces both in the SG and SC version, as expected. All other choices lead to similar, however much more accurate, results, with TD being the best space for Galerkin method and SM the best for Collocation.

We notice that different choices of collocation points for SC-SM (Gauss vs. Clenshaw Curtis) lead to similar results (see Fig. 2 (right)). Therefore from now on we will only use SC-SM with Gauss points.

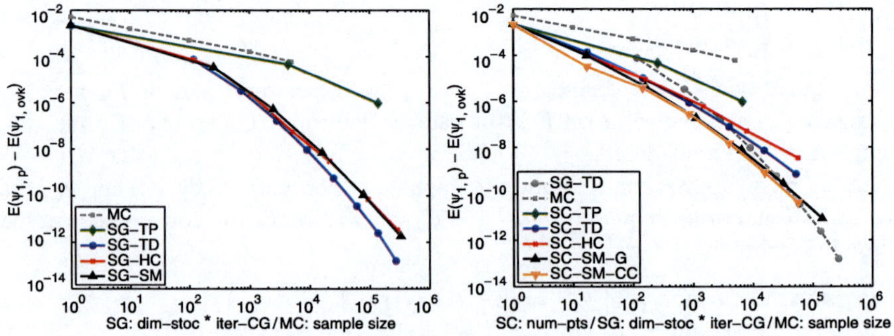

Fig. 2 Error $\varepsilon_{\text{mean}}\left[\psi_1\right]$ vs. estimated computational cost. *Left*: comparison between SG methods and Monte Carlo. *Right*: comparison between SC methods and SG-TD

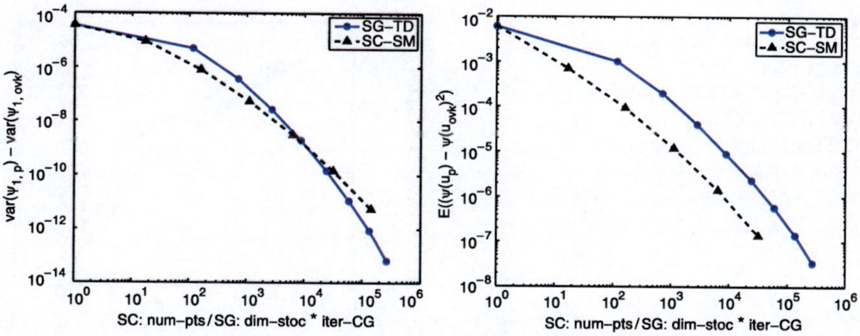

Fig. 3 Convergence curves for $\varepsilon_{\mathrm{var}}[\psi_1]$ (*left*) and $\varepsilon_{\mathrm{norm}}[\psi_1]$ (*right*) with respect to the computational cost. Comparison between SG-TD and SC-SM methods

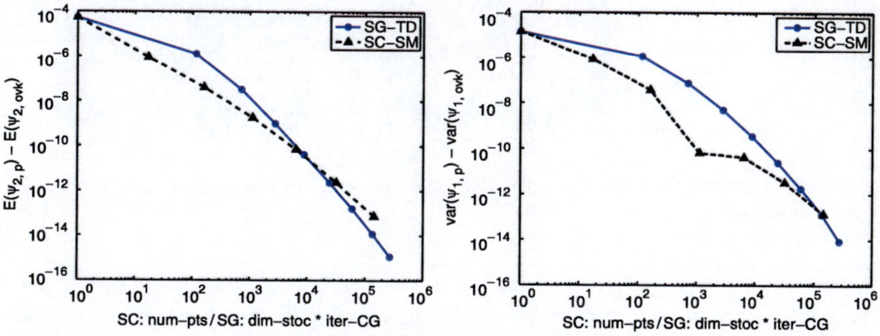

Fig. 4 Convergence curves for $\varepsilon_{\mathrm{mean}}[\psi_2]$ (*left*) and $\varepsilon_{\mathrm{var}}[\psi_2]$ (*right*) with respect to the computational cost. Comparison between SG-TD and SC-SM methods

From Fig. 2 (right) we conclude that the SC method is the best method with respect to the computational cost, at least for "practical" tolerances, while, for very small tolerances ($\leq 10^{-10}$), SG is a better choice. The same happens also for the other error indicators $\varepsilon_{\mathrm{var}}[\psi_1]$ and $\varepsilon_{\mathrm{norm}}[\psi_1]$, (see Fig. 3), as well as for the quantity ψ_2 (see Fig. 4).

We should point out that the plots may not represent a completely fair comparison. Actually, the solution of the global linear system for SG method is performed through preconditioned conjugate gradient iterations, with a fixed tolerance ($\epsilon = 10^{-12}$); this clearly over-resolves the system when the error in the stochastic dimension is much larger than ϵ. The performance of SG may be therefore improved by tuning the tolerance of the PCG method to an *a posteriori* estimation of the stochastic error. However, we have observed that running the same SG simulations with tolerance $\epsilon = 10^{-8}$ changes only slightly the results, so we can say that the choice of the tolerance for the PCG method is not deeply affecting our performance/cost analysis.

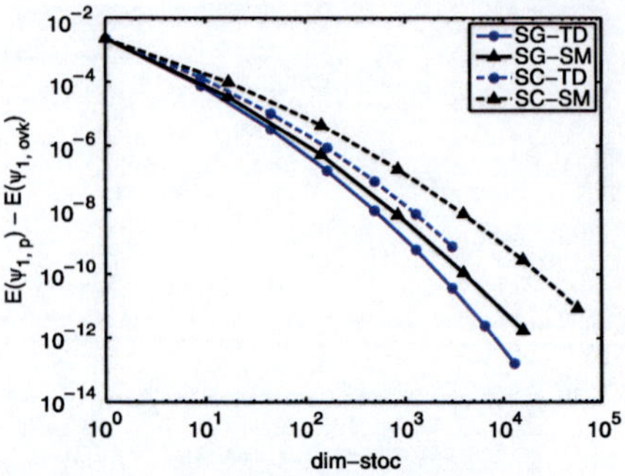

Fig. 5 Convergence curves for $\varepsilon_{\text{mean}}[\psi_1]$ with respect to the dimension of the stochastic space. Comparison between SG and SC methods with TD and SM polynomial spaces

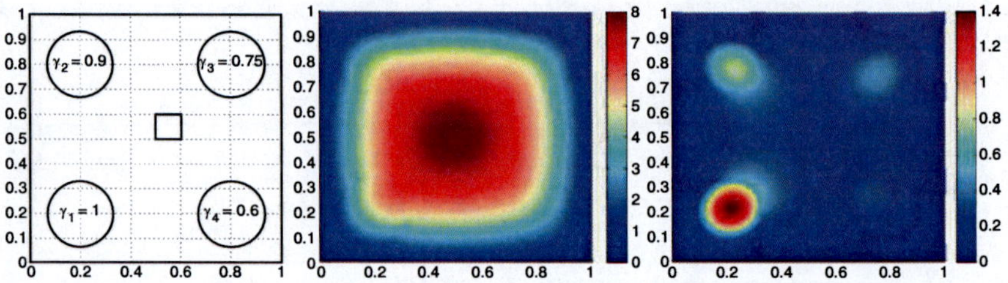

Fig. 6 *Left*: geometry for test case 2. *Middle*: expected value of the solution. *Right*: standard deviation of the solution

It is also instructive to look at the convergence plots of the error vs. the dimension of the stochastic space (Fig. 5). As expected from L^2 optimality, for a given polynomial space the Galerkin solution is more accurate than the collocation solution. We remind once more, however, that the computational cost in the two cases is quite different and the convergence plots in Fig. 2 give a more complete picture of the performances of the two methods.

4.2 Test Case 2: Anisotropic Problem

In this test we consider an anisotropic problem in which different random variables contribute differently to the total variability of the solution, in order to study the advantages of the anisotropic version of the SC and SG methods. We take the geometry and problem definition similar to test case 1; however, since our focus is on anisotropy, we consider only four inclusions (the ones in the corners, cf. Fig. 6 (left)) so that we can test many different choices of the weights that define the anisotropic spaces (10)–(13). Nonetheless, the anisotropic setting is particularly meant to be used in high dimensional spaces (see e.g. [15]). For convenience we consider a forcing term uniformly distributed on the whole domain and we look just at $\varepsilon_{\text{mean}}[\psi_1]$.

The random coefficient is $a(\omega, \mathbf{x}) = 1 + \sum_{n=1}^{4} \gamma_n y_n(\omega)\chi_n(\mathbf{x})$, with $y_n(\omega) \sim \mathcal{U}(-0.99, 0)$ and $\gamma_n \leq 1$. The values of the coefficients γ_n are shown in Fig. 6 (left). Notice that these values give different importance to the four random variables. In particular, the inclusion in the bottom-left corner has the largest variance and we expect it to contribute the most to the total variance of the solution. It is therefore intuitively justified to use polynomial degrees higher in the corresponding direction of the stochastic multidimensional space rather than in the other ones. Figure 6 also shows the mean value (middle) and the standard deviation (right) of the solution.

Our goal is to assess the performances of anisotropic polynomial spaces in comparison with their isotropic counterpart. For this we need to estimate the weights to be used in the construction of the anisotropic polynomial space.

We follow closely the argument in [15]. The overall random conductivity coefficient in the n-th inclusion Ω_n is a uniform random variable $\mathcal{U}(a_n, b_n)$ with $a_n = 1 - 0.99\gamma_n$ and $b_n = 1$. This can be rewritten as

$$a(\omega, \mathbf{x})_{|\Omega_n} = \frac{a_n + b_n}{2} + \frac{b_n - a_n}{2}\hat{y}_n, \quad \text{with } \hat{y}_n \sim \mathcal{U}(-1, 1).$$

It is easy to show that the solution $u = u(\cdot, \hat{y}_n)$ admits an analytic continuation in the complex region $\Sigma_n = \{z \in \mathbb{C} : \Re e\,(z) > -w_n\}$ with $w_n = \frac{a_n+b_n}{b_n-a_n} = \frac{2-0.99\gamma_n}{0.99\gamma_n}$, which contains, in particular, the interior of the ellipse

$$\mathcal{E}_{\rho_n} = \left\{ z \in \mathbb{C} : \Re e\,(z) = \frac{\rho_n + \rho_n^{-1}}{2}\cos\phi,\ \Im m\,(z) = \frac{\rho_n - \rho_n^{-1}}{2}\sin\phi,\ \phi \in [0, 2\pi) \right\}$$

with $\rho_n = w_n + \sqrt{w_n^2 - 1}$.

Standard spectral approximation analysis (see e.g. [6]) allows us to say that interpolation of $u(\cdot, \hat{y}_n)$ in $p_n + 1$ Gauss-Legendre points converges exponentially fast with rate $e^{-g_n p_n}$, with $g_n = \log \rho_n = \log(w_n + \sqrt{w_n^2 - 1})$.

Therefore the *theoretical estimate* (a priori choice) of the weight to be used for the n-th variable is $\alpha_n = g_n$. The larger γ_n, the smaller the corresponding weight α_n. In practice, we have renormalized the weights by dividing them by the smallest one. Notice that the spaces (10)–(13) remain unchanged by this normalization. The corresponding theoretical weights are in this case $\alpha^{th} = [1,\ 3.5,\ 5.5,\ 7.5]$. To assess the effectiveness of the proposed theoretical estimate, we also consider the weights $\alpha = [1, 2, 3, 4]$ (nearly half the theoretical estimate) and $\alpha = [1, 7, 11, 15]$ (twice the theoretical estimate). Finally, we have also considered an experimental (a posteriori) estimate of the coefficients (as suggested in [15]), where the exponential decay $e^{-g_n p_n}$ is estimated numerically by increasing the approximation level in only one direction at a time; the resulting weights are $\alpha^{exp} = [1,\ 2.5,\ 4,\ 5.5]$.

In this example we consider only SG methods in anisotropic TD spaces as they seem to be the most appropriate for this type of problem. Similarly, we restrict our study only to SC methods in the same ATD spaces, so they are directly comparable with the corresponding Galerkin version. The use of SC-ASM methods is expected to give even better results.

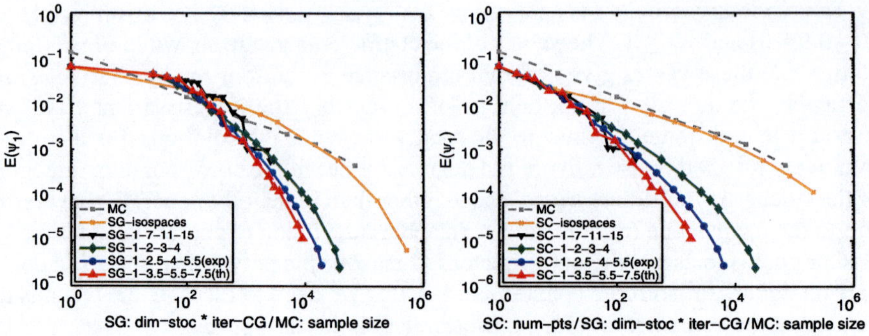

Fig. 7 Performance of SG-ATD (*left*) and SC-ATD (*right*) methods with different choices of weights, in the computation of $\mathbb{E}[\psi_1]$. Error $\varepsilon_{\text{mean}}[\psi_1]$ vs. computational cost

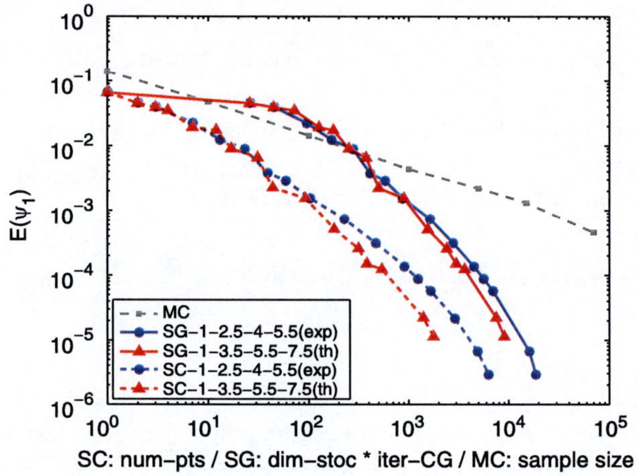

Fig. 8 Comparison between SG-ATD and SC-ATD methods with best weights in the computation of $\mathbb{E}[\psi_1]$. Error $\varepsilon_{\text{mean}}[\psi_1]$ vs. computational cost

We have computed the SG-ATD and SC-ATD with the different choices of weights up to level $w = 21$ and compared them with an overkill solution computed by SG-TD isotropic method at level $w = 22$. This solution has about 14000 sdof . In comparison, the SG-ATD solution has 837 sdof with weights $\alpha = [1, 2, 3, 4]$, 434 sdof with the experimental weights $\alpha^{exp} = [1, 2.5, 4, 5.5]$, 220 sdof with the theoretical weights $\alpha^{th} = [1, 3.5, 5.5, 7.5]$, and 68 sdof with the weights $\alpha = [1, 7, 11, 15]$. We observe that the level $w = 22$ isotropic TD space contains all the ATD spaces with level $w < 22$, therefore our overkill solution is much more accurate than the other ones considered here.

Figure 7 shows the error in computing $\mathbb{E}[\psi_1]$ vs. the estimated computational cost when using the SG-ATD (left) or SC-ATD (right) methods. For reference purposes we have also added the convergence plot for MCS.

First, we observe that SC and SG outperform the standard MCS. Figure 7 also shows that the theoretical estimate of the weights performs better than all other choices and seems to be very close to optimum for both SC and SG methods, while the a posteriori choice gives slightly worse results although the convergence curve is smoother.

In Fig. 8 we compare the performances of the SG-ATD and SC-ATD methods with the theoretical and experimental choices of the weights. In this test, the collocation method seems to be superior to the Galerkin one, even for very small tolerances.

Acknowledgements The second and third authors have been supported by the Italian grant FIRB-IDEAS "Advanced numerical techniques for uncertainty quantification in engineering and life science problems".

The authors would also like to recognize the support of the PECOS center at ICES, University of Texas at Austin (Project Number 024550, Center for Predictive Computational Science). Support from the VR project "Effektiva numeriska metoder för stokastiska differentialekvationer med tillämpningar" and King Abdullah University of Science and Technology (KAUST) is also acknowledged.

References

1. K. I. Babenko. Approximation by trigonometric polynomials in a certain class of periodic functions of several variables. *Soviet Math. Dokl.*, 1:672–675, 1960
2. I. M. Babuška, R. Tempone, and G. E. Zouraris. Galerkin finite element approximations of stochastic elliptic partial differential equations. *SIAM J. Numer. Anal.*, 42(2):800–825, 2004
3. I. Babuška, F. Nobile, and R. Tempone. A stochastic collocation method for elliptic partial differential equations with random input data. *SIAM J. Numer. Anal.*, 45(3):1005–1034, 2007
4. V. Barthelmann, E. Novak, and K. Ritter. High dimensional polynomial interpolation on sparse grids. *Adv. Comput. Math.*, 12(4):273–288, 2000
5. M. Bieri, R. Andreev, and C. Schwab. Sparse tensor discretization of elliptic spdes. SAM-Report 2009-07, Seminar für Angewandte Mathematik, ETH, Zurich, 2009
6. P.J. Davis. *Interpolation and approximation*. Dover, New York, 1975. Republication, with minor corrections, of the 1963 original, with a new preface and bibliography
7. O. G. Ernst, C. E. Powell, D. J. Silvester, and E. Ullmann. Efficient solvers for a linear stochastic Galerkin mixed formulation of diffusion problems with random data. *SIAM J. Sci. Comput.*, 31(2):1424–1447, 2008/09
8. P. Frauenfelder, C. Schwab, and R. A. Todor. Finite elements for elliptic problems with stochastic coefficients. *Comput. Methods Appl. Mech. Eng.*, 194(2–5):205–228, 2005
9. B. Ganapathysubramanian and N. Zabaras. Sparse grid collocation schemes for stochastic natural convection problems. *J. Comput. Phys.*, 225(1):652–685, 2007
10. W. Gautschi. *Orthogonal Polynomials: Computation and Approximation*. Oxford University Press, Oxford, 2004
11. R. G. Ghanem and P. D. Spanos. *Stochastic Finite Elements: A Spectral Approach*. Springer, New York, 1991
12. O. P. Le Maître, H. N. Najm, R. G. Ghanem, and O. M. Knio. Multi-resolution analysis of Wiener-type uncertainty propagation schemes. *J. Comput. Phys.*, 197(2):502–531, 2004
13. H. G. Matthies and A. Keese. Galerkin methods for linear and nonlinear elliptic stochastic partial differential equations. *Comput. Methods Appl. Mech. Eng.*, 194(12-16):1295–1331, 2005

14. F. Nobile and R. Tempone. Analysis and implementation issues for the numerical approximation of parabolic equations with random coefficients. MOX-Report 22-2008, Department of Mathematics, Politecnico di Milano, Italy, 2008. Int. J. Num. Methods Eng., Published online, June 12, 2009, DOI 10.1002/nme.2656
15. F. Nobile, R. Tempone, and C.G. Webster. An anisotropic sparse grid stochastic collocation method for partial differential equations with random input data. *SIAM J. Numer. Anal.*, 46(5):2411–2442, 2008
16. F. Nobile, R. Tempone, and C. G. Webster. A sparse grid stochastic collocation method for partial differential equations with random input data. *SIAM J. Numer. Anal.*, 46(5):2309–2345, 2008
17. A. Nouy, A. Clément, F. Schoefs, and N. Moës. An extended stochastic finite element method for solving stochastic partial differential equations on random domains. *Comput. Methods Appl. Mech. Eng.*, 197(51-52):4663–4682, 2008
18. M. F. Pellissetti and R. G. Ghanem. Iterative solution of systems of linear equations arising in the context of stochastic finite elements. *Adv. Eng. Software*, 31:607–616, 2000
19. C. E. Powell and H. C. Elman. Block-diagonal preconditioning for spectral stochastic finite-element systems. *IMA J. Numer. Anal.*, 29(2):350–375, 2009
20. A. Quarteroni, R. Sacco, and F. Saleri. *Numerical mathematics*, volume 37 of *Texts in Applied Mathematics*. Springer, Berlin, second edition, 2007
21. J. Shen and L-L. Wang. Sparse spectral approximations of high-dimensional problems based on hyperbolic cross. *SIAM J. Numer. Anal.*, 48(3):1087–1109, 2010
22. S. A. Smolyak. Quadrature and interpolation formulas for tensor products of certain classes of functions. *Dokl. Akad. Nauk SSSR*, 4:240–243, 1963
23. G. Stefanou. The stochastic finite element method: past, present and future. *Comput. Methods Appl. Mech. Eng.*, 198:1031–1051, 2009
24. R. A. Todor and C. Schwab. Convergence rates for sparse chaos approximations of elliptic problems with stochastic coefficients. *IMA J Numer Anal*, 27(2):232–261, 2007
25. X. Wan and G. E. Karniadakis. Multi-element generalized polynomial chaos for arbitrary probability measures. *SIAM J. Sci. Comput.*, 28(3):901–928, 2006
26. D. Xiu and J. S. Hesthaven. High-order collocation methods for differential equations with random inputs. *SIAM J. Sci. Comput.*, 27(3):1118–1139, 2005
27. D. Xiu and G. E. Karniadakis. The Wiener-Askey polynomial chaos for stochastic differential equations. *SIAM J. Sci. Comput.*, 24(2):619–644, 2002

Hybridizable Discontinuous Galerkin Methods

N.C. Nguyen, J. Peraire, and B. Cockburn

Abstract We present an overview of recent developments of HDG methods for numerically solving partial differential equations in fluid mechanics.

1 Background

In recent years, discontinuous Galerkin (DG) finite element methods have emerged as a competitive alternative for solving nonlinear hyperbolic systems of conservation laws. The advantages of the DG methods over classical finite difference and finite volume methods are well-documented in the literature: the DG methods work well on arbitrary meshes, result in stable high-order accurate discretizations of the convective and diffusive operators, allow for a simple and unambiguous imposition of boundary conditions and are very flexible to parallelization and adaptivity. Despite all these advantages, DG methods have not yet made a significant impact for practical applications. This is largely due to the high computational cost associated to them when compared to finite differences or finite volume schemes.

The hybridizable discontinuous Galerkin (HDG) methods were recently introduced to try to address this issue. In this paper, we present an overview of the recent developments of these methods with *implicit time-marching integration* as applied to some basic models in fluid mechanics.

The HDG methods retain the advantages of standard DG methods and result in a significantly reduced degree of freedom count, therefore allowing for a substantial reduction in the computational cost and memory storage. Hybridizable DG methods were initially developed for elliptic problems [4, 5, 9, 10, 12, 13] and have already

N.C. Nguyen and J. Peraire (✉)
Department of Aeronautics, MIT, 77 Massachusetts Avenue, Cambridge, MA 02139, USA
e-mail: cuongng@mit.edu, peraire@mit.edu

B. Cockburn
Department of Mathematics, University of Minnesota, Minneapolis, MN 55455, USA
e-mail: cockburn@math.umn.edu

J.S. Hesthaven and E.M. Rønquist (eds.), *Spectral and High Order Methods for Partial Differential Equations*, Lecture Notes in Computational Science and Engineering 76,
DOI 10.1007/978-3-642-15337-2_4,

been developed and demonstrated for linear and nonlinear convection-diffusion problems [6, 20, 21], linear elasticity [27], and incompressible flow [7, 11, 22–24].

The HDG methods we consider have the following main advantages over many existing discontinuous Galerkin methods:

- **Reduced number of globally coupled degrees of freedom** Unlike many other DG methods (analyzed in [1]) which result in a final system involving all the degrees of freedom of the approximate field variables, the HDG methods produce a final system in terms of the degrees of freedom of the *approximate traces* of the field variables. Since the approximate traces are defined on the element borders only, the HDG methods have significantly less the globally coupled unknowns as other DG methods. In fact, a variant of the HDG method – the Embedded DG method (EDG) [12, 17]) – has the same number of globally coupled unknowns than a standard continuous Galerkin method. This large reduction in the degrees of freedom can lead to significant savings for both computational time and memory storage.
- **Superconvergence** For convection-diffusion problems, the HDG methods provide *optimal convergence* for the approximation of the gradient – a special convergence property of the HDG methods for diffusion problems – whereas, for all of the DG methods studied in [1], as well as the standard continuous Galerkin approach the approximate gradient converges *suboptimally*. For incompressible flows, the approximate velocity, pressure, velocity gradient, and vorticity converge with the optimal order. This has to be contrasted with the fact that *all* the other DG methods display the suboptimal order of convergence for the approximate pressure, velocity gradient, and vorticity. Moreover, the HDG methods have superconvergence properties for the numerical traces and the average of the approximate variables.
- **Local postprocessing** Based on the optimal convergence and superconvergence of the HDG methods, local postprocessing can be developed to increase by one the spatial order of convergence of the numerical solution. For incompressible flows, local postprocessing can be employed to obtain a new approximation of the velocity which is exactly divergence-free, $\boldsymbol{H}(\text{div})$-conforming, and converges with an additional order. For time-dependent problems, postprocessing only needs to be done at those time levels for which a more accurate result is desired. Moreover, since the postprocessing is performed at the element level, it is less expensive than the solution procedure.
- **Geometric flexibility and mesh adaptation** The HDG methods can be implemented on general unstructured meshes and are well suited to handle h/p adaptivity since grid refinement or coarsening can be achieved without taking into account the continuity restrictions typical of conforming methods, and since different order of approximations can be used on different elements/subdomains. Adaptivity is of particular importance in compressible flow given the complexity of the solution structure and geometries involved.
- **Parallelization** The HDG methods remain highly parallelizable even when implicit time integration is used since the local problems are formulated at the

element or subdomain level, they can be solved independently for each of the subdomain blocks. For the global problem, the iterative techniques with p-multigrid and block ILU preconditioning developed for DG methods can also be applied here [26].

We attempt to give an overview of recent developments of the HDG methods for fluid dynamics. In Sect. 2 we describe the basic ideas of HDG methodology for a convection-diffusion model equation: a mixed formulation of the model equation, a characterization of the numerical solution in terms of the approximate trace, relationship between the HDG method and the standard DG methods, the choice of the stabilization parameter, and the local postprocessing to improve the order of convergence. In Sect. 3 we show how the main ideas can be extended to time-dependent and nonlinear convection-diffusion problems, Stokes flows, and incompressible Navier–Stokes equations. In Sect. 4 we present numerical results for fluid dynamics to demonstrate the performance and accuracy of the HDG method. Finally, in Sect. 5, we end the paper with some concluding remarks on future developments.

2 The HDG Method

2.1 The Convection-Diffusion Model Equation

We will describe the main ideas behind the hybridized discontinuous Galerkin method using the linear convection-diffusion equation as a model problem

$$\nabla \cdot (\boldsymbol{c} u) - \nabla \cdot (\kappa \nabla u) = f, \quad \text{in } \Omega, \tag{1}$$

with boundary conditions

$$\begin{aligned} u &= g_D, \text{ on } \Gamma_D, \\ (-\kappa \nabla u + \boldsymbol{c} u) \cdot \boldsymbol{n} &= g_N, \text{ on } \Gamma_N. \end{aligned} \tag{2}$$

Here u is the field variable, $\boldsymbol{c}$ and $\kappa > 0$ are constant and f, g_D and g_N are given (see [20] for additional details).

We introduce the auxiliary variable $\boldsymbol{q} = -\kappa \nabla u$ and rewrite the above equation as a first order system of equations

$$\begin{aligned} \boldsymbol{q} + \kappa \nabla u &= 0, \text{ in } \Omega, \\ \nabla \cdot (\boldsymbol{c} u + \boldsymbol{q}) &= f, \text{ in } \Omega, \end{aligned} \tag{3}$$

with boundary conditions

$$\begin{aligned} u &= g_D, \text{ on } \Gamma_D, \\ (\boldsymbol{q} + \boldsymbol{c} u) \cdot \boldsymbol{n} &= g_N, \text{ on } \Gamma_N. \end{aligned} \tag{4}$$

Next, we introduce the notation necessary for the description of the HDG method.

2.2 Mesh and Trace Operators

Let $\mathcal{T}_h$ be a collection of disjoint elements that partition Ω. We denote by $\partial\mathcal{T}_h$ the set $\{\partial K : K \in \mathcal{T}_h\}$. For an element K of the collection $\mathcal{T}_h$, $F = \partial K \cap \partial\Omega$ is the boundary face if the $d-1$ Lebesgue measure of F is nonzero. For two elements K^+ and K^- of the collection $\mathcal{T}_h$, $F = \partial K^+ \cap \partial K^-$ is the interior face between K^+ and K^- if the $d-1$ Lebesgue measure of F is nonzero. Let $\mathcal{E}_h^o$ and $\mathcal{E}_h^\partial$ denote the set of interior and boundary faces, respectively. We denote by $\mathcal{E}_h$ the union of $\mathcal{E}_h^o$ and $\mathcal{E}_h^\partial$.

Let $\boldsymbol{n}^+$ and $\boldsymbol{n}^-$ be the outward unit normals of ∂K^+ and ∂K^-, respectively, and let $(\boldsymbol{q}^\pm, u^\pm)$ be the traces of $(\boldsymbol{q}, u)$ on F from the interior of $K^\pm$. Then, we define the mean values $\{\cdot\}$ and jumps $[\![\cdot]\!]$ as follows. For $F \in \mathcal{E}_h^o$, we set

$$\begin{aligned} \{\boldsymbol{q}\} &= (\boldsymbol{q}^+ + \boldsymbol{q}^-)/2 & \{u\} &= (u^+ + u^-)/2, \\ [\![\boldsymbol{q}\cdot\boldsymbol{n}]\!] &= \boldsymbol{q}^+\cdot\boldsymbol{n}^+ + \boldsymbol{q}^-\cdot\boldsymbol{n}^- & [\![u\boldsymbol{n}]\!] &= u^+\boldsymbol{n}^+ + u^-\boldsymbol{n}^-. \end{aligned}$$

For $F \in \mathcal{E}_h^\partial$, the set of boundary edges on which $\boldsymbol{q}$ and u are singled value, we set

$$\begin{aligned} \{\boldsymbol{q}\} &= \boldsymbol{q} & \{u\} &= u, \\ [\![\boldsymbol{q}\cdot\boldsymbol{n}]\!] &= \boldsymbol{q}\cdot\boldsymbol{n} & [\![u\boldsymbol{n}]\!] &= u\boldsymbol{n}. \end{aligned}$$

Note that the jump in u is a vector, but the jump in $\boldsymbol{q}$ is a scalar. Furthermore, the jumps will be zero for a continuous function.

2.3 Approximation Spaces

Let $\mathcal{P}_m(D)$ denote the set of polynomials of degree at most m on a domain D. We introduce discontinuous finite element spaces

$$W_h = \{w \in L^2(\Omega) : w|_K \in \mathcal{P}_k(K),\ \forall K \in \mathcal{T}_h\},$$

and

$$\boldsymbol{V}_h = \{\boldsymbol{v} \in (L^2(\Omega))^d : \boldsymbol{v}|_K \in (\mathcal{P}_k(K))^d,\ \forall K \in \mathcal{T}_h\}.$$

Here $L^2(D)$ is the space of square integrable functions on D. In addition, we introduce a traced finite element space

$$M_h = \{\mu \in L^2(\mathcal{E}_h) : \mu|_F \in \mathcal{P}_k(F),\ \forall F \in \mathcal{E}_h\}.$$

We also set $M_h(g_D) = \{\mu \in M_h : \mu = \mathsf{P} g_D \text{ on } \Gamma_D\}$, where P denotes the L^2-projection into the space $\{\mu|_{\partial\Omega}\ \forall \mu \in M_h\}$. Note that M_h consists of functions which are continuous inside the faces (or edges) $F \in \mathscr{E}_h$ and discontinuous at their borders.

For functions $\boldsymbol{w}$ and $\boldsymbol{v}$ in $(L^2(D))^d$, we denote $(\boldsymbol{w}, \boldsymbol{v})_D = \int_D \boldsymbol{w} \cdot \boldsymbol{v}$. For functions u and v in $L^2(D)$, we denote $(u, v)_D = \int_D uv$ if D is a domain in $\mathbb{R}^d$ and $\langle u, v\rangle_D = \int_D uv$ if D is a domain in $\mathbb{R}^{d-1}$. We finally introduce

$$(w, v)_{\mathscr{T}_h} = \sum_{K \in \mathscr{T}_h} (w, v)_K, \ \langle \zeta, \rho\rangle_{\partial \mathscr{T}_h} = \sum_{K \in \mathscr{T}_h} \langle w, v\rangle_{\partial K}, \ \langle \mu, \eta\rangle_{\mathscr{E}_h} = \sum_{F \in \mathscr{E}_h} \langle \mu, \eta\rangle_F,$$

for functions w, v defined on $\mathscr{T}_h$, ζ, ρ defined on $\partial \mathscr{T}_h$, and μ, η defined on $\mathscr{E}_h$.

2.4 HDG Formulation

We seek an approximation $(\boldsymbol{q}_h, u_h) \in \boldsymbol{V}_h \times W_h$ such that for all $K \in \mathscr{T}_h$,

$$\begin{aligned} (\kappa^{-1}\boldsymbol{q}_h, \boldsymbol{v})_K - (u_h, \nabla \cdot \boldsymbol{v})_K + \langle \widehat{u}_h, \boldsymbol{v} \cdot \boldsymbol{n}\rangle_{\partial K} &= 0, & \forall\, \boldsymbol{v} \in (\mathscr{P}_k(K))^d, \\ -(\boldsymbol{c}u_h + \boldsymbol{q}_h, \nabla w)_K + \langle (\widehat{\boldsymbol{c}u}_h + \widehat{\boldsymbol{q}}_h) \cdot \boldsymbol{n}, w\rangle_{\partial K} &= (f, w)_K, & \forall\, w \in \mathscr{P}_k(K). \end{aligned} \tag{5}$$

Here, the numerical traces $\widehat{\boldsymbol{c}u}_h + \widehat{\boldsymbol{q}}_h$ and $\widehat{u}_h$ are approximations to $\boldsymbol{c}u - \kappa \nabla u$ and u over ∂K, respectively. Next, we express $(\boldsymbol{q}_h, u_h)$ in terms of $\widehat{u}_h$ only. To this end, we consider numerical traces $\widehat{\boldsymbol{c}u}_h + \widehat{\boldsymbol{q}}_h$ of the form

$$\widehat{\boldsymbol{c}u}_h + \widehat{\boldsymbol{q}}_h = \boldsymbol{c}\widehat{u}_h + \boldsymbol{q}_h + \tau(u_h - \widehat{u}_h)\boldsymbol{n}, \quad \text{on } \partial K. \tag{6}$$

Here, τ is the so-called *local stabilization parameter*; it has an important effect on both the stability and accuracy of the resulting scheme. The selection of the value of the parameter τ will be described below. Note that both $\widehat{\boldsymbol{c}u}_h$ and $\boldsymbol{c}\widehat{u}_h$ are different approximations to the same quantity $\boldsymbol{c}u$ and that the former is defined in terms of the latter.

We next express $\widehat{u}_h$ in terms of the boundary data g_D and a new variable $\lambda_h \in M_h(0)$ as

$$\widehat{u}_h = \begin{cases} \mathsf{P} g_D, & \text{on } \mathscr{E}_h \cap \Gamma_D, \\ \lambda_h, & \text{on } \mathscr{E}_h \backslash \Gamma_D. \end{cases}$$

By adding the contributions of (5) over all the elements and enforcing the continuity of the normal component of the numerical flux, we arrive at the following problem: find an approximation $(\boldsymbol{q}_h, u_h, \lambda_h) \in \boldsymbol{V}_h \times W_h \times M_h(0)$ such that

$$\begin{aligned} (\kappa^{-1}\boldsymbol{q}_h, \boldsymbol{v})_{\mathscr{T}_h} - (u_h, \nabla \cdot \boldsymbol{v})_{\mathscr{T}_h} + \langle \lambda_h, \boldsymbol{v} \cdot \boldsymbol{n}\rangle_{\partial \mathscr{T}_h} &= -\langle g_D, \boldsymbol{v} \cdot \boldsymbol{n}\rangle_{\Gamma_D}, & \forall \boldsymbol{v} \in \boldsymbol{V}_h, \\ -(\boldsymbol{c}u_h + \boldsymbol{q}_h, \nabla w)_{\mathscr{T}_h} + \langle (\widehat{\boldsymbol{c}u}_h + \widehat{\boldsymbol{q}}_h) \cdot \boldsymbol{n}, w\rangle_{\partial \mathscr{T}_h} &= (f, w)_{\mathscr{T}_h}, & \forall w \in W_h, \\ \langle [\![(\widehat{\boldsymbol{c}u}_h + \widehat{\boldsymbol{q}}_h) \cdot \boldsymbol{n}]\!], \mu\rangle_{\mathscr{E}_h} &= \langle g_N, \mu\rangle_{\Gamma_N}, & \forall \mu \in M_h(0). \end{aligned} \tag{7}$$

Note that the Dirichlet boundary condition has been enforced by requiring that $\widehat{u}_h = \mathsf{P}g_D$ on $\mathscr{E}_h \cap \Gamma_D$, whereas the continuity of the normal component of $\widehat{cu}_h + \widehat{\boldsymbol{q}}_h$ is enforced explicitly by the last equation.

We observe that λ_h is uniquely defined over each edge since λ_h belongs to M_h. Furthermore, if $[\![(\widehat{cu}_h + \widehat{\boldsymbol{q}}_h) \cdot \boldsymbol{n}]\!]$ belongs to M_h, then the last equation (7) simply states that $[\![(\widehat{cu}_h + \widehat{\boldsymbol{q}}_h) \cdot \boldsymbol{n}]\!] = 0$ pointwise over $\mathscr{E}_h \backslash \Gamma_N$ and that $(\widehat{cu}_h + \widehat{\boldsymbol{q}}_h) \cdot \boldsymbol{n} = \mathsf{P}g_N$ on Γ_N; in other words, the normal component of the numerical trace $\widehat{cu}_h + \widehat{\boldsymbol{q}}_h$ is single-valued. Hence, both λ_h and $\widehat{cu}_h + \widehat{\boldsymbol{q}}_h$ are conservative fluxes according to the definition in [1]. Note that our numerical traces remain conservative even when the diffusion coefficient κ is discontinuous at the interior element interface.

We note that, due to the discontinuous nature of both $\boldsymbol{V}_h$ and W_h, the first two equations in (6) can be used to eliminate both $\boldsymbol{q}_h$ and u_h to obtain a weak formulation in terms of λ_h only and thus a global system of equations involving the degrees of freedom of λ_h, as described below.

2.5 Characterization of the Numerical Trace

We first introduce the so-called local solver which associate to each function $(\mathsf{m}, f) \in M_h \times L^2(\Omega)$, the pair $(\boldsymbol{q}_h^{\mathsf{m},f}, u_h^{\mathsf{m},f})$ on Ω whose restriction to each element K is in $(\mathscr{P}_k(K))^d \times \mathscr{P}_k(K)$ and satisfies

$$(\kappa^{-1}\boldsymbol{q}_h^{\mathsf{m},f}, \boldsymbol{v})_K - (u_h^{\mathsf{m},f}, \nabla \cdot \boldsymbol{v})_K = -\langle \mathsf{m}, \boldsymbol{v} \cdot \boldsymbol{n} \rangle_{\partial K}, \tag{8a}$$

$$-(\boldsymbol{c}u_h^{\mathsf{m},f} + \boldsymbol{q}_h^{\mathsf{m},f}, \nabla w)_K + \left\langle (\widehat{\boldsymbol{c}u}_h^{\mathsf{m},f} + \widehat{\boldsymbol{q}}_h^{\mathsf{m},f}) \cdot \boldsymbol{n}, w \right\rangle_{\partial K} = (f, w)_K, \tag{8b}$$

for all $(\boldsymbol{v}, w) \in (\mathscr{P}_k(K))^d \times \mathscr{P}_k(K)$, where

$$\widehat{\boldsymbol{c}u}_h^{\mathsf{m},f} + \widehat{\boldsymbol{q}}_h^{\mathsf{m},f} = \boldsymbol{c}\mathsf{m} + \boldsymbol{q}_h^{\mathsf{m},f} + \tau(u_h^{\mathsf{m},f} - \mathsf{m})\boldsymbol{n}. \tag{8c}$$

It is now clear, see (7), that the approximate solution $(\boldsymbol{q}_h, u_h) \in \boldsymbol{V}_h \times W_h$ satisfies

$$1\boldsymbol{q}_h = \boldsymbol{q}_h^{\lambda_h, f}, \qquad u_h = u_h^{\lambda_h, f}, \tag{9a}$$

where $\lambda_h \in M_h(0)$ is such that

$$\left\langle [\![(\widehat{\boldsymbol{c}u}_h^{\lambda_h, f} + \widehat{\boldsymbol{q}}_h^{\lambda_h, f}) \cdot \boldsymbol{n}]\!], \mu \right\rangle_{\mathscr{E}_h} = \langle g_N, \mu \rangle_{\Gamma_N}, \quad \forall \mu \in M_h(0). \tag{9b}$$

We next show that we can eliminate $\boldsymbol{q}_h$ and u_h from the above equations to obtain a weak formulation in terms of λ_h only.

Let $(\boldsymbol{q}_h^{\mathsf{m},0}, u_h^{\mathsf{m},0})$ (respectively, $(\boldsymbol{q}_h^{0,f}, u_h^{0,f})$) solve (8) when we set $f = 0$ (respectively, $\mathsf{m} = 0$). If, for all η and $\mu \in M_h$, we set

$$a_h(\eta,\mu) = -\left\langle [\![(\widehat{cu}_h^{\eta,0} + \widehat{\boldsymbol{q}}_h^{\eta,0}) \cdot \boldsymbol{n}]\!], \mu \right\rangle_{\mathscr{E}_h}, \tag{10a}$$

$$b_h(\mu) = \left\langle [\![(\widehat{cu}_h^{0,f} + \widehat{\boldsymbol{q}}_h^{0,f}) \cdot \boldsymbol{n}]\!], \mu \right\rangle_{\mathscr{E}_h}, \tag{10b}$$

we have from (9b) and linearity of the problem (8) that the function $\lambda_h \in M_h(0)$ is the solution of the variational formulation

$$a_h(\lambda_h,\mu) = b_h(\mu) - \langle g_N, \mu \rangle_{\Gamma_N}, \quad \forall \mu \in M_h(0). \tag{11}$$

The existence and uniqueness of the numerical trace λ_h is presented in [20].

The above weak formulation gives rise to a matrix system of the form

$$\mathbb{K}\,\Lambda = \mathbb{F}, \tag{12}$$

where Λ is the vector of degrees of freedom of λ_h, $\mathbb{K}$ is the matrix associated with the bilinear form $a_h(\cdot,\cdot)$, and $\mathbb{F}$ the vector associated with the linear form $b_h(\cdot) - \langle g_N, \cdot \rangle_{\Gamma_N}$. Note that since

$$a_h(\eta,\mu) = -\left\langle (\widehat{cu}_h^{\eta,0} + \widehat{\boldsymbol{q}}_h^{\eta,0}) \cdot \boldsymbol{n}, \mu \right\rangle_{\partial\mathscr{T}_h},$$

we can easily deduce that if the support of η is the interior face $F = \partial K^+ \cap \partial K^-$, or the boundary face $F = \partial K \cap \partial\Omega$, then $a_h(\eta,\mu) = 0$ when the support of μ does not intersect $\partial K^+ \cup \partial K^-$, or ∂K, respectively. Thus, the matrix $\mathbb{K}$ has a block-structure of blocks of square matrices of order dim $\mathscr{P}_k$. In each block-row or block-column, there are at most five non-zero blocks when the elements are triangles, and at most seven non-zero blocks in three space dimension.

The construction of the matrix system (12) can be carried out in two steps. In the first step, we solve the local problem (8) for every element $K \in \mathscr{T}_h$. In the second step, we evaluate the face integrals (10) by using the standard finite element quadrature rule and assembly. This procedure can be implemented for arbitrary polynomial degrees. The detailed implementation discussed in [20] is omitted here to save space.

2.6 *Relation to Other DG Methods*

In order to derive an explicit expression for the numerical traces in terms of $(u_h, \boldsymbol{q}_h)$, we proceed as follows. Since the conservativity condition implies $[\![(\widehat{cu_h} + \widehat{\boldsymbol{q}}_h) \cdot \boldsymbol{n}]\!] = 0$ pointwise, we have, using expression (6), that

$$[\![\boldsymbol{q}_h \cdot \boldsymbol{n}]\!] + \tau^+ u_h^+ + \tau^- u_h^- - (\tau^+ + \tau^-)\lambda_h = 0, \qquad \text{on } \mathscr{E}_h^o.$$

Solving for λ_h and inserting the result into the expression for $\widehat{cu}_h + \widehat{\boldsymbol{q}}_h$ (6), we obtain on $\mathscr{E}_h^o$

$$
\begin{aligned}
\lambda_h &= \frac{\tau^+}{\tau^+ + \tau^-} u_h^+ + \frac{\tau^-}{\tau^+ + \tau^-} u_h^- + \left(\frac{1}{\tau^+ + \tau^-} \right) [\![\boldsymbol{q}_h \cdot \boldsymbol{n}]\!], \\
\widehat{\boldsymbol{c}u}_h + \widehat{\boldsymbol{q}}_h &= \boldsymbol{c}\lambda_h + \frac{\tau^-}{\tau^+ + \tau^-} \boldsymbol{q}_h^+ + \frac{\tau^+}{\tau^+ + \tau^-} \boldsymbol{q}_h^- + \left(\frac{\tau^+ \tau^-}{\tau^+ + \tau^-} \right) [\![u_h \boldsymbol{n}]\!].
\end{aligned}
\tag{13}
$$

These expressions for the numerical traces highlight the relationship between the HDG method and the more standard DG methods, as discussed below.

In the convective limit we have $\kappa = 0$ and consequently $\boldsymbol{q}_h = 0$. In this case, the expressions (13) become

$$
\begin{aligned}
\lambda_h &= \frac{\tau^+}{\tau^+ + \tau^-} u_h^+ + \frac{\tau^-}{\tau^+ + \tau^-} u_h^-, \\
\widehat{\boldsymbol{c}u}_h \cdot \boldsymbol{n}^+ &= \frac{\tau^+}{\tau^+ + \tau^-} (\boldsymbol{c} \cdot \boldsymbol{n}^+ + \tau^-) u_h^+ + \frac{\tau^-}{\tau^+ + \tau^-} (\boldsymbol{c} \cdot \boldsymbol{n}^+ - \tau^+) u_h^-.
\end{aligned}
\tag{14}
$$

In the diffusive limit $\boldsymbol{c} = \boldsymbol{0}$, expressions (13) become

$$
\begin{aligned}
\lambda_h &= \frac{\tau^+}{\tau^+ + \tau^-} u_h^+ + \frac{\tau^-}{\tau^+ + \tau^-} u_h^- + \left(\frac{1}{\tau^+ + \tau^-} \right) [\![\boldsymbol{q}_h \cdot \boldsymbol{n}]\!], \\
\widehat{\boldsymbol{q}}_h &= \frac{\tau^-}{\tau^+ + \tau^-} \boldsymbol{q}_h^+ + \frac{\tau^+}{\tau^+ + \tau^-} \boldsymbol{q}_h^- + \left(\frac{\tau^+ \tau^-}{\tau^+ + \tau^-} \right) [\![u_h \boldsymbol{n}]\!].
\end{aligned}
\tag{15}
$$

This case has been originally studied in [3]; see also [4, 9, 13].

By rearranging terms these expressions can be transformed into the more standard form considered in [3],

$$
\begin{aligned}
\widehat{\boldsymbol{q}}_h &= \{\boldsymbol{q}_h\} + C_{11} [\![u_h \boldsymbol{n}]\!] + \boldsymbol{C}_{12} [\![\boldsymbol{q}_h \cdot \boldsymbol{n}]\!], \\
\lambda_h = \hat{u}_h &= \{u_h\} - \boldsymbol{C}_{12} \cdot [\![u_h \boldsymbol{n}]\!] + C_{22} [\![\boldsymbol{q}_h \cdot \boldsymbol{n}]\!] .
\end{aligned}
\tag{16}
$$

where,

$$
C_{11} = \left(\frac{\tau^+ \tau^-}{\tau^+ + \tau^-} \right), \quad \boldsymbol{C}_{12} = \frac{1}{2} \left(\frac{[\![\tau \boldsymbol{n}]\!]}{\tau^+ + \tau^-} \right), \quad C_{22} = \left(\frac{1}{\tau^+ + \tau^-} \right) .
$$

It is interesting to note that for the simple choice of $\tau^\pm$ of order unity everywhere HDG methods yield optimal convergence rate of $k + 1$ for both the scalar variable and the flux, and that they display superconvergence properties of the scalar variable [4, 6, 13].

We point out that in the Local DG method [15], the trace λ_h is chosen to be independent of $\boldsymbol{q}_h$, that is $C_{22} = 0$. This has the advantage of allowing the degrees of freedom associated with the $\boldsymbol{q}_h$ to be locally eliminated and a global system involving only the degrees of freedom associated to u_h is thus solved. However, using $C_{22} = 0$ yields suboptimal convergence for the approximate gradient. It is shown in [13] that the superconvergent schemes require that C_{22} be non-zero.

While this presents a serious inconvenience for LDG methods, for HDG methods this represents no difficulty.

2.7 The Local Stabilization Parameter τ

To account for the diffusion and convection effects our local stabilization parameter τ will take the following form

$$\tau = \tau_d + \tau_c$$

where τ_d and τ_c are the local stabilization parameters related to the diffusion and convection, respectively. This allows us to write each component of the numerical trace $\widehat{\boldsymbol{q}}_h + \widehat{\boldsymbol{c}u_h}$ as

$$\begin{aligned}\widehat{\boldsymbol{q}}_h &= \boldsymbol{q}_h + \tau_d(u_h - \lambda_h)\boldsymbol{n},\\ \widehat{\boldsymbol{c}u_h} &= \boldsymbol{c}\lambda_h + \tau_c(u_h - \lambda_h)\boldsymbol{n}.\end{aligned}$$

A suitable expression for τ_c and τ_d is to take on each edge $\tau_c^+ = \tau_c^- = \eta_c$ and $\tau_d^+ = \tau_d^- = \eta_d$, where

$$\eta_c = |\boldsymbol{c} \cdot \boldsymbol{n}|, \qquad \eta_d = \frac{\kappa}{\ell}, \tag{17}$$

where ℓ denotes a representative diffusive length scale which is typically of unity order and independent of the mesh size h. In this case, the expressions for the numerical traces becomes

$$\begin{aligned}\lambda_h &= \{u_h\} + \frac{1}{2\tau}[\![\boldsymbol{q}_h \cdot \boldsymbol{n}]\!],\\ \widehat{\boldsymbol{c}u_h} + \widehat{\boldsymbol{q}}_h &= \boldsymbol{c}\,\lambda_h + \{\boldsymbol{q}_h\} + \frac{\tau}{2}[\![u_h \boldsymbol{n}]\!].\end{aligned}$$

It can be shown that the HDG method is well-defined with the above choice of the stabilization parameter. Alternative forms for τ_c and τ_d can be found in [20].

Numerical experiment and theory (see [6, 20]) confirm that the above choice of the stabilization parameter is optimal in the sense that both the approximate scalar variable and gradient converge with the optimal order $k + 1$. We point out that our stabilization parameter is independent of the polynomial degree and the mesh size. This is different from some DG methods such as the interior penalty DG method which typically select stabilization parameter to depend on the mesh size.

2.8 Local Postprocessing

We first show that we can postprocess the total approximate flux $\boldsymbol{q}_h^T = \boldsymbol{q}_h + \boldsymbol{c}u_h$ and its numerical trace $\widehat{\boldsymbol{q}}_h^T = \widehat{\boldsymbol{q}}_h + \widehat{\boldsymbol{c}u_h}$ with an element-by-element procedure to obtain an approximation of $\boldsymbol{q}+c u$, denoted $\boldsymbol{q}_h^{T*}$ that belongs to $H(\mathrm{div}, \Omega)$ and also converges in an optimal fashion [2, 6, 14]. On each simplex $K \in \mathscr{T}_h$, we define the new total flux $\boldsymbol{q}_h^{T*}$ as the only element of $(\mathscr{P}_k(K))^d + \boldsymbol{x}\mathscr{P}_k(K)$ satisfying, for $k \geq 0$,

$$\begin{aligned} \langle(\boldsymbol{q}_h^{T*} - \widehat{\boldsymbol{q}}_h^{T}) \cdot \boldsymbol{n}, \mu\rangle_F &= 0, \quad \forall \mu \in \mathscr{P}_k(F), \forall F \in \partial K, \\ (\boldsymbol{q}_h^{T*} - \boldsymbol{q}_h^{T}, \boldsymbol{v})_K &= 0, \quad \forall \boldsymbol{v} \in (\mathscr{P}_{k-1}(K))^d \quad \text{if } k \geq 1. \end{aligned} \tag{18}$$

It is clear that the function $\boldsymbol{q}_h^{T*}$ belongs to $H(\mathrm{div}, \Omega)$, thanks to the singlevaluedness of the normal component of the numerical trace $\widehat{\boldsymbol{q}}_h + \widehat{c u}_h$.

Next, we consider postprocessing u_h, $\boldsymbol{q}_h$, and $\widehat{\boldsymbol{q}}_h$ to obtain the new approximate scalar variable u_h^* of u. Towards this end, we find $(u_h^*, \boldsymbol{q}_h^*, \lambda_h^*) \in \mathscr{P}_{k^*}(K) \times (\mathscr{P}_{k^*}(K))^d \times (\mathscr{P}_{k^*}(F))^{d+1}$ for $k^* = k+1$ on the simplex $K \in \mathscr{T}_h$ such that

$$\begin{aligned} \left(\kappa^{-1}\nabla \boldsymbol{q}_h^*, \boldsymbol{v}\right)_K - \left(u_h^*, \nabla\cdot\boldsymbol{v}\right)_K + \left\langle\lambda_h^*, \boldsymbol{v}\cdot\boldsymbol{n}\right\rangle_{\partial K} &= 0 \\ -\left(\boldsymbol{q}_h^* + c u_h^*, \nabla w\right)_K + \left\langle(\widehat{\boldsymbol{q}}_h^* + \widehat{c u}_h^*)\cdot\boldsymbol{n}, w\right\rangle_{\partial K} &= \left(\nabla\cdot\boldsymbol{q}_h^{T*}, w\right)_K, \\ \left\langle(\widehat{\boldsymbol{q}}_h^* + \widehat{c u}_h^*)\cdot\boldsymbol{n}, \mu\right\rangle_{\partial K} &= \left\langle\boldsymbol{q}_h^{T*}\cdot\boldsymbol{n}, \mu\right\rangle_{\partial K}, \\ \left(u_h^*, 1\right)_K &= (u_h, 1)_K, \end{aligned} \tag{19}$$

for all $(\boldsymbol{v}, w, \mu) \in (\mathscr{P}_{k^*}(K))^d \times \mathscr{P}_{k^*}(K) \times (\mathscr{P}_{k^*}(F))^{d+1}$, where

$$\widehat{\boldsymbol{q}}_h^* + \widehat{c u}_h^* = \boldsymbol{q}_h^* + c\lambda_h^* + \tau(u_h^* - \lambda_h^*)\boldsymbol{n}.$$

We note that this local postprocessing is nothing but the HDG discretization *at the element level* of the following convection-diffusion Neumann problem

$$\begin{aligned} \nabla\cdot(-\kappa\nabla u + c u) &= \nabla\cdot\boldsymbol{q}_h^{T*}, \quad &&\text{in } K, \\ (-\kappa\nabla u + c u)\cdot\boldsymbol{n} &= \boldsymbol{q}_h^{T*}\cdot\boldsymbol{n}, \quad &&\text{on } \partial K, \\ (u, 1)_K &= (u_h, 1)_K. \end{aligned} \tag{20}$$

Therefore, the new approximation u_h^* is even much less expensive to compute than the original approximation u_h. This is because the local problems (19) have very few degrees of freedom and also because they can be solved independently of each other.

Our postprocessing procedure relies on the optimal convergence of $\boldsymbol{q}_h^{T*}$ and its divergence $\nabla\cdot\boldsymbol{q}_h^{T*}$, and on the superconvergence of the average of the approximate scalar variable u_h. In fact, these properties for the HDG method have been theoretically analyzed and confirmed by numerical experiments for the steady symmetric diffusion case in [4, 13]: both $\boldsymbol{q}_h^{T*}$ and $\nabla\cdot\boldsymbol{q}_h^{T*}$ converge with order $k+1$, while $(u_h, 1)_K$ superconverges with order $k+2$. We may thus expect that the scalar variable u_h^* converges with order $k+2$.

3 Extensions of the Basic Algorithm

In this section, we present several extensions of the basic algorithm described in the previous section.

3.1 *Time-Dependent Convection-Diffusion Problems*

We consider the time-dependent convection-diffusion model written as a system of first-order equations

$$\begin{aligned} \boldsymbol{q} + \kappa \nabla u &= 0, && \text{in } \Omega \times (0,T], \\ \frac{\partial u}{\partial t} + \nabla \cdot (\boldsymbol{c} u + \boldsymbol{q}) &= f, && \text{in } \Omega \times (0,T], \\ u &= g_D, && \text{on } \Gamma_D \times (0,T], \\ (\boldsymbol{q} + \boldsymbol{c} u) \cdot \boldsymbol{n} &= g_N, && \text{on } \Gamma_N \times (0,T], \\ u &= u_0, && \text{in } \Omega \text{ for } t = 0. \end{aligned} \tag{21}$$

The HDG method of lines for the above problem seeks an approximation $(\boldsymbol{q}_h, u_h) \in \boldsymbol{V}_h \times W_h$ such that for all $K \in \mathscr{T}_h$,

$$\begin{aligned} (\kappa^{-1}\boldsymbol{q}_h, \boldsymbol{v})_K - (u_h, \nabla \cdot \boldsymbol{v})_K + \langle \widehat{u}_h, \boldsymbol{v} \cdot \boldsymbol{n} \rangle_{\partial K} &= 0, \\ (\frac{\partial u_h}{\partial t}, w)_K - (\boldsymbol{c} u_h + \boldsymbol{q}_h, \nabla w)_K + \langle (\widehat{\boldsymbol{c} u}_h + \widehat{\boldsymbol{q}}_h) \cdot \boldsymbol{n}, w \rangle_{\partial K} &= (f, w)_K, \end{aligned} \tag{22}$$

for all $(\boldsymbol{v}, w) \in (\mathscr{P}_k(K))^d \times \mathscr{P}_k(K)$ and for all $t \in (0,T]$. Here, the numerical traces $\widehat{\boldsymbol{c} u}_h + \widehat{\boldsymbol{q}}_h$ and $\widehat{u}_h$ are approximations to $\boldsymbol{c} u - \kappa \nabla u$ and u over ∂K, respectively.

The above HDG formulation (22) can then be discretized in time using an appropriate time-stepping scheme. Here we consider backward difference formulaes (BDF) for the discretization of the time derivative. For instance, using the Backward-Euler scheme at time-level t^n with timestep Δt^n the HDG method then seeks an approximation $(\boldsymbol{q}_h^n, u_h^n, \lambda_h^n) \in \boldsymbol{V}_h \times W_h \times M_h(0)$ such that

$$\begin{aligned} (\kappa^{-1}\boldsymbol{q}_h^n, \boldsymbol{v})_{\mathscr{T}_h} - (u_h^n, \nabla \cdot \boldsymbol{v})_{\mathscr{T}_h} + \langle \lambda_h^n, \boldsymbol{v} \cdot \boldsymbol{n} \rangle_{\partial \mathscr{T}_h} &= -\langle g_D, \boldsymbol{v} \cdot \boldsymbol{n} \rangle_{\Gamma_D}, \\ \frac{1}{\Delta t^n} \left(u_h^n, w\right)_{\mathscr{T}_h} - (\boldsymbol{c} u_h^n + \boldsymbol{q}_h^n, \nabla w)_{\mathscr{T}_h} & \\ + \langle (\widehat{\boldsymbol{c} u}_h^n + \widehat{\boldsymbol{q}}_h^n) \cdot \boldsymbol{n}, w \rangle_{\partial \mathscr{T}_h} &= (f, w)_{\mathscr{T}_h} + \frac{1}{\Delta t^n} \left(u_h^{k-1}, w\right)_{\mathscr{T}_h}, \\ \langle [\![(\widehat{\boldsymbol{c} u}_h^n + \widehat{\boldsymbol{q}}_h^n) \cdot \boldsymbol{n}]\!], \mu \rangle_{\mathscr{E}_h} &= \langle g_N, \mu \rangle_{\Gamma_N}, \end{aligned} \tag{23}$$

for all $(\boldsymbol{v}, w, \mu) \in \boldsymbol{V}_h \times W_h \times M_h(0)$, where, as we did for the steady-state case, we choose $\widehat{\boldsymbol{c} u}_h^n + \widehat{\boldsymbol{q}}_h^n$ of the form

$$\widehat{\boldsymbol{c} u}_h^n + \widehat{\boldsymbol{q}}_h^n = \boldsymbol{c}\, \widehat{u}_h^n + \boldsymbol{q}_h^n + \tau (u_h^n - \widehat{u}_h^n) \boldsymbol{n}, \quad \text{on } \partial K.$$

This discrete system has a similar form as the system (7) for the steady-state case. Hence, we can apply exactly the same solution procedure described earlier for the steady-state case to the time-dependent case at every time step.

Of course, a similar procedure can be applied to treat any higher-order BDF method such as the widely used second-order and third-order BDF schemes. The HDG method can also work with other implicit time-stepping methods such as the fully implicit Runge–Kutta methods and DG methods in time.

The post-processing method described for the steady state convection diffusion problem can also be applied in the time-dependent case with identical results. That is, both $\boldsymbol{q}_h^T$ and $\nabla\cdot\boldsymbol{q}_h^T$ converge spatially with order $p+1$, while $(u_h, 1)_K$ superconverges in space with order $p+2$. This means that it is then possible to reconstruct, at any desired time level, a new scalar variable, u_h^*, which superconverges with order $p+2$ (see [20] for additional details).

3.2 Nonlinear Convection-Diffusion Problems

Here, we describe the HDG method for steady-state nonlinear convection-diffusion equations presented in [21]. Consider a nonlinear convection-diffusion equation of the form

$$\begin{aligned} -\nabla\cdot(\kappa\nabla u) + \nabla\cdot\boldsymbol{F}(u) &= f, \quad \text{in } \Omega, \\ u &= g_D, \text{ on } \partial\Omega. \end{aligned} \tag{24}$$

We rewrite the above equation as a first order system of equations

$$\begin{aligned} \boldsymbol{q} + \kappa\nabla u &= 0, && \text{in } \Omega, \\ \nabla\cdot(\boldsymbol{q} + \boldsymbol{F}(u)) &= f, && \text{in } \Omega, \\ u &= g_D, && \text{on } \partial\Omega. \end{aligned} \tag{25}$$

Here, $\boldsymbol{F} \in (L^\infty(\Omega))^d$ are vector-valued nonlinear functions of the scalar variable u.

Multiplying the first two equations of (25) by test functions, integrating by parts, and enforcing the continuity of the normal component of the total numerical flux, we obtain the following problem: find an approximation $(\boldsymbol{q}_h, u_h, \widehat{u}_h) \in \boldsymbol{V}_h \times W_h \times M_h(g_D)$ such that

$$\begin{aligned} \left(\kappa^{-1}\boldsymbol{q}_h, \boldsymbol{v}\right)_{\mathscr{T}_h} - (u_h, \nabla\cdot\boldsymbol{v})_{\mathscr{T}_h} + \langle\widehat{u}_h, \boldsymbol{v}\cdot\boldsymbol{n}\rangle_{\partial\mathscr{T}_h} &= 0, \\ -(\boldsymbol{q}_h + \boldsymbol{F}(u_h), \nabla w)_{\mathscr{T}_h} + \left\langle\left(\widehat{\boldsymbol{q}}_h + \widehat{\boldsymbol{F}}_h\right)\cdot\boldsymbol{n}, w\right\rangle_{\partial\mathscr{T}_h} &= (f, w)_{\mathscr{T}_h}, \\ \left\langle(\widehat{\boldsymbol{q}}_h + \widehat{\boldsymbol{F}}_h)\cdot\boldsymbol{n}, \mu\right\rangle_{\partial\mathscr{T}_h} &= 0, \end{aligned} \tag{26}$$

for all $(\boldsymbol{v}, w, \mu) \in \boldsymbol{V}_h \times W_h \times M_h(0)$, where

$$\widehat{\boldsymbol{q}}_h + \widehat{\boldsymbol{F}}_h = \boldsymbol{q}_h + \boldsymbol{F}(\widehat{u}_h) + \tau(u_h, \widehat{u}_h)(u_h - \widehat{u}_h)\boldsymbol{n}, \quad \text{on } \mathscr{E}_h. \tag{27}$$

This completes the definition of the general form of the HDG method. This nonlinear system of equations is solved by the Newton–Raphson method as described in [21]. Here we observe that, at each Newton iteration, we recover the HDG structure of the linear problem (7), and thus solve for the degrees of freedom of $\widehat{u}_h$ only.

The choice of the numerical flux $\widehat{\boldsymbol{q}}_h + \widehat{\boldsymbol{F}}_h$ is an extension of the expression for the numerical flux used for the linear case. The main difference is that, due to the nonlinearity of the convection, the *stabilization function* $\tau(\cdot,\cdot) : \partial\mathcal{T}_h \to \mathbb{R}$ can now be a nonlinear function of u_h and $\widehat{u}_h$. This implies that the last equation (26) *cannot* force the normal component of the total flux $\widehat{\boldsymbol{q}}_h + \widehat{\boldsymbol{F}}_h$ to be single valued on all interior faces $e \in \mathcal{E}_h^o$; it only forces its L^2*-projection* into $M_h(0)$ to be single valued. This is enough to guarantee the local conservativity of the method, as we can see from the second term of the left-hand side of the second equation (26).

Suitable expressions for the stabilization function and the associated entropy inequality as well as the extension to nonlinear time dependent problems and the postprocessing procedure are described in [21].

3.3 Stokes Flows

We describe here a hybridizable discontinuous Galerkin (HDG) method for the Stokes system [22]

$$\begin{aligned} -\nu\Delta\boldsymbol{u} + \nabla p &= \boldsymbol{f}, && \text{in } \Omega, \\ \nabla\cdot\boldsymbol{u} &= 0, && \text{in } \Omega, \\ \boldsymbol{u} &= \boldsymbol{g}, && \text{on } \partial\Omega. \end{aligned} \tag{28}$$

We rewrite the above equation as the following first order system of equations

$$\begin{aligned} \mathrm{L} - \nabla\boldsymbol{u} &= 0, && \text{in } \Omega \\ -\nu\nabla\cdot\mathrm{L} + \nabla p &= \boldsymbol{f}, && \text{in } \Omega, \\ \nabla\cdot\boldsymbol{u} &= 0, && \text{in } \Omega, \\ \boldsymbol{u} &= \boldsymbol{g} && \text{on } \partial\Omega. \end{aligned} \tag{29}$$

As usual we assume that $\boldsymbol{g}$ satisfies the compatibility condition $\int_{\partial\Omega} \boldsymbol{g}\cdot\boldsymbol{n} = 0$.

We first introduce discontinuous finite element approximation spaces for the gradient, velocity, and pressure as

$$\begin{aligned} \mathrm{G}_h &= \{\mathrm{G} \in (L^2(\mathcal{T}_h))^{d\times d} &&: \mathrm{G}|_K \in (\mathcal{P}_k(D))^{d\times d},\ \forall K \in \mathcal{T}_h\}, \\ \boldsymbol{V}_h &= \{\boldsymbol{v} \in (L^2(\mathcal{T}_h))^d &&: \boldsymbol{v}|_K \in (\mathcal{P}_k(K))^d,\ \forall K \in \mathcal{T}_h\}, \\ P_h &= \{q \in L^2(\mathcal{T}_h) &&: q|_K \in \mathcal{P}_k(K),\ \forall K \in \mathcal{T}_h\}. \end{aligned}$$

In addition, we introduce a finite element approximation space for the approximate trace of the velocity

$$\boldsymbol{M}_h = \{\boldsymbol{\mu} \in (L^2(\mathcal{E}_h))^d \ :\ \boldsymbol{\mu}|_F \in (\mathcal{P}_k(F))^d,\ \forall F \in \mathcal{E}_h\}.$$

We also set

$$\boldsymbol{M}_h(\boldsymbol{g}) = \{\boldsymbol{\mu} \in \boldsymbol{M}_h \; : \; \boldsymbol{\mu} = \mathsf{P}\boldsymbol{g} \text{ on } \partial\Omega\},$$

where P denotes the L^2-projection into the space $\{\boldsymbol{\mu}|_{\partial\Omega} \; \forall \; \boldsymbol{\mu} \in \boldsymbol{M}_h\}$. We further denote by $\overline{\Psi}_h$ the set of functions in $L^2(\partial\mathcal{T}_h)$ that are constant on each ∂K for all elements K

$$\overline{\Psi}_h = \{r \in L^2(\partial\mathcal{T}_h) \; : \; r \in \mathcal{P}_0(\partial K), \; \forall K \in \mathcal{T}_h\}.$$

The mean of our approximate pressure will belong to this space. For a function q in $L^2(\partial\mathcal{T}_h)$, the mean of q on the element boundary ∂K of an element K is defined as

$$\overline{q}|_{\partial K} = \frac{1}{|\partial K|}\int_{\partial K} q.$$

Obviously, we have $\overline{q} = q$ for any q in $\overline{\Psi}_h$.

We next define various inner products for our finite element spaces as

$$(r,q)_{\mathcal{T}_h} = \sum_{K\in\mathcal{T}_h}(r,q)_K, \quad (\boldsymbol{w},\boldsymbol{v})_{\mathcal{T}_h} = \sum_{K\in\mathcal{T}_h}(\boldsymbol{w},\boldsymbol{v})_K, \quad (\mathrm{H},\mathrm{G})_{\mathcal{T}_h} = \sum_{K\in\mathcal{T}_h}(\mathrm{H},\mathrm{G})_K,$$

for $r,q \in L^2(\mathcal{T}_h)$, $\boldsymbol{w},\boldsymbol{v} \in (L^2(\mathcal{T}_h))^d$, and $\mathrm{H},\mathrm{G} \in (L^2(\mathcal{T}_h))^{d\times d}$. We also define the boundary inner products as

$$\langle r,q\rangle_{\partial\mathcal{T}_h} = \sum_{K\in\mathcal{T}_h}\langle r,q\rangle_{\partial K}, \; \langle \boldsymbol{w},\boldsymbol{v}\rangle_{\partial\mathcal{T}_h} = \sum_{K\in\mathcal{T}_h}\langle \boldsymbol{w},\boldsymbol{v}\rangle_{\partial K}, \; \langle \mathrm{H},\mathrm{G}\rangle_{\partial\mathcal{T}_h} = \sum_{K\in\mathcal{T}_h}\langle \mathrm{H},\mathrm{G}\rangle_{\partial K},$$

for $r,q \in L^2(\mathcal{E}_h)$, $\boldsymbol{w},\boldsymbol{v} \in (L^2(\mathcal{E}_h))^d$, and $\mathrm{H},\mathrm{G} \in (L^2(\mathcal{E}_h))^{d\times d}$. Recall the standard notation $(\mathrm{H},\mathrm{G})_D = \int_D \mathrm{tr}(\mathrm{H}^T\mathrm{G})$, where tr is the trace operator.

The HDG method then seeks an approximation $(\mathrm{L}_h, \boldsymbol{u}_h, p_h, \widehat{\boldsymbol{u}}_h, \overline{\rho}_h) \in \mathrm{G}_h \times \boldsymbol{V}_h \times P_h \times \boldsymbol{M}_h(\boldsymbol{g}) \times \overline{\Psi}_h$ such that

$$\begin{aligned}
(\mathrm{L}_h,\mathrm{G})_{\mathcal{T}_h} + (\boldsymbol{u}_h, \nabla\cdot\mathrm{G})_{\mathcal{T}_h} - \langle\widehat{\boldsymbol{u}}_h, \mathrm{G}\cdot\boldsymbol{n}\rangle_{\partial\mathcal{T}_h} &= 0,\\
(\nu\mathrm{L}_h - p_h\mathrm{I}, \nabla\boldsymbol{v})_{\mathcal{T}_h} + \left\langle(-\nu\widehat{\mathrm{L}}_h + \widehat{p}_h\mathrm{I})\cdot\boldsymbol{n}, \boldsymbol{v}\right\rangle_{\partial\mathcal{T}_h} &= (\boldsymbol{f},\boldsymbol{v})_{\mathcal{T}_h},\\
-(\boldsymbol{u}_h, \nabla q)_{\mathcal{T}_h} + \langle\widehat{\boldsymbol{u}}_h\cdot\boldsymbol{n}, q-\overline{q}\rangle_{\partial\mathcal{T}_h} &= 0,\\
\overline{p}_h - \overline{\rho}_h &= 0,\\
\left\langle(-\nu\widehat{\mathrm{L}}_h + \widehat{p}_h\mathrm{I})\cdot\boldsymbol{n}, \boldsymbol{\mu}\right\rangle_{\partial\mathcal{T}_h} &= 0,\\
\langle\widehat{\boldsymbol{u}}_h\cdot\boldsymbol{n}, \overline{\psi}\rangle_{\partial\mathcal{T}_h} &= 0,\\
(p_h, 1)_{\mathcal{T}_h} &= 0,
\end{aligned} \tag{30}$$

for all $(\mathrm{G}, \boldsymbol{v}, q, \boldsymbol{\mu}, \overline{\psi}) \in \mathrm{G}_h \times \boldsymbol{V}_h \times P_h \times \boldsymbol{M}_h(\boldsymbol{0}) \times \overline{\Psi}_h$, where

$$-\nu\widehat{\mathrm{L}}_h + \widehat{p}_h\mathrm{I} = -\nu\mathrm{L}_h + p_h\mathrm{I} + \mathrm{S}(\boldsymbol{u}_h - \widehat{\boldsymbol{u}}_h)\otimes\boldsymbol{n}. \tag{31}$$

Here S is the second-order tensor consisting of stabilization parameters and I is the second-order identity tensor. Note also that the Dirichlet boundary condition has been enforced by requiring that $\widehat{\boldsymbol{u}}_h \in \boldsymbol{M}_h(\boldsymbol{g})$.

The first four equations of (30) define the local solver which can be used to eliminate all the variables L_h, $\boldsymbol{u}_h$, and p_h by inserting them into the last three equations of (30), thereby obtaining a linear system in terms of $(\widehat{\boldsymbol{u}}_h, \overline{p}_h)$ only. Since $\widehat{\boldsymbol{u}}_h$ is defined on the element faces and $\overline{p}_h$ has one degree of freedom per element, the HDG method reduces significantly the number of the globally coupled unknowns. In practice, we implement the HDG method by using the augmented Lagrangian approach [16]; see [22] for a detailed discussion.

Finally, we use the element-by-element postprocessing proposed in [11] obtain a new approximate velocity which is exactly divergence-free, $\boldsymbol{H}(div)$-conforming, and converges with the order $k+2$. In the three dimensional case, we define the postprocessed approximate velocity $\boldsymbol{u}_h^\star$ on the tetrahedron $K \in \mathscr{T}_h$ as the element of $(\mathscr{P}_{k+1}(K))^d$ such that

$$\langle(\boldsymbol{u}_h^\star - \widehat{\boldsymbol{u}}_h)\cdot\boldsymbol{n}, \mu\rangle_F = 0 \quad \forall\, \mu \in \mathscr{P}_k(F), \tag{32a}$$

$$\langle(\boldsymbol{n}\times\nabla)(\boldsymbol{u}_h^\star\cdot\boldsymbol{n}) - \boldsymbol{n}\times(\{\mathrm{L}_h^t\}\boldsymbol{n}), (\boldsymbol{n}\times\nabla)\mu\rangle_F = 0 \quad \forall\, \mu \in \mathscr{P}_{k+1}(F)^\perp, \tag{32b}$$

for all faces F of K, and such that

$$(\boldsymbol{u}_h^\star - \boldsymbol{u}_h, \nabla w)_K = 0 \quad \forall\, w \in \mathscr{P}_k(K), \tag{32c}$$

$$(\nabla\times\boldsymbol{u}_h^\star - \mathrm{w}_h, (\nabla\times\boldsymbol{v})\,\mathrm{B}_K)_K = 0 \quad \forall\, \boldsymbol{v} \in \mathscr{S}_k(K). \tag{32d}$$

Here

$$\mathscr{P}_{k+1}(F)^\perp := \{\mu \in \mathscr{P}_{k+1}(F) : \langle\mu, \widetilde{\mu}\rangle_F = 0, \quad \forall \widetilde{\mu} \in \mathscr{P}_k(F)\},$$

and

$$\mathrm{w}_h := (\mathrm{L}_{32}^h - \mathrm{L}_{23}^h, \mathrm{L}_{13}^h - \mathrm{L}_{31}^h, \mathrm{L}_{21}^h - \mathrm{L}_{12}^h)$$

is the approximation to the vorticity. Furthermore, B_K is the so-called *symmetric bubble matrix* introduced in [8], namely,

$$\mathrm{B}_K := \sum_{\ell=0}^{3} \lambda_{\ell-3}\lambda_{\ell-2}\lambda_{\ell-1}\nabla\lambda_\ell \otimes \nabla\lambda_\ell,$$

where λ_i are the barycentric coordinates associated with the tetrahedron K. Finally, $\mathscr{S}_k(K) := \{\boldsymbol{p} \in \boldsymbol{N}_k : (\boldsymbol{p}, \nabla\phi)_K = 0 \text{ for all } \phi \in \mathscr{P}_{k+1}(K)\}$, where $\boldsymbol{N}_k = \mathscr{P}_{k-1}(K) \oplus \boldsymbol{S}_k$ and $\boldsymbol{S}_\ell$ is the space of vector-valued homogeneous polynomials $\boldsymbol{v}$ of degree ℓ such that $\boldsymbol{v}\cdot\boldsymbol{x} = 0$; see [18, 19].

In the two dimensional case, the postprocessing is defined by the above equations if (32d) is replaced by

$$(\nabla \times \boldsymbol{u}_h^\star - \mathsf{w}_h, w\, b_K)_K = 0 \quad \forall\, w \in \mathscr{P}_{k-1}(K),$$

where $b_K := \lambda_0\lambda_1\lambda_2$ and $\mathsf{w}_h := \mathrm{L}_{21}^h - \mathrm{L}_{12}^h$.

We refer the reader to [11] for a proof of the fact that $\boldsymbol{u}_h^\star$ is a divergence-free velocity in $\boldsymbol{H}(div, \Omega)$ and converges with the order $k+2$ in the L^2-norm.

3.4 Incompressible Navier–Stokes Equations

Let us extend the HDG method described above to the steady incompressible Navier–Stokes equations written in conservative form

$$\begin{aligned} \mathrm{L} - \nabla \boldsymbol{u} &= 0, && \text{in } \Omega \\ -\nu \nabla \cdot \mathrm{L} + \nabla p + \nabla \cdot (\boldsymbol{u} \otimes \boldsymbol{u}) &= \boldsymbol{f}, && \text{in } \Omega, \\ \nabla \cdot \boldsymbol{u} &= 0, && \text{in } \Omega, \\ \boldsymbol{u} &= \boldsymbol{g}, && \text{on } \partial\Omega. \end{aligned} \tag{33}$$

The Navier–Stokes system differs from the Stokes one due to the presence of the nonlinear convective term $\nabla \cdot (\boldsymbol{u} \otimes \boldsymbol{u})$.

The HDG method for the above system seeks an approximation $(\mathrm{L}_h, \boldsymbol{u}_h, p_h, \widehat{\boldsymbol{u}}_h, \overline{p}_h) \in \mathrm{G}_h \times \boldsymbol{V}_h \times P_h \times \boldsymbol{M}_h(\boldsymbol{g}) \times \overline{\Psi}_h$ such that

$$\begin{aligned} (\mathrm{L}_h, \mathrm{G})_{\mathscr{T}_h} + (\boldsymbol{u}_h, \nabla \cdot \mathrm{G})_{\mathscr{T}_h} - \langle \widehat{\boldsymbol{u}}_h, \mathrm{G}\boldsymbol{n} \rangle_{\partial\mathscr{T}_h} &= 0, \\ (\nu \mathrm{L}_h - p_h \mathrm{I} - \boldsymbol{u}_h \otimes \boldsymbol{u}_h, \nabla \boldsymbol{v})_{\mathscr{T}_h} & \\ + \left\langle (-\nu \widehat{\mathrm{L}}_h + \widehat{p}_h \mathrm{I} + \widehat{\boldsymbol{u}}_h \otimes \widehat{\boldsymbol{u}}_h)\boldsymbol{n}, \boldsymbol{v} \right\rangle_{\partial\mathscr{T}_h} &= (\boldsymbol{f}, \boldsymbol{v})_{\mathscr{T}_h}, \\ -(\boldsymbol{u}_h, \nabla q)_{\mathscr{T}_h} + \langle \widehat{\boldsymbol{u}}_h \cdot \boldsymbol{n}, q - \overline{q} \rangle_{\partial\mathscr{T}_h} &= 0, \\ \overline{p}_h - \overline{p}_h &= 0, \\ \left\langle (-\nu \widehat{\mathrm{L}}_h + \widehat{p}_h \mathrm{I} + \widehat{\boldsymbol{u}}_h \otimes \widehat{\boldsymbol{u}}_h)\boldsymbol{n}, \boldsymbol{\mu} \right\rangle_{\partial\mathscr{T}_h} &= 0, \\ \langle \widehat{\boldsymbol{u}}_h \cdot \boldsymbol{n}, \overline{\psi} \rangle_{\partial\mathscr{T}_h} &= 0, \\ (p_h, 1)_{\mathscr{T}_h} &= 0, \end{aligned} \tag{34}$$

for all $(\mathrm{G}, \boldsymbol{v}, q, \boldsymbol{\mu}, \overline{\psi}) \in \mathrm{G}_h \times \boldsymbol{V}_h \times P_h \times \boldsymbol{M}_h(\boldsymbol{0}) \times \overline{\Psi}_h$, where

$$\left(-\nu \widehat{\mathrm{L}}_h + \widehat{p}_h \mathrm{I}\right) \boldsymbol{n} = (-\nu \mathrm{L}_h + p_h \mathrm{I})\, \boldsymbol{n} + \boldsymbol{s}_h(\boldsymbol{u}_h, \widehat{\boldsymbol{u}}_h). \tag{35}$$

Here $\boldsymbol{s}_h(\boldsymbol{u}_h, \widehat{\boldsymbol{u}}_h)$ is the *stabilization vector-valued function* the choice of which is crucial since it does have an important effect on both the stability and accuracy of the method. We consider an extension of the expression for $\boldsymbol{s}_h(\boldsymbol{u}_h, \widehat{\boldsymbol{u}}_h)$ proposed in [11, 22] for the Stokes system as follows

$$\boldsymbol{s}_h(\boldsymbol{u}_h, \widehat{\boldsymbol{u}}_h) = \mathrm{S}(\boldsymbol{u}_h, \widehat{\boldsymbol{u}}_h)(\boldsymbol{u}_h - \widehat{\boldsymbol{u}}_h), \tag{36}$$

where $\mathrm{S}(\boldsymbol{u}_h, \widehat{\boldsymbol{u}}_h)$ is the *stabilization tensor* which may depend on $\boldsymbol{u}_h$ and $\widehat{\boldsymbol{u}}_h$.

Substituting (35) into (34) we obtain that $(\mathrm{L}_h, \boldsymbol{u}_h, p_h, \widehat{\boldsymbol{u}}_h, \overline{\rho}_h) \in \mathrm{G}_h \times \boldsymbol{V}_h \times P_h \times \boldsymbol{M}_h(\boldsymbol{g}) \times \overline{\Psi}_h$ is the solution of

$$
\begin{aligned}
(\mathrm{L}_h, \mathrm{G})_{\mathscr{T}_h} + (\boldsymbol{u}_h, \nabla\cdot\mathrm{G})_{\mathscr{T}_h} - \langle \widehat{\boldsymbol{u}}_h, \mathrm{G}\boldsymbol{n}\rangle_{\partial\mathscr{T}_h} &= 0,\\
(\nabla\cdot(-\nu\mathrm{L}_h + p_h\mathrm{I}), \boldsymbol{v})_{\mathscr{T}_h} - (\boldsymbol{u}_h\otimes\boldsymbol{u}_h, \nabla\boldsymbol{v})_{\mathscr{T}_h} &\\
+ \langle(\widehat{\boldsymbol{u}}_h\otimes\widehat{\boldsymbol{u}}_h)\boldsymbol{n} + \boldsymbol{s}_h(\boldsymbol{u}_h, \widehat{\boldsymbol{u}}_h), \boldsymbol{v}\rangle_{\partial\mathscr{T}_h} &= (\boldsymbol{f}, \boldsymbol{v})_{\mathscr{T}_h},\\
-(\boldsymbol{u}_h, \nabla q)_{\mathscr{T}_h} + \langle\widehat{\boldsymbol{u}}_h\cdot\boldsymbol{n}, q - \overline{q}\rangle_{\partial\mathscr{T}_h} &= 0,\\
\overline{p}_h - \overline{\rho}_h &= 0,\\
\langle(-\nu\mathrm{L}_h + p_h\mathrm{I} + \widehat{\boldsymbol{u}}_h\otimes\widehat{\boldsymbol{u}}_h)\boldsymbol{n} + \boldsymbol{s}_h(\boldsymbol{u}_h, \widehat{\boldsymbol{u}}_h), \boldsymbol{\mu}\rangle_{\partial\mathscr{T}_h} &= 0,\\
\langle\widehat{\boldsymbol{u}}_h\cdot\boldsymbol{n}, \overline{\psi}\rangle_{\partial\mathscr{T}_h} &= 0,\\
(p_h, 1)_{\mathscr{T}_h} &= 0,
\end{aligned}
\tag{37}
$$

for all $(\mathrm{G}, \boldsymbol{v}, q, \boldsymbol{\mu}, \overline{\psi}) \in \mathrm{G}_h \times \boldsymbol{V}_h \times P_h \times \boldsymbol{M}_h(\boldsymbol{0}) \times \overline{\Psi}_h$.

The above nonlinear system of equations is solved by the Newton–Raphson method: Given the mth current iterate $(\mathrm{L}_h^m, \boldsymbol{u}_h^m, p_h^m, \widehat{\boldsymbol{u}}_h^m, \overline{\rho}_h^m)$, we find an increment $(\delta\mathrm{L}_h^m, \delta\boldsymbol{u}_h^m, \delta p_h^m, \delta\widehat{\boldsymbol{u}}_h^m, \delta\overline{\rho}_h^m) \in \mathrm{G}_h \times \boldsymbol{V}_h \times P_h \times \boldsymbol{M}_h(0) \times \overline{\Psi}_h$ such that

$$
\begin{aligned}
(\delta\mathrm{L}_h^m, \mathrm{G})_{\mathcal{T}_h} + (\delta\boldsymbol{u}_h^m, \nabla\cdot\mathrm{G})_{\mathcal{T}_h} - \langle\delta\widehat{\boldsymbol{u}}_h^m, \mathrm{G}\boldsymbol{n}\rangle_{\partial\mathcal{T}_h} &= r_1(\mathrm{G}),\\
(\nabla\cdot(-\nu\delta\mathrm{L}_h^m + \delta p_h^m\mathrm{I}), \boldsymbol{v})_{\mathcal{T}_h} - (\delta\boldsymbol{u}_h^m\otimes\boldsymbol{u}_h^m + \boldsymbol{u}_h^m\otimes\delta\boldsymbol{u}_h^m, \nabla\boldsymbol{v})_{\mathcal{T}_h} &\\
+\langle(\delta\widehat{\boldsymbol{u}}_h^m\otimes\widehat{\boldsymbol{u}}_h^m + \widehat{\boldsymbol{u}}_h^m\otimes\delta\widehat{\boldsymbol{u}}_h^m)\boldsymbol{n} + \partial_1\boldsymbol{s}_h(\boldsymbol{u}_h^m, \widehat{\boldsymbol{u}}_h^m)\delta\boldsymbol{u}_h^m + \partial_2\boldsymbol{s}_h(\boldsymbol{u}_h^m, \widehat{\boldsymbol{u}}_h^m)\delta\widehat{\boldsymbol{u}}_h^m, \boldsymbol{v}\rangle_{\partial\mathcal{T}_h} &= r_2(\boldsymbol{v}),\\
-(\delta\boldsymbol{u}_h^m, \nabla q)_{\mathcal{T}_h} + \langle\delta\widehat{\boldsymbol{u}}_h^m\cdot\boldsymbol{n}, q - \overline{q}\rangle_{\partial\mathcal{T}_h} &= r_3(q),\\
\delta\overline{p}_h^m - \delta\overline{\rho}_h^m &= r_4,\\
\langle(-\nu\delta\mathrm{L}_h^m + \delta p_h^m\mathrm{I} + \delta\widehat{\boldsymbol{u}}_h^m\otimes\widehat{\boldsymbol{u}}_h^m + \widehat{\boldsymbol{u}}_h^m\otimes\delta\widehat{\boldsymbol{u}}_h^m)\boldsymbol{n}\rangle_{\partial\mathcal{T}_h} &\\
+\langle\partial_1\boldsymbol{s}_h(\boldsymbol{u}_h^m, \widehat{\boldsymbol{u}}_h^m)\delta\boldsymbol{u}_h^m + \partial_2\boldsymbol{s}_h(\boldsymbol{u}_h^m, \widehat{\boldsymbol{u}}_h^m)\delta\widehat{\boldsymbol{u}}_h^m, \boldsymbol{\mu}\rangle_{\partial\mathcal{T}_h} &= r_5(\boldsymbol{\mu}),\\
\langle\delta\widehat{\boldsymbol{u}}_h^m\cdot\boldsymbol{n}, \overline{\psi}\rangle_{\partial\mathcal{T}_h} &= r_6(\overline{\psi}),\\
(\delta p_h^m, 1)_{\mathcal{T}_h} &= r_7,
\end{aligned}
\tag{38}
$$

for all $(\mathrm{G}, \boldsymbol{v}, q, \boldsymbol{\mu}, \overline{\psi}) \in \mathrm{G}_h \times \boldsymbol{V}_h \times P_h \times \boldsymbol{M}_h(\boldsymbol{0}) \times \overline{\Psi}_h$. Note here that the right-hand side residuals are evaluated from (37) at the current iterate.

We observe that the above system (38) has a similar structure as the HDG system (30) for the Stokes flow except that there are some additional terms due to the convective nonlinearity. Therefore, it can be solved in a similar manner by means of the hybridization technique. This leads to a linear system of algebraic equations involving the degrees of freedom of $(\delta\widehat{\boldsymbol{u}}_h^m, \delta\overline{\rho}_h^m)$ only. Alternatively, we may apply the augmented Lagrangian approach to the nonlinear system (37) and then use the hybridization technique to obtain a system in terms of $\delta\widehat{\boldsymbol{u}}^m$ only [23].

Although our discussion has focused primarily on the steady-state case, the same HDG method can be applied to the time-dependent problem with using an implicit time-stepping method; see Sect. 3.1 for further details. Finally, we emphasize that the postprocessing procedure described for Stokes flow can be used for both the steady and unsteady Navier–Stokes problems.

4 Numerical Results

In this section, we present numerical results for a benchmark problem in fluid dynamics. We would like to refer the readers to the previous work [4, 6, 7, 11, 20–23, 27] for many other examples which demonstrate the performance and accuracy of the HDG methods described in this paper.

The Taylor vortex problem is a well-known example of the unsteady incompressible Navier–Stokes equations. The problem has an exact solution of the form

$$\begin{aligned} u_x &= -\cos(\pi x)\sin(\pi y)\exp\left(\tfrac{-2\pi^2 t}{Re}\right), \\ u_y &= \sin(\pi x)\cos(\pi y)\exp\left(\tfrac{-2\pi^2 t}{Re}\right), \\ p &= -\tfrac{1}{4}(\cos(2\pi x)+\cos(2\pi y))\exp\left(\tfrac{-4\pi^2 t}{Re}\right), \end{aligned}$$

where $Re = 1/\nu$ is the Reynolds number. We consider the above problem on $\Omega = (0,1)^2$ with Reynolds number $Re = 20$ and final time $T = 1$. We take the Dirichlet boundary condition for the velocity as the restriction of the exact solution to the domain boundary and the initial condition as an instantiation of the exact solution at $t = 0$.

In our experiments, we consider triangular meshes that are obtained by splitting a regular $n \times n$ Cartesian grid into a total of $2n^2$ triangles, giving uniform element sizes of $h = 1/n$. On these meshes, we consider polynomials of degree k to represent all the approximate variables using a nodal basis within each element, with the nodes uniformly distributed. We use the third-order backward difference formula (BDF3) for the temporal discretization. The stabilization tensor S is chosen as

$$\mathrm{S} = \begin{pmatrix} \tau & 0 \\ 0 & \tau \end{pmatrix},$$

where τ is equal to 1 on $\mathcal{E}_h$.

We first look at the convergence and accuracy in terms of both k and h refinements. For this purpose, we select a small timestep of $\Delta t = 0.005$, so that the spatial error is dominant and the temporal error is negligible. We present in Table 1 the history of convergence of the HDG method at the final time $t = 1$. We observe that the approximate velocity, pressure, and velocity gradient converge with the optimal order $k + 1$ for $k = 1, 2, 3$. The fact that the HDG method yields optimal convergence for both the approximate pressure and velocity gradient is a very important advantage since many other DG methods provide suboptimal convergence of order k for the approximate pressure and velocity gradient. Moreover, we observe that all the approximate variables converge *exponentially* with the polynomial degree k as depicted in Fig. 1. We emphasize that these results are obtained with τ being set to 1 and thus independent of both k and h.

Equally important is the fact that the postprocessed velocity $\boldsymbol{u}_h^*$ converges with the order $k + 2$, which is one order higher than the original approximate velocity

Table 1 History of convergence of the HDG method for the Taylor vortex problem with $Re = 20$

Degree	Mesh	$\|\boldsymbol{u} - \boldsymbol{u}_h\|_{\mathcal{T}_h}$		$\|p - p_h\|_{\mathcal{T}_h}$		$\|\mathrm{L} - \mathrm{L}_h\|_{\mathcal{T}_h}$		$\|\boldsymbol{u} - \boldsymbol{u}_h^\star\|_{\mathcal{T}_h}$	
k	1/h	Error	Order	Error	Order	Error	Order	Error	Order
	2	4.73e–2	–	3.44e–2	–	3.29e–1	–	3.40e–2	–
	4	1.27e–2	1.89	8.59e–3	2.00	1.26e–1	1.39	8.04e–3	2.08
1	8	2.94e–3	2.11	2.14e–3	2.01	3.85e–2	1.71	1.34e–3	2.59
	16	6.95e–4	2.08	5.38e–4	1.99	1.07e–2	1.84	1.89e–4	2.82
	32	1.70e–4	2.03	1.36e–4	1.99	2.85e–3	1.91	2.50e–5	2.92
	2	1.14e–2	–	6.67e–3	–	1.04e–1	–	8.35e–3	–
	4	1.26e–3	3.17	8.43e–4	2.98	1.72e–2	2.60	6.12e–4	3.77
2	8	1.51e–4	3.06	1.07e–4	2.98	2.60e–3	2.73	4.07e–5	3.91
	16	1.87e–5	3.01	1.33e–5	3.00	3.64e–4	2.84	2.70e–6	3.91
	32	2.33e–6	3.00	1.67e–6	3.00	4.85e–5	2.91	1.76e–7	3.94
	2	1.81e–3	–	1.00e–3	–	2.01e–2	–	1.22e–3	–
	4	1.08e–4	4.06	7.00e–5	3.84	1.72e–3	3.54	4.67e–5	4.70
3	8	6.59e–6	4.04	4.33e–6	4.01	1.29e–4	3.74	1.63e–6	4.84
	16	4.08e–7	4.01	2.68e–7	4.01	8.92e–6	3.85	5.48e–8	4.89
	32	2.55e–8	4.00	1.67e–8	4.00	5.88e–7	3.92	1.82e–9	4.91

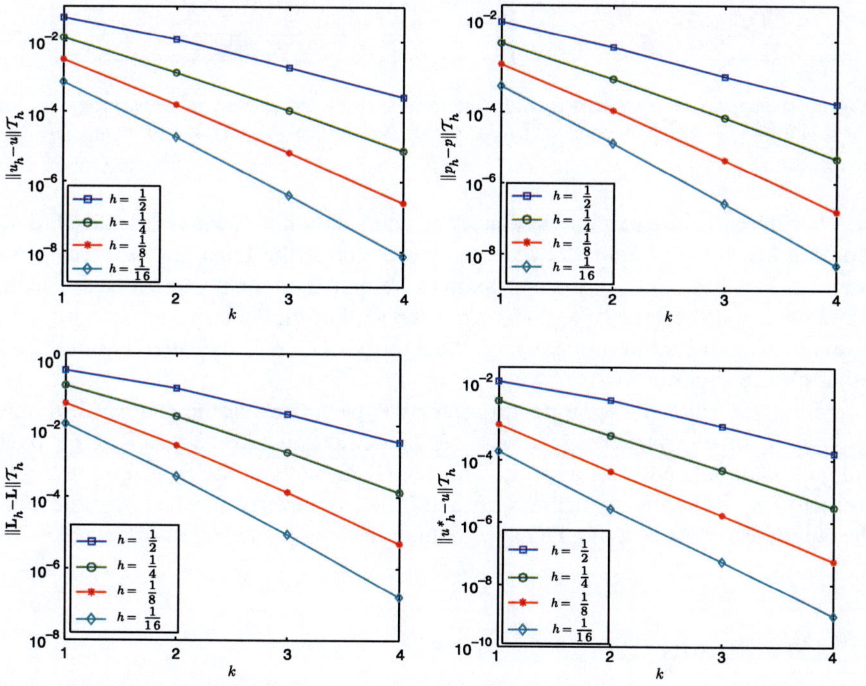

Fig. 1 The L^2 error in log scale as a function of h and k for $\boldsymbol{u}_h$ (*top left*), p_h (*top right*), $\mathbf{L}_h$ (*bottom left*), and $\boldsymbol{u}_h^*$ (*bottom right*)

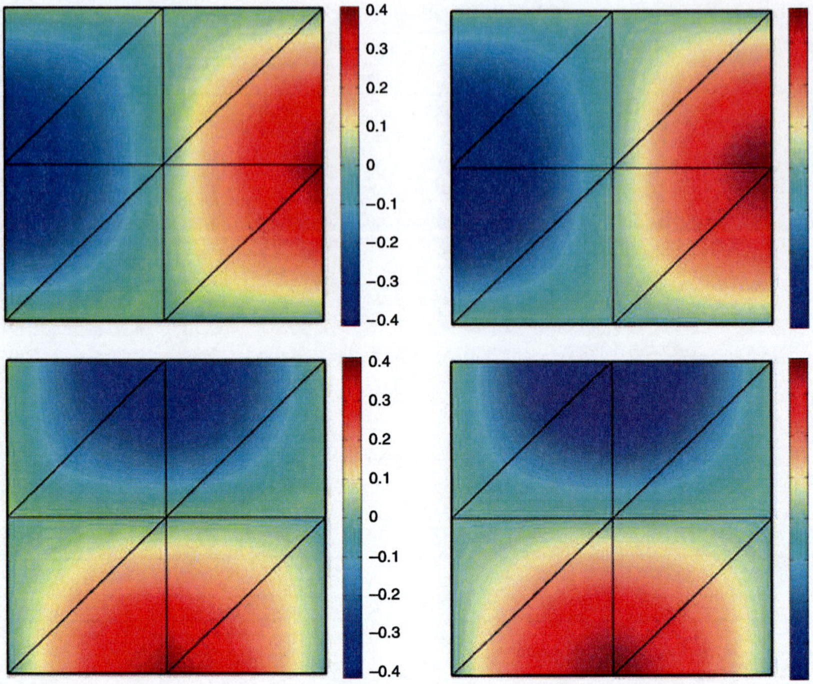

Fig. 2 The approximate velocity $\boldsymbol{u}_h$ (*left*) and the postprocessed velocity $\boldsymbol{u}_h^*$ (*right*) for $k = 2$ on the grid $h = 1/2$, with horizontal velocity at the *top* and vertical velocity at the *bottom*

$\boldsymbol{u}_h$. Furthermore, we emphasize that $\boldsymbol{u}_h^*$ is an exactly divergence-free and $\boldsymbol{H}(\text{div})$-conforming velocity field. To visualize the effect of the local postprocessing, we show in Fig. 2 the plots of the approximate velocity and the postprocessed velocity for $k = 2$ on the grid $h = 1/2$. We observe that the local postprocessing does provide a significant improvement in the approximation of the velocity field, since $\boldsymbol{u}_h^*$ is clearly superior to $\boldsymbol{u}_h$.

Moreover, since the local postprocessing is performed at the element level and only at the timestep where higher accuracy is desired, it adds very little to the overall computational cost. As a result, with the HDG method, the $(k+2)$-convergent velocity, $(k+1)$-convergent pressure, and $(k+1)$-convergent velocity gradient can be computed at the cost of a DG approximation using polynomials of degree k.

5 Conclusions

We present an overview of recent developments of HDG methods for numerically solving partial differential equations in fluid mechanics. The main philosophy of the HDG methodology includes the following steps:

- Identify the globally coupled unknowns as the numerical traces of the field variables associated with the essential boundary condition.
- Enforce explicitly the continuity of the normal component of the numerical fluxes associated with the Neumann boundary condition. This is called the conservativity condition.
- Define the local solver by applying the HDG method to the governing equations at the element level.
- Substitute all the volumetric unknowns from the local solver into the conservativity condition to obtain a final system in terms of the numerical traces only.
- Apply the local postprocessing to obtain an improved approximation of the field variables.

The above guidelines are very general and applicable beyond problems considered in this paper. Indeed, based on this general framework we have successfully developed HDG methods for the compressible Euler and Navier–Stokes equations [25]. Inspired by the simplicity and generality of this new DG methodology, our current research effort focuses on devising HDG methods for multi-physics applications.

References

1. D. N. Arnold, F. Brezzi, B. Cockburn, and L. D. Marini. Unified analysis of discontinuous Galerkin methods for elliptic problems. *SIAM J. Numer. Anal.*, 39(5):1749–1779, 2001
2. P. Bastian and B. Rivière. Superconvergence and $H(\text{div})$ projection for discontinuous Galerkin methods. *Int. J. Numer. Methods Fluids*, 42:1043–1057, 2003
3. P. Castillo, B. Cockburn, I. Perugia, and D. Schötzau. An a priori error analysis of the local discontinuous Galerkin method for elliptic problems. *SIAM J. Numer. Anal.*, 38(5):1676–1706, 2001
4. B. Cockburn, B. Dong, and J. Guzmán. A superconvergent LDG-hybridizable Galerkin method for second-order elliptic problems. *Math. Comp.*, 77:1887–1916, 2008
5. B. Cockburn and J. Gopalakrishnan. A characterization of hybridized mixed methods for second order elliptic problems. *SIAM J. Numer. Anal.*, 42(1):283–301, 2004
6. B. Cockburn, B. Dong, J. Guzmán, M. Restelli, and R. Sacco. An hybridizable discontinuous Galerkin method for steady-state convection-diffusion-reaction problems. *SIAM J. Sci. Comput.*, 31(5):3827–3846, 2009
7. B. Cockburn and J. Gopalakrishnan. The derivation of hybridizable discontinuous Galerkin methods for Stokes flow. *SIAM J. Numer. Anal.*, 47:1092–1125, 2009
8. B. Cockburn, J. Gopalakrishnan, and J. Guzmán. A new elasticity element made for enforcing weak stress symmetry. *Math. Comp.*, 79:1331–1349, 2009
9. B. Cockburn, J. Gopalakrishnan, and R. Lazarov. Unified hybridization of discontinuous Galerkin, mixed and continuous Galerkin methods for second-order elliptic problems. *SIAM J. Numer. Anal.*, 47:1319–1365, 2009
10. B. Cockburn, J. Gopalakrishnan, and F.-J. Sayas. A projection-based error analysis of HDG methods. *Math. Comp.*, 79:1351–1367, 2010
11. B. Cockburn, J. Gopalakrishnan, N. C. Nguyen, J. Peraire, and F.-J. Sayas. Analysis of an HDG method for Stokes flow. *Math. Comp.* DOI: 10.1090/S0025-5718-2010-02410-X

12. B. Cockburn, J. Guzmán, S.-C. Soon, and H. K. Stolarski. An analysis of the embedded discontinuous Galerkin method for second-order elliptic problems. *SIAM J. Numer. Anal.*, 47(4):2686–2707, 2009
13. B. Cockburn, J. Guzmán, and H. Wang. Superconvergent discontinuous Galerkin methods for second-order elliptic problems. *Math. Comp.*, 78:1–24, 2009
14. B. Cockburn, G. Kanschat, and D. Schötzau. A locally conservative LDG method for the incompressible Navier–Stokes equations. *Math. Comp.*, 74:1067–1095, 2005
15. B. Cockburn and C. W. Shu. The local discontinuous Galerkin method for convection-diffusion systems. *SIAM J. Numer. Anal.*, 35:2440–2463, 1998
16. M. Fortin, and R. Glowinski. *Augmented Lagrangian methods*, volume 15 of *Studies in Mathematics and its Applications*. North-Holland Publishing Co., Amsterdam, 1983. Applications to the numerical solution of boundary value problems, Translated from the French by B. Hunt and D. C. Spicer
17. S. Güzey, B. Cockburn, and H. Stolarski. The embedded discontinuous Galerkin methods: Application to linear shells problems, *Int. J. Numer. Methods Eng.*, 70:757–790, 2007
18. J.-C. Nédélec. Mixed finite elements in $\mathbf{R}^3$. *Numer. Math.*, 35:315–341, 1980
19. J.-C. Nédélec. A new family of mixed finite elements in $\mathbf{R}^3$. *Numer. Math.*, 50:57–81, 1986
20. N. C. Nguyen, J. Peraire, and B. Cockburn. An implicit high-order hybridizable discontinuous Galerkin method for linear convection-diffusion equations. *J. Comput. Phys.*, 228:3232–3254, 2009
21. N. C. Nguyen, J. Peraire, and B. Cockburn. An implicit high-order hybridizable discontinuous Galerkin method for nonlinear convection-diffusion equations. *J. Comput. Phys.*, 228:8841–8855, 2009
22. N. C. Nguyen, J. Peraire, and B. Cockburn. A hybridizable discontinuous Galerkin method for Stokes flow. *Comput. Methods Appl. Mech. Engrg.*, 199:582–597, 2010
23. N. C. Nguyen, J. Peraire, and B. Cockburn. An implicit high-order hybridizable discontinuous Galerkin method for the incompressible Navier–Stokes equations. Submitted to Journal of Computational Physics, 2009.
24. N. C. Nguyen, J. Peraire, and B. Cockburn. A hybridizable discontinuous Galerkin method for the incompressible Navier–Stokes equations. In Proceedings of the 48th AIAA Aerospace Sciences Meeting and Exhibit, Orlando, Florida, January 2010. AIAA-2010-362
25. J. Peraire, N. C. Nguyen, and B. Cockburn. A hybridizable discontinuous Galerkin method for the compressible Euler and Navier–Stokes Equations. In Proceedings of the 48th AIAA Aerospace Sciences Meeting and Exhibit, Orlando, Florida, January 2010. AIAA-2010-363
26. P. O. Persson and J. Peraire. Newton-GMRES preconditioning for discontinuous Galerkin discretizations of the Navier–Stokes equations *SIAM J. Sci. Comput.*, 30(6):2709–2733, 2008
27. S.-C. Soon, B. Cockburn, and H. K. Stolarski. A hybridizable discontinuous Galerkin method for linear elasticity. *Int. J. Numer. Methods Eng.*, 80(8):1058–1092, 2009

Multivariate Modified Fourier Expansions

Ben Adcock and Daan Huybrechs

Abstract In this paper, we review recent advances in the approximation of multivariate functions using eigenfunctions of the Laplace operator subject to homogeneous Neumann boundary conditions. Such eigenfunctions are known explicitly on a variety of domains, including the d-variate cube, equilateral triangle and numerous other higher dimensional simplices. Practical construction of truncated expansions is achieved using a mixture of asymptotic and classical quadratures. Moreover, by exploiting the hyperbolic cross, the number of expansion coefficients need only grow mildly with dimension.

Despite converging uniformly throughout the domain, the rate of convergence of such expansions may be slow. We review two techniques to accelerate convergence. The first smoothes the function by interpolating certain derivatives of the function evaluated on the boundary of the domain. The second numerically computes a smooth, periodic extension of the function on a larger domain.

1 Introduction

The subject of this paper is the approximation of a multivariate function $f : \Omega \to \mathbb{R}$ in eigenfunctions of the Laplace operator subject to homogeneous Neumann boundary conditions:

$$-\Delta\phi(x) = \mu\phi(x), \quad x \in \Omega, \quad \frac{\partial \phi}{\partial n}(x) = 0, \quad x \in \partial\Omega. \tag{1}$$

B. Adcock (✉)
DAMTP, Centre for Mathematical Sciences, University of Cambridge, Wilberforce Rd, Cambridge CB3 0WA, UK
e-mail: b.j.s.adcock@damtp.cam.ac.uk

D. Huybrechs
Department of Computer Science, Katholieke Universiteit Leuven, Celestijnenlaan 200A, 3001 Leuven, Belgium
e-mail: daan.huybrechs@cs.kuleuven.be

J.S. Hesthaven and E.M. Rønquist (eds.), *Spectral and High Order Methods for Partial Differential Equations*, Lecture Notes in Computational Science and Engineering 76, DOI 10.1007/978-3-642-15337-2_5,

The one-dimensional case of (1) was proposed and studied in [15] and has led to a rapidly growing list of papers on numerical aspects of eigenfunction expansions and related topics, which we will briefly review throughout this text.

We assume that the domain $\Omega \subset \mathbb{R}^d$ is non-empty, bounded and simply connected with piecewise smooth boundary. The eigenvalues of (1), $\mu_n \geq 0$, $n \in \mathbb{N}$, are real, countable and have no finite limit point in $\mathbb{R}$. The corresponding eigenfunctions, $\phi_n(x)$, are orthogonal. Aside from the zero eigenvalue $\mu_0 = 0$ with eigenfunction $\phi_0(x) \equiv 1$, all eigenvalues are positive.

Let $(\cdot,\cdot)$ be the standard Euclidean inner product on Ω with associated norm $\|\cdot\|$. A function f may be expanded in the infinite series

$$f(x) \sim \sum_{n=0}^{\infty} \frac{1}{\|\phi_n\|^2} \hat{f}_n \phi_n(x), \tag{2}$$

where $\hat{f}_n = (f, \phi_n)$. Standard spectral theory confirms density of the set $\{\phi_n : n \in \mathbb{N}\}$ in $L^2(\Omega)$ [11], thereby verifying convergence of this expansion in the $L^2(\Omega)$ norm.

In the unit interval $\Omega = (-1, 1)$, trivial calculations establish that

$$f(x) \sim \frac{1}{2}\hat{f}_0^{[0]} + \sum_{n=1}^{\infty} \left\{ \hat{f}_n^{[0]} \phi_n^{[0]}(x) + \hat{f}_n^{[1]} \phi_n^{[1]}(x) \right\}, \tag{3}$$

where $\phi_n^{[0]}(x) = \cos n\pi x$ and $\phi_n^{[1]}(x) = \sin(n - \frac{1}{2})\pi x$. Such expansion bears a striking resemblance to the classical Fourier expansion. For this reason we refer to (2) as a *modified Fourier expansion*. However, unlike the former, the series (3) converges uniformly on $[-1, 1]$. If this series is truncated after N terms, the uniform error is $\mathcal{O}\left(N^{-1}\right)$, whereas away from the endpoints the pointwise error is $\mathcal{O}\left(N^{-2}\right)$ [19]. In comparison, the truncated Fourier sum exhibits an $\mathcal{O}\left(N^{-1}\right)$ error away from the endpoints, but uniform convergence is lacking (the Gibbs phenomenon).

The improvement in convergence stems from the Neumann boundary conditions. Suppose that $\mu \neq 0$ is a Laplace–Neumann eigenvalue with corresponding eigenfunction ϕ. Then, for $f \in H^2(\Omega)$, two applications of Stokes' theorem yields

$$\begin{aligned}(f, \phi) &= -\frac{1}{\mu}(f, \Delta\phi) \\ &= -\frac{1}{\mu}\int_{\partial\Omega} f(x)\frac{\partial\phi(x)}{\partial n}dx + \frac{1}{\mu}\int_{\partial\Omega}\frac{\partial f(x)}{\partial n}\phi(x)dx - \frac{1}{\mu}(\Delta f, \phi).\end{aligned}$$

After substituting the boundary conditions, we obtain

$$(f, \phi) = \frac{1}{\mu}\int_{\partial\Omega}\frac{\partial f(x)}{\partial n}\phi(x)dx - \frac{1}{\mu}(\Delta f, \phi).$$

In the univariate setting, for example, the coefficients $\hat{f}_n^{[i]} = \mathcal{O}(n^{-2})$, ensuring uniform convergence of the expansion. In fact, iterating the above process, we obtain

$$\hat{f}_n^{[i]} = \sum_{r=0}^{k-1} \frac{(-1)^{n+i}\left[f^{(2r+1)}(1) + (-1)^{i+1} f^{(2r+1)}(-1)\right]}{[(n-\frac{i}{2})\pi]^{2r+2}} + \mathcal{O}\left(n^{-2k-2}\right), \quad (4)$$

for any $k \in \mathbb{N}_+ = \mathbb{N}\backslash\{0\}$. In comparison, the classical Fourier sine coefficient

$$\int_{-1}^{1} f(x) \sin n\pi x \, dx = \mathcal{O}\left(n^{-1}\right).$$

Non-uniform convergence of the Fourier expansion is now apparent.

Modified Fourier expansions possess great generality and relative simplicity. However, in the tensor product setting at least, they are usually cast aside in favour of spectrally convergent approximations comprised of orthogonal polynomials.

Nonetheless, such expansions enjoy a number of advantages. First, the coefficients $\hat{f}_n$ can be computed efficiently using a variety of numerical quadratures. The mainstay of this is the observation that the eigenfunctions ϕ_n become increasingly oscillatory for large n, thus facilitating the use of efficient computational schemes for highly oscillatory integrals. These enable the computation of coefficients one-by-one at a fixed cost per coefficient. In this manner, any M coefficients can be computed in $\mathcal{O}(M)$ operations. This approach is also completely adaptive: increasing M does not require recalculation of existing values. When looking at the computational cost, such a scheme compares favourably over alternative approaches, in particular the FFT. On the other hand, unlike in the FFT, the accuracy of each computed coefficient is not necessarily coupled to the total number of coefficients. Still, due to this adaptivity, modified Fourier expansions can successfully exploit tools such as the hyperbolic cross to greatly reduce the number of approximation coefficients. We consider this further in Sect. 3. For a discussion of the particular quadratures used in modified Fourier expansions, we refer the reader to [14–16].

Modified Fourier series offer a number of benefits with regard to applications. First, they lead to considerably better conditioned matrices than polynomial-based spectral methods for differential equations [2,3]. Second, they allow for cheaper and faster calculation of spectra of highly oscillatory Fredholm operators [8].

Modified Fourier expansions possess at least one other virtue: Laplace–Neumann eigenfunctions are known explicitly in a number of non-tensor product domains, including the equilateral triangle [14]. We consider this further in Sect. 5.

Unfortunately the convergence rate of such expansions may be slow. In Sect. 4 we introduce two techniques to accelerate convergence.

2 The d-Variate Cube

Univariate modified Fourier expansions were introduced in [15]. Their extension to the d-variate cube $\Omega = (-1,1)^d$, studied in [16], is obtained by Cartesian products. If $n = (n_1, \ldots, n_d) \in \mathbb{N}^d$ and $i = (i_1, \ldots, i_d) \in \{0,1\}^d$ are multi-indices then

$$f(x) \sim \sum_{i \in \{0,1\}^d} \sum_{n \in \mathbb{N}^d} \hat{f}_n^{[i]} \phi_n^{[i]}(x), \tag{5}$$

where $\phi_n^{[i]}(x) = \phi_{n_1}^{[i_1]}(x_1) \ldots \phi_{n_d}^{[i_d]}(x_d)$. Convergence of the expansion (5) has been addressed in [2]. As described, this is best studied in so-called *Sobolev spaces of dominating mixed smoothness*:

$$H^q_{\text{mix}}(\Omega) = \{f \in L^2(\Omega) : \partial^{\beta_1}_{x_1} \ldots \partial^{\beta_d}_{x_d} f \in L^2(\Omega),\ 0 \le \beta_1, \ldots, \beta_d \le q\}, \quad q \in \mathbb{N}.$$

Observe that $H^{qd}(\Omega) \subset H^q_{\text{mix}}(\Omega) \subset H^q(\Omega)$, where $H^q(\Omega)$ is the qth classical Sobolev space. Such spaces occur in a range of applications, including sparse grids [9] and hyperbolic cross approximations [20, 21]. In relation to modified Fourier expansions, we have the following result:

Theorem 1 (Adcock [2]). *The set $\{\phi_n^{[i]} : n \in \mathbb{N}^d, i \in \{0,1\}^d\}$ is dense and orthogonal in $H^1_{\text{mix}}(-1,1)^d$.*

This theorem highlights the advantage over the Fourier basis. Uniform convergence of the expansion (5) follows immediately from the continuous embedding $H^1_{\text{mix}}(\Omega) \hookrightarrow C(\bar{\Omega})$ [2].

For numerical computations we must truncate the expansion (5) suitably. To do so, we include only those coefficients $n \in I_N$, where $I_N \subset \mathbb{N}^d$ is finite. Standard intuition leads to the full index set

$$I_N = \{n \in \mathbb{N}^d : 0 \le n_1, \ldots, n_d \le N\}. \tag{6}$$

Provided the function f is sufficiently smooth, convergence rates of $\mathcal{O}\left(N^{-2}\right)$ and $\mathcal{O}\left(N^{-1}\right)$ are observed inside the domain and on the boundary respectively [2, 19]. Unfortunately $|I_N| = \mathcal{O}\left(N^d\right)$, rendering such expansions expensive to construct in two or more dimensions. However, as we now consider, this figure can be significantly reduced without affecting convergence rates unduly.

3 The Hyperbolic Cross

The majority of coefficients $\hat{f}_n^{[i]}$ with indices n in the set (6) have negligible contribution to the overall approximation. An alternative criterion to define I_N is to include only those coefficients with absolute value greater than some tolerance ϵ.

Due to the simple tensor product setting, the coefficients $\hat{f}_n^{[i]}$ are $\mathcal{O}\left(n_1^{-2} \dots n_d^{-2}\right)$ for large $n_1, \dots, n_d$, provided $f \in H^2_{\text{mix}}(\Omega)$. In fact, by applying the univariate expansion (4) in each variable, we readily obtain a multivariate analogue [2, 16]. Returning to the construction of I_N and setting the tolerance $\epsilon = N^{-2}$ we obtain the *hyperbolic cross* index set

$$I_N = \{n \in \mathbb{N}^d : \bar{n}_1 \dots \bar{n}_d \leq N\}, \tag{7}$$

where $\bar{m} = \max\{m, 1\}$ for $m \in \mathbb{N}$. A simple calculation (see [13]) verifies that $|I_N| = \mathcal{O}\left(N(\log N)^{d-1}\right)$. This figure grows much more mildly with dimension than the corresponding value of $\mathcal{O}\left(N^d\right)$ for the index set (6).

The application of the hyperbolic cross (7) to modified Fourier expansions was introduced in [13]. As the following theorem demonstrates, this greatly increases their effectiveness without deteriorating the convergence rate unduly:

Theorem 2 (Adcock [2]). *Suppose that f_N is the truncated modified Fourier expansion of f based on (7). Then $f(x) - f_N(x)$ is $\mathcal{O}\left(N^{-2}(\log N)^{d-1}\right)$ for $x \in \Omega$ and $\mathcal{O}\left(N^{-1}(\log N)^{d-1}\right)$ for $x \in \partial\Omega$.*

4 Accelerating Convergence

As demonstrated, the convergence rate of modified Fourier expansion is typically slow. In this section we discuss two approaches to accelerate convergence.

4.1 The Lanczos Representation and Its Computation

Consider the univariate setting. The quality of the approximation f_N is improved if f satisfies

$$f^{(2r+1)}(\pm 1) = 0, \quad r = 0, \dots, k-1, \tag{8}$$

for some $k = 1, 2, \dots$ (the analogue of periodicity for modified Fourier expansions). This is manifested in a number of ways. First, recalling (4), we observe more rapid decay of the coefficients $\hat{f}_n^{[i]} = \mathcal{O}\left(n^{-2k-2}\right)$. Moreover, the expansion f_N converges to f in the $H^{2k+1}(-1, 1)$ norm [3], and the pointwise convergence rate is $\mathcal{O}\left(N^{-2k-2}\right)$ in $(-1, 1)$ and $\mathcal{O}\left(N^{-2k-1}\right)$ at the endpoints [19]. If f does not satisfy (8) a standard approach is to seek a smooth function p such that

$$p^{(2r+1)}(\pm 1) = f^{(2r+1)}(\pm 1), \quad r = 0, \dots, k-1, \tag{9}$$

and decompose f as $(f - p) + p$. This is referred to as the *Lanczos representation* [18]. Since $f - p$ obeys (8), the new approximation defined by $g_N = f_N - p_N + p$ converges at the faster rates prescribed above. Usually the function p is chosen to be a polynomial of degree $2k$. In view of this fact, this process is often referred to as *polynomial subtraction* [17].

Unfortunately, this process requires exact derivatives. Such values (or functions of $(d-1)$ variables when $d \geq 2$) are unknown in general. However, if approximated to sufficient accuracy, the convergence rate of g_N need not deteriorate. One approach to accomplish this is Eckhoff's method [10].

The approximation g_N (a linear combination of a finite modified Fourier sum and a polynomial) with p satisfying (9) also satisfies $\widehat{g_N}_n^{[i]} = \hat{f}_n^{[i]} + \mathcal{O}\left(n^{-2k-2}\right)$. To avoid the use of derivatives, we construct a new function g_N that satisfies this condition approximately: $\widehat{g_N}_n^{[i]} = \hat{f}_n^{[i]}$, $n = 0, \ldots, N+k$, $i = 0, 1$. With g_N defined in this manner, it can be shown that the convergence rate does not deteriorate in comparison to polynomial subtraction [4].

The extension of this method to the d-variate cube is achieved by formulating an approximation g_N satisfying

$$\widehat{g_N}_n^{[i]} = \hat{f}_n^{[i]}, \quad 0 \leq n_1, \ldots, n_d \leq N + k, \quad i \in \{0,1\}^d.$$

In this setting, g_N consists of Cartesian products of univariate polynomials and modified Fourier eigenfunctions. Analysis of this approximation was presented in [1].

4.2 The Fourier Extension Problem

An alternative to the explicit subtraction of a smooth function interpolating boundary conditions is to enlarge the initial function space. A particular example that leads to interesting results and analysis is to consider *both Laplace–Neumann and Laplace–Dirichlet eigenfunctions*, as pursued in [12]. This combination of two orthogonal bases is no longer a basis itself, but a frame. It is overcomplete and therefore many representations of f may exist. The approach taken in [12] to single out a useful representation is through a least squares criterion, following earlier results in [6, 7]. It is shown that exponential convergence is achieved when f is analytic.

Laplace–Dirichlet eigenfunctions on $[-1,1]$ are $\cos(n-\frac{1}{2})\pi x$ and $\sin n\pi x$, $n = 1, 2, \ldots$. Combined with (3) and rearranging terms, this leads to an approximation $g_N(x)$ of the form

$$f(x) \approx g_N(x) = \frac{1}{2}a_0 + \sum_{n=1}^{N} \left(a_n \cos n\frac{\pi}{2}x + b_n \sin n\frac{\pi}{2}x \right).$$

Thus one is looking for a periodic function on $[-2,2]$, to represent f on $[-1,1]$. This function has exactly the form of a classical Fourier series, hence the name *Fourier extension*. It is shown in [12] that the least squares problem can be converted

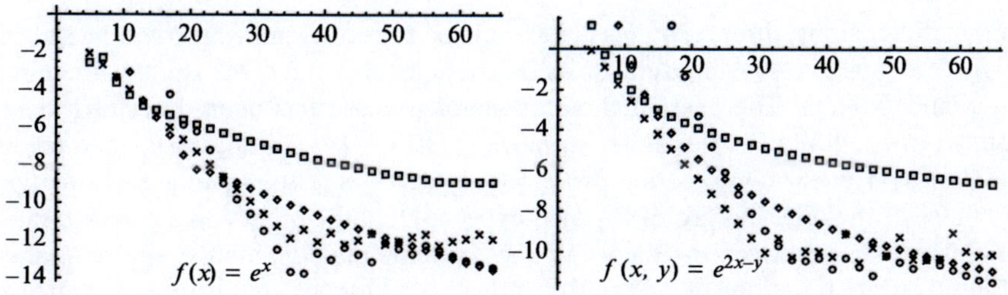

Fig. 1 Log uniform error $\log_{10} \|f - g_N\|_\infty$ where g_N is Eckhoff's approximation with $k = 2$ (*squares*), $k = 4$ (*diamonds*), $k = 8$ (*circles*) or the Fourier extension approximation (*crosses*)

into a polynomial approximation problem. Convergence theory and fast algorithms then follow from existing results on orthogonal polynomials. In particular, assuming sufficient analyticity of f near $[-1, 1]$, we have (see [12, Theorem 3.14])

$$f(x) - g_N(x) \sim (3 + 2\sqrt{2})^{-N}, \qquad x \in [-1, 1].$$

The extension of this approach to multivariate functions is straightforward in practice (see [7]), but delicate in theory. The generalisation of the theoretical analysis in [12] to many dimensions is a topic of current research. Preliminary results are along the lines of the theory described in Sect. 5.

In Fig. 1 we compare this approach and Eckhoff's method for univariate and bivariate examples. Exponential convergence of the former is verified. Despite offering only algebraic convergence, when $k = 8$ Eckhoff's approach yields similar results.

5 The Non-Tensor Product Case

The identification of modified Fourier expansions with eigenfunctions of the Laplace operator provides a useful link with geometry. The scope of multivariate modified Fourier expansions can be extended to include all domains for which eigenfunctions of the Laplace operator are known explicitly. A large and interesting family of domains consists of the so-called *fundamental regions* of *root systems*. These are simplical domains with special symmetry properties such that $\mathbb{R}^d$ can be tiled by reflecting the domain across all its sides and repeating the process indefinitely. It was first observed in [5] that eigenfunctions of the Laplace operator are obtained by symmetrizing classical multivariate Fourier series with respect to the set of symmetries described by a root system. The simplest example is the one-dimensional case, related to even symmetry in the root system A_1:

$$\cos n\pi x = \frac{1}{2}e^{in\pi x} + \frac{1}{2}e^{-in\pi x}.$$

In two dimensions, three triangles are associated to root systems: the equilateral triangle (root system A_2), the right isosceles triangle (B_2) and the triangle with angles $\frac{\pi}{2}$, $\frac{\pi}{3}$ and $\frac{\pi}{6}$ (C_2). The case of the equilateral triangle has been described in the context of modified Fourier series in more detail in [14]. In this case, each eigenfunction is a linear combination of six plane waves, and the symmetries involved are those of the dihedral group D_3. It is expected that the other cases can be treated similarly and, moreover, the theory of root systems may enable the approximation of multivariate functions on a long list of three- and higher dimensional simplices.

References

1. Adcock, B.: Convergence acceleration of modified Fourier series in one or more dimensions, Math. Comp. DOI: 10.1090/S0025-5718-2010-02393-2
2. Adcock, B.: Multivariate modified Fourier series and application to boundary value problems, Num. Math. **115**(4), 511–552 (2010)
3. Adcock, B.: Univariate modified Fourier methods for second order boundary value problems, BIT, **49**, 249–280 (2009)
4. Barkhudaryan, A., Barkhudaryan, R., and Poghosyan, A.: Asymptotic behavior of Eckhoffs method for Fourier series convergence acceleration, Anal. Theor. Appl., **23**, 228–242 (2007)
5. Bérard, P.: Spectres et groupes cristallographiques I: domaines euclidiens, Inv. Math., **58**, 179–199 (1980)
6. Boyd, J.: A comparison of numerical algorithms for Fourier extension of the first, second and third kinds, J. Comp. Phys., **178**, 118–160 (2002)
7. Bruno, O., Han, Y., and Pohlman, M.: Accurate high-order representation of complex three-dimensional surfaces via Fourier continuation analysis, J. Comp. Phys., **227**, 1094–1125 (2007)
8. Brunner, H., Iserles, A., and Nørsett, S. P.: The computation of the spectra of highly oscillatory Fredholm integral operators, J. Int. Eqn. Appl., (to appear)
9. Bungartz, H.-J. and Griebel, M.: Sparse grids, Acta Numerica, **13**, 147–269 (2004)
10. Eckhoff, K. S.: On a high order numerical method for functions with singularities. Math. Comp., **67**, 1063–1087 (1998)
11. Evans, L. C.: Partial Differential Equations. AMS, RI (1998)
12. Huybrechs, D.: On the Fourier extension of non-periodic functions, SIAM J. Numer. Anal., **47**(6), 4326–4355 (2010)
13. Huybrechs, D., Iserles, A. and Nørsett, S. P.: From high oscillation to rapid approximation IV: Accelerating convergence, IMA J. Num. Anal., (to appear)
14. Huybrechs, D., Iserles, A., and Nørsett, S. P.: From high oscillation to rapid approximation V: The equilateral triangle, Technical Report NA2009/04, DAMTP, University of Cambridge (2009)
15. Iserles, A. and Nørsett, S. P.: From high oscillation to rapid approximation I: Modified Fourier expansions, IMA J. Num. Anal., **28**, 862–887 (2008)
16. Iserles, A. and Nørsett, S. P.: From high oscillation to rapid approximation III: Multivariate expansions, IMA J. Num. Anal., **29**, 882–916 (2009)
17. Lanczos, C.: Discourse on Fourier Series. Hafner, New York (1966)
18. Lyness, J. N.: Computational techniques based on the Lanczos representation, Math. Comp., **28**, 81–123 (1974)
19. Olver, S.: On the convergence rate of a modified Fourier series, Math. Comp., **78**, 862–887 (2008)
20. Schmeißer, H.-J. and Triebel, H.: Topics in Fourier Analysis and Function Spaces. Wiley, NY (1987)
21. Temlyakov, V.: Approximation of Periodic Functions. Nova Science Publishers, New York (1993)

Constraint Oriented Spectral Element Method

E. Ahusborde, M. Azaïez, and R. Gruber

Abstract An original polynomial approximation to solve partial differential equations is presented. This spectral element version takes into account the underlying nature of the corresponding physical problem. For different types of operators, this approach allows to all terms in a variational form to be represented by the same functional dependence and by the same regularity, thus eliminating regularity constraints imposed by standard numerical methods. This method satisfies automatically different type of constraints, such as occur for the **grad**(div) and **curl**(curl) operators, and this for any geometry. It can be applied to a wide range of physical problems [Physical Review E, **75**(5), 056704 (2007)], including fluid flows, electromagnetism, material sciences, ideal linear magnetohydrodynamic stability analysis, and Alfvèn wave heating of fusion plasmas [Communications in Computational Physics, **5**(2–4), 413–425 (2009)].

1 Introduction

A wide range of physical phenomena can be described by mathematical models based on a set of coupled partial differential equations. Some operators pose significant problems, in particular those that are restricted by physical constraints. For example, for an incompressible flow, velocity $\mathbf{u}$ must satisfy the incompressibility constraint $\nabla \cdot \mathbf{u} = 0$. For some operators, such as **grad**(div) and **curl**(curl) operators, the solution is restricted by constraints that are an integral part of the solution. For example, in the case of the **grad**(div) operator, there are two classes of eigensolutions. The solenoidal modes are infinitely degenerate with $\lambda^2 = 0$ and $\nabla \cdot \mathbf{u} = 0$.

M. Azaïez (✉) and E. Ahusborde
TREFLE, ENSCPB, 33607 Pessac, France
e-mail: ahusborde@enscpb.fr, azaiez@enscpb.fr

R. Gruber
EPFL, 1015 Lausanne, Switzerland
e-mail: gruber@epfl.ch

J.S. Hesthaven and E.M. Rønquist (eds.), *Spectral and High Order Methods for Partial Differential Equations*, Lecture Notes in Computational Science and Engineering 76,
DOI 10.1007/978-3-642-15337-2_6, © Springer-Verlag Berlin Heidelberg 2011

The second class of irrotational modes are represented by a discrete spectrum with eigenmodes satisfying $\nabla \times \mathbf{u} = 0$. If these strong internal conditions cannot be satisfied precisely, so-called spectral pollution [1] appears and the numerical approach does not stably converge to the physical solution.

Our goal is to propose an approach taking into account the nature of the physical problem and avoiding the generation of nonphysical solutions.

2 Constraint Oriented Polynomial Approximation

Let Ω be a bounded connected open set in $I\!R^2$, with a Lipschitz-continuous boundary $\partial\Omega$. To describe the method, we introduce a partition of Ω without overlap: $\overline{\Omega} = \cup_{k=1}^{N} \overline{\Omega}_k$ and $\Omega_k \cap \Omega_k^{'} = \emptyset$, $1 \le k < k^{'} \le N$. For simplification, we consider only rectilinear elements with edges collinear to the x, y- axis, that is: $\Omega_k =]c_k, c_k^{'}[\times]d_k, d_k^{'}[$.

For each integer p, let $I\!P_p(\Omega_k)$ be the space of restrictions to Ω_k of polynomials with two variables and degree less or equal to p with respect to each variable.

We recall the standard quadrature formula: let $\Sigma_{GLL} = \{(\xi_i, \rho_i);\ 0 \le i \le p\}$ and $\Sigma_{GL} = \{(\zeta_i, \omega_i);\ 1 \le i \le p\}$ respectively denote the sets of Gauss–Lobatto–Legendre and Gauss–Legendre quadrature nodes and weights associated to polynomials of degree p (see [2]).

The canonical polynomial interpolation basis $h_i(x) \in I\!P_p(]-1,+1[)$ built on Σ_{GLL} is given by the relationships:

$$h_i(x) = -\frac{1}{p(p+1)}\frac{1}{L_p(\xi_i)}\frac{(1-x^2)\,L_p'(x)}{(x-\xi_i)}, \qquad -1 \le x \le +1, \quad 0 \le i \le p. \tag{1}$$

Denoting by F_k the mapping which sends the square $]-1,+1[^2$ onto Ω_k, we define the discrete product, for any u and v continuous on $\overline{\Omega}$: $(u,v)_h = \sum_{k=1}^{N} \frac{|\Omega_k|}{4} \sum_{i=0}^{p} \sum_{j=0}^{p} u \circ F_k(\xi_i, \xi_j) v \circ F_k(\xi_i, \xi_j) \rho_i \rho_j$. $\mid \Omega_k \mid$ is the measure of Ω_k.

Finally, we introduce the Lagrange interpolation operator I_h. For any continuous function φ on $\overline{\Omega}$, $I_h\, \varphi_{|\Omega_k}$ belongs to $I\!P_p(\Omega_k)$ and is equal to φ at all nodes $F_k(\xi_i, \xi_j)$, $0 \le i, j \le p$.

We use the standard notation for the Sobolev spaces $H^1(\Omega)$, provided with the corresponding norms.

2.1 Definition and Properties

We consider the set of functions $g_i(x)$ associated to the canonical basis (1) through the relationships:

$$g_i(x) = h_i(x) - \beta_i L_p(x), \qquad 0 \le i \le p, \tag{2}$$

where the constants β_i are such that all $g_i(x) \in I\!P_{p-1}(]-1,+1[)$. The functions $g_i(x)$ have the following properties:

1. Their moments up to order $(p-1)$ are equal to those of their corresponding element in the GLL canonical basis, i.e.,: *For* $0 \le i \le p$,

$$\int_{-1}^{+1} (g_i(x) - h_i(x))\, x^j \, dx = 0, \qquad \forall\, j, \quad 0 \le j \le (p-1). \tag{3}$$

   The difference $(g_i(x) - h_i(x))$ being proportional to $L_p(x)$ is orthogonal to all polynomials of degree less or equal to $(p-1)$.
2. Interpolation of their corresponding element in the canonical basis at the GL nodes, i.e.,: *For* $0 \le i \le p$,

$$g_i(\zeta_j) = h_i(\zeta_j), \qquad \forall\, j, \quad 1 \le j \le p. \tag{4}$$

3. The constants β_i can be obtained through a series expansion of (1) and one gets:

$$\beta_i = \frac{1}{(p+1)\, L_p(\xi_i)}, \qquad 0 \le i \le p. \tag{5}$$

The set of $(p+1)$ functions $g_i(x) \in I\!P_{p-1}(]-1,+1[)$ is linearly dependent. However, the preceding properties ensure that any combination of p elements in the list is linearly independent. We therefore arbitrarily discard one element in the set $\{g_i(x)\}_{i=0}^{p}$ (for, instance the first one, $g_0(x)$) and use the remaining elements to span the $I\!P_{p-1}(]-1,+1[)$ space.

For any continuous function φ on $]-1,+1[$ we consider two polynomial approximations. The Lagrange one, based on the h_j basis is given by:

$$\varphi_p(x) = \sum_{j=0}^{p} \varphi(\xi_j) h_j(x), \tag{6}$$

and a second one uses the new basis (g_j), $j = 1, \cdots p$:

$$\varphi_{p-1}^{(0)}(x) = \sum_{j=1}^{p} \varphi_j^{(0)} g_j(x). \tag{7}$$

The coefficients in expansion (7) can be derived from those of (6) thanks to the p relationships:

$$\int_{-1}^{1} \left(\varphi_p(x) - \varphi_{p-1}^{(0)}(x)\right) g_i(x)\, dx = 0, \quad \forall\, i, \quad 1 \le i \le p. \tag{8}$$

One can easily verify:

Theorem 1.

$$\forall i, \quad 1 \le i \le p, \quad \varphi_i^{(0)} = \varphi(\xi_i) + \alpha_i \varphi(\xi_0), \tag{9}$$

where the coefficients α_i *are the unique solutions of:*

$$\forall i, \quad 1 \le i \le p, \qquad \sum_{j=1}^{p} (g_j, g_i)\, \alpha_j = (g_0, g_i)\,. \tag{10}$$

Let's introduce the projection operator $\pi_h^{(0)}$ defined on Ω by: for any function f, $\pi_h^{(0)} f_{|\Omega_k}$ belongs to $I\!P_{p-1}(\Omega_k)$ and satisfies:

$$\int_{\Omega_k} \left(f_{|\Omega_k} - \pi_h^{(0)} f_{|\Omega_k} \right) g_i \, dx = 0, \quad \forall i, \quad 1 \le i \le p. \tag{11}$$

2.1.1 First Numerical Result

The first experiment is made on $]-1,1[$ divided into N equal elements. Its purpose is to test the efficiency of the new basis to approximate any given function. For instance, we consider the function $f(x) = \sin\left((x^2-1) \times (x+3)\right)$.

Figure 1 shows the efficiency of $\pi_h^{(0)} f$ to approximate the function f since the expected algebraic $\mathcal{O}(N^{-p})$ (see [2, 3]) decrease is observed. $I_h f$ is a piecewise-polynomial function of degree p on each Ω_k and continuous across element borders. It can be used to be derived. The function $\pi_h^{(0)} f$ is a piecewise-polynomial function of degree $p-1$ less regular than $I_h f$ and discontinuous across element borders when $p = 1$. It can be used to represent variations in directions without derivatives. By consequent, the key of our approach is the possibility to use both approximations $I_h f$ or $\pi_h^{(0)} f$ depending if the function is derived or not.

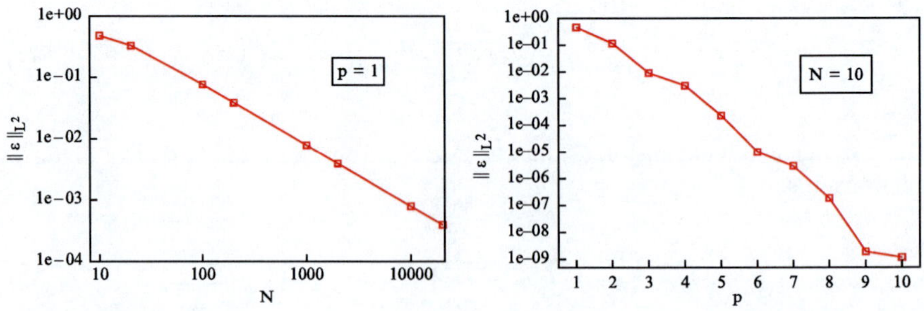

Fig. 1 The $L^2(\Omega)$–norm $||f - \pi_h^{(0)} f||$ error behavior. *Left*: as a function of N with $p = 1$ on a logarithmic scale. *Right*: as a function of p with $N = 10$ on a semi-logarithmic scale

2.2 Extension to Multidimensional Case

Extension to the multidimensional case is almost straightforward. For a given vector field $\mathbf{u} = (u_x, u_y)$, we consider three different possibilities of approximation:

$$u_r^{(0)}(x,y) = \sum_{e=1}^{N_x} \sum_{f=1}^{N_y} \sum_{i=\delta_{e1}}^{p} \sum_{j=\delta_{f1}}^{p} \overline{u}_{ij}^{ref}\, g_i(x_{ef}) g_j(y_{ef}), \tag{12}$$

$$u_r^{(1)}(x,y) = \sum_{e=1}^{N_x} \sum_{f=1}^{N_y} \sum_{i=\delta_{e1}}^{p} \sum_{j=\delta_{f1}}^{p} \overline{u}_{ij}^{ref}\, h_i(x_{ef}) g_j(y_{ef}), \tag{13}$$

$$u_r^{(2)}(x,y) = \sum_{e=1}^{N_x} \sum_{f=1}^{N_y} \sum_{i=\delta_{e1}}^{p} \sum_{j=\delta_{f1}}^{p} \overline{u}_{ij}^{ref}\, g_i(x_{ef}) h_j(y_{ef}). \tag{14}$$

Here, δ_{e1} and δ_{f1} are the Kronecker symbols and the index r denotes x or y. N_r is the number of elements on the r-direction. The piecewise-polynomial functions $u_r^{(s)}$ $(s = 0, 1, 2)$ have different regularity and different local polynomial degree.

Remark 1. Following the relation (9), one can verify that if the vector field belongs to $(H_0^1(\Omega))^2$, all the coefficients in the expansions (12)–(14) are the same while the functions are different. The approximation of the vector field $\mathbf{u} = (u_x, u_y)$ can be achieved using any of the following collocative expressions according to the functional dependence and the regularity required:

$$u_r^{(0)}(x,y) = \sum_{e=1}^{N_x} \sum_{f=1}^{N_y} \sum_{i=\delta_{e1}}^{p} \sum_{j=\delta_{f1}}^{p} u^r(\xi_i^{ef}, \xi_j^{ef})\, g_i(x_{ef}) g_j(y_{ef}), \tag{15}$$

$$u_r^{(1)}(x,y) = \sum_{e=1}^{N_x} \sum_{f=1}^{N_y} \sum_{i=\delta_{e1}}^{p} \sum_{j=\delta_{f1}}^{p} u^r(\xi_i^{ef}, \xi_j^{ef})\, h_i(x_{ef}) g_j(y_{ef}), \tag{16}$$

$$u_r^{(2)}(x,y) = \sum_{e=1}^{N_x} \sum_{f=1}^{N_y} \sum_{i=\delta_{e1}}^{p} \sum_{j=\delta_{f1}}^{p} u^r(\xi_i^{ef}, \xi_j^{ef})\, g_i(x_{ef}) h_j(y_{ef}). \tag{17}$$

The approximation $u_r^{(0)}$ will be used when u_r is not derived whereas the approximations $u_r^{(1)}$ and $u_r^{(2)}$ will be used respectively when u_r is derived in the directions x and y.

3 The Constraint Oriented Effect

To illustrate the capability of our approach to satisfy different constraints without changing the definition of the spectral element, let's consider *a formal test problem* which consists in solving an eigenvalue problem written on the square $\Omega := [-1,+1]^2$ cut into N^2 elements. The variational form of our problem consists in: *Finding* $\mathbf{u} \in (H_0^1(\Omega))^2$ *and* λ *such that*:

$$\int_\Omega \nabla\mathbf{u}\cdot\nabla\mathbf{v}d\mathbf{x} + \alpha\int_\Omega \nabla\cdot\mathbf{u}\;\nabla\cdot\mathbf{v}\,d\mathbf{x} + \beta\int_\Omega \nabla\times\mathbf{u}\cdot\nabla\times\mathbf{v}d\mathbf{x} = \lambda^2\int_\Omega \mathbf{u}\cdot\mathbf{v}\,d\mathbf{x}.$$

When $\beta = 0$ and α is large enough, the resulting eigenvalues are those of Stokes and the associated eigenvectors are divergence-free. For $\alpha = 0$ and β large enough, the eigenvalues remain the same but the associated eigenvectors become curl-free.

To provide a stable element we suggest to replace in (18) the two penalty bilinear forms by a stable approximation using the new basis. The discrete version is then: *Find* $\mathbf{u}_h = (u_x, u_y) \in \mathbf{X}_h$ *and* λ *such that:*

$$\mathscr{A}_h(\mathbf{u}_h,\mathbf{v}_h) + \alpha\,\mathscr{B}_h(\mathbf{u}_h,\mathbf{v}_h) + \beta\,\mathscr{C}_h(\mathbf{u}_h,\mathbf{v}_h) = \lambda^2\int_\Omega \mathbf{u}_h\cdot\mathbf{v}_h\,d\mathbf{x}, \quad \forall\mathbf{v}_h \in \mathbf{X}_h, \tag{18}$$

where:

$$\mathscr{A}_h(\mathbf{u}_h,\mathbf{v}_h) = (\nabla\mathbf{u}_h, \nabla\mathbf{v}_h)_h, \tag{19}$$

$$\mathscr{B}_h(\mathbf{u}_h,\mathbf{v}_h) = \left(\frac{\partial u_x^{(1)}}{\partial x} + \frac{\partial u_y^{(2)}}{\partial y}, \frac{\partial v_x^{(1)}}{\partial x} + \frac{\partial v_y^{(2)}}{\partial y}\right)_h, \tag{20}$$

$$\mathscr{C}_h(\mathbf{u}_h,\mathbf{v}_h) = \left(\frac{\partial u_y^{(1)}}{\partial x} - \frac{\partial u_x^{(2)}}{\partial y}, \frac{\partial v_y^{(1)}}{\partial x} - \frac{\partial v_x^{(2)}}{\partial y}\right)_h. \tag{21}$$

$\mathbf{X}_h$ is the space of continuous piecewise-polynomial functions vanishing on $\partial\Omega$.

We remark that all terms in the right-hand side of (20) and (21) are piecewise-polynomial functions with a local degree equal to $p-1$, a basic requirement to ensure a stable approximation for **grad**(div) and **curl**(curl) operators.

3.1 Numerical Results

This section discusses some numerical results related to the problem (18) showing its numerical efficiency in comparison with classical approaches. We start with case $\beta = 0$ and $\alpha = 10^5$.

The eigenvalue problem (18) gives $2(Np-1)^2$ eigenvalues and associated eigenvectors corresponding to the degrees of freedom in $\mathbf{X}_h$. Among these eigenvalues,

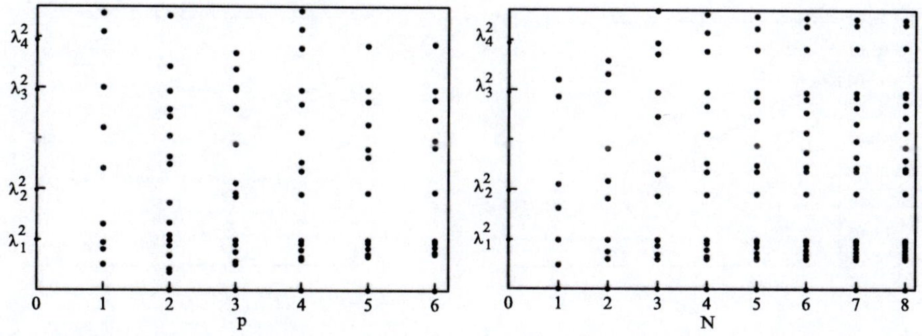

Fig. 2 The dependence of the **grad**(div) spectrum computed using a standard *hp* method with *p* for fixed $N = 4$ (*left*), and with *N* for fixed $p = 4$ (*right*)

there are the Stokes eigenvalues and the non-zero eigenvalues of the **grad**(div) operator multiplied by α. We are particularly interested in predicting the number of Stokes eigenvalues $\mathcal{N}_S$. This number corresponds to the size of the kernel of the discretized **grad**(div) operator, *ie* to the number of zero eigenvalues. It can be prooved that $\mathcal{N}_S = (Np-2)^2$. Consequently, the resolution of the problem (18) must lead to $(Np-2)^2$ Stokes eigenmodes. The $(Np)^2 - 2$ remaining eigenmodes are those of the **grad**(div) operator multiplied by α.

Thus, the main difficulty to solve the problem (18) consists in providing a stable discretization to the **grad**(div) operator. If a standard approach is chosen, the solution to the problem (18) is searched in $\mathbf{X}_h = I\!P_p^0(\Omega)^2$. With such a choice the discretization of the **grad**(div) operator is unstable and so-called spectral pollution appears [1]. Indeed, Fig. 2 represents the spectrum of the **grad**(div) operator computed using a standard *hp* method as a function of *p* (left) and *N* (right).

In both cases, besides the expected eigenvalues $\lambda_i (1 \leq i \leq 4)$, the class of divergence-free eigensolutions expands to an unphysical discrete spectrum. Consequently, in the resolution of the problem (18) the number of Stokes eigenvalues is false and besides the non-zero eigenvalues of the **grad**(div) operator, we can notice the presence of spurious eigenvalues.

Our approach offers a stable discretization of **grad**(div) operator and consequently we obtain $(Np-2)^2$ Stokes eigenvalues and no spurious modes.

Figure 3 illustrates the convergence of the difference ϵ between the four lowest Stokes eigenvalues computed by our method with those produced by Leriche et al. in [4]. The left part of the figure exhibits the error convergence as a function of p for $N = 2$ on a semi-logarithmic scale. The error is exponentially decreasing. The right part of the figure shows the same error convergence as a function of N for $p = 1$ on a log-log diagram. Here, we observe an algebraic decrease.

The second experiment concerns the case $\alpha = 0$ and $\beta = 10^5$. Here the main difficulty consists in providing a stable discretization of the **curl**(curl) operator. The numerical conclusion is almost the same than for the previous study. We limit ourself to the production of the graphs (Fig. 4).

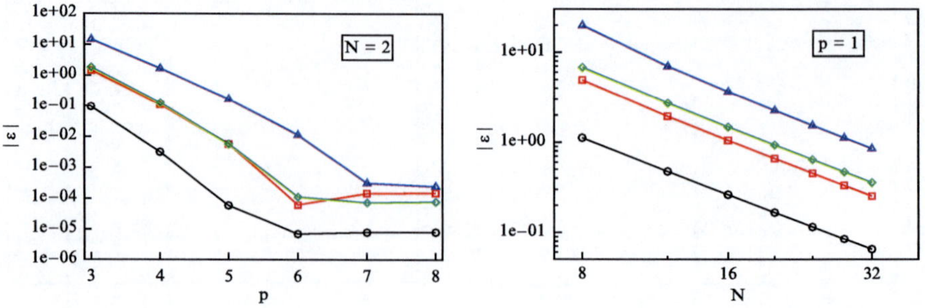

Fig. 3 Convergence plots obtained using the penalty method for the four lowest divergence-free modes (*circle* = λ_1, *square* = λ_2, *diamond* = λ_3, *triangle* = λ_4). *Left*: as a function of p with fixed $N = 2$. *Right*: as a function of N with fixed $p = 1$

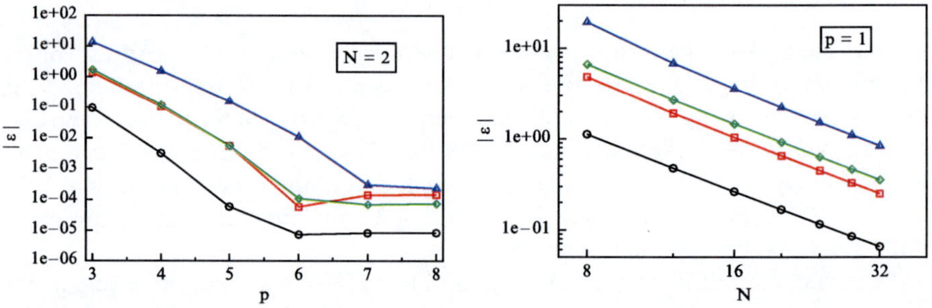

Fig. 4 Convergence plots obtained using the penalty method for the four lowest curl-free modes (*circle* = λ_1, *square* = λ_2, *diamond* = λ_3, *triangle* = λ_4). *Left*: as a function of p with fixed $N = 4$. *Right*: as a function of N with fixed $p = 1$

The similarity between the two results proves the efficiency of our approach to solve two different constraints without a significant modification of the spectral element. The same tools remain valid to ensure the stability and expected convergence.

References

1. Gruber R., Rappaz J.: Finite element methods in linear ideal MHD. Springer Series in Computational Physics. Springer, Berlin (1985)
2. Bernardi C., Maday Y., Rapetti F.: Discrétisation variationnelles de problèmes aux limites ellitiques. Springer, Paris (2000)
3. Canuto C., Hussaini M. Y., Quarteroni A., Zang T. A.: Spectral methods. Evolution to complex geometries and applications to fluid dynamics. Scientific Computation. Springer, Berlin (2007)
4. Leriche E., Labrosse G.: Stokes eigenmodes in a square domain and the stream function-velocity correlation. Journal of Computational Physics, **200**, 489–511 (2004)
5. Appert K., Azaiez M., Gruber R.: Modes of a plasma-filled waveguide determined by a numerical hp method. Communications in Computational Physics, **5**(2–4), 413–425 (2009)
6. Ahusborde E., Gruber R., Azaiez M., Sawley M. L.: Physics-conforming constraints-oriented numerical method. Physical Review E, **75**(5), 056704 (2007)

Convergence Rates of Sparse Tensor GPC FEM for Elliptic sPDEs

Marcel Bieri, Roman Andreev, and Christoph Schwab

Abstract We propose a novel class of *sparse tensor algorithms* for the numerical solution of stochastic elliptic PDEs. The methods are based on a hierarchic discretization in both, physical and probability space. The discretization spaces are then intertwined in a sparse tensor product fashion, leading to algorithms of log-linear complexity. We will present this idea in the framework of the stochastic Galerkin method. Theoretical results as well as numerical examples indicate the superiority of this sparse tensor product algorithms, compared to the full tensor product approaches used so far.

1 Introduction

Most, if not all, engineering models are subject to significant uncertainties, due to either variabilities in the model parameters or a fundamental lack of knowledge of the processes and quantities identified with the model. Neglecting the latter type of model uncertainties, we will only consider inherent parameter uncertainties modeled as random fields. Our model problem under consideration, is the following stochastic diffusion problem

$$\begin{cases} -\operatorname{div}(a(\omega,\mathbf{x})\nabla u(\omega,\mathbf{x})) = f(\mathbf{x}) & \text{in } D, \\ u(\omega,\mathbf{x})|_{\mathbf{x}\in\partial D} = 0, \end{cases} \qquad P\text{-a.e. } \omega \in \Omega \tag{1}$$

where $D \subset \mathbb{R}^d$ is a Lipschitz domain, (Ω, Σ, P) a complete probability space and the diffusivity a a random field. To guarantee existence and uniqueness of a solution, we assume $f \in H^{-1}(D)$ and, by Lax-Milgram,

M. Bieri (✉), R. Andreev, and C. Schwab
Seminar for Applied Mathematics, ETH Zürich, Switzerland
e-mail: mbieri@math.ethz.ch, andreevr@math.ethz.ch, schwab@math.ethz.ch

J.S. Hesthaven and E.M. Rønquist (eds.), *Spectral and High Order Methods for Partial Differential Equations*, Lecture Notes in Computational Science and Engineering 76,
DOI 10.1007/978-3-642-15337-2_7,

$$P\left\{\omega \in \Omega : a_{\min} \leq \operatorname*{ess\,inf}_{\mathbf{x}\in D} a(\omega,\mathbf{x}) \wedge \operatorname*{ess\,sup}_{\mathbf{x}\in D} a(\omega,\mathbf{x}) \leq a_{\max}\right\} = 1, \tag{2}$$

i.e. the diffusion coefficient is bounded from above and below.

All numerical schemes to approximate the solution u to (1), e.g. [1,2,8,9,11,12], consist of a sequence of stochastic approximations, e.g. polynomial chaos (PC) or collocation interpolation operators, to the law of the random solution and a spatial approximation, e.g. by finite elements, chosen independently of the stochastic approximant. Thus, they exhibit an overall complexity, i.e. total number of degrees of freedom, of $O(N_D \times N_\Omega)$, where N_D denotes the number of degrees of freedom of the spatial discretization and N_Ω the number of stochastic degrees of freedom, e.g. number of terms in the PC expansion. This is prohibitive, especially if a fine resolution of the spatial behavior or a description of the random input by a large number of variables is required, e.g. due to short correlation lengths in the input random fields. The main idea of this work is to choose suitable hierarchic approximations in space and random parameter domain and combine them in a sparse tensor product fashion, leading to algorithms of $O(N_D \log N_\Omega + N_\Omega \log N_D)$ overall complexity, and hence a considerable reduction in computation time and memory requirement.

The outline of this work is as follows: In Sect. 2, we will derive a parametrized form of the model problem (1) and associated variational formulations. In Sect. 3, we will then present the sparse tensor stochastic Galerkin method, along with theoretical results on the convergence rates. Finally, in Sect. 4, we will show a numerical example to underline the superiority of this novel approach, compared to existing ones.

We note here, that the present work is based on [4,5] and the theoretical results presented here, can be found therein. Moreover, we mention that the same idea is applicable in the framework of stochastic collocation algorithms, and has been thoroughly investigated in [3,4].

2 Parametrization of the Model Problem

The sparse tensor stochastic Galerkin method presented here, relies on a separation of spatial and stochastic parts in the input parameter a. Assuming so, one can then derive the parametrized variational formulation of the model problem, on which the Galerkin scheme is based.

2.1 Separation of Stochastic and Deterministic Variables

Assumption 1. *We assume that the diffusion coefficient is given by a series of the form*

$$a(\omega, \mathbf{x}) = \mathbb{E}_a(\mathbf{x}) + \sum_{m \geq 1} \psi_m(\mathbf{x}) Y_m(\omega), \tag{3}$$

where $\mathbb{E}_a$ denotes the mean field, the ψ_m are $L^2(D)$-orthonormalized functions in $L^\infty(D)$, satisfying $\{\|\psi_m\|_{L^\infty}\}_{m\in\mathbb{N}} \in \ell_1$, and $Y_m : \Omega \to \mathbb{R}$ are pairwise uncorrelated random variables.

Such a series can e.g. be derived from a so-called Karhunen-Loève (KL) expansion of the random field a. The KL expansion is guaranteed to exist if we assume that a has finite second moments, i.e. the mean field $\mathbb{E}_a$ and covariance C_a exist. In this case, the covariance operator $\mathscr{C}_a : L^2(D) \longrightarrow L^2(D)$, defined through $(\mathscr{C}_a u)(\mathbf{x}) := \int_D C_a(\mathbf{x}, \mathbf{x}') u(\mathbf{x}')\, d\mathbf{x}'$, has, under some general assumptions on C_a, a countable sequence of eigenpairs $(\lambda_m, \varphi_m)_{m\in\mathbb{N}}$. Setting $\psi = \sqrt{\lambda_m}\varphi_m$, the series (3) is in fact the KL series of a.

To be able to numerically handle the series (3), it is truncated after M terms, i.e. we define

$$a_M(\omega, \mathbf{x}) = \mathbb{E}_a(\mathbf{x}) + \sum_{m=1}^{M} \psi_m(\mathbf{x}) Y_m(\omega), \tag{4}$$

where M is usually determined in the course of the discretization process.

2.2 Parametric Deterministic Problem

We make the following assumption on the random variables Y_m and deterministic functions ψ_m in the series representation (3) of a.

Assumption 2. *(a) The family $(Y_m)_{m\geq 1} : \Omega \to \mathbb{R}$ is* independent,

(b) With each random variable $Y_m(\omega)$ in (3), is associated a complete probability space $(\Omega_m, \Sigma_m, P_m)$, with

(i) The range of Y_m, $\Gamma_m :=$ Ran$(Y_m) \subseteq \mathbb{R}$, is compact and, after eventually rescaling, equal to $[-1, 1]$ for all m,

(ii) The probability measure P_m admits a uniform probability density function $\rho_m = \frac{1}{2}$ such that $dP_m(\omega) = \rho_m(y_m) dy_m$, $m \in \mathbb{N}$, $y_m \in \Gamma_m$, and

(c) The deterministic functions ψ_m are decaying at least algebraically to zero, i.e. $\|\psi_m\|_{L^\infty} \leq Cm^{-s}$ with $s > 1$.

Due to the independency, one can view the Y_m as different coordinates in probability space, and we therefore set $y_m := Y_m(\omega) \in \Gamma_m$. Furthermore, we define

$$\Gamma := \Gamma_1 \times \cdots \times \Gamma_M, \quad \mathbf{y} = (y_1, \ldots, y_M) \in \Gamma, \quad \rho(\mathbf{y}) = \prod_{m=1}^{M} \rho_m(y_m).$$

Hence, the (truncated) diffusion coefficient can equivalently be stated as

$$a_M(\mathbf{y},\mathbf{x}) = \mathbb{E}_a(\mathbf{x}) + \sum_{m=1}^{M} \psi_m(\mathbf{x}) y_m. \tag{5}$$

By $L^2_\rho(\Gamma; H^1_0(D))$, we denote the Bochner space of functions $v : \Gamma \to H^1_0(D)$, for which $\|v(\mathbf{y},\cdot)\|_{H^1_0(D)} : \Gamma \to \mathbb{R}$ belongs to $L^2_\rho(\Gamma)$. It holds

$$L^2_\rho(\Gamma; H^1_0(D)) \cong L^2_\rho(\Gamma) \otimes H^1_0(D), \tag{6}$$

where $\otimes$ denotes the tensor product between separable Hilbert spaces, see e.g. [10, Chap. 1].

Replacing the random coefficient a in (1) by its parametrized version (5) and multiplying with a test function v, gives rise to the parametric deterministic variational formulation: find $u \in L^2_\rho(\Gamma; H^1_0(D))$, s.t.

$$b(u,v) = l(v), \qquad \forall\, v \in L^2_\rho(\Gamma; H^1_0(D)), \tag{7}$$

where the bilinear and linear form b and l, respectively, are given by

$$b(u,v) = \mathbb{E}\left[\int_D a_M(\mathbf{y},\mathbf{x}) \nabla u(\mathbf{y},\mathbf{x}) \cdot \nabla v(\mathbf{y},\mathbf{x}) d\mathbf{x}\right],\ l(v) = \mathbb{E}\left[\int_D f(\mathbf{x}) v(\mathbf{y},\mathbf{x}) d\mathbf{x}\right]. \tag{8}$$

The unique solvability of (7) is a direct consequence of (2).

3 Sparse Tensor Stochastic Galerkin Method

3.1 *Sparse Tensor Galerkin Formulation*

In stochastic Galerkin finite element methods (sGFEM), we discretize the variational formulation (7) by Galerkin projection onto a hierarchic sequence of finite dimensional subspaces of $L^2_\rho(\Gamma; H^1_0(D))$ in (6), i.e.

$$V_0^\Gamma \subset V_1^\Gamma \subset \cdots \subset L^2_\rho(\Gamma) \quad \text{and} \quad V_0^D \subset V_1^D \subset \cdots H^1_0(D) \tag{9}$$

We introduce *detail spaces* $W^\Gamma_{l_1}$ and $W^D_{l_2}$ such that

$$V^\Gamma_{l_1} = V^\Gamma_{l_1-1} \oplus W^\Gamma_{l_1} \quad \text{and} \quad V^D_{l_2} = V^D_{l_2-1} \oplus W^D_{l_2} \text{ for } l_1, l_2 = 1, 2, \ldots \tag{10}$$

where l_1, l_2 denote the stochastic and spatial *level of refinement*, respectively. Further, we set $W^\Gamma_0 := V^\Gamma_0$ and $W^D_0 := V^D_0$. The sums in (10) are direct, so that the (finite-dimensional) approximation spaces V^Γ_L and V^D_L admit a multilevel decomposition

$$V_L^\Gamma = \bigoplus_{l_1=0}^{L} W_{l_1}^\Gamma \quad \text{and} \quad V_L^D = \bigoplus_{l_2=0}^{L} W_{l_2}^D .$$

We denote by

$$V_L^\Gamma \otimes V_L^D = \bigoplus_{0 \le l_1, l_2 \le L} W_{l_1}^\Gamma \otimes W_{l_2}^D \subset L_\rho^2(\Gamma) \otimes H_0^1(D) \tag{11}$$

the (full) tensor product space of the finite dimensional component subspaces V_L^D and V_L^Γ, respectively. However, we approximate the parametric deterministic problem (7) by Galerkin projection onto the sparse tensor product space

$$V_L^\Gamma \hat{\otimes} V_L^D := \bigoplus_{0 \le l_1 + l_2 \le L} W_{l_1}^\Gamma \otimes W_{l_2}^D . \tag{12}$$

Hence, the sparse tensor sGFEM discretization can be written as, in variational form: find $\hat{u} \in V_L^\Gamma \hat{\otimes} V_L^D$

$$b(\hat{u}, v) = l(v) \quad \forall\, v \in V_L^\Gamma \hat{\otimes} V_L^D , \tag{13}$$

where b and l are given by (8).

3.2 *Hierarchic Discretization in* $L_\rho^2(\Gamma)$

Here, we propose a best-N-term approximation of $u : \Gamma \to H_0^1(D)$, based on an expansion in Legendre polynomials.

By $\mathbb{N}_0^M$ denote the set of all multiindices of length M. If $\alpha \in \mathbb{N}_0^M$, denote by

$$\mathscr{L}_\alpha(\mathbf{y}) := L_{\alpha_1}(y_1) \cdots L_{\alpha_M}(y_M)$$

the tensorized Legendre polynomial of degree α. Hence, u can be represented in terms of Legendre polynomials as

$$u(\mathbf{y}, \mathbf{x}) = \sum_{\alpha \in \mathbb{N}_0^M} u_\alpha(\mathbf{x}) \mathscr{L}_\alpha(\mathbf{y}), \tag{14}$$

with 'coefficients' $u_\alpha \in H_0^1(D)$. For a parameter $\gamma > 0$ and a level $l_1 \in \mathbb{N}_0$, define the index sets

$$\Lambda_\gamma(l_1) := \underset{\substack{\Lambda \subset \mathbb{N}_0^M \\ |\Lambda| = \lceil 2^{\gamma l_1} \rceil}}{\arg\max} \left(\sum_{\alpha \in \Lambda} \|u_\alpha\|_{H_0^1} \right) \subset \mathbb{N}_0^M, \qquad l_1 = 0, 1, 2, \ldots \tag{15}$$

and the truncated Legendre expansion

$$u_{\Lambda_\gamma(l_1)} = \sum_{\alpha \in \Lambda_\gamma(l_1)} u_\alpha(\mathbf{x})\mathscr{L}_\alpha(\mathbf{y}) \tag{16}$$

Given the space $\Lambda_\gamma(l_1)$, the stochastic approximation spaces in (9) are defined by

$$V_{l_1}^\Gamma := \{v \in L^2_\rho(\Gamma) : v = \sum_{\alpha \in \Lambda_\gamma(l_1)} v_\alpha \mathscr{L}_\alpha, v_\alpha \in \mathbb{R}\}.$$

The index sets $\Lambda_\gamma(l_1)$ are in general not computationally available. In Sect. 4, however, we will present an algorithm to find index sets $\tilde{\Lambda}_\gamma(l_1)$, which prove to be close to optimal in numerical examples. With the L^2_ρ-projection $P_{l_1}^\Gamma : L^2_\rho(\Gamma) \to V_{l_1}^\Gamma$, defined by $P_{l_1}^\Gamma u := u_{\Lambda_\gamma(l_1)}$ as in (16), the following approximation properties hold:

Proposition 1. *Let $s > 1$ be the decay rate of $\psi_m(\mathbf{x})$ as assumed in Assumption 2. If u solves (7), then, for each $0 < r < s - \frac{3}{2}$ there exists a constant $C(r)$, s.t. for every $\gamma > 0$ and for the sequence of projections $P_{l_1}^\Gamma$, corresponding to the index sets $\Lambda_\gamma(l_1)$ in (15), it holds*

$$\|u - P_{l_1}^\Gamma u\|_{L^2_\rho(\Gamma;H^1_0(D))} \leq C(r)(N_{l_1}^\Gamma)^{-r} |u|_{\mathscr{A}_r(H^1_0(D))}, \tag{17}$$

where $N_{l_1}^\Gamma := |\Lambda_\gamma(l_1)|$ or, equivalently,

$$\|u - P_{l_1}^\Gamma u\|_{L^2_\rho(\Gamma;H^1_0(D))} \leq C(r)2^{-l_1\gamma r} |u|_{\mathscr{A}_r(H^1_0(D))}. \tag{18}$$

In the above proposition, $\mathscr{A}_r(H^1_0)$ denotes the space of functions in $L^2_\rho(\Gamma; H^1_0(D))$, which can be best-$N$-term approximated at rate r. For details and a proof, we refer to [5, Sect. 3.2] and [4, Sect. 7.1].

3.3 Hierarchic Discretization in D

As a basis for the spatial approximation spaces $V_{l_2}^D$, we choose linear finite element wavelets, based on a nested sequence $\{\mathscr{T}_{l_2}\}_{l_2 \geq 0}$ of regular simplicial triangulations of D, i.e.

$$V_{l_2}^D := S^1(D, \mathscr{T}_{l_2}) = \left\{u \in H^1_0(D) : u|_T \in \mathscr{P}_1(T) \text{ for } T \in \mathscr{T}_{l_2}\right\}.$$

For the construction of such wavelet bases, we refer to [5, Sect. 3.3] and, more generally, to [6]. They satisfy the approximation property

$$\|u - P_{l_2}^D u\|_{H^1(D)} \leq C 2^{-l_2 t} \|u\|_{H^{1+t}(D)}, \quad t \in [0, 1], \tag{19}$$

or, w.r.t. $N_{l_2}^D$,

$$\|u - P_{l_2}^D u\|_{H^1(D)} \leq C(N_{l_2}^D)^{-t/d} \|u\|_{H^{1+t}(D)}, \quad t \in [0,1], \tag{20}$$

where $P_{l_2}^D : L^2(D) \to V_{l_2}^D$ denotes the H^1-projection onto $V_{l_2}^D$.

3.4 Convergence Rates of Sparse Tensor sGFEM

In the present section, we state our main result, that the sparse tensor sGFEM converges algebraically in terms of the total numbers of degrees of freedom. We define the sparse tensor product projection operator $\hat{P}_L : L^2_\rho(\Gamma) \otimes H^1_0(D) \to V_L^\Gamma \hat{\otimes} V_L^D$ by

$$(\hat{P}_L v)(\mathbf{y}, \mathbf{x}) := \sum_{0 \leq l_1 + l_2 \leq L} \left(P_{l_1}^\Gamma - P_{l_1 - 1}^\Gamma\right) \otimes \left(P_{l_2}^D - P_{l_2-1}^D\right) v(\mathbf{y}, \mathbf{x}).$$

Proposition 2. *Let the solution u to the model problem (1) satisfy*

$$u \in \mathscr{A}_r((H^{1+t} \cap H^1_0)(D)) \quad \text{for some} \quad 0 < r < s - \frac{3}{2}, \; 0 < t \leq 1 \tag{21}$$

with $s > \frac{3}{2}$ as in Assumption 2. Let $\hat{u}$ denote the sGFEM solution to the problem (13) w.r.t. the sparse tensor product spaces sequence $V_L^\Gamma \hat{\otimes} V_L^D$, defined in (12). Then, there exists a constant $C(\beta) > 0$, independent of L and M, such that

$$\|u - \hat{u}\|_{L^2_\rho(\Gamma; H^1_0(D))} \leq C(\beta) L^{1+\beta} \hat{N}_L^{-\beta} \|u\|_{\mathscr{A}_r(H^{1+t}(D))}, \tag{22}$$

where $\hat{N}_L := \dim(V_L^\Gamma \hat{\otimes} V_L^D)$ and $\beta = \min(r, t/d)$.

Hence, by using the sparse tensor formulation of the sGFEM, we retrieve the less of the two convergence rates r and $\frac{t}{d}$, stemming from the stochastic (17) and spatial (20) discretization, respectively. In a full tensor approach (11), on the other hand, the convergence rates can be shown to be

$$\|u - P_L^D \otimes P_L^\Gamma u\| \leq C(N_L)^{-\bar{\beta}} \|u\|_{\mathscr{A}_r((H^1_0 \cap H^{p+1})(D))}, \tag{23}$$

with $\bar{\beta} = (d/t + 1/r)^{-1}$, see [5, Remark 3.7].

4 Implementation and Numerical Examples

In this final section, we will first provide an algorithm to identify a quasi-optimal set $\tilde{\Lambda}_\gamma(l_1)$ and then show a numerical example which confirms the theoretical results presented in Sect. 3.4.

4.1 Localization of Quasi-Best-N-Term Coefficients

The identification of $\tilde{\Lambda}_\gamma(l_1)$ is based on the observation, proved in [5], that the decay of the coefficients ψ_m of the input random field determines the decay of the coefficients u_α in the expansion (14). Precisely, one can show that

$$\|u_\alpha\|_{\mathscr{W}(D)} \lesssim \eta^{-\alpha} := \prod_{m\geq 1} \eta_m^{-\alpha_m}$$

where $\eta_m^{-1} \lesssim \|\psi_m\|_{L^\infty} m^{1+\delta}$ and $\eta_m^{-1} \lesssim m^{-s+(1+\delta)}$, with s given as in Assumption 2. Hence, the index sets $\tilde{\Lambda}_\gamma(L)$ consist simply of the $\lceil 2^{\gamma L}\rceil$ largest upper bounds $\eta^{-\nu}$, i.e.

$$\tilde{\Lambda}_\gamma(L) := \underset{\substack{\Lambda\subset\mathbb{N}_0^M\\ |\Lambda|=\lceil 2^{\gamma L}\rceil}}{\arg\max}\left(\sum_{\alpha\in\Lambda}\eta^{-\alpha}\right) \subset \mathbb{N}_0^M, \qquad l_1 = 0,1,2,\ldots \tag{24}$$

It has been shown in [5], that the computation of these index sets can be done linear in time and memory requirement, w.r.t. to their size.

4.2 Numerical Example

We consider a problem of the form (1) on the unit square $D = [-1,1]^2$ with a diffusion coefficient a given by a series as in (3). To verify the convergence result provided by Proposition 2, we choose $\psi_{mn}(\mathbf{x}) = \sqrt{\lambda_m\lambda_n}\varphi_m(x_1)\varphi_n(x_2)$ where

$$\lambda_m = \frac{8^{5/2}}{(\pi(2m-1))^5}, \quad \varphi_m(x_i) = \sin\left(\frac{x_i+1}{\sqrt{2\lambda_m}}\right), \quad i = 1,2,$$

hence the sequence $\|\psi_m\|_{L^\infty(D)}$ exhibits an algebraic decay with rate $s = \frac{5}{2}$, which, by Proposition 1, in turn implies a stochastic rate of $r = 1$. Furthermore, we assume $\mathbb{E}_a(\mathbf{x}) = x_1 + 5$ and $f \equiv 1$. Using piecewise linear wavelets in space corresponds to a spatial approximation rate $\frac{t}{d} = \frac{1}{2}$, provided the solution is accordingly regular. Hence, by Proposition 2, we expect a sparse tensor rate of $\beta = \frac{1}{2}$ while as in the full

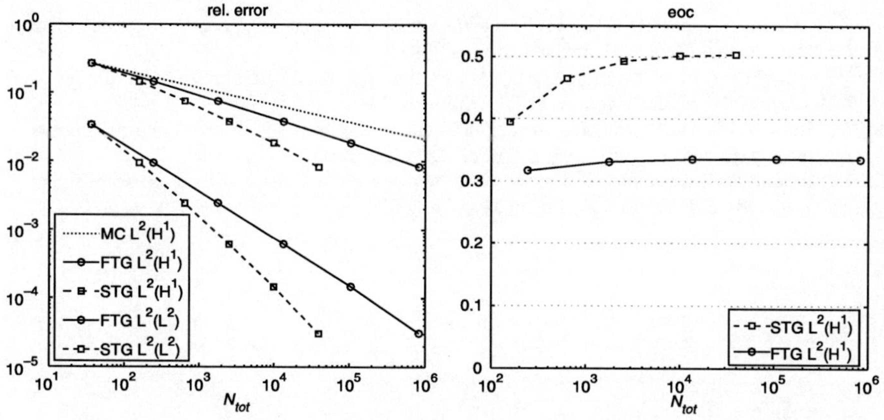

Fig. 1 *Left*: Convergence of the solution computed by a sparse tensor Galerkin (STG), full tensor Galerkin (FTG) and Monte Carlo (MC) method. *Right*: Corresponding estimated orders of convergence

tensor case we expect a rate $\bar{\beta} = \frac{1}{3}$, see (23). Those rates are exactly retrieved by the numerical experiment as one can see in Fig. 1. There, we plot the relative error of the computed solutions in the $L^2_\rho(\Gamma; H^1_0(D))$- and $L^2_\rho(\Gamma; L^2(D))$-norm, respectively, the expected convergence of the Monte Carlo method and the estimated order of convergence, computed between consecutive data points.

Acknowledgements The work of the author Marcel Bieri was supported in part by the Swiss National Science Foundation under Grant No. 200021-120290/1.

References

1. I. Babuška, R. Tempone, and G.E. Zouraris. Galerkin finite element approximations of stochastic elliptic partial differential equations. *SIAM J. Num. Anal.*, 42(2):800–825, 2002
2. I. Babuška, F. Nobile, and R. Tempone. A stochastic collocation method for elliptic partial differential equations with random input data. *SIAM J. Num. Anal.*, 45(3):1005–1034, 2007
3. M. Bieri. A sparse composite collocation finite element method for elliptic sPDEs. Technical Report 2009-08, Seminar for Applied Mathematics, ETH Zürich, 2009 (submitted)
4. M. Bieri. *Sparse tensor discretizations of elliptic PDEs with random input data*. PhD thesis, ETH Zürich, 2009 (in preparation)
5. M. Bieri, R. Andreev, and Ch. Schwab. Sparse tensor discretization of elliptic sPDEs. *SIAM J. Sci. Comput.*, 31(6):4281–4304, 2009
6. A. Cohen. *Numerical analysis of wavelet methods*, volume 32 of *Studies in Mathematics and its Applications*. Elsevier, Amsterdam, 2003
7. A. Cohen, R. DeVore, and Ch. Schwab. Convergence rates of best N-term stochastic Galerkin FE-approximations for a class of elliptic sPDEs. Technical Report 2009-02, Seminar for Applied Mathematics, ETH Zürich, 2009 (submitted)
8. P. Frauenfelder, Ch. Schwab, and R.-A. Todor. Finite elements for elliptic problems with stochastic coefficients. *Comp. Meth. Appl. Mech. Engrg.*, 194:205–228, 2005

9. R.G. Ghanem and P.D. Spanos. *Stochastic Finite Elements, a Spectral Approach*. Dover Publications Inc., New York, revised edition, 2003
10. W. A. Light and E. W. Cheney. *Approximation theory in tensor product spaces*, volume 1169 of *Lecture Notes in Mathematics*. Springer, Berlin, 1985
11. D. Xiu and J.S. Hesthaven. High-order collocation methods for differential equations with random inputs. *SIAM J. Sci. Comp.*, 27(3):1118–1139, 2005
12. D. Xiu and G.E. Karniadakis. The wiener-askey polynomial chaos for stochastic differential equations. *SIAM J. Sci. Comp.*, 24(2):619–644, 2002

A Conservative Spectral Element Method for Curvilinear Domains

Mick Bouman, Artur Palha, Jasper Kreeft, and Marc Gerritsma

Abstract This paper describes a mimetic spectral element method on curvilinear grids applied to the Poisson equation. The Poisson equation is formulated in terms of differential forms. The spectral basis functions in which the differential forms are expressed lead to a metric free discrete representation of the gradient and the divergence operator. Using the fact that the pullback operator commutes with the wedge product and the exterior derivative leads to a mimetic spectral element formulation on curvilinear grids which displays exponential convergence and satisfies the divergence exactly. The robustness of the proposed scheme will be demonstrated for a sample problem for which exponential convergence is obtained.

1 Introduction

Mimetic discretization schemes aim to reformulate partial differential equations in discrete form such that its structure is preserved as much as possible. Many invariants/symmetries can be described in terms of algebraic topology, where these operators can be defined without any reference to metric. The continuous counterpart of Algebraic Topology is Differential Geometry, expressing the variables in terms of differential forms. The continuous description is necessary for the expression of the metric dependent part. The relation between the continuous formulation in terms of differential forms and algebraic topology can be found in [1–5, 9, 12]. Recently, these ideas have also been introduced for spectral methods, [7, 10, 11]. In this paper these ideas are extended to curvilinear domains.

M. Gerritsma (✉), Mick Bouman, Artur Palha, and Jasper Kreeft
Delft University of Technology, Kluyverweg 1, 2629 HS Delft, The Netherlands
e-mail: M.I.Gerritsma@tudelft.nl

J.S. Hesthaven and E.M. Rønquist (eds.), *Spectral and High Order Methods for Partial Differential Equations*, Lecture Notes in Computational Science and Engineering 76,
DOI 10.1007/978-3-642-15337-2_8,

2 The Poisson Equation in Terms of Differential Forms

Consider

$$\mathbf{div}\, q = f, \quad u = \mathbf{grad}\, \phi, \quad q = u, \tag{1}$$

in a domain Ω, with $n = \dim(\Omega)$. Along the boundary a combination of ϕ and $q \cdot n$ is prescribed, where n is the outward unit normal to $\partial\Omega$. An equivalent representation in terms of differential forms is given by

$$dq^{(n-1)} = f^{(n)}, \quad d\phi^{(0)} = u^{(1)}, \quad q^{(n-1)} = \star u^{(1)}. \tag{2}$$

Here $\phi^{(0)}$ is a 0-form, $u^{(1)}$ a 1-form, $q^{(n-1)}$ a $(n-1)$-form and $f^{(n)}$ a n-form in $T^*\Omega$, the cotangent bundle of Ω. d is the exterior derivative which maps k-forms onto $(k+1)$-forms and $\star$ is the Hodge-$\star$ operator which establishes an isometry between k-forms and $(n-k)$-forms.

The wedge product, $\wedge$, is an anti-symmetric, multi-linear operator which maps k-forms and m-forms onto $(k+m)$-forms, i.e.,

$$\wedge \,:\, \Lambda^k(\Omega) \times \Lambda^m(\Omega) \mapsto \Lambda^{k+m}(\Omega), \tag{3}$$

with

$$\alpha^{(k)} \wedge \alpha^{(k)} \equiv 0^{(2k)}, \quad \forall \alpha^{(k)} \in \Lambda^k(\Omega), \tag{4}$$

where $\Lambda^k(\Omega)$ is the space of k-forms on Ω. If we define an inner-product, $(\cdot,\cdot)$ for k-forms in terms of its vector proxies, the Hodge-$\star$ operator maps

$$\star \,:\, \Lambda^k(\Omega) \mapsto \Lambda^{(n-k)}(\Omega), \tag{5}$$

defined by

$$\zeta^{(k)} \wedge \star\eta^{(k)} := \left(\zeta^{(k)}, \eta^{(k)}\right) \omega^{(n)}, \tag{6}$$

where $\omega^{(n)}$ is the normalized n-form which satisfies $\omega^{(n)} \wedge \star\omega^{(n)} = \omega^{(n)}$. We have that

$$\star\star\eta^{(k)} = (-1)^{k(n-k)}\, \eta^{(k)}, \tag{7}$$

for a metric with positive signature.

A mapping $\Phi \,:\, \Omega' \mapsto \Omega$ induces a map $\Phi^* \,:\, \Lambda^k(\Omega) \mapsto \Lambda^k(\Omega')$, called the *pullback operator*. The pullback operator maps differential forms over Ω to differential forms over Ω', such that for all $\alpha^{(k)} \in \Lambda^k(\Omega)$ we have

$$\int_{\Phi(\Omega')} \alpha^{(k)} \equiv \int_{\Omega'} \Phi^*\alpha^{(k)}. \tag{8}$$

For the spectral method on arbitrary domains, we rely on two important properties of the pullback operator:

1. *The pullback operator commutes with the exterior derivative*: For all $\alpha^{(k)} \in \Lambda^k(\Omega)$

$$\Phi^*\left(d\alpha^{(k)}\right) = d\left(\Phi^*\alpha^{(k)}\right); \tag{9}$$

2. *The pullback operator is an algebra homomorphism*: For all $\alpha^{(k)} \in \Lambda^k(\Omega)$ and $\beta^{(m)} \in \Lambda^m(\Omega)$

$$\Phi^*\left(\alpha^{(k)} \wedge \beta^{(m)}\right) = \left(\Phi^*\alpha^{(k)}\right) \wedge \left(\Phi^*\beta^{(m)}\right). \tag{10}$$

Using these properties of the pullback operator we can transform (2) from an arbitrary curvilinear domain Ω to an orthogonal domain Ω'

$$d\Phi^*q^{(n-1)} = \Phi^*f^{(n)}\ ,\ d\Phi^*\phi^{(0)} = \Phi^*u^{(1)}\ ,\ \Phi^*q^{(n-1)} = \Phi^* \star \left(\Phi^*\right)^{-1}\Phi^*u^{(1)}. \tag{11}$$

Introducing the transformed variables $\tilde{\alpha}^{(k)} = \Phi^*\alpha^{(k)}$, this gives

$$d\tilde{q}^{(n-1)} = \tilde{f}^{(n)},\quad d\tilde{\phi}^{(0)} = \tilde{u}^{(1)},\quad \tilde{q}^{(n-1)} = \tilde{\star}\tilde{u}^{(1)}, \tag{12}$$

where

$$\tilde{\star} := \left(\Phi^* \star \left(\Phi^*\right)^{-1}\right)\ :\ \Lambda^k\left(\Omega'\right) \longrightarrow \Lambda^{(n-k)}\left(\Omega'\right). \tag{13}$$

Note that (12) represents the Poisson equation mapped onto the orthonormal domain, Ω'.

Now that we have transformed our problem form curvilinear coordinates in Ω to orthogonal coordinates in Ω', we can use the mimetic spectral element scheme described in the next section. Once we have solved (12) for $\tilde{\phi}^{(0)}$ and $\tilde{q}^{(n-1)}$, we retrieve the solution in physical space by pre-multiplication by $(\Phi^*)^{-1}$, such as

$$\phi^{(0)} = \left(\Phi^*\right)^{-1}\tilde{\phi}^{(0)},\quad q^{(n-1)} = \left(\Phi^*\right)^{-1}\tilde{q}^{(n-1)}. \tag{14}$$

For a more extensive discussion on differential forms see, for instance, [1,6].

3 Discretization of the Transformed Poisson Equation

Gauss-Lobatto Grid Let $n = 2$ and $\Omega' = [-1,1]^2$ with coordinates (ξ,η), then the k-forms can be expanded in terms of Lagrange and edge functions. Consider the Gauss-Lobatto-Legendre (GLL) nodes in the interval $[-1,1]$ given by the $N+1$ zeros, ξ_i, of the polynomial $\left(1-\xi^2\right)L'_N(\xi)$. Let $h_i(\xi)$ be the Lagrange polynomials through the GLL-nodes,

$$h_i(\xi_j) = \begin{cases} 1 & \text{if } i = j \\ 0 & \text{if } i \neq j \end{cases},\quad i,j = 0,\ldots,N. \tag{15}$$

The edge functions, $e_i(\xi)$, are expressed in terms of the Lagrange functions as

$$e_i(\xi) = -\sum_{k=0}^{i-1} dh_k(\xi) = -\sum_{k=0}^{i-1} \frac{dh_k}{d\xi} d\xi\,, \quad i = 1, \ldots, N. \tag{16}$$

The edge functions satisfy

$$\int_{\xi_{k-1}}^{\xi_k} e_i(\xi) = \begin{cases} 1 & \text{if } i = k \\ 0 & \text{if } i \neq k \end{cases}, \quad i, j = 0, \ldots, N. \tag{17}$$

See [7] for a derivation and properties of the edge functions.

These basis functions can be seen as differential forms, where $h_i(\xi)$ is a 0-form and $e_i(\xi)$ a 1-form. Now any k-form $(0 \leq k \leq n)$ can be constructed by selecting one of the two for each direction, as will be shown below.

Expand the normal flux $\tilde{q}^{(1)}$ in

$$\left(\tilde{q}^{(1)}\right)^N (\xi, \eta) = -\sum_{i=1}^{N}\sum_{j=0}^{N} q^{\eta}_{i,j} e_i(\xi) h_j(\eta) + \sum_{i=0}^{N}\sum_{j=1}^{N} q^{\xi}_{i,j} h_i(\xi) e_j(\eta), \tag{18}$$

where $q^{\xi}_{i,j}$ and $q^{\eta}_{i,j}$ are the normal flux components as shown in Fig. 1. Let

$$\left(\tilde{f}^{(2)}\right)^N (\xi, \eta) = \sum_{i=1}^{N}\sum_{j=1}^{N} f_{i,j} e_i(\xi) e_j(\eta), \tag{19}$$

and insert this in the equation $d\tilde{q}^{(1)} = \tilde{f}^{(2)}$, we obtain

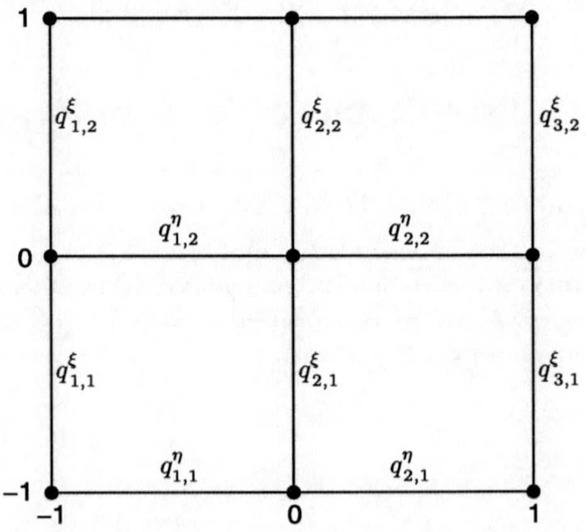

Fig. 1 Gauss-Lobatto grid and locations of $q^{\xi}_{i,j}$ and $q^{\eta}_{i,j}$ for $N = 2$

$$\sum_{i=1}^{N}\sum_{j=1}^{N}\left[q^{\xi}_{i,j}-q^{\xi}_{i-1,j}+q^{\eta}_{i,j}-q^{\eta}_{i,j-1}-f_{i,j}\right]e_i(\xi)e_j(\eta)=0. \tag{20}$$

Since, the basis functions, $e_i(\xi)$ are linearly independent, this can only be satisfied when

$$q^{\xi}_{i,j}-q^{\xi}_{i-1,j}+q^{\eta}_{i,j}-q^{\eta}_{i,j-1}-f_{i,j}=0\,,\quad i,j=1,\ldots,N. \tag{21}$$

The result is a finite-volume formulation for the divergence relation. It can be shown that this is an exact, metric-free discretization, [7]. The discrete divergence equation becomes

$$Dq=f, \tag{22}$$

where matrix D consists only of the values -1, 0 and 1.

Extended Gauss Grid The gradient equation, $d\tilde{\phi}^{(0)}=\tilde{u}^{(1)}$, will be discretized on a second grid which is dual to the GLL grid. The reason for the two dual grids stems from the fact that the variables in the divergence equation are externally oriented with respect to the geometric objects they are associated with, whereas the variables in the gradient equation are internally oriented with respect to the underlying geometric elements, see [9, 12] for a lucid explanation of this difference. The dual grid consists of the Gauss points plus boundary points, which are the zeros of the Legendre polynomial $L_N(\xi_i)$ complemented with nodes on the boundary, see Fig. 2. This grid will be referred to as the *extended Gauss grid*. Let $\tilde{h}_i(\xi)$ be the one-dimensional Lagrange polynomial through the extended Gauss points, then the expansion of $\tilde{\phi}^{(0)}$ is given by

$$\left(\tilde{\phi}^{(0)}\right)^N(\xi,\eta)=\sum_{i=1}^{N}\sum_{j=1}^{N}\phi_{i,j}\tilde{h}_i(\xi)\tilde{h}_j(\eta)\quad\text{for }\xi,\eta\in\Omega', \tag{23}$$

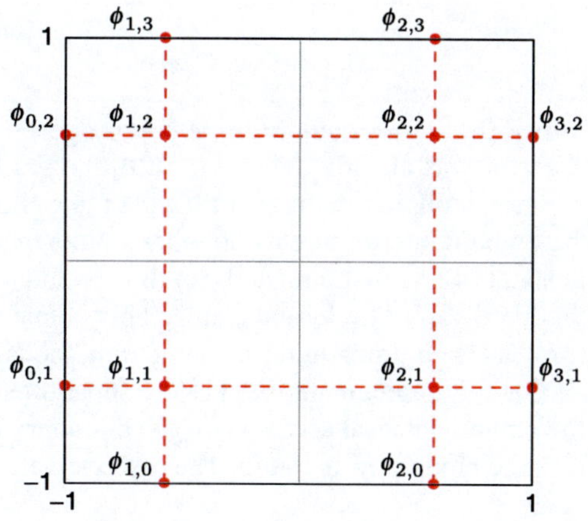

Fig. 2 Extended Gauss grid and the locations of $\phi^{(0)}_{i,j}$ for $N=2$

for the interior part of domain Ω' and

$$\left(\tilde{\phi}^{(0)}\right)^N(\xi,\eta) = \begin{cases} \phi(\xi,-1) = \sum_{i=1}^N \phi_{i,0}\tilde{h}_i(\xi) \\ \phi(\xi,1) \quad = \sum_{i=1}^N \phi_{i,N+1}\tilde{h}_i(\xi) \\ \phi(-1,\eta) = \sum_{j=1}^N \phi_{0,j}\tilde{h}_j(\eta) \\ \phi(1,\eta) \quad = \sum_{j=1}^N \phi_{N+1,j}\tilde{h}_j(\eta) \end{cases} \tag{24}$$

for the boundary of the domain, $\partial\Omega'$.

If we discretize $\tilde{u}^{(1)}$ along the edges of the dual grid, one can show that the gradient equation is also metric-free and exact, see [7]. The discrete gradient operator on the extended Gauss grid is directly related to the divergence matrix defined on the Gauss-Lobatto grid (22), as

$$G = -D^T. \tag{25}$$

Hodge-⋆ Operator The discrete Hodge-⋆ operator maps k-forms on the extended Gauss grid to $(n-k)$-forms on the GLL grid, thus establishing the connection between the two dual meshes. This connection is based on the definition of the Hodge-⋆ operator (6), integrated over the domain Ω, and is written as

$$\begin{aligned} \int_\Omega \left(q^{(1)}, \star d\phi^{(0)}\right)\omega^{(n)} &= \int_\Omega \left(\left(\Phi^*\right)^{-1}\tilde{q}^{(1)}, \star d\left(\Phi^*\right)^{-1}\tilde{\phi}^{(0)}\right)\left(\Phi^*\right)^{-1}\tilde{\omega}^{(n)} \\ &= \int_\Omega \left(\Phi^*\right)^{-1}\tilde{q}^{(1)} \wedge \star\star\left(\Phi^*\right)^{-1} d\tilde{\phi}^{(0)} \\ &= \int_{\Omega'} \tilde{q}^{(1)} \wedge \Phi^* \star\star\left(\Phi^*\right)^{-1} d\tilde{\phi}^{(0)} \\ &= -\int_{\Omega'} \tilde{q}^{(1)} \wedge d\tilde{\phi}^{(0)} \\ &= -\int_{\partial\Omega'} \tilde{q}^{(1)} \wedge \tilde{\phi}^{(0)} + \int_{\Omega'} d\tilde{q}^{(1)} \wedge \tilde{\phi}^{(0)}, \quad \forall \tilde{q}^{(1)}. \end{aligned} \tag{26}$$

This is the *support operator method* proposed by Hyman et al., [8]. The divergence operator was already defined in (20), the gradient operator and Hodge-⋆ operator, $\star d$, are implicitly defined in (26). At the righthand side a division is made between the domain interior and its boundary. Note that the complete righthand side is independent of any mapping and can thus be integrated on the standard domain Ω'. All the metric is in the lefthand side. This connection relation is given in a variational form, as is common in finite element methods.

For the implementation, (18) is substituted in (26) for $\tilde{q}^{(n-1)}$, (23) is used for the domain integral and (24) for the boundary integral. By numerical integration the discrete Hodge, H is found. The discrete values ϕ and $\mathbf{q}$ are found by solving

$$\begin{bmatrix} H & D^T \\ D & 0 \end{bmatrix} \begin{bmatrix} \mathbf{q} \\ \phi \end{bmatrix} = \begin{bmatrix} 0 \\ \mathbf{f} \end{bmatrix}. \tag{27}$$

4 Results

In this section we want to establish that $\mathbf{div}\, q = f$ is satisfied exactly on curvilinear grids (implying that the scheme is conservative) and that the convergence rate toward the exact solution is exponential in the L^2-norm. Consider therefore the exact solution given by

$$\phi(x, y) = \sin(\pi x) \sin(\pi y), \quad (x, y) \in [-1, 1]^2. \tag{28}$$

Curvilinear coordinates are obtained by the mapping $\Phi(\xi, \eta) = (x, y)$ given by

$$\begin{aligned} x(\xi, \eta) &= \xi + c \sin(\pi \xi) \sin(\pi \eta), \\ y(\xi, \eta) &= \eta + c \sin(\pi \xi) \sin(\pi \eta), \qquad \xi, \eta \in [-1, 1]. \end{aligned} \tag{29}$$

The shape parameter c is used to create the curvilinear coordinate system, see Fig. 3. Figure 4 shows the convergence of the error in the L^2-norm of $\phi^{(0)}$ and $q^{(1)}$ as a function of the polynomial degree N. Also depicted in this figure is the interpolation error of the exact solution, denoted by $\|\phi_{ex} - \phi\|_{L^2}$. This figure shows that exponential convergence is retained in curvilinear coordinates even for the self-overlapping case $c = 0.6$. Figure 4 also demonstrates that the divergence equation is satisfied up to machine precision for all grids and all polynomial degrees N in the L^1- and L^∞ norm, which is a direct consequence of the fact that (21) is metric-free and exact.

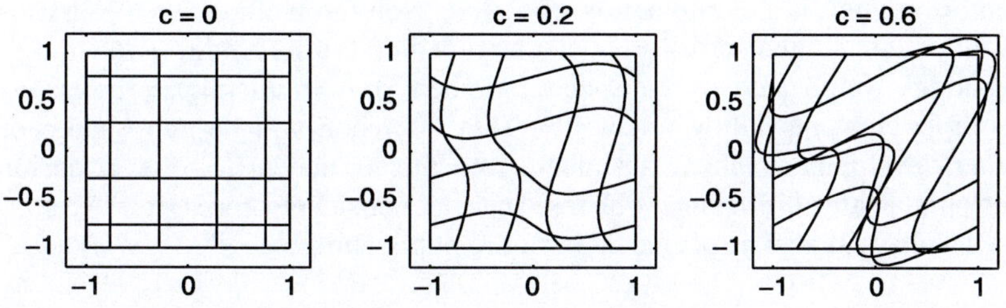

Fig. 3 Three grids with, from *left* to *right*, increasing shape parameter c

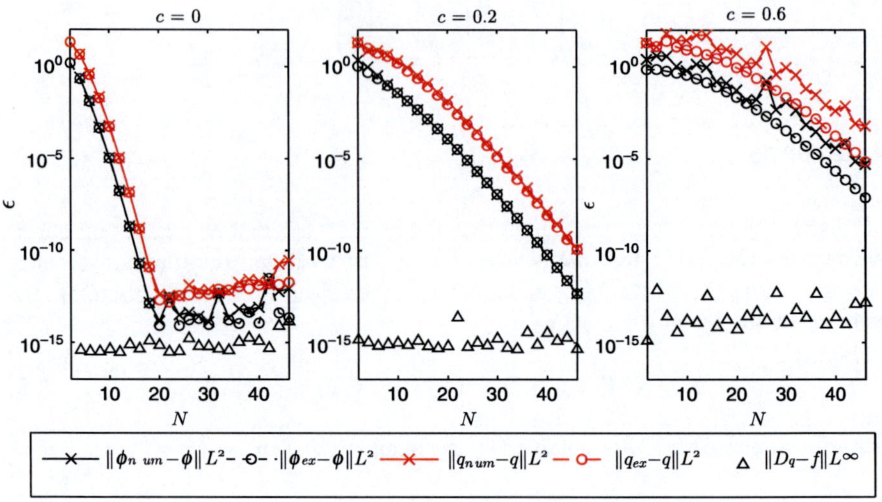

Fig. 4 Convergence of L^2 error for $\phi^{(0)}$ and $q^{(1)}$, and the L^∞ error of the divergence equation, as a function of the polynomial degree for the three grids

5 Concluding Remarks

In this paper a framework is described that leads to a finite-volume like metric-free discretization of the divergence and gradient operators. These equations are satisfied exactly, independent of the shape of the grid and the polynomial degree used in the approximation. This is a consequence of the use of the edge functions. The metric dependent part of the Poisson equation is in the Hodge-$\star$ operator, for which the support operator method is introduced. Therefore the conservative spectral element method presented is a combination of a finite volume method for the first-order derivatives and a higher-order finite element method for the Hodge-$\star$ relation.

Although we considered one spectral element only in this paper, the extension to multiple elements follows naturally. This extension will be published shortly. Another development which fits nicely into this framework is the extension to anisotropic steady diffusion problems, such as considered by Hyman et al., [8]. These results will also be presented in future publications.

References

1. Bochev, P.B., Hyman, J.M.: Principles of Mimetic Discretizations of Differential Equations. In: D. Arnold, P. Bochev, R. Lehoucq, R. Nicolaides and M. Shashkov (eds.), IMA Volume 142, Springer, New York, 2006
2. Bossavit, A.: On the geometry of electromagnetism. Journal of Japanese Society of Applied Electromagnetics and Mechanics 6, 17–28 (no 1), 114–123 (no 2), 233–240 (no 3), 318–326 (no 4), 1998

3. Bossavit, A.: Computational electromagnetism and geometry: Building a finite-dimensional "Maxwell's house". Journal of Japanese Society of Applied Electromagnetics and Mechanics 7, 150–159 (no 1), 294–301 (no 2), 401–408 (no 3), 1999 and 8, 102–109 (no 4), 203–209 (no 5), 372–377 (no 6), 2000
4. Desbrun, M., Hirani, A.N., Leok, M., Marsden, J.E.: Discrete Exterior Calculus. arXiv:math/0508341v2, 18 Aug 2005
5. Desbrun, M., Kanso, E., Tong, Y.: Chapter 7: Discrete Differential Forms for Computational Modeling. ACM SIGGRAPH ASIA 2008 Courses, SIGGRAPH Asia'08, art. no. 15, 2008
6. Flanders, H.: Differential Forms with Applications to the Physical Sciences. Academic Press, New York, 1963
7. Gerritsma, M.I.: Edge Functions for Spectral Elements. Submitted to the proceedings of ICOSAHOM 2009 (this issue)
8. Hyman, J., Shaskov, M., Steinberg, S.: The numerical solution of diffusion problems in strongly heterogeneous non-isotropic materials. Journal of Computational Physics 132, 130–148, 1997
9. Mattiussi, C.: The Finite Volume, Finite Difference, and Finite Elements Methods as Numerical Methods for Physical Field Problems. In: Advances in Imaging and Electron Physics, vol. 113, pp. 1–146, 2000
10. Palha, A., Gerritsma, M.I.: Spectral Element Approximation of the Hodge operator in Curved Domains. Submitted to the proceedings of ICOSAHOM 2009 (this issue)
11. Robidoux, N.: Polynomial Histopolation, Superconvergent Degrees of Freedom and Pseudospectral Discrete Hodge Operators. Unpublished: http://www.cs.laurentian.ca/nrobidoux/prints/super/histogram.
12. Tonti, E.: On the Mathematical Structure of a Large Class of Physical Theories. Accademia Nazionale dei Lincei, Estratto dai Rendiconti Della Classe di Scienze Fisiche, Matematiche e Naturali, Serie VIII, Vol. LII, fasc. 1, Gennaio, 1972

An Efficient Control Variate Method for Parametrized Expectations

Sébastien Boyaval

Abstract Two new variance reduction approaches have been recently introduced in [*A variance reduction method for parametrized stochastic differential equations using the reduced basis paradigm*, Commun. Math. Sci. 8, 2010], to speed up the Monte-Carlo evaluations of the expectations of many parametrized random variables at many values of the parameter, when the random variables are scalar functional of parametrized Itô stochastic processes.

The two approaches make an original use of ideas previously developed in the certified reduced-basis method initiated by [*Reliable real-time solution of parametrized partial differential equations: Reduced-basis output bounds methods*, JFE, 124(1):7–80, 2002]. Indeed, although the reduced-basis method is now developed by many contributors and allows one to fast compute many parametrized solutions to a large variety of partial differential equations, it accelerates only deterministic computational methods. The variance reduction viewpoint allows one to extend the use of some essential reduced-basis ideas to probabilistic computational methods. (We refer to the recent review [*Reduced Basis Techniques for Stochastic Problems*, Archives of Computational Methods in Engineering, December 2009] and the bibliography therein for an overview of the reduced-basis capabilities.)

In this work, we concentrate on one of the two variance reduction approaches. We first briefly recall the motivations for, and the principles of, our accelerated variance reduction technique. Then, the parallel with ideas underpinning the reduced-basis method is emphasized, to explain why and how we expect our variance reduction approach to achieve computational reductions. Last, we present two open problems about better understanding and better using this very recent variance reduction approach (some previously unpublished numerical results give a partial answer to the second question). Our goal is to encourage further studies about this method.

S. Boyaval (✉)
Université Paris-Est, Laboratoire d'hydraulique Saint-Venant, Ecole des ponts ParisTech, 6&8 avenue Blaise Pascal, Cité Descartes, 77455 Marne-la-Vallée Cedex 2, France and INRIA, MICMAC project, Domaine de Voluceau, BP. 105, Rocquencourt, 78153 Le Chesnay Cedex, France
e-mail: sebastien.boyaval@inria.fr

J.S. Hesthaven and E.M. Rønquist (eds.), *Spectral and High Order Methods for Partial Differential Equations*, Lecture Notes in Computational Science and Engineering 76, DOI 10.1007/978-3-642-15337-2_9,

1 A Control Variate Method for Parametrized Expectations

1.1 Setting of the Problem

Let $(\Omega, \mathscr{F}, \mathbb{P})$ be a probability space. We denote $\mathbf{E}_{\mathbb{P}}(Z)$ the expectation of a d-dimensional integrable random variable Z (with respect to $\mathbb{P}$), $d \in \mathbb{N}_{>0}$. There are many applications where one manipulates parametrized collections of square-integrable random variables $Z^\lambda \in L^2_{\mathbb{P}}(\Omega)$, using a parameter $\lambda \in \Lambda$ which belongs to a bounded subset Λ of $\mathbb{R}^P$, $P \in \mathbb{N}_{>0}$, and where one has to compute $\mathbf{E}_{\mathbb{P}}(Z^\lambda)$ for many values of the parameter λ. For instance, this is the case in some optimization algorithms for parameter estimation, where the parameter λ is calibrated in order to fit observed data, or in segregated algorithms simulating a system of equations, where $\mathbf{E}_{\mathbb{P}}(Z^\lambda)$ is used by one group of equations and λ is defined by another equation. We refer to [1] for specific examples in finance and rheology of these so-called *many-query* frameworks, where many expectations $\mathbf{E}_{\mathbb{P}}(Z^\lambda)$ have to be computed for many values of λ (see also the references therein for details about the probabilistic framework that we use here).

We are interested in situations where $\mathbf{E}_{\mathbb{P}}(Z^\lambda)$ is approximated by a Monte-Carlo (MC) method. The simpler MC method reads as such: one directly simulates the law of Z^λ using (pseudo-)random numbers, and by virtue of the law of large numbers, $\mathbf{E}_{\mathbb{P}}(Z^\lambda)$ is approximated as one realization of the random variable $\mathrm{E}_M(Z^\lambda) := \frac{1}{M}\sum_{m=1}^{M} Z^\lambda_m$ using M independant copies Z^λ_m, $m = 1, \ldots, M$, of Z^λ. Unfortunately, the previous simple MC approach converges very slowly with respect to M. Indeed, by virtue of the Chebyshev inequality:

$$\forall \epsilon > 0\,,\ \mathbb{P}\left(\left|\mathrm{E}_M\left(Z^\lambda\right) - \mathbf{E}_{\mathbb{P}}\left(Z^\lambda\right)\right| \geq \epsilon\right) \leq \frac{\mathbf{Var}_{\mathbb{P}}\left(Z^\lambda\right)}{M\epsilon^2}\,, \tag{1}$$

so the control we have on the error committed in the approximation of $\mathbf{E}_{\mathbb{P}}(Z^\lambda)$ by $\mathrm{E}_M(Z^\lambda)$ decreases slowly with M. Similarly, the quantitative estimates of the probabilities of the events "the statistical error is below a given level" which are suggested by the Central Limit Theorem (CLT) when one expects to be close to the asymptotic regime $M \to \infty$ are also scaled by the same ratio $\mathbf{Var}_{\mathbb{P}}(Z^\lambda)/M$:

$$\forall a > 0\,,\ \mathbb{P}\left(\left|\mathrm{E}_M\left(Z^\lambda\right) - \mathbf{E}_{\mathbb{P}}\left(Z^\lambda\right)\right| \leq a\sqrt{\frac{\mathbf{Var}_{\mathbb{P}}\left(Z^\lambda\right)}{M}}\right) \underset{M\to\infty}{\longrightarrow} \int_{-a}^{a} \frac{e^{-x^2/2}}{\sqrt{2\pi}}dx\,. \tag{2}$$

Now, a remedy to slow convergence is to develop more refined MC approaches. In particular, since the variance $\mathbf{Var}_{\mathbb{P}}(Z^\lambda)$ also enters the MC error bounds (1) and (2), one possibility for more elaborate MC approaches is typically to invoke the law of large numbers for another random variable than Z^λ, which has less variance (a i.e., *reduced variance* compared to Z^λ) but still permits the computation of $\mathbf{E}_{\mathbb{P}}(Z^\lambda)$ in the end. Among many possible approaches to variance reduction [3], we next concentrate on the *control variate* method (see Sect. 1.2). We focus in particular on the many-query frameworks which are so computationally demanding that even a refined MC approach using a variance reduction method would still be untractable for a large number of queries (that is, many parameter values $\lambda \in \Lambda$). More precisely, assuming that two random variables Z^{λ_1} and Z^{λ_2} are correlated one another for any $\lambda_1 \neq \lambda_2$ in Λ, our goal is to use the control variate method in combination with ideas from the Reduced-Basis (RB) method [2, 4, 5] for speeding up the numerous reiterated computations involved in a refined MC approach of $\mathbf{E}_{\mathbb{P}}\left(Z^{\lambda}\right)$ for many queries $\lambda \in \Lambda$.

1.2 The Control Variate Method

Let $\lambda \in \Lambda$ be fixed, the control variate method aims at evaluating $\mathbf{E}_{\mathbb{P}}\left(Z^{\lambda}\right)$ as

$$\mathbf{E}_{\mathbb{P}}\left(Z^{\lambda}\right) = \mathbf{E}_{\mathbb{P}}\left(Z^{\lambda} - Y^{\lambda}\right) + \mathbf{E}_{\mathbb{P}}\left(Y^{\lambda}\right)$$

using a so-called *control variate* $Y^{\lambda} \in L^2_{\mathbb{P}}(\Omega)$, which is chosen such that $\mathbf{E}_{\mathbb{P}}\left(Y^{\lambda}\right)$ is easily evaluated and the expectation $\mathbf{E}_{\mathbb{P}}\left(Z^{\lambda} - Y^{\lambda}\right)$ is easily approximated by Monte-Carlo estimations (that is $\mathbf{Var}_{\mathbb{P}}\left(Z^{\lambda}\right) \gg \mathbf{Var}_{\mathbb{P}}\left(Z^{\lambda} - Y^{\lambda}\right)$). Here, we will next choose Y^{λ} such that $\mathbf{E}_{\mathbb{P}}\left(Y^{\lambda}\right) = 0$, that is $\mathbf{E}_{\mathbb{P}}\left(Z^{\lambda}\right) = \mathbf{E}_{\mathbb{P}}\left(Z^{\lambda} - Y^{\lambda}\right)$. Then the choice $Y^{\lambda}_{o} \equiv Z^{\lambda} - \mathbf{E}_{\mathbb{P}}\left(Z^{\lambda}\right)$ is clearly optimal in the sense that $\mathbf{Var}_{\mathbb{P}}\left(Z^{\lambda} - Y^{\lambda}_{o}\right) = 0$, but it is of course idealistic since one needs $\mathbf{E}_{\mathbb{P}}\left(Z^{\lambda}\right)$ for the computation of Y^{λ}_{o}.

In practice, a first possible choice is to take $Y^{\lambda}_{M_1} = Z^{\lambda} - \mathrm{E}_{M_1}\left(Z^{\lambda}\right)$ as a control variate approximating the optimal choice Y^{λ}_{o}, where the MC estimator $\mathrm{E}_{M_1}\left(Z^{\lambda}\right)$ uses a large fixed number $M_1 \in \mathbb{N}$ of independent copies of Z^{λ}. Then, the *residual variance* obtained after using the control variate is indeed reduced and reads:

$$\mathbf{Var}_{\mathbb{P}}\left(Z^{\lambda} - Y^{\lambda}_{M_1}\right) = \mathbf{Var}_{\mathbb{P}}\left(\mathrm{E}_{M_1}\left(Z^{\lambda}_{m}\right)\right) = \frac{1}{M_1}\mathbf{Var}_{\mathbb{P}}\left(Z^{\lambda}\right). \tag{3}$$

But of course there is no gain if, for each $\lambda \in \Lambda$, one computes $Y^{\lambda}_{M_1}$ and then $\mathrm{E}_{M}\left(Z^{\lambda} - Y^{\lambda}_{M_1}\right)$ to evaluate $\mathbf{E}_{\mathbb{P}}\left(Z^{\lambda} - Y^{\lambda}_{M_1}\right) = \mathbf{E}_{\mathbb{P}}\left(Z^{\lambda}\right)$, since this in fact requires $M \times M_1$ independent copies of Z^{λ}: it has the same variance and the same computational cost as the simpler MC estimator $\mathrm{E}_{M \times M_1}\left(Z^{\lambda}\right)$.

A second practical approach is to take, as control variate, a linear combination $Y^{\lambda}_{M_1,N} = \sum_{n=1}^{N} \alpha_n(\lambda) Y^{\lambda_n(\lambda)}_{M_1}$ using N elements $Y^{\lambda_n(\lambda)}_{M_1}$ to be chosen in $\mathcal{Y}_{M_1} = \mathbf{Span}\left(Y^{\lambda}_{M_1},\ \lambda \in \Lambda\right)$, with coefficients $\{\alpha_n(\lambda) \in \mathbb{R}\ ,\ 1 \leq n \leq N\}$ to be determined with a view to minimizing the variance $\mathbf{Var}_{\mathbb{P}}\left(Z^{\lambda} - Y^{\lambda}_{M_1,N}\right)$. The latter choice reduces the variance at least as much as the former since $Y^{\lambda}_{M_1} \in \mathcal{Y}_{M_1}$

is a possible value for the linear combination $Y^{\lambda}_{M_1,N}$. Moreover, although the computations for one $Y^{\lambda}_{M_1,N}$ are more expensive than those for $Y^{\lambda}_{M_1}$, the choice $Y^{\lambda}_{M_1,N}$ in fact suggests a third possible choice which yields a similar reduction of variance, plus interesting computational reductions in the many-query limit of many parameter values $\lambda \in \Lambda$.

Our choice is a third possible control variate $Y^{\lambda}_{N,M_1} = \sum_{n=1}^{N} \alpha_n(\lambda) Y^{\lambda_n}_{M_1}$ with *fixed* λ_n, $n = 1, \cdots, N$, in the linear combination Y^{λ}_{N,M_1}. With such λ_n chosen once, for all $\lambda \in \Lambda$ (in contrast with $Y^{\lambda}_{M_1,N}$), the choice Y^{λ}_{N,M_1} can still yield a good reduction of variance if the variations of $Y^{\lambda}_{M_1}$ in $L^2_{\mathbb{P}}(\Omega)$ with respect to $\lambda \in \Lambda$ are smooth. In the sequel, we recall from [1] how to use the control variate Y^{λ}_{N,M_1} in practice, based on a parallel with the RB method [2, 4, 5]. Taking profit by the many-query setting, a practical methodology can indeed be developed to efficiently compute *many* control variates Y^{λ}_{N,M_1} at *many* parameter values $\lambda \in \Lambda$ with a fixed precomputed set $\{Y^{\lambda_n}_{M_1}, n = 1, \dots, N\}$, and with coefficients $\{\alpha_n(\lambda) \in \mathbb{R}, n = 1, \dots, N\}$ chosen at each $\lambda \in \Lambda$ to minimize the variance

$$\mathbf{Var}_{\mathbb{P}}\left(Z^{\lambda} - Y^{\lambda}_{N,M_1}\right) = \mathbf{E}_{\mathbb{P}}\left(|Y^{\lambda}_{o} - Y^{\lambda}_{N,M_1}|^2\right). \tag{4}$$

One point is how to identify a fixed N-dimensional subset $\mathcal{Y}_{N,M_1} \subset \mathcal{Y}_{M_1}$, so that the *worst*[1] residual variance for $\lambda \in \Lambda$ (after reduction) is controlled ($\equiv$ minimized).

1.3 A Practical Approach of the Control Variate Method Deduced from Parallels with the Standard Reduced-Basis Method

The standard RB method has been developed for applications where one has to compute many μ-parametrized functions $u(\mu) \in X$ in a Hilbert space X, for many values of the input parameter $\mu \in \Lambda$. In the many-query frameworks of application (real-time simulation, parameter optimization, multiscale computation,...), $u(\mu)$ is typically the solution to a (well-posed, e.g. coercive and continuous) μ-parametrized elliptic Boundary Value Problem (BVP) with variational formulation:

$$a(u(\mu), v; \mu) = l(v), \ \forall v \in X.$$

Then, usual linear discrete subspaces $X_{\mathcal{N}} \subset X$ (e.g. finite-element spaces) which yield accurate Galerkin approximations $u_{\mathcal{N}}(\mu) \in X_{\mathcal{N}}$ at any fixed $\mu \in \Lambda$ typically have very large dimension $\mathcal{N}$. So it is expensive to compute $u_{\mathcal{N}}(\mu)$ for many μ in

[1] The *worst* is to be understood as the largest residual variance in Λ as opposed to e.g. a L^2 *mean* of the residual variances for instance provided Λ is endowed with a topology; the latter would have suggested a parallel with Proper Orthogonal Decomposition (POD) methods rather than RB. We make this choice on purpose, to follow RB ideas.

a many-query framework. The RB method aims at rigorously reducing the total cost of computations, based on the fact that if $u_{\mathcal{N}}(\mu)$ is a smooth function of μ, then the set $\{u_{\mathcal{N}}(\mu),\ \mu \in \Lambda\}$ is only a thin portion of $X_{\mathcal{N}}$ close to a small-dimensional linear subspace of $X_{\mathcal{N}}$ (say with dimension $N \ll \mathcal{N}$).

Assume $a(\cdot,\cdot;\mu)$ is bilinear and symmetric for the sake of simplicity, on noting

$$u_{\mathcal{N}}(\mu) = \operatorname*{arginf}_{v \in X_{\mathcal{N}}} \sqrt{a(u(\mu)-v, u(\mu)-v;\mu)}\,,\ \forall \mu \in \Lambda\,, \tag{5}$$

the RB method suggests to approximate $u_{\mathcal{N}}(\mu)$ by $u_{\mathcal{N},N}(\mu) \in X_{\mathcal{N},N}$ solution to a similar least-squares problem in a N-dimensional vector subspace $X_{\mathcal{N},N} \subset X_{\mathcal{N}}$

$$u_{\mathcal{N},N}(\mu) = \operatorname*{arginf}_{v \in X_{\mathcal{N},N}} \sqrt{a(u_{\mathcal{N}}(\mu)-v, u_{\mathcal{N}}(\mu)-v;\mu)}\,,\ \forall \mu \in \Lambda\,. \tag{6}$$

In practice, it is still computationally expensive to find $X_{\mathcal{N},N}$ such that the worst approximation error $\|u_{\mathcal{N}}(\mu) - u_{\mathcal{N},N}(\mu)\|_X$ for $\mu \in \Lambda$ is small. The RB method suggests a pratical approach to that difficulty, which has proved feasible and computatially profitable in most of the many-query frameworks where it has been applied (provided sufficiently many queries in parameter values $\mu \in \Lambda$ compensate for the expensive construction of $X_{\mathcal{N},N}$). It combines inexpensive a posteriori error estimators for $\|u_{\mathcal{N}}(\mu) - u_{\mathcal{N},N}(\mu)\|_X$ with an iterative construction of $X_{\mathcal{N},N}$ (that is, such that $X_{\mathcal{N},N} \subset X_{\mathcal{N},N+1}$) through a *greedy algorithm* (see e.g. [2,4]).

We now present at the same time the main ideas of the RB computational strategy to address the standard μ-parametrized elliptic BVP above and the λ-parametrized MC problem with control-variates of Sect. 1.2. The parallel between an efficient practical approach for reiterated least-squares problems (4) and the standard RB approach for reiterated Galerkin approximations (5) should thus be clear. We assume M_1 is such that the reduced variance $\mathbf{Var}_{\mathbb{P}}\left(Z^\lambda - Y^\lambda_{M_1}\right)$ is small enough for all $\lambda \in \Lambda$.

1. The RB approximation spaces are spanned by "snapshot" solutions, hence:

$$X_{\mathcal{N},N} = \mathbf{Span}\left(u_{\mathcal{N}}(\mu_n^N), n = 1,\dots,N\right) \text{ and } \mathcal{Y}_{M_1,N} = \mathbf{Span}\left(Y_{M_1}^{\lambda_n^N}, n = 1,\dots,N\right),$$

   invoking N accurate approximations: $u_{\mathcal{N}}(\mu_n^N)$ for $u(\mu_n^N)$ and $Y_{M_1}^{\lambda_n^N}$ for $Y_o^{\lambda_n^N}$, at N well-chosen parameter values, resp. $(\mu_n^N)_{n=1,\dots,N} \in \Lambda^{\mathbb{N}}$ and $(\lambda_n^N)_{n=1,\dots,N} \in \Lambda^{\mathbb{N}}$.
2. The linear approximation spaces $X_{\mathcal{N},N}$ and $\mathcal{Y}_{M_1,N}$ are constructed in an *offline* stage as approximations to the "optimal" spaces that minimize (resp.)[2]

[2] The fact that the energy norm $\sqrt{a(\cdot,\cdot;\mu)}$ depends on μ in (5,7) and not in (4,8) is rather an *additional* difficulty for the standard RB method, and does not preclude comparisons.

$$\min_{\{\mu_n^N, n=1,\dots,N\}\subset\Lambda} \sup_{\mu\in\Lambda} \min_{v\in X_{\mathcal{N},N}} \sqrt{a(u_{\mathcal{N}}(\mu)-v, u_{\mathcal{N}}(\mu)-v;\mu)}\,, \tag{7}$$

$$\min_{\{\lambda_n^N, n=1,\dots,N\}\subset\Lambda} \sup_{\lambda\in\Lambda} \min_{\{\alpha_n(\lambda), n=1,\dots,N\}\subset\Lambda} \mathbf{Var}_{\mathbb{P}}\left(Y_o^\lambda - \sum_{n=1}^{N}\alpha_n(\lambda)Y_{M_1}^{\lambda_n^N}\right), \tag{8}$$

using a *greedy* algorithm. Hence we choose iteratively, until sufficiently large N:

$$\mu_n^N \equiv \mu_n \qquad \lambda_n^N \equiv \lambda_n \qquad n = 0,\dots,N-1\,,$$

by (a) solving successive one-dimensional optimization problems rather than by solving straightforwardly the N-dimensional minimization problems (7, 8) and (b) replacing the suprema over the whole parameter range Λ in (7, 8) by suprema in a discrete training sample $\Lambda_{\text{trial}} \subset \Lambda$ (see details in e.g. [2, 4]).

3. RB approximations $u_{\mathcal{N},N}(\mu) \in X_{\mathcal{N},N} \subset X_{\mathcal{N}}$ and $Y^\lambda_{N,M_1,M_2} \in \mathcal{Y}_{M_1,N} \subset \mathcal{Y}_{M_1}$ are next computed *online* for many queries in Λ. For a given μ, the computations are fast and the RB error can be *certified*: $u_{\mathcal{N},N}(\mu)$ is solution to a *small* N-dimenionsal linear system and the RB error is evaluated *fast* using the same a posteriori estimator as offline. For a given λ, coefficients $\alpha^{N,M_1}_{n,M_2}(\lambda) \in \mathbb{R}$ in $Y^\lambda_{N,M_1,M_2} = \sum_{n=1}^N \alpha^{N,M_1}_{n,M_2}(\lambda) Y^{\lambda_n}_{M_1}$ are computed *fast*, as minimizers of a *cheap* MC estimation of the $L^2_{\mathbb{P}}(\Omega)$ distance between Y_o^λ and $\mathcal{Y}_{M_1,N}$ function of some $\alpha_n(\lambda)$:

$$\min_{Y^\lambda\in\mathcal{Y}_{M_1,N}} \mathbf{Var}_{\mathbb{P}}\left(Z^\lambda - Y^\lambda\right) = \min_{Y^\lambda\in\mathcal{Y}_{M_1,N}} \mathbf{E}_{\mathbb{P}}\left(|Y_o^\lambda - Y^\lambda|^2\right). \tag{9}$$

The MC estimation of (9) should use a *small* number M_2 of copies for Z^λ and $Y^{\lambda_n}_{M_1}$, $n = 1,\dots,N$; and after minimization in $\alpha_n(\lambda)$, the resulting minimum $V^\lambda \equiv \mathrm{Var}_{M_2}\left(Z^\lambda - \sum_{n=1}^N \alpha^{N,M_1}_{n,M_2}(\lambda)Y^{\lambda_n}_{M_1}\right)$ can be used to evaluate the statistical error in the MC estimation $E^\lambda \equiv \mathrm{E}_{M_2}\left(Z^\lambda - \sum_{n=1}^N \alpha^{N,M_1}_{n,M_2}(\lambda)Y^{\lambda_n}_{M_1}\right)$ of the output $\mathbf{E}_{\mathbb{P}}\left(Z^\lambda\right)$. (Note that in the two latter MC estimators the coefficients $\alpha^{N,M_1}_{n,M_2}(\lambda)$ are fixed.)

To sum up our RB algorithm will compute fast many control variates in two stages:

(a) (*Offline pre-computations*) a greedy algorithm selects iteratively N parameter values λ_n, $n = 1,\dots,N$, in a training sample Λ_{trial}: for $n = 1,\dots,N-1$, it chooses some $\lambda_{n+1} \in \operatorname{argmax}\left\{\mathbf{Var}_{\mathbb{P}}\left(Z^\lambda - Y^\lambda_{n,M_1}\right), \lambda \in \Lambda_{\text{trial}}\right\}$ and computes $Y^{\lambda_{n+1}}_{M_1}$ (in fact, only an accurate MC estimation $\mathrm{E}_{M_1}\left(Z^{\lambda_{n+1}}\right) \simeq \mathbf{E}_{\mathbb{P}}\left(Z^{\lambda_{n+1}}\right)$ with $M_1 \gg 1$);

(b) (*Online queries*) for each query $\lambda \in \Lambda$, a control variate Y^{λ}_{N,M_1,M_2} is computed as an approximate minimizer of (9). That is, the variance in (9) is computed by a MC estimator using M_2 copies and as a function of unknown coefficients $\alpha_n(\lambda)$, then coefficients $\alpha^{N,M_1}_{n,M_2}(\lambda)$ are compute to minimize the MC estimator. Finally, we obtain a *certified* estimation of the output as $\mathbf{E}_{\mathbb{P}}\left(Z^{\lambda}\right) \simeq E^{\lambda} \pm O(V^{\lambda})$.

Like in the applications of the standard RB method, our practical approach is interesting when the number of online queries is large enough to compensate for the offline computations. Unlike the standard RB method, the evaluation of the error due to reduction is not fully rigorous, in particular because the evaluation of the output E^{λ} and its error bound V^{λ} invokes MC estimations.

Remark 1. In the offline stage, the greedy algorithm also needs to numerically evaluate variances: $\mathbf{Var}_{\mathbb{P}}\left(Z^{\lambda} - Y^{\lambda}_{n,M_1}\right)$, $\forall \lambda \in \Lambda_{\text{trial}}$ and $n = 0, \dots, N-1$. This also necessitates approximations, which in turn influence the selection by the greedy algorithm of the λ_n in Y^{λ}_{N,M_1,M_2}. In practice, to evaluate those variances, we use the same procedure as online for V^{λ}: at each step $n = 0, \dots, N-1$, for all $\lambda \in \Lambda_{\text{trial}}$, $\text{Var}_{M_2}\left(Z^{\lambda} - \sum_{m=1}^{n} \alpha^{n,M_1}_{m,M_2}(\lambda) Y^{\lambda_p}_{M_1}\right)$ is approximated with the same number $M_2 \ll M_1$ of copies. So the approximation error is fast, and consistent throughout the procedure with that in the V^{λ} used online as a bound of the error $|\mathbf{E}_{\mathbb{P}}\left(Z^{\lambda}\right) - E^{\lambda}|$.

2 Open Questions

2.1 Rigorous Certification of the Variance Reduction?

Proposition 1. *Let Y^{λ}_{o} depend smoothly on $\lambda \in \Lambda$ (in the sense of the assumptions of Proposition 1 in [1]). Then one can choose $\{\lambda_n\ ,\ n \in \mathbb{N}_{>0}\} \subset \Lambda$ such that $Y^{\lambda}_{o,N} = \sum_{n=1}^{N} \beta_n(\lambda) Y^{\lambda_n}_{o} \underset{N\to\infty}{\overset{L^2_{\mathbb{P}}(\Omega)}{\longrightarrow}} Y^{\lambda}_{o}$ for all $\lambda \in \Lambda$, with coefficients $\beta_n(\lambda)$ minimizing $\mathbf{Var}_{\mathbb{P}}\left(Z^{\lambda} - Y^{\lambda}_{o,N}\right)$. If in addition, $\forall N, M_1, M_2 \in \mathbb{N}_{>0}$, the $N \times N$ matrices with entries $\text{E}_{M_2}\left(Y^{\lambda_i}_{M_1} Y^{\lambda_j}_{M_1}\right)$, $1 \le i, j \le N$, are almost surely (a.s.) non-singular (where $M_1 + M_2$ independent copies are used for the MC estimations), then it holds for all $\lambda \in \Lambda$*

$$\lim_{N\to\infty} \lim_{M_2\to\infty} \lim_{M_1\to\infty} \text{E}_{M_2}\left(Z^{\lambda} - \sum_{n=1}^{N} \alpha^{N,M_1}_{n,M_2}(\lambda) Y^{\lambda_n}_{M_1}\right) \overset{a.s.}{=} \mathbf{E}_{\mathbb{P}}\left(Z^{\lambda}\right) \quad \textit{and}$$

$$\lim_{N\to\infty} \lim_{M_2\to\infty} \lim_{M_1\to\infty} \text{Var}_{M_2}\left(Z^{\lambda} - \sum_{n=1}^{N} \alpha^{N,M_1}_{n,M_2}(\lambda) Y^{\lambda_n}_{M_1}\right) \overset{a.s.}{=} 0\,.$$

Proof. For fixed N, M_2, first apply the strong law of large numbers when $M_1 \to \infty$. Then, by virtue of the strong law of large numbers when $M_2 \to \infty$, and on noting $\lim_{M_2\to\infty} \lim_{M_1\to\infty} \alpha_{n,M_2}^{N,M_1}(\lambda) \overset{a.s.}{=} \beta_n(\lambda)$ (because the MC estimations of the covariance matrices are a.s. non-singular so we can pass to the limit in Cramers' rule), we obtain

$$\lim_{M_2\to\infty} \lim_{M_1\to\infty} \mathrm{Var}_{M_2}\left(Z^\lambda - \sum\nolimits_{n=1}^{N} \alpha_{n,M_2}^{N,M_1}(\lambda) Y_{M_1}^{\lambda_n}\right) \overset{a.s.}{=} \mathbf{Var}_{\mathbb{P}}\left(Z^\lambda - Y_{o,N}^\lambda\right) \text{ and}$$

$$\lim_{M_2\to\infty} \lim_{M_1\to\infty} \mathrm{E}_{M_2}\left(Z^\lambda - \sum\nolimits_{n=1}^{N} \alpha_{n,M_2}^{N,M_1}(\lambda) Y_{M_1}^{\lambda_n}\right) \overset{a.s.}{=} \mathbf{E}_{\mathbb{P}}\left(Z^\lambda\right).$$

Last, we get the result for $N \to \infty$ by Proposition 1 in [1]. □

The Proposition 1 above is only M_1-asymptotic, so there are still interesting questions: conditionally to the M_1 offline copies, do the random variables $\mathrm{Var}_{M_2}\left(Z^\lambda - \sum_{n=1}^{N} \alpha_{n,M_2}^{N,M_1}(\lambda) Y_{M_1}^{\lambda_n}\right)$ and $\mathrm{E}_{M_2}\left(Z^\lambda - \sum_{n=1}^{N} \alpha_{n,M_2}^{N,M_1}(\lambda) Y_{M_1}^{\lambda_n}\right)$ converge as $M_2 \to \infty$? Does a CLT hold and allow one to derive rigorous error bounds?

2.2 *Computational Efficiency: Optimize MC Estimations?*

Let δt denote the marginal time needed for computing one realization of Z^λ, for a given $\lambda \in \Lambda$. The total cost of computations in the procedure of Sect. 1.2 is then

- *(offline)* $O\left(|\Lambda_{\text{trial}}|(\delta t + M_2^2) + N M_1 \delta t\right)$ plus
- *(online)* $O\left(J(\delta t + M_2^2)\right)$ where J is the number of λ values queried online, and the same small number M_2 of copies as offline is used for the MC estimations.

This cost has to be compared to $O(J M_1 M_2 \delta t)$ for a direct one-stage MC approach with a similar variance (hence, with a similar statistical error) in the end. So the gain depends on how large J and M_1 are, M_2 being as small as possible. Now, as sets of M_2 independent copies are first repeatedly used offline (at each step $n = 0, \dots, N-1$ of the greedy algorithm) for MC estimations of variances $\mathrm{Var}_{M_2}\left(Z^\lambda - \sum_{m=1}^{n} \alpha_{m,M_2}^{n,M_1}(\lambda) Y_{M_1}^{\lambda_p}\right)$ when $\lambda \in \Lambda_{\text{trial}}$, before M_2 copies are intensively used online ("step" $n = N$) for MC estimations of the similar variances at any $\lambda \in \Lambda$, one might want to optimize the choice of those few M_2 copies step after step during the greedy in Λ_{trial}, so that the selection of M_2 "good" copies (using e.g. concepts from Quasi-Monte-Carlo (QMC) methods [3]) yields online MC estimators Var_{M_2} and E_{M_2} that are as accurate as possible $\forall \lambda \in \Lambda$ (M_2 being fixed *small*).

We show in Fig. 1 the results of a first test about how the variance reduction achieved by our RB approach depends on the M_2 copies of the collection $\{Z^\lambda, \lambda \in \Lambda_{\text{trial}}\}$ that are used online to compute both control variates and MC estimations V^λ for residual variances. The results are for the same FENE-dumbbell

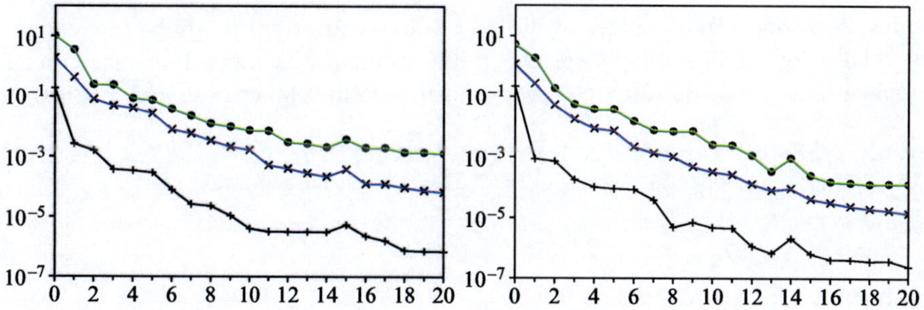

Fig. 1 Two figures (*left/right*): different choices for the M_2 copies used online ($\equiv$ different from one another, and also different from the offline M_2 copies, in contrast to Fig. 4.7 in [1]). In each figure, three curves: minimum $+$, mean $\times$ and maximum $\circ$ of a realization of residual variances $\text{Var}_{M_2}(Z^\lambda - Y^\lambda_{N,M_1,M_2})$ (y-axis) for a sample of parameter values $\lambda \in [-1,1]^4$. Number of precomputed expectations used for the control variates Y^λ_{N,M_1,M_2}: $N = 0, \dots, 20$ (x-axis)

test-problem as in [1]. The output is $Z^\lambda = X_1^\lambda(T)^2$ where $X_1^\lambda(T)$ denotes the first component at final time T of a 2-d vector stochastic process $X^\lambda(t)$ solution on $t \in [0,T]$ to $dX^\lambda(t) = \left(\lambda\, X^\lambda(t) - F(X^\lambda(t))\right) dt + dB(t)$, where $(B(t))$ is a 2-d Brownian motion, $F(X) = X/(1-|X|^2/b)$ with $b = 9$ is the Finitely-Extensible Nonlinear Elastic (FENE) force and λ is a 2×2 traceless matrix with entries in $[-1,1]^4$. Computing viscoelastic flows with the FENE-dumbbell model indeed defines a challenging many-query parametric framework that has useful applications to the simulation of dilute polymer flows in rheology (X models polymer molecules).

For this test-problem, the Fig. 4.7 in [1] show results obtained when the M_2 online copies of the stochastic processes $(X^\lambda(t), \lambda \in \Lambda_{\text{trial}})$, which are used to both compute control variates Y^λ_{N,M_1,M_2} and then evaluate residual variances V^λ at each λ, were exactly the *same* as the M_2 offline copies used during the greedy algorithm. (In fact, M_2 copies of the *same* Brownian motion are used for all λ.) Here in Fig. 1, we test different choices for the M_2 copies of the stochastic processes $(X^\lambda(t), \lambda \in \Lambda_{\text{trial}})$ that are used online. We use for the online stage either M_2 online copies that are different and independent from the M_2 offline one (left), or only the first half of the same M_2 copies as offline, hence only $M_2/2$ copies in fact (right). Notice that not only the MC estimations of the residual variances change online, but also the coefficients $\alpha^{N,M_1}_{n,M_2}(\lambda)$ in the linear combinations Y^λ_{N,M_1,M_2} used as control variates.

Little difference is observed here whatever the online choice for M_2: the results in fact seem hardly sensitive to the choice of the M_2 copies! This is good news for the robustness of our approach. But the question whether this is quite a general fact or only due to the FENE model is still open. As a con to going further in optimizing the M_2 online copies, the analysis of the residual statistical error could complicate, in particular because no simple CLT holds for MC estimators where the M_2 optimally selected realizations are not independent (this is the case in QMC). Yet, in some cases, optimized choices of the M_2 online realizations might be interesting, especially for applications where cheap but accurate MC estimations are crucial (applications where the variance *is* very sensitive to the choice of the M_2 copies!).

Acknowledgements Thanks to A.T. Patera (MIT) and T. Lelièvre (Université Paris-Est, CERMICS) for past collaborations, ongoing discussions and interesting suggestions.

References

1. S. Boyaval and T. Lelièvre. A variance reduction method for parametrized stochastic differential equations using the reduced basis paradigm. *Commun. Math. Sci.*, 8(3):735–762, 2010
2. S. Boyaval, C. Le Bris, T. Lelièvre, Y. Maday, N.C. Nguyen, and A.T. Patera, *Reduced Basis Techniques for Stochastic Problems*, Archives of Computational Methods in Engineering, submitted, December 2009. Preprint available on http://augustine.mit.edu
3. B. Jourdain. Adaptive variance reduction techniques in finance. Radon Series Comp. Appl. Math 8. De Gruyter, 2009
4. A.T. Patera and G. Rozza, Reduced Basis Approximation and A Posteriori Error Estimation for Parametrized Partial Differential Equations, Version 1.0, Copyright MIT 2006–2007, (tentative rubric) MIT Pappalardo Graduate Monographs in Mechanical Engineering (to appear)
5. C. Prud'homme, D. Rovas, K. Veroy, Y. Maday, A. T. Patera, and G. Turinici. Reliable real-time solution of parametrized partial differential equations: Reduced-basis output bounds methods. *Journal of Fluids Engineering*, 124(1):70–80, 2002

A Proof, Based on the Euler Sum Acceleration, of the Recovery of an Exponential (Geometric) Rate of Convergence for the Fourier Series of a Function with Gibbs Phenomenon

John P. Boyd

Abstract When a function $f(x)$ is singular at a point x_s on the real axis, its Fourier series, when truncated at the Nth term, gives a pointwise error of only $O(1/N)$ over the entire real axis. Boyd and Moore [Summability methods for Hermite functions. Dyn. Atmos. Oceans **10**, 51–62 (1986)] and Boyd [A lag-averaged generalization of Euler's method for accelerating series. Appl. Math. Comput. **72**, 146–166 (1995)] proved that it is possible to recover an exponential rate of convegence at all points away from the singularity in the sense that $|f(x) - f_N^{\sigma}(x)| \sim O(\exp(-\mu(x)N))$ where $f_N^{\sigma}(x)$ is the result of applying a summability method to the partial sum $f_N(x)$ and $\mu(x)$ is a proportionality constant that is a function of $d(x) \equiv |x - x_s|$, the distance from x to the singularity. Here we improve these earlier results and give an elementary proof of great generality using conformal mapping in a dummy variable z, which is the Euler acceleration. We show $\exp(\mu(x)) \approx \min(2, \cos(d(x)/2))$ for the Euler filter when the Fourier period is 2π and $f(x)$ has no off-axis singularities very close to the real axis. We correct recent claims that only a root-exponential rate of convergence can be recovered.

1 Introduction

Let $f(x)$ denote the sum of the Fourier series

$$f(x) = \sum_{n=-\infty}^{\infty} c_n \exp(inx) \tag{1}$$

with a convergence-slowing singularity on the real axis.

John P. Boyd
Department of Atmospheric, Oceanic and Space Science, University of Michigan, Ann Arbor, MI, USA
e-mail: jpboyd@umich.edu

J.S. Hesthaven and E.M. Rønquist (eds.), *Spectral and High Order Methods for Partial Differential Equations*, Lecture Notes in Computational Science and Engineering 76, DOI 10.1007/978-3-642-15337-2_10,

For simplicity, we shall focus attention on functions for which there is just a single such singularity which without loss of generality we shall assume is at the origin. Our goal is to show that by elementary means that it is possible to recover a geometric rate of convergence, that is, an error proportional to $\exp(-\mu(x)N)$, everywhere except at the singularity itself by using the Euler acceleration.

The Euler acceleration is a *local* filter in the sense that it accelerates the series at a given x without using information from different x. The Gibbs reprojection filter developed by Gottleib and Shu accelerates the Fourier series by replacing it by a polynomial approximation at all x simultaneously, thus treating different x *collectively* [8]. Such nonlocal filters are very powerful, but also more complicated, and share the defect of spatial non-uniformity, failing at the singularities themselves. Because of page limits, comparisons of local and non-local filters must be left for another time.

Although recovery-of-geometric-convergence by the Euler acceleration was analytically demonstrated in 1984 by Boyd and Moore [6] and again by Boyd in [1], articles in 2002 and 2005 displayed filters which recovered only "root-exponential" convergence, which is coefficients falling proportional to $\exp(q'\sqrt{n})$ for some positive constant q' [13, 14], These "adaptive filters" are described in the first half of Tadmor's review [12]. Our goal here is to remove some of this confusion, and strengthen the conclusions of our earlier papers by replacing the *qualitative* assertion of geometric convergence by a new theorem that gives the *quantitative* rate of convergence.

The central result is the following.

Theorem 1. *Let $f(x)$ be a 2π-periodic analytic function which is singular only at $x = x_0 = 0$ and also $x_j = \sigma_j + i\tau_j$ where the number of off-axis singularities x_j may be zero, finite or infinite. Let $f_N^\sigma(x;N)$ denote its Euler-accelerated partial sum:*

$$f_N^\sigma \equiv \sum_{n=-N}^{N} \sigma_E(n/(N+1);N)\, c_n \exp(inx) \tag{2}$$

where $\sigma_E(0) = 1$, $\sigma_E\left(\frac{|j|}{N+1}\right) = \sum_{k=j}^{N} \mu_{Nk}$, $|j| = 1,2,\ldots M$, $\mu_{Nk} = \frac{N!}{2^N}\frac{1}{k!(N-k)!}$.

Then

$$|f(x) - f_N^\sigma(x)| \le p(x)\exp(-\mu(x)N) \tag{3}$$

$$\exp(\mu(x)) = \rho(x) = \begin{cases} \min(2, |\zeta_1(x)|, |\zeta_2(x)|, \ldots), |x| \ge (2/3)\pi \\ \min(|\zeta_0(x)|, |\zeta_1(x)|, \ldots), |x| < (2/3)\,\pi \end{cases} \tag{4}$$

where, denoting $r_j \equiv \exp(|\Im(x_j)|)$,

$$|\zeta_j| = \frac{2r_j}{\sqrt{1 + r_j^2 + 2r_j\cos(x - \Re(x_j))}} \tag{5}$$

which is greater than one – geometric convergence – for all $|\Im(x_j)| > 0$, *and*

$$|\zeta_0(x)| = \frac{1}{\cos(x/2)} \tag{6}$$

which is also always greater than one except for $x = 0$.

Furthermore, if the singularity on the real axis is no stronger than a Dirac delta-function, the N*-independent factor in (3)*$|p(x)| \leq C/|x|$.

2 Acceleration by Conformal Mapping

2.1 Abel Extension and Conformal Mapping

The Euler acceleration is a particular case of a general acceleration strategy:

1. Inflate the series to a function of a dummy variable z by multiplying the nth term of the series by z^n
2. Apply a conformal mapping by replacing z by a new ζ where

$$z = \mathscr{Z}(\zeta) \tag{7}$$

   with the mapping chosen to be an analytic function such that, (a) $\mathscr{Z}(1) = 1$ and (b) $\mathscr{Z}(z) \propto \zeta$ for small ζ.
3. Expand the Nth partial sum of the inflated series as a power series in ζ; because of the requirement that $z \propto \zeta + O(\zeta^2)$ for small ζ, the first N terms in the ζ series are completely determined by only the first N terms in the original series.

The accelerated approximation is then just the partial sum of the ζ series evaluated at $\zeta = z = 1$. The inflated series is dubbed the "Abel extension" of the sum S in [6].

Reviews of acceleration-by-conformal-mapping include [9–11].

To understand the magic of the transformation from a power series in z to the power series in ζ, we need to recall the following.

Theorem 2 (Convergence of a power series). *Suppose that the disk* $|\zeta| < \rho$ *is the largest disk centered on the origin in the complex* ζ*-plane which is free of singularities of an analytic function* $S(\zeta)$. *Then the power series of* $S(\zeta)$ *converges everywhere within the disk and* ρ *is the "radius of convergence" of the series. Let* $S_N(\zeta)$ *denote the partial sum of the power series up to and including the Nth term. Then*

$$|S(\zeta) - S_N(\zeta)| \leq \textit{constant} \exp(-N \log(\rho - \epsilon)) \tag{8}$$

for all N *with a proportionality constant independent of* N *where* $\epsilon > 0$ *is a constant that may be arbitrarily small.*

2.2 Möbius Transformation and Euler Acceleration

Euler acceleration is the choice of the simple rational transformation

$$z = \frac{\zeta}{2-\zeta} \qquad \leftrightarrow \zeta = \frac{2z}{1+z} \tag{9}$$

To motivate this, consider an example which is an alternating series,

$$S^{log2} \equiv \log(2) = \sum_{n=1}^{\infty} (-1)^{n+1}/n \tag{10}$$

$$S^{log2,Abel}(z) = \log(1+z) = \sum_{n=1}^{\infty} \frac{(-1)^{n+1}}{n} z^n \tag{11}$$

The Abel extension is *logarithmically singular* at $z = -1$. Consequently, its power series in z has a *unit* radius of convergence. At $z = 1$, the point corresponding to the original series for log(2), the power series converges very slowly because this point is right on the boundary of the disk of convergence. And yet the inflated function $S^{log2,Abel}(z)$ is *not* singular at $z = 1$, the only value of z that we really care about, but only at $z = -1$.

The Euler mapping takes the singular point, $z = -1$, to $\zeta = \infty$.

$$S^{log2,Abel}(z(\zeta)) = \log(1 + \zeta/(2-\zeta)) = -\log(1-\zeta/2)$$

$$= \sum_{n=1}^{\infty} \frac{(-1)^{n+1}}{n} \left(\frac{\zeta}{2-\zeta}\right)^n \tag{12}$$

$$= -\sum_{n=1}^{\infty} \frac{2}{n} \left(\frac{1}{2}\right)^n \zeta^n \tag{13}$$

Evaluating the last series at $\zeta = 1$ gives a sum which converges geometrically with each term smaller than its predecessor by a factor of one-half. Because of the singularity in the mapping itself at $\zeta = 2$, the radius of convergence in ζ is $\rho = 2$ so that $\zeta = 1$ is only one-half of the distance to the boundary of the disk of convergence.

A rational map which is linear in both the numerator and denominator polynomials is also known as a "Möbius transformation." It has the important property that all Möbius transformations map *circles* to *circles*. The nth term decays proportionally to $\exp(-\log(2)|n|)$. Six-digit accuracy requires about a million terms of the original series, but only 20 ($= -\log_2(10^{-6})$) terms of the Eulerized expansion.

3 Accelerating a Fourier Series

There is no loss of generality in assuming a period of 2π since a general period P in y may be normalized by the trivial dilation $x = (2\pi/P)y$. The most significant restriction we shall impose is that the Fourier series is slowly converging because of a single singularity on $x \in [-\pi, \pi]$. We shall *not* exclude the possibility that $f(x)$ has singularities off the real axis at $x = x_k, k = 1, 2, \ldots, n_s$ where n_s is any integer between zero and infinity and $|\Im(x_k)|$ is nonzero for all k. Without loss of generality, we shall assume that the singularity on the real axis is located at $x = 0$ (and, by periodicity, at $x = \pm 2\pi m$ where m is an integer). A singularity at a point $w = x_0$ in a coordinate w can be translated into our standard form by defining $x = w - x_s$.

To inflate $f(x)$ to a power series in z, it is helpful to write

$$f(x) = f_1(x) + f_2(x) \tag{14}$$

where

$$f_1(x) = \sum_{n=0}^{\infty} c_n \exp(inx); \qquad f_2(x) = \sum_{n=1}^{\infty} c_{-n} \exp(-inx) \tag{15}$$

$$f^{Abel}(x, z) = f_1^{Abel}(x, z) + f_2^{Abel}(x, z) \tag{16}$$

$$f_1^{Abel}(x, z) = \sum_{n=0}^{\infty} c_n z^n \exp(inx); \; f_2^{Abel}(x, z) = \sum_{n=1}^{\infty} c_{-n} z^n \exp(-inx) \tag{17}$$

To determine the x-dependent rate of convergence, we need to determine (1) the images of the singularities of $f(x)$ in the z-plane and (2) the images of these same singularities in the ζ-plane. This geometry-of-singularities is described the last three theorems of this section.

To prove the first, we need the following.

Theorem 3. *If the Fourier coefficients satisfy*

$$|c_n| \leq constant \exp(-q|n|)n^{2m} \tag{18}$$

for some positive constant q and constant m, then the Fourier series is ANALYTIC *everywhere within the strip*

$$|\Im(x)| < q \tag{19}$$

Noted on p. 271 of [7].

Theorem 4. *The function $f_1(x) \equiv \sum_{n=0}^{\infty} c_n \exp(inx)$ has no singularities in the upper half-plane $\Im(x) > 0$ while $f_2(x) \equiv \sum_{n=1}^{\infty} c_{-n} \exp(-inx)$ is free of singularities in the lower half-plane.*

Proof. Let $x = \sigma + i\tau$. Then

$$f_1(\sigma + i\tau) = \sum_{n=0}^{\infty} c_n \exp(-n\tau) \exp(in\sigma) \tag{20}$$

When $\tau = \Im(x) > 0$, the series converges geometrically. Theorem 3 then implies that $f_1(x)$ must be analytic for all σ and similarly for f_2.

Theorem 5 (Location of Singularities in z**).** *Suppose $f(x)$ has a singularity at*

$$x_j = \sigma_j + i\tau_j, \qquad j = 0, 1, 2, \ldots \tag{21}$$

Then the corresponding singularity of $f_1^{Abel}(x, z)$ is at

$$\Im(x_j) < 0, \qquad |z_J| = \exp(|\tau_j|), \qquad arg(z_j(x)) = \Re(x_j) - x$$

The image of the singularity at $x = 0$ is

$$|z| = 1, \qquad arg(z_0(x)) = -x \tag{22}$$

Those of f_2^{Abel} are the same except $arg(z_j(x)) = x - \Re(x_j)$ and $arg(z_0(x)) = x$.

Proof. With $x = \sigma + i\tau$ and $z = r\exp(i\theta)$,

$$f_1^{Abel}(x, z) = \sum_{n=0}^{\infty} c_n \exp(n[\log(r) - \tau]) \exp(in[\theta + \sigma]) \tag{23}$$

Thus, $f_1^{Abel}(x, z)$ does not depend on r and τ separately nor on θ and σ separately, but only on the combinations $\log(r) - \tau$ and $\theta + \sigma$. Similar remarks apply to $f_2^{Abel}(x, z)$. It follows that if $f_1(x) = f_1^{Abel}(x, z = 1)$ has a singularity at $x = \sigma + i\tau$ where, as already proved, $\tau_j \leq 0$, then

$$\log(|z_j|(\tau)) = \Im(x) - \tau_j \qquad \& \qquad \arg(z_j)(x) = \Re(x_j) - \Re(x) \tag{24}$$

When x is real, these specialize to the theorem.

Theorem 6. *If $f(z)$ is singular at $z = r_j \exp(i\theta_j)$, then $f(z(\zeta))$ is singular at*

$$|\zeta_j| = \frac{2r_j}{\sqrt{1 + r_j^2 + 2r_j \cos(\theta_j)}} \tag{25}$$

Proof. Let $\bar{z}$ denotes the complex conjugate of z. Recall that $\zeta = 2z/(1+z)$. Then multiplying both sides by $\bar{\zeta}$ yield

$$|\zeta|^2 = \frac{4|z|^2}{1+z+\bar{z}+|z|^2} \tag{26}$$

Substituting $z = r_j \exp(i\theta_j)$ and $z+\bar{z} = 2\cos(\arg(z))$ gives the theorem. The proof that the constant C is independent of x is omitted.

Theorem 1 then follows from combining Theorems 2, 4–6. At a given x, the Fourier sum converges at a rate proportional to $(1/\rho(x))^N$ where $\rho(x)$ is the radius of convergence of the ζ power series. This is the simply $|\zeta_j|$ for whichever singularity is closest to $\zeta = 0$.

4 Numerical Illustration of Geometric Convergence

A useful example is the "shifted sawtooth" function, [1]

$$\mathrm{Sws}(x) \equiv \begin{cases} x - \pi & x \in [0, 2\pi] \\ \mathrm{Sws}(x + 2\pi m) & \text{otherwise}, m = \text{integer} \end{cases} \tag{27}$$

for which $c_0 = 0, c_n = i, c_{-n} = -i$ for $n = 1, 2, \ldots$.

Figure 1 shows the errors for the Euler-accelerated Fourier series. On a log-linear plot, a geometric rate of convergence appears as a straight line. The errors in the Fourier series fluctuate with N, but the *envelope* displays the predicted rate of convergence.

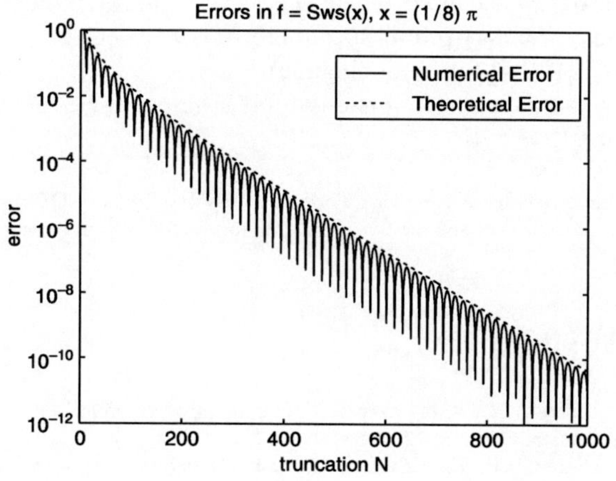

Fig. 1 The *solid curve* is the error of the Euler-accelerated series for the function $f(x) = \mathrm{Sws}(x)$ for $x = \pi/8$. The *dotted line* is the theoretical prediction of the envelope of the errors, $2\exp(-\mu(x)N)/N$ where $\mu(x) = -\log(\cos(x/2))$

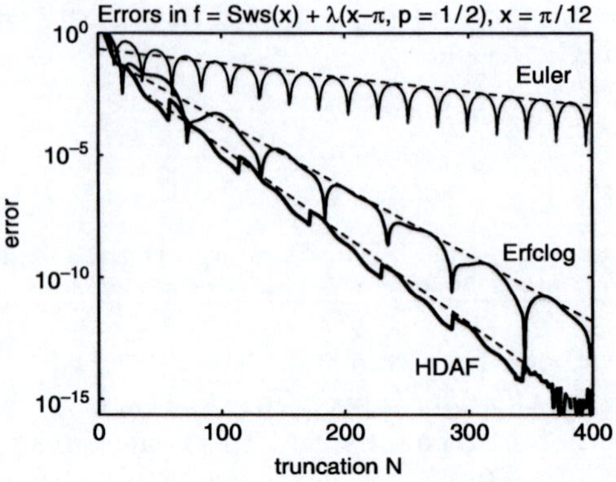

Fig. 2 Errors at $x = \pi/12$ vs. the truncation N for $f(x) = \text{Sws}(x) + \Lambda$ where $\Lambda = (1 - p^2)/(1 + p^2) - 2\,p\,\cos(x - \pi)$ as accelerated by three different filters. The *dashed* guidelines show that the rate of convergence is geomeric – proportional to $\exp(-\mu N)$ for some constant μ – for all three filters. The function Λ has poles on the imaginary axis; such off-axis singularities do not destroy the efficacy of filters

5 Summary

Tanner's [15] Hermite Distributed Approximating Function (HDAF) and Boyd's ErfcLog filter [2] are x-adaptive in the sense that contain an order parameter p, and both are best applied with *spatially varying order*. Both are more accurate than the Euler acceleration as shown in Fig. 2. The point is not that the Euler acceleration is best, but rather than it has the simplest theory. In work in progress, we show that a generalized adaptive Euler method is better than the HDAF and ErfcLog filters.

Recovery of spectral accuracy in the presence of shocks and other singularities has become a "big business." A partial list of other efforts to accelerate Fourier series through local filters include [1,3–6,12–14,16]. It therefore seems worthwhile to give an elementary proof "spectral recovery." Theorem 1 improves on [1] by quantifying the effects of off-axis singularities.

A much longer version of this article may be found on ArXiv at http://arxiv.org/abs/1003.5263.

Acknowledgements This work was supported by NSF OCE0451951 and ATM 0723440. I thank both reviewers and the Scientific Committee for helpful comments.

References

1. Boyd, J.P.: A lag-averaged generalization of Euler's method for accelerating series. Appl. Math. Comput. **72**, 146–166 (1995)
2. Boyd, J.P.: The Erfc-Log filter and the asymptotics of the Vandeven and Euler sequence accelerations. In: A.V. Ilin, L.R. Scott (eds.) Proceedings of the 3rd International Conference on Spectral and High Order Methods, pp. 267–276. Houston J. Mathematics, Houston (1996)

3. Boyd, J.P.: Trouble with Gegenbauer reconstruction for defeating Gibbs' phenomenon: Runge phenomenon in the diagonal limit of Gegenbauer polynomial approximations. J. Comput. Phys. **204**(1), 253–264 (2005)
4. Boyd, J.P.: Exponentially accurate Runge-free approximation of non-periodic functions from samples on an evenly-spaced grid. Appl. Math. Lett. **20**(9), 971–975 (2007)
5. Boyd, J.P.: Acceleration of algebraically-converging Fourier series when the coefficients have series in powers of $1/n$. J. Comput. Phys. **228**(5), 1401–1411 (2008)
6. Boyd, J.P., Moore, D.W.: Summability methods for Hermite functions. Dyn. Atmos. Oceans **10**, 51–62 (1986)
7. Canuto, C., Hussaini, M.Y., Quarteroni, A., Zang, T.A.: Spectral Methods: Fundamentals in Single Domain, p. 558. Springer, New York (2006)
8. Gottlieb, D., Shu, C.W.: On the Gibbs phenomenon and its resolution. SIAM Rev. **39**(4), 644–668 (1997)
9. Guttmann, A.J.: Asymptotic analysis of power-series expansions. In: C. Domb, J.L. Lebowitz (eds.) Phase Transitions and Critical Phenomena, no. 13 in Phase Transitions and Critical Phenomena, pp. 1–234. Academic Press, New York (1989)
10. Morse, P.M., Feshbach, H.: Methods of Theoretical Physics (two volumes). McGraw-Hill, New York (1953)
11. Pearce, C.J.: Transformation methods in the analysis of series for critical properties. Adv. Phys. **27**(1), 89–148 (1978)
12. Tadmor, E.: Filters, mollifiers and the computation of Gibbs phenomenon. In: A. Iserles (ed.) Acta Numerica, 16, pp. 305–378. Cambridge University Press, Cambridge (2007)
13. Tadmor, E., Tanner, J.: Adaptive mollifiers for high resolution recovery of piecewise smooth data from its spectral information. Foundations Comput. Math. **2**(2), 155–189 (2002)
14. Tadmor, E., Tanner, J.: Adaptive filters for piecewise smooth spectral data. IMA J. Numer. Anal. **25**(4), 635–647 (2005)
15. Tanner, J.: Optimal filter and mollifier for piecewise smooth spectral data. Math. Comput. **75**(254), 767–790 (2006)
16. Vandeven, H.: Family of spectral filters for discontinuous problems. J. Sci. Comput. **6**, 159–192 (1991)

A Seamless Reduced Basis Element Method for 2D Maxwell's Problem: An Introduction

Yanlai Chen, Jan S. Hesthaven, and Yvon Maday

Abstract We present a reduced basis element method (RBEM) for the time-harmonic Maxwell's equation. The RBEM is a Reduced Basis Method (RBM) with parameters describing the geometry of the computational domain, coupled with a domain decomposition method. The basic idea is the following. First, we decompose the computational domain into a series of subdomains, each of which is deformed from some reference domain. Then, we associate with each reference domain precomputed solutions to the same governing partial differential equation, but with different choices of deformations. Finally, one seeks the approximation on a new domain as a linear combination of the corresponding precomputed solutions on each subdomain. Unlike the work on RBEM for thermal fin and fluid flow problems, we do not need a mortar type method to "glue" the various local functions. This "gluing" is done "automatically" thanks to the use of a discontinuous Galerkin method. We present the rationale for the method together with numerical results showing exponential convergence for the simulation of a metallic pipe with both ends open.

Keywords Discontinuous Galerkin method · Domain Decomposition · Reduced basis element method · Reduced basis method · Reduced order model · Maxwell's equations

---

Y. Chen (✉), and J.S. Hesthaven
Division of Applied Mathematics, Brown University, 182 George St, Providence, RI 02912, USA
e-mail: Yanlai_Chen@Brown.edu, Jan.Hesthaven@Brown.edu

Y. Maday
Université Pierre et Marie Curie-Paris 6, UMR 7598, Laboratoire J.-L. Lions, 75005 Paris, France and Division of Applied Mathematics, Brown University, 182 George St, Providence, RI 02912, USA
e-mail: maday@ann.jussieu.fr

J.S. Hesthaven and E.M. Rønquist (eds.), *Spectral and High Order Methods for Partial Differential Equations*, Lecture Notes in Computational Science and Engineering 76, DOI 10.1007/978-3-642-15337-2_11,

1 Introduction

There is a need to rapidly, perhaps even in real time, and accurately predict some quantities of interest under the variation of a set of parameters in applications such as computational optimization, control and design, the development of efficient ways to quantify uncertainties and their impact. In such cases, an output of interest, here denoted by s^e, is defined by a functional applied to the solution of a parameterized partial differential equation (PDE). Let us write the problem in weak form as

$$\left| \begin{array}{l} \text{For an input } \nu \in \mathcal{D} \subset \mathbb{R}^p \text{ the output is defined by} \\ \qquad s^e(\nu) := l(u^e(\nu);\nu) \in \mathbb{C}, \\ \text{where } u^e(\nu) \in X^e \text{ is the exact solution of} \\ \qquad a(u^e(\nu), v; \nu) = f(v;\nu), \quad \forall\, v \in X^e, \end{array} \right. \tag{1}$$

where a and f are bilinear and linear forms, respectively, associated to the PDE, and X^e is the space of the exact solution u^e.

To approximate its solution $u^{\mathcal{N}}(\nu) \simeq u^e(\nu)$, one could use the following finite element (FE) discretization: Given $\nu \in \mathcal{D} \subset \mathbb{R}^P$, find $u^{\mathcal{N}}(\nu) \in X^{\mathcal{N}}$ satisfying $a^{\mathcal{N}}(u^{\mathcal{N}}(\nu), v; \nu) = f^{\mathcal{N}}(v;\nu)$, $\forall v \in X^{\mathcal{N}}$. Here $X^{\mathcal{N}}$ is the finite element space approximating X^e with $\dim(X^{\mathcal{N}}) \equiv \mathcal{N}$, and $a^{\mathcal{N}}(\cdot,\cdot;\cdot)$ and $f^{\mathcal{N}}(\cdot;\cdot)$ are computable approximations of $a(\cdot,\cdot;\cdot)$ and $f(\cdot;\cdot)$, respectively. We assume $u^{\mathcal{N}}$ provides a *reference* solution (called the truth approximation) that is accurate enough for all $\nu \in \mathcal{D}$. For that purpose, one usually must choose a very large $\mathcal{N}$. This makes it time-consuming to solve for the truth approximation, in particular when the solution is needed for many instances of ν.

The Reduced Basis Method (RBM) is particularly well suited for this "many-query" scenario to improve the simulation efficiency. It was introduced in [6, 13]. See [4, 7, 14, 17] for recent developments including rigorous a posteriori error estimators and greedy algorithm to form the global approximation spaces. See also the review paper [16] and the extensive reference therein. The first theoretical a priori convergence result for a 1D parametric space problem is presented in [10] where exponential convergence of the reduced basis approximation is confirmed.

The key idea of the RBM is to store the solutions of the PDE for a specific set of parameters, and then find the reduced basis approximation for a new parameter as a linear combination of these precomputed solutions. The fundamental observation is that the parameter dependent solution $u^e(\nu)$ is not simply an arbitrary member of the infinite-dimensional space associated with the PDE, but rather that it evolves on a lower-dimensional manifold induced by the parametric dependence. Under this assumption we can expect that as ν ($\in \mathcal{D} \subset \mathbb{R}^q$) varies, the set of all solutions $u^e(\nu)$ can be well approximated by a finite and low dimensional vector space. How to choose the initial set of parameters used to compute the basis functions is crucial to the method. It is guided by the *a posteriori* error estimators, which are also used to certify the quality of the approximations.

In this paper, we are going to concentrate on the case of ν in (1) describing the geometries. This special parameter allows us to combine domain decomposition ideas with RBM to obtain a method called *reduced basis element method*. It was introduced in [11] and later applied to a thermal fin problem [12] and the Stokes problem [9]. We shall here, for the first time, apply it to Maxwell's equation. The first ingredient of reduced basis element method (RBEM) is a domain decomposition approach to generalize a discretization method, originally designed over a simple geometry, to a complicated one. The second ingredient is the RBM mentioned above. The extension from RBM to RBEM is like the one from a single element method to a multi-element method. We first decompose the computational domain into a series of subdomains that are deformed from several generic, *reference* building blocks. As a precomputation, we associate with each reference domain solutions to the same problem, but with different choices of the parameters describing the deformations. Finally, we seek the approximation on a new shape as a linear combination of the precomputed solutions mapped from the reference block onto each particular subdomain of interest.

We focus on the time-harmonic Maxwell's problem. The motivation is the design of waveguides, where pipes of different shapes have to be connected together. For the RBEM introduced in [9, 11, 12], a mortar type method is needed to "glue" the local functions on the subdomains due to the non-conformity of the method. This "gluing" is not necessary for our method. Since it is formulated in such a way that the reduced basis element space is a subspace of the finite element space, and the connection between neighboring elements are taken care of by the numerical fluxes.

The paper is organized as follows. In Sect. 2, we formulate the RBEM. Some numerical results are given in Sect. 3 to show the superior convergence toward the truth approximation with a seamless "gluing" of the decomposed subdomains. Finally, concluding remarks are presented in Sect. 4.

2 Reduced Basis Element Method

In this section, we discuss the reduced basis element method for electromagnetics. We first formulate the RBM with geometry as a parameter and lay out the numerical scheme for the truth approximation. Then, we study the RBEM. Finally, we discuss the a posteriori error estimate for the RBEM.

2.1 Reduced Basis Method with Geometry As a Parameter

We are seeking the frequency-domain solution of the 2D Maxwell's equations in normalized differential form,

$$\begin{cases} -\epsilon\omega^2 \widehat{E}_\xi + \frac{1}{\mu}\frac{\partial}{\partial\eta}\left(\frac{\partial \widehat{E}_\eta}{\partial\xi} - \frac{\partial \widehat{E}_\xi}{\partial\eta}\right) = i\omega J_\xi, \\ -\epsilon\omega^2 \widehat{E}_\eta - \frac{1}{\mu}\frac{\partial}{\partial\xi}\left(\frac{\partial \widehat{E}_\eta}{\partial\xi} - \frac{\partial \widehat{E}_\xi}{\partial\eta}\right) = i\omega J_\eta. \end{cases} \tag{2}$$

As shown by Fig. 1, we want to solve the problem on a domain, Ω^a, which consists of three subdomains, denoted from left to right by Ω_1^a, Ω_2^a and Ω_3^a. A dipole antenna is located in Ω_1^a. To the left of the antenna is Bérenger's perfectly matched layer (PML), see [3, 5], also [1], which is also used in the symmetric part of Ω_3^a. A Silver–Müller condition is enforced on the exterior boundary of the PML. The other boundaries are assumed to be perfectly electrically conducting (PEC) metallic wall, i.e., boundary condition $\widehat{E}_\xi \hat{n}_\eta - \widehat{E}_\eta \hat{n}_\xi = 0$ is enforced. Here, $(\hat{n}_\xi, \hat{n}_\eta)$ denotes the unit outward normal. This models a metallic pipe with segments of varying shapes and both ends open.

As the parameter $\nu := (\alpha, \beta, \theta)$ change, it appears that we need to change the computational domain. In fact, the computation is always done on a reference domain, Ω, as shown by Fig. 2. The deformation of the i-th subdomain is denoted by $\mathcal{F}_i$, i.e., $\Omega_i^a = \mathcal{F}_i(\Omega_i)$. We denote by $\widehat{K}$ an element in Ω^a corresponding to a reference domain element K. The map $\mathcal{F}_i$, shown in Fig. 3, is defined as follows

$$\mathcal{F}_i\left(\begin{pmatrix} x \\ y \end{pmatrix}\right) = \begin{pmatrix} \frac{a_i}{L_i}\sin\theta_i & 0 \\ \frac{a_i}{L_i}\cos\theta_i & 1 \end{pmatrix}\begin{pmatrix} x - x_i \\ y \end{pmatrix} + \begin{pmatrix} \xi_i \\ \eta_i \end{pmatrix}$$

with $(\xi_i, \eta_i)^T$ defined recursive by

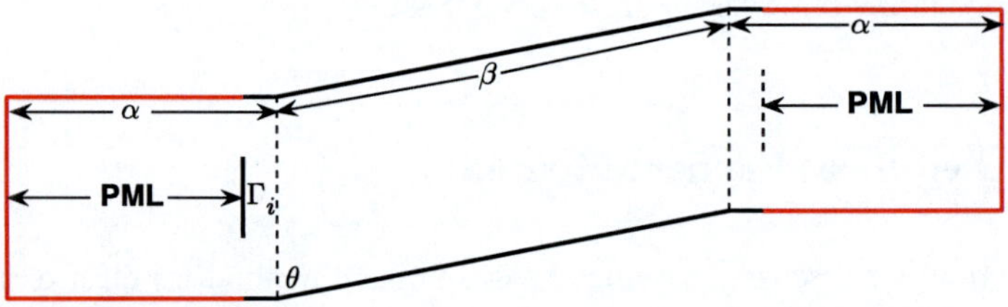

Fig. 1 Actual decomposed domain

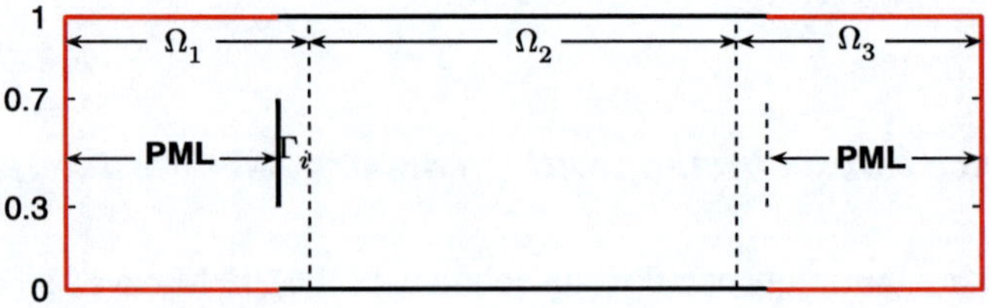

Fig. 2 Decomposed reference domain

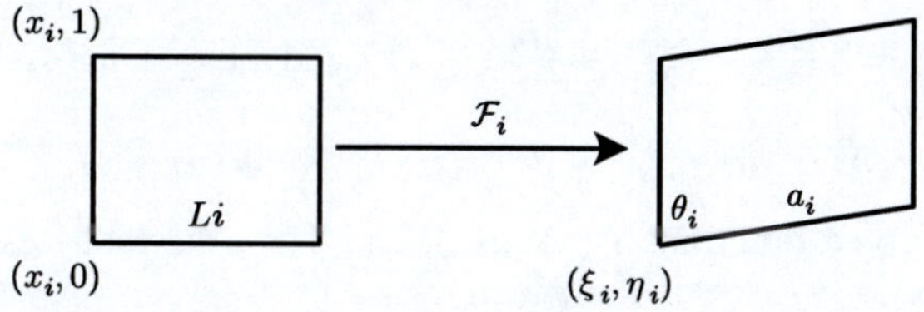

Fig. 3 A generic mapping on one subdomain

$$\begin{pmatrix}\xi_1\\ \eta_1\end{pmatrix} = \begin{pmatrix}0\\ 0\end{pmatrix}, \quad \begin{pmatrix}\xi_{i+1}\\ \eta_{i+1}\end{pmatrix} = \mathcal{F}_i\left(\begin{pmatrix}x_i + L_i\\ 0\end{pmatrix}\right).$$

To maintain tangential components and map $H(curl, \widehat{K})$ to $H(curl, K)$, we apply the Piola transform (see [2] and [15] for a more complete presentation) to $(\widehat{E}_\eta, -\widehat{E}_\xi)^T$:

$$\begin{pmatrix}\widehat{E}_\eta\\ -\widehat{E}_\xi\end{pmatrix} = \mathcal{P}_i\left(\begin{pmatrix}E_y\\ -E_x\end{pmatrix}\right) := \frac{1}{|J\mathcal{F}_i|} J\mathcal{F}_i \begin{pmatrix}E_y\\ -E_x\end{pmatrix},$$

where $J\mathcal{F}_i$ is the Jacobian matrix of the map $\mathcal{F}_i$. After the application of the Piola transform $\mathcal{P}_i$, we obtain, on the reference domain Ω_i, the following system of equations

$$\begin{cases} i\omega\mu H_z + \frac{L_i}{a_i \sin\theta_i}\left(\frac{\partial E_y}{\partial x} - \frac{\partial E_x}{\partial y}\right) = 0, \\ i\epsilon\omega\left(\frac{L_i}{a_i \sin\theta_i} E_x - (\cot\theta_i) E_y\right) - \frac{1}{\mu}\frac{\partial H_z}{\partial y} = J_x, \\ i\epsilon\omega E_y + \frac{1}{\mu}\left(\frac{L_i}{a_i \sin\theta_i}\frac{\partial H_z}{\partial x} - (\cot\theta_i)\frac{\partial H_z}{\partial y}\right) = J_y. \end{cases} \tag{3}$$

Next, we account for the PML by modifying the system as follows (see [3,5] for details).

$$\begin{cases} (i\omega\mu + \frac{\mu\sigma}{\epsilon})H_z + \frac{L_i}{a_i \sin\theta_i}\left(\frac{\partial E_y}{\partial x} - \frac{\partial E_x}{\partial y}\right) = 0, \\ i\epsilon\omega\left(\frac{L_i}{a_i \sin\theta_i} E_x - \cot\theta_i (1 - \frac{i\sigma}{\epsilon\omega}) E_y\right) - \frac{1}{\mu}(1 - \frac{i\sigma}{\epsilon\omega})\frac{\partial H_z}{\partial y} = (1 - \frac{i\sigma}{\epsilon\omega}) J_x, \\ i\epsilon\omega(1 - \frac{i\sigma}{\epsilon\omega}) E_y + \frac{1}{\mu}\left(\frac{L_i}{a_i \sin\theta_i}\frac{\partial H_z}{\partial x} - (\cot\theta_i)\frac{\partial H_z}{\partial y}\right) = J_y. \end{cases} \tag{4}$$

Here, σ is a piecewise quadratic C^1-function of x (constant along the y-axis). It is identically zero in the non-PML region and monotonically increasing from the PML/non-PML interfaces to both boundaries on the left end and right end. If we let

$$\mathbf{u} = \begin{pmatrix}H_z\\ E_x\\ E_y\end{pmatrix} \qquad F(\mathbf{u}) = \begin{pmatrix}e_3 \times E\\ e_1 \times H\\ e_2 \times H\end{pmatrix} \qquad \mathcal{S} = \begin{pmatrix}0\\ (1 - \frac{i\sigma}{\epsilon\omega}) J_x\\ J_y\end{pmatrix},$$

we have $\nabla \cdot F(\mathbf{u}) = \left(\frac{\partial E_x}{\partial y} - \frac{\partial E_y}{\partial x}, -\frac{\partial H_z}{\partial y}, \frac{\partial H_z}{\partial x}\right)^T$ and the system (4) can be rewritten as

$$\mathrm{A}(L_i, a_i, \theta_i)\,\mathbf{u} + \mathrm{B}(L_i, a_i, \theta_i)\,\nabla \cdot F(\mathbf{u}) = \mathcal{S},$$

where, for any given functions $L(\mathbf{x})$, $a(\mathbf{x})$, and $\theta(\mathbf{x})$, the matrices $\mathrm{A}(L(\mathbf{x}), a(\mathbf{x}), \theta(\mathbf{x}))$ and $\mathrm{B}(L(\mathbf{x}), a(\mathbf{x}), \theta(\mathbf{x}))$ are defined by

$$\mathrm{A} = \begin{pmatrix} \mu i\omega + \frac{\mu}{\epsilon}\sigma & 0 & 0 \\ 0 & \frac{i\epsilon\omega L(\mathbf{x})}{a(\mathbf{x})\sin\theta(\mathbf{x})} & -\cot\theta(\mathbf{x})(i\epsilon\omega + \sigma) \\ 0 & 0 & i\epsilon\omega + \sigma \end{pmatrix},$$

$$\mathrm{B} = \begin{pmatrix} -\frac{L(\mathbf{x})}{a(\mathbf{x})\sin\theta(\mathbf{x})} & 0 & 0 \\ 0 & \frac{1}{\mu}(1 - \frac{i\sigma}{\epsilon\omega}) & 0 \\ 0 & \frac{1}{\mu}\cot\theta(\mathbf{x}) & \frac{L(\mathbf{x})}{\mu a(\mathbf{x})\sin\theta(\mathbf{x})} \end{pmatrix}.$$

Now, we can state the equation on the domain $\Omega := \Omega_1 \bigcup \Omega_2 \bigcup \Omega_3$ as,

$$\mathrm{A}(L(\mathbf{x}), a(\mathbf{x}), \theta(\mathbf{x}))\,\mathbf{u} + \mathrm{B}(L(\mathbf{x}), a(\mathbf{x}), \theta(\mathbf{x}))\,\nabla \cdot F(\mathbf{u}) = \mathcal{S}, \tag{5}$$

where, $L(\mathbf{x})$ is a piecewise function defined to be equal to the width of Ω_i on Ω_i, $a(\mathbf{x})$ is equal to α on $\Omega_1 \bigcup \Omega_3$ and β on Ω_2, $\theta(\mathbf{x})$ is equal to $\frac{\pi}{2}$ on $\Omega_1 \bigcup \Omega_3$ and θ on Ω_2. Given a mesh $\mathcal{T}_h$, we define the following finite element space

$$X^{\mathcal{N}} = \{\mathbf{v} \in \left(L^2(\mathcal{T}_h)\right)^3 : \text{ for all elements } K \in \mathcal{T}_h, \mathbf{v}|K \in (P_k(K))^3\},$$

and use a discontinuous Galerkin method [8] to solve (5) as in [4] with the same numerical fluxes, but without the elimination of H_z since this cannot be done with PML and Silver–Müller boundary condition present. As a result, we obtain the following system

$$a^{\mathcal{N}}(\mathbf{u}^{\mathcal{N}}(\nu), \mathbf{v}; \alpha, \beta, \theta) = f^{\mathcal{N}}(\mathbf{v}), \quad \forall\ \mathbf{v} \in X^{\mathcal{N}}. \tag{6}$$

Here,

$$\begin{aligned} a^{\mathcal{N}}(\mathbf{u}^{\mathcal{N}}(\nu), \mathbf{v}; \alpha, \beta, \theta) = (\mathrm{A}\,\mathbf{u}^{\mathcal{N}}(\nu), \mathbf{v})_{\mathcal{T}_h} &+ \langle \mathrm{B}\widehat{F}(\mathbf{u}^{\mathcal{N}}(\nu))\mathbf{n}, \mathbf{v}\rangle_{\partial\mathcal{T}_h} \\ &-(\mathrm{B}F(\mathbf{u}^{\mathcal{N}}(\nu)), \nabla\mathbf{v})_{\mathcal{T}_h}, \end{aligned}$$

and $f^{\mathcal{N}}(\mathbf{v}) = (\mathcal{S}, \mathbf{v})_{\mathcal{T}_h}$, where $\widehat{F} = (e_3 \times \widehat{E},\ e_1 \times \widehat{H},\ e_2 \times \widehat{H})^T$, $\mathbf{n}$ the unit outward normal vector, $(\cdot,\cdot)_{\mathcal{T}_h}$ denotes the broken inner product on the elements and $(\cdot,\cdot)_{\partial\mathcal{T}_h}$ the broken inner product on the set of faces of all the elements. The standard RBM is then applied to build the reduced basis space. The accuracy of the reduced basis solution is certified by the residual-based a posteriori error estimate. This error estimate is cheap to obtain on-line. It also guides the selection of the parameters and

Fig. 4 A general domain decomposed into $D + 2$ subdomains

thus the building of the reduced basis space in the greedy algorithm. See, e.g., [4,16] for details.

2.2 *Reduced Basis Element Method: Formulation*

In this subsection, we discuss in detail how we apply RBEM to our problem to enable a highly efficient simulation on more complicated domains. We are going to concentrate on domains consisting of pipes such as that shown in Fig. 4. The standard discontinuous Galerkin FE method for the full problem would result in system:

$$a^{\mathcal{N}}\left(u^{\mathcal{N}}(\nu), v; (\alpha; \beta_1, \theta_1; \cdots; \beta_D, \theta_D)\right) = f^{\mathcal{N}}(v), \quad \forall v \in X^{\mathcal{N}}. \tag{7}$$

We observe, see, e.g., Fig. 7, that the global solution has a certain amount of "repetitiveness" in the middle domains. The two ends have different behavior because of the PML material. Moreover, there is a dipole antenna in the left end. This motivates us to treat the three parts differently, to decompose the domain, and to make assumptions in the following way: We can subdivide the middle part into many blocks. Since the solution has similar patterns on all these blocks, we can mimic RBM, and only seek the solution on each block in a space spanned by a certain set of precomputed functions instead of the (very rich) FE space. Moreover, these spaces on all the blocks can be assumed to be the same.

The natural choice of the elementary block is, of course, rectangles in our case. So we denote the subdomains from left to right by $\mathrm{S}_0, \mathrm{S}_1, \cdots, \mathrm{S}_D, \mathrm{S}_{D+1}$. Obviously, S_0 corresponds to Ω_1^a in Fig. 1, S_{D+1} to Ω_3^a and all the remaining to Ω_2^a.

Next, we need to identify an appropriate set of functions as our basic patterns. We take the basis functions in the reduced basis space obtained in Sect. 2.1. Note that we have built a reduced basis space that is spanned by N solutions (on the domain as in Fig. 1) to (6) at N judiciously chosen parameters [4, 14, 17]:

$$W^N = \mathrm{span}\,\{u^{\mathcal{N}}(\nu^1), \cdots, u^{\mathcal{N}}(\nu^N)\}.$$

We can cut each of these solutions into three pieces to obtain three sets of basis functions to be used on S_0, $\{\mathrm{S}_1, \cdots, \mathrm{S}_D\}$, S_{D+1}. That is, we define a space $\mathcal{Y}_N$ of dimension $N\,(D+2)$,

$$\begin{aligned}\mathcal{Y}_N = \Big\{v \in X^{\mathcal{N}} \text{s. t. } v|_{\mathrm{S}_0} &\in \operatorname{span}\left\{u^{\mathcal{N}}(\nu^1)|_{\Omega_1}, \cdots, u^{\mathcal{N}}(\nu^N)|_{\Omega_1}\right\} \\ v|_{\mathrm{S}_{D+1}} &\in \operatorname{span}\left\{u^{\mathcal{N}}(\nu^1)|_{\Omega_3}, \cdots, u^{\mathcal{N}}(\nu^N)|_{\Omega_3}\right\} \qquad (8) \\ v|_{\mathrm{S}_i} &\in \operatorname{span}\left\{u^{\mathcal{N}}(\nu^1)|_{\Omega_2}, \cdots, u^{\mathcal{N}}(\nu^N)|_{\Omega_2}\right\} \text{ for } i = 1, \cdots, D\Big\}.\end{aligned}$$

The RBEM is nothing but to seek a solution in the space $\mathcal{Y}_N$, i.e., the RBEM solution is the solution of the following problem:

$$\begin{aligned}&\text{Find} \quad u^N(\nu) \in \mathcal{Y}_N \quad \text{such that} \\ &a^{\mathcal{N}}\left(u^N(\nu), v; (\alpha; \beta_1, \theta_1; \cdots; \beta_D, \theta_D)\right) = f^{\mathcal{N}}(v), \quad \forall v \in \mathcal{Y}_N. \qquad (9)\end{aligned}$$

Note that no "gluing" is necessary since our original space $X^{\mathcal{N}}$ consists of discontinuous functions and the new space $\mathcal{Y}_N$ is naturally a subspace of $X^{\mathcal{N}}$. This dramatic difference from the method in [9, 11, 12] motivates the name of our method – *seamless reduced basis element method.*

2.3 *Reduced Basis Element Method: Error Estimate*

To end this section, we briefly remark on the error estimate. Since the space $\mathcal{Y}_N$ is a subspace of $X^{\mathcal{N}}$, the error estimate can be done in exactly the same ways as the RBM. The parameter $(\alpha; \beta_1, \theta_1; \cdots; \beta_D, \theta_D)$ has a higher dimension than for the RBM used to build $\mathcal{Y}_N$. However, this is not much of a problem for relatively small D.

3 Numerical Results

In this paper, we set $D = 1$ and thus only consider problems with three subdomains. Next, we show numerical results for a two-parameter case and then a three-parameter case.

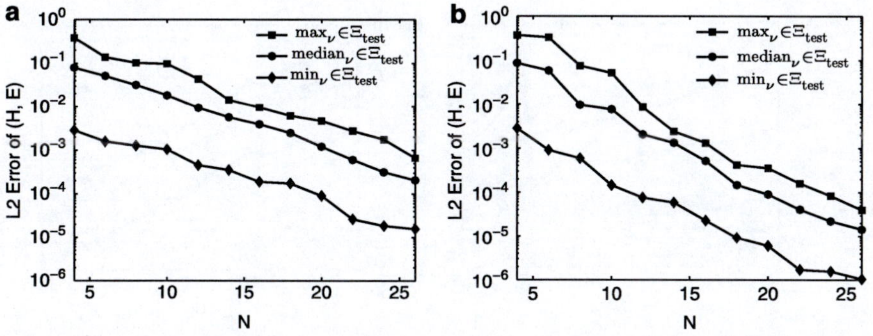

Fig. 5 History of convergence for **(a)** RBM and **(b)** RBEM. Three pieces and two parameters

3.1 *Two-Parameter Case*

We choose two parameters: the length of Ω_1^a (i.e. Ω_3^a), α, and the length of Ω_2^a, β and let $\theta = \frac{\pi}{2}$. The parameter domain is set to be $(\alpha, \beta) \in [0.48, 1.00]\times[0.84, 1.75]$. We set $J_x = 0$, $J_y = \cos(\omega(y - \frac{1}{2}))\delta_{\Gamma_i}$.

First, we perform a reduced basis analysis on $\Omega := \Omega_1 \bigcup \Omega_2 \bigcup \Omega_3$. We use the standard reduced basis method, see, e.g., [4, 16], to obtain 26 bases. To test the validity of our basis, we solve for the reduced basis solution for parameters in a set Ξ_{test} containing 250 randomly selected points. The history of convergence of the reduced basis solution toward the truth approximation is shown in Fig. 5(a). Clearly, exponential convergence is observed.

Then, we test RBEM on the same set Ξ_{test}. The maximum, median and minimum of the error between the reduced basis element solution and the truth approximation is plotted in Fig. 5(b). The exponential convergence with respect to the number of bases used in the RBEM is observed. This clearly shows that the RBEM is working as expected. It is not surprising that RBEM is providing much more accurate solutions than RBM since the RBEM solution are sought in a higher dimensional ($3N$) space. Note that for a given RB dimension, the RBM is performing better. However the construction of $3N$ basis functions for the RBEM requires only N FE solutions, which considerably reduces the offline effort. Moreover, with the same settings for α, β and $D > 1$ (addressed in a future paper), RBEM can solve problems on a domain having length up to $D \times 1.75 + 2.0$, but RBM can reach at most 3.75.

3.2 *Three-Parameter Case*

Here, we vary θ in $[\frac{3\pi}{7}, \frac{4\pi}{7}]$. A set of 50 bases are generated through the reduced basis analysis. The RBM and RBEM are tested on a set consisting of 500 randomly

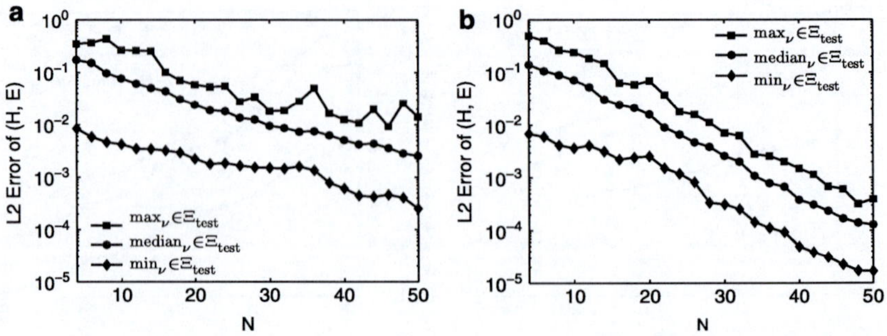

Fig. 6 History of convergence for (**a**) RBM and (**b**) RBEM. Three pieces and three parameters

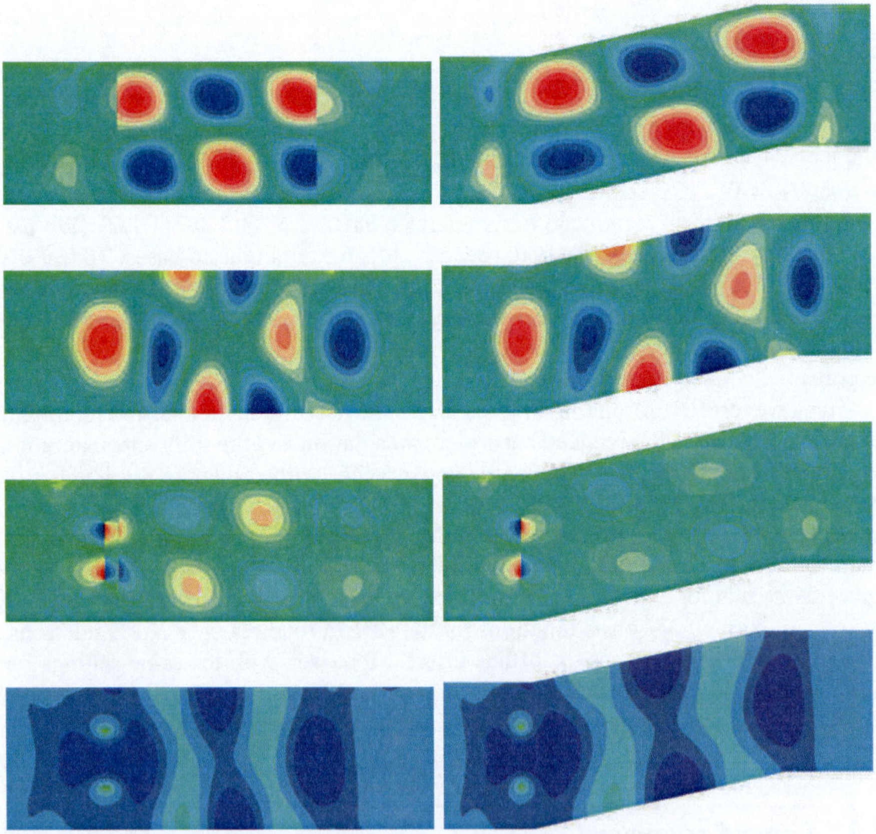

Fig. 7 RBEM solution with $(\alpha, \beta, \theta) = (0.6, 1.75, \frac{3\pi}{7})$: the *left column* is the solution to the parametrized PDE (E_x, E_y), and the *right column* is the solution on the actual domain $(\widehat{E}_\xi, \widehat{E}_\eta)$; from *top* to *bottom* is real part of E_x $(\widehat{E}_\xi)$, real part of E_y $(\widehat{E}_\eta)$, imaginary part of E_x $(\widehat{E}_\xi)$ and imaginary part of E_y $(\widehat{E}_\eta)$

chosen points in the parameter domain $[0.48, 1.00] \times [0.84, 1.75] \times [\frac{3\pi}{7}, \frac{4\pi}{7}]$. The convergence results, shown in Fig. 6, are similar to the two-parameter case.

Moreover, we plot in Fig. 7 the RBEM solutions for the case with the subdomain lengths being 0.6, 1.75 and 0.6 with $\theta = \frac{3\pi}{7}$. We see that the discontinuity (due to the piecewise Piola transform) in E_x is clearly captured by our method, and then recovered nicely (see $\widehat{E}_\xi$) after the (piece-wise) application of the inverse Piola transform $\mathcal{P}_i^{-1}$. The solutions on three patches are "glued" together seamlessly. Note that the Piola transform for E_y in our particular case is identity.

4 Concluding Remarks

In this paper, we have formulated a reduced basis element method to simulate electromagnetic wave propagation in a domain consisting of pipes of different shapes. High efficiency and accuracy of the method is confirmed. The second part of this series is going to be devoted to the study of the multi-element case, i.e., $D \geq 2$.

Acknowledgement Research supported by AFOSR Grant FA9550-07-1-0425.

References

1. S. Abarbanel, D. Gottlieb, and J.S. Hesthaven. *Well-posed perfectly matched layers for advective acoustics.* J. Comput. Phys., 154 (1999) 266–283
2. F. Brezzi and M. Fortin *Mixed and hybrid Finite Element Methods.* Springer, Berlin 1991
3. J.P. Bérenger, *A perfectly matched layer for the absorption of electromagnetics waves.* J. Comput. Phys., 114 (1994) 185–200
4. Y. Chen, J.S. Hesthaven, Y. Maday, and J. Rodríguez, *Certified reduced basis methods and output bounds for the harmonic Maxwell's equations.* SIAM J. Sci. Comput., 32(2):970–996, 2010
5. F. Collino, P.B. Monk, *Optimizing the perfectly matched layer.* Comput. Methods Appl. Mech. Eng., 164 (1998) 157–171
6. J.P. Fink and W.C. Rheinboldt, *On the error behavior of the reduced basis technique for nonlinear finite element approximations.* Z. Angew. Math. Mech., 63(1):21–28, 1983
7. M.A. Grepl, Y. Maday, N.C. Nguyen, and A.T. Patera, *Efficient reduced-basis treatment of nonaffine and nonlinear partial differential equations.* Math. Model. Numer. Anal., 41(3):575–605, 2007
8. J.S. Hesthaven and T. Warburton, *Nodal Discontinuous Galerkin Methods: Algorithms, Analysis, and Applications.* Springer Texts in Applied Mathematics 54, Springer, New York, 2008
9. A.E. Løvgren, Y. Maday, and E.M. Rønquist, *A reduced-basis element method for the steady Stokes problem.* ESAIM: M^{2}AN 40 (2006) 529–552
10. Y. Maday, A.T. Patera, and G. Turinici, *Global a priori convergence theory for reduced-basis approximations of single-parameter symmetric coercive elliptic partial differential equations.* C. R. Acad. Sci. Paris, Ser. I, 335 (2002) 289–294
11. Y. Maday and E.M. Rønquist, *A reduced-basis element method.* J. Sci. Comput. 17 (2002) 447–459

12. Y. Maday and E.M. Rønquist, *The reduced-basis element method: Application to a thermal fin problem.* SIAM J. Sci. Comput. 26 (2004) 240–258
13. A.K. Noor and J.M. Peters, *Reduced basis technique for nonlinear analysis of structures.* AIAA, 18(4):455–462, 1980
14. C. Prud'homme, D. Rovas, K. Veroy, Y. Maday, A.T. Patera, and G. Turinici, *Reliable real-time solution of parametrized partial differential equations: Reduced-basis output bound methods.* J. Fluids Eng., 124(1):70–80, 2002
15. P.A. Raviart and J.M. Thomas, *A mixed finite element method for second order elliptic problems.* In: I. Galligani and E. Magenes, editors, Mathematical Aspects of Finite Element Methods, Lecture Notes in Mathematics, Vol. 606. Springer, New York, 1977
16. G. Rozza, D.B.P. Huynh, and A.T. Patera, *Reduced basis approximation and a posteriori error estimation for affinely parametrized elliptic coercive partial differential equations: Application to transport and continuum mechanics.* Arch. Comput. Methods Eng., 15(3):229–275, 2008
17. K. Veroy, C. Prud'homme, D.V. Rovas, and A.T. Patera, *A posteriori error bounds for reduced-basis approximation of parametrized noncoercive and nonlinear elliptic partial differential equations.* AIAA, 2003

An hp-Nitsche's Method for Interface Problems with Nonconforming Unstructured Finite Element Meshes

Alexey Chernov and Peter Hansbo

Abstract In this paper we propose an hp-Nitsche's method for the finite element solution of interface elliptic problems using non-matched unstructured meshes of triangles and parallelograms in $\mathbb{R}^2$ and tetrahedra and parallelepipeds in $\mathbb{R}^3$. We obtain an explicit lower bound for the penalty weighting function in terms of the local inverse inequality constant. We prove a priori error estimates which are explicit in the mesh size h and in the polynomial degree p. The error bound is optimal in h and suboptimal in polynomial degree by $p^{1/2}$.

1 Introduction

Let us consider a bounded open domain in $\Omega \subset \mathbb{R}^d$, $d = 2, 3$ with polygonal (polyhedral) boundary, decomposed into two mutually disjoint subdomains $\overline{\Omega} = \overline{\Omega^1} \cup \overline{\Omega^2}$. We assume that the interface $\Gamma = \overline{\Omega^1} \cap \overline{\Omega^2}$ is a polygonal curve (polyhedral surface). Γ is allowed to be open or closed, and in the latter case $\partial\Omega \cap \partial\Omega^i = \emptyset$, $\Gamma = \partial\Omega^i$ and $\partial\Omega^{3-i} = \partial\Omega \cup \Gamma$ for $i = 1$ or 2. For any function v on Ω we abbreviate $v^i := v|_{\Omega^i}$. We want to solve for u the diffusion equation, $i = 1, 2$

$$
\begin{aligned}
-\nabla \cdot (\kappa^i \nabla u^i) &= f && \text{in } \Omega^i, && (1)\\
u^i &= 0 && \text{on } \partial\Omega \cap \partial\Omega^i, && (2)\\
u^1 - u^2 &= 0 && \text{on } \Gamma && (3)\\
\kappa^1 \nabla u^1 \cdot \mathbf{n}^1 + \kappa^2 \nabla u^2 \cdot \mathbf{n}^2 &= 0 && \text{on } \Gamma. && (4)
\end{aligned}
$$

A. Chernov (✉)
Hausdorff Center for Mathematics and Institute for Numerical Simulation, University of Bonn, Endenicher Allee 62, 53115 Bonn, Germany
e-mail: chernov@hcm.uni-bonn.de

P. Hansbo
Department of Mathematical Sciences, Chalmers University of Technology, 41262 Göteborg, Sweden
e-mail: peter.hansbo@chalmers.se

J.S. Hesthaven and E.M. Rønquist (eds.), *Spectral and High Order Methods for Partial Differential Equations*, Lecture Notes in Computational Science and Engineering 76, DOI 10.1007/978-3-642-15337-2_12,

Here $f \in L_2(\Omega)$; $\kappa \in L_\infty(\Omega)$ and $\kappa_- \le \kappa(\mathbf{x}) \le \kappa_+$ for almost all $\mathbf{x} \in \Omega$; $\mathbf{n}^i$ is the outward normal vector field on $\partial\Omega^i$, $i = 1, 2$. We utilize the standard Sobolev spaces and abbreviate

$$V := V^1 \times V^2, \quad \text{where} \quad V^i := \{v \in H^1(\Omega^i) : v|_{\partial\Omega \cap \partial\Omega^i} = 0\}. \tag{5}$$

This problem has already been proposed and analyzed in [2] in the h-refinement setting. We here extend it to the hp-setting. In this context the proofs require an auxiliary formulation with the bilinear form A_{hp} (23) using lifting operators. This is done only for technical purposes and formulation (22), similar to that of [2] (and equivalent to (23) on the discrete level), should be used for implementation due to its simpler form.

2 Discretization and Notations

By assumption $\partial\Omega^i$, $i = 1, 2$ have a piecewise flat boundary. We assume that we are given a conforming partition $\mathcal{T}_h^i$ of $\partial\Omega^i$ into triangles and parallelograms (tetrahedra or parallelepipeds) $\Omega^i = \cup_{K \in \mathcal{T}_h^i} K$ and set $\mathcal{T}_h := \mathcal{T}_h^1 \cup \mathcal{T}_h^2$. In order to define the finite element spaces, we assume that every $K \in \mathcal{T}_h$ is an affine image of the standard simplex $S_d = \{x_i \ge 0, \sum_{i=1}^d x_i \le 1\}$ or of the standard cube $[0,1]^d$:

$$\forall K \in \mathcal{T}_h \quad \exists F_K : \begin{cases} \hat{K} \to K \\ \hat{\mathbf{x}} \mapsto A\hat{\mathbf{x}} + \mathbf{x}_0 \end{cases}, \qquad \hat{K} = S_d \text{ or } [0,1]^d. \tag{6}$$

We define the local polynomial spaces by

$$\mathcal{R}_p(\hat{K}) := \begin{cases} \text{polynomials of total degree } \le p \text{ on } S_d & \text{if } \hat{K} = S_d, \\ \text{polynomials of degree } \le p_i \text{ in } \hat{x}_i \text{ on } [0,1]^d & \text{if } \hat{K} = [0,1]^d. \end{cases} \tag{7}$$

In the latter case p is understood as a multiindex $p = (p_1, \dots, p_d)$. We utilize the global finite element spaces of piecewise polynomials

$$V_{hp} = V_{hp}^1 \times V_{hp}^2, \quad V_{hp}^i = \{v \in V^i : v|_K \circ F_k \in \mathcal{R}_p(\hat{K})\}, \tag{8}$$

$$W_{hp} = W_{hp}^1 \times W_{hp}^2, \quad W_{hp}^i = \{v \in L_2(\Omega^i) : v|_K \circ F_k \in \mathcal{R}_p(\hat{K})\}. \tag{9}$$

Note that V_{hp}^i consists of continuous and W_{hp}^i consists of possibly discontinuous functions and $\nabla V_{hp} \subset W_{hp}^d$. We utilize the standard scalar- and vector-valued *jump operators*, cf. e.g., [1,4]. For $v \in V$, $\mathbf{v} \in V^d$ we define on Γ

$$[v] := v^1 - v^2, \qquad [\![v]\!] := v^1\mathbf{n}^1 + v^2\mathbf{n}^2. \tag{10}$$

Note that $[u][v] = [\![u]\!] \cdot [\![v]\!]$ almost everywhere on Γ, and the transmission conditions (3) and (4) reduce to

$$[u] = 0 \quad \Leftrightarrow \quad [\![u]\!] = 0; \qquad [\![\kappa \nabla u]\!] = 0. \tag{11}$$

Let $\chi_A(\mathbf{x}) = 1$ for $\mathbf{x} \in A \subset \mathbb{R}^d$ and zero otherwise. For $v \in V \cap (C^0(\overline{\Omega^1}) \times C^0(\overline{\Omega^2}))$ and some fixed $0 \leq \alpha \leq 1$ we define a *weighting operator*

$$\{v\}_\alpha := \alpha v^1 \chi_{\overline{\Omega^1}} + (1-\alpha) v^2 \chi_{\overline{\Omega^2}}. \tag{12}$$

Note that $\{v\}_\alpha|_\Gamma = \alpha v^1 + (1-\alpha) v^2|_\Gamma$. We carry out the error analysis in the following mesh dependent norm, cf. e.g., [2–4, 8], which mimics the energy norm for (24). Suppose $\gamma \in L_\infty(\Gamma)$ is a uniformly positive function of local mesh parameters. For every $v \in V$ we define the norm

$$\|v\|_{hp,\gamma} := \left(\int_\Omega \kappa |\nabla v|^2 + \int_\Gamma \gamma [v]^2 \right)^{1/2}, \tag{13}$$

where the piecewise gradient operator is understood as $\nabla v|_{\Omega^i} := \nabla v^i$ for $i = 1, 2$.

Remark 1. Note that $\|\cdot\|_{hp,\gamma}$ is an equivalent norm on V if $\gamma(\mathbf{x}) \geq \gamma_-$ where γ_- does not depend on the discretization parameters. In the case $\partial\Omega \cap \partial\Omega^i \neq \emptyset$, $i = 1, 2$ this follows by the standard Poincare inequality. The argument is more elaborate if Γ does not intersect $\partial\Omega$, cf. [3, 5] for more details.

For technical purposes we introduce a lifting operator $\mathscr{L} : V \to W_{hp}^d$ elementwise

$$\forall K \in \mathscr{T}_h \quad \int_K \mathscr{L}(v)|_K \cdot \mathbf{w} = \int_{\partial K \cap \Gamma} [\![v]\!] \cdot \mathbf{w}, \quad \forall \mathbf{w} \in \mathscr{R}_p(K)^d. \tag{14}$$

Lifting operators of this type is often used in the analysis of Discontinuous Galerkin methods, cf. [1, 7]. Note that $\mathscr{L}(v)$ vanishes on K if none of its faces is a subset of Γ and with (12) we have

$$\int_\Omega \mathscr{L}(v) \cdot \{\mathbf{w}\}_\alpha = \int_\Gamma [\![v]\!] \cdot \{\mathbf{w}\}_\alpha, \quad \forall \mathbf{w} \in W_{hp}^d \tag{15}$$

Lemma 1 (Inverse inequality). *Let $K \in \mathscr{T}_h$, J be a face of K and $w \circ F_K \in \mathscr{R}_p(\hat{K})$.*

$$\int_J w^2 \leq \mathscr{G}_K \int_K w^2, \tag{16}$$

$$\text{where} \quad \mathscr{G}_K = \begin{cases} \dfrac{(p+1)(p+1)}{d} \dfrac{\text{volume}(J)}{\text{volume}(K)}, & \text{if } \hat{K} = S_d \\ (p_{J\perp}+1)^2 \dfrac{\text{volume}(J)}{\text{volume}(K)}, & \text{if } \hat{K} = [0,1]^d \end{cases} \tag{17}$$

where $p_{J\perp}$ is the polynomial degree of $w \circ F_K$ in the direction orthogonal to the face $F^{-1}(J)$.

Proof. The proof for $\hat{K} = S_d$ is given in [9, Theorem 5]. The proof for $\hat{K} = [0,1]^d$ follows similarly, cf. [3].

From now on we assume for simplicity that $\kappa|_K = \kappa_K$ is constant on $K \in \mathscr{T}_h$. We write similarly $\mathscr{G}|_K := \mathscr{G}_K$ and $\mathscr{G}^i := \mathscr{G}|_{\Omega^i}$.

Lemma 2. *For all $v \in V$ there holds*

$$\int_{\Omega^i} \kappa^i |\mathscr{L}(v)|^2 \leq \int_\Gamma \kappa^i \mathscr{G}^i [v]^2, \qquad i = 1, 2. \tag{18}$$

Proof. For arbitrary $K \in \mathscr{T}_h$ with a face $J = K \cap \Gamma$ and $\mathbf{w} \circ F_K \in \mathscr{R}_p(\hat{K})^d$ there holds

$$\int_K \mathscr{L}(v)|_K \cdot \mathbf{w} \overset{(14)}{=} \int_J [\![v]\!] \cdot \mathbf{w} \leq \left(\int_J [v]^2\right)^{1/2} \left(\int_J |\mathbf{w}|^2\right)^{1/2} \tag{19}$$

$$\overset{(16)}{\leq} \left(\int_J [v]^2\right)^{1/2} \left(\mathscr{G}_K \int_K |\mathbf{w}|^2\right)^{1/2}, \tag{20}$$

hence we have $\int_K |\mathscr{L}(v)|^2 \leq \mathscr{G}_K \int_J [v]^2$ and thus

$$\int_{\Omega^i} \kappa^i |\mathscr{L}(v)|^2 \leq \sum_{K \in \mathscr{T}_h^i} \kappa_K^i \mathscr{G}_K^i \int_{K \cap \Gamma} [v]^2 = \int_\Gamma \kappa^i \mathscr{G}^i [v]^2 \tag{21}$$

3 hp-Nitsche's Method

For a fixed discretization V_{hp} and $\{\kappa \nabla u\}_\alpha = \alpha \kappa^1 \nabla u^1 + (1 - \alpha) \kappa^2 \nabla u^2$ we define

$$a_{hp}(u, v) := \int_\Omega \kappa \nabla u \cdot \nabla v - \int_\Gamma [\![u]\!] \cdot \{\kappa \nabla v\}_\alpha - \int_\Gamma \{\kappa \nabla u\}_\alpha \cdot [\![v]\!] + \int_\Gamma \sigma [u][v], \tag{22}$$

$$A_{hp}(u, v)$$

$$:= \int_\Omega \kappa \nabla u \cdot \nabla v - \int_\Omega \mathscr{L}(u) \cdot \{\kappa \nabla v\}_\alpha - \int_\Omega \{\kappa \nabla u\}_\alpha \cdot \mathscr{L}(v) + \int_\Gamma \sigma [u][v], \tag{23}$$

Consider the problem of finding $u_{hp} \in V_{hp}$:

$$a_{hp}(u, v) = \int_\Omega f v \quad \text{or} \quad A_{hp}(u, v) = \int_\Omega f v \qquad \forall v \in V_{hp} \tag{24}$$

The above formulations are equivalent, since $a_{hp} \equiv A_{hp}$ on $V_{hp} \times V_{hp}$ due to (15), but (22) does not include the lifting operator and therefore is easier to implement. We shall use (23) in the error analysis, see [3] for more details. In order to quantify inconsistency of the second formulation we define a *residual operator*

$$R_{hp,\gamma}(w) := \sup_{0 \neq v \in V_{hp}} \frac{\int_\Omega f v - A_{hp}(w,v)}{\|v\|_{hp,\gamma}}. \tag{25}$$

We denote by $\Pi_{hp} : L_2(\Omega^1)^d \times L_2(\Omega^2)^d \to W_{hp}^d$ the L_2 projection. Note that W_{hp} consists of in general discontinuous piecewise polynomials, thus Π_{hp} is a local projection.

Theorem 1 (Consistency error). *Suppose $u \in V$ is a weak solution of (1)–(4), then*

$$R_{hp,\gamma}(u) \leq \left(\int_\Gamma \frac{1}{\gamma} \left| \{\kappa(\Pi_{hp}(\nabla u) - \nabla u)\}_\alpha \right|^2 \right)^{1/2} \tag{26}$$

for arbitrary $\gamma \in L_\infty(\Gamma)$ such that $\gamma(\mathbf{x}) \geq \gamma_- > 0$ almost everywhere on Γ.

Proof. Recalling the transmission conditions (11) and (14) we get $\mathcal{L}(u) \equiv 0$,

$$\int_\Gamma \{\kappa \nabla u\}_\alpha \cdot [\![v]\!] = \int_\Gamma \kappa^1 \nabla u^1 \cdot \mathbf{n}^1 v^1 + \int_\Gamma \kappa^2 \nabla u^2 \cdot \mathbf{n}^2 v^2 \tag{27}$$

and by partial integration over Ω^1 and Ω^2

$$\int_\Omega f v - A_{hp}(u,v) = \int_\Omega f v - \int_\Omega \kappa \nabla u \cdot \nabla v + \int_\Omega \{\kappa \nabla u\}_\alpha \cdot \mathcal{L}(v) \tag{28}$$

$$= - \int_\Gamma \{\kappa \nabla u\}_\alpha \cdot [\![v]\!] + \int_\Omega \{\kappa \nabla u\}_\alpha \cdot \mathcal{L}(v). \tag{29}$$

Recall that κ is piecewise constant on $\mathcal{T}_h$, hence $\{\kappa \nabla u\}_\alpha, \mathcal{L}(v) \in W_{hp}^d$ yielding

$$\int_\Omega \{\kappa \nabla u\}_\alpha \cdot \mathcal{L}(v) = \int_\Omega \{\kappa \Pi_{hp}(\nabla u)\}_\alpha \cdot \mathcal{L}(v) \overset{(15)}{=} \int_\Gamma \{\kappa \Pi_{hp}(\nabla u)\}_\alpha \cdot [\![v]\!] \tag{30}$$

thus

$$\int_\Omega f v - A_{hp}(u,v) = \int_\Gamma \{\kappa(\Pi_{hp}(\nabla u) - \nabla u)\}_\alpha \cdot [\![v]\!] \tag{31}$$

$$\leq \left(\int_\Gamma \frac{1}{\gamma} \left| \{\kappa(\Pi_{hp}(\nabla u) - \nabla u)\}_\alpha \right|^2 \right)^{1/2} \left(\int_\Gamma \gamma [v]^2 \right)^{1/2} \tag{32}$$

and (26) follows for any uniformly positive and bounded γ.

4 Quasi-Optimal Convergence

Lemma 3 (Continuity). *For* $\forall v, w \in V$ *and* $\tilde{\alpha} = \max(\alpha, (1-\alpha))$ *there holds*

$$A_{hp}(w, v) \le (1+\tilde{\alpha})\|w\|_{hp,\{\sigma+\kappa\mathcal{G}\}_\alpha} \|v\|_{hp,\{\sigma+\kappa\mathcal{G}\}_\alpha}. \tag{33}$$

Proof. We bound all terms in $A_{hp}(w, v)$ by the Cauchy–Schwarz inequality. In particular we have $\int_{\Omega^i} \kappa^i \nabla w^i \cdot \nabla v^i \le \left(\int_{\Omega^i} \kappa^i |\nabla w^i|^2\right)^{1/2} \left(\int_{\Omega^i} \kappa^i |\nabla v^i|^2\right)^{1/2}$;

$$\int_\Omega \{\kappa \nabla w\}_\alpha \cdot \mathcal{L}(v) \le \alpha \left(\int_{\Omega^1} \kappa^1 |\nabla w^1|^2\right)^{1/2} \left(\int_{\Omega^1} \kappa^1 |\mathcal{L}(v)|^2\right)^{1/2} \tag{34}$$

$$+(1-\alpha) \left(\int_{\Omega^2} \kappa^2 |\nabla w^2|^2\right)^{1/2} \left(\int_{\Omega^2} \kappa^2 |\mathcal{L}(v)|^2\right)^{1/2} \tag{35}$$

and

$$\int_\Gamma \sigma[w][v] \le \left(\int_\Gamma \sigma[w]^2\right)^{1/2} \left(\int_\Gamma \sigma[v]^2\right)^{1/2}. \tag{36}$$

We use Lemma 1 and obtain the bound $A_{hp}(w, v) \le T(w)T(v)$ where

$$T(v)^2 = (1+\tilde{\alpha}) \int_\Omega \kappa |\nabla v|^2 + \int_\Gamma (\alpha(\kappa^1 \mathcal{G}^1) + (1-\alpha)(\kappa^2 \mathcal{G}^2) + \sigma)[v]^2 \tag{37}$$

and the assertion follows, since $T(v) \le \sqrt{1+\tilde{\alpha}}\|v\|_{hp,\sigma+\{\kappa\mathcal{G}\}_\alpha}$.

Lemma 4 (Coercivity). *For* $\forall v \in V_{hp}$ *and* $\tilde{\alpha} = \max(\alpha, 1-\alpha)$ *there holds*

$$A_{hp}(v, v) \ge \frac{\delta - \tilde{\alpha}}{\delta} \|v\|^2_{hp,\sigma - \delta\{\kappa\mathcal{G}\}_\alpha} \tag{38}$$

provided $\delta > \tilde{\alpha}$ *and* $\sigma > \delta\{\kappa\mathcal{G}\}_\alpha$ *almost everywhere on* Γ.

Proof. We have

$$A_{hp}(v, v) = \int_\Omega \kappa |\nabla v|^2 - 2\int_\Omega \{\kappa \nabla v\}_\alpha \cdot \mathcal{L}(v) + \int_\Gamma \sigma[v]^2. \tag{39}$$

We use the arithmetic-geometric mean inequality $2ab \le \delta^{-1}a^2 + \delta b^2$ for $\delta > 0$ and Lemma 1 to obtain

$$2\int_{\Omega}\{\kappa\nabla v\}_\alpha \cdot \mathscr{L}(v) \le \alpha\left(\frac{1}{\delta}\int_{\Omega^1}\kappa^1|\nabla v^1|^2 + \delta\int_\Gamma \kappa^1\mathscr{G}^1[v]^2\right) \tag{40}$$

$$+(1-\alpha)\left(\frac{1}{\delta}\int_{\Omega^2}\kappa^2|\nabla v^2|^2 + \delta\int_\Gamma \kappa^2\mathscr{G}^2[v]^2\right) \tag{41}$$

$$\le \frac{\tilde{\alpha}}{\delta}\int_\Omega \kappa|\nabla v|^2 + \int_\Gamma \delta\{\kappa\mathscr{G}\}_\alpha[v]^2. \tag{42}$$

Inserting this in (39) we obtain

$$A_{hp}(v,v) \ge \frac{\delta-\tilde{\alpha}}{\delta}\int_\Omega \kappa|\nabla v|^2 + \int_\Gamma(\sigma - \delta\{\kappa\mathscr{G}\}_\alpha)[v]^2 \tag{43}$$

and (38) follows.

We remark that the norms $\|\cdot\|_{hp,(\sigma+\{\kappa\mathscr{G}\}_\alpha)}$ and $\|\cdot\|_{hp,(\sigma-\delta\{\kappa\mathscr{G}\}_\alpha)}$ are equivalent on the finite dimensional space V_{hp} and equivalence constants are independent of the discretization parameters if the penalty weighting function σ is chosen such that $\sigma/\{\kappa\mathscr{G}\}_\alpha > \sigma_0 > \delta$. The Lax-Milgram Lemma guarantees unique solvability of discrete formulations (24).

Theorem 2 (Quasi-optimal convergence). *Suppose $u \in V$ is a weak solution of (1)–(4) and $u_{hp} \in V_{hp}$ is a solution of (24) with the penalty weighting function $\sigma := \sigma_0\{\kappa\mathscr{G}\}_\alpha$ with a fixed constant $\sigma_0 > \delta > \tilde{\alpha}$. Then*

$$\|u-u_{hp}\|_{hp,\{\kappa\mathscr{G}\}_\alpha} \le ((1+\tilde{\alpha})c_{\tilde{\alpha},\delta,\sigma_0}+1)\inf_{v\in V_{hp}}\|u-v\|_{hp,\{\kappa\mathscr{G}\}_\alpha} + \frac{c_{\tilde{\alpha},\delta,\sigma_0}}{\sigma_0+1}R_{hp,\{\kappa\mathscr{G}\}_\alpha}(u) \tag{44}$$

with a positive constant

$$c_{\tilde{\alpha},\delta,\sigma_0} = \frac{\delta}{\delta-\tilde{\alpha}}\frac{\sigma_0+1}{\min(\sigma_0-\delta,1)}. \tag{45}$$

Proof. For any $v \in V$, $\gamma(\mathbf{x}) \ge \gamma_- > 0$ and $a > 0$ there holds

$$\min(a,1)\|v\|^2_{hp,\gamma} \le \|v\|^2_{hp,a\gamma} = \int_\Omega \kappa|\nabla v|^2 + a\int_\Gamma \gamma[v]^2 \le \max(a,1)\|v\|^2_{hp,\gamma}, \tag{46}$$

hence Lemma 4 gives for arbitrary $v \in V_{hp}$

$$\frac{(\delta-\tilde{\alpha})\min(\sigma_0-\delta,1)}{\delta}\|u_{hp}-v\|^2_{hp,\{\kappa\mathscr{G}\}_\alpha} \le A_{hp}(u_{hp}-v,u_{hp}-v) \tag{47}$$

$$= A_{hp}(u-v,u_{hp}-v) + \int_\Omega fv - A_{hp}(u,u_{hp}-v). \tag{48}$$

We use continuity (33) and definition (25) of the residual and obtain

$$A_{hp}(u-v,u_{hp}-v) \le (1+\tilde{\alpha})(\sigma_0+1)\|u-v\|_{hp,\{\kappa\mathcal{G}\}_\alpha}\|u_{hp}-v\|_{hp,\{\kappa\mathcal{G}\}_\alpha} \quad (49)$$

$$\int_\Omega fv - A_{hp}(u,u_{hp}-v) \le R_{hp,\{\kappa\mathcal{G}\}_\alpha}(u)\|u_{hp}-v\|_{hp,\{\kappa\mathcal{G}\}_\alpha} \quad (50)$$

hence

$$\|u_{hp}-v\|_{hp,\{\kappa\mathcal{G}\}_\alpha} \le \frac{1+\tilde{\alpha}}{1-\tilde{\alpha}\delta^{-1}}\frac{\sigma_0+1}{\min(\sigma_0-\delta,1)}\|u-v\|_{hp,\{\kappa\mathcal{G}\}_\alpha} \quad (51)$$

$$+\frac{1}{(1-\tilde{\alpha}\delta^{-1})\min(\sigma_0-\delta,1)}R_{hp,\{\kappa\mathcal{G}\}_\alpha}(u). \quad (52)$$

Finally, we use triangle inequality

$$\|u-u_{hp}\|_{hp,\{\kappa\mathcal{G}\}_\alpha} \le \|u-v\|_{hp,\{\kappa\mathcal{G}\}_\alpha} + \|u_{hp}-v\|_{hp,\{\kappa\mathcal{G}\}_\alpha} \quad (53)$$

and obtain (44) taking infimum over all $v \in V_{hp}$.

The following theorem gives the convergence estimate for quasiuniform discretization, see [3] for more details.

Theorem 3. *Suppose $u \in V \cap (H^s(\Omega^1) \times H^s(\Omega^2))$ with $s \ge 2$ is a weak solution of (1)–(4) and $u_{hp} \in V_{hp}$ is a solution of (24) based on quasiuniform shape regular mesh $\mathcal{T}_h$, uniform polynomial degree p and the penalty weighting function $\sigma := \sigma_0\{\kappa\mathcal{G}\}_\alpha$, $0 \le \alpha \le 1$ with a constant $\sigma_0 > \max(\alpha, 1-\alpha)$. Let*

$$h := \max_{K\in\mathcal{T}_h} \operatorname{diam}(K), \qquad p := \min_{K\in\mathcal{T}_h} p_K, \quad (54)$$

then $\exists C > 0$ independent of the discretization parameters such that

$$\sum_{i=1}^{2}\|u^i - u^i_{hp}\|_{H^1(\Omega^i)} \le C\frac{h^{\min\{s-1,p\}}}{p^{s-3/2}}\sum_{i=1}^{2}\|u^i\|_{H^s(\Omega^i)}. \quad (55)$$

Remark 2. As Theorem 3 shows, the convergence rate is optimal in h and suboptimal in p by $p^{1/2}$, which agrees with related results for DG-FEM cf. e.g., [6, 7].

Remark 3. As Theorem 3 shows, convergence is achieved for every $0 \le \alpha \le 1$. We remark that a better sparsity pattern of the stiffness matrix is achieved if $\alpha = 0$ or 1. This becomes more important if at least one of Ω^i is discretized with boundary elements, cf. [3] for this generalization and more details.

References

1. D. N. Arnold, F. Brezzi, B. Cockburn, L. D. Marini, *Unified analysis of discontinuous Galerkin methods for elliptic problems*, SIAM J. Numer. Anal., 39 (2001/02), 1749–1779 (electronic)
2. R. Becker, P. Hansbo, R. Stenberg, *A finite element method for domain decomposition with non-matching grids*, M2AN Math. Model. Numer. Anal., 37 (2003), 209–225
3. A. Chernov, P. Hansbo, *An hp-Nitsche's method for interface problems with non-matching finite element and boundary element discretizations* (in preparation)
4. A. Hansbo, P. Hansbo, M. G. Larson, *A finite element method on composite grids based on Nitsche's method*, M2AN Math. Model. Numer. Anal., 37 (2003), 495–514
5. P. Hansbo, C. Lovadina, I. Perugia, G. Sangalli, *A Lagrange multiplier method for the finite element solution of elliptic interface problems using non-matching meshes*, Numer. Math., 100 (2005), 91–115
6. P. Houston, C. Schwab, E. Süli, *Discontinuous hp-finite elements for advection-diffusion-reaction problems*, SIAM J. Numer. Anal., 39 (2002), 2133–2163
7. D. Schötzau, C. Schwab, A. Toselli, *Mixed hp-DGFEM for incompressible flows*, SIAM J. Numer. Anal., 40 (2002), 2171–2194 (electronic) (2003)
8. D. Schötzau, T. P. Wihler, *Exponential convergence of mixed hp-DGFEM for Stokes flow in polygons*, Numer. Math., 96 (2003), 339–361
9. T. Warburton, J. S. Hesthaven, *On the constants in hp-finite element trace inverse inequalities*, Comput. Meth. Appl. Mech. Eng., 192 (2003), 2765–2773

Hybrid Explicit–Implicit Time Integration for Grid-Induced Stiffness in a DGTD Method for Time Domain Electromagnetics

Victorita Dolean, Hassan Fahs, Loula Fezoui, and Stéphane Lanteri

Abstract In the recent years, there has been an increasing interest in discontinuous Galerkin time domain (DGTD) methods for the numerical modeling of electromagnetic wave propagation. Such methods most often rely on explicit time integration schemes which are constrained by a stability condition that can be very restrictive on highly refined meshes. In this paper, we report on some efforts to design a hybrid explicit–implicit DGTD method for solving the time domain Maxwell equations on locally refined simplicial meshes. The proposed method consists in applying an implicit time integration scheme locally in the refined regions of the mesh while preserving an explicit time scheme in the complementary part.

1 Introduction

Nowadays, a variety of methods exist for the numerical treatment of the time domain Maxwell equations, ranging from the well established and still prominent finite difference time domain (FDTD) methods based on Yee's scheme to the more recent finite element time domain (FETD) and discontinuous Galerkin time domain (DGTD) methods. The latter are very well adapted to local mesh refinement but at the expense of a restrictive time step in order to preserve the stability of the explicit time integration schemes. In the first one, a local time stepping strategy is combined to an explicit time integration scheme, while the second approach relies on the use of an implicit or a hybrid explicit–implicit time integration scheme. In the present work, we consider the second approach.

V. Dolean (✉)
University of Nice-Sophia Antipolis, J.A. Dieudonné Lab., CNRS UMR 6621, 06108 Nice Cedex, France
e-mail: dolean@unice.fr

H. Fahs, L. Fezoui, and S. Lanteri
INRIA, 2004 Route des Lucioles, BP 93, 06902 Sophia Antipolis Cedex, France
e-mail: Stephane.Lanteri@inria.fr

J.S. Hesthaven and E.M. Rønquist (eds.), *Spectral and High Order Methods for Partial Differential Equations*, Lecture Notes in Computational Science and Engineering 76, DOI 10.1007/978-3-642-15337-2_13,

Explicit–implicit methods for the solution of the system of Maxwell equations have been studied by several authors with the shared goal of designing numerical methodologies able to deal with hybrid structured-unstructured meshes. For example, a stable hybrid FDTD–FETD method is considered by Rylander and Bondeson in [8], while Degerfeldt and Rylander [3] propose a FETD method with stable hybrid explicit–implicit time stepping working on brick-tetrahedral meshes that do not require an intermediate layer of pyramidal elements. In [6], the authors study the application of explicit–implicit Runge–Kutta (so-called IMEX-RK) methods in conjunction with high order discontinuous Galerkin discretizations on unstructured triangular meshes, in the framework of unsteady compressible flow problems (i.e., the numerical solution of Euler or Navier–Stokes equations).

This study is concerned with the design of a non-dissipative hybrid explicit–implicit DGTD method for solving the time domain Maxwell equations on unstructured simplicial meshes. The hybrid explicit–implicit DGTD method considered here has been initially introduced by Piperno [7]. However, to our knowledge, this hybrid explicit–implicit DGTD method has not been investigated numerically so far for the simulation of realistic electromagnetic wave propagation problems. The rest of the paper is organized as follows: in Sect. 2, we state the initial and boundary value problem to be solved; the discretization in space by a discontinuous Galerkin method is discussed in Sect. 3 while the integration in time is considered in Sect. 4; numerical results and conclusions are respectively reported in Sect. 5.

2 Continuous Problem

We consider the Maxwell equations in three space dimensions for linear isotropic media with no source. The electric and magnetic fields $\mathbf{E}(\mathbf{x},t)$ and $\mathbf{H}(\mathbf{x},t)$) verify:

$$\varepsilon\partial_t\mathbf{E} - \mathrm{curl}\mathbf{H} = -\mathbf{J}, \quad \mu\partial_t\mathbf{H} + \mathrm{curl}\mathbf{E} = 0, \tag{1}$$

where $\mathbf{J}(\mathbf{x},t)$ is a current source term. These equations are set on a bounded polyhedral domain Ω of $\mathbb{R}^3$. The permittivity $\varepsilon(\mathbf{x})$ and the magnetic permeability tensor $\mu(\mathbf{x})$ are varying in space, time-invariant and both positive functions. Our goal is to solve system (1) in a domain Ω with boundary $\partial\Omega = \Gamma_a \cup \Gamma_m$, where we impose the following boundary conditions:

$$\begin{cases} \mathbf{n}\times\mathbf{E} = 0 \text{ on } \Gamma_m, \\ \mathbf{n}\times\mathbf{E} - \sqrt{\dfrac{\mu}{\varepsilon}}\mathbf{n}\times(\mathbf{H}\times\mathbf{n}) = \mathbf{n}\times\mathbf{E}_{\mathrm{inc}} - \sqrt{\dfrac{\mu}{\varepsilon}}\mathbf{n}\times(\mathbf{H}_{\mathrm{inc}}\times\mathbf{n}) \text{ on } \Gamma_a. \end{cases} \tag{2}$$

Here $\mathbf{n}$ denotes the unit outward normal to $\partial\Omega$ and $(\mathbf{E}_{\mathrm{inc}}, \mathbf{H}_{\mathrm{inc}})$ is a given incident field. The first boundary condition is called *metallic* (referring to a perfectly conducting surface) while the second condition is called *absorbing* and takes here the form of the Silver–Müller condition which is a first order approximation of the exact absorbing boundary condition. This absorbing condition is applied on Γ_a which represents an artificial truncation of the computational domain. Finally, system (1) is supplemented with initial conditions: $\mathbf{E}_0(\mathbf{x}) = \mathbf{E}(\mathbf{x}, t)$ and $\mathbf{H}_0(\mathbf{x}) = \mathbf{H}(\mathbf{x}, t)$.

3 Discretization in Space

We consider a partition $\mathscr{T}_h$ of Ω into a set of tetrahedra τ_i of size h_i with boundary $\partial\tau_i$ such that $h = \max_{\tau_i \in \mathscr{T}_h} h_i$. For each τ_i, V_i denotes its volume, and ε_i and μ_i are respectively the local electric permittivity and magnetic permeability of the medium, which are assumed constant inside the element τ_i. For two distinct tetrahedra τ_i and τ_k in $\mathscr{T}_h$, the intersection $\tau_i \cap \tau_k$ is a triangle a_{ik} which we will call interface. For a given partition $\mathscr{T}_h$, we seek approximate solutions to (1) in the finite dimensional subspace $V_p(\mathscr{T}_h) = \{\mathbf{v} \in L^2(\Omega)^3 : v_{k|\tau_i} \in \mathbb{P}_p(\tau_i), \text{for } k = 1,3 \text{ and } \forall \tau_i \in \mathscr{T}_h\}$ where $\mathbb{P}_p(\tau_i)$ denotes the space of nodal polynomial functions of degree at most p inside the element τ_i. Following the discontinuous Galerkin approach, the electric and magnetic fields inside each finite element are searched for as linear combinations $(\mathbf{E}_i, \mathbf{H}_i)$ of linearly independent basis vector fields $\boldsymbol{\varphi}_{ij}$, $1 \le j \le d$, where d denotes the local number of degrees of freedom inside τ_i. The discretization in space yields the following system of ODEs:

$$M_i^{\varepsilon}\frac{d\mathbf{E}_i}{dt} = K_i\mathbf{H}_i - \sum_{k\in\mathscr{V}_i} S_{ik}\mathbf{H}_k, \; M_i^{\mu}\frac{d\mathbf{H}_i}{dt} = -K_i\mathbf{E}_i + \sum_{k\in\mathscr{V}_i} S_{ik}\mathbf{E}_k, \tag{3}$$

where the symmetric positive definite mass matrices M_i^{σ} (σ stands for ε or μ), the symmetric stiffness matrix K_i and the symmetric interface matrix S_{ik} (all of size $d \times d$) are given by:

$$(M_i^{\sigma})_{jl} = \sigma_i \int_{\tau_i} {}^t\boldsymbol{\varphi}_{ij} \cdot \boldsymbol{\varphi}_{il}, (S_{ik})_{jl} = \frac{1}{2}\int_{a_{ik}} {}^t\boldsymbol{\varphi}_{ij} \cdot (\boldsymbol{\varphi}_{kl} \times \mathbf{n}_{ik}),$$
$$(K_i)_{jl} = \frac{1}{2}\int_{\tau_i} {}^t\boldsymbol{\varphi}_{ij} \cdot \mathrm{curl}\boldsymbol{\varphi}_{il} + {}^t\boldsymbol{\varphi}_{il} \cdot \mathrm{curl}\boldsymbol{\varphi}_{ij}.$$

4 Time Discretization

The choice of the time discretization method is a crucial step for the global efficiency of the numerical method. Then, a possible alternative is to combine the strengths of explicit (easy to implement, greater accuracy with less computational effort) and implicit schemes (unconditional stability) applying an implicit time integration scheme locally in the refined regions of the mesh while preserving an explicit time scheme in the complementary part, resulting in an hybrid explicit–implicit (or locally implicit) time integration strategy. The set of local system of ordinary differential equations for each τ_i (3) can be formally transformed in a global system. To this end, we suppose that all electric (resp. magnetic) unknowns are gathered in a column vector $\mathbb{E}$ (resp. $\mathbb{H}$) of size $d_g = N_{\mathscr{T}_h} d$ where $N_{\mathscr{T}_h}$ stands for the number of elements in $\mathscr{T}_h$. Then system (3) can be rewritten as (we set $\mathbb{S} = \mathbb{K} - \mathbb{A} - \mathbb{B}$):

$$\mathbb{M}^{\varepsilon} \frac{d\mathbb{E}}{dt} = \mathbb{K}\mathbb{H} - \mathbb{A}\mathbb{H} - \mathbb{B}\mathbb{H} = \mathbb{S}\mathbb{H}, \quad \mathbb{M}^{\mu} \frac{d\mathbb{H}}{dt} = -\mathbb{K}\mathbb{E} + \mathbb{A}\mathbb{E} - \mathbb{B}\mathbb{E} = -{}^t\mathbb{S}\mathbb{E}. \tag{4}$$

where we have the following definitions and properties:

- $\mathbb{M}^{\varepsilon}, \mathbb{M}^{\mu}$ and $\mathbb{K}$ are $d_g \times d_g$ block diagonal matrices with diagonal blocks equal to M_i^{ε}, M_i^{μ} and K_i respectively.
- $\mathbb{A}$ is also a $d_g \times d_g$ block sparse matrix, whose non-zero blocks are equal to S_{ik} when a_{ik} is an internal interface of the mesh.
- $\mathbb{B}$ is a $d_g \times d_g$ block diagonal matrix, whose non-zero blocks are associated to the numerical treatment of the boundary conditions (2).

4.1 Explicit and Implicit Time Schemes

The system (4) can be time integrated using a second-order Leap–Frog scheme as:

$$\mathbb{M}^{\varepsilon} \left(\frac{\mathbb{E}^{n+1} - \mathbb{E}^{n}}{\Delta t} \right) = \mathbb{S}\mathbb{H}^{n+\frac{1}{2}}, \quad \mathbb{M}^{\mu} \left(\frac{\mathbb{H}^{n+\frac{3}{2}} - \mathbb{H}^{n+\frac{1}{2}}}{\Delta t} \right) = -{}^t\mathbb{S}\mathbb{E}^{n+1}. \tag{5}$$

The resulting fully explicit DGTD-$\mathbb{P}_p$ method is analyzed in [5] where it is shown that the method is non-dissipative, conserves a discrete form of the electromagnetic energy and is stable under the CFL-like condition:

$$\Delta t \le \frac{2}{d_2}, \quad \text{with } d_2 = \| (\mathbb{M}^{-\mu})^{\frac{1}{2}} \, {}^t\mathbb{S} \, (\mathbb{M}^{-\varepsilon})^{\frac{1}{2}} \|, \tag{6}$$

where $\|.\|$ denote the canonical norm of a matrix ($\forall X, \|AX\| \le \|A\|\|X\|$), and the matrix $(\mathbb{M}^{-\sigma})^{\frac{1}{2}}$ is the inverse square root of $\mathbb{M}^{\sigma}$. Alternatively, (4) can be time integrated using a second-order Crank–Nicolson scheme as:

$$\begin{cases} \mathbb{M}^{\varepsilon} \left(\dfrac{\mathbb{E}^{n+1} - \mathbb{E}^{n}}{\Delta t} \right) = \mathbb{S} \left(\dfrac{\mathbb{H}^{n} + \mathbb{H}^{n+1}}{2} \right), \\ \mathbb{M}^{\mu} \left(\dfrac{\mathbb{H}^{n+1} - \mathbb{H}^{n}}{\Delta t} \right) = -{}^t\mathbb{S} \left(\dfrac{\mathbb{E}^{n} + \mathbb{E}^{n+1}}{2} \right). \end{cases} \tag{7}$$

Such a fully implicit DGTD-$\mathbb{P}_p$ method is considered in [2] for the solution of the 2D Maxwell equations. In particular, the resulting method is unconditionally stable.

4.2 Hybrid Explicit–Implicit Time Scheme

We consider here a method of this kind that was recently proposed by Piperno in [7]. The set of elements τ_i of the mesh is now assumed to be partitioned into two subsets: one made of the smallest elements and the other one gathering the remaining elements. In the following, these two subsets are respectively referred as $\mathscr{S}_i$ and $\mathscr{S}_e$. In the proposed hybrid time scheme, the small elements are handled using a Crank–Nicolson scheme while all other elements are time advanced using a variant of the classical Leap–Frog scheme known as the Verlet method. Then, starting from the values of the fields at time $t^n = n\Delta t$, the proposed hybrid explicit–implicit time integration scheme consists in three sub-steps:

1. The components of $\mathbb{H}$ and $\mathbb{E}$ associated to the set $\mathscr{S}_e$ are time advanced from t^n to $t^{n+\frac{1}{2}}$ with time step $\Delta t/2$ using a pseudo-forward Euler scheme,
2. The components of $\mathbb{H}$ and $\mathbb{E}$ associated to the set $\mathscr{S}_i$ are time advanced from t^n to t^{n+1} with time step Δt using the Crank-Nicolson scheme,
3. The components of $\mathbb{H}$ and $\mathbb{E}1$ associated to the set $\mathscr{S}_e$ are time advanced from $t^{n+\frac{1}{2}}$ to t^{n+1} with time step $\Delta t/2$ using the reversed pseudo-forward Euler scheme.

In order to further describe this scheme, the problem unknowns are reordered such that sub-vectors with an e subscript (respectively, an i subscript) are associated to the elements of the set $\mathscr{S}_e$ (respectively, the set $\mathscr{S}_i$). Thus, the global system of ordinary differential equations (4) can be split into two systems:

$$\begin{cases} \mathbb{M}_e^{\varepsilon}\dfrac{d\mathbb{E}_e}{dt} = \mathbb{S}_e\mathbb{H}_e - \mathbb{A}_{ei}\mathbb{H}_i, \\ \mathbb{M}_e^{\mu}\dfrac{d\mathbb{H}_e}{dt} = -{}^t\mathbb{S}_e\mathbb{E}_e + \mathbb{A}_{ei}\mathbb{E}_i, \end{cases} \qquad \begin{cases} \mathbb{M}_i^{\varepsilon}\dfrac{d\mathbb{E}_i}{dt} = \mathbb{S}_i\mathbb{H}_i - \mathbb{A}_{ie}\mathbb{H}_e, \\ \mathbb{M}_i^{\mu}\dfrac{d\mathbb{H}_i}{dt} = -{}^t\mathbb{S}_i\mathbb{E}_i + \mathbb{A}_{ie}\mathbb{E}_e. \end{cases} \tag{8}$$

Then, the proposed hybrid explicit–implicit algorithm consists in the following steps:

$$\text{Step 1} \;:\; \begin{cases} \mathbb{M}_e^{\mu}\left(\dfrac{\mathbb{H}_e^{n+\frac{1}{2}} - \mathbb{H}_e^n}{\Delta t/2}\right) = -{}^t\mathbb{S}_e\mathbb{E}_e^n + \mathbb{A}_{ei}\mathbb{E}_i^n, \\ \mathbb{M}_e^{\varepsilon}\left(\dfrac{\mathbb{E}_e^{n+\frac{1}{2}} - \mathbb{E}_e^n}{\Delta t/2}\right) = \mathbb{S}_e\mathbb{H}_e^{n+\frac{1}{2}} - \mathbb{A}_{ei}\mathbb{H}_i^n. \end{cases} \tag{9}$$

$$\text{Step 2} \;:\; \begin{cases} \mathbb{M}_i^{\varepsilon}\left(\dfrac{\mathbb{E}_i^{n+1} - \mathbb{E}_i^n}{\Delta t}\right) = \mathbb{S}_i\left(\dfrac{\mathbb{H}_i^{n+1} + \mathbb{H}_i^n}{2}\right) - \mathbb{A}_{ie}\mathbb{H}_e^{n+\frac{1}{2}}, \\ \mathbb{M}_i^{\mu}\left(\dfrac{\mathbb{H}_i^{n+1} - \mathbb{H}_i^n}{\Delta t}\right) = -{}^t\mathbb{S}_i\left(\dfrac{\mathbb{E}_i^{n+1} + \mathbb{E}_i^n}{2}\right) + \mathbb{A}_{ie}\mathbb{E}_e^{n+\frac{1}{2}}. \end{cases} \tag{10}$$

$$\text{Step 3}\ :\ \begin{cases} \mathbb{M}_e^{\varepsilon}\left(\dfrac{\mathbb{E}_e^{n+1}-\mathbb{E}_e^{n+\frac{1}{2}}}{\Delta t/2}\right) & = \quad \mathbb{S}_e\mathbb{H}_e^{n+\frac{1}{2}} - \mathbb{A}_{ei}\mathbb{H}_i^{n+1}, \\ \mathbb{M}_e^{\mu}\left(\dfrac{\mathbb{H}_e^{n+1}-\mathbb{H}_e^{n+\frac{1}{2}}}{\Delta t/2}\right) & = -{}^t\mathbb{S}_e\mathbb{E}_e^{n+1} + \mathbb{A}_{ei}\mathbb{E}_i^{n+1}. \end{cases} \tag{11}$$

In [7], the author shows that the hybrid explicit–implicit scheme (9)-(11) for time integration of the semi-discrete system (4) associated to the DGTD-$\mathbb{P}_p$ method exactly conserves the following quadratic form of the numerical unknowns $\mathbb{E}_e^n$, $\mathbb{E}_i^n$, $\mathbb{H}_e^n$ and $\mathbb{H}_i^n$:

$$\mathscr{E}^n = \mathscr{E}_e^n + \mathscr{E}_i^n + \mathscr{E}_h^n \quad \text{with} \quad \begin{cases} \mathscr{E}_e^n = {}^t\mathbb{E}_e^n\mathbb{M}_e^{\varepsilon}\mathbb{E}_e^n + {}^t\mathbb{H}_e^{n+\frac{1}{2}}\mathbb{M}_e^{\mu}\mathbb{H}_e^{n-\frac{1}{2}}, \\ \mathscr{E}_i^n = {}^t\mathbb{E}_i^n\mathbb{M}_i^{\varepsilon}\mathbb{E}_i^n + {}^t\mathbb{H}_i^n\mathbb{M}_i^{\mu}\mathbb{H}_i^n, \\ \mathscr{E}_h^n = -\dfrac{\Delta t^2}{4}\,{}^t\mathbb{H}_i^n\,{}^t\mathbb{A}_{ei}(\mathbb{M}_e^{\varepsilon})^{-1}\mathbb{A}_{ei}\mathbb{H}_i^n, \end{cases} \tag{12}$$

as far as $\Gamma_a = \emptyset$. However, the condition under which $\mathscr{E}^n$ is a positive definite quadratic form and thus represents a discrete form of the electromagnetic energy is not given. In the following we state such a condition on the global time step Δt.

Lemma 1. *The discrete electromagnetic energy $\mathscr{E}^n$ given by (12) is a positive definite quadratic form of the numerical unknowns $\mathbb{E}_e^n$, $\mathbb{E}_i^n$, $\mathbb{H}_e^n$ and $\mathbb{H}_i^n$ if:*

$$\Delta t \le \frac{2}{\alpha_e + \max(\beta_{ei}, \gamma_{ei})} \quad \textit{with} \quad \begin{cases} \alpha_e = \| (\mathbb{M}_e^{\varepsilon})^{-\frac{1}{2}}\mathbb{S}_e(\mathbb{M}_e^{\mu})^{-\frac{1}{2}} \|, \\ \beta_{ei} = \| (\mathbb{M}_e^{\varepsilon})^{-\frac{1}{2}}\mathbb{A}_{ei}(\mathbb{M}_i^{\mu})^{-\frac{1}{2}} \|, \\ \gamma_{ei} = \| (\mathbb{M}_e^{\mu})^{-\frac{1}{2}}\mathbb{A}_{ei}(\mathbb{M}_i^{\varepsilon})^{-\frac{1}{2}} \|, \end{cases} \tag{13}$$

where $\| \,.\, \|$ denotes a matrix norm and the matrix $(\mathbb{M}^{\sigma}_{e/i})^{-\frac{1}{2}}$ is the inverse of the square root of the matrix $\mathbb{M}^{\sigma}_{e/i}$ (σ stands for ε or μ).

The proof can be found in [4]. In summary, (13) states that the stability of the hybrid explicit–implicit DGTD-$\mathbb{P}_p$ method is deduced from a criterion which is essentially the one obtained for the fully explicit method here restricted to the subset of explicit elements $\mathscr{S}_e$, augmented by two terms involving elements of the implicit subset $\mathscr{S}_i$ associated to hybrid internal interfaces (i.e., interfaces a_{ik} such that $\tau_i \in \mathscr{S}_e$ and $\tau_k \in \mathscr{S}_i$).

5 Numerical Results

In this section we apply the proposed hybrid explicit–implicit DGTD-$\mathbb{P}_p$ method to the simulation of a 3D problem involving the scattering of a plane wave (F = 1 GHz) by a perfectly conducting sphere with wall thickness $e = 5\ 10^{-3}$ m and radius

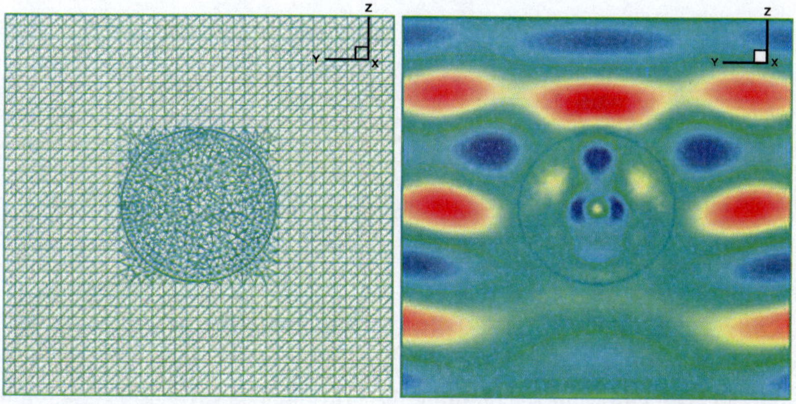

Fig. 1 Scattering of plane wave by a spherical mesh cavity with a hole. Geometry setting and unstructured mesh (*left*), contour lines of E_x for the hybrid explicit–implicit DGTD-$\mathbb{P}_2$ method (*right*)

$R = 0.2$ m with a hole of radius $r = 2.5\ 10^{-2}$ m at one of its pole (see Fig. 1 left for a view of the geometry and the unstructured mesh in a selected plane). The computational domain is artificially bounded by a cubic surface on which the Silver–Müller boundary condition is applied. The underlying tetrahedral mesh consists of 56,482 vertices and 301,116 tetrahedra. The contour lines of E_x for a physical simulation time corresponding to 10 periods of the incident wave are shown on Fig. 1 right. The definition of the subsets $\mathscr{S}_i$ and $\mathscr{S}_e$ relies on the geometric criterion $c_g(\tau_i) = 4 \min_{j \in \mathscr{V}_i} \frac{V_i V_j}{P_i P_j}$. In the present case, the threshold value $2.5\ 10^{-3}$ m has been selected resulting in $|\mathscr{S}_e| = 300{,}526$ and $|\mathscr{S}_i| = 590$ (i.e., only 0.2% of the mesh elements are treated implicitly). The time steps used in the simulations are the following: 0.34 (2.8) picosec for the explicit (hybrid) DGTD-$\mathbb{P}_1$ method and 0.17 (1.4) picosec for the explicit (hybrid) DGTD-$\mathbb{P}_2$ method. Numerical simulations have been conducted on a cluster of Intel Xeon 2.33 GHz based nodes interconnected by a high performance Myrinet network. Each node consists of a dual processor quad core board sharing 16 GB of RAM memory. The parallelization of the hybrid explicit–implicit DGTD-$\mathbb{P}_p$ method relies on a SPMD (Single Program Multiple Data) strategy which combines a partitioning of the tetrahedral mesh with a message passing programming using the MPI interface. Performance results for the simulations based on the DGTD-$\mathbb{P}_1$ and DGTD-$\mathbb{P}_2$ methods are summarized in Table 1 where "RAM (LU)" is the maximum per-processor memory overhead for computing and storing the sparse L and U factors (after an AMD reordering for the minimization of the bandwidth), while "Time (LU)" gives the maximum factors construction time. The direct solver used is MUMPS (see [1]) The results of Table 1 show that the memory overhead associated to the construction and the storage of the L and U factors of the implicit matrix is acceptable while the gain in computing time is roughly equal to 4.4 for both the $\mathbb{P}_1$ and $\mathbb{P}_2$ interpolation methods.

Table 1 Scattering of plane wave by a spherical mesh cavity. Performance results ($N_s = 8$ processing units)

Method	RAM (LU)	Time (LU)	Total time
Explicit DGTD-$\mathbb{P}_1$	–	–	44 mn
Hybrid explicit–implicit DGTD-$\mathbb{P}_1$	2 MB	<1 s	10 mn
Explicit DGTD-$\mathbb{P}_2$	–	–	4 h 24 mn
Hybrid explicit–implicit DGTD-$\mathbb{P}_2$	8 MB	<1 s	56 mn

6 Conclusions

We have presented some preliminary results of the development of a hybrid explicit–implicit DGTD method for overcoming the grid-induced stiffness in time domain electromagnetics. The proposed method allows to reduce notably the overall computing time as compared to a fully explicit method, when a rather small number of the mesh elements are treated implicitly (typically a few percent) which is often the case in practical situations involving locally refined simplicial meshes. Future works will follow several directions: (a) improvement of the temporal accuracy by studying the combination of a high order Leap-Frog scheme with a high order implicit time scheme, (b) design of an auto-adaptive solution strategy for the selection of the reference time step minimizing dispersion error and, (c) treatment of load balancing issues raised by the separation of mesh elements into two subsets in order to obtained a scalable hybrid explicit–implicit DGTD method.

References

1. P.R. Amestoy, I.S. Duff, J.-Y. L'Excellent, Multifrontal parallel distributed symmetric and unsymmetric solvers, Comput. Meth. App. Mech. Engng. 184 (2000) 501–520
2. A. Catella, V. Dolean, S. Lanteri, An implicit discontinuous Galerkin time-domain method for two-dimensional electromagnetic wave propagation, COMPEL, 29 (3) (2010) 602–625
3. D. Degerfeldt, T. Rylander, A brick-tetrahedron finite-element interface with stable hybrid explicit-implicit time-stepping for Maxwell's equations, J. Comput. Phys. 220 (1) (2006) 383–393
4. V. Dolean, H. Fahs, L. Fezoui, S. Lanteri, Locally implicit discontinuous Galerkin method for time domain electromagnetics, J. Comput. Phys. (2009). http://dx.doi.org/10.1016/j.jcp.2009.09.038
5. L. Fezoui, S. Lanteri, S. Lohrengel, S. Piperno, Convergence and stability of a discontinuous Galerkin time-domain method for the 3D heterogeneous Maxwell equations on unstructuredmeshes, ESAIM: M2AN 39 (6) (2005) 1149–1176
6. A. Kanevsky, M. Carpenter, D. Gottlieb, J. Hesthaven, Application of implicit-explicit high order Runge-Kutta methods to discontinuous Galerkin schemes, J. Comput. Phys. 225 (2) (2007) 1753–1781
7. S. Piperno, Symplectic local time stepping in non-dissipative DGTD methods applied to wave propagation problem, ESAIM: M2AN 40 (5) (2006) 815–841
8. T. Rylander, A. Bondeson, Stability of explicit-implicit hybrid time-stepping schemes for Maxwell's equations, J. Comput. Phys. 179 (2) (2002) 426–438

High-Order Quasi-Uniform Approximation on the Sphere Using Fourier-Finite-Elements

T. Dubos

Abstract Solving transport equations on the whole sphere using an explicit time stepping and a Eulerian formulation on a latitude-longitude grid is relatively straightforward but suffers from the pole problem: due to the increased zonal resolution near the pole, numerical stability requires unacceptably small time steps. Commonly used workarounds such as near-pole zonal filters affect the qualitative properties of the numerical method. Rigorous solutions based on spherical harmonics have a high computational cost. The numerical method we propose to avoid this problem is based on a Galerkin formulation in a subspace of a Fourier-finite element spatial discretization, providing quasi-uniform resolution and high-order accuracy. For N^2 degrees of freedom, the computational cost is $O(N^2 \log N)$, intermediate between finite-difference or finite-volume methods and spherical harmonics methods. We present experimental results and standard benchmarks demonstrating the accuracy and stability of the method applied to the rotating shallow-water equations.

1 Introduction

Global weather and climate modelling require the numerical solution of partial differential equations on the whole sphere. A difficult part of this task is the discretization of the dynamical core, dealing with the transport of mass, momentum and various species. Ideally a numerical scheme should be accurate, stable and computationally efficient. In the context of climate studies, a crucial additional requirement is that it be conservative: the discretized system should enforce the exact conservation of a discrete approximation of the total mass, momentum and, if possible, energy and enstrophy. For an in-depth review of the evolution of dynamical cores, the reader is referred to [17]. Although a single optimal scheme has not emerged yet, most dynamical cores in use today use either one of two methods: the finite

T. Dubos
IPSL, Laboratoire de Météorologie Dynamique, École Polytechnique, Palaiseau, France
e-mail: dubos@lmd.polytechnique.fr

J.S. Hesthaven and E.M. Rønquist (eds.), *Spectral and High Order Methods for Partial Differential Equations*, Lecture Notes in Computational Science and Engineering 76,
DOI 10.1007/978-3-642-15337-2_14, © Springer-Verlag Berlin Heidelberg 2011

difference method [1, 2, 13] or the spectral transform method [9, 14], both using a structured latitude-longitude grid.

With explicit time-stepping, the temporal stability is limited by a Courant–Friedrichs–Lewy (CFL) condition. With finite-difference methods in latitude-longitude coordinates, zonal grid intervals near the poles are $\sim a/N^2$ with a the Earth radius, much smaller than the grid interval $\sim a/N$ near the Equator. By representing the solution as a combination of spherical harmonics, the spectral transform method elegantly removes the singularity at the pole introduced by spherical coordinates. This comes at a fairly high computational cost. Furthermore because temporal integration is usually of low order, the overall accuracy is typically no better than third order.

Much current research on numerical schemes for dynamical cores has abandoned the latitude-longitude grid and focuses on the use of quasi-uniform grids with less severe singularities [11, 17]. In [6] instead, a new numerical method formulated on the familiar latitude-longitude grid is designed which is more accurate than finite differences and more efficient than the spectral transform, borrowing from the two approaches to achieve comparable stability and conservation properties. Constructing the method boils down to designing the functional space used for the approximation of the dynamical fields. Adhering to the Galerkin framework then guarantees the conservation of linear and quadratic invariants.

[6] is restricted to non-divergent flows while the minimal testbed for atmospheric applications is the compressible Saint-Venant model. This restriction was motivated by the fact that exact conservation of linear and quadratic integral invariants can be generically obtained within the Galerkin framework. The energy and enstrophy of the Saint-Venant model are not quadratic invariants, and will not be exactly conserved. Nevertheless the method developed in [6] can be readily extended to compressible flows. For this we use here the stream function – velocity potential representation of the velocity field, as in spectral models.

In Sect. 2 the functional space used in [6] is described. Zonal Fourier discretization brings spectral accuracy, zonal invariance and fast transform. Latitudinal finite elements provide adjustable accuracy, exact quadrature and spatial locality. Latitude-dependant zonal truncation controlled by the largest eigenvalue of the Laplacian operator brings quasi-uniform resolution, overcoming the pole problem. In Sect. 3 the rotating shallow-water equations are solved using this functional space. A standard benchmark is implemented and the properties of accuracy and conservation of the method are discussed.

2 Quasi-Uniform Approximation of Scalar Fields by Fourier-Finite Elements

Let (x, y, z) be a set of Cartesian coordinates, and (λ, ϕ, r) the associated longitude-latitude-radius coordinates, i.e., $x = r\cos\lambda\cos\phi$, $y = r\sin\lambda\cos\phi$ and $z = r\sin\phi$. Here $\lambda \in [-\pi, \pi]$ is the longitude and $\phi \in [-\pi/2, \pi/2]$ is the latitude. We note $(\mathbf{e}_\lambda, \mathbf{e}_\phi, \mathbf{e}_r)$ the local orthonormal basis associated to the longitude-latitude-radius coordinates. Smoothness of a scalar function f across the poles $\phi = \pm\pi/2$ imposes that the zonal Fourier modes $\hat{f}_m(\phi)$ decay like $\cos^{|m|}\phi$ as $\phi \to \pm\pi/2$:

$$\hat{f}_m(\phi) = \frac{1}{2\pi}\int f(\lambda,\phi)\, e^{-im\lambda} \mathrm{d}\lambda \sim \cos^{|m|}\phi \qquad \phi \to \pm\pi/2. \tag{1}$$

These conditions hold for $m \le k$ if f is k times continuously differentiable. We consider the space $\mathscr{S}'$ spanned by basis functions F_{ml} where

$$F_{2ml} = B_l(\sin\phi)\, e^{2mi\lambda} \qquad F_{2m+1\,l} = \cos\phi\, B_l(\sin\phi)\, e^{(2m+1)i\lambda}$$

where the B_l are B–splines of degree d relative to nodes $-1 = z_0 < z_1 < \cdots < z_N = 1$ subdividing the latitudinal interval in N elements. The nodes z_k can be chosen arbitrarily. We choose equally spaced latitudes: $z_k = \cos(\pi k/N)$. We have not tried to improve the accuracy or stability of the numerical method by adjusting the nodes z_k. Notice that other families of piecewise polynomials can replace B–splines. The multiplicative factor $\cos\phi$ in front of odd zonal modes is consistent with the decay rule (1) and ensures that exact quadrature rules exist for products of functions belonging to $\mathscr{S}'$. Indeed the integral

$$\langle f \rangle = \int f \mathrm{d}\lambda \cos\phi \mathrm{d}\phi = 2\pi \int \hat{f}_0 \mathrm{d}z \tag{2}$$

can be computed in two steps from the values of f on a regular latitude-longitude grid, by applying first equal-weight quadrature in the zonal direction then element-wise Gaussian quadrature in the latitudinal direction [6].

When integrating a transport equation with an explicit temporal scheme, the time the step τ must satisfy the CFL criterion

$$\tau U \le c\delta. \tag{3}$$

where c is a constant depending on the details of the temporal scheme, U is the maximum velocity, and

$$\delta^{-2} = \|S\|_{\mathscr{S}'} = \sup_{g\in\mathscr{S}'} \frac{S(g,g)}{\|g\|^2} \qquad \text{where } S(f,g) = \langle \nabla f^* \cdot \nabla g \rangle. \tag{4}$$

The effective grid scale δ entering (3) and controlling the time step is therefore defined from the largest eigenvalue of the Laplacian operator. Within the functional space $\mathscr{S}'$, δ is controlled by the near-pole zonal resolution, much finer than near the Equator. This fine resolution is wasted since the discretization error made near the equator eventually propagates to the whole sphere under the effect of advection. This excess resolution is now removed by restricting the Galerkin formulation to a subspace $\mathscr{S}''$ of $\mathscr{S}'$.

For this we discard for each zonal mode m a number $L(m)$ of near-pole degrees of freedom and define $\mathscr{S}''$ as the space spanned by the $(F_{ml})_{(m,l)\in K}$ with $K = \{-M \le m \le M$ and $L(m) \le l < N + d - L(m)$. Since the Laplacian does not couple the different zonal modes,

$$\|S\|_{\mathscr{S}''} = \max_{0\le m\le M} \|S\|_{\mathscr{S}_m^{L(m)}} \tag{5}$$

where $\mathscr{S}_m^L$ the space spanned by the basis functions (F_{ml}) with $L \le l < N+d-L$. Since the basis functions for $-1 \le m \le 1$ have the correct near-pole decay, we decide that $L(m) = 0$ for $-1 \le m \le 1$. We then define $L(m)$ for $l > 1$ as the smallest number L such that $\|S\|_{\mathscr{S}_m^L} \le \delta^{-2}$ where

$$\delta^{-2} = \max\left(\|S\|_{\mathscr{S}_0^0}, \|S\|_{\mathscr{S}_1^0}\right) \tag{6}$$

and $\|S\|_{\mathscr{S}_m^L}$ is defined as in (4). The resolution δ is entirely determined by the latitudinal resolution, e.g., by the number N of latitudinal intervals and by the positions of the nodes z_k. Notice that $\|S\|_{\mathscr{S}_m^L} \ge m^2$, which also bounds the number M of zonal modes for a given N. We now set $N = M$. By analogy with spectral truncation, a specific choice of M is called in the sequel the "truncation M," for instance T42 in the case $M = N = 42$. Calculations show that δ is roughly 1/3 the zonal grid size at the Equator [6].

We now show that, despite near-pole zonal truncation, the functional space $\mathscr{S}''$ provides an approximation of the same order as $\mathscr{S}'$, i.e., of order $d + 1$. For this we pick a point (λ_0, ϕ_0) on the sphere and consider two scalar functions, a Gaussian:

$$f(\lambda,\phi) = \exp\frac{\cos\alpha - 1}{\alpha_c^2} \tag{7}$$

and a cosine-bell:

$$f(\lambda,\phi) = 1 + \cos\left(\pi \min\left(1, \alpha/\alpha_c\right)\right) \tag{8}$$

where α is the geodesic angle between (λ,ϕ) and (λ_0,ϕ_0) and $\alpha_c = \pi/8$. We compute an approximation $\tilde{f} \in \mathscr{S}''$ from the value of f at the quadrature points then the maximal pointwise error, as well as the largest pointwise error in the approximation of the gradient:

$$\varepsilon(N,d) = \max\left|f(\lambda_i,\phi_j) - \tilde{f}(\lambda_i,\phi_j)\right|, \tag{9}$$

$$\varepsilon_\nabla(N,d) = \max\left\|\nabla f(\lambda_i,\phi_j) - \nabla\tilde{f}(\lambda_i,\phi_j)\right\|. \tag{10}$$

We repeat the process for 100 random values of (λ_0,ϕ_0) and retain the largest errors. Figure 1 displays $\varepsilon(N,d)$ and $\varepsilon_\nabla(N,d)$ as a function of the zonal grid size at the Equator $360/N$, for finite elements of degree $d = 1, 2, 3$ (circles, crosses,

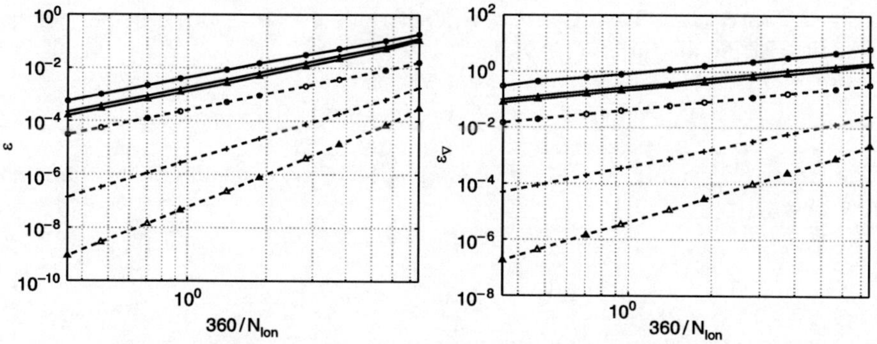

Fig. 1 Discretization errors $\varepsilon(N,d) = \max\left|f(\lambda_i,\phi_j) - \tilde{f}(\lambda_i,\phi_j)\right|$ (*left*) and $\varepsilon_\nabla(N,d) = \max\left\|\nabla f(\lambda_i,\phi_j) - \nabla\tilde{f}(\lambda_i,\phi_j)\right\|$ (*right*) as a function of the zonal grid size $360/N_{lon}$ (*degrees*) for finite elements of degree $d = 1, 2, 3$ (*circles, crosses, triangles*), for the Gaussian (*dashed line*) and for the cosine bell (*solid line*)

triangles), for the Gaussian (solid line) and for the cosine bell (dashed line). For the cosine bell, the error scales like $\varepsilon \sim N^{-2}$, indicating second-order accuracy. This is consistent with the fact that the cosine bell is only continuously differentiable with a bounded second derivative. For the Gaussian however, the error scales like $\varepsilon \sim N^{-(d+1)}$ demonstrating that the formal order of accuracy is indeed achieved in practice. As expected for a finite-element method, the order of the approximation of the gradients is one less than the order of the pointwise approximation.

3 Rotating Shallow-Water Equations

The rotating shallow-water equations describe the quasi-horizontal motion of a thin fluid layer over a spherical Earth of radius a rotating at an angular rate Ω. The flow state is described by the velocity field $\mathbf{u}$ and the geopotential p, equal to the hydrostatic pressure divided by the fluid density, and to the fluid layer thickness multiplied by the acceleration of gravity. Motion results from the Coriolis force and hydrostatic pressure. Among several possible equivalent formulations, we use the so-called vector-invariant form:

$$\partial_t p + \text{div}\, p\mathbf{u} = 0 \tag{11}$$

$$\partial_t \mathbf{u} + (f + \zeta)\mathbf{e}_r \times \mathbf{u} + \nabla H = 0 \tag{12}$$

$$\text{where } H = p + \frac{\mathbf{u}\cdot\mathbf{u}}{2}, \tag{13}$$

$f = 2\Omega \sin\phi$ is the local Coriolis parameter and $\zeta = \mathbf{e}_r \cdot \text{curl}\,\mathbf{u}$ is the relative vorticity. Taking the curl of (12) yields:

$$\partial_t (f + \zeta) + \text{div}(f + \zeta)\mathbf{u} = 0. \tag{14}$$

Combining (11) and (12) yields the conservation of axial angular momentum and total energy:

$$\partial_t L_z = 0 \qquad L_z = \mathbf{e}_z \cdot \langle \mathbf{x} \times p\mathbf{u} \rangle \tag{15}$$

$$\partial_t E = 0 \qquad E = \frac{1}{2} \langle p^2 + p\mathbf{u} \cdot \mathbf{u} \rangle . \tag{16}$$

Combining (11) and (14) yields the conservation of total enstrophy:

$$\partial_t Z = 0 \qquad Z = \frac{1}{2} \left\langle \frac{(f + \zeta)^2}{p} \right\rangle . \tag{17}$$

We discretize (11)–(13) by the Galerkin method. The velocity field derives from a stream function ψ and a potential π:

$$\mathbf{u} = \nabla\pi + \mathbf{e}_r \times \nabla\psi. \tag{18}$$

The unknown, time-dependent scalar fields π, ψ, p, H belong to the finite-dimensional space $\mathcal{S}''$. Equation (11)–(13) is tested against test functions $\hat{H}, \hat{p}, \hat{\mathbf{u}} = \nabla\hat{\pi} + \mathbf{e}_r \times \nabla\hat{\psi}$:

$$\forall \hat{H} \qquad \langle \hat{H} H \rangle = \left\langle \hat{H} \left(p + \frac{\mathbf{u} \cdot \mathbf{u}}{2} \right) \right\rangle \tag{19}$$

$$\forall \hat{p} \qquad \langle \hat{p} \partial_t p \rangle - \langle \nabla\hat{p} \cdot p\mathbf{u} \rangle = 0 \tag{20}$$

$$\forall \hat{\mathbf{u}} \qquad \langle \hat{\mathbf{u}} \partial_t \mathbf{u} \rangle + \langle \hat{\mathbf{u}} (f + \zeta) \times \mathbf{u} \rangle + \langle \hat{\mathbf{u}} \cdot \nabla H \rangle = 0. \tag{21}$$

We run Williamson's Rossby–Haurwitz test case 3.6 [16] at coarse resolutions T21 and T42. Let the Courant number be $C = \tau\sqrt{gh_{max}}/\delta$ where $g = 9.80616$ is the acceleration of gravity and $h_{max} = 10{,}350$ m is the maximum initial fluid layer thickness. We find that a leap-frog temporal scheme is stable up to $C \simeq 0.98$, corresponding at resolution $T42$ to a time step of 5 min. The results shown below are obtained with a low-storage, three-step, third-order Runge-Kutta temporal scheme [18] which was found stable up to $C \simeq 1.7$, corresponding at resolution $T42$ to a time step of 8 min.

We display in Fig. 2 the relative drift with time of the integral invariants. Mass is conserved to machine accuracy and not plotted. Energy and angular momentum are very well conserved (about 10^{-8} relative drift at T42). The relative drift in enstrophy is larger (about 10^{-5} at T42) but still small. Since enstrophy involves vorticity, a gradient quantity, this is consistent with the accuracy results shown in Fig. 1.

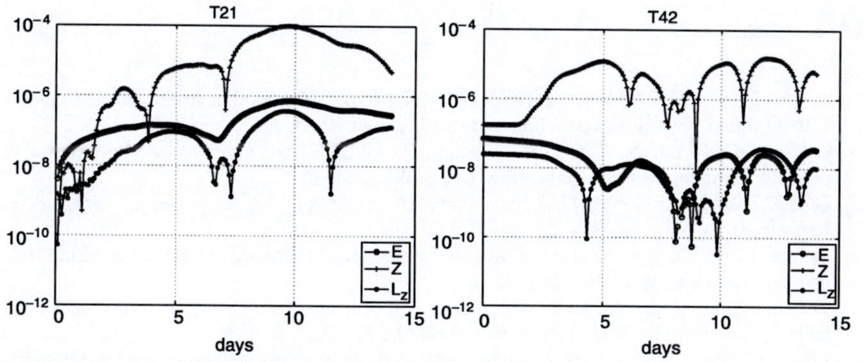

Fig. 2 Relative drift of the integral invariants during the integration of Williamson Rossby–Haurwitz' wave test case; at resolutions T21 and T42

4 Discussion

We have designed a conservative, accurate and stable method for the solution of scalar PDEs on the whole sphere. Attractive features of the method are its simplicity associated to the latitude-longitude grid and its efficiency, intermediate between finite difference and spectral methods. Conservativity of the method relies on the Galerkin framework and exact quadrature of nonlinear terms. The pole problem is overcome by varying the zonal resolution near the poles. [6] discusses relationships with the spectral transform method, other finite-element methods [5], the spectral element method [3, 15], zonal filters [4, 10] and finite-volume methods [7, 8, 12].

We have used the functional spaces developed in [6] to solve the Saint-Venant equations in vorticity-streamfunction formulation. The energy and enstrophy of the Saint-Venant model are not quadratic invariants, and are not exactly conserved even with exact quadrature. Nevertheless good to very good conservation of the integral invariants is found in the widely accepted Rossby–Haurwitz wave test case.

An alternative to using the vorticity-streamfunction formulation would be to design functional spaces to represent the zonal and latitudinal wind components, with their specific near-pole behavior. This may be slightly more economical since the required smoothness, hence polynomial degree, is less. More importantly, the next step towards a full-blown dynamical core is to implement a multi-layer shallow-water model and submit it to baroclinic benchmarks. Work is under way towards this goal.

Acknowledgements The author would like to thank Catherine Mavriplis for supporting the presentation of this work to ICOSAHOM 2009. This work was initiated during a research visit to the Applied Physics Laboratory, University of Washington, Seattle (USA) supported by the Dáélégation Générale à l'Armement, grant ERE 0760027.

References

1. Arakawa, A., 1966: Computational design for long-term numerical integration of the equations of fluid motion: Two-dimensional incompressible flow. Part I. *J. Comput. Phys.*, **1**(1), 119–143
2. Arakawa, A. and Lamb, V. R., 1981: A potential enstrophy and energy conserving scheme for the shallow water equations. *Mon. Wea. Rev.*, **109**(1), 18–36
3. Baer, F., Wang, H., Tribbia, J. J., and Fournier, A., 2006: Climate modeling with spectral elements. *Mon. Wea. Rev.*, **134**(12), 3610–3624
4. Cheong, H. B., 2000: Application of double fourier series to the shallow-water equations on a sphere. *J. Comput. Phys.*, **165**, 261–287
5. Côté, J. and Staniforth, A., 1990: An accurate and efficient finite-element global model of the shallow-water equations. *Mon. Wea. Rev.*, **118**(12), 2707–2717
6. Dubos, T., 2009: A conservative fourier-finite-element method for solving partial differential equations on the whole sphere. *Quat. J. Roy. Met. Soc.*, **135**(644), 1877–1889
7. Hourdin, F. and Armengaud A., 1999: The use of finite-volume methods for atmospheric advection of trace species. Part I: Test of various formulations in a general circulation model. *Mon. Wea. Rev.*, **127**(5), 822–837
8. Lin, S.-J. and Rood, R. B., 1996: Multidimensional flux-form semi-lagrangian transport schemes. *Mon. Wea. Rev.*, **124**(9), 2046–2070
9. Orszag, S. A., 1970: Transform method for the calculation of vector-coupled sums: Application to the spectral form of the vorticity equation. *J. Atmos. Sci.*, **27**, 890–895
10. Purser, R. J., 1988: Degradation of numerical differencing caused by Fourier filtering at high-latitudes. *Mon. Wea. Rev.*, **116**(5), 1057–1066
11. Rančić, M., Zhang, H., and Savic-Jovcic, V., 2008: Nonlinear advection schemes on the octagonal grid. *Mon. Wea. Rev.*, **136**(12), 4668–4686
12. Rasch, P. J. and Williamson, D. L., 1990: Computational aspects of moisture transport in global models of the atmosphere. *Quat. J. Roy. Met. Soc.*, **116**(495), 1071–1090
13. Sadourny, R., 1975: Compressible model flows on the sphere. *J. Atmos. Sci.*, **32**(11), 2103–2110
14. Swarztrauber, P. N., 1996: Spectral transform methods for solving the shallow-water equations on the sphere. *Mon. Wea. Rev.*, **124**, 730–744
15. Taylor, M., 1997: The spectral element method for the shallow water equations on the sphere. *J. Comput. Phys.*, **130**(1), 92–108
16. Williamson, D., Drake, J., Hack, J., Jakob, R., and Swarztrauber, P., 1992: A standard test set for numerical approximations to the shallow water equations in spherical geometry. *J. Comput. Phys.*, **102**(1), 211–224
17. Williamson, D. L., 2007: The evolution of dynamical cores for global atmospheric models. *J. Meteor. Soc. Jpn.*, **85B**, 241–269
18. Yoh, J. J. and Zhong, X. L., 2004: New hybrid Runge-Kutta methods for unsteady reactive flow simulation. *AIAA J.*, **42**(8), 1593–1600

An *hp* Certified Reduced Basis Method for Parametrized Parabolic Partial Differential Equations

Jens L. Eftang, Anthony T. Patera, and Einar M. Rønquist

Abstract We extend previous work on a parameter multi-element *hp* certified reduced basis method for elliptic equations to the case of parabolic equations. A POD (in time)/Greedy (in parameter) sampling procedure is invoked both in the partitioning of the parameter domain (h-refinement) and in the construction of individual reduced basis approximation spaces for each parameter subdomain (p-refinement). The critical new issue is proper balance between additional POD modes and additional parameter values in the initial subdivision process. We present numerical results to compare the computational cost of the new approach to the standard (p-type) reduced basis method.

1 Introduction

The reduced basis (RB) method is a model-order reduction framework for rapid evaluation of functional outputs – such as surface temperatures or fluxes – for partial differential equations which depend on an input parameter vector – such as geometric factors or material properties. Given *any* parameter vector from a predefined parameter domain, the field variable is approximated as a Galerkin-optimal linear combination of accurately pre-computed "truth" finite element (FE) snapshots of the solution at judiciously selected parameter values [2,6]; assuming that the field depends smoothly on the parameters, a RB approximation can be obtained with very few snapshots. Moreover, rigorous a posteriori upper bounds for the error in the RB approximation with respect to the truth discretization can be developed [4,9].

J.L. Eftang (✉) and E.M. Rønquist
Department of Mathematical Sciences, Norwegian University of Science and Technology, 7049 Trondheim, Norway
e-mail: eftang@math.ntnu.no, ronquist@math.ntnu.no

A.T. Patera
Department of Mechanical Engineering, Massachusetts Institute of Technology, Cambridge, MA 02139, USA
e-mail: patera@mit.edu

J.S. Hesthaven and E.M. Rønquist (eds.), *Spectral and High Order Methods for Partial Differential Equations*, Lecture Notes in Computational Science and Engineering 76, DOI 10.1007/978-3-642-15337-2_15,

The RB equation formation, solution, *and* error estimation can be made very efficient in the case of (perhaps approximate) "affine" parameter dependence through an offline-online procedure [8, 9]; the method is computationally attractive in two important engineering contexts – "real-time" and "many-query".

For many problems, the field variable may be quite different in different regions of the parameter domain, and hence a snapshot from one region may be of little value in approximating the solution in another region: the RB space is thus in some sense too large. In [3], an *hp* reduced basis method is introduced for linear elliptic equations: we adaptively subdivide the original parameter domain into smaller regions; we then build individual RB approximation spaces spanned by snapshots restricted to parameter vectors within each parameter subdomain. The RB approximation associated with any new parameter vector is then constructed as a linear (Galerkin) combination of snapshots from the parameter subdomain in which the new parameter vector resides. We thus expect the dimension of the (local) approximation space, and thus the online computational cost, to be very low: every basis function contributes significantly to the RB approximation. An alternative parameter-element reduced-order "interpolation" approach is introduced in [1].

In this paper, we extend the work in [3] to linear parabolic equations through a POD (in time)/Greedy (in parameter) sampling approach [5, 7]. This procedure determines the partition of the parameter domain *and* the construction of the individual RB approximation spaces for each subdomain. The elliptic machinery from [3] readily extends to the parabolic case since we only subdivide the parameter (and not the temporal) domain. The critical new issue is proper balance between additional POD modes and additional parameter values in the initial subdivision process.

Let $\Omega \subset \mathbf{R}^2$, define $L^2(\Omega) = \{v : \int_\Omega v^2 \, d\Omega < \infty\}$, $H^1(\Omega) = \{v : |\nabla v| \in L^2(\Omega)\}$, $H_0^1(\Omega) = \{v \in H^1(\Omega) : v|_{\partial\Omega} = 0\}$, and introduce $X^e \equiv X^e(\Omega)$ such that $H_0^1(\Omega) \subseteq X^e(\Omega) \subset H^1(\Omega)$. The admissible parameter domain is $\mathcal{D} \subset \mathbf{R}^P$. For each $\mu \in \mathcal{D}$, $a(\cdot, \cdot; \mu)$ is a coercive and continuous bilinear form, $b(\cdot, \cdot; \mu)$ is an $L^2(\Omega)$ inner-product, and $f(\cdot; \mu)$ is a linear and bounded functional. We assume that a, b and f admit affine expansions in at most Q terms in the sense that (for example) $a(\cdot, \cdot; \mu) = \sum_{q=1}^{Q_a} \Theta^q(\mu) a^q(\cdot, \cdot)$, where the Θ^q are μ-dependent functions and the $a^q(\cdot, \cdot)$ are μ-independent bilinear forms, $1 \le q \le Q_a (\le Q)$; similar expansions in $Q_b \le Q$ and $Q_f \le Q$ terms apply for b and f, respectively. Let $\bar{\mu} \in \mathcal{D}$ be a fixed "reference parameter", and denote the symmetric part of a by a_s; we then define the X-inner-product and X-norm by $a_s(\cdot, \cdot; \bar{\mu})$ and $\|\cdot\|_X = (a_s(\cdot, \cdot; \bar{\mu}))^{1/2}$, respectively.

We shall consider problems already discretized in time with the Euler Backward (EB) method. Let $[0, T]$ be the time interval and introduce $K + 1$ discrete time-values $t^k = k\Delta t$, $0 \le k \le K$, where $\Delta t = T/K$ is the step-size. Our "exact" (hence ^{e}) problem then reads: Given any $\mu \in \mathcal{D}$, find $u^e(t^k, \mu) \in X^e$, $1 \le k \le K$, such that

$$\frac{1}{\Delta t} b\big(u^e(t^k; \mu) - u^e(t^{k-1}; \mu), v; \mu\big) + a\big(u^e(t^k; \mu), v; \mu\big) = f(v; \mu), \quad \forall v \in X^e; \tag{1}$$

we assume zero initial conditions, $u^e(t^0;\mu) = 0$. The output of interest can now be evaluated as a functional of the field variable; in this paper however, for simplicity of exposition, we consider no particular outputs of interest. Note that since our problem is linear time-invariant (LTI), we may also readily treat time-dependent (offline unknown) control functions through an impulse approach [4].

The RB approximation will be built upon truth FE approximations to the "exact" solution; let $X \equiv X^{\mathcal{N}}(\Omega) \subset X^e(\Omega)$ denote a FE space of dimension $\mathcal{N}$. We shall assume that X is rich enough that the error between the truth and "exact" solutions is in practice negligible. The truth discretization then reads: Given any $\mu \in \mathcal{D}$, find $u^k(\mu) \equiv u(t^k;\mu) \in X$, $1 \le k \le K$, such that

$$\frac{1}{\Delta t} b(u^k(\mu) - u^{k-1}(\mu), v; \mu) + a(u^k(\mu), v; \mu) = f(v;\mu), \quad \forall v \in X; \qquad (2)$$

for initial condition $u^0(\mu) = 0$.

In Sect. 2, we formulate the *hp* RB method for parabolic problems, review the POD/greedy sampling procedure from [5], and discuss the new parameter domain partitioning approach. In Sect. 3, we present numerical results and discuss the computational cost of the new approach relative to the standard method.

2 The *hp* Reduced Basis Method

Reduced Basis Approximation Assume that $\mathcal{D}$ is divided into M parameter subdomains $\mathcal{V}_m \subset \mathcal{D}$, $1 \le m \le M$. The partitioning procedure is briefly reviewed below; see [3] for further details. Each subdomain has an associated set of nested RB approximation spaces $X_{N,m} \subset X$, $1 \le N \le N_{\max,m}$, (where $\dim(X_{N,m}) = N$) constructed by the POD/Greedy sampling procedure. The parameter domain partitioning, the POD/Greedy sampling, and the computation of the truth snapshots are all effected in an offline computational stage; this stage may be rather expensive in terms of computational cost, but is carried out only once as a pre-processing step.

Given any new $\mu \in \mathcal{D}$ in the online stage, the algorithm first determines which subdomain $\mathcal{V}_{m^*} \subset \mathcal{D}$ contains μ, and then selects the associated approximation space X_{N,m^*} from a database of offline-constructed spaces. Once m^* ($1 \le m^* \le M$) is determined, the RB approximation reads: Given any N and any $\mu \in \mathcal{D}$, find $u_N^k(\mu) \equiv u^k_{\hat{N},m^*}(\mu) \in X_{\hat{N},m^*}$, $1 \le k \le K$, such that

$$\frac{1}{\Delta t} b(u_N^k(\mu) - u_N^{k-1}(\mu), v; \mu) + a(u_N^k(\mu), v; \mu) = f(v;\mu), \quad \forall v \in X_{\hat{N},m^*}, \qquad (3)$$

subject to $u_N^0(\mu) = 0$; here $\hat{N} = \min\{N, N_{\max,m^*}\}$. The offline-online decoupling and associated computational procedures are explained in detail in [8, 9]. In particular, the online computational cost and storage requirements are independent of $\mathcal{N}$ – the dimension of the truth FE space – thanks to our "affine" assumption on the parameter dependence.

A Posteriori Error Estimation For each $\mu \in \mathcal{D}$, denote by $\alpha_{\mathrm{LB}}(\mu) < \alpha(\mu) = \inf_{v \in X} a(v, v; \mu)/\|v\|_X^2$ a lower bound for the coercivity constant of $a(\cdot, \cdot; \mu)$. We then define the "energy norm" for $w^k \in X$, $1 \le k \le K$,

$$|||w^k||| = \left(b(w^k, w^k; \mu) + \Delta t \sum_{k'=1}^{k} a_s(w^{k'}, w^{k'}; \mu) \right)^{1/2}, \quad 1 \le k \le K. \tag{4}$$

Given an RB approximation for $\mu \in \mathcal{V}_m \subset \mathcal{D}$, $u_N^k(\mu)$, $1 \le k \le K$, we write the residual as $r_N^k(v; \mu) = f(v; \mu) - b(u_N^k(\mu) - u_N^{k-1}(\mu), v; \mu)/\Delta t - a(u_N^k(\mu), v; \mu)$ and denote by $\epsilon_N^k(\mu) = \sup_{v \in X} r_N^k(v; \mu)/\|v\|_X$ the residual dual norm. The energy norm of the RB error $e_N^k(\mu) = u^k(\mu) - u_N^k(\mu)$, $1 \le k \le K$, is bounded by

$$\Delta_N^k(\mu) \equiv \left(\Delta t \sum_{k'=1}^{k} (\epsilon_N^{k'}(\mu))^2 \Big/ \alpha_{\mathrm{LB}}(\mu) \right)^{1/2} \ge |||e_N^k(\mu)|||. \tag{5}$$

For a proof of (5) and the associated (offline-online) computational procedures for the dual norm of the residuals and the coercivity lower bound, see [4, 8, 9].

POD/Greedy Sampling In order to determine the parameter domain partitioning (h-refinement) and, associated with each subdomain, individual RB approximation spaces (p-refinement), we invoke the POD/Greedy sampling procedure introduced in [5] (see also [7]). We first describe in this section the standard p-type POD/Greedy procedure applied to the entire parameter domain $\mathcal{D}$. We then consider in the next section the application of the POD/Greedy procedure in the hp context.

Let the function POD($\{w^k \in X, 1 \le k \le K\}$, R) return $R \le K$ X-orthonormal functions $\{\chi^i \in X, 1 \le i \le R\}$ such that $\mathcal{P}_R = \mathrm{span}\{\chi^i, 1 \le i \le R\}$ satisfies the optimality property

$$\mathcal{P}_R = \arg \inf_{Y \subset \mathrm{span}\{w^k, 1 \le k \le K\}} \left(\frac{1}{K} \sum_{k=1}^{K} \inf_{w \in Y} \|w^k - w\|_X^2 \right)^{1/2}. \tag{6}$$

To obtain the set $\{\chi^i, 1 \le i \le R\}$ – the first R *POD modes* of $\mathrm{span}\{w^1, \ldots, w^K\}$ – we first solve the eigenvalue problem $C\psi^i = \lambda^i \psi^i$ for ($\psi^i \in \mathbf{R}^K, \lambda^i \in \mathbf{R}$) associated with the R largest eigenvalues of C, where $C_{ij} = (w^i, w^j)_X/K$, $1 \le i, j \le K$; we then compute $\chi^i = \sum_{k=1}^{K} \psi_k^i w^k$ for $1 \le i \le R$.

Let $\Xi \subset \mathcal{D}$ be a (typically very rich) finite training sample over $\mathcal{D}$. We initialize the POD/Greedy(R, L) algorithm by choosing (randomly, say) $\mu^* \in \mathcal{D}$ and setting $N = 0$, $X_N = \{0\}$. Then, while $N < L$, we first compute the projection error

$e^k_{N,\mathrm{proj}}(\mu^*) = u^k(\mu^*) - \mathrm{proj}_{X_N}(u^k(\mu^*))$, $1 \le k \le K$, where $\mathrm{proj}_{X_N}(w)$ denotes the X-orthogonal projection of $w \in X$ onto X_N. Next, we define R (nested) RB spaces as $X_{N+i} \equiv X_N \oplus \mathrm{span}\{\mathrm{POD}(\{e^k_{N,\mathrm{proj}}(\mu^*), 1 \le k \le K\}, i)\}$, $1 \le i \le R$, and set $N \leftarrow N + R$. Finally, the next parameter vector is chosen greedily over Ξ based on the a posteriori error estimator at the final time: $\mu^* \leftarrow \arg\max_{\mu \in \Xi} \Delta^K_N(\mu)$.

Parameter Domain Partitioning Since we subdivide only the parameter (and not the temporal) domain, the *hp* reduced basis framework described in detail for elliptic problems in [3] also applies to the parabolic context of this paper. The "parabolic" algorithm developed here differs from the "elliptic" algorithm of [3] in the definition of the error bound and in particular in the choice of the parameter sampling procedure: care must be taken to properly balance additional POD modes and additional parameter values in partitioning the parameter domain.

The parameter domain partition is determined in the offline stage. We start from the original domain $\mathcal{D}$, choose $\mu^* = \mu^*_0 \in \mathcal{D}$, and perform the POD/Greedy(R, L) algorithm over $\mathcal{D}$ with $R = R_1 \ge 1$ and $L = R_1$ (such that we perform only a single POD). We denote the resulting (nested) approximation spaces as $X_{N,1}$, $1 \le N \le R_1$, and the next parameter vector as μ^*_1. Based on proximity (e.g. Euclidian distance) to the two parameter *anchor points* μ^*_0 and μ^*_1, we can now divide $\mathcal{D}$ into two new subdomains $\mathcal{V}_0 \subset \mathcal{D}$, $\mathcal{V}_1 \subset \mathcal{D}$, respectively. We now repeat the procedure within each subdomain for $\mu^* = \mu^*_0$ and $\mu^* = \mu^*_1$ as the initial parameter vectors, respectively; note that one of the two "child" subdomains inherits the parameter anchor point, and thus the associated approximation space, from its "parent." In Fig. 1, we illustrate the partitioning algorithm with two levels of refinement; we proceed recursively until the error bound at the final time is less than $\epsilon^1_{\mathrm{tol}}$ (over train samples) over each subdomain.

We must comment on the tuning parameter R_1, which is crucial to the convergence of the h-refinement stage of the algorithm. In particular, R_1 must be chosen large enough such that the RB error bound associated with the (R_1-dimensional) RB approximation at the final time is less than $\epsilon^1_{\mathrm{tol}}$ in a neighborhood of μ^*. Otherwise, the procedure would not converge since the tolerance would not be reached. Note it is not sufficient that the tolerance is satisfied only at μ^*, since then the tolerance might not be satisfied at any point arbitrarily close to μ^*, and the partitioning algorithm might yield arbitrarily small subdomains.

In particular, we shall require that the error bound associated with the RB approximation of $u^K(\mu^*)$ based on R_1 POD modes is less than $\epsilon^1_{\mathrm{tol}}/\rho_1$ with $\rho_1 > 1$. This requirement ensures that the RB error bound is smaller than $\epsilon^1_{\mathrm{tol}}$ in a neighborhood of μ^*; the refinement algorithm will then converge since eventually a finite subdomain containing μ^* will be included in this neighborhood. Note that choosing $\rho_1 > 1$ too

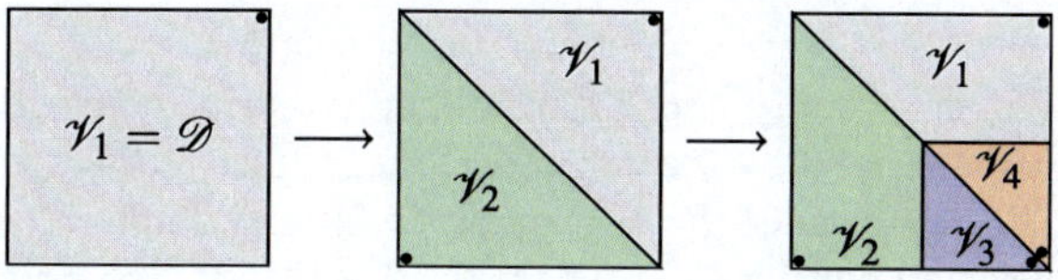

Fig. 1 Hierarchical partitioning of the parameter domain based on proximity to greedily chosen parameter anchor points

small would lead to a large number of subdomains, while large ρ_1 will require more POD modes to be included in the RB space; in the limit $\rho_1 \to \infty$, we would need to include all K POD modes in the RB space in order to achieve a zero RB error (bound) at μ^* at the final time – as in the elliptic case, there would thus perforce be a neighborhood around μ^* where the RB error bound would be very small and in particular less than ϵ^1_{tol}.

It remains to determine R_1 automatically. Towards that end, we note that the POD norm defined in (6) is similar to the energy norm defined in (4); since $e^k_{0,\text{proj}}(\mu^*) = u^k(\mu^*)$, the POD error is thus closely related to the associated RB error at μ^*. As an initial guess, we thus choose R_1 such that the POD error at μ^* – realized as the square root of the sum of the eigenvalues λ^i, $i = R_1 + 1, \dots, K$ – is less than $\epsilon^1_{\text{tol}}/(\rho_1\rho_2)$, where we choose $\rho_2 \geq \alpha_{\text{LB}}(\mu^*)^{1/2}$ because the POD error is a lower bound for the RB error (and thus RB error bound) divided by $\alpha_{\text{LB}}(\mu^*)^{1/2}$. Next, we compute the RB error bound associated with the RB approximation of $u^k(\mu^*)$, $1 \leq k \leq K$, based on R_1 POD modes: if the error bound is smaller than $\epsilon^1_{\text{tol}}/\rho_1$, we conclude that R_1 is sufficiently large; if not, we successively set $R_1 \leftarrow R_1 + 1$, increase the number of POD modes, and compute a new RB error bound – until the tolerance is satisfied.

This h-refinement results in a total of M subdomains $\mathscr{V}_m \subset \mathscr{D}$, $1 \leq m \leq M$. The next step is p-refinement: we expand the approximation spaces associated with each subdomain $\mathscr{V}_m$, $m = 1, \dots, M$, by application of the POD/Greedy(R, L) sampling procedure (but not initialized; hence $N = R_1$) for $R = R_2$ and $L > R_1$ "specified"; in actual practice, we terminate the POD/Greedy in subdomain m for $L \equiv N_{\max,m}(\epsilon^2_{\text{tol}})$ such that the error bound is less than a second tolerance $\epsilon^2_{\text{tol}} < \epsilon^1_{\text{tol}}$ (over the training sample) over the subdomain – the final approximation spaces $X_{N,m}$, $1 \leq N \leq N_{\max,m}$, $1 \leq m \leq M$, will thus in general have different dimensions. We typically choose $R_2 = 1$; note that $R_2 > 1$ will lead to improved offline performance but worse online performance.

We now turn to the online stage: for every new $\mu \in \mathscr{D}$, the algorithm first determines which approximation space to invoke, and then computes the RB approximation and associated RB error bound. Note that since the subdomains are constructed hierarchically based on proximity to the parameter anchor points associated with each subdomain, we can determine the subdomain containing μ in an efficient (typically negligible) $\mathscr{O}(\log_2 M)$-operations binary search. In particular, once $\mathscr{V}_{m^*} \subset \mathscr{D}$ containing μ is found, we solve (3) for the RB space X_{N,m^*}, and compute the error bound (5); the total cost is $\mathscr{O}(N^3 + Q^2N^2)$, as described in more detail shortly.

3 A Convection-Diffusion Model Problem

We now apply the hp RB method to a convection-diffusion model problem parametrized by the angle and magnitude of the specified velocity field: let $\mu = (\mu_1, \mu_2)$ (hence $P = 2$ parameters) and define $\mathbf{V}(\mu) = [\mu_2 \cos \mu_1, \mu_2 \sin \mu_1]^{\text{T}}$; we shall consider $\mu \in \mathscr{D} = [0, \pi] \times [1, 10]$. The physical domain is $\Omega = \{(x, y) :$

$x^2 + y^2 < 2\}$; the final time is $T = 1$ and the timestep is $\Delta t = 0.05$ such that $K = 20$. The "exact" field $\tilde{u}^e(t, \mu)$ satisfies $(\tilde{u}^e(t^k; \mu) - \tilde{u}^e(t^{k-1}; \mu))/\Delta t - \nabla^2 \tilde{u}^e(t^k; \mu) + \mathbf{V}(\mu) \cdot \nabla \tilde{u}^e(t^k; \mu) = 10, 1 \le k \le K$; we apply homogeneous Dirichlet boundary conditions; we consider an *inhomogeneous* initial condition (hence the tilde) $\tilde{u}^e(t^0) = g$, where g satisfies $-\nabla^2 g = 10$ in Ω.

We now reduce our equation to the desired form (1). We first write $\tilde{u}^e = u^e + g$ such that u^e now satisfies homogeneous initial conditions. We then define $b(w, v; \mu) = \int_\Omega wv \, d\Omega$, $a(w, v; \mu) = \int_\Omega \nabla w \cdot \nabla v \, d\Omega + \int_\Omega (\mathbf{V}(\mu) \cdot \nabla w) v \, d\Omega$, and $f(v; \mu) = 10 \int_\Omega v \, d\Omega - a(g, v; \mu)$; note that a_s is μ-independent and in fact we may choose $\alpha_{\text{LB}}(\mu) = 1$. Thus u^e satisfies (1) (and homogeneous initial conditions) with $Q_a = 3$, $Q_b = 1$, and $Q_f = 4$. We next introduce a truth space $X \equiv X^{\mathcal{N}}(\Omega)$: five spectral elements each of polynomial order 10. Figure 2 depicts the truth solution at $t = 0, 0.1, 0.25$ for the parameter value $\mu = (\pi, 10)$. As the parameters vary, the solution changes dramatically – a good candidate for *hp* treatment.

We now apply the POD/Greedy procedure to partition $\mathcal{D}$ into M parameter subdomains; the resulting *hp* RB approximation can then be written in the form (3). In Fig. 3, we show the partition of the parameter domain for $M = 97$ and $M = 2{,}258$ subdomains corresponding to $\epsilon^1_{\text{tol}} = 5$ and $\epsilon^1_{\text{tol}} = 1$, respectively; we choose $\rho_1 = 2$ and $\rho_2 = 1$. We also report, for each of the two partitions shown, the maximum of the error bound over the training samples over all subdomains as a function of N; we include the standard p-type RB approximation ($M = 1$) as well. Clearly, with smaller subdomains we need fewer basis functions for each approximation space.

We summarize in Table 1 for different error tolerances ϵ^2_{tol} the offline and online performance of the *hp* approach *relative to that of the standard p-type RB method.* We report the number of truth solves (effectively, parameters visited in the POD/Greedy); the number of operations for online evaluation of $u^k_N(\mu)$, $1 \le k \le K$, and $\Delta^k_N(\mu)$; and the online storage. The values in the table are based on the *theoretical* operation count and storage requirement. For N basis functions the online operation count (for our LTI system) is roughly $2N^3/3 + 2KN^2$ operations

Fig. 2 Example solutions for the convection-diffusion problem at $t = 0, 0.1, 0.25$ for the parameter value $\mu = (\pi, 10)$

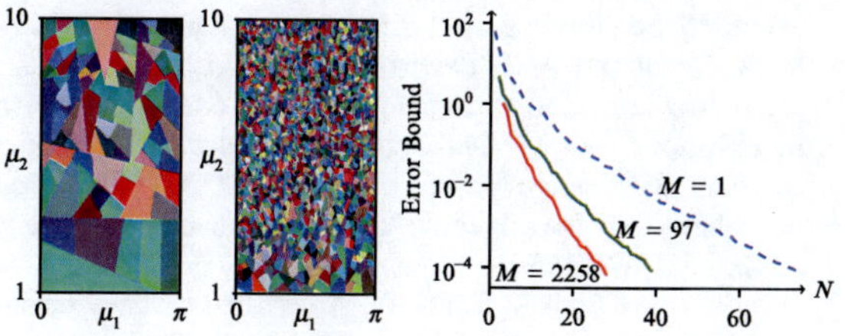

Fig. 3 Partition of $\mathcal{D}$ into $M = 97$ and $M = 2{,}258$ subdomains ($\epsilon_{\text{tol}} = 5$ and $\epsilon^1_{\text{tol}} = 1$, respectively); maximum error bound as a function of the RB approximation space dimension

Table 1 Offline and online effort relative to the standard (p-type) RB method for the two partitions $M = 97$ subdomains (*left*) and $M = 2{,}258$ subdomains (*right*) for different tolerances $\epsilon_{\mathrm{tol}}^2$

Tolerance, $\epsilon_{\mathrm{tol}}^2$	10^{-2}	10^{-3}	10^{-4}
Truth solves	39.2	40.7	40.6
Online $u_N^k(\mu)$	0.20	0.22	0.21
Online $\Delta_N^k(\mu)$	0.25	0.28	0.28
Online storage	16.7	17.7	17.5

Tolerance, $\epsilon_{\mathrm{tol}}^2$	10^{-2}	10^{-3}	10^{-4}
Truth solves	597	660	659
Online $u_N^k(\mu)$	0.08	0.09	0.09
Online $\Delta_N^k(\mu)$	0.11	0.13	0.14
Online storage	166	200	197

for the RB solution (and, in practice, output), and $\mathcal{O}(Q^2N^2 + KN^2)$ operations for the RB error bound (see [4, 8] for details); we neglect the $\mathcal{O}(QN^2)$ cost of forming the RB system and the $\mathcal{O}(\log_2 M)$ cost of finding the correct subdomain. For each space (subdomain) the storage requirement is $\mathcal{O}(Q^2N^2)$.

The new method is admittedly more expensive in terms of the *offline* cost – the number of truth solves. However, significant computational savings are achieved in the *online* computation of the RB solution and RB error bound; note that for modest Q the costs of the RB solution and RB error bound are comparable. For real-time or many-query applications the online cost is often our main concern and the *hp* approach is thus very attractive. Note however, that p-type refinement plays a crucial role in controlling the offline cost, in particular in higher parameter dimensions.

Future work on the *hp* approach will focus on quadratically nonlinear problems: in these cases the online operation count is $\mathcal{O}(N^4)$ and thus computational performance can greatly benefit from the (further) dimension reduction afforded by the *hp* approach.

Acknowledgements We acknowledge helpful discussions with D. Knezevic, N. C. Nguyen, D. B. P. Huynh, and S. Boyaval. The work has been supported by the Norwegian University of Science and Technology, AFOSR Grant FA 9550-07-1-0425, and OSD/AFOSR Grant FA 9550-09-1-0613.

References

1. D. Amsallem, J. Cortial, and C. Farhat. On-demand CFD-based aeroelastic predictions using a database of reduced-order bases and models. In: 47th AIAA Aerospace Sciences Meeting (2009)
2. B. O. Almroth, P. Stern, and F. A. Brogan. Automatic choice of global shape functions in structural analysis. *AIAA J.*, **16**, 525–528, 1978
3. J. L. Eftang, A. T. Patera, and E. M. Rønquist. An *hp* certified reduced basis method for parametrized elliptic partial differential equations. *SIAM J. Sci. Comput.*, accepted 2010
4. M. A. Grepl and A. T. Patera. A posteriori error bounds for reduced-basis approximations of parametrized parabolic partial differential equations. *M2AN*, **39**, 157–181, 2005
5. B. Haasdonk and M. Ohlberger. Reduced basis method for finite volume approximations of parametrized linear evolution equations. *M2AN Math. Model. Numer. Anal.*, **42**, 277–302, 2008
6. A. K. Noor and J. M. Peters. Reduced basis technique for nonlinear analysis of structures. *AIAA J.*, **18**, 455–462, 1980
7. D. J. Knezevic and A. T. Patera. A certified reduced basis method for the Fokker-Planck equation of dilute polymeric fluids: FENE dumbbells in extensional flow. *SIAM J. Sci. Comput.*, 32(2):793–817, 2010

8. N. C. Nguyen, G. Rozza, D. B. P. Huynh, and A. T. Patera. Reduced Basis Approximation and A Posteriori Error Estimation for Paramtrized Parabolic PDEs; Application to Real-Time Bayesian Parameter Estimation. In: Biegler, et al. Computational Methods for Large Scale Inverse Problems and Uncertainty Quantification. Wiley, London (2009)
9. G. Rozza, D. B. P. Huynh, and A. T. Patera. Reduced Basis Approximation and A Posteriori Error Estimation for Affinely Parametrized Elliptic Coercive Partial Differential Equations. *Arch. Comput. Methods Eng.*, **15**, 229–275, 2008

Highly Accurate Discretization of the Navier–Stokes Equations in Streamfunction Formulation

D. Fishelov, M. Ben-Artzi, and J.-P. Croisille

Dedicated to the memory of Professor David Gottlieb for his Wisdom and Generosity

Abstract A discrete version of the pure streamfunction formulation of the Navier–Stokes equation is presented. The proposed scheme is fourth order in both two and three spatial dimensions.

1 Fourth Order Scheme for the Navier–Stokes Equations in Two Dimensions

We consider the Navier–Stokes equations in pure streamfunction form, which in the two-dimensional case leads to the scalar equation

$$\begin{cases} \partial_t \Delta\psi + \nabla^{\perp}\psi \cdot \nabla\Delta\psi - \nu\Delta^2\psi = f(x,y,t), \\ \psi(x,y,t) = \psi_0(x,y). \end{cases} \tag{1}$$

Recall that $\nabla^{\perp}\psi = (-\partial_y\psi, \partial_x\psi)$ is the velocity vector. The no-slip boundary condition associated with this formulation is

$$\psi = \frac{\partial\psi}{\partial n} = 0, \quad (x,y) \in \partial\Omega, \quad t > 0 \tag{2}$$

D. Fishelov (✉)
Afeka-Tel-Aviv Academic College for Engineering, 218 Bnei-Efraim St. Tel-Aviv 69107, Israel
e-mail: daliaf@post.tau.ac.il

M. Ben-Artzi
Institute of Mathematics, The Hebrew University, Jerusalem 91904, Israel
e-mail: mbartzi@math.huji.ac.il

J.-P. Croisille
Department of Mathematics, University of Metz, France,
e-mail: croisil@poncelet.univ-metz.fr

J.S. Hesthaven and E.M. Rønquist (eds.), *Spectral and High Order Methods for Partial Differential Equations*, Lecture Notes in Computational Science and Engineering 76,
DOI 10.1007/978-3-642-15337-2_16,

and the initial condition is

$$\psi(x,y,0)=\psi_0(x,y),\quad (x,y)\in\Omega. \tag{3}$$

The spatial derivatives in Equation (1) are discretized as we describe next. The fourth order discrete Laplacian $\tilde{\Delta}_h\psi$ and biharmonic $\tilde{\Delta}_h^2\psi$ operators introduced in [4] are perturbations of the second order operators $\Delta_h\psi = (\delta_x^2+\delta_y^2)\psi$ and $\Delta_h^2\psi = (\delta_x^4+\delta_y^4+2\delta_x^2\delta_y^2)\psi$. They are designed as follows.

$$\tilde{\Delta}_h\psi_{i,j} = 2\Delta_h\psi_{i,j} - (\delta_x(\psi_x)_{i,j}+\delta_y(\psi_y)_{i,j}) = (\Delta\psi)_{i,j} + O(h^4). \tag{4}$$

Here, ψ_x, ψ_y are the fourth-order Hermitian approximations to $\partial_x\psi, \partial_y\psi$ described as

$$\begin{cases} \sigma_x\psi_x = \dfrac{1}{6}(\psi_x)_{i-1,j} + \dfrac{2}{3}(\psi_x)_{i,j} + \dfrac{1}{6}(\psi_x)_{i+1,j} = \delta_x\psi_{i,j}, & 1\le i,j\le N-1 \\ \sigma_y\psi_y = \dfrac{1}{6}(\psi_y)_{i,j-1} + \dfrac{2}{3}(\psi_y)_{i,j} + \dfrac{1}{6}(\psi_y)_{i,j+1} = \delta_y\psi_{i,j}, & 1\le i,j\le N-1. \end{cases} \tag{5}$$

We use the standard central difference operators $\delta_x, \delta_y, \delta_x^2, \delta_y^2$.

The fourth-order approximation to the biharmonic operator $\Delta^2\psi$ is

$$\tilde{\Delta}_h^2\psi = \delta_x^4\psi + \delta_y^4\psi + 2\delta_x^2\delta_y^2\psi - \frac{h^2}{6}(\delta_x^4\delta_y^2\psi + \delta_y^4\delta_x^2\psi) = \Delta^2\psi + O(h^4), \tag{6}$$

where δ_x^4 and δ_y^4 are the compact approximations of ∂_x^4 and ∂_y^4, respectively.

$$\delta_x^4\psi_{i,j} = \frac{12}{h^2}\left((\delta_x\psi_x)_{i,j} - \delta_x^2\psi_{i,j}\right),\quad \delta_x^4\psi = \partial_x^4\psi - \frac{1}{720}h^4\partial_x^8\psi + O(h^6), \tag{7}$$

$$\delta_y^4\psi_{i,j} = \frac{12}{h^2}\left((\delta_y\psi_y)_{i,j} - \delta_y^2\psi_{i,j}\right),\quad \delta_y^4\psi = \partial_y^4\psi - \frac{1}{720}h^4\partial_y^8\psi + O(h^6). \tag{8}$$

The convective term in (1) is $C(\psi) = -\partial_y\psi\Delta(\partial_x\psi) + \partial_x\psi\Delta(\partial_y\psi)$. Its fourth-order approximation needs special care. The mixed derivative $\partial_x\partial_y^2\psi$ may be approximated to fourth-order accuracy by $\tilde{\psi}_{yyx}$ using a suitable combination of lower order approximations.

$$\tilde{\psi}_{yyx} = \delta_y^2\psi_x + \delta_x\delta_y^2\psi - \delta_x\delta_y\psi_y = \partial_x\partial_y^2\psi + O(h^4). \tag{9}$$

For the pure third order derivative $\partial_x^3\psi$ we note that if ψ is smooth then

$$\psi_{xxx} = \frac{3}{2h^2}\left(10\delta_x\psi - h^2\delta_x^2\partial_x\psi - 10\partial_x\psi\right)_{i,j} + O(h^4). \tag{10}$$

One needs to approximate $\partial_x \psi$ to sixth-order accuracy in order to obtain from (10) a fourth-order approximation for $\partial_x^3 \psi$. Denoting this approximation by $\tilde{\psi}_x$, we invoke the Pade formulation [5], having the following form.

$$\frac{1}{3}(\tilde{\psi}_x)_{i+1,j} + (\tilde{\psi}_x)_{i,j} + \frac{1}{3}(\tilde{\psi}_x)_{i-1,j} = \frac{14}{9}\frac{\psi_{i+1,j} - \psi_{i-1,j}}{2h} + \frac{1}{9}\frac{\psi_{i+2,j} - \psi_{i-2,j}}{4h}. \tag{11}$$

At near-boundary points we apply a special treatment as in [5]. Carrying out the same procedure for $\partial_y \psi$, which yields the approximate value $\tilde{\psi}_y$, and combining with all other mixed derivatives, a fourth order approximation of the convective term is

$$\begin{aligned}\tilde{C}_h(\psi) &= -\psi_y\big(\Delta_h \tilde{\psi}_x + \frac{5}{2}(6\frac{\delta_x\psi - \tilde{\psi}_x}{h^2} - \delta_x^2\tilde{\psi}_x) + \delta_x\delta_y^2\psi - \delta_x\delta_y\tilde{\psi}_y\big) \qquad (12)\\ &+ \psi_x\big(\Delta_h \tilde{\psi}_y + \frac{5}{2}(6\frac{\delta_y\psi - \tilde{\psi}_y}{h^2} - \delta_y^2\tilde{\psi}_y) + \delta_y\delta_x^2\psi - \delta_y\delta_x\tilde{\psi}_x\big)\\ &= C(\psi) + O(h^4).\end{aligned}$$

Our implicit–explicit time-stepping scheme is of the Crank–Nicholson type as follows.

$$\frac{(\tilde{\Delta}_h\psi_{i,j})^{n+1/2} - (\tilde{\Delta}_h\psi_{i,j})^n}{\Delta t/2} = -\tilde{C}_h\psi^{(n)} + \frac{\nu}{2}[\tilde{\Delta}_h^2\psi_{i,j}^{n+1/2} + \tilde{\Delta}_h^2\psi_{i,j}^n] \tag{13}$$

$$\frac{(\tilde{\Delta}_h\psi_{i,j})^{n+1} - (\tilde{\Delta}_h\psi_{i,j})^n}{\Delta t} = -\tilde{C}_h\psi^{(n+1/2)} + \frac{\nu}{2}[\tilde{\Delta}_h^2\psi_{i,j}^{n+1} + \tilde{\Delta}_h^2\psi_{i,j}^n]. \tag{14}$$

Due to stability reasons we have chosen an Explicit–Implicit time stepping scheme. It is possible however to use an explicit time-stepping scheme if one can afford a small time step in order to advance the solution in time. The set of linear equations is solved via a FFT solver using the Sherman–Morrison formula (see [2]). This solver is of $O(N^2 logN)$ operations, where N is the number of grid points in each spatial direction. For the application of the pure streamfunction formulation on an irregular domain see [3].

2 The Pure Streamfunction Formulation in Three Dimensions

Let Ω be a bounded domain in R^3. The three-dimensional Navier–Stokes equations in vorticity-velocity formulation is

$$
\begin{aligned}
&\boldsymbol{\omega}_t + \nabla \times (\boldsymbol{\omega} \times \mathbf{u}) - \nu \Delta \boldsymbol{\omega} = \nabla \times \mathbf{f}, \quad \text{in} \quad \Omega \\
&\boldsymbol{\omega} = \nabla \times \mathbf{u}, \ \ \nabla \cdot \mathbf{u} = 0, \quad \text{in} \quad \Omega \\
&\mathbf{u} = \mathbf{0} \quad \text{on } \partial\Omega \\
&\boldsymbol{\omega}(\mathbf{x}, 0) = \boldsymbol{\omega}_0(\mathbf{x}) := \nabla \times \mathbf{u}_0, \quad \text{in} \quad \Omega.
\end{aligned}
\tag{15}
$$

where $\boldsymbol{\omega} = \nabla \times \mathbf{u}$ and the no-slip boundary condition has been imposed. The pure streamfunction formulation for this system is obtained by introducing a streamfunction $\boldsymbol{\psi}(\mathbf{x}, t) \in R^3$, such that

$$\mathbf{u} = -\nabla \times \boldsymbol{\psi}. \tag{16}$$

This is always possible since $\nabla \cdot \mathbf{u} = \mathbf{0}$. Thus,

$$\boldsymbol{\omega} = \nabla \times \mathbf{u} = \Delta \boldsymbol{\psi} - \nabla(\nabla \cdot \boldsymbol{\psi}). \tag{17}$$

Imposing a gauge condition

$$\nabla \cdot \boldsymbol{\psi} = 0, \tag{18}$$

yields

$$\omega = \Delta \boldsymbol{\psi}. \tag{19}$$

The system (15) can now be rewritten as

$$\frac{\partial \Delta \boldsymbol{\psi}}{\partial t} - \nabla \times (\Delta \boldsymbol{\psi} \times (\nabla \times \boldsymbol{\psi})) = \nu \Delta^2 \boldsymbol{\psi} + \nabla \times \mathbf{f}, \quad \text{in} \quad \Omega. \tag{20}$$

The boundary conditions $\mathbf{u} = 0$ translates to $\nabla \times \boldsymbol{\psi} = 0$ on $\partial\Omega$. We require that

$$\mathbf{n} \times \boldsymbol{\psi} = \mathbf{0}, \quad \mathbf{n} \times (\nabla \times \boldsymbol{\psi}) = \mathbf{0}, \quad \text{on} \quad \partial\Omega. \tag{21}$$

The condition $\mathbf{n} \times \boldsymbol{\psi} = \mathbf{0}$ means that $\boldsymbol{\psi}$ is parallel to $\mathbf{n}$, hence the normal component of the velocity vector is zero on the boundary. Adding the condition $\mathbf{n} \times (\nabla \times \boldsymbol{\psi}) = \mathbf{0}$ ensures that the full velocity vector vanishes on the boundary. The requirements in (21) are equivalent to four scalar conditions, namely the vanishing of the two tangential components of $\boldsymbol{\psi}$ and $\nabla \times \boldsymbol{\psi}$.

Turning now to the gauge condition $\nabla \cdot \boldsymbol{\psi} = 0$, we add the condition

$$\frac{\partial(\boldsymbol{\psi} \cdot \mathbf{n})}{\partial n} = 0, \quad \text{on} \quad \partial\Omega. \tag{22}$$

Together with the vanishing of the tangential components of $\boldsymbol{\psi}$, it implies that $\nabla \cdot \boldsymbol{\psi} = 0$ on $\partial\Omega$.

Equations (21) and (22) consist of five scalar conditions for $\boldsymbol{\psi}$ on the boundary. We can still add one more scalar boundary condition, as the equations for the 3-component streamfunction $\boldsymbol{\psi}$ contain the fourth order biharmonic operator. The sixth scalar boundary condition that we choose to add is

$$\Delta(\nabla \cdot \boldsymbol{\psi}) = 0, \quad \text{on} \quad \partial\Omega. \tag{23}$$

We thus obtain

$$\nabla \cdot \psi = 0, \quad \Delta(\nabla \cdot \psi) = 0, \quad \text{on} \quad \partial\Omega. \tag{24}$$

We assume that the initial value $\psi(\mathbf{x}, 0)$ satisfies $(\nabla \cdot \psi)(\mathbf{x}, 0) = 0$. Taking the divergence of (20) we obtain an evolution equation for $\nabla \cdot \psi$.

$$\frac{\partial \Delta(\nabla \cdot \psi)}{\partial t} = \nu \Delta^2 (\nabla \cdot \psi), \quad \text{in} \quad \Omega. \tag{25}$$

Equations (24) and (25) together with the assumption that $\nabla \cdot \psi = 0$ initially ensure that $\nabla \cdot \psi = 0$ for all $t > 0$. See also [1, 6, 7]. Finally, we have the following three-dimensional pure streamfunction formulation

$$\begin{cases} \frac{\partial \Delta \psi}{\partial t} - \nabla \times (\Delta \psi \times (\nabla \times \psi)) = \nu \Delta^2 \psi + \nabla \times \mathbf{f}, & \text{in} \quad \Omega \\ \mathbf{n} \times \psi = \mathbf{0}, \ \frac{\partial (\psi \cdot \mathbf{n})}{\partial n} = 0, \quad \text{on} \quad \partial\Omega \\ \mathbf{n} \times (\nabla \times \psi) = \mathbf{0}, \quad \Delta(\nabla \cdot \psi) = 0, & \text{on} \quad \partial\Omega. \end{cases} \tag{26}$$

3 The Numerical Scheme

Our numerical scheme is based on the approximation of the following equation

$$\frac{\partial \Delta \psi}{\partial t} - ((\nabla \times \psi) \cdot \nabla)\Delta \psi + (\Delta \psi \cdot \nabla)(\nabla \times \psi) - \nu \Delta^2 \psi = \nabla \times \mathbf{f}, \quad \text{in} \quad \Omega, \tag{27}$$

assuming that $\psi \in H_0^2(\Omega)$. For the vector function ψ we construct a fourth-order approximation to the biharmonic operator as follows. The pure fourth-order derivatives are approximated by $\delta_x^4, \delta_y^4, \delta_z^4$ as in (7) and (8).

The mixed terms ψ_{xxyy}, ψ_{yyzz} and ψ_{zzxx} are approximated by

$$\begin{cases} \tilde{\delta}_{xy}^2 \psi_{i,j,k} = 3\delta_x^2 \delta_y^2 \psi_{i,j,k} - \delta_x^2 \delta_y \psi_{y,i,j,k} - \delta_y^2 \delta_x \psi_{x,i,j,k} = \partial_x^2 \partial_y^2 \psi_{i,j,k} + O(h^4) \\ \tilde{\delta}_{yz}^2 \psi_{i,j,k} = 3\delta_y^2 \delta_z^2 \psi_{i,j,k} - \delta_y^2 \delta_z \psi_{z,i,j,k} - \delta_z^2 \delta_y \psi_{y,i,j,k} = \partial_y^2 \partial_z^2 \psi_{i,j,k} + O(h^4) \\ \tilde{\delta}_{zx}^2 \psi_{i,j,k} = 3\delta_z^2 \delta_x^2 \psi_{i,j,k} - \delta_z^2 \delta_x \psi_{z,i,j,k} - \delta_x^2 \delta_z \psi_{x,i,j,k} = \partial_z^2 \partial_x^2 \psi_{i,j,k} + O(h^4). \end{cases} \tag{28}$$

A fourth order approximation of the biharmonic operator is then obtained as

$$\tilde{\Delta}_h^2 \psi = \delta_x^4 \psi + \delta_y^4 \psi + \delta_z^4 \psi + 2\tilde{\delta}_{xy}^2 \psi + 2\tilde{\delta}_{yz}^2 \psi + 2\tilde{\delta}_{zx}^2 \psi. \tag{29}$$

The approximate derivatives ψ_x, ψ_y and ψ_z are related to ψ via the Hermitian derivatives as in (5).

Equation (29) provides a fourth order compact operator for $\Delta^2 \psi$, which involves values of ψ, ψ_x, ψ_y and ψ_z at (i, j, k) and at its 26 nearest neighbors. The Laplacian operator is approximated by a fourth order operator via

$$\tilde{\Delta}_h \psi = 2\Delta_h \psi - (\delta_x \psi_x + \delta_y \psi_y + \delta_z \psi_z). \tag{30}$$

The nonlinear part in (27) consists of two terms, the convective term and the stretching term. We design a fourth-order scheme which approximates the convective term. The convective term in the three-dimensional case is

$$C(\psi) = -((\nabla \times \psi) \cdot \nabla)\Delta\psi = u\Delta\partial_x \psi + v\Delta\partial_z \psi + w\Delta\partial_z \psi. \tag{31}$$

Here $(u, v, w) = \mathbf{u} = -\nabla \times \psi$ is the velocity vector, whose components contain first order derivatives of the streamfunction, and thus may be approximated to fourth-order accuracy. The terms $\Delta\partial_x \psi$, $\Delta\partial_z \psi$, $\Delta\partial_z \psi$ may be approximated as in the two-dimensional case. The term $\Delta\partial_x \psi$, for example, may be written as

$$\Delta\partial_x \psi = \partial_x^3 \psi + \partial_x \partial_y^2 \psi + \partial_x \partial_z^2 \psi. \tag{32}$$

Here, the pure and mixed type derivatives may be approximated as in the two-dimensional Navier–Stokes equations (see (9) and (10)). We denote the approximation to the convective term by $\tilde{C}_h(\psi)$.

Now, we construct a fourth-order approximation to the stretching term $S = (\boldsymbol{\omega} \cdot \nabla)\mathbf{u} = -(\Delta\psi \cdot \nabla)(\nabla \times \psi)$. Note that the stretching term contains $\Delta\psi$ and mixed second order derivatives of the streamfunction. The Laplacian of ψ may be approximated to fourth-order accuracy, as in (30). The second order mixed terms, such as $\partial_x \partial_y \psi$, may be approximated using a Hermitian approximation of the type

$$(\sigma_x \sigma_y)(\psi_{xy})_{i,j,k} = \delta_x \delta_y \psi_{i,j,k}. \tag{33}$$

Hence,

$$(I + \frac{h^2}{6}\delta_x^2)(I + \frac{h^2}{6}\delta_y^2)(\psi_{xy})_{i,j,k} = \delta_x \delta_y \psi_{i,j,k} \quad , 1 \le i, j, k \le N - 1 \tag{34}$$

is an implicit equation for ψ_{xy}. We denote the approximation of the stretching term by $\tilde{S}_h(\psi)$. For the approximation in time, we apply a Crank–Nicholson scheme (see the comment after (13) and (14)).

We obtain the following scheme

$$\frac{(\tilde{\Delta}_h \psi_{i,j,k})^{n+1/2} - (\tilde{\Delta}_h \psi_{i,j,k})^n}{\Delta t/2} = -\tilde{C}_h \psi_{i,j,k}^{(n)} + \tilde{S}_h \psi_{i,j,k}^{(n)} + \frac{\nu}{2}[\tilde{\Delta}_h^2 \psi_{i,j,k}^{n+1/2} + \tilde{\Delta}_h^2 \psi_{i,j,k}^n] \tag{35}$$

$$\frac{(\tilde{\Delta}_h \psi_{i,j,k})^{n+1} - (\tilde{\Delta}_h \psi_{i,j,k})^n}{\Delta t} = -\tilde{C}_h \psi_{i,j,k}^{(n+1/2)} + \tilde{S}_h \psi_{i,j,k}^{(n+1/2)} + \frac{\nu}{2}[\tilde{\Delta}_h^2 \psi_{i,j}^{n+1} + \tilde{\Delta}_h^2 \psi_{i,j,k}^n]. \tag{36}$$

At present, a direct solver is invoked to solve the linear set of equations (35) and (36).

Some preliminary MATLAB computations with coarse grids confirm the fourth order accuracy of the scheme. We first show numerical results for the time-dependent

Stokes equations

$$\frac{\partial \Delta \psi}{\partial t} = \nu \Delta^2 \psi + \mathbf{f}, \quad \text{in} \quad \Omega. \tag{37}$$

We have picked the exact solution ψ

$$\psi^T(\mathbf{x}, t) = -\frac{1}{4} e^{-t} \left(z^4, x^4, y^4 \right) \tag{38}$$

in the cube $\Omega = (0,1)^3$. Here, $\mathbf{f}$ is chosen such that ψ in (38) satisfied (37) exactly. Infg the numerical results shown here we have chosen the time step Δt of order h^2 in order to retain the overall fourth-order accuracy of the scheme. In practice, if we are interested mainly in the steady state solution, a larger time step, which is independent of h, may be used. In Table 1 we show results for the Stokes problem with $\Delta t = 0.1h^2$ and $t = 0.00625$. Here e is the error in the l_h^2 norm, i.e.,

$$e^2 = \sum_i \sum_j \sum_k (\psi_3(x_i, y_j, z_k) - \tilde{\psi}_3(x_i, y_j, z_k))^2 h^3,$$

where ψ_3 is the z component of the exact solution and $\tilde{\psi}_3$ is the z component of the approximate solution. e_y is the l_h^2 in the y derivative of ψ_3. In Table 2 we display the results for $t = 0.0625$ using $\Delta t = h^2$.

Next we show results for the Navier–Stokes Equations

$$\frac{\partial \Delta \psi}{\partial t} - ((\nabla \times \psi) \cdot \nabla) \Delta \psi + (\Delta \psi \cdot \nabla)(\nabla \times \psi) - \nu \Delta^2 \psi = \nabla \times \mathbf{f}, \quad \text{in} \quad \Omega \tag{39}$$

in the cube $\Omega = (0,1)^3$. Here, the source term $\mathbf{g} = \nabla \times \mathbf{f}$ is chosen such that $\psi^T(\mathbf{x}, t) = -\frac{1}{4} e^{-t} \left(z^4, x^4, y^4 \right)$ is an exact solution of (39). In Table 3 we present results for $t = 0.00625$ using $\Delta t = 0.1h^2$.

Table 1 Stokes equations for $t = 0.00625$ using $\Delta t = 0.1h^2$

	Grid $5 \times 5 \times 5$	Rate	Grid $9 \times 9 \times 9$	Rate	Grid $17 \times 17 \times 17$
e	2.5460(−9)	3.82	1.8017(−10)	3.98	1.1443(−11)
e_y	7.7417(−9)	3.73	5.8037(−10)	3.96	3.7391(−11)
div (ψ)	1.3409(−8)	3.74	1.0052(−9)	3.96	6.4621(−11)

Table 2 Stokes equations with $\Delta t = h^2$ for $t = 0.0625$

	Grid $5 \times 5 \times 5$	Rate	Grid $9 \times 9 \times 9$	Rate	Grid $17 \times 17 \times 17$
e	9.6461(−7)	4.41	4.5309(−8)	4.00	2.8291(−9)
e_y	3.0293(−6)	4.33	1.5049(−7)	3.99	9.4269(−9)
div (ψ)	5.2470(−6)	4.33	2.6066(−7)	4.00	1.6328(−8)

Table 3 Navier–Stokes equations for $t = 0.00625$ using $\Delta t = 0.1h^2$

	Grid $5 \times 5 \times 5$	Rate	Grid $9 \times 9 \times 9$	Rate	Grid $17 \times 17 \times 17$
e	2.4497(−9)	3.86	1.6924(−10)	4.01	1.0473(−11)
e_y	7.6486(−9)	3.75	5.6845(−10)	3.98	3.5917(−11)
div (ψ)	1.2294(−8)	3.71	9.3619(−10)	3.92	6.1700(−11)

Table 4 Navier–Stokes equations for $t = 0.0625$ using $\Delta t = h^2$

	Grid $5 \times 5 \times 5$	Rate	Grid $9 \times 9 \times 9$	Rate	Grid $17 \times 17 \times 17$
e	9.4418(−7)	4.46	4.2709(−8)	4.04	2.5934(−9)
e_y	2.9836(−6)	4.38	1.4334(−7)	4.03	8.7800(−9)
div (ψ)	5.0471(−6)	4.40	2.3944(−7)	4.02	1.4778(−8)

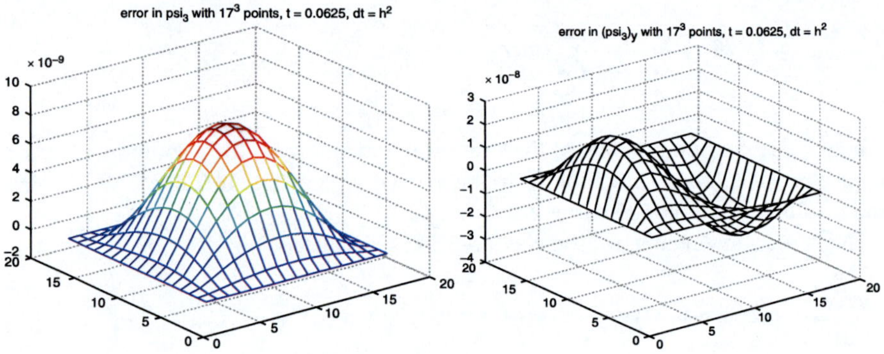

Fig. 1 Navier–Stokes : Errors in **(a)** ψ_3 and **(b)** $(\psi_3)_y$ for $N = 17, t = 0.0625,\ dt = h^2$

In Table 4 we show results for the Navier–Stokes Equations with $\Delta t = h^2$ for $t = 0.0625$. In Fig. 1a, b we display the errors for Navier–Stokes equations in ψ_3 and $(\psi_3)_y$ at $t = 0.0625$ with $dt = h^2$ and a 17^3 grid.

References

1. M. Ben-Artzi, *Planar Navier–Stokes equations, vorticity approach*, Handbook of Mathematical Fluid Dynamics, Chapter 5, Vol. II (2003)
2. M. Ben-Artzi, J.-P. Croisille, D. Fishelov, *A fast direct solver for the biharmonic problem in a rectangular grid*, SIAM J. Sci. Computing, Vol. 31 (1), pp. 303–333 (2008)
3. M. Ben-Artzi, I. Chorev, J-P. Croisille, D. Fishelov, *A compact difference scheme for the biharmonic equation in planar irregular domains*, SIAM J. Numer. Anal., Vol. 47 (4), pp. 3087–3108 (2009)
4. M. Ben-Artzi, J.-P. Croisille, D. Fishelov, *A High Order Compact Scheme for the Pure-Streamfunction Formulation of the Navier-Stokes Equations*, J. Sci. Computing, Vol. 42 (2), pp. 216–250 (2010)

5. M. H. Carpenter, D. Gottlieb, S. Abarbanel *The stability of numerical boundary treatments for compact high-order schemes finite difference schemes*, J. Comput. Phys., Vol. 108, pp. 272–295 (1993)
6. V. Ruas, L. Quartapelle, *Uncouples finite element solutions of biharmonic problems for vector potentials*, Inter. J. Numer. Methods in Fluids, Vol. 11, pp. 811–822 (1990)
7. A. Rubel, G. Volpe, *Biharmonic vector stream function formulation and multigrid solutions for a three-dimensional driven-cavity Stokes flow*, AIAA Computational Fluid Dynamics Conference, 9th, Buffalo, NY, June 13–15, 1989, AIAA, pp. 380–388 (1989). Inter. J. Numer. Methods in Fluids, Vol. 11, pp. 811–822 (1990)

Edge Functions for Spectral Element Methods

Marc Gerritsma

Abstract It is common practice in finite element methods to expand the unknowns in nodal functions. The discretization of the gradient, curl and divergence operators requires H^1, $H(\text{curl})$ and $H(\text{div})$ function spaces and their discrete representation. Especially in mixed formulations this involved quite some mathematical machinery which can be avoided once we recognize that not all unknowns are associated with point-values. In this short paper higher order basis functions will be presented which have the property that conservation laws become independent of the basis functions. The basis functions proposed in this paper yield a discrete representation of **grad**, **curl** and **div** which are exact and completely determined by the topology of the grid. The discretization of these vector operators is invariant under general C^1 transformations.

1 Introduction

Mimetic discretization schemes aim to preserve symmetries of the physical system to be modeled. If we are able to represent such symmetries in a discrete setting, we satisfy the associated conservation law in the discrete sense.

Mimetic discretizations are based on the strong analogy between differential geometry and algebraic topology. The global, metric-free description can be rephrased without error in terms of cochains, while the local description requires differential field reconstructions. For an introduction to the interplay between differential forms and cochains the reader is referred to [1–3, 5, 6, 8, 12, 15].

A key ingredient in mimetic methods is to re-establish the explicit connection between physical variables and the geometric objects these variables are associated with. The operation that connects the physical variable with its associated geometric object is *integration*, where the geometry, $\mathscr{C}$, enters the integral as the domain of

M. Gerritsma
TU Delft, Delft, The Netherlands
e-mail: M.I.Gerritsma@TUDelft.nl

J.S. Hesthaven and E.M. Rønquist (eds.), *Spectral and High Order Methods for Partial Differential Equations*, Lecture Notes in Computational Science and Engineering 76, DOI 10.1007/978-3-642-15337-2_17, © Springer-Verlag Berlin Heidelberg 2011

integration and the physical variable, Φ, appears as the integrand.

$$\int_{\mathscr{C}} \Phi \, dC = \langle \mathscr{C}, \Phi \rangle \in \mathbb{R}. \tag{1}$$

Equation (1) expresses the fact that geometric integration is in fact duality pairing between geometry, $\mathscr{C}$, and physical variables, Φ, since integration is a bilinear operation.

In [9, 10, 13] this approach is applied to spectral element methods. The main ingredient in this approach is the use of spectral basis functions which are associated with points, lines, surfaces and volumes. This paper focuses on the construction of the basis function associated with line segments, the so-called *edge functions*. Since we will consider quadrilateral elements only and employ tensor products to form the spectral element basis, the higher-dimensional basis functions are formed naturally by applying tensor products. For instance, the surface element is the tensor product of two edge functions and one nodal function, whereas the volume basis function is the tensor product of three edge functions.

The outline of the paper is as follows: In Sect. 2 the so-called *edge functions* will be derived for a simple one-dimensional equation. In Sect. 3 the edge functions will be used to discretize the differential operators, **grad**, **curl** and **div**. In Sect. 4 the transformation of differential forms is presented. Concluding remarks can be found in Sect. 5.

2 The Edge Functions

Consider the one-dimensional equation

$$u(\xi) = \frac{d\phi}{d\xi}, \quad \xi \in [a, b]. \tag{2}$$

Let $a = \xi_0 < \xi_1 < \cdots < \xi_{N-1} < \xi_N = b$ be a partitioning of the interval $[a, b]$, then the function $\phi(\xi)$ can be expanded in nodal basis functions

$$\phi(\xi) = \sum_{i=0}^{N} \phi_i h_i(\xi), \tag{3}$$

in which $h_i(\xi)$ are Lagrange basis functions through the points ξ_i, $i = 0, \ldots, N$ and $\phi_i = \phi(\xi_i)$. The traditional way to discretize (2) is to expand $u(\xi)$ in the *same* Lagrangian basis functions $h_i(\xi)$, i.e.,

$$u(\xi) = \sum_{i=0}^{N} u_i h_i(\xi). \tag{4}$$

If we insert the expansions (3) and (4) in (2) we obtain

$$\sum_{i=0}^{N} u_i h_i(\xi) = \sum_{i=0}^{N} \phi_i \frac{dh_i}{d\xi}. \tag{5}$$

There are however a few objections to this approach: First, a polynomial of degree N on the left hand side is equated to a polynomial of degree $N-1$ on the right hand side. Second, this formulation is *not* invariant under general C^1 coordinate transformations. These shortcomings can be attributed to the fact that *u cannot be associated with nodes*. In terms of differential geometry: If ϕ is a 0-form, then $u = d\phi$ is a 1-form which is associated with line segments. On a more engineering level we have that

$$\phi(p) = \phi(q) + \int_q^p u(x)\,dx, \tag{6}$$

i.e., the point-wise evaluation of ϕ in two arbitrary points p and q is associated with the integral of u over the interval (p,q). Let us therefore define the integral quantities

$$\overline{u}_i = \int_{\xi_{i-1}}^{\xi_i} u(x)\,dx, \quad i = 1,\ldots,N. \tag{7}$$

Note that by defining $\overline{u}_i$ as an integral quantity instead of the value in particular points, we exactly satisfy $\overline{u}_i = \phi_i - \phi_{i-1}$, which is the discrete analogue of the integral relation (6). Interpolation of integral quantities is called *histopolation*, see Robidoux [14].

We have that

$$\begin{aligned}
u^N(\xi) &= \sum_{i=0}^{N} \phi_i\, dh_i(\xi) \\
&= \sum_{i=0}^{N} (\phi_i - \phi_k)\, dh_i(\xi) \\
&= \sum_{i=0}^{N} \left[-\sum_{j=i+1}^{k} \overline{u}_j + \sum_{j=k+1}^{i} \overline{u}_j \right] dh_i(\xi) \\
&= -\sum_{i=0}^{k-1} dh_i(\xi) \sum_{j=i+1}^{k} \overline{u}_j + \sum_{i=k+1}^{N} dh_i(\xi) \sum_{j=k+1}^{i} \overline{u}_j \\
&= -\sum_{j=1}^{k} \left(\sum_{i=0}^{j-1} dh_i(\xi) \right) \overline{u}_j + \sum_{j=k+1}^{N} \left(\sum_{i=j}^{N} dh_j(\xi) \right) \overline{u}_j
\end{aligned} \tag{8}$$

The first line states that the discrete approximation (histopolation) of u, denoted by $u^N(\xi)$, can be exactly expressed as the derivative of the discrete interpolation of

$\phi(\xi)$. In the second line we use

$$\sum_{i=0}^{N} h_i(\xi) \equiv 1 \quad \Longrightarrow \quad \sum_{i=0}^{N} dh_i(\xi) = d \sum_{i=0}^{N} h_i(\xi) \equiv 0, \tag{9}$$

and in the third line we used $\overline{u}_i = \phi_i - \phi_{i-1}$ repeatedly. In the remaining lines we re-arrange the summations. Since (8) is true for all $k = 0, \ldots, N$, we can eliminate k by averaging over all k

$$\begin{aligned}
u^N(\xi) &= \frac{1}{N+1} \sum_{k=0}^{N} u^N(\xi) \\
&= -\frac{1}{N+1} \sum_{k=0}^{N} \sum_{j=1}^{k} \left(\sum_{i=0}^{j-1} dh_i(\xi) \right) \overline{u}_j + \frac{1}{N+1} \sum_{k=0}^{N} \sum_{j=k+1}^{N} \left(\sum_{i=j}^{N} dh_j(\xi) \right) \overline{u}_j \\
&= \frac{1}{N+1} \sum_{j=1}^{N} \left[-(N+1-j) \overline{u}_j \sum_{i=0}^{j-1} dh_i(\xi) + j \overline{u}_j \sum_{i=j}^{N} dh_i(\xi) \right] \\
&= \frac{1}{N+1} \sum_{j=1}^{N} \left[-(N+1) \overline{u}_j \sum_{i=0}^{j-1} dh_i(\xi) + j \overline{u}_j \sum_{i=0}^{N} dh_i(\xi) \right] \\
&= -\sum_{j=1}^{N} \overline{u}_j \sum_{i=0}^{j-1} dh_i(\xi).
\end{aligned} \tag{10}$$

If we now define the basis functions

$$e_j(\xi) = -\sum_{i=0}^{j-1} dh_i(\xi), \quad j = 1, \ldots, N, \tag{11}$$

we can express u in terms of the integral quantities $\overline{u}_i$ as

$$u^N(\xi) = \sum_{i=1}^{N} \overline{u}_i e_i(\xi). \tag{12}$$

The basis functions $e_i(\xi)$ can be interpreted as polynomial indicator functions, Fig. 1, because they satisfy

$$\int_{\xi_{k-1}}^{\xi_k} e_i(\xi) = \delta_{i,k} = \begin{cases} 1 & \text{if } i = k \\ 0 & \text{if } i \neq k \end{cases}. \tag{13}$$

Compare this with the nodal interpolation where we have that $h_i(\xi_j) = \delta_{i,j}$. The basis functions $e_i(\xi)$ correspond to higher order Whitney forms, see [3, 6, 19]. Note

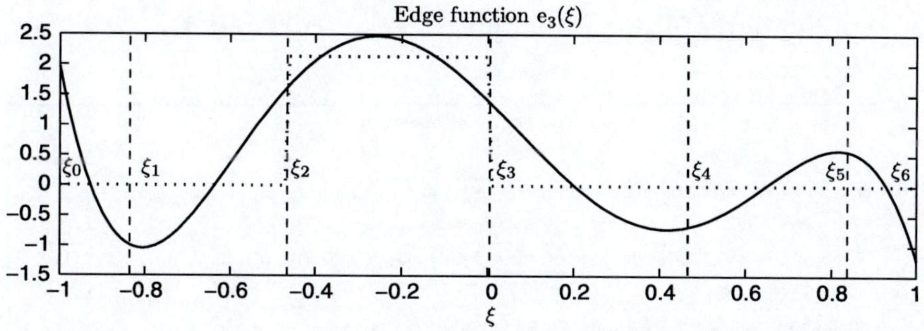

Fig. 1 Example of an edge function: Partitioning of the interval [−1,1] with Gauss–Lobatto nodes and the edge function $e_3(\xi)$

that we have $de_j(\xi) = -d \circ d \sum h_i(\xi) \equiv 0$, see for instance Flanders [7]. This property will be used repeatedly in the next section. If we insert the expansion of u in terms of edge functions into our one-dimensional model problem, we obtain

$$\begin{aligned}\sum_{i=1}^{N} \overline{u}_i e_i(\xi) &= \sum_{i=0}^{N} \phi_i dh_i(\xi) \\ &= -\sum_{i=1}^{N} (\phi_i - \phi_{i-1}) \sum_{j=0}^{i-1} dh_j(\xi) \\ &= \sum_{i=1}^{N} (\phi_i - \phi_{i-1})\, e_i(\xi) \end{aligned} \tag{14}$$

This shows that there is strict equality: The polynomial degrees on both sides are the same and this relation remains valid under arbitrary C^1 transformations, since the basis functions on both sides of the equality sign transform in the same way. Because the basis functions $e_i(\xi)$ are linearly independent, we in fact have

$$\sum_{i=1}^{N} [\overline{u}_i - (\phi_i - \phi_{i-1})]\, e_i(\xi) = 0 \quad \Longrightarrow \quad \overline{u}_i - (\phi_i - \phi_{i-1}) = 0. \tag{15}$$

This is a purely topological, metric-free relation because all the metric properties are encoded in the basis functions and its form is solely determined by the topology of the grid. Once we know which nodes form the boundary of a given line segment, we can set up this relation. The definitions of $\overline{u}_i$, (7), and ϕ_i show that (15) is *exact*; no approximations are involved.

The metric-free form, (15), resembles a finite volume discretization, see for instance [11, 16–18] of the sample problem. In the next section discrete representations of vector operators in terms of the edge functions will be addressed. The resulting discrete equations also resemble finite volume discretizations.

3 Application of Edge Functions to grad, curl and div

The gradient operator

Consider $u = \textbf{grad}\phi.$ (16)

Let ϕ be expanded as a tensor product of nodal functions in the coordinates (ξ, η, ζ)

$$\phi(\xi,\eta,\zeta) = \sum_{i=0}^{N}\sum_{j=0}^{N}\sum_{k=0}^{N} \phi_{i,j,k} h_i(\xi) h_j(\eta) h_k(\zeta), \tag{17}$$

then it can be shown by straightforward calculation that

$$\overline{u}^{\xi}_{i,j,k} = \phi_{i,j,k} - \phi_{i-1,j,k}, \quad \overline{u}^{\eta}_{i,j,k} = \phi_{i,j,k} - \phi_{i,j-1,k} \quad \text{and}$$
$$\overline{u}^{\zeta}_{i,j,k} = \phi_{i,j,k} - \phi_{i,j,k-1}, \tag{18}$$

with

$$u^{\xi}(\xi,\eta,\zeta) = \sum_{i=1}^{N}\sum_{j=0}^{N}\sum_{k=0}^{N} \overline{u}^{\xi}_{i,j,k} e_i(\xi) h_j(\eta) h_k(\zeta),$$
$$u^{\eta}(\xi,\eta,\zeta) = \sum_{i=0}^{N}\sum_{j=1}^{N}\sum_{k=0}^{N} \overline{u}^{\eta}_{i,j,k} h_i(\xi) e_j(\eta) h_k(\zeta),$$
$$u^{\zeta}(\xi,\eta,\zeta) = \sum_{i=0}^{N}\sum_{j=0}^{N}\sum_{k=1}^{N} \overline{u}^{\zeta}_{i,j,k} h_i(\xi) h_j(\eta) e_k(\zeta). \tag{19}$$

Equation (18) is exact, coordinate free and invariant under C^1 transformations.

The curl operator

Let u be defined along edges, (19), then $\omega = \textbf{curl}\, u$ is given by

$$\overline{\omega}^{\xi}_{i,j,k} = \overline{u}^{\zeta}_{i,j,k} - \overline{u}^{\zeta}_{i,j-1,k} - \overline{u}^{\eta}_{i,j,k} + \overline{u}^{\eta}_{i,j,k-1},$$
$$\overline{\omega}^{\eta}_{i,j,k} = \overline{u}^{\xi}_{i,j,k} - \overline{u}^{\xi}_{i,j,k-1} - \overline{u}^{\zeta}_{i,j,k} + \overline{u}^{\zeta}_{i-1,j,k},$$
$$\overline{\omega}^{\zeta}_{i,j,k} = \overline{u}^{\eta}_{i,j,k} - \overline{u}^{\eta}_{i-1,j,k} - \overline{u}^{\xi}_{i,j,k} + \overline{u}^{\xi}_{i,j-1,k}, \tag{20}$$

with

$$\omega^{\xi}(\xi,\eta,\zeta) = \sum_{i=0}^{N}\sum_{j=1}^{N}\sum_{k=1}^{N} \overline{\omega}^{\xi}_{i,j,k} h_i(\xi) e_j(\eta) e_k(\zeta),$$
$$\omega^{\eta}(\xi,\eta,\zeta) = \sum_{i=1}^{N}\sum_{j=0}^{N}\sum_{k=1}^{N} \overline{\omega}^{\eta}_{i,j,k} e_i(\xi) h_j(\eta) e_k(\zeta),$$
$$\omega^{\zeta}(\xi,\eta,\zeta) = \sum_{i=1}^{N}\sum_{j=1}^{N}\sum_{k=0}^{N} \overline{\omega}^{\zeta}_{i,j,k} e_i(\xi) e_j(\eta) h_k(\zeta).$$

If u is a gradient, (18), then $\omega^{\xi} = \omega^{\eta} = \omega^{\zeta} \equiv 0$, which implies

$$u = \textbf{grad}\,\phi \quad \Longleftrightarrow \quad \textbf{curl}\, u \equiv 0. \tag{21}$$

Equation (20) is exact, metric-free and invariant under C^1 transformations.

The divergence operator

Consider the divergence equation

$$a = \mathbf{div}\, \boldsymbol{f}. \tag{22}$$

Given fluxes defined over surfaces. Let the flux vector be expanded as

$$f^{\xi}(\xi,\eta,\zeta) = \sum_{i=0}^{N}\sum_{j=1}^{N}\sum_{k=1}^{N} \overline{f}^{\xi}_{i,j,k} h_i(\xi) e_j(\eta) e_k(\zeta),$$

$$f^{\eta}(\xi,\eta,\zeta) = \sum_{i=1}^{N}\sum_{j=0}^{N}\sum_{k=1}^{N} \overline{f}^{\eta}_{i,j,k} e_i(\xi) h_j(\eta) e_k(\zeta),$$

$$f^{\zeta}(\xi,\eta,\zeta) = \sum_{i=1}^{N}\sum_{j=1}^{N}\sum_{k=0}^{N} \overline{f}^{\zeta}_{i,j,k} e_i(\xi) e_j(\eta) h_k(\zeta). \tag{23}$$

If a is expanded in terms of volume basis functions

$$a(\xi,\eta,\zeta) = \sum_{i=1}^{N}\sum_{j=1}^{N}\sum_{k=1}^{N} \overline{a}_{i,j,k} e_i(\xi) e_j(\eta) e_k(\zeta), \tag{24}$$

the divergence equation reduces to

$$\overline{a}_{i,j,k} = \overline{f}^{\xi}_{i,j,k} - \overline{f}^{\xi}_{i-1,j,k} + \overline{f}^{\eta}_{i,j,k} - \overline{f}^{\eta}_{i,j-1,k} + \overline{f}^{\zeta}_{i,j,k} - \overline{f}^{\zeta}_{i,j,k-1}. \tag{25}$$

Again we see that the divergence equation reduces to a topological equation which is independent of the basis functions. And therefore these relations will remain unchanged under general coordinate transformations. The unknowns $\overline{a}_{i,j,k}$ and $\overline{f}^{\xi}_{i,j,k}$ represent

$$\overline{a}_{i,j,k} = \int_{\xi_{i-1}}^{\xi_i}\int_{\eta_{j-1}}^{\eta_j}\int_{\zeta_{k-1}}^{\zeta_k} a(\xi,\eta,\zeta)\, d\xi d\eta d\zeta, \quad \overline{f}^{\xi}_{i,j,k} = \int_{\eta_{j-1}}^{\eta_j}\int_{\zeta_{k-1}}^{\zeta_k} f^{\xi}(\xi,\eta,\zeta)\, d\eta d\zeta, \tag{26}$$

it follows that (25) is an exact representation of the divergence equation. No numerical approximations are involved. If the fluxes f are the curl of a vector ω, i.e., $f = \mathbf{curl}\,\omega$ then also in the discrete setting we have

$$\mathbf{div}\, f = 0 \iff f = \mathbf{curl}\,\omega \tag{27}$$

4 Transformations

Let V^p be a p-dimensional, compact oriented submanifold of M^n, $\dim(M^n) = n$ and let $F : M^n \to W^m$ be a C^1 map into a manifold W^m, $\dim(W^m) = m$. The image of $F(V)$ in W^m need not be a submanifold. It might even have self-intersections and all sorts of pathologies. Still, if β^p is a differential form on W^m, it makes sense to talk about the integral of β^p over $F(V)$ and in fact we have

$$\int_{F(V^p)} \beta^p = \int_{V^p} F^\star \beta^p, \tag{28}$$

where $F^\star$ is the pull-back operator associated with F, which is a linear map from the cotangent bundle $T^\star W^m$ onto the cotangent bundle $T^\star M^n$, [8, p. 155].

Let M^n be the domain $[-1,1]^n$ and F a map onto a curvilinear domain $\hat{\Omega}$, then any differential form β^p on $\hat{\Omega}$ is mapped onto a p-form $F^\star \beta^p$ on the parent domain, M^n. Due to the fact that (28) holds and the unknowns in our system are integral quantities (so-called *co-chains*), we can expand $F^\star \beta^p$ in a tensor product as described in Sect. 3. Let $\tilde{e}_j$ be an edge function along a curvilinear contour bounded by $\tilde{\xi}_{i-1} = F(\xi_{i-1})$ and $\tilde{\xi}_i = F(\xi_i)$, then

$$\int_{\tilde{\xi}_{i-1}}^{\tilde{\xi}_i} \tilde{e}_j = \int_{F(\xi_{i-1})}^{F(\xi_i)} \tilde{e}_j = \int_{\xi_{i-1}}^{\xi_i} F^\star \tilde{e}_j = \int_{\xi_{i-1}}^{\xi_i} e_j = \delta_{i,j}, \tag{29}$$

by (13). All expansion coefficients are integral quantities (co-chains) which remain invariant due to (28). Furthermore, the pullback operator commutes with the exterior derivative, $F^\star \circ d \equiv d \circ F^\star$. These properties demonstrate that the relations in Sect. 3 are valid under a wide class of transformations. This is what one would expect, since the relations in Sect. 3, representing the *generalized Stokes' Theorem*, are purely topological and cannot depend on the particular coordinate system or polynomial representation. An example of this invariance is used in [4].

5 Concluding Remarks

In this paper the edge functions $e_i(x)$ were derived, representing basis functions along line segments. Using tensor products, these edge functions can be used to represent variables defined over surfaces and volumes. The extension to higher order dimensions is straightforward. Using these basis functions the discrete representation of the gradient, curl and divergence are purely topological and independent of the basis functions. These relations are exact; no numerical approximation is involved. Since these operations are metric-free, they are preserved under C^1 mappings and in this sense they extend the Thomas-Raviart and Nédélec elements, which are only invariant under affine transformations. Although an arbitrary partitioning was considered, for spectral element methods usually the Gauss–Lobatto nodes are taken. An application of these edge function for partial differential equations in curvilinear coordinates can be found in Bouman [4].

References

1. Bochev, P.B.: A discourse on variational and geometric aspects of stability of discretizations. 33rd Computational Fluid Dynamics Lecture Series, VKI LS 2003-05, edited by H. Deconinck, ISSN0377-8312. Von Karman Institute for Fluid Dynamics, Belgium, 2005
2. Bochev, P.B., Hyman, J.M.: Principles of mimetic discretizations of differential equations. IMA Volume 142, edited by D. Arnold, P. Bochev, R. Lehoucq, R. Nicolaides, M. Shashkov. Springer, Berlin, 2006
3. Alain Bossavit's Japanese papers: http://butler.cc.tut.fi/~bossavit/Books/IEEEJapan.html
4. Bouman, M.P., Palha da Silva Clérigo, A., Kreeft, J.J., Gerritsma, M.I.: A conservative spectral element method for arbitrary domains. Proceedings of ICOSAHOM 2009 (this issue)
5. Desbrun, M., Hirani, A.N., Leok, M., Marsden, J.E.: Discrete Exterior Calculus. arXiv:math/0508341v2, 18 Aug 2005
6. Desbrun, M., Kanso, E., Tong, Y.: Chapter 7: Discrete differential forms for computational modeling. ACM SIGGRAPH ASIA 2008 Courses, SIGGRAPH Asia'08, Art. no. 15, 2008
7. Flanders, H.: Differential forms with applications to the physical sciences. Academic Press, New York, 1963
8. Frankel, Th.: The geometry of physics – an introduction, 2nd edition. Cambridge University Press, London, 2004
9. Gerritsma, M.I., Bouman, M.P., Palha da Silva Clérigo, A.: Least-squares spectral element method on a staggered grid. Lecture Notes on Computer Science, Large-Scale Scientific Computing, 7th International Conference, LSSC 2009, Sozopol, Bulgaria, June 2009, pp. 653–661. Springer, Berlin, 2010
10. Gerritsma, M.I.: An introduction to a compatible spectral discretization method. Mechanics of Advanced Materials and Structures, 2009 (Submitted)
11. LeVeque, R.J.: Finite volume methods for hyperbolic problems. Cambridge University Press, London, 2002
12. Mattiussi, C.: The finite volume, finite difference, and finite elements methods as numerical methods for physical field problems. In: Advances in Imaging and Electron Physics, volume 113, pp. 1–146, 2000
13. Palha, A., Gerritsma, M.I.: Mimetic least-squares spectral/*hp* finite element method for the Poisson equation. Lecture Notes on Computer Science, Large-Scale Scientific Computing, 7th International Conference, LSSC 2009, Sozopol, Bulgaria, June 2009, pp. 662–670. Springer, Berlin, 2010
14. Robidoux, N.: Polynomial histopolation, superconvergent degrees of freedom and pseudospectral discrete hodge operators. Unpublished: http://www.cs.laurentian.ca/nrobidoux/prints/super/histogram.pdf
15. Tonti, E.: On the mathematical structure of a large class of physical theories. Accademia Nazionale dei Lincei, estratto dai Rendiconti della Classe di Scienze fisiche, matematiche e naturali, Serie VIII, volume LII, fasc. 1, Gennaio, 1972
16. Toro, E.J.: Riemann solvers and numerical methods for fluid dynamics – a practical introduction. Springer, Berlin, 1999
17. Versteeg, H.K., Malalasekera, W.: An introduction to computational fluid dynamics – the finite volume method. Prentice Hall, NJ, 1995
18. Vinokur, M.: An analysis of finite difference and finite volume formulations of conservation laws. Journal of Computational Physics, 81, 1–52, 1989
19. Whitney, H.: Geometric integration. Dover, NY, 2000. ISBN 0486445836

Modeling Effects of Electromagnetic Waves on Thin Wires with a High-Order Discontinuous Galerkin Method

N. Gödel, T. Warburton, and M. Clemens

Abstract An efficient method for modeling a strong coupling between electromagnetic fields and currents in a thin wire is presented. The Discontinuous Galerkin Finite Element Method (DG-FEM) is used to discretize both, Maxwell's equations and the wire equations in the time domain. Suitable tests for investigation of the accuracy of the model and its implementation are provided.

1 Introduction

Since electrical devices are getting more and more sophisticated, the modeling of these devices, especially the treatment of small and detailed parts is challenging the development of simulation codes. In general, DG-FEM is able to treat small parts, i.e., electrical wires and harnesses in enclosures simply by meshing them with very small elements. This results in high geometric aspect ratios and, consequently, severe time step restrictions for explicit timestepping schemes. One method to improve timestepping efficiency is the implementation of a multirate timestepping method as described in [4].

In this work, a field-wire coupling formulation is implemented using a thin wire discretization, where the wire is not discretized with volume elements, but along a curve defined inside the computational domain. There are two possible implementations of this model: the curve can be defined along the tetrahedral edges or alternatively arbitrarily in space. The former solution leads to an easier tetrahedra-wire coupling implementation but includes severe constraints for the tetrahedral

N. Gödel (✉) and M. Clemens
Chair for Theory of Electrical Engineering and Computational Electromagnetics, Faculty of Electrical Engineering, Helmut-Schmidt-University of the Federal Armed Forces Hamburg, P.O. Box 700822, 22008 Hamburg, Germany
e-mail: Nico.Goedel@hsu-hh.de

T. Warburton
Computational and Applied Mathematics, Rice University, 6100 Main Street MS-134, Houston, TX, USA

J.S. Hesthaven and E.M. Rønquist (eds.), *Spectral and High Order Methods for Partial Differential Equations*, Lecture Notes in Computational Science and Engineering 76, DOI 10.1007/978-3-642-15337-2_18, © Springer-Verlag Berlin Heidelberg 2011

mesh generation algorithm. The latter results in a curve definition that is completely independent from tetrahedral mesh generation. However, the coupling of fields on the wire and fields inside the tetrahedrons is more sophisticated.

Having realistic applications with complex 3D geometries in mind, this work focuses on the second option of defining the wire geometry. Since CAD data management and mesh generation is still challenging and often time consuming, the guideline of this work is to avoid any additional mesh generation constraints and to make the wire discretization completely independent from the volume mesh generation.

Thin wire modeling was first presented by Holland et al. in [7] and discretized using a finite difference method. Edelvik et al. used this formulation for a continuous FEM approach in [2] and Volpert et al. suggested a first DG-FEM thin wire model in [9]. In this work, a DG-FEM formulation with upwind fluxes is used to discretize both, Maxwell's equations as well as the thin wire equations. A full coupling between the electromagnetic fields and the wire currents is implemented and first benchmarks are presented.

The paper is organized as follows: In Sect. 2, the DG-FEM discretization of Maxwell's equations in the time domain is briefly introduced. Section 3 contains the thin wire equations and their discretization. In Sect. 4, the coupling mechanism of the field to wire coupling is numerically analyzed. Section 5 describes the effects of a radiating wire to the ambient field and Sect. 6 highlights the proposed formulation of the full coupling algorithm.

2 DG-FEM Discretization of Maxwell's Equations

For the discretization of Maxwell's equations in the time domain, the DG-FEM has proven to be an efficient and suitable formulation [3,6]. The two main characteristics of DG-FEM are its parallel efficiency due to the elementwise FEM approach and the ability of using explicit time integration schemes. Both characteristics together allow for a highly parallel implementation on specialized hardware such as graphics processing units (GPU), which provide high memory bandwidth and floating point performance [5, 8].

A variational formulation of Maxwell's equations in the time domain with a vectorial testfunction $\vec{\phi}$ defined in the computational domain Ω and its boundary $\Gamma = \partial\Omega$ is derived in [6] with

$$(\vec{\phi}, \mu \frac{\mathrm{d}\vec{H}}{\mathrm{dt}} + \nabla \times \vec{E})_{\Omega} \quad + \quad (\vec{\phi}, \vec{n} \times (\vec{E}^{*} - \vec{E}))_{\Gamma=\partial\Omega} \quad = \quad 0, \tag{1}$$

$$(\vec{\phi}, \varepsilon \frac{\mathrm{d}\vec{E}}{\mathrm{dt}} - \nabla \times \vec{H})_{\Omega} \quad - \quad (\vec{\phi}, \vec{n} \times (\vec{H}^{*} - \vec{H}))_{\Gamma=\partial\Omega} \quad = \quad -(\vec{\phi}, \vec{J})_{\Omega}, \tag{2}$$

where $\vec{E}$ and $\vec{H}$ denote the electric and magnetic field variables, respectively, and $\vec{J}$ the electric current density. This formulation can be solved elementwise with the boundary terms ensuring the connection between the elements with help of the numerical fluxes $\vec{H}^*$ and $\vec{E}^*$. The electric permittivity is identified by ε and the magnetic permeability by μ. The vector $\vec{n}$ is the outward pointing normal vector on the elemental boundaries.

As presented in [6], (1) and (2) can be decomposed into a discretized system of six equations for the 3D components of $\vec{E}$ and $\vec{H}$, leading to an elementwise formulation on 3D simplices. In this work, nodal Lagrange polynomials are used as basis functions and a tetrahedral mesh is used as geometric discretization.

3 Thin Wire Equations and DG-FEM Discretization

The current I and the charges per unit length q' on a perfectly conducting thin wire can be expressed with help of the Holland formulation published in [7]

$$L'\left(\frac{\partial}{\partial t}I + v^2\frac{\partial}{\partial s}q'\right) = E_s, \tag{3}$$

$$\frac{\partial}{\partial t}q' + \frac{\partial}{\partial s}I = 0. \tag{4}$$

Here, v denotes the wave speed and L' the wire inductance per unit length. The partial derivative $\frac{\partial}{\partial s}$ is a directional derivative in wire direction. The right-hand side reflects the excitation term with E_s being the projected electric field onto the wire. A variational formulation of (3) and (4) with a testfunction ψ reads

$$\left(\psi, L'\left(\frac{\mathrm{d}I}{\mathrm{dt}} + v^2\frac{\mathrm{d}q'}{\mathrm{d}s}\right)\right)_\Omega + \left(\psi, L'v^2(f_I^* - f_I)\right)_{\Gamma=\partial\Omega} = (\psi, E_s)_\Omega, \tag{5}$$

$$\left(\psi, \frac{\mathrm{d}q'}{\mathrm{dt}} + \frac{\mathrm{d}I}{\mathrm{d}s}\right)_\Omega + \left(\psi, (f_{q'}^* - f_{q'})\right)_{\Gamma=\partial\Omega} = 0. \tag{6}$$

Here, f_I^* and $f_q'^*$ denote the flux terms for the current and charge evaluation, being

$$f_I^* - f_I = (q'^* - q') + \alpha(I^* - I) \tag{7}$$

$$f_{q'}^* - f_{q'} = (I^* - I) + \alpha(q'^* - q'). \tag{8}$$

The parameter α is 0 for a central flux and 1 for an upwind flux. For modeling currents and charges in a thin wire, the penalization of jumps in the current is suitable, since there is no physical explanation of jumps in a current density. However, although it is not considered in this work, the quantity of charges along a straight wire can jump in case of jumps in the conductivity. In this case, penalty terms for the charges would be less suitable. With a wire oriented in x-direction, a DG-FEM discretization on each wire element is given by

$$\frac{\mathrm{d}}{\mathrm{d}t}\mathbf{I} = -v^2\mathbf{D}_x\mathbf{q} + \mathbf{M}^{-1}\mathbf{F}\left(\mathbf{f}_I^* - \mathbf{f}_I\right) + \frac{\mathbf{E}_l}{L}, \tag{9}$$

$$\frac{\mathrm{d}}{\mathrm{d}t}\mathbf{q} = -\mathbf{D}_x\mathbf{I} + \mathbf{M}^{-1}\mathbf{F}\left(\mathbf{f}_q^* - \mathbf{f}_q\right), \tag{10}$$

with $\mathbf{D}_x$ being the differentiation matrix in x-direction and $\mathbf{M}, \mathbf{F}$ representing the DG-FEM mass- and fluxmatrices on the wire, respectively. The term $\frac{E_l}{L}$ is responsible for the excited currents on the wire. The wire inductance L

$$L = \frac{\mu_0}{2\pi}\log\frac{r_0 + a}{2a}, \tag{11}$$

has to be chosen according to the problem geometry where a is the wire radius and r_0 the radius within electric fields have effects on the wire.

4 Field to Wire Coupling

As a first benchmark, a receiving dipole antenna is simulated. The antenna is a wire of $l = 41$ m length as used in [2] and a radius of 10 mm is used to model the thin wire inductance in (11). The wire is excited by a broadband electric pulse presented in Fig. 1, where the time signal as well as the spectral properties are presented. The excitation signal is coupled into the thin wire equation through the electric field term of the right-hand side of (9), where the electric field forces the charges in the wire to move along the wire resulting in an electric current. With this broadband excitation different modes of the wire current can be excited. The analytic modes

$$f_n = n \cdot \frac{v}{l/2}, \quad n \in \mathbb{N} \tag{12}$$

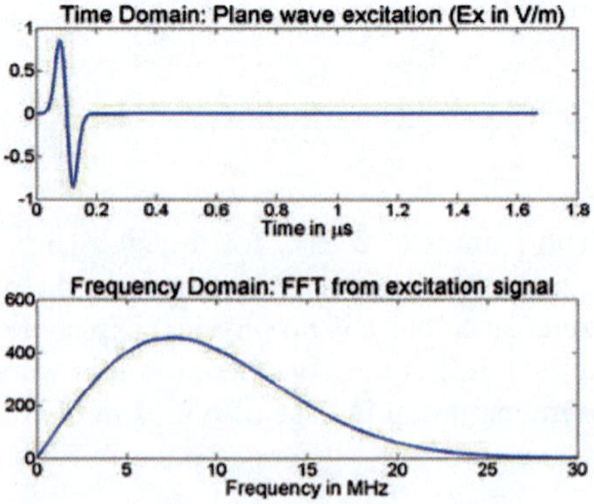

Fig. 1 Excitation signal in time and frequency domain

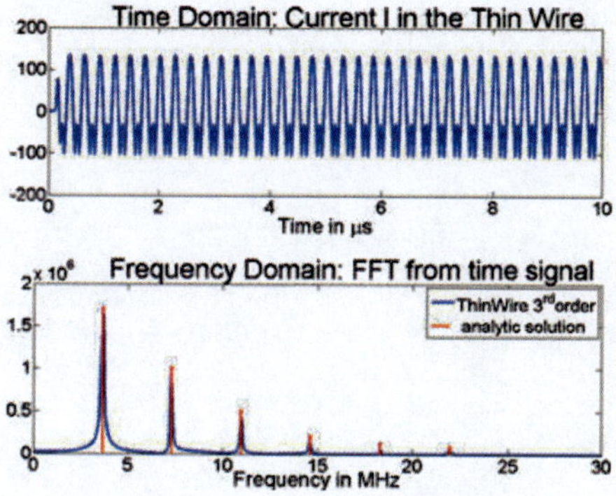

Fig. 2 Excited current in time and frequency domain

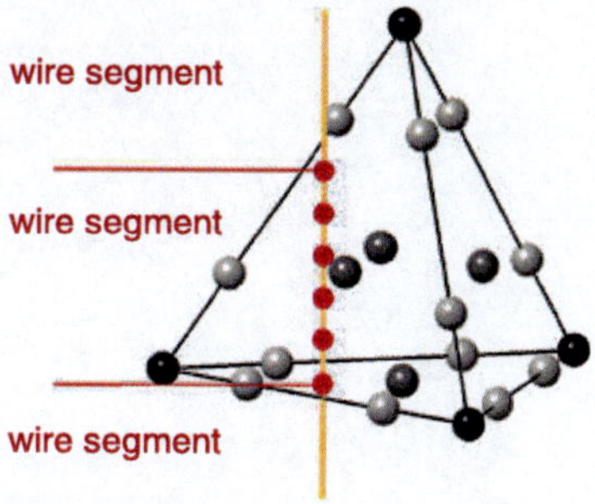

Fig. 3 Wire to tetrahedra connectivity

as well as the computed modes and the time signal are presented in Fig. 2. Once the electric pulse has passed the wire, different modes are excited. A fourier transformation of the time signal provides the numerical modes of the wire, which are almost exactly matching the analytic modes (12). Since there is no backcoupling enabled into Maxwell's equations at this point, the energy coupled into the wire is maintained. For a coupling into Maxwell's equations, the wire current I has to treated as a input value into Ampère's Law as a current density.

5 Wire to Field Coupling

In this section the coupling of a wire current into Maxwell's equations is investigated. Since the current in the wire is a global quantity, it has to be divided by the wire cross section to get the current density, which couples into the DG discretization of Ampre's Law. As highlighted in Fig. 3, the current on the wire is discretized in a 1D polynomial space. The electromagnetic fields in the tetrahedra are approximated by 3D polynomial spaces. Consequently, for the coupling, the wire current has to be extended to a 3D current-density.

Figure 3 describes the wire element decomposition into wire segments with respect to the intersections with the tetrahedral boundaries. On each wire segment,

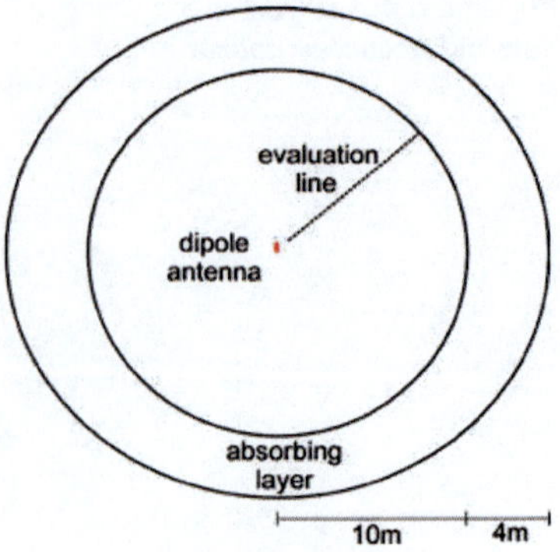

Fig. 4 Geometry of a radiating dipole antenna with evaluation line

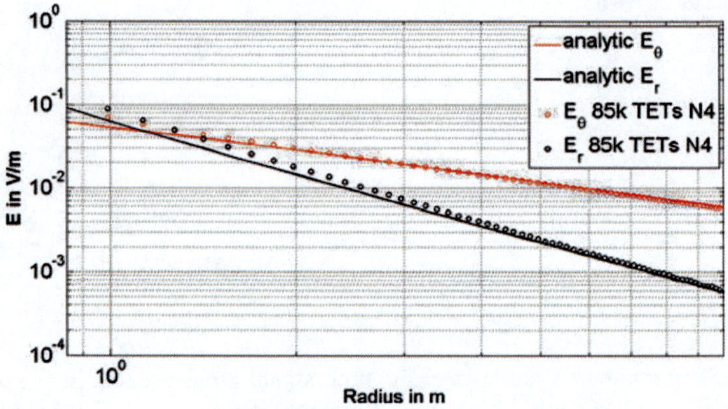

Fig. 5 Electric field components of a radiating dipole antenna compared to the analytic solution of an ideal dipole

Gauss nodes are defined and the wire current is interpolated at these Gauss nodes. The wire node data is lifted onto the volume node data at these Gauss nodes with help of a dirac function which reflects the lowest energy approximation of the wire field by the volume data.

To test this algorithm, a transmitting dipole antenna is simulated with help of this formulation. The wire is 0.5 m long with a radius of 0.2 mm and is situated in free space as presented in Fig. 4. The vacuum space is discretized with 85,235 tetrahedra and the fields are approximated using fourth order polynomials. A sinusoidal excitation current with a wavelength of 3 m is provided and the radiating electric field is computed along the evaluation line. Figure 5 shows the envelope of the electric field along the evaluation line compared to an analytic solution. Since analytic solutions to this kind of problem are only available for the ideal dipole of zero length, the comparison in Fig. 5 has to be treated carefully, especially in the ultra-near field of the antenna.

The radial component of the electric field should decay with a $1/r^2$ behaviour and the propagating azimuthal component E_θ with $1/r$ characteristic. To investigate

Table 1 Convergence of the singular solution with respect to the mesh size

No. of neighboring tetrahedra	L_2-error
12	4.5466
36	0.0072
43	2.8598e-4
109	2.8596e-4
575	2.6975e-4
3,056	2.5954e-4

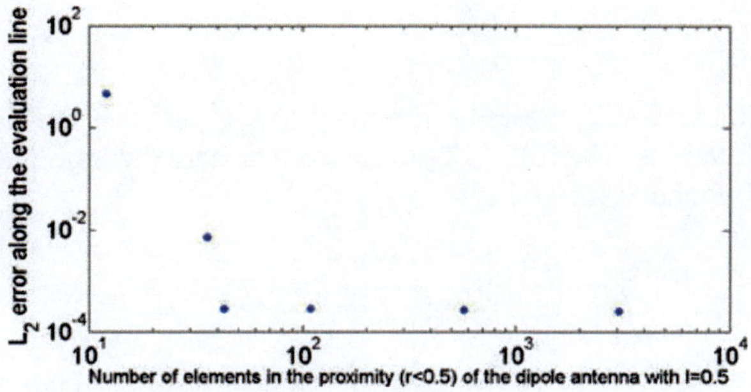

Fig. 6 Convergence of the singular solution

the convergence of the singular solutions of the azimuthal and radial components at the wire location, effects of a mesh refinement in the vicinity of the wire is analyzed. The L_2-error of the azimuthal component is computed along the evaluation line with radius $2.687 \leq r \leq 9.0$. Table 1 lists the L_2-error for different numbers of tetrahedra within a sphere of radius $= 0.5$. In Fig. 6, the convergence is presented. It can be seen that the solution converges with decreasing mesh size to a minimum L_2-error of order 10^{-4}. The field computation is executed on a Nvidia Tesla GPU with single precision accuracy. In earlier investigations published in [8], the effects of the single precision accuracy within GPU computations is of order 10^{-6}, which can be confirmed for normalized cavity simulations with simple geometries. Here, the singularity of the wire to field coupling influences the accuracy of the computation by the largest field value approximating the singularity. The highest values can be found at the wire ends and are of order 10^2 in the performed simulation.

6 Full Field to Wire coupling

This section combines the two presented interactions between electromagnetic waves and wires, i.e., a current excited by an external electric field also leads to a radiation of the wire itself. The fully coupled field wire system can be described with help of the variational formulation

$$(\vec{\phi}, \mu \frac{\mathrm{d}\vec{H}}{\mathrm{dt}} + \nabla \times \vec{E})_\Omega \quad + \quad (\vec{\phi}, \vec{n} \times (\vec{E}^* - \vec{E}))_{\Gamma=\partial\Omega} \quad = 0 \tag{13}$$

$$(\vec{\phi}, \varepsilon \frac{\mathrm{d}\vec{E}}{\mathrm{dt}} - \nabla \times \vec{H})_\Omega \quad - \quad (\vec{\phi}, \vec{n} \times (\vec{H}^* - \vec{H}))_{\Gamma=\partial\Omega} \quad = -(\vec{\phi}, \frac{I\delta_w}{A}))_\Omega, \tag{14}$$

$$(\psi, L'\left(\frac{\mathrm{dI}}{\mathrm{dt}} + v^2 \frac{\mathrm{dq}'}{\mathrm{d}s}\right)) \quad + \quad (\psi, L'v^2(f_I'^* - f_I)) \quad = \quad (\psi, E_s) \tag{15}$$

$$(\psi, \frac{\mathrm{dq}'}{\mathrm{dt}} + \frac{\mathrm{dI}}{\mathrm{d}s}) \quad + \quad (\psi, (f_q^* - f_q)) \quad = 0, \tag{16}$$

where the right-hand sides describe the interchange between the field equations and the wire equations. The fully coupled system is stable with the following semi discrete energy equation inequality

$$\frac{\mathrm{d}}{\mathrm{dt}}(\frac{\mu}{2}||\mathrm{H}||^2_\Omega + \frac{\varepsilon}{2}||\mathrm{E}||^2_\Omega + \frac{\mu}{4\pi}||\mathrm{I}||^2_{\Omega_w} + \frac{1}{4\pi\varepsilon}||\mathrm{q}'||^2_{\Omega_w}) \tag{17}$$
$$= -\sum_{k=1}^{K} \sum_{f=1}^{\#faces} (\frac{\mu}{2}||\hat{n} \times [\mathrm{H}]||^2 + \frac{\varepsilon}{2}||\hat{n} \times [\mathrm{E}]||^2 + \frac{\mu}{4\pi}||[\mathrm{I}]||^2 + \frac{1}{4\pi\varepsilon}||[\mathrm{q}']||^2) <= 0,$$

as long as the coupling terms show symplectic behaviour for the field energy W_f and the wire energy W_w transitions, i.e.,

$$\frac{\mathrm{d}}{\mathrm{dt}}(W_\mathrm{f} + W_\mathrm{w}) = 0 \tag{18}$$

resulting in a equivalent power loss and gain of the two systems with

$$P_E = \int E \frac{I}{A} \delta_w \mathrm{d}\Omega = \int EI \mathrm{d}s = P_I, \tag{19}$$

which is ensured by the variational formulation of the Holland equations.

The consistency of the proposed method is subject to further investigation. In [1], Cockburn and Guzmán provided fundamental analysis and a consistency proof of a field problem with discontinuous initial data resulting in oscillatory behaviour. These oscillations are shown in [1] to remain localized within a region whose extent depends on the mesh size, time step size, and simulation time.

In order to test the fully coupled field wire solver, a thin wire is situated in a box as presented in Fig. 7. The cylinder surrounding the wire is defined for meshing purpose. An inflow boundary condition with a Gaussian pulse is defined at the coloured face of the box. The pictures in Fig. 8 are highlighting the resulting fields for different time steps and with different scalings. The two left pictures show the incoming wave hitting the wire such that the field in the vicinity of the wire is disturbed. With this scaling, no back-scattering can be observed. The right picture has a different scaling showing a maximum of 2% of the excitation signal. Here, the back-scattering of the thin wire can be observed.

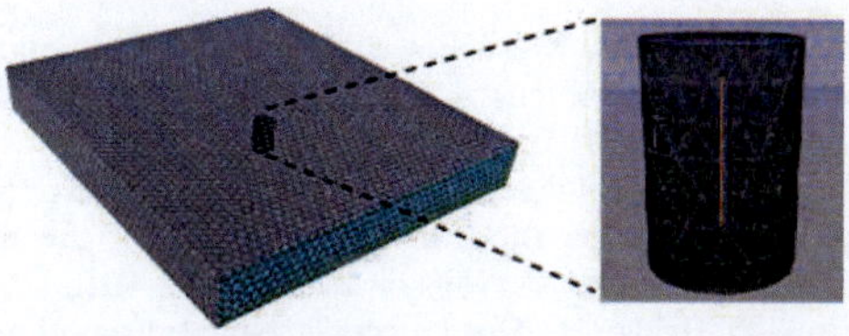

Fig. 7 Computational domain of the full field wire coupling test case

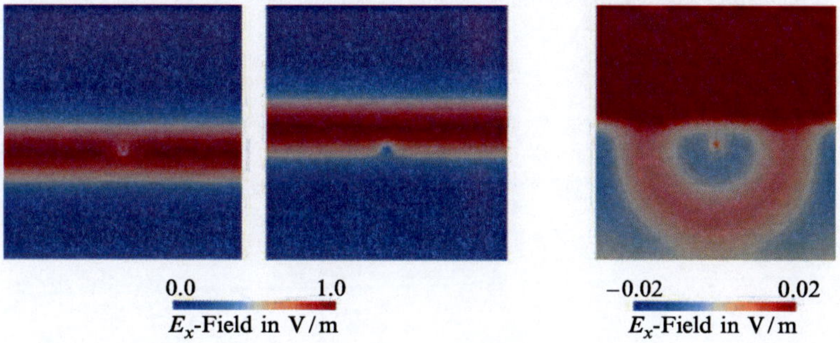

Fig. 8 Electric field components of a radiating dipole antenna compared to the analytic solution of an ideal dipole

7 Conclusion and Outlook

A strong coupling between Maxwell's equations and Holland's thin wire equations in the time domain has been presented. The field to wire coupling has been investigated by computing the modes of a receiving dipole antenna. The wire to field coupling has been tested by simulating a transmitting dipole antenna. For the full coupling, a back-scattering of a thin wire in free space has been computed.

References

1. Cockburn, B., Guzmán, J.: Error estimates for the Runge–Kutta discontinuous Galerkin method for the transport equation with discontinuous initial data. SIAM J. Numer. Anal. **46**(3), 1364–1398 (2008)
2. Edelvik, F., Ledfelt, G., Lotstedt, P., Riley, D.: An unconditionally stable subcell model for arbitrarily oriented thin wires in the fetd method. Antennas and Propagation, IEEE Transactions on **51**(8), 1797–1805 (2003)
3. Gödel, N., Lange, S., Clemens, M.: Time domain discontinuous Galerkin method with efficient modelling of boundary conditions for simulations of electromagnetic wave propagation. IEEE CNF APEMC pp. 594–597 (2008)
4. Gödel, N., Schomann, S., Warburton, T., Clemens, M.: Local timestepping discontinuous Galerkin methods for electromagnetic RF field problems. In: Antennas and Propagation, 2009. EuCAP 2009. 3rd European Conference on pp. 2149–2153 (2009)

5. Gödel, N., Warburton, T., Clemens, M.: GPU accelerated discontinuous Galerkin FEM for electromagnetic radio frequency problems. Antennas and Propagation Society International Symposium, 2009. APS-URSI '09. IEEE pp. 1–4 (2009)
6. Hesthaven, J.S., Warburton, T.: Nodal discontinuous Galerkin methods. Springer, Berlin (2008)
7. Holland, R., Simpson, L.: Implementation and optimization of the thin-strut formalism in THREDE. Nuclear Science, IEEE Transactions on **27**(6), 1625–1630 (1980)
8. Klöckner, A., Warburton, T., Bridge, J., Hesthaven, J.: Nodal discontinuous Galerkin methods on graphics processors. J. Comput. Phys. **228**(21), 7863 – 7882 (2009)
9. Volpert, T., Ferrieres, X., Pecqueux, B., Cohen, G.: Introduction of composite material and thin wire formalism in a discontinuous galerkin time domain method. ACES – preprint (2008)

A Hybrid Method for the Resolution of the Gibbs Phenomenon

Jae-Hun Jung

Abstract For the resolution of the Gibbs phenomenon, the inverse polynomial and the statistical filter methods were proposed independently. In this paper, we show how these two methods are different and similar, both mathematically and numerically. After comparing these methods, we propose a hybrid inverse polynomial and statistical filter method for the resolution of the Gibbs phenomenon.

1 Introduction

If a function to be approximated is analytic and periodic, its Fourier approximation yields a fast convergence. If not, then the Fourier approximation is highly oscillatory near the domain boundaries and the local jump discontinuity resulting in the deterioration of spectral accuracy. This is known as the Gibbs phenomenon. Since Gottlieb and his coworkers developed the *Gegenbauer reconstruction method* to recover spectral accuracy in the Fourier reconstruction contaminated by the Gibbs phenomenon in 1992 [3, 4], several other methods have also been developed to resolve the Gibbs phenomenon in the Fourier or polynomial approximations of discontinuous functions (for review, see Boyd's paper and references therein [2]).

In this paper, we consider two Gibbs-defeating methods: (1) the inverse polynomial reconstruction method (IPRM) [7, 10], and the statistical filter (SF) method [11]. The purpose of this paper is to compare these two methods mathematically and propose a Gibbs-defeating hybrid method to remedy the weaknesses of these methods, which include the ill-conditioning, slow convergence and high dimensionality of the transformation matrix.

J.-H. Jung
Department of Mathematics, The State University of New York at Buffalo, Buffalo, NY 14260-2900, USA
e-mail: jaehun@buffalo.edu.

J.S. Hesthaven and E.M. Rønquist (eds.), *Spectral and High Order Methods for Partial Differential Equations*, Lecture Notes in Computational Science and Engineering 76, DOI 10.1007/978-3-642-15337-2_19,

Since the IPRM was developed, several issues related to the IPRM have been addressed including the uniqueness, convergence, ill-posedness [7] and the truncation method for the removal of round-off errors [8]. The method was also applied to two-dimensional applications [1]. More rigorous proof of the existence of the IPRM was provided by Krebs in 2007 [9], and the generalized IPRM was recently proposed by Hrycak and Gröchenig in 2010 [5]. Although the SF method was published in 1995, it has not been recognized in the literature until recently.

These two methods are mathematically very similar although their numerical performances differ. No previous research has been done to study the mathematical relation between these two methods. As the numerical results in this paper show, the proposed hybrid method exploits strengths of these two methods and, consequently, is more robust and efficient than the IPRM and SF methods. Furthermore, the paper demonstrates how the recent development of the IPRM, such as the recent work by Hrycat and Gröchenig [5], could be generalized in the SF method.

Section 2 briefly explains the IPRM and the SF method. In Sect. 3, mathematical comparisons are made. In Sect. 4, a hybrid method is proposed and numerical results are given; and in Sect. 5, a brief summary is provided.

2 The Inverse and Statistical Filter Methods

We assume that the unknown function $f(x) \in L^2[-1,1]$ is analytic but not necessarily periodic and can be represented as a polynomial $f(x) = \sum_{l=0}^{\infty} g_l L_l(x)$, where $L_l(x)$ are the orthogonal polynomials such as the Legendre polynomials and g_l the corresponding expansion coefficients with the proper inner product $(\cdot,\cdot)$. We also assume that the finite Fourier data $\{\hat{f}_k\}_{-N}^{N}$ of $f(x)$ is given a priori. Let the expansion coefficient vector in the Legendre polynomials be $\mathbf{g} = (g_0, g_1, \cdots)^T$ and the Fourier coefficient vector $\hat{\mathbf{f}}$ be $\hat{\mathbf{f}} = (\hat{f}_{-N}, \cdots, \hat{f}_N)^T$. The Legendre inner product $(\cdot,\cdot)_L$ and the Fourier inner product $(\cdot,\cdot)_F$ are used for the expansion coefficients, $g_l = (f(x), L_l(x))_L := \frac{2l+1}{2}\int_{-1}^{1} f(x)L_l(x)dx$, and $\hat{f}_k = (f(x), \exp(ik\pi x))_F := \frac{1}{2}\int_{-1}^{1} f(x)\exp(-ik\pi x)dx$, respectively. We define the *transformation matrix* (or *connection matrix*) $A \in C^{M\times\infty}$ as

$$A_{kl} = (L_l(x), \exp(ik\pi x))_F, \tag{1}$$

where $M = 2N + 1$. Then the Fourier coefficients are obtained by $\hat{\mathbf{f}} = A\mathbf{g}$.

2.1 *Inverse Polynomial Reconstruction Method*

The IPRM seeks a reconstruction $\tilde{f}(x)$ such that $\tilde{f}(x)$ is a polynomial of degree at most m as $\tilde{f}(x) = \sum_{l=0}^{m} \tilde{g}_l \psi_l(x)$, where $\psi_l(x)$ is the polynomial of degree l

and $\tilde{g}_l$ is the corresponding expansion coefficient. Here the basis polynomials $\psi_l(x)$ are not necessarily to be orthogonal [7]. The reconstruction is unique whether it is expanded by the orthogonal or non-orthogonal polynomials. If the function $f(x)$ is a polynomial of finite degree, the reconstruction $\tilde{f}(x)$ is exact, i.e., $\tilde{f}(x) = f(x)$, and the IPRM yields spectral accuracy [7, 9]. The unknown expansion coefficients with the IPRM are found by minimizing the reconstruction in the Galerkin sense, that is, $(f(x) - \tilde{f}(x)) \perp F_N$, where F_N is the given Fourier space of $dim = 2N + 1$ spanned by $\{\exp(ik\pi x)\}_{-N}^{N}$ and the symbol $\perp$ denotes that the residue between the reconstruction and the function is orthogonal to the Fourier space, F_N in the sense of the Fourier inner product. Then the expansion coefficients $\{\tilde{g}_l\}_{l=0}^{m}$ are determined by $\mathbf{W} \cdot \tilde{\mathbf{g}} = \hat{\mathbf{f}}$, where the transformation matrix is given by $W_{kl} = (\psi_l(x), \exp(ik\pi x))_F$.

2.2 *Statistical Filter Method*

The reconstruction $\tilde{f}(x)$ by the SF method is given by the infinite sum of polynomials. For example, $\tilde{f}(x) = \sum_{l=0}^{\infty} \tilde{g}_l L_l(x)$, where $\tilde{g}_l$ are the expansion coefficients and $L_l(x)$ are the Legendre polynomials. Let $\tilde{\mathbf{g}}$ be the expansion coefficient vector. The SF method determines the expansion coefficients by minimizing $\| \tilde{\mathbf{g}} - P\hat{\mathbf{f}} \|^2$ and the minimizing matrix P is given by the pseudo-inversion [11] $P = C(AC)^{\div}$, where $\div$ denotes the pseudo-inversion. The matrix C is the *covariance matrix* diagonal elements of which are the expectation values of the square of g_l [11]. Thus for any analytic function, the diagonal elements C_{ii} decay when the reconstruction is sought with orthogonal polynomials and all the off-diagonal elements of C vanish. Since we seek a reconstruction by the linear sum of orthogonal polynomials, we use C whose non-zero elements are only diagonal elements. Then the expansion coefficients are given by $\tilde{\mathbf{g}} = C(AC)^{\div}\hat{\mathbf{f}}$.

Spectral convergence of the SF method has not been proven yet in the literature. To see spectral convergence, we define the error function $E(x)$

$$E(x) := \tilde{f}(x) - f(x) = \sum_{l=0}^{\infty} (\tilde{g}_l - g_l) L_l(x). \tag{2}$$

Notice that $\tilde{\mathbf{g}}$ is not given by $A^{\div}\hat{\mathbf{f}}$ but by $C(AC)^{\div}\hat{\mathbf{f}}$ using the covariance matrix. This is because the linear system $A\tilde{\mathbf{g}} = \hat{\mathbf{f}}$ is under-determined possibly yielding infinitely many solutions for $\mathbf{g}$ and the direct pseudo-inverse of A does not necessarily guarantee the convergence of the method. By using the covariance matrix C, the SF method seeks the convergence and the condition on C is crucial for the successful performance of the SF method.

We remark that the generalized IPRM was proposed recently by Hrycak and Gröchenig [5] that uses a similar formulation as the SF method but does not require the covariance matrix. Instead, the generalized method uses much larger matrix (e.g., $N \geq m^2$) [5]. However, our proposed hybrid method can reduce the matrix size significantly by using the covariance matrix in the generalized IPRM framework.

3 Convergence, Accuracy and Exactness

3.1 Convergence

For spectral convergence of the IPRM, let the truncation error function $TE(x)$ be $TE(x) = f_m(x) - \tilde{f}_m(x)$, where $f_m(x)$ is the truncated sum of $f(x)$ as $f_m(x) = \sum_{l=0}^{m} g_l L_l(x)$. The recovery of spectral accuracy can be shown by proving that the L_∞ norm of the projection of $TE(x)$ to the Fourier space decays with N, such that

$$\max_{-1\leq x\leq 1} |TE_N(x)| \leq C_1 A(\rho) q^N, \tag{3}$$

where $TE_N(x)$ is the Fourier approximation of $TE(x)$, C_1 a positive constant independent of N, A the function of ρ only, ρ the distance from the singularity of $f(x)$ in complex plane and $0 < q < 1$.

As for the SF method, note that the SF method seeks $\tilde{f}(x)$ in the same space where $f(x)$ resides and no truncation or regularization errors are introduced in the error analysis. Using $\tilde{\mathbf{g}} = C(AC)^{\div}\hat{\mathbf{f}}$,

$$\max_{-1\leq x\leq 1} |E(x)| = \max_{-1\leq x\leq 1} |\sum_{l=0}^{\infty}(\tilde{g}_l - g_l)L_l(x)| \leq |\sum_{l=0}^{\infty}(\tilde{g}_l - g_l)| \leq \| C(AC)^{\div}\hat{\mathbf{f}} - \mathbf{g} \|_1 .$$

Thus using $\hat{\mathbf{f}} = A\mathbf{g}$, we have

$$\begin{aligned}\max_{-1\leq x\leq 1} |E(x)| &\leq \| C(AC)^{\div}ACC^{-1} - I \|_1 \| \mathbf{g} \|_1 \leq D_1 \| C(AC)^{\div}ACC^{-1} - I \|_1 \\ &= D_1 \| C((AC)^{\div}AC - I)C^{-1} \|_1, \end{aligned} \tag{4}$$

where D_1 is a constant independent of N and we use the Parseval's theorem for $f(x) \in L^2$. $\| \cdot \|_1$ denotes the matrix 1−norm. Since the rank of A is only $2N + 1$, $(AC)^{\div}AC \neq I$ and the error vanishes only when $N \to \infty$, i.e., $\lim_{N\to\infty}(AC)^{\div}AC = I$. Let U and V be such that $AC = U[\Sigma_N \;\; \mathbf{0}]V^T$, where Σ_N is the diagonal matrix composed of the singular values of AC and U and V are unitary vectors. Then

$$(AC)^{\div}AC = V\begin{bmatrix} I_N & \mathbf{0} \\ \mathbf{0} & \mathbf{0} \end{bmatrix}V^T, \tag{5}$$

where I_N is the identity matrix, $I^{(2N+1)\times(2N+1)}$ and $\mathbf{0}$ the null matrix. The convergence of the SF method depends on how fast $(AC)^{\div}AC$ converges to I. Since V depends on the choice of C, convergence of the SF depends on C. This implies that how to choose the covariance matrix C is the critical question to obtain spectral convergence for the SF method while the IPRM does not necessarily need such condition.

3.2 Covariance Matrix

In order to suggest the improved SF method, we consider special cases for which C is not necessarily a covariance matrix,

$$C = \begin{bmatrix} \Delta & \mathbf{0} \\ \mathbf{0} & \mathbf{0} \end{bmatrix}, \quad \Delta \in \mathbf{R}^{M\times M}.$$

Case I ($M = 2N+1$)**:** First consider the case that Δ is a diagonal matrix and its diagonal elements are *arbitrary*;

$$\Delta = \begin{bmatrix} d_{11} & & \mathbf{0} \\ & \ddots & \\ \mathbf{0} & & d_{MM} \end{bmatrix}. \tag{6}$$

Then we have $C(AC)^{\div} = C\left([A_l A_r]C\right)^{\div} = \begin{pmatrix} A_l^{-1} \\ \mathbf{0} \end{pmatrix}$, where $A_l \in C^{M\times M}$, $A_r \in C^{M\times\infty}$, $A = [A_l \;\; A_r]$ and $\mathbf{0}$ the null matrix with M columns. By definition, this case is the same as the IPRM. Notice that no particular conditions for the elements of Δ were used. Δ is not necessarily the identity matrix or its diagonal elements are not necessarily decaying. For any choice of Δ the formulation is the same. Following the error analysis for the IPRM, the error is given by

$$\max_{-1\le x\le 1} |E(x)| \le \| A_l^{-1} A_r \mathbf{g}^{\perp} \|_1 + \| \mathbf{g}^{\perp} \|_1,$$

where $\mathbf{g}^{\perp} = (g_{M+1}, \cdots)^T$. Convergence is determined by how $A_l^{-1}A_r$ increases or decreases. The error is determined solely by the properties of the transformation matrix and the matrix C is irrelevant. Although the SF method is equivalent to the IPRM for this case, this formulation yields different results *numerically*. For example, consider the simple case that $C = \Delta$. Then the SF method is $\tilde{\mathbf{g}} = \Delta(A_l\Delta)^{\div}\hat{\mathbf{f}}$. Mathematically, we obtain $\tilde{\mathbf{g}} = \Delta(A_l\Delta)^{\div}\hat{\mathbf{f}} = A_l^{-1}\hat{\mathbf{f}}$, but numerically we obtain two different set of $\tilde{\mathbf{g}}$ due to round-off errors. By using the truncated singular value decomposition (SVD) method for the pseudo-inversion, the SF method yields more robust method than the IPRM.

Case II: We consider the case that Δ has the rank $m < M$. Then

$$C(AC)^{\div} = \begin{pmatrix} \Delta(A_l\Delta)^{\div} \\ \mathbf{0} \end{pmatrix},$$

where $A_l \in C^{M\times m}$. Since the reconstruction has a polynomial order m, this method yields the over-determined problem. This case is close to the IPRM with $A_l \in C^{m\times m}$ and the inversion is carried out by the pseudo-inverse.

For both cases I and II, we use any arbitrary diagonal matrix Δ and the results are equal or close to the inverse or pseudo-inverse methods respectively. The matrix C is independent of the regularity of $f(x)$ for these cases. In particular when A_l is the square matrix, C is completely arbitrary. In this aspect, the SF method is different from the IPRM. That is, the method is generalized to

$$\mathbf{g} = \begin{pmatrix} A_l^{\div} \\ \mathbf{0} \end{pmatrix} \hat{\mathbf{f}}. \qquad (7)$$

3.3 Spectral Accuracy and Exactness

As explained in the previous section, the matrix C has to be prescribed properly for the SF method. Due to the analyticity of the function, it is reasonable to have $C_{ii} = q^{2(i-1)}$, for $i = 1, 2, \cdots$, where $q < 1$ and $C_{ij} = 0$ if $i \neq j$. The free parameter q was chosen to be $q \sim \frac{1}{2}$ in [11]. This implies that (1) the SF method is not exact for a polynomial $f(x)$, (2) if $q \ll 1$, the SF method is close to the low order IPRM and thus the method does not yield exponential convergence after a certain polynomial order, and (3) if $q \sim 1$, the SF method is very slowly convergent. As the SF method is not exact, the exactness is only achieved when Δ is adopted, that is, the exactness is recovered only when the IPRM is used. Suppose that the unknown function is indeed a polynomial of order m and $m < M = 2N + 1$; $f(x) = \sum_{l=0}^{m} g_l L_l(x)$. Then we have $\hat{\mathbf{f}} = A_l(g_0, \cdots, g_m, 0, 0, \cdots)^T$, and $\tilde{\mathbf{g}} = C(AC)^{\div} A_l(g_0, \cdots, g_m, 0, 0, \cdots)^T$, where $A_l \in C^{M\times m}$. For the exactness, $\tilde{\mathbf{g}} = \mathbf{g}$, we need $I - C(AC)^{\div} A_l = \mathbf{0}$. This is only possible when Δ is a unit matrix whose size is $m \times m$. Then by the definition of the pseudo-inverse, $I - C(AC)^{\div} A_l = I - A_l^{\div} A_l = I - I = \mathbf{0}$. This is exactly the same as the IPRM of order m.

3.4 Numerical Convergence with Round-Off Errors

The numerical convergence of the IPRM or the SF method is affected by round-off errors [7, 11]. The source of round-off errors for the SF method are singular values that are much smaller than machine accuracy. By truncating such small singular values, the SF method can be less sensitive to round-off errors. Let ε_t be the tolerance level that all the singular values smaller than ε_t are truncated, such that $\Sigma^+ = diag(1/\sigma_1, \cdots, 1/\sigma_{N_t}, 0, \cdots, 0)$, where Σ^+ is the pseudo-inverse of the singular value matrix, σ_i the singular values, and N_t the index for $\forall \sigma_i \leq \varepsilon_t, i > N_t$. Based on the truncated SVD, two improvements can be made: (1) use $A_l \in C^{M \times m}$ with $m < M$ and $g = A_l^{\div} f$, for which no priori assumption for C is necessary. A similar idea was proposed and the rigorous proof of the bounded condition number ($N \geq m^2$) was given in [5]. And (2) use the IPRM with the preconditioner C and pseudo-inversion, i.e., $\tilde{\mathbf{g}} = C(A_l C)^{\div} f$. This makes the IPRM less sensitive to round-off errors. Recently a similar idea was applied to the polynomial reconstruction for the resolution of the Runge phenomenon [6].

4 A Hybrid IPRM and SF Method: Numerical Results

The proposed hybrid method is that the IPRM is used with the square matrix A for given N and A is preconditioned by C. Then the pseudo-inverse is used for the reconstruction.

For numerical experiments, first we consider $f(x) = \sin(0.4\pi x)$. The left figure of Fig. 1 shows the pointwise errors of the IPRM for $N = 64$, and $A \in C^{(129)\times(129)}$ with various $q = 0.95, 0.9$, and 0.8. The MATLAB command *pinv* is used for the pseudo-inverse which uses the truncated SVD. The MATLAB computation of the condition number of A is 2.6848×10^{17}. The figure shows that the improved IPRM (blue, green and purple solid lines) improves a lot the original IPRM (red solid line).

The right figure of Fig. 1 shows a convergence of the SF method with different values of q. We use the matrix $A \in C^{(2N+1)\times(10N)} = C^{(129)\times(640)}$, that is, $N = 64$ and A is truncated at the $10N = 640$th column. The figure shows how the convergence of the SF method can be affected by q. The best performance of the SF method was obtained with $q = 0.8$. Our numerical results (not included in the figure for graphical clarity) also shows that almost similar errors were obtained with $q = 0.7$ and 0.5. The red solid line in the right figure shows the hybrid method with $q = 0.8$. The figure shows that the hybrid method achieves almost similar results as the SF method with $q = 0.8$. But the hybrid method uses much smaller matrix.

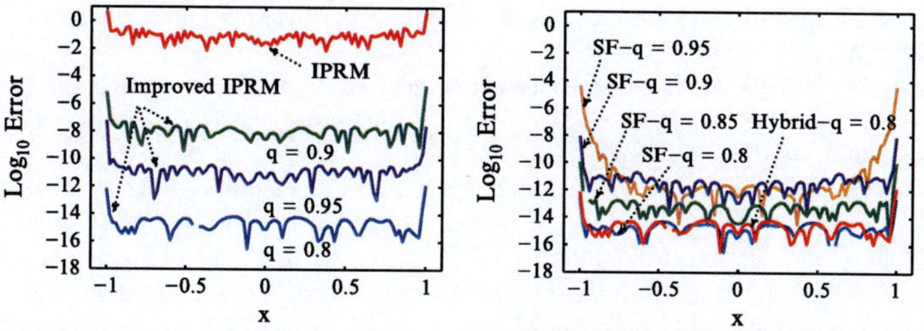

Fig. 1 The pointwise errors in logarithmic scale. *Left*: the IPRM (*red*) and improved IPRM reconstructions with $q = 0.8$ (*blue*), $q = 0.9$ (*green*), and $q = 0.95$ (*purple*). *Right*: the SF methods ($q = 0.95, 0.9, 0.85, 0.8$ (*brown, purple, green, blue*) and the hybrid method $q = 0.8$ (*red*)

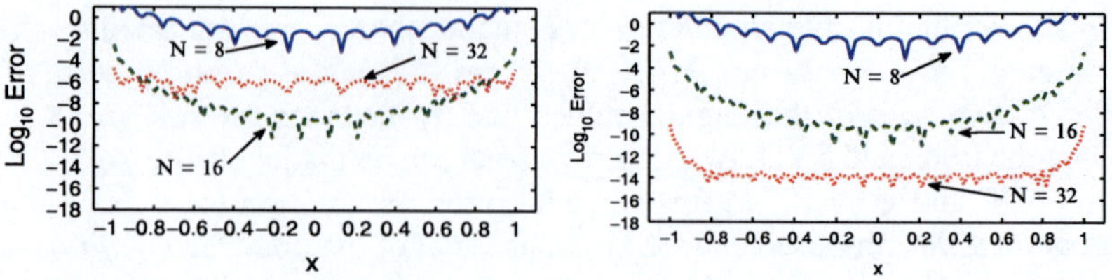

Fig. 2 The pointwise errors in logarithmic scale for the IPRM (*left*) and the hybrid method (*right*). $N = 8, 16, 32$

We also consider $f(x) = \sin(5.5\pi x)$. Figure 2 shows the pointwise errors for the IPRM and the hybrid method with $N = 8, 16, 32$. For the hybrid method, we use $q = 0.7$. As shown in the figure, the hybrid method performs much better and yields better convergence than the direct IPRM.

5 Conclusions

In this paper, the IPRM and the SF method are mathematically compared. Based on this comparison, we propose a hybrid IPRM and SF method. The baseline hybrid method is the IPRM with the covariance matrix defined in the SF method, which significantly reduces round-off errors and the size of the connection matrix. Future research should center around the optimization of parameter q.

Acknowledgements The author acknowledges the useful discussion with Alex Solomonoff and also thanks Tomasz Hrycak for allowing me to cite his unpublished manuscript, which is now published [5]. This work has been supported by the NSF under Grant No. DMS-0608844. The author would like to dedicate this paper to the late Prof. David Gottlieb, whose inspiration helped to make much of this work.

References

1. A. Abdi, M. Hosseini, An investigation of resolution of 2-variate Gibbs phenomenon, Appl. Math. Comput. 203 (2008), 714–732
2. J.P. Boyd, Jun Rong Ong, Exponentially-convergent strategies for defeating the Runge phenomenon for the approximation of non-periodic functions, part I: Single-interval schemes, Commun. Comput. Phys. 5 (2009), 484–497
3. D. Gottlieb, C.-W Shu, On the Gibbs phenomenon and its resolution, SIAM Rev. 39 (1997), 644–668
4. D. Gottlieb, C.-W. Shu, A. Solomonoff, H. Vandeven, On the Gibbs phenomenon I: Recovering exponential accuracy from the Fourier partial sum of a nonperiodic analytic function, J. Comput. Appl. Math. 43 (1992), 81–92
5. T. Hrycak, K. Gröchenig, Pseudospectral Fourier reconstruction with the modified inverse polynomial reconstruction method, J. Comput. Phys. 229 (2010), 933–946

6. J.-H. Jung, W. Stefan, A simple regularization of the polynomial interpolation for the resolution of the Runge phenomenon, J. Sci. Comput. DOI: 10.1007/s10915-010-9397-7
7. J.-H. Jung, B. D. Shizgal, Generalization of the inverse polynomial reconstruction method in the resolution of the Gibbs phenomena, J. Comput. Appl. Math. 172 (2004), 131–151
8. J.-H. Jung, B. D. Shizgal, On the numerical convergence with the inverse polynomial reconstruction method for the resolution of the Gibbs phenomenon, J. Comput. Phys. 2241 (2007), 477–488
9. M. Krebs, Reduktion des Gibbs-Phänomens in der Magnetresonanztomographie, Master's thesis, Technical University of Dortmund, 2007
10. B. D. Shizgal, J.-H Jung, Towards the resolution of the Gibbs phenomena, J. Comput. Appl. Math. 161 (2003), 41–65
11. A. Solomonoff, Reconstruction of a discontinuous function from a few Fourier coefficients using bayesian estimation, J. Sci. Comput. 10 (1995), 29–80

Numerical Simulation of Fluid–Structure Interaction in Human Phonation: Verification of Structure Part

Martin Larsson and Bernhard Müller

Abstract A high order finite-difference method has been developed to model fluid–structure interaction during phonation in the human larynx. The motion of the vocal folds is obtained by solving the elastic equations while the airflow is modeled by solving the compressible Navier–Stokes equations. In this paper, we address the problem of obtaining time-stable solutions for the linear elastic equations.

1 Introduction

Fluid–structure interaction in the human larynx generates our voice [5, 9]. We have developed a high order difference method to simulate the interaction of compressible flow in the larynx with the elastic structure of the vocal folds [3]. This paper deals with obtaining time-stable solutions for the linear elastic wave equation in a first-order formulation, which form the basis for more advanced structure models.

When written as a system of first order equations, the stability theory which is well developed for hyperbolic systems, applies directly. The disadvantage compared with a second order formulation in time is the increased computational effort required for the additional variables. Our main motivation for using a first order formulation is, however, related to the application of fluid–structure interaction where the traction boundary condition dictates the stresses on the elastic body. In the first order formulation, the traction boundary condition can be easily formulated as a simple Dirichlet condition for a subset of the solution variables. Dirichlet boundary conditions are not at all straight-forward to impose in a second order formulation [6].

M. Larsson (✉) and B. Müller
Norwegian University of Science and Technology (NTNU), Department of Energy and Process Engineering (EPT), 7491 Trondheim, Norway
e-mail: martin.larsson@ntnu.no, bernhard.muller@ntnu.no

J.S. Hesthaven and E.M. Rønquist (eds.), *Spectral and High Order Methods for Partial Differential Equations*, Lecture Notes in Computational Science and Engineering 76, DOI 10.1007/978-3-642-15337-2_20, © Springer-Verlag Berlin Heidelberg 2011

2 Theory

The 2D linear elastic wave equation in first order form are

$$
\begin{aligned}
u_t &= (1/\rho) f_x + (1/\rho) g_y \\
v_t &= (1/\rho) g_x + (1/\rho) h_y \\
f_t &= (\lambda + 2\mu) u_x + \lambda v_y \\
g_t &= \mu v_x + \mu u_y \\
h_t &= \lambda u_x + (\lambda + 2\mu) v_y
\end{aligned} \tag{1}
$$

where u, v are the velocity components and f, g, h are the three independent components of the symmetric Cauchy stress tensor. The Lamé parameters λ, μ and the density ρ are here taken to be constant in space and time.

Introducing the solution vector q and coefficient matrices A and B allows us to write the linear elastic wave equation (1) as a first order hyperbolic system

$$
q_t = A q_x + B q_y, \tag{2}
$$

where $q = (u, v, f, g, h)^{\mathrm{T}}$. The wave speeds of the system are $c_s = \sqrt{\mu/\rho}$ and $c_p = \sqrt{(\lambda + 2\mu)/\rho}$, referred to as secondary (or shear) and primary wave velocity, respectively. For convenience, we also define the parameter $\alpha = (\lambda + 2\mu)/\lambda = c_p^2/(c_p^2 - 2c_s^2)$.

In order to obtain simultaneous approximation (SAT) terms (to be explained below) for the system, we need to transform the system to characteristic variables. This can indeed be done, since the system (2) is hyperbolic. Thus, there exists an invertible matrix $T(k)$ for all directions k in 2D such that $T^{-1}(k)P(k)T(k) = \Lambda(k)$, where $P(k) = k_1 A + k_2 B$. The diagonal real eigenvalue matrix $\Lambda(k)$ can be chosen such that the eigenvalues occur in descending order. For the x- and y-directions, we get the following characteristic variables for the system (2).

$$
u^{(x)} = T_x^{-1} q = \frac{1}{2}\begin{bmatrix} \lambda u/c_p + f/\alpha \\ v + g/(\rho c_s) \\ -2f/\alpha + 2h \\ v - g/(\rho c_s) \\ -\lambda u/c_p + f/\alpha \end{bmatrix}, \quad u^{(y)} = T_y^{-1} q = \frac{1}{2}\begin{bmatrix} \lambda v/c_p + h/\alpha \\ u + g/(\rho c_s) \\ 2f - 2h/\alpha \\ u - g/(\rho c_s) \\ -\lambda v/c_p + h/\alpha \end{bmatrix}. \tag{3}
$$

Note that we use the symbol u to refer to both the vector of characteristic variables and the first velocity component. The meaning of u should be clear from context. The transformation back to flow variables is given by $q = Tu$.

3 Summation by Parts Operators

The idea behind the summation by parts technique [8] is to construct a difference operator Q which satisfies a discrete analogue to the continuous integration by parts property. This is called a summation by parts (SBP) property and by the energy method (cf. e.g., [2]), the discrete problem then satisfies the same energy estimate as the continuous problem.

For diagonal norm matrices H, there exist difference operators Q accurate to order $\mathcal{O}(h^{2s})$ in the interior and $\mathcal{O}(h^s)$ near the boundaries for $s = 1, 2, 3$ and 4. These operators have an effective order of accuracy $\mathcal{O}(h^{s+1})$ in the entire domain. Explicit forms of such operators Q and norm matrices H have been derived by Strand [8]. For this study, we use an SBP operator based on the central sixth order explicit finite difference operator ($s = 3$) which has been modified near the boundaries in order to satisfy the SBP property giving an effective $\mathcal{O}(h^4)$ order of accuracy in the whole domain.

With the injection method, numerical solutions with stable schemes can still exhibit a nonphysical growth in time which is not explained by the continuous equation. Simultaneous approximation terms (SAT) were devised to obtain time-stable solutions [1]. In this approach, a linear combination of the boundary condition and the differential equation is solved at the boundary instead of injecting the value at the end of each Runge–Kutta stage. This leads to a weak imposition of the physical boundary conditions. The imposition of SAT boundary conditions is accomplished by adding a term to the derivative operator, proportional to the difference between the value of the discrete solution u_N and the boundary condition to be fulfilled.

The strictly stable SAT method for a hyperbolic system in one space dimension with diagonal coefficient matrices was derived in [1] and is the basis for this work. The continuous 1D model problem is $u_t = \Lambda u_x$, $0 \leq x \leq 1$, with r unknowns and r equations and the coefficient matrix Λ is chosen such that the eigenvalues are in descending order, i.e., $\lambda_1 > \lambda_2 > \ldots > \lambda_k > 0 > \lambda_{k+1} > \ldots > \lambda_r$. We split the solution vector into two parts corresponding to positive and negative eigenvalues $\mathbf{u}^{\mathrm{I}} = (\mathbf{u}^{(1)}, \ldots, \mathbf{u}^{(k)})^{\mathrm{T}}$ and $\mathbf{u}^{\mathrm{II}} = (\mathbf{u}^{(k+1)}, \ldots, \mathbf{u}^{(r)})^{\mathrm{T}}$. For the variables in $\mathbf{u}^{\mathrm{I}}$ (each a grid function of length $N+1$) we have boundary conditions at $x = 1$, and for $\mathbf{u}^{\mathrm{II}}$ we need to specify boundary conditions at $x = 0$, as this is required for well-posedness.

Since we are here dealing with characteristic variables, we need to transform our physical boundary conditions to boundary conditions for the characteristic variables. This is done through the boundary functions $\mathbf{g}^{\mathrm{I}}(t) = (g^{(1)}(t), \ldots, g^{(k)}(t))$, $\mathbf{g}^{\mathrm{II}}(t) = (g^{(k+1)}(t), \ldots, g^{(r)}(t))$ and the coupling matrices R and L defined by

$$\mathbf{u}^{\mathrm{I}}(1,t) = R\mathbf{u}^{\mathrm{II}}(1,t) + \mathbf{g}^{\mathrm{I}}(t), \quad \mathbf{u}^{\mathrm{II}}(0,t) = L\mathbf{u}^{\mathrm{I}}(0,t) + \mathbf{g}^{\mathrm{II}}(t) \tag{4}$$

The SAT method is then:

$$\begin{aligned} \frac{d\mathbf{u}^{(i)}}{dt} &= \lambda_i Q\mathbf{u}^{(i)} - \lambda_i \tau \mathbf{S}^{(i)}(\mathbf{u}_N^{(i)} - (R\mathbf{u}^{\mathrm{II}})_N^{(i)} - g^{(i)}(t)), \quad 1 \leq i \leq k \\ \frac{d\mathbf{u}^{(i)}}{dt} &= \lambda_i Q\mathbf{u}^{(i)} + \lambda_i \tau \mathbf{S}^{(i)}(\mathbf{u}_0^{(i)} - (L\mathbf{u}^{\mathrm{I}})_0^{(i-k)} - g^{(i)}(t)), \quad k+1 \leq i \leq r \end{aligned} \tag{5}$$

where $\mathbf{S}^{(i)} = H^{-1}(0, 0, \ldots, 1)^{\mathrm{T}}$ for $1 \leq i \leq k$ and $\mathbf{S}^{(i)} = H^{-1}(1, 0, \ldots, 0)^{\mathrm{T}}$ for $k + 1 \leq i \leq r$. Regarding the notation, $(R\mathbf{u}^{\mathrm{II}})_N^{(i)}$ should be interpreted as follows: $\mathbf{u}^{\mathrm{II}}$ is an $(r-k) \times 1$ vector where each component is a grid function of length $N+1$. Multiplying R (being a $k \times (r-k)$ matrix) with $\mathbf{u}^{\mathrm{II}}$ yields a new vector of grid functions ($k \times 1$ vector). Take the (i)th grid function in this vector and finally the Nth component in the resulting grid function. As shown in [1], the SAT method is both stable and time stable provided that

$$\frac{1 - 1\sqrt{1 - |R||L|}}{|R||L|} \leq \tau \leq \frac{1 + 1\sqrt{1 - |R||L|}}{|R||L|} \tag{6}$$

with the additional constraint that $|R||L| \leq 1$, where the matrix norm is defined as $|R| = \rho(R^T R)^{1/2}$ and ρ is the spectral radius.

4 Application to Elastic Wave Equation

Now, we shall apply the general method outlined above to derive SAT expressions for boundary conditions on the velocity components. The vector of characteristic variables in the x-direction is given in (3), but henceforth we drop the superscript $^{(x)}$. The derivation for the y-direction is analogous.

We let the grid functions $\mathbf{u}$ and $\mathbf{v}$ in 2D with points labeled $0 \leq i \leq N$ and $0 \leq j \leq M$ in the x- and y-directions, respectively, correspond to the solution variables u and v in the discretization of the linear elastic wave equation. We label the boundary $i = 0$ "left," $i = N$ "right," $j = 0$ "bottom," $j = M$ "top."

The boundary conditions for the velocity components in 2D are of the form $u(x = 0, y, t) = u_{\mathrm{left}}(y, t)$, i.e., a given function of time, which we write for the discrete variables as $u_{0,j}(t) = u_{\mathrm{left},j}(t)$, with similar notation for the other edges and the other solution variables. The SAT expressions, one for each spatial direction and for each solution variable, will also be grid functions.

We split the vector of characteristic variables into two smaller vectors corresponding to the positive and negative eigenvalues, omitting the characteristic with zero eigenvalue as the corresponding SAT expression will be zero,

$$u^{\mathrm{I}} = \frac{1}{2}\begin{bmatrix} (\lambda/c_p)u + (1/\alpha)f \\ v + (1/c_p\rho)g \end{bmatrix}, \quad u^{\mathrm{II}} = \frac{1}{2}\begin{bmatrix} v - (1/c_p\rho)g \\ (-\lambda/c_p)u + (1/\alpha)f \end{bmatrix}$$

and define the coefficient matrices $\Lambda^{\mathrm{I}} = \mathrm{diag}(c_p, c_s)$, $\Lambda^{\mathrm{II}} = \mathrm{diag}(-c_s, -c_p)$. The boundary functions g^{I}, g^{II}, and the matrices L and R are defined by the relations (4).

If we impose the boundary condition in the x-direction $u(x = 1, t) = u_{\mathrm{right}}(t)$, $v(x = 1, t) = v_{\mathrm{right}}(t)$, $u(x = 0, t) = u_{\mathrm{left}}(t)$,$v(x = 0, t) = v_{\mathrm{left}}(t)$, then the boundary matrices and functions are given by

$$R = \begin{bmatrix} 0 & 1 \\ -1 & 0 \end{bmatrix}, \; L = \begin{bmatrix} 0 & -1 \\ 1 & 0 \end{bmatrix}, \; g^{\mathrm{I}} = \begin{bmatrix} \lambda u_{\mathrm{right}}(t)/c_p \\ v_{\mathrm{right}}(t) \end{bmatrix}, \; g^{\mathrm{II}}(t) = \begin{bmatrix} v_{\mathrm{left}}(t) \\ -\lambda u_{\mathrm{left}}(t)/c_p \end{bmatrix}.$$

In the y-direction, dictating $u(y = 1, t) = u_{\mathrm{top}}(t)$, $v(y = 1, t) = v_{\mathrm{top}}(t)$, $u(y = 0, t) = u_{\mathrm{bottom}}(t)$, $v(y = 0, t) = v_{\mathrm{bottom}}(t)$, we get, likewise

$$R = \begin{bmatrix} 0 & 1 \\ -1 & 0 \end{bmatrix}, \; L = \begin{bmatrix} 0 & -1 \\ 1 & 0 \end{bmatrix}, \; g^{\mathrm{I}} = \begin{bmatrix} \lambda v_{\mathrm{top}}(t)/c_p \\ u_{\mathrm{top}}(t) \end{bmatrix}, \; g^{\mathrm{II}}(t) = \begin{bmatrix} u_{\mathrm{bottom}}(t) \\ -\lambda v_{\mathrm{bottom}}(t)/c_p \end{bmatrix}.$$

Corresponding expressions can be derived for boundary conditions on the stress components, but these are omitted here due to space limitations. We note that, as the spectral radius of both L and R is 1 in each case, the inequalities (6) lend no other choice than $\tau = 1$ for time-stability.

Using the definition (5) with the expressions above, the SAT terms are first obtained for the characteristic variables, and then for the flow variables by applying the transformation matrices T for the x- and y-directions. The resulting expressions are

$$\begin{aligned}
\overline{\mathrm{SAT}}^x_{i,j,1} &= -c_p \tau h_{00}^{-1}\left[\delta_{iN}\left(u_{N,j} - u_{\mathrm{right},j}(t)\right) + \delta_{i0}\left(u_{0,j} - u_{\mathrm{left},j}(t)\right)\right] \\
\overline{\mathrm{SAT}}^x_{i,j,2} &= -c_s \tau h_{00}^{-1}\left[\delta_{iN}\left(v_{N,j} - v_{\mathrm{right},j}(t)\right) + \delta_{i0}\left(v_{0,j} - v_{\mathrm{left},j}(t)\right)\right] \\
\overline{\mathrm{SAT}}^x_{i,j,3} &= -(\lambda + 2\mu)\tau h_{00}^{-1}\left[\delta_{iN}\left(u_{N,j} - u_{\mathrm{right},j}(t)\right) - \delta_{i0}\left(u_{0,j} - u_{\mathrm{left},j}(t)\right)\right] \\
\overline{\mathrm{SAT}}^x_{i,j,4} &= -\mu \tau h_{00}^{-1}\left[\delta_{iN}\left(v_{N,j} - v_{\mathrm{right},j}(t)\right) - \delta_{i0}\left(v_{0,j} - v_{\mathrm{left},j}(t)\right)\right] \\
\overline{\mathrm{SAT}}^x_{i,j,5} &= -\lambda \tau h_{00}^{-1}\left[\delta_{iN}\left(u_{N,j} - u_{\mathrm{right},j}(t)\right) - \delta_{i0}\left(u_{0,j} - u_{\mathrm{left},j}(t)\right)\right]
\end{aligned} \tag{7}$$

$$\begin{aligned}
\overline{\mathrm{SAT}}^y_{i,j,1} &= -c_s \tau h_{00}^{-1}\left[\delta_{jM}\left(u_{i,M} - u_{\mathrm{top},i}(t)\right) + \delta_{j0}\left(u_{i,0} - u_{\mathrm{bottom},i}(t)\right)\right] \\
\overline{\mathrm{SAT}}^y_{i,j,2} &= -c_p \tau h_{00}^{-1}\left[\delta_{jM}\left(v_{i,M} - v_{\mathrm{top},i}(t)\right) + \delta_{j0}\left(v_{i,0} - v_{\mathrm{bottom},i}(t)\right)\right] \\
\overline{\mathrm{SAT}}^y_{i,j,3} &= -\lambda \tau h_{00}^{-1}\left[\delta_{jM}\left(v_{i,M} - v_{\mathrm{top},i}(t)\right) - \delta_{j0}\left(v_{i,0} - v_{\mathrm{bottom},i}(t)\right)\right] \\
\overline{\mathrm{SAT}}^y_{i,j,4} &= -\mu \tau h_{00}^{-1}\left[\delta_{jM}\left(u_{i,M} - u_{\mathrm{top},i}(t)\right) - \delta_{j0}\left(u_{i,0} - u_{\mathrm{bottom},i}(t)\right)\right] \\
\overline{\mathrm{SAT}}^y_{i,j,5} &= -(\lambda + 2\mu)\tau h_{00}^{-1}\left[\delta_{jM}\left(v_{i,M} - v_{\mathrm{top},i}(t)\right) - \delta_{j0}\left(v_{i,0} - v_{\mathrm{bottom},i}(t)\right)\right],
\end{aligned} \tag{8}$$

where δ_{ij} is the Dirac delta function, i.e., $\delta_{ij} = 1$ if $i = j$ and 0 otherwise. The overline stands for physical (flow) variables.

5 Discretization

We consider the mapping $x = x(\xi, \eta)$, $y = y(\xi, \eta)$ and introduce an equidistant computational grid with coordinates $\xi_i, i = 0, \ldots, N$, $\eta_j, j = 0, \ldots, M$. The Jacobian determinant of the transformation is given by $J^{-1} = x_\xi y_\eta - x_\eta y_\xi$. The linear elastic wave equation can then be written

$$\hat{q}_t = (\hat{A}\hat{q})_\xi + (\hat{B}\hat{q})_\eta \tag{9}$$

where the hats signify that the quantities are in transformed coordinates, i.e., $\hat{q} = J^{-1}q$, $\hat{A} = \xi_x A + \xi_y B$ and $\hat{B} = \eta_x A + \eta_y B$.

Introduce a vector $\hat{\mathbf{q}} = (\hat{q}_{ijk})^{\mathrm{T}} = (\hat{q}_{001}, \ldots, \hat{q}_{005}, \hat{q}_{101}, \ldots, \hat{q}_{105}, \ldots, \hat{q}_{NM5})^{\mathrm{T}}$ where the three indices i, j and k represent the ξ-coordinate, η-coordinate and the solution variable, respectively. We shall define our discretization in terms of Kronecker products. This formulation is convenient because it mimics the finite difference implementation. Let $\mathbf{Q}_\xi = Q_\xi \otimes I_\eta \otimes I_5$ and $\mathbf{Q}_\eta = I_\xi \otimes Q_\eta \otimes I_5$ where Q_ξ is the 1D difference operator in the ξ-direction and I_ξ is the unit matrix of size $(N+1) \times (N+1)$. In the other direction, Q_η and I_η are $(M+1) \times (M+1)$ matrices. The computation of the spatial derivatives of $\hat{\mathbf{q}}$ can then be seen as operating on $\hat{\mathbf{q}}$ with one of the Kronecker products, i.e., $\mathbf{Q}_\eta \hat{\mathbf{q}}$ operates on the second index and yields a vector of the same size as $\hat{\mathbf{q}}$ representing the first derivative in the η-direction. To express the semi-discrete linear elastic wave equation, we also need to define $\hat{\mathbf{A}} = I_\xi \otimes I_\eta \otimes \hat{A}$ and $\hat{\mathbf{B}} = I_\xi \otimes I_\eta \otimes \hat{B}$. Note that these products are never actually explicitly formed as they are merely theoretical constructs to make the notation more compact. Using the Kronecker products defined above, the semidiscrete linear elastic wave equation including the SAT term can be written

$$\frac{d\hat{\mathbf{q}}}{dt} = \mathbf{Q}_\xi(\hat{\mathbf{A}}\hat{\mathbf{q}}) + \mathbf{Q}_\eta(\hat{\mathbf{B}}\hat{\mathbf{q}}) + \widehat{\mathbf{SAT}}. \tag{10}$$

This system of ordinary differential equations, including the SAT expression is solved using the classical 4th order explicit Runge–Kutta method. No injection is needed to impose boundary conditions, as this is taken care of by the SAT method. A 6th order explicit filter [7] is used to suppress unresolved modes.

Equations (7) and (8) give the expressions in Cartesian coordinates. However, we need the SAT expression for curvilinear coordinates. These can be obtained by considering the system $\hat{q}_t = ((k_x A + k_y B)\hat{q})_k$ for the two spatial directions $k = \xi, \eta$. As the expressions become quite long, they are omitted here (cf. [4]).

6 Numerical Experiment

We consider now a simple test for our 2D discretization: a square domain occupies the region $-1\,\mathrm{m} \le x \le 1\,\mathrm{m}$, $-1\,\mathrm{m} \le y \le 1\,\mathrm{m}$. At $t = 0\,\mathrm{s}$, we give an initial condition for the stress component $g(x, y, t = 0) = g_0(x, y)$, while all other variables are initially zero. The initial condition is defined by $g_0(x, y) = s(2r_1 + 0.5) - s(2r_2 + 0.5)$, where $r_1^2 = (x - 0.5\,\mathrm{m})^2 + y^2$, $r_2^2 = (x + 0.5\,\mathrm{m})^2 + y^2$ and

$$s(r) = \begin{cases} \exp(-1/r - 1/(1-r) + 4)\,\mathrm{kg}/(\mathrm{m}\cdot\mathrm{s}^2), & \text{if } 0\,\mathrm{m} < r < 1\,\mathrm{m} \\ 0, & \text{otherwise.} \end{cases} \tag{11}$$

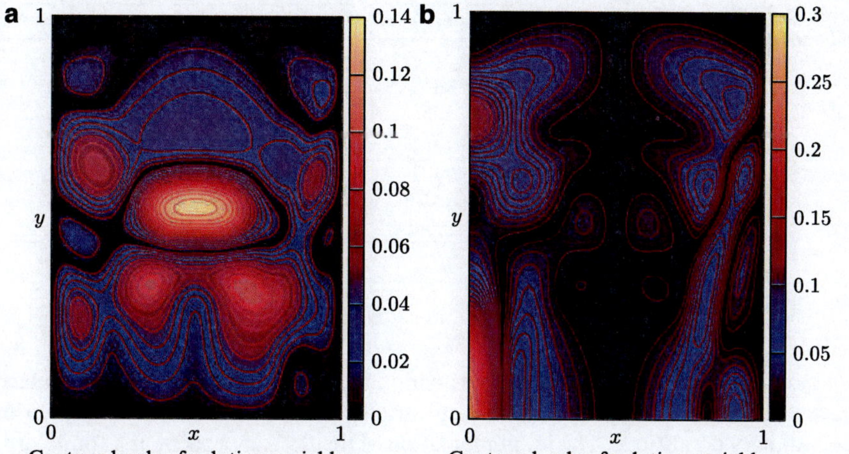

Fig. 1 Contour plots showing the absolute value of the velocity components u and v with contour levels spaced 0.01 m s^{-1} apart

Thus, $g_0(x, y)$ is a smooth function with infinitely many derivatives and compact support representing two sources located at $(\pm 0.5, 0)$ m. The material parameters are $\lambda = \mu = 1.0\,\mathrm{kg}/(\mathrm{m}\cdot\mathrm{s}^2)$, $\rho = 1.0\,\mathrm{kg\,m}^{-3}$. We integrate the solution with CFL number 0.8 and impose homogeneous Dirichlet boundary conditions for u and v using the SAT approach (3). We plot the solution evaluated at time $t = 0.5$ s in Fig. 1a, b. Since the largest wave speed $c_p = \sqrt{(\lambda + 2\mu)/\rho} = \sqrt{3}\,\mathrm{m\,s}^{-1}$, at $t = 0.5$ s the P-wave will have reached and been reflected from the left and right boundaries which are situated a distance 0.5 m from the sources. As the solution is symmetric with respect to the center lines $x = 0$ and $y = 0$, only the first quadrant is shown. The value of the velocity components at the boundary is zero, as enforced by the SAT term.

For the same set of parameters and initial/boundary conditions, we compute the solution at different grid resolutions and consider the solution at the finest grid to be exact. We can then calculate the error at each grid level and thus determine the rate of convergence. We define the 2-norm of the error at any grid level k as

$$e_2^{(k)} = \left[\frac{1}{NM} \sum_{\phi \in \{u,v,f,g,h\}} \sum_{i=0}^{N} \sum_{j=0}^{M} \left| \phi_{i,j}^{(k)} - \phi_{\mathrm{exact},i,j}^{(k)} \right|^2 \right]^{1/2}, \qquad (12)$$

where $\phi_{\mathrm{exact}}^{(k)}$ is the restriction of the solution $\phi^{(0)}$ on the finest grid to the grid on level k. As can be seen in Table 1, the order in the 2-norm approaches 4 as N and M increase, which is what we expected.

Table 1 2-norms of error and rates of convergence

$N \times M$	k	$e_2^{(k)}$	$\log_2(e_2^{(k+1)}/e_2^{(k)})$
32×32	6	3.646×10^{-2}	–
64×64	5	6.800×10^{-3}	2.423
128×128	4	8.521×10^{-4}	2.996
256×256	3	7.421×10^{-5}	3.521
512×512	2	4.558×10^{-6}	4.025
$1{,}024 \times 1{,}024$	1	2.710×10^{-7}	4.072
$2{,}048 \times 2{,}048$	0	0	–

7 Conclusions

We have derived simultaneous approximation terms (SAT) for the 2D linear elastic wave equation in first-order form to yield strictly stable schemes for general Dirichlet boundary conditions. The implementation of the SAT approach for a fourth order difference scheme has proved that the convergence rate is indeed fourth order for a test case with smooth data. The advantage of our approach is that Dirichlet boundary conditions can easily be imposed for either the velocity or the stress components which is required for fluid-structure interaction. In the future, we plan to apply this approach to the nonlinear elastic equations based on a Neo-Hookean model.

Acknowledgements The authors thank Bjørn Skallerud, Paul Leinan and Victorien Prot at the Department of Structural Engineering, NTNU for valuable discussions on the structure model and for Abaqus support. The current research has been funded by the Swedish Research Council under the project "Numerical Simulation of Respiratory Flow."

References

1. M.H. Carpenter, D. Gottlieb, and S. Abarbanel. Time-stable boundary conditions for finite-difference schemes solving hyperbolic systems: Methodology and application to high-order compact schemes. *J. Comput. Phys.*, 111:220–236, 1994
2. B. Gustafsson. *High order difference methods for time-dependent PDE.* Springer, Berlin, 2008
3. M. Larsson and B. Müller. Numerical simulation of fluid-structure interaction in human phonation: Application. In *Proceedings of ENUMATH 2009 Eighth European Conference on Numerical Mathematics and Advanced Applications*, Uppsala, Sweden, 2009 (to be published by Springer)
4. M. Larsson and B. Müller. Strictly stable high order difference method for the linear elastic wave equation. *Commun. Comput. Phys.* (Submitted)
5. H. Luo, R. Mittal, X. Zheng, S.A. Bielamowicz, R.J. Walsh, and J.K. Hahn. An immersed-boundary method for flow – structure interaction in biological systems with application to phonation. *J. Comput. Phys.*, 227:9303–9332, 2008
6. K. Mattsson, F. Ham, and G. Iaccarino. Stable boundary treatment for the wave equation on second-order form. *J. Sci. Comput.*, 41:366–383, 2009
7. B. Müller. High order numerical simulation of aeolian tones. *Comput. Fluid*, 37(4):450–462, 2008
8. B. Strand. Summation by parts for finite difference approximations for d/dx. *J. Comput. Phys.*, 110:47–67, 1994
9. I.R. Titze. *Principles of voice production.* National Center for Voice and Speech, 2000

A New Spectral Method on Triangles

Youyun Li, Li-Lian Wang, Huiyuan Li, and Heping Ma

Abstract We propose in this note a spectral method on triangles based on a new rectangle-to-triangle mapping, which leads to more reasonable grid distributions and efficient implementations than the usual approaches based on the collapsed transform. We present the detailed implementation for spectral approximations on a triangle and discuss the extension to spectral-element methods and three dimensions.

1 Introduction

Spectral element methods, which are capable of extending the standard spectral methods to complex geometries, have become an important tool for simulations of fluid dynamics, atmospheric modeling and many other phenomena. Since the seminal work [1], a large body of literature has been devoted to the tensor-based quadrilateral/hexahedral element methods (QSEM) (see, e.g., [2]). Recently, some progress has also been made in the triangular/tetrahedral spectral/hp element methods (TSEM), and the current approaches are mainly based on (1) the Koornwinder-Dubiner polynomials [3, 4]; (2) non-polynomial on triangular

L.-L. Wang (✉)
Division of Mathematical Sciences, School of Physical and Mathematical Sciences, Nanyang Technological University, Singapore 637371
e-mail: lilian@ntu.edu.sg

Y. Li
College of Mathematics and Computing Science, Changsha University of Science and Technology, Hunan 410004, China
e-mail: liyouyun8@hotmail.com

H. Li
Institute of Software, Chinese Academy of Sciences, Beijing 100190, China
e-mail: hynli@mail.rdcps.ac.cn

H. Ma
Department of Mathematics, College of Sciences, Shanghai University, Shanghai 200444, China
e-mail: hpma@shu.edu.cn

J.S. Hesthaven and E.M. Rønquist (eds.), *Spectral and High Order Methods for Partial Differential Equations*, Lecture Notes in Computational Science and Engineering 76, DOI 10.1007/978-3-642-15337-2_21,

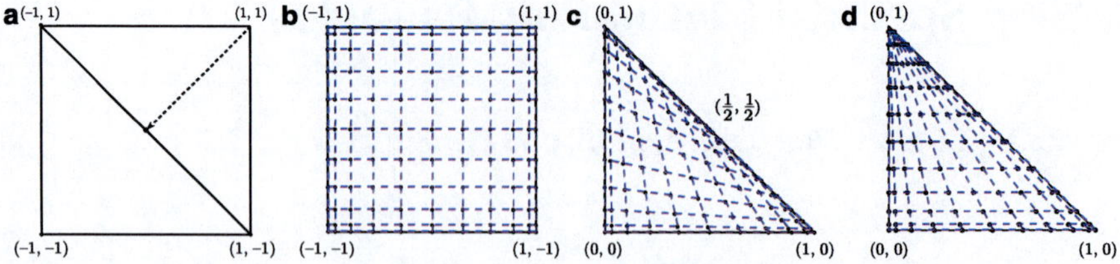

Fig. 1 (**a**) Illustration of the mapping (1) from the square **Q** onto the triangle $\{(x, y) : -1 \leq x, y; x + y \leq 0\}$; (**b**) tensorial Legendre–Gauss–Lobatto (LGL) grids on **Q**; (**c**) mapped LGL grids on **T** using the mapping (3); (**d**) mapped LGL grids on **T** using Duffy's transform in [4]

elements [5,6]; or (3) special nodal points [7,8]. In the first approach, the collapsed mapping (i.e., the Duffy's transform) is used to generate warped tensorial orthogonal polynomials on triangles/tetrahedra from the tensorial polynomial bases on rectangles/hexahedra. The second technique is also based on such a mapping to generate rational basis functions rather than polynomials. It is known that the Duffy's transform collapses one edge/face of the reference rectangle/hexahedron into a vertex of the triangle/tetrahedron, so the computational grids are severely clustered near the singular vertex.

This note aims to introduce a new rectangle-to-triangle mapping, which pulls one edge (at the middle point) of the triangle to two edges of the reference rectangle (cf. Fig. 1a). In contrast with the collapsed mapping, such a mapping is one-to-one, and leads to a more reasonable distribution on the triangle (cf. Fig. 1c, d). Most importantly, with a slight modification of the nodal Lagrange polynomial basis on the reference rectangle, we can derive a nodal basis (formed by irrational functions) on the triangles, which allows for an efficient implementation as with the QSEM. In a nutshell, we can view a triangular element as a deformed rectangular element, and demonstrate that the numerical issues induced by the deformation can be handled effectively. Significantly, this provides a great flexibility for the mesh generation and improves the performance of QSEM. Typically, allowing the elements being triangles along the boundaries, one can handle more complex computational domains with more regular meshes.

In this note, we first introduce the mapping and the nodal basis, then consider the implementation on a triangle, followed by the extensions to three dimensions and spectral-element methods.

2 Rectangle-to-Triangle Mapping and Nodal Basis

Hereafter, (x, y) is the Cartesian coordinate of a generic point in a triangle, while (ξ, η) represents the Cartesian coordinate of a point in the reference square: $\mathbf{Q} = (-1, 1)^2$. Given the vertex coordinates $\{(x_A, y_A), (x_B, y_B), (x_D, y_D)\}$ of a triangle $\triangle_{ABD}$, we can one-to-one map the square **Q** to the triangle region through

$$(x, y) = (x_A, y_A)\frac{(1-\xi)(1-\eta)}{4} + (x_B, y_B)\frac{(1+\xi)(3-\eta)}{8} + (x_D, y_D)\frac{(3-\xi)(1+\eta)}{8}. \tag{1}$$

Under this mapping, the vertices $(-1,-1), (1,-1)$ and $(-1,1)$ of $\mathbf{Q}$ correspond to the vertices A, B, D of $\triangle_{ABD}$, respectively, while the middle point C of the edge BD is the image of the vertex $(1,1)$ of $\mathbf{Q}$. Hence, this mapping deforms two edges ($\xi = 1$ and $\eta = 1$) of $\mathbf{Q}$ into one single edge (BD) of $\triangle_{ABD}$. An illustration of such a one-to-one correspondence is depicted in Fig. 1.

To be more specific, we confine ourselves to the special triangle:

$$\mathbf{T} := \{(x,y) : 0 < x, y < 1,\ 0 < x + y < 1\}, \tag{2}$$

and in this case, the mapping (1) (with BD being the hypotenuse) takes the form:

$$x = \frac{1}{8}(1+\xi)(3-\eta), \quad y = \frac{1}{8}(3-\xi)(1+\eta), \quad \forall\ (\xi,\eta) \in \mathbf{Q}, \tag{3}$$

with the inversion

$$\begin{cases} \xi = 1 + x - y - \sqrt{(x-y)^2 + 4(1-x-y)}, \\ \eta = 1 - x + y - \sqrt{(x-y)^2 + 4(1-x-y)}, \end{cases} \quad \forall\ (x,y) \in \mathbf{T}. \tag{4}$$

Under this mapping, we have

$$\frac{\partial x}{\partial \xi} = \frac{3-\eta}{8}, \quad \frac{\partial x}{\partial \eta} = -\frac{1+\xi}{8}, \quad \frac{\partial y}{\partial \xi} = -\frac{1+\eta}{8}, \quad \frac{\partial y}{\partial \eta} = \frac{3-\xi}{8}, \tag{5}$$

so the Jacobian determinant is given by

$$J = \det\left(\frac{\partial(x,y)}{\partial(\xi,\eta)}\right) = \frac{2-\xi-\eta}{16}. \tag{6}$$

In the sequel, we always associate a function u in $\mathbf{T}$ with a unction $\tilde{u}$ in $\mathbf{Q}$ via the mapping (3): $\tilde{u}(\xi,\eta) = u(x,y)$ and likewise for $\tilde{v}$ etc. One verifies that

$$\nabla u = (\partial_x u,\ \partial_y u) = \frac{2}{2-\xi-\eta}\Big((3-\xi)\partial_\xi \tilde{u} + (1+\eta)\partial_\eta \tilde{u},\ (1+\xi)\partial_\xi \tilde{u} + (3-\eta)\partial_\eta \tilde{u}\Big) := \widetilde{\nabla} \tilde{u}, \tag{7}$$

and

$$\begin{aligned} \iint_{\mathbf{T}} \nabla u \cdot \nabla v \, dxdy &= \iint_{\mathbf{Q}} \big(\widetilde{\nabla} \tilde{u} \cdot \widetilde{\nabla} \tilde{v}\big)\, J \, d\xi d\eta \\ &= \iint_{\mathbf{Q}} \Big(G_1(\xi)\partial_\xi \tilde{u} \partial_\xi \tilde{v} + G_2(\xi,\eta)\big(\partial_\xi \tilde{u} \partial_\eta \tilde{v} + \partial_\eta \tilde{u} \partial_\xi \tilde{v}\big) + G_1(\eta)\partial_\eta \tilde{u} \partial_\eta \tilde{v}\Big)\frac{1}{J}\, d\xi d\eta, \end{aligned} \tag{8}$$

where G_1 and G_2 are given by

$$G_1(z) = \frac{1}{64}\big((1+z)^2 + (3-z)^2\big), \quad G_2(\xi,\eta) = \frac{1}{8} - \frac{1}{32}(1-\xi)(1-\eta). \tag{9}$$

Consequently, the space $H^1(\mathbf{T})$ is mapped to the weighted space over $\mathbf{Q}$:

$$\widetilde{H}^1_\omega(\mathbf{Q}) := \left\{\tilde{u} \in L^2_\omega(\mathbf{Q}) \ : \widetilde{\nabla}\tilde{u} \in L^2_\omega(\mathbf{Q})\right\} \ \text{with} \ \omega = J, \tag{10}$$

and vice verse. We observe from (7) and (8) that if ∇u is well-defined at the middle point $(\frac{1}{2}, \frac{1}{2})$ of the hypotenuse of $\mathbf{T}$, then we have

$$\left(\frac{\partial \tilde{u}}{\partial \xi} + \frac{\partial \tilde{u}}{\partial \eta}\right)\Big|_{(1,1)} = 0. \tag{11}$$

This condition induced by the rectangle-to-triangle deformation can be viewed as an analogy of the pole condition in the polar and spherical coordinates. An essential point here is how to treat this condition effectively without loss of accuracy and implementation efficiency. For this purpose, we next construct a nodal basis for the finite-dimensional approximation space over $\mathbf{Q}$:

$$\widetilde{X}_N := \widetilde{H}^1_\omega(\mathbf{Q}) \cap [\mathscr{P}_N]^2 = \left\{\phi \in [\mathscr{P}_N]^2 \ : \ (\partial_\xi \phi + \partial_\eta \phi)\big|_{(1,1)} = 0\right\}, \tag{12}$$

where $\mathscr{P}_N$ is the set of all algebraic polynomials of degree $\le N$ in $(-1, 1)$. Let $\{z_j\}_{j=0}^N$ (with $z_0 = -1$ and $z_N = 1$) be the Legendre–Gauss–Lobatto points, i.e., the zeros of the polynomial $(1-z^2)L'_N(z)$, where L_N is the Legendre polynomial of degree N. Let $\{h_j\}_{j=0}^N$ be the Lagrange polynomial basis associated with $\{z_j\}_{j=0}^N$, and denote $d_{jk} = h'_k(z_j)$. Define

$$\tilde{h}_j(z) := h_j(z) - \frac{d_{Nj}}{2d_{NN}} h_N(z), \quad 0 \le j \le N-1. \tag{13}$$

It is clear that $h_j(z) \in \mathscr{P}_N$ and

$$\tilde{h}_j(z_k) = \delta_{kj}, \ \ \tilde{h}_j(1) = -\frac{d_{Nj}}{2d_{NN}}, \quad \tilde{h}'_j(1) = \frac{d_{Nj}}{2}, \quad 0 \le k, j \le N-1. \tag{14}$$

Setting

$$\psi_{ij}(\xi, \eta) = \begin{cases} h_i(\xi)h_j(\eta), & 0 \le i, j \le N-1, \\ \tilde{h}_i(\xi)h_N(\eta), & 0 \le i \le N-1; \ \ j = N \quad (\text{edge}: \xi = 1), \\ h_N(\xi)\tilde{h}_j(\eta), & i = N, \ \ 0 \le j \le N-1 \quad (\text{edge}: \eta = 1), \end{cases} \tag{15}$$

and

$$\Upsilon_N := \left\{(i, j) \ : \ 0 \le i, j \le N \ \ \text{but} \ \ (i, j) \ne (N, N)\right\},$$

we find from (14) that all the ψ_{ij} satisfy (11) and

$$\widetilde{X}_N = \text{span}\{\psi_{ij} : (i,j) \in \Upsilon_N\} \;\Rightarrow\; \dim(\widetilde{X}_N) = (N+1)^2 - 1. \tag{16}$$

It is seen that we modified the usual tensorial nodal basis $\{h_i(\xi)h_j(\eta)\}_{i,j=0}^N$ along the edges: $\xi = 1$ and $\eta = 1$ of $\mathbf{Q}$ so as to meet the condition (11) at the singular point. In view of (14), $\{\psi_{ij}\}_{i,j\in\Upsilon_N}$ forms a nodal basis of $\widetilde{X}_N$. More precisely, we have

$$\psi_{ij}(\xi_p, \eta_q) = \delta_{pi}\delta_{qj}, \quad \forall\,(i,j),\,(p,q) \in \Upsilon_N, \tag{17}$$

where $\{\xi_k = \eta_k = z_k\}_{k=0}^N$ are the LGL points as before.

The above nodal basis is complete in $\widetilde{H}^1_\omega(\mathbf{Q})$, but in order to enforce continuity across the elements, we need to define a nodal basis function at the singular vertex $(1,1)$. Define

$$\psi_{NN}(\xi,\eta) = \tilde{h}_N(\xi)\tilde{h}_N(\eta), \tag{18}$$

where $\tilde{h}_N(z) = \big(1 + d_{NN} - z d_{NN}\big)h_N(z)$. Observe that $\psi_{NN} \notin \widetilde{X}_N$ and satisfies

$$\Big(\frac{\partial \psi_{NN}}{\partial \xi} + \frac{\partial \psi_{NN}}{\partial \eta}\Big)\Big|_{(1,1)} = 0, \quad \psi_{NN}(\xi_p,\eta_q) = \delta_{pN}\delta_{qN}, \quad 0 \le p,q \le N. \tag{19}$$

Hence, ψ_{NN} must be linearly independent with the basis functions defined in (15). Hereafter, we update $\widetilde{X}_N$ by adding ψ_{NN} with dimensionality $(N+1)^2$.

Another important property of this basis is that the singularity induced by the transform is removable in the following sense.

Lemma 1. *For any $\tilde{u}, \tilde{v} \in \widetilde{X}_N$,*

$$\big\{(\widetilde{\nabla}\tilde{u}\cdot\widetilde{\nabla}\tilde{v})J\big\}\Big|_{(1,1)} = 0, \tag{20}$$

where J and $\widetilde{\nabla}$ are defined in (6) *and* (7)*, respectively.*

Proof. For any $\tilde{u} \in \widetilde{X}_N$, define

$$w(\xi,\eta) := (3-\xi)\partial_\xi \tilde{u} + (1+\eta)\partial_\eta \tilde{u},$$

and we have

$$\big(\partial_\xi \tilde{u} + \partial_\eta \tilde{u}\big)\big|_{(1,1)} = 0 \;\Rightarrow\; w(1,1) = 0.$$

Using Taylor expansion yields

$$w(\xi,\eta) = -(1-\xi)\partial_\xi w(1,1) - (1-\eta)\partial_\eta w(1,1) + O\Big((1-\xi)^2 + (1-\xi)(1-\eta) + (1-\eta)^2\Big).$$

It is obvious that

$$0 \le \frac{1-\xi}{(1-\xi)+(1-\eta)} \le 1, \quad 0 \le \frac{(1-\xi)^2}{(1-\xi)+(1-\eta)} \le 1-\xi, \quad \forall\,(\xi,\eta) \in \mathbf{Q},$$

and likewise for $1-\eta$ and other terms in big "O," so we have

$$\frac{w(\xi,\eta)}{2-\xi-\eta}\Big|_{(1,1)} = \text{constant}.$$

Consequently, $\widetilde{\nabla}\tilde{u}$ is well-defined at $(1,1)$, so is $\widetilde{\nabla}\tilde{v}$. As the determinant Jacobian J vanishes at $(1,1)$, (20) holds.

3 Implementations and Numerical Results

To test the approximation property of the foregoing nodal basis, we now implement the spectral methods for the elliptic equation in $\mathbf{T}$:

$$-\operatorname{div}\big(a \operatorname{grad} u\big) + b\,u = f \quad \text{in } \mathbf{T}; \quad u = 0 \quad \text{on } \Gamma_1; \quad \frac{\partial u}{\partial \mathrm{n}} = g \quad \text{on } \Gamma_2, \tag{21}$$

where a, b and f are given functions satisfying

$$a \in L^\infty(\mathbf{T}), \quad a(x,y) \geq a_0 > 0, \quad b(x,y) \geq 0, \quad \forall\,(x,y) \in \overline{\mathbf{T}}, \tag{22}$$

for certain constant a_0, Γ_1 (resp. Γ_2) consists of the edges $x = 0$ and $y = 0$ (resp. $x + y = 1$), and n is the unit outer normal vector along Γ_2. The weak formulation of (21) is to find $u \in H^1_{\Gamma_1}(\mathbf{T}) := \{u \in H^1(\mathbf{T}) : u|_{\Gamma_1} = 0\}$ such that

$$\mathscr{B}(u,v) = \big(a\nabla u, \nabla v\big)_{\mathbf{T}} + \big(bu, v\big)_{\mathbf{T}} = (f,v)_{\mathbf{T}} + (ag,v)_{\Gamma_2}, \quad \forall\, v \in H^1_{\Gamma_1}(\mathbf{T}), \tag{23}$$

where $(g,v)_{\Gamma_2} = \int_{\Gamma_2} gv d\gamma$.

We view $\mathbf{T}$ as a deformed triangle as a deformed quadrilateral element, and perform the numerical integration and differentiation on the reference element $\mathbf{Q}$. Define the discrete inner product associated with the usual tensorial LGL quadrature rule:

$$\langle u,v\rangle_{N,\mathbf{T}} = \sum_{0\leq p,q\leq N} \big(\tilde{u}\,\tilde{v}\,J\big)\big|_{(\xi_p,\eta_q)}\omega_p\omega_q := \big\langle \tilde{u}, \tilde{v}J\big\rangle_{N,\mathbf{Q}}, \quad \forall\, u,v \in C(\overline{\mathbf{T}}), \tag{24}$$

where $\{\omega_k\}$ are the LGL quadrature weights associated with LGL points $\{\xi_k = \eta_k\}$. Similarly, we can define the discrete rule, denoted by $\langle\cdot,\cdot\rangle_{N,\Gamma_2}$, along Γ_2, which sums the contributions from two edges $\xi = 1$ and $\eta = 1$.

The Galerkin approximation with numerical integration (GaNI) of (23) is to find $u_N \in V_N := \operatorname{span}\{\phi_{ij}(x,y) = \psi_{ij}(\xi,\eta) \in \widetilde{X}_N : 1 \leq i,j \leq N\}$ such that

$$\begin{aligned}\mathscr{B}_N(u_N,v_N) &= \big\langle a\nabla u_N, \nabla v_N\big\rangle_{N,\mathbf{T}} + \big\langle bu_N, v_N\big\rangle_{N,\mathbf{T}} \\ &= \langle f, v_N\rangle_{N,\mathbf{T}} + \langle ag, v_N\rangle_{N,\Gamma_2}, \quad \forall v_N \in V_N.\end{aligned} \tag{25}$$

Table 1 L^2-error, Max-error and the error at the middle point $(1/2, 1/2)$ for Example 1

	Without (18)			With (18)		
N	L^2	Max	$(1/2,1/2)$	L^2	Max	$(1/2,1/2)$
4	2.186e–3	5.624e–3	2.782e–4	2.186e–3	5.624e–3	3.538e–3
8	4.784e–7	3.693e–6	1.733e–6	4.784e–7	4.781e–6	4.781e–6
12	1.180e–10	1.486e–9	1.614e–10	1.180e–10	1.486e–9	2.070e–10
16	3.422e–14	3.457e–13	6.006e–14	3.422e–14	3.457e–13	1.267e–13
20	2.075e–14	9.892e–14	2.231e–14	2.075e–14	9.892e–14	2.120e–14
24	1.344e–13	6.971e–13	3.363e–14	1.344e–13	6.971e–13	3.386e–14
28	2.109e–13	1.000e–12	1.841e–13	2.109e–13	1.000e–12	1.874e–13
32	8.701e–14	3.211e–13	1.387e–13	8.701e–14	3.211e–13	1.371e–13

Some remarks are in order. Firstly, we could remove the extra basis function (18) at the singular point from V_N for a single triangle. Moreover, in view of Lemma 1, the physical values of the terms at the singular point vanish. The well-posedness of (23) and (25) can be proved by a standard argument.

We next present some examples to illustrate the approximability of the nodal basis.

Example 1. We consider (21) with $a(x, y) = x+2, b(x, y) = x + y$ and a smooth exact solution:

$$u(x, y) = e^{x+y-1} \sin\Big(3y\Big(y - \frac{\sqrt{3}}{2}x + \frac{\sqrt{3}}{4}\Big)\Big). \tag{26}$$

We tabulate in Table 1 the maximum pointwise and discrete L^2 errors on $\mathbf{T}$ for various N. Particularly, we single out the errors at the singular middle point $(1/2, 1/2)$, and list the numerical errors for the scheme (25) with or without the extra basis function (18). We observe an exponential decay of the errors with a convergence behavior similar to that of the quadrilateral element case using tensorial Lagrange polynomial basis (see, e.g., Fig. 2.17 in [9]). Moreover, the presence of the basis function (18) essentially does not affect the performance of the scheme (25).

Example 2. We consider (21) with $a = b = 1$ and test the exact solution with a finite regularity

$$u(x, y) = (1 - x - y)^{\frac{5}{2}}(e^{xy} - 1) \in H^{3-\epsilon}(\mathbf{T}), \quad \epsilon > 0. \tag{27}$$

We list in Table 2 the errors for various N, which indicates an algebraic decay of the errors with a convergence rate around $O(N^{-3})$. It is known that for a tensor-based spectral approximation on a rectangle, the theoretical order of convergence is $O(N^{-3+\varepsilon})$. Although we have not provided the analysis, the proposed scheme really enjoys a similar convergence behavior.

Table 2 L^2-error, Max-error and the error at the middle point $(1/2, 1/2)$ for Example 2

N	Without (18)			With (18)		
	L^2	Max	$(1/2, 1/2)$	L^2	Max	$(1/2, 1/2)$
15	2.866e–6	1.018e–5	5.895e–6	2.866e–6	1.018e–5	5.895e–6
30	3.410e–7	1.203e–6	7.045e–7	3.410e–7	1.203e–6	7.045e–7
45	9.940e–8	3.513e–7	2.054e–7	9.940e–8	3.513e–7	2.054e–7
60	4.158e–8	1.469e–7	8.600e–8	4.159e–8	1.468e–7	8.598e–8
75	2.101e–8	7.599e–8	4.757e–8	2.118e–8	7.486e–8	4.375e–8
90	1.221e–8	4.318e–8	2.533e–8	1.222e–8	4.316e–8	2.528e–8
105	7.669e–9	2.723e–8	1.620e–8	7.683e–9	2.706e–8	1.553e–8
120	1.075e–8	3.472e–8	2.632e–8	5.279e–9	1.942e–8	1.817e–8

4 Extensions and Discussions

A key point in the previous discussion is to one-to-one map a triangular element to the reference rectangle, and to view it as a deformed quadrilateral element. This provides some flexibility for mesh generation of QSEM. Typically, a hybrid spectral-element method can be constructed by using the triangular elements along the boundaries (with the singular edges facing the boundaries) and quadrilateral elements in the interior of the computational domains. This might lead to a more regular mesh and enhance the capability of QSEM for more complex geometries. On the other hand, the number of points on the singular edge is double of the points on the other two edges, so the singular edge should adjoin two quadrilateral elements and/or triangular elements (but share two nonsingular edges), or a triangular element (but share the singular edge). The availability of the aforementioned nodal basis makes the implementation of the hybrid spectral-element method almost as efficient as the usual QSEM.

We now discuss the extensions to tetrahedral elements. Let T be a tetrahedron with vertices A, B, C and D. Denote by Q the reference cube $\{(\xi, \eta, \zeta) : -1 < \xi, \eta, \zeta < 1\}$. The counterpart of (1) reads

$$\begin{aligned}(x, y, z) = (x_A, y_A, z_A)\frac{(1-\xi)(1-\eta)(1-\zeta)}{8} + (x_B, y_B, z_B)\frac{(1+\xi)(7-2\eta-2\zeta+\eta\zeta)}{24} \\ + (x_C, y_C, z_C)\frac{(1+\eta)(7-2\xi-2\zeta+\xi\zeta)}{24} + (x_D, y_D, z_D)\frac{(1+\zeta)(7-2\xi-2\eta+\xi\eta)}{24},\end{aligned}$$

which is one-to-one and maps the vertices $(-1, -1, -1)$, $(1, -1, -1)$, $(-1, 1, -1)$ and $(-1, -1, 1)$ of Q to the vertices of A, B, C and D of the tetrahedron T, respectively, while the images of vertices $(-1, 1, 1)$, $(1, -1, 1)$, $(1, 1, -1)$ and $(1, 1, 1)$ of Q are the middle points of the sides CD, DB, BC and the barycenter of the face ΔBCD, respectively. An illustration of this mapping is depicted in Fig. 2a.

In particular, for the specific tetrahedron

$$\mathsf{T} = \{(x, y, z) : 0 \le x, y, z; x + y + z \le 1\},$$

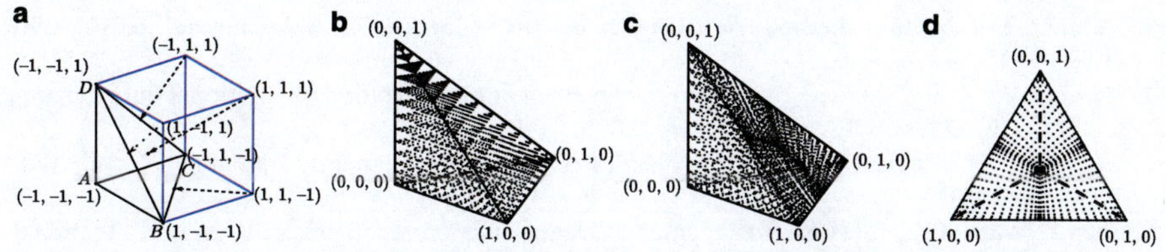

Fig. 2 (**a**) Illustration of the mapping from the cube $\mathsf{Q} = (-1, 1)^3$ and the tetrahedron $\{(x, y, z) : -1 < x, y, z; x + y + z < -1\}$; (**b**) mapped tensorial Legendre–Gauss–Lobatto (LGL) grids on T based on the Duffy's mapping in [4]; (**c**) mapped LGL grids on T using the mapping (28); (**d**) distribution of the grids on the singular face $x + y + z = 1$ of T

the mapping takes the form

$$\begin{cases} x = \frac{1}{24}(1+\xi)(7-2\eta-2\zeta+\eta\zeta), \\ y = \frac{1}{24}(1+\eta)(7-2\xi-2\zeta+\xi\zeta), \\ z = \frac{1}{24}(1+\zeta)(7-2\xi-2\eta+\xi\eta). \end{cases} \tag{28}$$

We plot in Fig. 2b, c the distributions of the mapped tensorial Legendre–Gauss–Lobatto grids on T based on the Duffy's mapping and the mapping (28). The Duffy's mapping collapses one face of Q into a vertex of T, so many collocation points cluster near the singular vertex, which turn out to be wasted. In contrast, the use of (28) leads to a more reasonable grid distribution. Like (11), similar conditions induced by the mapping should be imposed along the three lines that connect the barycenter and the middle points of three side of the singular face of T. Hence, the construction of the nodal basis is much more involved.

We shall report the numerical analysis and the applications of such spectral-element methods in a forthcoming paper.

Acknowledgements The work of the author Youyun Li was supported by Scientific Research Foundation for Returned Scholars, Ministry of Education of China. The work of the author Li-Lian Wang was supported by Singapore AcRF Tier 1 Grant RG58/08, Singapore MOE Grant T207B2202, Singapore NRF2007IDM-IDM002-010, and Leading Academic Discipline Project of Shanghai Municipal Education Commission Grant J50101. The work of the author Huiyuan Li was supported by National Natural Science Foundation of China (NSFC) Grants 10601056 and 10971212. The work of the author Heping Ma was supported by National Natural Science Foundation of China (NSFC) Grant 60874039.

References

1. Rønquist, M., Patera T.: A Legendre spectral element method for the incompressible Navier–Stokes equations. Brunswick, pp. 318–326. Friedrich Vieweg und Sohn, Germany (1988)
2. Deville, M.O., Fischer, P.F, Mund, E.H.: High-order methods for incompressible fluid flow. Cambridge University Press, London (2002)

3. Dubiner, M.: Spectral methods on triangles and other domains. J. Sci. Comput., **6**, 345–390 (1991)
4. Karniadakis, G.E., Sherwin, S.J.: Spectral/*hp* element methods for computational fluid dynamics. Oxford University Press, New York (2005)
5. Heinrichs, W., Loch, B.I., Spectral schemes on triangular elements. J. Comput. Phys., **173**, 279–301 (2001)
6. Shen J., Wang, L.L., Li H.Y.: A triangular spectral element method using fully tensorial rational basis functions. SIAM. J. Numer. Anal., **47**, 1619–1650 (2009)
7. Hesthaven, J.S.: From electrostatics to almost optimal nodal sets for polynomial interpolation in a simplex. SIAM J. Numer. Anal., **35**, 655–676 (1998)
8. Taylor, M.A., Wingate, B.A., Vincent, R.E.: An algorithm for computing Fekete points in the triangle. SIAM J. Numer. Anal., **38**, 1707–1720 (2000)
9. Canuto, C., Hussaini, M.Y., Quarteroni, A., and Zang, T.A.: Spectral methods: Fundamentals in single domains. Scientific Computation. Springer, Berlin, 2006

The Reduced Basis Element Method: Offline-Online Decomposition in the Nonconforming, Nonaffine Case

A.E. Løvgren, Y. Maday, and E.M. Rønquist

Abstract This work focuses on the reduced basis element method applied to the steady Stokes problem with geometric parameter dependence [2, 4]. We present a decoupling of the operators involved in the steady Stokes problem, which together with empirical interpolation [1] allows for complete separation of the offline-online complexity for the nonaffine case. We present numerical results from a hierarchical flow system in two dimensions, where both pipes and bifurcations are used as building blocks.

1 Introduction

We let Ω be a domain given by

$$\overline{\Omega} = \overline{\Phi(\widehat{\Omega})} = \bigcup_{k=1}^{K} \overline{\Omega^k}, \tag{1}$$

where each building block Ω^k is defined as a one-to-one mapping of one of several reference domains $\widehat{\Omega}$, i.e., $\Omega^k = \Phi^k(\widehat{\Omega})$. Corresponding to each reference domain

A.E. Løvgren (✉)
Center for Biomedical Computing, Simula Research Laboratory, P.O. Box 134, 1325 Lysaker, Norway
e-mail: emill@simula.no

Y. Maday
Laboratoire J.-L. Lions, Université Pierre et Marie Curie-Paris6, UMR 7598, Paris, 75005 France and Division of Applied Mathematics, Brown University 182 George Street, Providence, RI 02912, USA
e-mail: maday@ann.jussieu.fr

E.M. Rønquist
Department of Mathematical Sciences, Norwegian University of Science and Technology, 7491 Trondheim, Norway
e-mail: ronquist@math.ntnu.no

J.S. Hesthaven and E.M. Rønquist (eds.), *Spectral and High Order Methods for Partial Differential Equations*, Lecture Notes in Computational Science and Engineering 76, DOI 10.1007/978-3-642-15337-2_22,

there is a set of precomputed basis functions $\{(\hat{\mathbf{u}}_i, \hat{p}_i)\}_{i=1}^N$, found as the velocity–pressure solutions of the steady Stokes problem for different mappings Φ_i of the given reference domain; see Løvgren et al. [2] for details.

The reduced basis element approximation $(\mathbf{u}_N, p_N)$ is found by mapping the basis functions to the appropriate generic domains $\{\Omega^k\}_{k=1}^K$, and then solving the steady Stokes equations through a Galerkin projection. To ensure weak continuity of the reduced basis velocity across subdomain interfaces, Lagrange multipliers are imposed when solving the steady Stokes problem [2]. We note that the resulting reduced basis element approximation is nonconforming [3].

2 Offline-Online Decomposition

We use the viscous operator in the steady Stokes problem to illustrate the offline-online decomposition. On a generic domain $\Omega = \Phi(\widehat{\Omega})$, the computation of the reduced basis approximation involves the calculation of

$$a(\tilde{\mathbf{v}}, \tilde{\mathbf{w}}; \Phi) = \nu \int_{\Omega} \nabla\tilde{\mathbf{v}} \cdot \nabla\tilde{\mathbf{w}} d\Omega, \tag{2}$$

where $\tilde{\mathbf{v}}$ and $\tilde{\mathbf{w}}$ represent the velocity basis functions $\{\hat{\mathbf{u}}_i\}_{i=1}^N$ mapped to the generic domain through the inverse Piola transformation

$$\tilde{\mathbf{v}} = \Psi^{-1}(\hat{\mathbf{v}}, \Phi) = \frac{1}{|J|}\mathcal{J}(\hat{\mathbf{v}} \circ \Phi^{-1}). \tag{3}$$

Here $\mathcal{J}$ is the Jacobian of the map Φ, and J is the corresponding Jacobian determinant. We show that for $Q = 17$ we may write

$$a(\tilde{\mathbf{v}}, \tilde{\mathbf{w}}; \Phi) = \sum_{q=1}^{Q} a^q(\hat{\mathbf{v}}, \hat{\mathbf{w}}, g^q(\Phi)), \tag{4}$$

where the parameter dependent part of $g^q(\Phi)$ can be evaluated online by empirical interpolation [1]. In the following we set the viscosity $\nu = 1$.

The generic domain is not known in the offline stage, so we map the operator in (2) to the corresponding reference domain, and get

$$a(\tilde{\mathbf{v}}, \tilde{\mathbf{w}}; \Phi) = \int_{\widehat{\Omega}} \mathcal{J}^{-T}\widehat{\nabla}(\tilde{\mathbf{v}} \circ \Phi) \cdot \mathcal{J}^{-T}\widehat{\nabla}(\tilde{\mathbf{w}} \circ \Phi)|J| d\widehat{\Omega}, \tag{5}$$

where $\widehat{\nabla} = \mathcal{J}^T\nabla$. Using (3), we replace the velocities with their counterparts stored on the reference domain, and (5) gives

$$a(\tilde{\mathbf{v}}, \tilde{\mathbf{w}}; \Phi) = \int_{\widehat{\Omega}} \mathcal{J}^{-T}\widehat{\nabla}(\frac{1}{|J|}\mathcal{J}\hat{\mathbf{v}}) \cdot \mathcal{J}^{-T}\widehat{\nabla}(\frac{1}{|J|}\mathcal{J}\hat{\mathbf{w}})|J| d\widehat{\Omega}. \tag{6}$$

If we do not use the Piola transformation to map the velocities, we may collect the contributions from the transposed inverse Jacobians and the Jacobian determinant in a tensor T_{ij}, and use the elements of this tensor as the parameter dependent shape functions. This procedure is shown in detail in [4] and is sufficient as long as the inflow and outflow boundaries of the generic domain are undeformed relative to the reference domain. Since the Jacobian of the generic mapping is present in the gradient operator in (6), things get a little more complicated. We introduce the notation

$$\hat{\mathbf{v}} = \begin{bmatrix} \hat{v}_\xi \\ \hat{v}_\eta \end{bmatrix}, \quad \mathcal{J} = \begin{bmatrix} \mathcal{J}_{11} & \mathcal{J}_{12} \\ \mathcal{J}_{21} & \mathcal{J}_{22} \end{bmatrix}, \tag{7}$$

for the components of the velocities and the Jacobian. After multiplying the Jacobian with the reference velocities, we get the component form of (6)

$$\begin{aligned} &\int_{\widehat{\Omega}} \mathcal{J}^{-T}\widehat{\nabla}(\frac{1}{|J|}(\mathcal{J}_{11}\hat{v}_\xi + \mathcal{J}_{12}\hat{v}_\eta)) \cdot \mathcal{J}^{-T}\widehat{\nabla}(\frac{1}{|J|}(\mathcal{J}_{11}\hat{w}_\xi + \mathcal{J}_{12}\hat{w}_\eta))|J|\, d\widehat{\Omega} \\ &+ \int_{\widehat{\Omega}} \mathcal{J}^{-T}\widehat{\nabla}(\frac{1}{|J|}(\mathcal{J}_{21}\hat{v}_\xi + \mathcal{J}_{22}\hat{v}_\eta)) \cdot \mathcal{J}^{-T}\widehat{\nabla}(\frac{1}{|J|}(\mathcal{J}_{21}\hat{w}_\xi + \mathcal{J}_{22}\hat{w}_\eta))|J|\, d\widehat{\Omega}. \end{aligned} \tag{8}$$

Next we note that $\widehat{\nabla} u = [\frac{\partial u}{\partial \xi}, \frac{\partial u}{\partial \eta}]^T$, and use this to get the equivalent form

$$\begin{aligned} &\int_{\widehat{\Omega}} \mathcal{J}^{-T}\begin{bmatrix} \frac{\partial}{\partial \xi}(\frac{1}{|J|}(\mathcal{J}_{11}\hat{v}_\xi + \mathcal{J}_{12}\hat{v}_\eta)) \\ \frac{\partial}{\partial \eta}(\frac{1}{|J|}(\mathcal{J}_{11}\hat{v}_\xi + \mathcal{J}_{12}\hat{v}_\eta)) \end{bmatrix} \cdot \mathcal{J}^{-T}\begin{bmatrix} \frac{\partial}{\partial \xi}(\frac{1}{|J|}(\mathcal{J}_{11}\hat{w}_\xi + \mathcal{J}_{12}\hat{w}_\eta)) \\ \frac{\partial}{\partial \eta}(\frac{1}{|J|}(\mathcal{J}_{11}\hat{w}_\xi + \mathcal{J}_{12}\hat{w}_\eta)) \end{bmatrix} |J|\, d\widehat{\Omega} \\ &+ \int_{\widehat{\Omega}} \mathcal{J}^{-T}\begin{bmatrix} \frac{\partial}{\partial \xi}(\frac{1}{|J|}(\mathcal{J}_{21}\hat{v}_\xi + \mathcal{J}_{22}\hat{v}_\eta)) \\ \frac{\partial}{\partial \eta}(\frac{1}{|J|}(\mathcal{J}_{21}\hat{v}_\xi + \mathcal{J}_{22}\hat{v}_\eta)) \end{bmatrix} \cdot \mathcal{J}^{-T}\begin{bmatrix} \frac{\partial}{\partial \xi}(\frac{1}{|J|}(\mathcal{J}_{21}\hat{w}_\xi + \mathcal{J}_{22}\hat{w}_\eta)) \\ \frac{\partial}{\partial \eta}(\frac{1}{|J|}(\mathcal{J}_{21}\hat{w}_\xi + \mathcal{J}_{22}\hat{w}_\eta)) \end{bmatrix} |J|\, d\widehat{\Omega}. \end{aligned} \tag{9}$$

After multiplying each vector in (9) with $\mathcal{J}^{-T}$ and writing out the inner products, we get the sum of four products under each integral. The first of these four products is as follows,

$$\begin{aligned} &\frac{1}{|J|}\left(\mathcal{J}_{22}\frac{\partial}{\partial \xi}(\frac{1}{|J|}(\mathcal{J}_{11}\hat{v}_\xi + \mathcal{J}_{12}\hat{v}_\eta)) - \mathcal{J}_{21}\frac{\partial}{\partial \eta}(\frac{1}{|J|}(\mathcal{J}_{11}\hat{v}_\xi + \mathcal{J}_{12}\hat{v}_\eta))\right) \\ &* \left(\mathcal{J}_{22}\frac{\partial}{\partial \xi}(\frac{1}{|J|}(\mathcal{J}_{11}\hat{w}_\xi + \mathcal{J}_{12}\hat{w}_\eta)) - \mathcal{J}_{21}\frac{\partial}{\partial \eta}(\frac{1}{|J|}(\mathcal{J}_{11}\hat{w}_\xi + \mathcal{J}_{12}\hat{w}_\eta))\right). \end{aligned} \tag{10}$$

The other three products are similar, only the indices are different. To separate the elements of the Jacobian from the components of the reference velocity, we first differentiate with respect to ξ and η inside the four products. From the first product (10) we then get

$$
\begin{aligned}
&\frac{1}{|J|}\Bigg(\Big(\mathcal{J}_{22}\frac{\partial}{\partial\xi}\frac{\mathcal{J}_{11}}{|J|}-\mathcal{J}_{21}\frac{\partial}{\partial\eta}\frac{\mathcal{J}_{11}}{|J|}\Big)\hat{v}_\xi+\Big(\mathcal{J}_{22}\frac{\partial}{\partial\xi}\frac{\mathcal{J}_{12}}{|J|}-\mathcal{J}_{21}\frac{\partial}{\partial\eta}\frac{\mathcal{J}_{12}}{|J|}\Big)\hat{v}_\eta\\
&\quad+\frac{1}{|J|}\Big(\mathcal{J}_{22}(\mathcal{J}_{11}\frac{\partial\hat{v}_\xi}{\partial\xi}+\mathcal{J}_{12}\frac{\partial\hat{v}_\eta}{\partial\xi})-\mathcal{J}_{21}(\mathcal{J}_{11}\frac{\partial\hat{v}_\xi}{\partial\eta}+\mathcal{J}_{12}\frac{\partial\hat{v}_\eta}{\partial\eta})\Big)\Bigg)\\
&*\Bigg(\Big(\mathcal{J}_{22}\frac{\partial}{\partial\xi}\frac{\mathcal{J}_{11}}{|J|}-\mathcal{J}_{21}\frac{\partial}{\partial\eta}\frac{\mathcal{J}_{11}}{|J|}\Big)\hat{w}_\xi+\Big(\mathcal{J}_{22}\frac{\partial}{\partial\xi}\frac{\mathcal{J}_{12}}{|J|}-\mathcal{J}_{21}\frac{\partial}{\partial\eta}\frac{\mathcal{J}_{12}}{|J|}\Big)\hat{w}_\eta\\
&\quad+\frac{1}{|J|}\Big(\mathcal{J}_{22}(\mathcal{J}_{11}\frac{\partial\hat{w}_\xi}{\partial\xi}+\mathcal{J}_{12}\frac{\partial\hat{w}_\eta}{\partial\xi})-\mathcal{J}_{21}(\mathcal{J}_{11}\frac{\partial\hat{w}_\xi}{\partial\eta}+\mathcal{J}_{12}\frac{\partial\hat{w}_\eta}{\partial\eta})\Big)\Bigg).
\end{aligned}
\tag{11}
$$

After carrying out the multiplications for all four products, and collecting terms corresponding to the same velocity components, we find that we may decouple the viscous operator as in (4) for $Q = 17$. As an example of the operators in the decoupling, we let

$$
a^1(\hat{\mathbf{v}},\hat{\mathbf{w}},g^1(\Phi)) = \int_{\widehat{\Omega}}(\frac{\partial\hat{v}_\xi}{\partial\xi}\frac{\partial\hat{w}_\xi}{\partial\xi}+\frac{\partial\hat{v}_\eta}{\partial\eta}\frac{\partial\hat{w}_\eta}{\partial\eta})g^1(\Phi)d\widehat{\Omega}, \tag{12}
$$

with corresponding parameter dependent function

$$
g^1(\Phi) = \frac{1}{|J|^3}(\mathcal{J}_{11}^2+\mathcal{J}_{21}^2)(\mathcal{J}_{12}^2+\mathcal{J}_{22}^2). \tag{13}
$$

The rest of the operators in the decoupling with corresponding parameter functions can be found in Tables 1–3.

We use empirical interpolation to approximate the parameter functions as

$$
g^q(\Phi) \approx \sum_{m=1}^{M_q}\beta_m^q(\Phi)\tilde{g}_m^q, \quad q = 1,\ldots,Q, \tag{14}
$$

where each $\beta_m^q(\Phi)$ is a constant, and $\tilde{g}_m^q$ are modified versions of the parameter functions $g^q(\Phi)$, sampled at predefined mappings Φ_m.

In the offline stage we thus compute

$$
A_{ij}^{mq} = a^q(\hat{\mathbf{u}}_\mathbf{i},\hat{\mathbf{u}}_\mathbf{j},\tilde{g}_m^q), \tag{15}
$$

for all $i, j = 1,\ldots,N$, $q = 1,\ldots,Q$, and $m = 1,\ldots,M_q$. In the online stage the coefficients $\{\beta_m^q(\Phi)\}_{m=1}^{M_q}$ are found for each q by sampling the parameter function $g^q(\Phi)$ in M_q points and solving a lower triangular matrix. Once the coefficients are found, we assemble

$$
a(\tilde{\mathbf{u}}_\mathbf{i},\tilde{\mathbf{u}}_\mathbf{j};\Phi) \approx \sum_{q=1}^{Q}\sum_{m=1}^{M_q}\beta_m^q(\Phi)A_{ij}^{mq}. \tag{16}
$$

Table 1 The operators in the decoupling of the viscous operator

$$
\begin{aligned}
a^1(\hat{\mathbf{v}},\hat{\mathbf{w}},g^1(\Phi)) &= \int_{\hat\Omega}\Big(\frac{\partial \hat v_\xi}{\partial \xi}\frac{\partial \hat w_\xi}{\partial \xi} + \frac{\partial \hat v_\eta}{\partial \eta}\frac{\partial \hat w_\eta}{\partial \eta}\Big)g^1(\Phi)d\hat\Omega\\
a^2(\hat{\mathbf{v}},\hat{\mathbf{w}},g^2(\Phi)) &= \int_{\hat\Omega}\Big(\frac{\partial \hat v_\xi}{\partial \xi}\frac{\partial \hat w_\eta}{\partial \eta} + \frac{\partial \hat v_\eta}{\partial \xi}\frac{\partial \hat w_\xi}{\partial \eta} + \frac{\partial \hat v_\xi}{\partial \eta}\frac{\partial \hat w_\eta}{\partial \xi} + \frac{\partial \hat v_\eta}{\partial \eta}\frac{\partial \hat w_\xi}{\partial \xi}\Big)g^2(\Phi)d\hat\Omega\\
a^3(\hat{\mathbf{v}},\hat{\mathbf{w}},g^3(\Phi)) &= \int_{\hat\Omega}\Big(\frac{\partial \hat v_\eta}{\partial \xi}\frac{\partial \hat w_\eta}{\partial \xi}\Big)g^3(\Phi)d\hat\Omega\\
a^4(\hat{\mathbf{v}},\hat{\mathbf{w}},g^4(\Phi)) &= \int_{\hat\Omega}\Big(\frac{\partial \hat v_\xi}{\partial \eta}\frac{\partial \hat w_\xi}{\partial \eta}\Big)g^4(\Phi)d\hat\Omega\\
a^5(\hat{\mathbf{v}},\hat{\mathbf{w}},g^5(\Phi)) &= \int_{\hat\Omega}\Big(\frac{\partial \hat v_\xi}{\partial \xi}\frac{\partial \hat w_\eta}{\partial \xi} + \frac{\partial \hat v_\eta}{\partial \xi}\frac{\partial \hat w_\xi}{\partial \xi} - \frac{\partial \hat v_\eta}{\partial \xi}\frac{\partial \hat w_\eta}{\partial \eta} - \frac{\partial \hat v_\eta}{\partial \eta}\frac{\partial \hat w_\eta}{\partial \xi}\Big)g^5(\Phi)d\hat\Omega\\
a^6(\hat{\mathbf{v}},\hat{\mathbf{w}},g^6(\Phi)) &= \int_{\hat\Omega}\Big(\frac{\partial \hat v_\xi}{\partial \eta}\frac{\partial \hat w_\eta}{\partial \eta} + \frac{\partial \hat v_\eta}{\partial \eta}\frac{\partial \hat w_\xi}{\partial \eta} - \frac{\partial \hat v_\xi}{\partial \xi}\frac{\partial \hat w_\xi}{\partial \eta} - \frac{\partial \hat v_\xi}{\partial \eta}\frac{\partial \hat w_\xi}{\partial \xi}\Big)g^6(\Phi)d\hat\Omega\\[1ex]
a^7(\hat{\mathbf{v}},\hat{\mathbf{w}},g^7(\Phi)) &= \int_{\hat\Omega}(\hat v_\xi \hat w_\xi)g^7(\Phi)d\hat\Omega\\
a^8(\hat{\mathbf{v}},\hat{\mathbf{w}},g^8(\Phi)) &= \int_{\hat\Omega}(\hat v_\xi \hat w_\eta + \hat v_\eta \hat w_\xi)g^8(\Phi)d\hat\Omega\\
a^9(\hat{\mathbf{v}},\hat{\mathbf{w}},g^9(\Phi)) &= \int_{\hat\Omega}(\hat v_\eta \hat w_\eta)g^9(\Phi)d\hat\Omega\\[1ex]
a^{10}(\hat{\mathbf{v}},\hat{\mathbf{w}},g^{10}(\Phi)) &= \int_{\hat\Omega}\Big(\frac{\partial \hat v_\xi}{\partial \xi}\hat w_\xi + \hat v_\xi\frac{\partial \hat w_\xi}{\partial \xi}\Big)g^{10}(\Phi)d\hat\Omega\\
a^{11}(\hat{\mathbf{v}},\hat{\mathbf{w}},g^{11}(\Phi)) &= \int_{\hat\Omega}\Big(\frac{\partial \hat v_\eta}{\partial \xi}\hat w_\xi + \hat v_\xi\frac{\partial \hat w_\eta}{\partial \xi}\Big)g^{11}(\Phi)d\hat\Omega\\
a^{12}(\hat{\mathbf{v}},\hat{\mathbf{w}},g^{12}(\Phi)) &= \int_{\hat\Omega}\Big(\frac{\partial \hat v_\xi}{\partial \xi}\hat w_\eta + \hat v_\eta\frac{\partial \hat w_\xi}{\partial \xi}\Big)g^{12}(\Phi)d\hat\Omega\\
a^{13}(\hat{\mathbf{v}},\hat{\mathbf{w}},g^{13}(\Phi)) &= \int_{\hat\Omega}\Big(\frac{\partial \hat v_\eta}{\partial \xi}\hat w_\eta + \hat v_\eta\frac{\partial \hat w_\eta}{\partial \xi}\Big)g^{13}(\Phi)d\hat\Omega\\
a^{14}(\hat{\mathbf{v}},\hat{\mathbf{w}},g^{14}(\Phi)) &= \int_{\hat\Omega}\Big(\frac{\partial \hat v_\xi}{\partial \eta}\hat w_\xi + \hat v_\xi\frac{\partial \hat w_\xi}{\partial \eta}\Big)g^{14}(\Phi)d\hat\Omega\\
a^{15}(\hat{\mathbf{v}},\hat{\mathbf{w}},g^{15}(\Phi)) &= \int_{\hat\Omega}\Big(\frac{\partial \hat v_\eta}{\partial \eta}\hat w_\xi + \hat v_\xi\frac{\partial \hat w_\eta}{\partial \eta}\Big)g^{15}(\Phi)d\hat\Omega\\
a^{16}(\hat{\mathbf{v}},\hat{\mathbf{w}},g^{16}(\Phi)) &= \int_{\hat\Omega}\Big(\frac{\partial \hat v_\xi}{\partial \eta}\hat w_\eta + \hat v_\eta\frac{\partial \hat w_\xi}{\partial \eta}\Big)g^{16}(\Phi)d\hat\Omega\\
a^{17}(\hat{\mathbf{v}},\hat{\mathbf{w}},g^{17}(\Phi)) &= \int_{\hat\Omega}\Big(\frac{\partial \hat v_\eta}{\partial \eta}\hat w_\eta + \hat v_\eta\frac{\partial \hat w_\eta}{\partial \eta}\Big)g^{17}(\Phi)d\hat\Omega
\end{aligned}
$$

Together with simimlar contributions from the divergence operator in the steady Stokes problem, we build the reduced basis system matrix and solve the reduced basis steady Stokes problem.

3 *A Posteriori* Error Estimation

Following the ideas of Maday et al. [3], we propose a conforming correction of the reduced basis element approximation in order to compute bounds on the output of interest. This involves solving local Stokes problems on small domains related to the subdomain interfaces on the generic domain. The jump in the reduced basis element approximation across the interface is given as a boundary condition in the local problem, and the solution is added to the reduced basis element approximation in order to produce a conforming approximation. Using the conforming approximation we are able to compute the bound gap $s^+(\mathbf{u}_N;\Phi) - s^-(\mathbf{u}_N;\Phi)$ for the output of interest also in the multi-domain case. To decouple the computation of the conforming correction in an offline-online procedure, we introduce a basis for the jump across the interface, and precompute solutions to the local Stokes problem in the offline stage. In the online stage the jump in the reduced basis element

Table 2 The parameter functions $g^1, \ldots, g^{11}$ in the decoupling of the viscous operator

$$
\begin{aligned}
g^1(\Phi) &= \tfrac{1}{|J|^3}\left(\mathcal{J}_{11}^2 + \mathcal{J}_{21}^2\right)\left(\mathcal{J}_{12}^2 + \mathcal{J}_{22}^2\right)\\
g^2(\Phi) &= -\tfrac{1}{|J|^3}\left(\mathcal{J}_{11}\mathcal{J}_{12} + \mathcal{J}_{21}\mathcal{J}_{22}\right)^2\\
g^3(\Phi) &= \tfrac{1}{|J|^3}\left(\mathcal{J}_{12}^2 + \mathcal{J}_{22}^2\right)^2\\
g^4(\Phi) &= \tfrac{1}{|J|^3}\left(\mathcal{J}_{11}^2 + \mathcal{J}_{21}^2\right)^2\\
g^5(\Phi) &= \tfrac{1}{|J|^3}\left(\mathcal{J}_{12}^2 + \mathcal{J}_{22}^2\right)\left(\mathcal{J}_{11}\mathcal{J}_{12} + \mathcal{J}_{21}\mathcal{J}_{22}\right)\\
g^6(\Phi) &= \tfrac{1}{|J|^3}\left(\mathcal{J}_{11}^2 + \mathcal{J}_{21}^2\right)\left(\mathcal{J}_{11}\mathcal{J}_{12} + \mathcal{J}_{21}\mathcal{J}_{22}\right)\\
\\
g^7(\Phi) &= \tfrac{1}{|J|}\Bigg(\left(\mathcal{J}_{22}\tfrac{\partial}{\partial\xi}\tfrac{\mathcal{J}_{11}}{|J|} - \mathcal{J}_{21}\tfrac{\partial}{\partial\eta}\tfrac{\mathcal{J}_{11}}{|J|}\right)^2 + \left(\mathcal{J}_{12}\tfrac{\partial}{\partial\xi}\tfrac{\mathcal{J}_{11}}{|J|} - \mathcal{J}_{11}\tfrac{\partial}{\partial\eta}\tfrac{\mathcal{J}_{11}}{|J|}\right)^2\\
&\quad + \left(\mathcal{J}_{22}\tfrac{\partial}{\partial\xi}\tfrac{\mathcal{J}_{21}}{|J|} - \mathcal{J}_{21}\tfrac{\partial}{\partial\eta}\tfrac{\mathcal{J}_{21}}{|J|}\right)^2 + \left(\mathcal{J}_{12}\tfrac{\partial}{\partial\xi}\tfrac{\mathcal{J}_{21}}{|J|} - \mathcal{J}_{11}\tfrac{\partial}{\partial\eta}\tfrac{\mathcal{J}_{21}}{|J|}\right)^2\Bigg)\\
g^8(\Phi) &= \tfrac{1}{|J|}\Bigg(\left(\mathcal{J}_{22}\tfrac{\partial}{\partial\xi}\tfrac{\mathcal{J}_{11}}{|J|} - \mathcal{J}_{21}\tfrac{\partial}{\partial\eta}\tfrac{\mathcal{J}_{11}}{|J|}\right)\left(\mathcal{J}_{22}\tfrac{\partial}{\partial\xi}\tfrac{\mathcal{J}_{12}}{|J|} - \mathcal{J}_{21}\tfrac{\partial}{\partial\eta}\tfrac{\mathcal{J}_{12}}{|J|}\right)\\
&\quad + \left(\mathcal{J}_{12}\tfrac{\partial}{\partial\xi}\tfrac{\mathcal{J}_{11}}{|J|} - \mathcal{J}_{11}\tfrac{\partial}{\partial\eta}\tfrac{\mathcal{J}_{11}}{|J|}\right)\left(\mathcal{J}_{12}\tfrac{\partial}{\partial\xi}\tfrac{\mathcal{J}_{12}}{|J|} - \mathcal{J}_{11}\tfrac{\partial}{\partial\eta}\tfrac{\mathcal{J}_{12}}{|J|}\right)\\
&\quad + \left(\mathcal{J}_{22}\tfrac{\partial}{\partial\xi}\tfrac{\mathcal{J}_{21}}{|J|} - \mathcal{J}_{21}\tfrac{\partial}{\partial\eta}\tfrac{\mathcal{J}_{21}}{|J|}\right)\left(\mathcal{J}_{22}\tfrac{\partial}{\partial\xi}\tfrac{\mathcal{J}_{22}}{|J|} - \mathcal{J}_{21}\tfrac{\partial}{\partial\eta}\tfrac{\mathcal{J}_{22}}{|J|}\right)\\
&\quad + \left(\mathcal{J}_{12}\tfrac{\partial}{\partial\xi}\tfrac{\mathcal{J}_{21}}{|J|} - \mathcal{J}_{11}\tfrac{\partial}{\partial\eta}\tfrac{\mathcal{J}_{21}}{|J|}\right)\left(\mathcal{J}_{12}\tfrac{\partial}{\partial\xi}\tfrac{\mathcal{J}_{22}}{|J|} - \mathcal{J}_{11}\tfrac{\partial}{\partial\eta}\tfrac{\mathcal{J}_{22}}{|J|}\right)\Bigg)\\
g^9(\Phi) &= \tfrac{1}{|J|}\Bigg(\left(\mathcal{J}_{22}\tfrac{\partial}{\partial\xi}\tfrac{\mathcal{J}_{12}}{|J|} - \mathcal{J}_{21}\tfrac{\partial}{\partial\eta}\tfrac{\mathcal{J}_{12}}{|J|}\right)^2 + \left(\mathcal{J}_{12}\tfrac{\partial}{\partial\xi}\tfrac{\mathcal{J}_{12}}{|J|} - \mathcal{J}_{11}\tfrac{\partial}{\partial\eta}\tfrac{\mathcal{J}_{12}}{|J|}\right)^2\\
&\quad + \left(\mathcal{J}_{22}\tfrac{\partial}{\partial\xi}\tfrac{\mathcal{J}_{22}}{|J|} - \mathcal{J}_{21}\tfrac{\partial}{\partial\eta}\tfrac{\mathcal{J}_{22}}{|J|}\right)^2 + \left(\mathcal{J}_{12}\tfrac{\partial}{\partial\xi}\tfrac{\mathcal{J}_{22}}{|J|} - \mathcal{J}_{11}\tfrac{\partial}{\partial\eta}\tfrac{\mathcal{J}_{22}}{|J|}\right)^2\Bigg)\\
\\
g^{10}(\Phi) &= \tfrac{1}{|J|^2}\Bigg(\mathcal{J}_{22}\mathcal{J}_{11}\left(\mathcal{J}_{22}\tfrac{\partial}{\partial\xi}\tfrac{\mathcal{J}_{11}}{|J|} - \mathcal{J}_{21}\tfrac{\partial}{\partial\eta}\tfrac{\mathcal{J}_{11}}{|J|}\right)\\
&\quad + \mathcal{J}_{12}\mathcal{J}_{11}\left(\mathcal{J}_{12}\tfrac{\partial}{\partial\xi}\tfrac{\mathcal{J}_{11}}{|J|} - \mathcal{J}_{11}\tfrac{\partial}{\partial\eta}\tfrac{\mathcal{J}_{11}}{|J|}\right)\\
&\quad + \mathcal{J}_{22}\mathcal{J}_{21}\left(\mathcal{J}_{22}\tfrac{\partial}{\partial\xi}\tfrac{\mathcal{J}_{21}}{|J|} - \mathcal{J}_{21}\tfrac{\partial}{\partial\eta}\tfrac{\mathcal{J}_{21}}{|J|}\right)\\
&\quad + \mathcal{J}_{12}\mathcal{J}_{21}\left(\mathcal{J}_{12}\tfrac{\partial}{\partial\xi}\tfrac{\mathcal{J}_{21}}{|J|} - \mathcal{J}_{11}\tfrac{\partial}{\partial\eta}\tfrac{\mathcal{J}_{21}}{|J|}\right)\Bigg)\\
g^{11}(\Phi) &= \tfrac{1}{|J|^2}\Bigg(\mathcal{J}_{22}\mathcal{J}_{12}\left(\mathcal{J}_{22}\tfrac{\partial}{\partial\xi}\tfrac{\mathcal{J}_{11}}{|J|} - \mathcal{J}_{21}\tfrac{\partial}{\partial\eta}\tfrac{\mathcal{J}_{11}}{|J|}\right)\\
&\quad + \mathcal{J}_{12}\mathcal{J}_{12}\left(\mathcal{J}_{12}\tfrac{\partial}{\partial\xi}\tfrac{\mathcal{J}_{11}}{|J|} - \mathcal{J}_{11}\tfrac{\partial}{\partial\eta}\tfrac{\mathcal{J}_{11}}{|J|}\right)\\
&\quad + \mathcal{J}_{22}\mathcal{J}_{22}\left(\mathcal{J}_{22}\tfrac{\partial}{\partial\xi}\tfrac{\mathcal{J}_{21}}{|J|} - \mathcal{J}_{21}\tfrac{\partial}{\partial\eta}\tfrac{\mathcal{J}_{21}}{|J|}\right)\\
&\quad + \mathcal{J}_{12}\mathcal{J}_{22}\left(\mathcal{J}_{12}\tfrac{\partial}{\partial\xi}\tfrac{\mathcal{J}_{21}}{|J|} - \mathcal{J}_{11}\tfrac{\partial}{\partial\eta}\tfrac{\mathcal{J}_{21}}{|J|}\right)\Bigg)
\end{aligned}
$$

approximation then defines coefficients used when adding the precomputed conforming corrections to the reduced basis element approximation. All details on *a posteriori* error estimation and the basis for the jump across subdomain interfaces will be presented in future work.

4 Numerical Experiment

To illustrate the effect of the offline-online decomposition, we solve the steady Stokes problem on the domain Ω depicted in Fig. 1. Clearly, Ω consists of four subdomains, where each subdomain is a one-to-one map of either a reference pipe,

Table 3 The parameter functions $g^{12}, \ldots, g^{17}$ in the decoupling of the viscous operator

$$
\begin{aligned}
g^{12}(\Phi) = \tfrac{1}{|J|^2} \Big(& \mathscr{J}_{22}\mathscr{J}_{11}\Big(\mathscr{J}_{22}\tfrac{\partial}{\partial\xi}\tfrac{\mathscr{J}_{12}}{|J|} - \mathscr{J}_{21}\tfrac{\partial}{\partial\eta}\tfrac{\mathscr{J}_{12}}{|J|}\Big) \\
& + \mathscr{J}_{12}\mathscr{J}_{11}\Big(\mathscr{J}_{12}\tfrac{\partial}{\partial\xi}\tfrac{\mathscr{J}_{12}}{|J|} - \mathscr{J}_{11}\tfrac{\partial}{\partial\eta}\tfrac{\mathscr{J}_{12}}{|J|}\Big) \\
& + \mathscr{J}_{22}\mathscr{J}_{21}\Big(\mathscr{J}_{22}\tfrac{\partial}{\partial\xi}\tfrac{\mathscr{J}_{22}}{|J|} - \mathscr{J}_{21}\tfrac{\partial}{\partial\eta}\tfrac{\mathscr{J}_{22}}{|J|}\Big) \\
& + \mathscr{J}_{12}\mathscr{J}_{21}\Big(\mathscr{J}_{12}\tfrac{\partial}{\partial\xi}\tfrac{\mathscr{J}_{22}}{|J|} - \mathscr{J}_{11}\tfrac{\partial}{\partial\eta}\tfrac{\mathscr{J}_{22}}{|J|}\Big)\Big) \\
g^{13}(\Phi) = \tfrac{1}{|J|^2} \Big(& \mathscr{J}_{22}\mathscr{J}_{12}\Big(\mathscr{J}_{22}\tfrac{\partial}{\partial\xi}\tfrac{\mathscr{J}_{12}}{|J|} - \mathscr{J}_{21}\tfrac{\partial}{\partial\eta}\tfrac{\mathscr{J}_{12}}{|J|}\Big) \\
& + \mathscr{J}_{12}\mathscr{J}_{12}\Big(\mathscr{J}_{12}\tfrac{\partial}{\partial\xi}\tfrac{\mathscr{J}_{12}}{|J|} - \mathscr{J}_{11}\tfrac{\partial}{\partial\eta}\tfrac{\mathscr{J}_{12}}{|J|}\Big) \\
& + \mathscr{J}_{22}\mathscr{J}_{22}\Big(\mathscr{J}_{22}\tfrac{\partial}{\partial\xi}\tfrac{\mathscr{J}_{22}}{|J|} - \mathscr{J}_{21}\tfrac{\partial}{\partial\eta}\tfrac{\mathscr{J}_{22}}{|J|}\Big) \\
& + \mathscr{J}_{12}\mathscr{J}_{22}\Big(\mathscr{J}_{12}\tfrac{\partial}{\partial\xi}\tfrac{\mathscr{J}_{22}}{|J|} - \mathscr{J}_{11}\tfrac{\partial}{\partial\eta}\tfrac{\mathscr{J}_{22}}{|J|}\Big)\Big) \\
g^{14}(\Phi) = -\tfrac{1}{|J|^2} \Big(& \mathscr{J}_{21}\mathscr{J}_{11}\Big(\mathscr{J}_{22}\tfrac{\partial}{\partial\xi}\tfrac{\mathscr{J}_{11}}{|J|} - \mathscr{J}_{21}\tfrac{\partial}{\partial\eta}\tfrac{\mathscr{J}_{11}}{|J|}\Big) \\
& + \mathscr{J}_{11}\mathscr{J}_{11}\Big(\mathscr{J}_{12}\tfrac{\partial}{\partial\xi}\tfrac{\mathscr{J}_{11}}{|J|} - \mathscr{J}_{11}\tfrac{\partial}{\partial\eta}\tfrac{\mathscr{J}_{11}}{|J|}\Big) \\
& + \mathscr{J}_{21}\mathscr{J}_{21}\Big(\mathscr{J}_{22}\tfrac{\partial}{\partial\xi}\tfrac{\mathscr{J}_{21}}{|J|} - \mathscr{J}_{21}\tfrac{\partial}{\partial\eta}\tfrac{\mathscr{J}_{21}}{|J|}\Big) \\
& + \mathscr{J}_{11}\mathscr{J}_{21}\Big(\mathscr{J}_{12}\tfrac{\partial}{\partial\xi}\tfrac{\mathscr{J}_{21}}{|J|} - \mathscr{J}_{11}\tfrac{\partial}{\partial\eta}\tfrac{\mathscr{J}_{21}}{|J|}\Big)\Big) \\
g^{15}(\Phi) = -\tfrac{1}{|J|^2} \Big(& \mathscr{J}_{21}\mathscr{J}_{12}\Big(\mathscr{J}_{22}\tfrac{\partial}{\partial\xi}\tfrac{\mathscr{J}_{11}}{|J|} - \mathscr{J}_{21}\tfrac{\partial}{\partial\eta}\tfrac{\mathscr{J}_{11}}{|J|}\Big) \\
& + \mathscr{J}_{11}\mathscr{J}_{12}\Big(\mathscr{J}_{12}\tfrac{\partial}{\partial\xi}\tfrac{\mathscr{J}_{11}}{|J|} - \mathscr{J}_{11}\tfrac{\partial}{\partial\eta}\tfrac{\mathscr{J}_{11}}{|J|}\Big) \\
& + \mathscr{J}_{22}\mathscr{J}_{21}\Big(\mathscr{J}_{22}\tfrac{\partial}{\partial\xi}\tfrac{\mathscr{J}_{21}}{|J|} - \mathscr{J}_{21}\tfrac{\partial}{\partial\eta}\tfrac{\mathscr{J}_{21}}{|J|}\Big) \\
& + \mathscr{J}_{11}\mathscr{J}_{22}\Big(\mathscr{J}_{12}\tfrac{\partial}{\partial\xi}\tfrac{\mathscr{J}_{21}}{|J|} - \mathscr{J}_{11}\tfrac{\partial}{\partial\eta}\tfrac{\mathscr{J}_{21}}{|J|}\Big)\Big) \\
g^{16}(\Phi) = -\tfrac{1}{|J|^2} \Big(& \mathscr{J}_{21}\mathscr{J}_{11}\Big(\mathscr{J}_{22}\tfrac{\partial}{\partial\xi}\tfrac{\mathscr{J}_{12}}{|J|} - \mathscr{J}_{21}\tfrac{\partial}{\partial\eta}\tfrac{\mathscr{J}_{12}}{|J|}\Big) \\
& + \mathscr{J}_{11}\mathscr{J}_{11}\Big(\mathscr{J}_{12}\tfrac{\partial}{\partial\xi}\tfrac{\mathscr{J}_{12}}{|J|} - \mathscr{J}_{11}\tfrac{\partial}{\partial\eta}\tfrac{\mathscr{J}_{12}}{|J|}\Big) \\
& + \mathscr{J}_{21}\mathscr{J}_{21}\Big(\mathscr{J}_{22}\tfrac{\partial}{\partial\xi}\tfrac{\mathscr{J}_{22}}{|J|} - \mathscr{J}_{21}\tfrac{\partial}{\partial\eta}\tfrac{\mathscr{J}_{22}}{|J|}\Big) \\
& + \mathscr{J}_{11}\mathscr{J}_{21}\Big(\mathscr{J}_{12}\tfrac{\partial}{\partial\xi}\tfrac{\mathscr{J}_{22}}{|J|} - \mathscr{J}_{11}\tfrac{\partial}{\partial\eta}\tfrac{\mathscr{J}_{22}}{|J|}\Big)\Big) \\
g^{17}(\Phi) = -\tfrac{1}{|J|^2} \Big(& \mathscr{J}_{21}\mathscr{J}_{12}\Big(\mathscr{J}_{22}\tfrac{\partial}{\partial\xi}\tfrac{\mathscr{J}_{12}}{|J|} - \mathscr{J}_{21}\tfrac{\partial}{\partial\eta}\tfrac{\mathscr{J}_{12}}{|J|}\Big) \\
& + \mathscr{J}_{11}\mathscr{J}_{12}\Big(\mathscr{J}_{12}\tfrac{\partial}{\partial\xi}\tfrac{\mathscr{J}_{12}}{|J|} - \mathscr{J}_{11}\tfrac{\partial}{\partial\eta}\tfrac{\mathscr{J}_{12}}{|J|}\Big) \\
& + \mathscr{J}_{21}\mathscr{J}_{22}\Big(\mathscr{J}_{22}\tfrac{\partial}{\partial\xi}\tfrac{\mathscr{J}_{22}}{|J|} - \mathscr{J}_{21}\tfrac{\partial}{\partial\eta}\tfrac{\mathscr{J}_{22}}{|J|}\Big) \\
& + \mathscr{J}_{11}\mathscr{J}_{22}\Big(\mathscr{J}_{12}\tfrac{\partial}{\partial\xi}\tfrac{\mathscr{J}_{22}}{|J|} - \mathscr{J}_{11}\tfrac{\partial}{\partial\eta}\tfrac{\mathscr{J}_{22}}{|J|}\Big)\Big)
\end{aligned}
$$

or a reference bifurcation. We consider the vertical top-left boundary to be the inflow boundary, and we have four outflow boundaries in the bottom-right of the domain. The walls between the inflow and outflow are considered to be no-slip boundaries.

Following the procedure described in Løvgren et al. [2], we compute separate sets of basis functions for each reference domain and each type of inflow/outflow boundary condition. We use a spectral element code with polynomials of degree $\mathscr{N} = 20$ to compute the basis functions, and we also compute a reference solution of the steady Stokes problem on the entire domain. We are not interested in the error of the reduced basis approximation itself, but the absolute error of an output derived from this approximation. Our output of interest $s(\mathbf{u}; \Phi)$ is the volume flow rate across the inflow boundary, and for the reference solution we get $s(\mathbf{u}; \Phi) = 4.80 \cdot 10^{-4}$. We use $N_p = 15$ basis functions on the pipe domain in Fig. 1, and

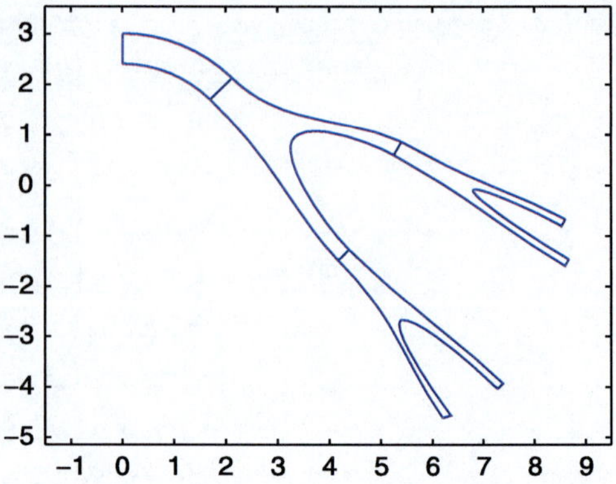

Fig. 1 The computational domain Ω

Table 4 Comparison of the computation time spent when solving the steady Stokes problem on the domain depicted in Fig. 1

Method	Time
Spectral element	4,746 s $\approx$ 1 h20 m
Reduced basis element	299 s $\approx$ 5 m
RB offline-online	38 s

$N_b = 30$ basis functions on the bifurcation domains. The mappings used to generate the basis functions are the same as in [2]. For the resulting reduced basis element approximation the lower bound for the output of interest is $s(\mathbf{u}; \Phi) - s^-(\mathbf{u}_N; \Phi) = 1.5 \cdot 10^{-7}$, both with and without the offline-online decoupling. The upper bound is more conservative, giving $s^+(\mathbf{u}_N; \Phi) - s(\mathbf{u}; \Phi) = 8.1 \cdot 10^{-5}$; see [2] for details.

In Table 4 we present the time spent to compute the reference solution and the reduced basis approximation with and without offline-online decoupling. The speed-up when applying the offline-online decoupling is almost a factor 10 relative to the reduced basis element method, and more than a factor 100 relative to the spectral element method.

Acknowledgements This work has been supported by the Research Council of Norway through a Centre of Excellence grant to the Center for Biomedical Computing.

References

1. M. Barrault, Y. Maday, N. C. Nguyen, and A. T. Patera. An empirical interpolation method: Application to efficient reduced-basis discretization of partial differential equations. *C. R. Acad. Sci. Paris, Serie I*, **339**, 667–672 (2004)
2. A. E. Løvgren, Y. Maday, and E. M. Rønquist. A reduced basis element method for the steady Stokes problem. *M2AN*, **40**(3), 529–552 (2006)
3. Y. Maday and E. M. Rønquist. The reduced-basis element method: Application to a thermal fin problem. *SIAM J. Sci. Comput.*, **26**(1), 240–258 (2004)
4. G. Rozza and A. Quarteroni. Numerical solution of parametrized Navier–Stokes equations by reduced basis methods. *Numer. Meth. for PDEs*, **23**(4), 923–948 (2007)

The Challenges of High Order Methods in Numerical Weather Prediction

Catherine Mavriplis

Abstract This paper reports on the communications made at the 2009 ICOSAHOM meeting minisymposium on the challenges of high order methods in numerical weather prediction, with contributions from mathematicians as well as atmospheric and ocean modelers. Motivation for an investment in high order method development for numerical weather prediction is given in terms of the potential payoff in light of the current challenges in the field. Among other issues, the implementation of physical parameterizations with high order methods stands out as a yet-unexplored and potentially difficult challenge to resolve. Adaptivity is also expected by some to significantly advance the state-of-the-art but no consensus seems to be reached that it will be feasible. Among the recommendations expressed at the workshop are the need for demonstrated efficiency comparisons between high order and low order methods for a desired level of accuracy in resolving waves.

1 Introduction

The following paper reflects the communications made at the 2009 ICOSAHOM meeting in the minisymposium with the same title organised by the author. The goal of this workshop was to bring together mathematicians and geoscientists to examine and discuss the challenges of introducing and sustaining the development of high order methods in the field of numerical weather prediction. While numerical weather prediction distinguishes itself from climate and other atmospheric modeling, it shares with these fields as well as similar efforts in ocean modeling some basic characteristics. The workshop therefore gathered ocean as well atmospheric modelers, from both the high and low order communities.

Traditionally, high order methods have been confined to more theoretical applications: for example, studies of isotropic turbulence in a cubical domain are routinely

C. Mavriplis
University of Ottawa, Ottawa, Canada
e-mail: Catherine.Mavriplis@uottawa.ca

J.S. Hesthaven and E.M. Rønquist (eds.), *Spectral and High Order Methods for Partial Differential Equations*, Lecture Notes in Computational Science and Engineering 76,
DOI 10.1007/978-3-642-15337-2_23,

studied with spectral methods. High order methods have excellent convergence properties, e.g., spectral methods exhibit exponential convergence to smooth solutions and have very low dispersion and dissipation errors. However, the smoothness implied tends to restrict their application to sets of problems that do not exhibit discontinuities (such as compressible aerodynamics) nor sharp gradients (such as sharp weather fronts in otherwise noisy fields). With the development of the Spectral Element (SE) method [1] however, and later, the Discontinuous Galerkin (DG) method [2] and others, high order methods have become more accessible to engineering calculations and modeling of real, complex phenomena. High order methods are also quite expensive in comparison with low order methods for relatively low accuracy. However as accuracy requirements increase there is a trade-off: high order methods converge much more quickly and hence become more efficient. Furthermore, with the advent of massively parallel computers the expense has become less restrictive since high order subdomain calculations make use of the power and speed of individual processors and the disjoint subdomain formulation decreases interprocessor communication, the bottleneck of parallel machines, e.g., [3,4].

The intent of the minisymposium and this paper is to broaden the applicability of high order methods to real, complex physical problems of grand challenge type. As stated above, the advantages and disadvantages of high order versus low order methods are intertwined and are not always clear to physical modelers. Perception and unfamiliarity with the methods that others use often limit progress. It is indeed difficult to introduce new techniques to those who have worked and dedicated themselves to tremendous advances with traditionally used methods because of the vast infrastructure development. Some progress has been made in several fields: whereas the first (1989) ICOSAHOM featured mostly classical fluid mechanics problems with simple geometries [5], the most recent meeting in Beijing (2007) featured minisymposia on aeroacoustics, plasma physics, biomechanics, electromagnetics and significantly more complex fluid mechanics [6]. In the atmospheric sciences, while spectral Fourier methods have enjoyed a relatively long history in global circulation modeling (using spherical harmonics in longitude), since the 1990s there has been steady development of more flexible high order models (e.g., the Spectral Element Atmospheric Model (SEAM) [7], the US National Center for Atmospheric Research (NCAR) High Order Method Modeling Environment (HOMME) [8], which now also contains a Discontinuous Galerkin model for global atmospheric modeling [9]).

However, in mesoscale meteorology where weather modeling and short term predictions are sought, finite difference (FD) methodologies have dominated modeling efforts 50 years. Most current operational mesoscale numerical weather prediction models employ fixed grid FD methods with high order FD in some areas. Across the globe weather modelers are striving to increase resolution, but, due to the fixed grid scheme, the only choice is to reduce grid spacing dx. Adaptivity in atmospheric modeling remains rare, as it does in most high order efforts in other fields as well. There are certainly good arguments for reluctance to move to high order methods in mesoscale meteorology: the microphysics of atmospheric modeling, such as evaporation, precipitation, radiation and chemistry, are quite complex and at a level of accuracy that is much lower than that of the fluid dynamics. On the other hand, the calculations of multiple passive scalars for the modeling of cloud processes and atmospheric chemistry require large amounts of computation and, if efficiency can be increased, significant gains can be made. Lastly, local storm modeling can be aided by higher resolution adaptive grid methodologies as localized structures such as tornadoes moving and developing quickly across large terrains are often poorly forecast by or missing from existing operational and research models [10].

The motivation is therefore clearly laid out to spur research in high order methods for numerical weather prediction. The challenges, however, are many and the workshop discussion of these is captured in this paper.

2 Overview of Atmospheric Modeling Challenges and Status

As summarized by Boyd at the workshop[1] [11], contemporary atmospheric models are made up of two basic components: the dynamics core which solves the fluid mechanics equation models of the desired atmospheric flows and the "physics", a comprehensive package to treat the complex associated physical phenomena such as radiative transfer, photochemistry, phase change, precipitation, etc. Atmospheric models also vary according to their intended use, with global circulation models (GCM) being used for long term climate studies, global weather forecasting (GWF) for large scale predictions and limited area models (LAM) for local or national weather prediction. While GCMs are run at low resolution for long times with arbitrary initial conditions to reach statistical equilibrium, GWFs run at high resolution for a handful of days starting with initial conditions derived from observation of the atmosphere. LAMs use even smaller computational domains at very high resolution, often nested in a global model. Data assimilation models are also used where observational data is input to the simulation at regular intervals to"corral" the simulation to reality. Obviously, none of these situations lends itself to a clean mathematical initial value problem statement reflecting the reality of the atmosphere. Initial conditions need to reflect the dynamical balances (quasi-geostrophic, i.e., between Coriolis and pressure gradient forces, and hydrostatic) that dominate atmospheric flows. With poorly defined initial and boundary conditions, high order models immediately run into problems as their high order accuracy and low dispersion and dissipation errors will create spurious phenomena, such as artificial gravity waves, that will persist in long time simulations.

Other peculiarities of atmospheric models have found work-around solutions in the atmospheric modeling community that have become so entrenched they are difficult to sort out when proposing a new approach such as high order methods. For one, there is a large difference in scale between the horizontal and vertical in atmospheric modeling, this difference also varying depending on the application. Indeed, for GCMs, a thin layer of the atmosphere is assumed, at its simplest using

[1] The workshop participants' names are listed in the Acknowledgement section of this paper.

the shallow water equations. For weather modeling, however, particularly in strong convection environments such as for storms, the vertical component is dominant. In NCAR's Community Climate Model, e.g., the vertical is treated by finite differences, while the horizontal is treated by spectral methods. Resolution in the horizontal is usually much coarser than in the vertical, creating a cell anisotropy that can be problematic for efficient schemes, e.g., O (25 km) horizontally versus O (0.5 km) vertically in a typical high resolution LAM run. Marginal horizontal resolutions have also led to fixes such as "convective parameterizations", wherein a local convection event, e.g., moisture rising from a mountain and creating local precipitation, must be sub-grid-triggered within a larger cell.

These different situations lead to different assumptions in the equation models from the outset. For example, while a hydrostatic assumption is the norm for GCMs and GWFs, the non-hydrostatic equations must be used for severe weather prediction. The hydrostatic assumption ignores vertical momentum terms other than pressure gradient and gravity. This effectively eliminates the meteorologically unimportant, but very rapidly propagating acoustic waves. However, with an increased interest in higher resolution, the need to treat the vertical direction more completely requires a non-hydrostatic assumption. In this case, as pointed out by Durran at the workshop [12], the treatment of the acoustic waves will have to be dealt with efficiently, since they do not contribute to the meteorological dynamics but do make the compressible equations stiff. Several methods are being used: (1) filtering the governing equations to remove the acoustic waves through the Boussinesq approximation (ignoring density variations in the mass conservation principle as well as in vertical momentum terms but including density in the buoyancy terms), the anelastic formulation (in which a reference density varying only with the vertical is included in mass conservation) or the pseudo-incompressible system (in which density perturbations are included in mass conservation but only depend on reference rather than perturbation pressure as well as temperature); (2) advancing the acoustic waves implicitly or on a separate shorter time step through either complete or partial operator splitting. In both of the latter cases, divergence generated by the operators is evaluated on the large time step and propagated by the system. In the complete splitting, this divergence accumulates over the series of steps taken. While partial operator splitting keeps the divergence changes small during the small steps, it has been shown to be unstable; and yet it is used with work-around solutions of filtering [13] or divergence damping [14].

In the above methodologies, conservation is compromised to differing degrees. For weather prediction, with strong convective motion, local conservation takes on a more crucial role than in long-term climate calculations, where global conservation is important. For this reason, many modelers are now considering finite volume, e.g., [15], discontinuous Galerkin, e.g., [16] and other flux-based methods as alternates to the familiar spectral and finite difference methods.

Lagrangian approaches are preferred to Eulerian by some groups in the community. A Lagrangian frame of reference eliminates the nonlinear advection terms, thereby alleviating restrictive advective time-stepping CFL (Courant–Friedrichs–Lewy stability) conditions. Additional backward trajectory calculations are needed however, and these may introduce errors: higher order interpolation is used to improve the accuracy of marginally resolved waves. This practice is sometimes also used in finite difference Eulerian schemes: the advection terms are approximated to a higher order. Semi-implicit semi-Lagrangian schemes are popular in global modeling as the time-splitting can resolve faster moving gravity waves with a larger time step, revealing the method to be more efficient than its Eulerian counterpart [12, 17].

Indeed, for Eulerian schemes as well, semi-implicit methods are popular for the same reason. Efficiently resolving the geophysical waves in all atmospheric modeling is challenging due to the different speeds of propagation of the acoustic, gravity and Rossby waves. While the advection terms are usually treated explicitly and the pressure gradient and divergence terms implicitly, the different wave speeds suggest time splitting or implicit treatment: the fastest moving waves may be treated implicitly in order to avoid drastically small time steps to accommodate those high speeds. At the same time, any nonlinear parts of the other terms may be treated explicitly. For high order methods, explicit schemes are expensive as the CFL-limited time steps can be quite restrictive due to the non-uniform collocation point distribution of element-based spectral basis functions. However, trade-offs between the savings in the number of degrees of freedom required to resolve waves and the increased computational expense need to be quantified to determine whether the shift to high order methods can be justified. Efficient time stepping schemes for high order methods will need to be demonstrated.

Another peculiarity of atmospheric modeling is the "pole problem": the fact that a latitude-longitude grid converges at the poles creating singularities. Many novel grids as well polar filtering have been proposed to alleviate this problem. The Lagrangian approach also circumvents this problem. Global spectral and subdomain-based high order methods sidestep the pole problem quite naturally either through spherical harmonics or the cubed sphere tiling approach, e.g., [18, 19]. Dubos presented a mixed Fourier-finite element method at the workshop [20], that offers conservation and elimination of the pole problem. For low order methods unstructured icosahedral grids are becoming more popular for this reason among others [21].

Orography, or the topography of the Earth's surface, also factors into the accuracy and stability of numerical atmospheric models. While terrain-following vertical coordinate systems have been in use for many years, terrain-intersecting grids are now being explored as a means of increasing resolution while maintaining numerical stability, e.g., [22] and have been tested in the Deutscher Wetterdienst COSMO-DE model. Lock [23] presented a high resolution method based on a terrain-intersecting grid for flow over very steep orography at the workshop.

As models become more powerful, in particular because of the increased computer power and storage, and hence the ability to use finer grid spacing, some of the entrenched modeling approximations are hitting their limits. For example, the hydrostatic assumption has conveniently served to filter out sound waves, that, while they exist, do not affect the meteorology. At high resolution, mathematical formulation of the governing equations should change to properly describe nonhydrostatic motions, but this change also allows the system to support gravity and sound waves, with large as well as small scale motions becoming three-dimensional, which can force a reduction in the maximum stable time step and significantly increase the computational burden (see, e.g., [24]). Many high-resolution models now use the mode-splitting or a even filtered equation set approach [25] to more efficiently integrate the non-hydrostatic equations. The convective parameterization mentioned above also becomes useless [26, 27] at finer resolutions, e.g., mysteriously absent thunderstorms in the refined portion of the grid within a coarse grid region of thunderstorm activity according to Boyd [11]. In such cases, the "fixes" implemented by the accumulated infrastructure may need to be rethought in the context of new methods, such as high order and adaptive methods. For a discussion of physical parameterizations in the context of new weather prediction models see [28].

Adaptivity seems to be a natural approach to modeling currently unresolved features, either calculated, such as waves, or modeled, such as orography. But as more and more features get resolved, more appear and the "physics" respond poorly because they have been built and tuned over many years for regular grids with relatively low resolution. This is an area ripe for investigation. Furthermore, because of this continual underresolution, areas of transition between adapted grids will probably trigger spurious waves, unless the adaptivity is done carefully. Again, standard work-around solutions exist such as a relaxation zone [29] between the coarse grid and fine LAM grid. Continuously-varying resolution reduces some of the reflections at least in the long waves. Recent work in adaptive mesh refinement with second order finite volume schemes was presented by Jablonowski at the workshop: Fig. 1a shows an adapted grid for the calculation of a shallow water model case where two cyclonic Rankine vortices initially placed near the Equator in the Southern Hemisphere merge. This work [30] has successfully shown that more meteorologically-important features can be resolved and tracked but has also uncovered a host of issues: such as wave reflections at coarse/fine grid interfaces and dispersion and diffusion on coarse cells. A comparison with spectral element adaptive calculations of St-Cyr's [31] showed that the spectral element method was better able to control global errors. Refinement criteria are also under investigation: flow-based criteria, such as the vorticity-based one used in the calculation of Fig. 1a, seem to be tailored to specific flow conditions. Numerically-based refinement criteria that are naturally implemented with high order methods [32] may be applied more generally. Adjoint-based techniques that are gaining ground in other computational arenas are also making an appearance in the geoscience community [33]. More investigation with complex atmospheric flows is needed here.

Some similarities can be drawn between ocean and atmospheric modeling. While the ocean dynamics equations are similar, much of the "physics" mentioned above can be omitted while tracking salt and biological influences becomes pertinent. In ocean modeling, the domain is perhaps more complex at least at the coarsest level: domains are multiply-connected basins with islands and continents creating gridding and boundary value problem challenges. Depth can also be approximated by shallow water assumptions in the simplest of models, but eventually becomes more important as model fidelity increases. Ocean modelers have very sparse observational data and forecasting is less common. The models are used mainly for long term prediction and circulation understanding. The current research trend leans

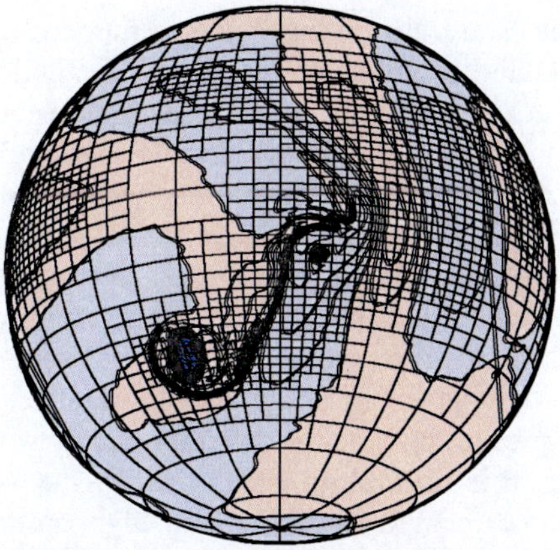

(a) Adaptive mesh refinement second order finite volume solution and grid for a shallow water model of two merging cyclonic Rankine vortices initially placed near the Equator in the Southern Hemisphere. A snapshot of the relative vorticity field at day 5 is depicted. *Blue* indicates a clockwise rotation, *red* indicates counterclockwise rotation. Each adapted block contains additional grid points that are not shown [30]. Courtesy of C. Jablonowski

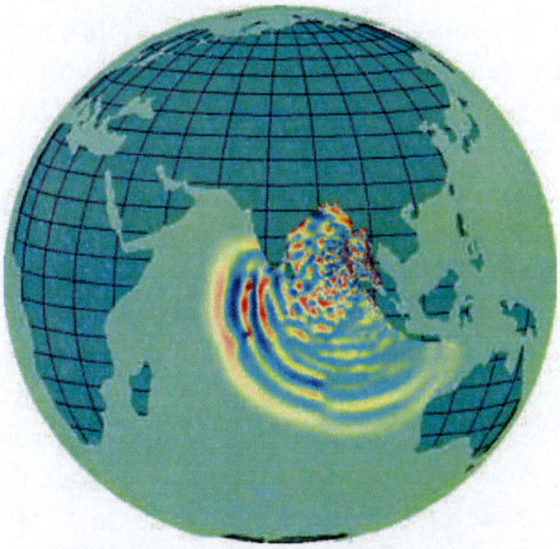

(b) Snapshot of a global simulation of the 2004 Great Andaman–Sumatra Tsunami, using an unstructured triangular mesh for accurate topography representation and a second order conforming/non-conforming finite element numerical scheme, developed at Alfred Wegener Institute and implemented in the operational tsunami model TsunAWI [34]. *Red* indicates height above mean sea level, *blue* below. Courtesy of J. Behrens/S. Harig

Fig. 1 Advanced low order atmosphere and ocean simulations

toward coupled ocean-atmosphere dynamics. A unified approach would obviously simplify modeling efforts, but many challenges remain in the understanding of the coupling and the disparity of scales.

One particular case where many parallels can be drawn, however, is in tsunami modeling, a topic addressed by Behrens [34] in the workshop. The problem of tsunami prediction is closely related to severe weather event prediction, such as hurricanes and tornadoes in particular. The short warning times of 23 mn for tsunamis and 18 mn for tornadoes preclude long simulation times in prediction codes. The detrimental effect of false warnings is also a shared characteristic. The interaction of these phenomena with inhabited areas is a complex problem: e.g., in the case of tsunamis: ragged coastlines with topography and more complex (debris-filled and obstacle impeded) flow in inundation areas. These requirements may preclude the use of expensive, more detailed high order methods at least for short-term prediction. An example result of a simulation of the 2004 Great Andaman–Sumatra Tsunami, performed with TsunAWI, an unstructured grid finite element operational tsunami model, is shown in Fig. 1b.

In more general ocean modeling, advances are also being made in numerical modeling: spectral element, discontinuous Galerkin and unstructured grid methods are being considered as alternatives to structured finite difference methods. Legat [35] has worked to implement the discontinuous Galerkin method in this field, and in particular to accurately represent complex boundaries (e.g., coastlines and curved manifolds) with suitable high fidelity grids that do not compromise the increased accuracy of a higher order DG method (e.g., fourth order).

In summary, there are many challenging characteristics to atmospheric and, in particular, numerical weather modeling. Several different methods (global spectral, finite difference, finite volume, discontinuous Galerkin and spectral element methods) have been used in the field, each presenting potential advantages. As we move forward, with advances on all fronts, it is difficult to predict which methods will be clear winners. In fact, few direct comparisons between high order and low order methods exist. Global spectral methods were compared to finite differences of up to sixth order in [36]: the spectral methods were shown to be more efficient for a prescribed high accuracy. A comparison of a new finite volume version of the Community Climate System Model (CCSM3) has been made with its previous spectral dynamics core showing significant improvements in some areas [37]. A recent comparison of finite volume and spectral element methods by Jablonowski and St-Cyr was presented at the workshop [31]. Spectral element and discontinuous Galerkin nonhydrostatic models were recently compared in [16]. A systematic comparison for both model problems such as those shown by Crowell [38] (Fig. 2) and Jablonowski [31] at the workshop can start to establish some metrics for comparison, as well as meaningful test cases, while uncovering areas in need of future development. Figure 2 shows the ability of the discontinuous Galerkin method to reduce errors around and provide better definition of complex structures in comparison with finite difference methods for the same CPU time.

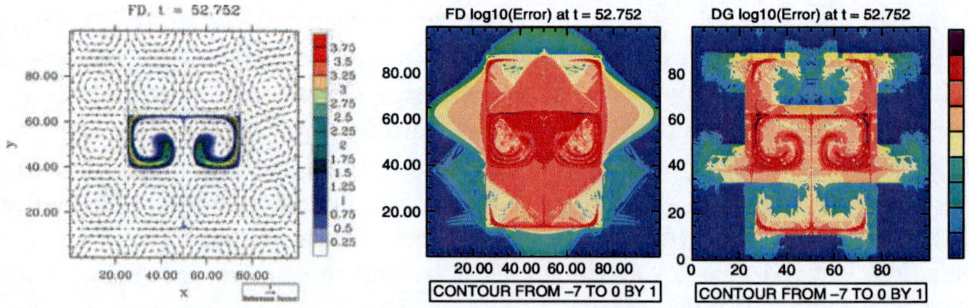

Fig. 2 Deformational flow: a tracer is advected by an array of vortices into fine structures. Comparison of finite difference (FD) and discontinuous Galerkin (DG) solutions for same CPU time. Representative solution (*left*) (FD–DG solution looks similar) and FD and DG $\log_{10}$ errors (*right*) for same CPU time. Resolutions: FD: sixth order [320 × 320], DG: tenth order [180 × 180]. The DG solution restricts errors to finer areas [38]

3 Challenges

Given this history and current status of atmospheric modeling and, in particular, numerical weather prediction, it is difficult to predict whether high order methods will have an immediate or future impact on the field. Nevertheless, many issues raised are ripe for a fresh look at them. With the recent progress in high order method implementations for model equations [7–9] and more complex systems [16], adaptive methodologies [30, 32] and high performance computing [3], it is clear that on-going efforts are needed to determine the possibility of success, perhaps uncovering difficulties to surmount, but hopefully also providing breakthroughs in some areas. From the workshop discussion, the following conclusions were drawn.

3.1 Where High Order Holds Promise

High order subdomain-based methods will ease gridding problems, and potentially prove as efficient as lower order methods on high performance computing platforms. It was not clear, however, how high an order would be best to pursue. While some participants felt that second order was high enough (i.e., that high order is not necessary), some thought fourth order would be more than adequate, in particular in the context of the "physics" being the limiting factor. High order methods will certainly help in resolving a wider range of waves with lower dissipation than low order methods. High order is viewed as attractive for adaptivity in particular for lossless transition between regions of varying grid density.

3.2 Where High Order Instills Doubts

High order methods will perhaps not ever be as efficient as lower order methods for atmospheric flows because of the complexity of the system and the continual marginal resolution capabilities. Some researchers question the ability of high order methods to give stable solutions in this underresolved regime, both for the dynamics and the physics, bringing up the need for examination of filters for high order methods in the atmospheric context. Of course the more applied problem solvers often feel it is better to get a "quick and dirty" approximate solution than wait for a really accurate one that bears little resemblance to reality. For these people, low order methods seem most promising to pursue. Lastly, the coupling with the physical parameterizations as they have been developed is overwhelmingly viewed as the greatest roadblock to acceptance of high order methods in this community.

3.3 Recommendations

The recommendations for further research therefore are to:

- Carry out more comprehensive comparisons of low and high order methods on both simple model problems and more sophisticated test cases providing a fair basis of comparison for storage, efficiency, and accuracy.
- Explore the behaviour and the stability limitations of high order methods in underresolved calculations including the use of adaptivity to resolve certain (but not all – how do we limit?) features.
- Tackle the problem of the physical parameterization schemes: either exploring how they could be coupled meaningfully and efficiently to a high order dynamical core or rethinking the framework of how they are linked to the dynamics.

Many of these suggestions mirror those distilled from a similar meeting on adaptive methodologies for atmospheric and ocean modeling held in Reading this year [39]. We recommend that the two efforts be combined, in particular that the high order community involve itself in the proposed Newton Institute program.

4 Conclusions

In conclusion, it is evident that atmospheric and ocean modeling, and in particular, numerical weather prediction are very complex fields, whose vast incremental development make them difficult to penetrate as modelers or mathematicians uninitiated to their many particularities. As such, the introduction of newer, high order methods is a challenge, but evidence of initial successes has emerged in the last decades. The task of rethinking the framework in the context of a new computational modeling tool is before us. It will require careful comparison testing, development of efficient schemes and a cooperative effort with physical modelers.

Acknowledgements The author wishes to acknowledge the input of all the workshop participants: Jörn Behrens (Alfred Wegener Institute for Polar and Marine Research), John Boyd (University of Michigan), Sean Crowell (University of Oklahoma), Thomas Dubos (École Polytechnique, France), Dale Durran (University of Washington), Christiane Jablonowski (University of Michigan), Vincent Legat (University of Catholique de Louvain), and Sarah-Jane Lock (University of Leeds). In particular, the careful revision of this manuscript by J. Boyd, D. Durran, C. Jablonowski, T. Dubos, J. Behrens and L. Wicker is appreciated. This work and the workshop participation were supported by a US National Science Foundation grant No. 0530820. We also wish to thank the ICOSAHOM scientific committee for accommodating this non-traditional workshop.

References

1. Patera, A.T.: A Spectral Element Method for Fluid Dynamics: Laminar Flow in a Channel Expansion. J. Comput. Phys. **54**, 468–488 (1984)
2. Cockburn, B. and Shu, C.-W.: Runge–Kutta Discontinuous Galerkin Methods for Convection-Dominated Problems. J. Sci. Comp. **16**, no. 3, 173–261 (2001)
3. Fischer, P.F., Lottes, J., Pointer, W.D. and Siegel, A.: Petascale Algorithms for Reactor Hydrodynamics. J. Phys. Conf. Series **125**, 012076 (2008)
4. Klockner, A., Warburton, T., Bridge, J. and Hesthaven, J.S.: Nodal Discontinuous Galerkin Methods on Graphics Processors. J. Comput. Phys. **228**, 7863–7882 (2009)
5. Canuto, C., Quarteroni, A., eds, *Spectral and high order methods for partial differential equations, Proceedings of the ICOSAHOM'89 Conference* (North-Holland, Amsterdam, 1990)
6. Chen, Z., Shi, Z.-C., Shi, C.-W., Tang, T., eds, Selected Papers from 7th International Conference on Spectral and High Order Methods (ICOSAHOM07) Chinese Academy of Sciences, Beijing, June 18–22, 2007. Commun. Comput. Phys. **5**, no. 2–4 (2009)
7. Taylor, M.A., Loft, R. and Tribbia, J.: Performance of a spectral element atmospheric model (SEAM) on the HP Exemplar SPP2000. NCAR TN 439 EDD (1998)
8. Dennis, J, Fournier, A., Spotz, W.F., St-Cyr, A., Taylor, M.A., Thomas, S.J. and Tufo, H.: High Resolution Mesh Convergence Properties and Parallel Efficiency of a Spectral Element Atmospheric Dynamical Core. Int. J. High Perf. Comp. Appl. **19**, no. 3, 225–235 (2005)
9. Dennis, J.M., Nair, R.D., Tufo, H.M., Levy, M. and Voran, T.: Development of a Scalable Global Discontinuous Galerkin Atmospheric Model. Int. J. Comp. Sci. Eng., to appear.
10. NOAA, NOAA Hazardous Weather Testbed (HWT) 2008 HWT Spring Experiment. http://hwt.nssl.noaa.gov/Spring_2008/ Cited Sep 16 2009 (2008)
11. Boyd, J.P.: Challenges and Controversies in Multiscale Fluid Dynamics Algorithms for Numerical Weather Prediction and Climate Modeling. SIAM News **41**, no. 9, 1-1 (2008)
12. Durran, D.R.: *Numerical Methods for Wave Equations in Geophysical Fluid Dynamics* (Springer, New York, 1999)
13. Tatsumi, Y.: An Economical Explicit Time Integration Scheme for a Primitive Model. J. Meteor. Soc. Japan **61**, 269–287 (1983)
14. Skamarock, W.C. and Klemp, J.B.: The Stability of Time-Split Numerical Methods for the Hydrostatic and Nonhydrostatic Elastic Equations. Mon. Wea. Rev. **120**, 2109–2127 (1992)
15. Lin, S.-J.: A "Vertically Lagrangian" Finite-Volume Dynamical Core for Global Models. Mon. Wea. Rev. **132**, 2293–2307 (2004)
16. Giraldo, F.X. and Restelli, M.: A Study of Spectral Element and Discontinuous Galerkin Methods for the Navier–Stokes Equations in Nonhydrostatic Mesoscale Atmospheric Modeling: Equation Sets and Test Cases. J. Comp. Phys. **227**, no. 8, 3849–3877 (2008)
17. St-Cyr, A. and Thomas, S.J.: Nonlinear Operator Integration Factor Splitting for the Shallow Water Equations. Appl. Num. Math. **52**, no. 4, 429–448 (2005)
18. Putnam, W.M. and Lin, S.J.: Finite Volume Transport on Various Cubed-Sphere Grids. J. Comp. Phys. **227**, 55–78 (2007)
19. Nair, R., Thomas, S.J. and Loft, R.: A Discontinuous Galerkin Transport Scheme on the Cubed Sphere. Mon. Wea. Rev. **133**, 814–828 (2005)
20. Dubos, T.: A Conservative Fourier-Finite-Element Method for Solving PDEs on the Whole Sphere. Quart. J. Roy. Met. Soc., accepted (2009)
21. Behrens, J., *Adaptive Atmospheric Modeling – Key techniques in grid generation, data structures, and numerical operations with applications*, LNCSE **54** (Springer, Heidelberg, 2006)
22. Steppeler, J., Bitzer, H., Minotte, M. and Bonaventura, L.: Nonhydrostatic Atmospheric Modeling Using a z-Coordinate Representation. Mon. Wea. Rev. **130**, 2143–2149 (2002)
23. Lock, S.: Development of a new numerical method for studying microscale atmsopheric dynamics. Ph.D. thesis, University of Leeds UK (2008)
24. Bartello, P. and Thomas, S. J.: The Cost-Effectiveness of Semi-Lagrangian Advection. Mon. Wea. Rev. **124**, 2883–2897 (1996)

25. Smolarkiewicz, P.K., Margolin, L.G. and Wyszogrodzki, A.A.: A Class of Nonhydrostatic Global Models. J. Atmos. Sci. **58**, 349–364 (2001)
26. Skamarock, W.C. and Klemp, J.B.: Adaptive Grid Refinement for Two-Dimensional and Three-Dimensional Nonhydrostatic Atmospheric Flow. Mon. Wea. Rev. **121**, 788–804 (1993)
27. Lorant, V. and Royer, J.F.: Sensitivity of Equatorial Convection to Horizontal Resolution in Aquaplanet Simulations with a Variable-Resolution GCM. Mon. Wea. Rev. **129**, no. 11, 2730–2745 (2001)
28. Lee, T.-Y. and Hong, S.-Y.: Physical Parameterization in Next-Generation NWP. Bull. Am. Met. Soc. **86**, no. 11, 1615–1618 (2005)
29. Davies, H.: A Lateral Boundary Formulation for Multi-Level Prediction Models. Quart. J. Roy. Met. Soc. **102**, no. 432, 405–418 (1976)
30. Jablonowski, C., Herzog, M., Penner, J. E., Oehmke, R. C., Stout, Q. F., van Leer, B. and Powell, K. G.: Block-Structured Adaptive Grids on the Sphere: Advection Experiments. Mon. Wea. Rev. **134**, 3691–3713 (2006)
31. St-Cyr, A., Jablonowski, C., Dennis, J.M., Tufo, H.M. and Thomas, S.J.: A Comparison of Two Shallow Water Models with Non-Conforming Adaptive Grids. Mon. Wea. Rev. **136**, 1898–1922 (2008)
32. Feng, H. and Mavriplis, C.: Adaptive Spectral Element Simulations of Thin Flame Sheet Deformation. J. Sci. Comp. **17**, no. 1–4, 385-395 (2002)
33. Power, P.W., Piggott, M.D. and Fang, F.: Adjoint Goal-Based Error Norms for Adaptive Mesh Ocean Modeling, Ocean Model. **15**, 3–38 (2006)
34. Harig, S., Chaeroni, C., Pranowo, W.S. and Behrens, J.: Tsunami Simulations on Several Scales: Comparison of Approaches with Unstructured Meshes and Nested Grids. Ocean Dyn. **58**, no. 5–6, 429–440 (2008)
35. Bernard, P.-E., Remacle, J.-F., Comblen, R., Legat, V. and Hillewaert, K.: High-Order Discontinuous Galerkin Schemes on General 2D Manifolds Applied to the Shallow Water Equations. J. Comp. Phys. **228**, 6514–6535 (2009)
36. Browning, G.L., Hack, J.J. and Swarztrauber, P.N.: A Comparison of Three Numerical Methods for Solving Differential Equations on The Sphere. Mon. Wea. Rev. **117**, 1058–1075 (1988)
37. Bala, G., Rood, R.B. et al.: Evaluation of a CCSM3 Simulation with a Finite Volume Dynamical Core for the Atmosphere at 1 Degrees Latitude × 1.25 Degrees Longitude Resolution. J. Clim. **21**, no. 7, 1467–1486 (2008)
38. Crowell, S., Williams, D., Mavriplis, C. and Wicker, L.: Comparison of Traditional and Novel Discretization Methods for Advection Models in Numerical Weather Prediction. Allen, G. et al. (Eds.): ICCS 2009, Part II, LNCS **5545**, 263–272 (2009)
39. Weller, H., Ringler, T., Piggott, M. and Wood, N.: Challenges Facing Adaptive Mesh Modelling of the Atmosphere and Ocean. Bull. Am. Met. Soc. **91**, 105–108 (2010)

GMRES for Oscillatory Matrix-Valued Differential Equations

Sheehan Olver

Abstract We investigate the use of Krylov subspace methods to solve linear, oscillatory ODEs. When we apply a Krylov subspace method to a properly formulated equation, we retain the asymptotic accuracy of the asymptotic expansion whilst converging to the exact solution. We demonstrate the effectiveness of this method by computing error and Mathieu functions.

1 Introduction

Our aim is to compute the fundamental solution to the differential equation

$$Y'(t) = (B(t) + \omega A(t))Y(t), \qquad t \in (a, b) \tag{1}$$

where A and B are $d \times d$ matrix-valued functions and ω is large. Applications of such equations include the computation of special functions (such as Airy, Bessel, hypergeometric and Mathieu functions [1]) the time-independent Schrödinger equation and semi-discretizations of the linear time-dependent Schrödinger equation.

When the eigenvalues of A are imaginary, the solutions to (1) become more oscillatory as $\omega \to \infty$. Thus traditional time-stepping methods are inefficient for large ω. The accuracy of modified Magnus expansions [5] does not degenerate as ω increases when used to compute (1). On the other hand, the approach we construct actually improves with accuracy as ω increases, and at an arbitrarily high asymptotic order. Moreover, we solve the equation globally, allowing us to compute over unbounded domains and to higher accuracy than a time stepping approach can achieve.

The simplest form of (1) is when $d = 1$, in which case we obtain the solution exactly:

S. Olver
Oxford University Mathematical Institute, 24-29 St Giles', Oxford, UK
e-mail: sheehan.olver@sjc.ox.ac.uk

J.S. Hesthaven and E.M. Rønquist (eds.), *Spectral and High Order Methods for Partial Differential Equations*, Lecture Notes in Computational Science and Engineering 76, DOI 10.1007/978-3-642-15337-2_24,

$$Y = \exp \int (B + \omega A) \mathrm{d}t.$$

The inhomogeneous form of (1) is the Levin differential equation [6] (changing notation to emphasize that these are scalar functions)

$$\mathscr{L} y = y' + \mathrm{i}\omega g' y = f. \tag{2}$$

A particular solution is

$$y = \mathrm{e}^{-\mathrm{i}\omega g} \int f \mathrm{e}^{\mathrm{i}\omega g} \, \mathrm{d}t.$$

In other words, solving (2) allows us to compute oscillatory integrals:

$$\int_a^b f \mathrm{e}^{\mathrm{i}\omega g} \, \mathrm{d}t = y(b) \mathrm{e}^{\mathrm{i}\omega g(b)} - y(a) \mathrm{e}^{\mathrm{i}\omega g(a)}.$$

Many methods have been developed in recent years for computing oscillatory integrals, with a recent review in [4]. One particular approach is to apply the GMRES method [10] directly to the differential equation [7–9], which we refer to as differential GMRES. By reformulating (2) as a shifted linear system

$$\mathscr{M} u = (\mathscr{M}_0 + \mathrm{i}\omega) u = f, \tag{3}$$

where $\mathscr{M}_0$ represents a linear operator, we achieve an asymptotic error in residual of

$$\mathscr{O}(\omega^{-n-1}),$$

where n is the number of GMRES iterations. This is the same asymptotic order as an asymptotic expansion, however, differential GMRES actually converges for fixed ω, subject to a condition on the growth of f in the complex plane. In Sect. 2, we review the details of this approach.

The goal, then, is to generalize this approach to the higher dimensional case of (1). This is accomplished by reforming the equation as a (matrix-valued) shifted linear system (3). By doing so, we obtain a method which also simultaneously achieves high asymptotic order and (based on numerical results) convergence.

Differential GMRES in its pure form requires taking derivatives (and integrals in the higher dimensional case) of the functions involved. In the general case, this is impractical. To avoid this, while we develop the framework in the infinite-dimensional setting, in practice we represent non-oscillatory functions as Chebyshev polynomials. This could be handled automatically and adaptive by the `chebfun` system [2], however, for concreteness and speed we use fixed order Chebyshev polynomials in our examples. More precisely, we represent functions by their values at Chebyshev–Lobatto points, and the fast cosine transform can be used to compute derivatives and anti-derivatives.

2 Oscillatory Integrals

GMRES [10] is an iterative algorithm originally developed for solving finite-dimensional linear systems

$$A\mathbf{v} = \mathbf{b} \qquad \text{for} \qquad A \in \mathbb{C}^{d\times d} \quad \text{and} \quad \mathbf{b} \in \mathbb{C}^{d}.$$

The Krylov subspace is defined as

$$\mathscr{K}_n[A, \mathbf{b}] = \text{span}\{\mathbf{b}, A\mathbf{b}, \ldots, A^{n-1}\mathbf{b}\}.$$

GMRES finds an element $\mathbf{v} \in \mathscr{K}_n[A, \mathbf{b}]$ such that the norm

$$||A\mathbf{v} - \mathbf{b}||_2$$

is minimized. This is accomplished through Arnoldi iteration [11], which constructs an orthonormal matrix $Q_n = (q_1, \ldots, q_n)$ whose columns span $\mathscr{K}_n[A, \mathbf{b}]$ and an upper Hessenberg (only zeros below the first subdiagonal) matrix $H \in \mathbb{C}^{(n+1)\times n}$ such that

$$AQ_n = Q_{n+1}H \qquad \text{and} \qquad q_1 = \frac{\mathbf{b}}{||\mathbf{b}||_2}.$$

Then $\mathbf{c} \in \mathbb{C}^n$ is chosen to minimize the norm

$$||H\mathbf{c} - ||\mathbf{b}||_2\, \mathbf{e}_1||_2\,. \tag{4}$$

The GMRES approximation is now $\mathbf{v} = Q_n\mathbf{c}$. Assuming that (4) decreases rapidly as n increases, $\mathbf{v}$ is indeed a good approximation to the true inverse:

$$||A\mathbf{v} - \mathbf{b}||_2 = ||AQ_n\mathbf{c} - ||\mathbf{b}||_2\, Q_n\mathbf{e}_1||_2 = ||Q_{n+1}[H\mathbf{c} - ||\mathbf{b}||_2\, \mathbf{e}_1]||_2 = (4).$$

In [7, 9], this was generalized for unbounded, infinite-dimensional operators such as the differentiation operator $\mathscr{D}$. In exactly the same manner as the finite-dimensional case, given the linear operator $\mathscr{L}$, function f and a semi-inner product $\langle\cdot,\cdot\rangle$, Arnoldi iteration determines a row-vector $\mathbf{q}_n = (q_1, \ldots, q_n)$ whose entries span the Krylov subspace $\mathscr{K}_n[\mathscr{L}, f]$ and an upper Hessenberg matrix $H \in \mathbb{C}^{(n+1)\times n}$ such that (where we use the convention $\mathscr{L}\mathbf{q}_n = (\mathscr{L}q_1, \ldots, \mathscr{L}q_n)$)

$$\mathscr{L}\mathbf{q}_n = \mathbf{q}_{n+1}H \qquad \text{and} \qquad q_1 = \frac{f}{||f||}.$$

Similarly, differential GMRES finds a function $v \in \mathscr{K}_n[\mathscr{L}, f]$ that minimizes the seminorm

$$||\mathscr{L}v - f||\,.$$

This is accomplished by finding $\mathbf{c} \in \mathbb{C}^n$ that minimizes the finite-dimensional norm

$$||H\mathbf{c} - ||\mathbf{b}||\,\mathbf{e}_1||_2,$$

thence $v = \mathbf{q}_n\mathbf{c}$.

Now consider the case of Arnoldi iteration applied to a shifted linear system of the form (3). We will denote the Hessenberg matrix produced by Arnoldi iteration for a particular value of ω by H_ω. A fact known from the finite-dimensional case which is also true in the infinite-dimensional case is that the orthonormal basis $\mathbf{q}_n$ is independent of ω and

$$H_\omega = H_0 + \mathrm{i}\omega I_{n,n+1}. \qquad [3]$$

In other words, we only need to compute the Arnoldi iteration for one choice of ω to determine the GMRES approximation for all choices of ω. Furthermore, GMRES satisfies the property that the error in residual is

$$||\mathscr{M}v - f|| = \mathscr{O}(w^{-n}). \qquad [9]$$

In our particular case we wish to solve the Levin differential equation (2). But $\mathscr{L}$ is not in the form of a shifted linear operator. If we assume that g' does not vanish, we can trivially put it into the required form:

$$\mathscr{M} = \mathscr{L}\frac{1}{g'} = \mathscr{D}\frac{1}{g'} + \mathrm{i}\omega.$$

We thus apply differential GMRES to $\mathscr{M}$, f and a suitable inner product to obtain v. Then $y = \frac{v}{g'}$ and hence we approximate the oscillatory integral by

$$\frac{v(b)}{g'(b)}\mathrm{e}^{\mathrm{i}\omega g(a)} - \frac{v(a)}{g'(a)}\mathrm{e}^{\mathrm{i}\omega g(a)}.$$

since we represent functions by their values, we use the standard dot product on the sample vector at Chebyshev–Lobatto points.

If the integral does contain a stationary point, i.e., g' vanishes in $[a,b]$, modifying the operator to take the form of a shifted linear operator is more complicated, and detailed in [8].

Consider the integral

$$\int_1^\infty \mathrm{e}^{\mathrm{i}\omega t^2}\mathrm{d}t = \frac{\sqrt{\pi}\,\mathrm{erfc}\sqrt{-\mathrm{i}\omega}}{2\sqrt{-\mathrm{i}\omega}} \quad \text{for} \quad \omega > 0.$$

Since the interval is unbounded, we represent functions by rational Chebyshev series – i.e., in terms of the basis $T_k(\frac{t-2}{t})$ – or, more precisely, by the values they take at the mapped Chebyshev–Lobatto points. In Fig. 1, we compute the absolute error of our approximation for several choices of ω. As can be seen, the rate of convergence as $n \to \infty$ increases with the frequency, and the number of

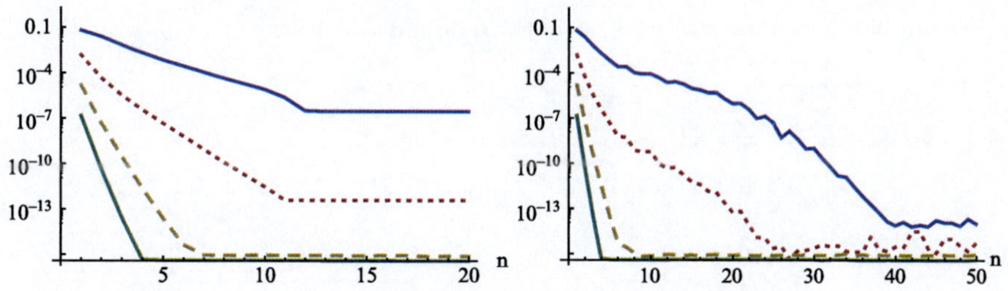

Fig. 1 The error in computing $\frac{\sqrt{\pi}\,\mathrm{erfc}\sqrt{-\mathrm{i}\omega}}{2\sqrt{-\mathrm{i}\omega}}$ using differential GMRES with 15 (*left figure*) and 50 (*right figure*) mapped Chebyshev–Lobatto points, for $\omega = 1$ (*plain*), 10 (*dotted*), 100 (*dashed*) and 1,000 (*thick*)

mapped Chebyshev–Lobatto points required to achieve machine precision accuracy decreases with the frequency.

3 Oscillatory Differential Equations

We will use the notation exp to denote the matrix exponential, though we only apply it to diagonal matrices, where

$$\exp \mathrm{diag}\,(d_1, \ldots, d_n) = \mathrm{diag}\,(\exp d_1, \ldots, \exp d_n).$$

We will use the indefinite integral notation to denote

$$\int A(t)\mathrm{d}t = \int_a^t A(t)\mathrm{d}t.$$

We now consider the solution of (1). Motivated by the preceding section, our goal is to transform this equation into a inhomogeneous shifted linear system. We will only consider the case where A is diagonalizable with distinct eigenvalues. Thus assume that there exists a matrix-valued function V that is nonsingular for all $t \in [a, b]$ and diagonal matrix-valued function $\Lambda = \mathrm{diag}\,(\lambda_1, \ldots, \lambda_d)$ where $\lambda_i \neq \lambda_j$ for any i and j, so that

$$AV = V\Lambda.$$

The requirement that V is smooth and nonsingular prevents application of this expansion to coalescing eigenvalues. This is similar to the case of stationary points for oscillatory integrals, and likewise outside the scope of this discussion. Note also that V is not determined uniquely, however, our approach works for any choice of V.

We apply the transformation $Y = VW$ to obtain

$$\begin{aligned}(VW)' &= (B + \omega A)VW \Leftrightarrow \\ V'W + VW' &= (BV + \omega V\Lambda)W \Leftrightarrow \\ W' &= (H + \omega \Lambda)W \qquad \text{for} \qquad H = V^{-1}BV - V^{-1}V'.\end{aligned}$$

We now apply the transformation

$$W = (I + U)\mathrm{e}^{\int (\operatorname{diag} H + \omega \Lambda)\mathrm{d}t} m$$

where $\operatorname{diag} H$ is the diagonal matrix whose entries are the diagonal of H. Therefore

$$U' + (I + U)(\operatorname{diag} H + \omega \Lambda) = (H + \omega \Lambda)(I + U).$$

We can rephrase this as

$$\mathscr{L}U = F \qquad \text{for} \qquad \mathscr{L}U = U' + U \operatorname{diag} H - HU + \omega[U, \Lambda],$$

where $[U, \Lambda]$ is the standard matrix commutator $[U, \Lambda] = U\Lambda - \Lambda U$ and $F = H - \operatorname{diag} H$.

Our goal now is to premultiply this operator by an inverse to the commutator operator. Because of our choice of transformations, we have ensured that F has zeros along the diagonal. Thus we can utilize the following inverse to the commutator:

Definition 1. For $\Lambda = \operatorname{diag}(\lambda_1, \ldots, \lambda_n)$ with distinct entries and M with zeros along the diagonal,

$$\mathscr{Q}M = \begin{pmatrix} 0 & \frac{m_{12}}{\lambda_2 - \lambda_1} & \cdots & \frac{m_{1n}}{\lambda_n - \lambda_1} \\ \frac{m_{21}}{\lambda_1 - \lambda_2} & 0 & \ddots & \vdots \\ \vdots & \ddots & \ddots & \frac{m_{(n-1)n}}{\lambda_n - \lambda_{n-1}} \\ \frac{m_{n1}}{\lambda_1 - \lambda_n} & \cdots & \frac{m_{n(n-1)}}{\lambda_{n-1} - \lambda_n} & 0 \end{pmatrix}$$

From inspection, it is clear then that $[\mathscr{Q}F, \Lambda] = \mathscr{Q}[F]\Lambda - \Lambda\mathscr{Q}[F] = F$.

However, the term

$$\mathscr{L}\mathscr{Q}F = (\mathscr{Q}F)' + (\mathscr{Q}F)\operatorname{diag} H - H\mathscr{Q}F + \omega F$$

will not necessarily have zeros along the diagonal, because of the term $H\mathscr{Q}F$. In other words, we cannot generate the Krylov subspace for $\mathscr{L}\mathscr{Q}$ and F. Fortunately, diagonal matrices lie in the kernel of the commutator. Hence we use the following, alternative commutator inverse:

$$[\cdot, \Lambda]^{-1}U = \mathscr{Q}U + \int \operatorname{diag}(H\mathscr{Q}U)\mathrm{d}t.$$

Then (using $D = \int \operatorname{diag}(H\mathscr{Q}F)\mathrm{d}t$)

$$\begin{aligned}\mathscr{M}F = \mathscr{L}[\cdot,\Lambda]^{-1}F = (\mathscr{Q}F)' + \operatorname{diag}(H\mathscr{Q}F) - H\mathscr{Q}F \\ + (\mathscr{Q}F)\operatorname{diag}H + D\operatorname{diag}H - HD + \omega F.\end{aligned}$$

Since $\mathscr{Q}F$ has zeros along the diagonal, the first and fourth terms also have zeros along the diagonal. The second term cancels the diagonal of the third term. Finally, since D is diagonal

$$D\operatorname{diag}H - HD = (\operatorname{diag}H - H)D$$

also has zeros on the diagonal. Thus $\mathscr{M}$ successfully maps the set of infinitely differentiable matrix-valued functions with zeros along the diagonal to itself.

Without modification, we can now construct a differential GMRES method for $\mathscr{M}$ and F, provided an appropriate semi-inner product is used. We will use the Frobenius inner product, where the dot product of two functions remains the dot product of their values at Chebyshev–Lobatto points. This returns a function G which satisfies

$$F \approx \mathscr{M}G = \mathscr{L}[\cdot,\Lambda]^{-1}G.$$

Therefore, $Y' \approx (B + \omega A)Y$ for

$$Y = V(I + [\cdot,\Lambda]^{-1}G)\mathrm{e}^{\int(\operatorname{diag}H + \omega\Lambda)\mathrm{d}t}.$$

The fundamental solution is then

$$Y_{\mathrm{F}}(t) = Y(t)Y(0)^{-1}.$$

4 Example: Mathieu Functions

Consider the system

$$Y' = (B_\alpha + \omega A)Y,$$

$$B_\alpha = \begin{pmatrix} 0 & 1 \\ \alpha & 0 \end{pmatrix} \quad \text{and} \quad A = \begin{pmatrix} 0 & 1 \\ -\cosh 2t & 0 \end{pmatrix}.$$

The eigenvalues of A are $\pm\mathrm{i}\sqrt{\cosh 2t}$, hence the solution is highly oscillatory. Indeed, as $t \to \infty$, it becomes exponentially more oscillatory. The fundamental solution to this equation can be written in terms of Mathieu functions [1] as

$$\begin{pmatrix} C(\alpha(1+\omega), \frac{1}{2}\omega(1+\omega), it) & -S(\alpha(1+\omega), \frac{1}{2}\omega(1+\omega), it) \\ \mathrm{i}\frac{C'(\alpha(1+\omega), \frac{1}{2}\omega(1+\omega), it)}{1+\omega} & -\mathrm{i}\frac{S'(\alpha(1+\omega), \frac{1}{2}\omega(1+\omega), it)}{1+\omega} \end{pmatrix}.$$

We apply our approach to approximate this function with $\alpha = 0$ and $\omega = 10$ over the interval $(0, 5)$. As can be seen in Fig. 2, our solution is equal to that of the built-in

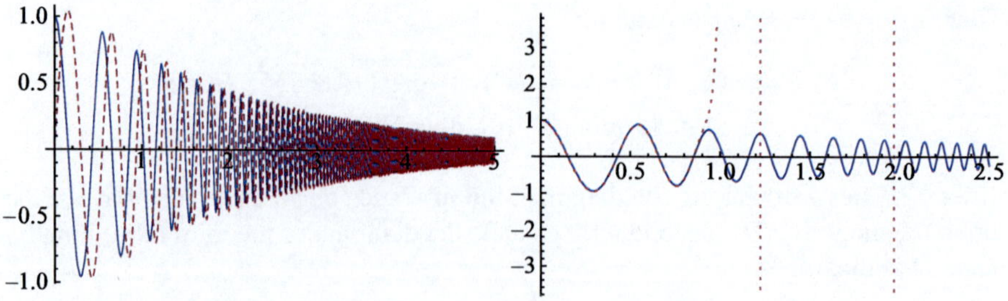

Fig. 2 In the *left* graph, the $(1,1)$-entry (*plain*) and the $(1,2)$-entry (*dotted*) of the differential GMRES approximation to the fundamental solution of the Mathieu equation with $\alpha = 0$ and $\omega = 10$. In the *right* graph, a comparison of the $(1,1)$-entry of the GMRES approximation (*solid*) to the real part of `Mathematica`'s built-in routine (*dotted*)

`Mathematica` routine (after scaling to obtain the fundamental solution) for computing Mathieu functions when t is small. As t increases, the `Mathematica` routine quickly explodes, whereas our method remains nicely behaved. Furthermore, the true solution must be real, as is our approximate solution, whilst the `Mathematica` routine grows a nonzero imaginary component. Comparison with a numerical ODE solver with a very small step size reveals that our solution is indeed the correct one. We omit a graph of the convergence rate for different values of ω, which is similar in behaviour to Fig. 1: the larger ω is, the faster the rate of convergence.

References

1. M. Abramowitz and I.A. Stegun. *Handbook of Mathematical Functions with Formulas, Graphs, and Mathematical Tables*. US Government Printing Office, Washington, DC, 1964
2. Z. Battles and L. Trefethen. An extension of matlab to continuous functions and operators. *SIAM J. Sci. Comput.*, 25:1743–1770, 2004
3. B. Datta and Y. Saad. Arnoldi methods for large sylvester-like observer matrix equations and an associated algorithm for partial pole assignment. *Lin. Algebra Appl.*, 154–156:225–244, 1991
4. D. Huybrechs and S. Olver. Highly oscillatory quadrature. ed. B. Engquist et al. *Highly oscillatory quadrature: Computation, Theory and Applications*. Cambridge University Press, Cambridge, 2008
5. A. Iserles. On the method of neumann series for highly oscillatory equations. *Bit Numer. Math.*, 44(3):473–488, 2004
6. D. Levin. Procedures for computing one and two-dimensional integrals of functions with rapid irregular oscillations. *Math. Comp.*, 38(158):531–538, 1982
7. S. Olver. Gmres for the differentiation operator. *SIAM J. Numer. Anal.*, 47:3359–3373, 2009
8. S. Olver. Fast, numerically stable computation of oscillatory integrals with stationary points. *BIT*, 50:149–171, 2010
9. S. Olver. Shifted gmres for oscillatory integrals. *Numer. Math.*, 114:607–628, 2010
10. Y. Saad and M. Schultz. Gmres: A generalized minimal residual algorithm for solving nonsymmetric linear systems. *SIAM J. Sci. Stat. Comput.*, 7:856–869, 1986
11. L. Trefethen and D. Bau III. *Numer. Lin. Algebra*. SIAM, 1997

Sensitivity Analysis of Heat Exchangers Using Perturbative Methods

J.C. Pacio, C.A. Dorao, and M. Fernandino

Abstract The solution of heat exchanger models is usually influenced by different parameters related to fluid properties, geometry and flow conditions. This implies that several simulations are usually required for design and optimization purposes. In this work, a sensitivity analysis using perturbative methods is presented as an alternative for reducing the computational cost of studying the sensitivity of the solution to different parameters. The method is based on the computation of the adjoint problem, and thus only one extra simulation is required for studying the sensitivity of an integral response, regardless of the number of parameters. Both the direct and the adjoint problem were solved using a least squares formulation.

1 Introduction

The design, optimization and scaling up of new processes require to take into account the most favorable and most disfavorable conditions in order to guarantee optimal performance and production under specifications. For that reason, it is important to quantify the influence of different parameters in the predictive solution by performing a sensitivity analysis.

Heat transfer is influenced by several parameters, such as mass flow rates, fluid properties, inlet pressure and temperature, geometry. The effect of variations in these parameters on the solution can be quantified by determining the sensitivity coefficients which can be computed by performing a set of simulations for different variations in the parameters. This approach is commonly referred to as construction of response surfaces and is widely used for reliability analysis. However, this approach can result in overwhelming computational costs if a lot of parameters are present. For a problem with eight parameters and considering five representative

J.C. Pacio (✉), C.A. Dorao, and M. Fernandino
Department of Energy and Process Engineering, Norwegian University of Science and Technology, 7491 Trondheim, Norway
e-mail: julio.pacio@ntnu.no, carlos.dorao@ntnu.no, maria.fernandino@ntnu.no

J.S. Hesthaven and E.M. Rønquist (eds.), *Spectral and High Order Methods for Partial Differential Equations*, Lecture Notes in Computational Science and Engineering 76,
DOI 10.1007/978-3-642-15337-2_25,

values for each one, $5^8 \approx 400.000$ simulations are required for obtaining the response surface.

An interesting alternative to the direct calculation is the methodology of Sensitivity Analysis using Generalized Perturbation Theory (GPT), widely used in reactors physics and thermal-hydraulics [2, 8, 9]. Sensitivity analysis consists on estimating the variation in the response due to a perturbation in the parameters.

GPT makes use of the solution in a point of interest, and requires one extra simulation for computing sensitivity coefficients. The main advantage is that it can be accomplished without previously choosing the parameter to be studied, and the calculations are faster and more efficient, since the system of equations that describes the physical behavior of the problems is solved only once. However, the main disadvantage of this methodology is that the answer is linearized around a point of interest. Such disadvantage can be minimized through the application of higher order GPT [4], but this escapes the limits of this work.

The GPT method results in the formulation of the adjoint problem, which is dependent on the integral response chosen. For a complete mathematical formulation, further reading of [3, 4, 9] is recommended.

In this work the GPT method is applied for studying the sensitivity of a heat exchanger model to variations in the model parameters. The main goal of the work is to analyze the advantages, accuracy and limitations of GPT. Although the GPT method is commonly applied for studying some thermo-hydraulic models, the application in the context of heat exchanger models is rather limited.

The final models, i.e. the direct and adjoint problems are solved numerically. Usually variations of the Finite Difference Method (FDM) or Finite Volume Method (FVM) are used for GPT, e.g. [3, 7]. In this work a least squares spectral element method (LSSEM) is used for solving all the governing equations. For the general description of the method, refer to [5, 6]. For smooth problems, this method shows a spectral convergence [1]. This means that each simulation can be solved faster and more accurately when compared to the traditional formulations.

The structure of this paper is as follows. The heat exchanger model used in this work is presented in Sect. 2. Section 3 presents the reference case. The GPT results are shown in Sect. 4, and their accuracy is studied in Sect. 5. Final conclusions and remarks are treated in Sect. 6.

2 The One–Dimensional Horizontal Heat Exchanger Problem

The scope of this work is limited to configurations that can be represented by a 1D analysis and operating conditions that fit the following approximations:

- Steady state
- Horizontal and constant cross section
- Two streams: single phase incompressible, constant physical properties
- Heat leakage, axial conduction and radiation heat transfer are negligible small.

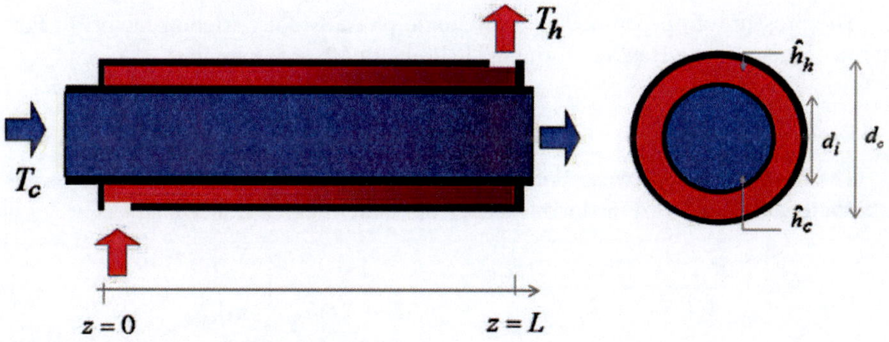

Fig. 1 Two–streams tube–in–tube heat exchanger

Table 1 Nomenclature used in this work

Symbol	Meaning	SI unit	Symbol	Meaning	SI unit
p	Pressure	Pa	T	Temperature	K
z	Position	m	f	Friction factor	–
$\dot{m}$	Mass flow rate	kg/s	G	Mass flux	$kg/(m^2s)$
ρ	Density	kg/m^3	c_p	Heat capacity	$J/(kgK)$
D_h	Hydraulic diameter	m	d_i	Inner diameter	m
t	Wall thickness	m	d_o	Outer diameter	m
λ	Thermal conductivity	$W/(Km)$	$\hat{h}$	Standard heat transfer coefficient	$W/(Km^2)$
Re	Reynolds number	–	UP	Linear heat transfer coefficient	$W/(Km)$
Pr	Prandtl number	–	Q	Heat duty	W
Nu	Nusselt number	–	$\dot{W}_p$	Pumping power	W

Considering all these constraints, the geometry chosen for study was a tube–in–tube heat exchanger, as shown in Fig. 1. This configuration is widely used for small scale and laboratory applications, since it's the cheapest and simplest option to build and maintain [11].

With these approximations, 1D momentum and energy balance equations are formulated for each stream j as follows (Table 1 summarizes the nomenclature):

$$\mp \frac{\partial p_j}{\partial z} = \frac{f_j}{2} \frac{G_j^2}{\rho_i D_{h_j}} \tag{1}$$

$$\pm \dot{m}_j c_{p_j} \frac{\partial T_j}{\partial z} = \frac{f_j}{2} \frac{\dot{m}_j G_j^2}{\rho_i^2 D_{h_j}} + UP(T_k - T_j) \tag{2}$$

with proper inlet boundary conditions. The proper sign should be chosen for both streams flowing in opposite (counterflow) or the same direction (co-current flow, as in Fig. 1). As a consequence of the previous approximations, there are no acceleration or gravitational terms in (1) and (2).

The pressure drop is modeled with single phase Moody friction factor f. For turbulent regime the Haaland formula [12] for smooth tubes applies:

$$f = [1.8ln(6.9/Re)]^{-2} \tag{3}$$

The heat transfer between both streams is represented by a linear heat transfer coefficient (HTC) UP. For steady–state, from a thermal resistance analysis.

$$\frac{1}{UP} = \frac{ln(1 + 2t/d_i)}{2\pi\lambda_{wall}} + \frac{1}{\pi(d_i + 2t)\hat{h}_h} + \frac{1}{\pi d_i \hat{h}_c} \tag{4}$$

Generally, an empirical or semi–empirical correlation is used to compute $\hat{h}$. In this case, the Dittus-Bolter correlation is appropriate for the inner tube (cold), and the annulus flow (hot) is represented by the Pethukov and Roizen formula [10] as follows:

$$\hat{h}_c = \frac{Nu_c\lambda_c}{D_{h_c}} = 0.023Re_c^{0.8}Pr_c^{0.3}\frac{\lambda_c}{D_{h_c}} \tag{5}$$

$$\hat{h}_h = \frac{Nu_h\lambda_h}{D_{h_h}} = 0.020Re_h^{0.8}Pr_h^{0.4}\frac{\lambda_h}{D_{h_h}}\left(\frac{d_o}{d_i + 2t}\right)^{0.86} \tag{6}$$

The physical properties in (5) and (6) are evaluated at inlet conditions.

From a design perspective, some integral results (FOM = figures of merit) may be more interesting that the exact distribution of pressure and temperatures, such as:

- Heat duty: $Q = \int_{z=0}^{L} UP(T_h - T_c)dz$
- Pressure drop on stream i: $\Delta p_i = \pm[p_i(z = 0) - p_i(z = L)]$
- Pumping power: $\dot{W}_p = \frac{\dot{m}_h}{\rho_h}\Delta p_h + \frac{\dot{m}_c}{\rho_c}\Delta p_c$

In the following sections, a reference case is defined and the sensitivity of these FOM is studied around this working point.

3 Reference Case

For a given wall material (stainless steel), the wall thermal conductivity λ_{wall} can be assumed constant; and for given working fluids (both water), their thermophysical properties depend only on the inlet pressure and temperature. Therefore, the problem is completely defined by only nine independent parameters

$$\mathbf{p} = \left[p_{c_0}, p_{h_0}, T_{c_0}, T_{h_0}, \dot{m}_c, \dot{m}_h, d_i, t, d_o\right] \tag{7}$$

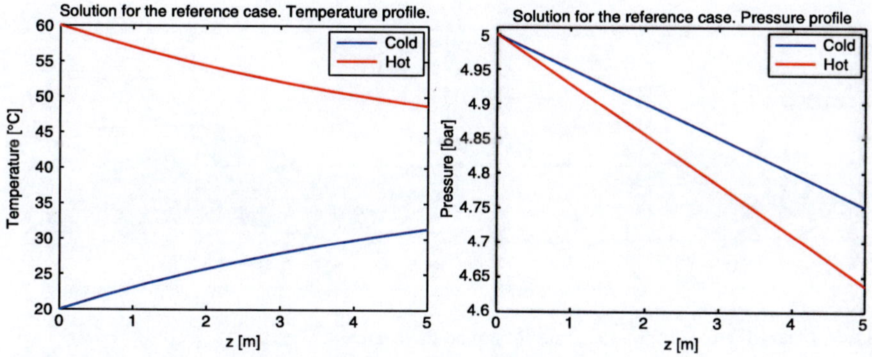

Fig. 2 Reference solution for the temperature and pressure profiles

The reference case is in parallel flow arrangement, and the parameters are fixed in their reference values.[1] Figure 2 shows the solution for this reference case. These plots were obtained with a coarse mesh (only one element) and a high order approximation (order 10). This was chosen because of the spectral convergence of the method for smooth problems. This yields a least-squares error of $\mathscr{J} \approx 10^{-8}$, in a short CPU time ($t < 10^{-2} s$).

4 Sensitivity Analysis Results

With one extra simulation, GPT gives the derivative of the response against all parameters, i.e. the gradient, which is the best direction for optimization. Figure 3 shows the response surface of Q and $\dot{W}_p$ for perturbations on the inner diameter (d_i) and the hot mass flow rate ($\dot{m}_h$) up to $\pm 20\%$ around their reference values. Underneath every point of this surface, the GPT results is plotted as the gradient vector (normal to constant level lines), pointing towards the optimum.

The response surface was constructed for 40 values of each parameter, i.e. $40^2 = 1{,}600$ simulations were required. This is an example with only two parameters, a complete analysis should consider all the nine parameters listed in $\mathbf{p}_2$, but this could not be plotted. Therefore, $40^9 \approx 2 \times 10^{14}$ would be required for computing the response surface. When lots of parameters are present, computing the response surface is practically impossible. With GPT, local information is obtained in each point with only one extra simulation, pointing towards the optimum, and the number of simulations required for optimization can be reduced significantly.

[1] $p_{c0} = p_{h0} = 5 bar$, $T_{c0} = 20°C$, $T_{h0} = 60°C$, $\dot{m}_c = \dot{m}_h = 1 \frac{kg}{s}$, $d_i = 20$ mm, $t = 0.15$ mm, $d_o = 30$ mm.

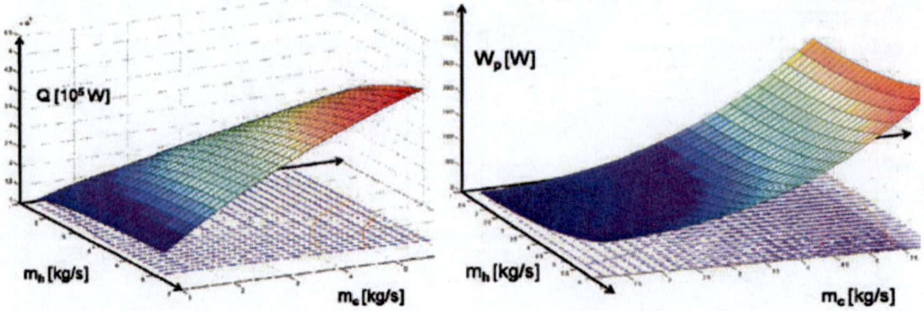

Fig. 3 Q and W_p for different $\dot{m}_h$ and d_i around their reference values

Table 2 Important parameters for every response

Integral response	Important parameters
Q	T_{h_in}, T_{c_in}, $\dot{m}_c$, $\dot{m}_h$, d_o
Δp_c	$\dot{m}_c$, d_i
Δp_h	$\dot{m}_h$, d_i, d_o
W_p	$\dot{m}_c$, $\dot{m}_h$, d_i, d_o

With GPT we can also identify which parameters are important for design. If a $\pm 10\%$ perturbation on a parameter produces an effect smaller than $\pm 1\%$ on the response, it is considered negligible. Table 2 summarizes these results.

The inlet pressure is not relevant, since their effect on physical properties is very small (the HTC only changes 0.003% and c_p a 0.007% for $\pm 20\%$ perturbation on p_{c_0} or p_{h_0}). For the same reason, the inlet temperatures are only relevant to the heat duty, mainly given by the temperature difference.

5 Sensitivity Analysis Accuracy

An open question arises concerning the limitations of the method and the need for higher order GPT. The mathematical formulation of GPT [2, 9] does not introduce any approximation, and therefore the sensitivity coefficients obtained are exact. But this is a first–order approximation for estimating variations in the response; therefore the accuracy of this approximation is related to dependence of the response on the parameters. An accuracy analysis for $\pm 10\%$ perturbations on all the parameters for the four integral responses is presented in Fig. 4.

As expected, in all cases a better accuracy is obtained for smaller perturbations. This accuracy is acceptable for perturbations as large as $\pm 10\%$, except in the case of the influence of diameters on pressure drops and pumping power. The reason for this is that the dependence of these responses on these parameters is far away from being linear, as can be seen in Fig. 5.

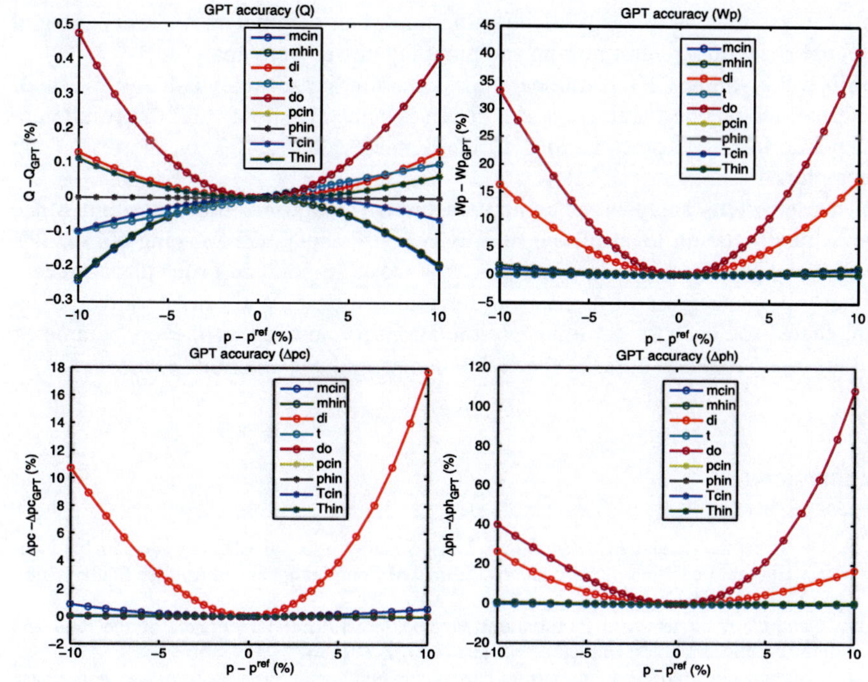

Fig. 4 GPT accuracy for Q, W_p, Δp_c and Δp_h for $\pm 10\%$ on all the parameters

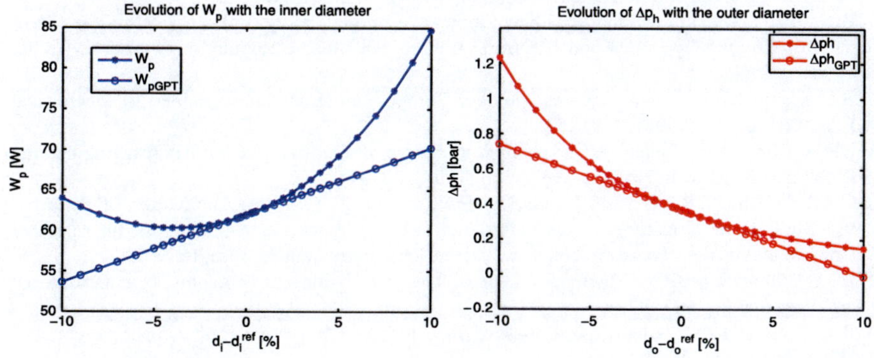

Fig. 5 The effect of diameters on Δp_h and W_p is not linear

6 Conclusions

In this work first–order GPT was used for a sensitivity analysis on single–phase one–dimensional heat exchangers. With only one extra simulation, GPT gives the sensitivity of an integral response to all parameters around a reference point.

This analysis allows to easily identify the important parameters for every integral response (heat duty, pressure drop and pumping power were analysed).

Since first–order GPT is a linear approximation, its accuracy is better for small perturbations on the parameters. An accuracy analysis shows that the results are acceptable for $\pm 5\%$ perturbations on the diameters and $\pm 10\%$ on the rest of the parameters.

This sensitivity analysis can be implemented in an optimization algorithm. Since one extra simulation gives all the sensitivity coefficients on a working point, GPT gives the 'best direction' for finding a new working point in order to optimize a properly defined figure of merit as an integral response. With this iterative procedure, there is no need to perform new simulations for new values of every parameter, and the computational cost is expected to be reduced considerably.

References

1. B. De Maerschalck and M.I. Gerritsma, Least-Squares Spectral Element Method for Non-Linear Hyperbolic Differential Equations, Journal of Computational and Applied Mathematics, 215 (2008), 357–367
2. A. Gandini, A Generalized Perturbation Method for Bilinear Functionals of the Real and Adjoint Neutrons Fluxes, Journal of Nuclear Energy, 21 (1967), 755–765
3. A. Gandini, Generalized Perturbation Theory for Nonlinear Systems from the Importance Conservation Principle, Nuclear Science and Engineering, 77 (1981), 316
4. A. Gandini, Generalized Perturbation Theory (GPT) Methods. A Heuristic Approach, Advances in Nuclear Science and Technology, 19 (1987), 205–380
5. B.N. Jiang, The Least-Squares Finite Element Method, Theory and Applications in Computational Fluid Dynamics and Electromagnetics, Scientific Computation, Springer, Berlin, 1998
6. B.N. Jiang, On the Least Squares Method, Computational Methods Applied to Mechanical Engineering, 152 (1998), 239–257
7. F.R.A. Lima, et al., Recent Advances in Perturbative Methods Applied to Nuclear Engineering Problems, Progress in Nuclear Energy, 33 (1998), 23–97
8. C.A.B.O. Lira, F.R. Andrade Lima, S.V. Freitoza and A. Gandini, Application of Perturbation Methods for Sensitivity Analysis for Nuclear Power Plant Steam Generators, International Conference on New Trends in Nuclear Systems Thermohydraulics, Pisa, Italy, 1994
9. E.M. Oblow, Sensitivity Theory for Reactor Thermal-Hydraulics Problems, Nuclear Science and Engineering, 68 (1978), 322–337
10. R.K. Shah and M.S. Bhatti, Handbook of Single-Phase Convective Heat Transfer, chapter 3, Wiley, New York, 1987
11. R.K. Shah and D.P. Sekulic, Fundamentals of Heat Exchanger Design, Wiley, New Jersey, 2003
12. F. White, Fluid Mechanics, McGraw-Hill Series in Mechanical Engineering, New York, 1986

Spectral Element Approximation of the Hodge-$\star$ Operator in Curved Elements

Artur Palha and Marc Gerritsma

Abstract Mimetic approaches to the solution of partial differential equations (PDE's) produce numerical schemes which are compatible with the structural properties – conservation of certain quantities and symmetries, for example – of the systems being modelled. Least Squares (LS) schemes offer many desirable properties, most notably the fact that they lead to symmetric positive definite algebraic systems, which represent an advantage in terms of computational efficiency of the scheme. Nevertheless, LS methods are known to lack proper conservation properties which means that a mimetic formulation of LS, which guarantees the conservation properties, is of great importance. In the present work, the LS approach appears in order to minimize the error between the dual variables, implementing weakly the material laws, obtaining an optimal approximation for both variables. The application to a 2D Poisson problem and a comparison will be made with a standard LS finite element scheme, see, for example, Cai et al. (J. Numer. Anal. 34:425–454, 1997).

1 Introduction

Numerical schemes are an essential tool for solving partial differential equations (PDE's). These schemes, being a model reduction, inherently lead to loss of information of the system being modeled, namely on its structure, e.g., conservation of certain quantities – mass, momentum, energy, etc. – and symmetries, which are embedded into the PDE's as a result of the geometrical properties of the differential operators. It is known today, see [3, 13, 15], that the well-posedness of many PDE problems reflects geometrical, algebraic topological and homological structures underlying the problem. It is, therefore, important for the numerical scheme to

A. Palha (✉) and M. Gerritsma
Department of Aerodynamics, Faculty of Aerospace Engineering, TUDelft, Kluyverweg 2, 2629 HT Delft, The Netherlands
e-mail: apalhadasilvaclerigo@tudelft.nl

J.S. Hesthaven and E.M. Rønquist (eds.), *Spectral and High Order Methods for Partial Differential Equations*, Lecture Notes in Computational Science and Engineering 76,
DOI 10.1007/978-3-642-15337-2_26,

be compatible with these structures (the physics), i.e., to mimic them. The goal of mimetic methods is to satisfy exactly, or as good as possible, the structural properties of the continuous model. It is becoming clear that in doing so, one obtains stable schemes. Additionally, a clear separation between the processes of discretization and approximation arises, the latter only take place in the constitutive relations.

Least Squares (LS) schemes offer many desirable properties, most notably the fact that they lead to symmetric positive definite algebraic systems, which represent an advantage in terms of computational efficiency of the scheme. Nevertheless, LS methods fail to satisfy the conservation laws [14]. A mimetic LS formulation will satisfy the conservation law exactly. In the current paper, the LS approach is used to minimize the error between the dual variables, obtaining an optimal approximation. This LS approximation is known in the literature as the implementation of Weak Material Laws, as proposed by Bochev and Hyman in [1].

2 Mimetic Approaches for the 2D Poisson Equation

The introduction of mimetic approaches to the solution of PDE's relies on a prior knowledge of differential geometry, mainly of the concepts of k-differential forms, of the wedge product, $\wedge$, of the exterior derivative, d, of the inner product, $(\cdot,\cdot)$, and of the Hodge-$\star$ operator, $\star$. It is out of the scope of this work to give an introduction to this theory, for that, the reader is referred to the works by Flanders [7], Burke [5] and Bossavit [3]. For a very short introduction with the same notation followed in this work, the reader is referred to the prior works of the authors [8, 11]. One can rewrite the Poisson equation using the framework of differential geometry, obtaining a system of first order PDE's:[1]

$$\begin{cases} -\nabla\phi = \mathbf{u} \\ \nabla\cdot\mathbf{u} = f \end{cases} \Leftrightarrow \begin{cases} -\mathrm{d}\phi^0 = u^1 \\ \mathrm{d}\tilde{v}^1 = \tilde{f}^2 \\ \tilde{v}^1 = \star u^1 \end{cases}, \tag{1}$$

where one clearly sees that the u^1 that appears in the first equation is not equal to the $\tilde{v}^1$ (a twisted 1-form) that appears in the second equation, but rather it is related to it through a material constitutive relation. This is not explicit when using standard vector calculus and, as it will be seen later (Sects. 3 and 4), it plays an important role in the accurate numerical solution of the Poisson equation.

In order to numerically solve this problem, different discretization approaches may be implemented. The approach followed in this work satisfies exactly the equilibrium equations between unknown physical quantities but relaxes the constitutive equation, being enforced weakly. Hence, discretization of the problem in appropriate function spaces leads to the following Tonti diagram to be solved (dotted lines

[1] Here f^2 denotes a 2-form. Not to be confused with the square of f.

represent weakly imposed relations):

$$
\begin{array}{ccc}
\phi_h^0 & & \tilde{f}_h^2 \\
\Big\downarrow{\scriptstyle d} & & \Big\uparrow{\scriptstyle d} \\
u_h^1 & \underset{\star}{\dashrightarrow} & \tilde{v}_h^1
\end{array}
\tag{2}
$$

3 Weak Material Laws: The Role of Least-Squares

As proposed by Bochev and Hyman [1] and Bochev and Gunzburger [2], a way of defining the Hodge-$\star$ operator and hence the constitutive equation in a weak sense is by using a least-squares minimization process that penalizes the discrepancy between the dual physical quantities. The exact equilibrium equation appears as a linear constraint that must be satisfied by the minimizers of the functional. Hence the problem is reduced to a constrained minimization problem:

$$
\begin{gathered}
\text{Seek } (\phi_h^0, u_h^1, \tilde{v}_h^1) \text{ in } \Lambda_h^0 \times \Lambda_h^1 \times \tilde{\Lambda}_h^1 \text{ such that} \\
\mathscr{I}(\phi_h^0, u_h^1, \tilde{v}_h^1) = \frac{1}{2}\left(\| \star \tilde{v}_h^1 + u_h^1 \|_0^2 + \| \mathrm{d}\tilde{v}_h^1 - \tilde{f}^2 \|_0^2 \right)
\end{gathered}
\tag{3}
$$

$$
\text{subject to: } - \mathrm{d}\phi_h^0 = u_h^1 \tag{4}
$$

The choice for an L^2 inner product lies on the fact that, using such an inner product, it is possible to demonstrate an optimal error estimate, as can be seen in Bochev et al. [2, Sect. 3.2]. If the subspaces Λ_h^0, Λ_h^1 and Λ_h^2, and the twisted ones, are chosen in such a way that they constitute a de Rham complex:

$$
\begin{array}{ccccccccc}
\mathbb{R} & \longrightarrow & \Lambda_h^0 & \xrightarrow{\mathrm{d}} & \Lambda_h^1 & \xrightarrow{\mathrm{d}} & \Lambda_h^2 & & \\
 & & \Big\downarrow{\scriptstyle \star} & & \Big\downarrow{\scriptstyle \star} & & \Big\downarrow{\scriptstyle \star} & & \\
 & & \tilde{\Lambda}_h^2 & \underset{\mathrm{d}}{\longleftarrow} & \tilde{\Lambda}_h^1 & \underset{\mathrm{d}}{\longleftarrow} & \tilde{\Lambda}_h^0 & \longleftarrow & \mathbb{R}
\end{array}
\tag{5}
$$

then (4) is satisfied exactly and one can substitute u_h^1 by $-\mathrm{d}\phi_h^0$ without any approximation involved. In this way the constrained minimization problem is reduced to a simple minimization problem only on two variables, ϕ_h^0 and $\tilde{v}_h^1$:

$$
\begin{gathered}
\text{Seek } (\phi_h^0, \tilde{v}_h^1) \text{ in } \Lambda_h^0 \times \tilde{\Lambda}_h^1 \text{ such that} \\
\mathscr{I}(\phi_h^0, \tilde{v}_h^1) = \frac{1}{2}\left(\| \star \tilde{v}_h^1 - \mathrm{d}\phi^0 \|_0^2 + \| \mathrm{d}\tilde{v}_h^1 - \tilde{f}^2 \|_0^2 \right)
\end{gathered}
\tag{6}
$$

In this way, the Hodge-$\star$ operator is implemented as L^2 projections between the different dual spaces.

4 Application to the 2D Poisson Equation

4.1 Straight Elements

To apply the above mentioned scheme to the 2D Poisson equation first the appropriate subspaces Λ^0_h, Λ^1_h and Λ^2_h, and the associated twisted form spaces, must be specified. Since one will use a spectral/*hp* LS method, these spaces are defined as:

$$\Lambda^0_{h,p} = \text{span}\left\{h_i^p(\xi)h_j^p(\eta)\right\},\ i = 0,\dots,p \quad j = 0,\dots,p$$

$$\Lambda^1_{h,p} = \text{span}\left\{e_i^p(\xi)h_j^p(\eta) \otimes h_n^p(\xi)e_m^p(\eta)\right\},\ i,m = 1,\dots,p \quad j,n = 0,\dots,p$$

$$\Lambda^2_{h,p} = \text{span}\left\{e_i^p(\xi)e_j^p(\eta)\right\},\ i = 1,\dots,p \quad j = 1,\dots,p$$

where $h_i^p(\xi)$ is the i-th Lagrange interpolant of order p over Gauss–Lobatto–Legendre points and $e_i^p(\xi)$ is the i-th edge interpolant of order p, introduced in [9]. We see that with this choice the degrees of freedom associated with these bases of the discrete subspaces are located where they should be: at nodal points (for 0-forms), at edges (for 1-forms) and at volumes (for 2-forms). In this way, the resulting reconstructed physical quantities will have different continuity properties: continuity across elements (0-forms), tangential (or normal) continuity along edges (1-forms) and no continuity across elements (2-forms). It is possible to show that these subspaces constitute a de Rham complex, as in (5) and hence they are suitable to be used to represent the unknown degrees of freedom. Kopriva [12], employed a similar use of staggered spectral element grids.

In order to assess the above described method, it will be applied to the solution of the 2D Poisson equation with a particular right hand side and Dirichlet boundary conditions in order to obtain the following analytical solution: $\phi(x,y) = \cos(x^2) + y^2$, $(x,y) \in [-1,1] \times [-1,1]$.

In Fig. 2 (left), one sees the convergence of the error in ϕ and $\mathbf{u}$ with the mesh size. In the mimetic approach one obtains again the optimal convergence rate of $p+1$, in $\mathbf{u}$, that was lost in the standard LS case. This result confirms what was stated in Bochev et al. [2, Sect. 3.2]. In Fig. 2 (right), one sees that the convergence of the error of $\nabla \times \mathbf{u}$ is higher for the mimetic case.

4.2 Curved Elements

The extension to curved elements requires some care, since the usual operations like the inner product of k-forms and the Hodge-$\star$ behave in a more sophisticated way, since, as it will be shown, the canonical space, $\widehat{\Omega}$, of local coordinates (ξ,η) is not Euclidean as the physical space of each element, Ω, of coordinates (x,y). This

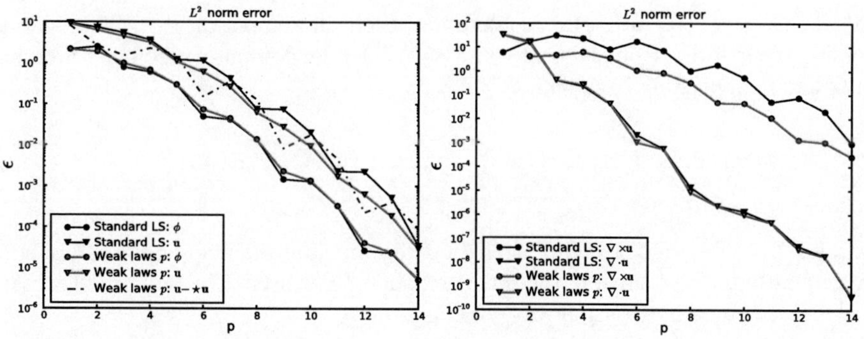

Fig. 1 L^2 convergence for mimetic and standard LS (*left*) and $\nabla \times \mathbf{u}$ for mimetic LS and standard LS (*right*), as a function of the order p of the approximation space

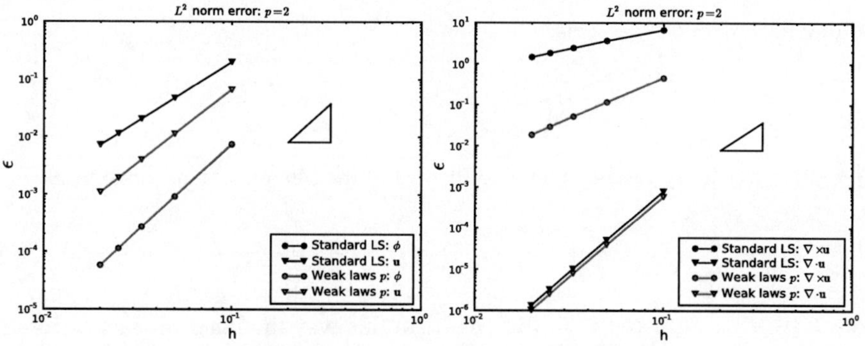

Fig. 2 L^2 convergence for mimetic and standard LS (*left*) and $\nabla \times \mathbf{u}$ for mimetic LS and standard LS (*right*), as a function of mesh size h

is induced by the transformation of coordinates implied to map from the physical space to the canonical space, where all basis k-forms are defined.

4.2.1 The Inner Product

A mapping $\Phi : \widehat{\Omega} \longrightarrow \Omega$ induces a map $\Phi^* : T^*(\Omega) \longrightarrow T^*(\widehat{\Omega})$, called the *pullback operator*, and its inverse $((\Phi^*)^{-1})$. The pullback operator maps differential forms over Ω to differential forms over $\widehat{\Omega}$, such that for all $\alpha^k \in \Lambda^k(\Omega)$ we have

$$\Phi^* : \alpha^k(x, y) \longrightarrow \left[\Phi^* \alpha^k\right](\xi, \eta) = \widehat{\alpha^k}(\xi, \eta). \tag{7}$$

As stated above, $\widehat{\Omega}$ is not like $\mathbb{R}^2$, in the sense that it does not have a metric identical to the identity, which means that one can no longer compute the inner products as in $\mathbb{R}^2$, that is Ω: $\langle \mathrm{d}x^i, \mathrm{d}x^j \rangle = g^{ij} = \delta^{ij}$. For $\widehat{\Omega}$ one will have in general:

$\langle \mathrm{d}\xi^i, \mathrm{d}\xi^j \rangle = \widehat{g}^{ij} \neq \delta^{ij}$. The problem lies, then, in computing $\widehat{g}^{ij}$. This is done passing the k-forms from $\widehat{\Omega}$ to Ω with $((\Phi^*)^{-1})$, then computing the inner product in Ω and then pushing back with Φ^*. Hence

$$\mathrm{d}\xi = \xi_x \mathrm{d}x + \xi_y \mathrm{d}y = \left((\Phi^*)^{-1} y_\eta \mathrm{d}x - (\Phi^*)^{-1} x_\eta \mathrm{d}y \right) \frac{1}{(\Phi^*)^{-1} J} \tag{8}$$

where $\xi^i_{x^j} = \frac{\partial \xi^i}{\partial x^j}$ and J is the Jacobian determinant of the mapping Φ. To obtain this equality the derivative of the inverse function was used. The inner product can now be computed in the usual way, yielding:

$$\widehat{(\mathrm{d}\xi, \mathrm{d}\xi)} = \frac{1}{(\Phi^*)^{-1} J^2} \left((\Phi^*)^{-1} y_\eta^2 + (\Phi^*)^{-1} x_\eta^2 \right). \tag{9}$$

Pulling back with Φ^* one gets:

$$(\mathrm{d}\xi, \mathrm{d}\xi) = \frac{1}{J^2} (y_\eta^2 + x_\eta^2). \tag{10}$$

The same procedure can be applied to the rest of the inner products, obtaining:

$$(\mathrm{d}\eta, \mathrm{d}\eta) = \frac{1}{J^2} \left(y_\xi^2 + x_\xi^2 \right), \quad (\mathrm{d}\eta, \mathrm{d}\xi) = (\mathrm{d}\xi, \mathrm{d}\eta) = -\frac{1}{J^2} \left(x_\xi x_\eta + y_\xi y_\eta \right) \tag{11}$$

which gives the metric $\widehat{g}^{ij} = (\mathrm{d}\eta^i, \mathrm{d}\eta^j)$. In this way the inner product of forms $\alpha = \sum a_i \mathrm{d}\eta i$ and $\beta = \sum b_i \mathrm{d}\eta i$ is given by:

$$(\alpha, \beta) = \sum a_i b_j \widehat{g}^{ij} \tag{12}$$

4.2.2 The Hodge-$\star$ Operator

The Hodge-$\star$ operator can be defined in the following way, see [6],

$$\widehat{\star}\mathrm{d}\xi^i = \widehat{g}^{ik} \widehat{\epsilon}_{kj} \mathrm{d}\xi^j \tag{13}$$

where $\widehat{g}^{ij}$ is the metric tensor and $\widehat{\epsilon}_{ij}$ is the usual Levi–Civita tensor, see [6]. In this way one has:

$$\widehat{\star}\mathrm{d}\xi = \frac{1}{J} \left[\left(y_\eta y_\xi + x_\eta y_\xi \right) \mathrm{d}\xi + \left(x_\eta^2 + y_\eta^2 \right) \mathrm{d}\eta \right] \tag{14}$$

$$\widehat{\star}\mathrm{d}\eta = \frac{1}{J} \left[- \left(x_\xi^2 + y_\xi^2 \right) \mathrm{d}\xi - \left(y_\eta y_\xi + x_\eta y_\xi \right) \mathrm{d}\eta \right] \tag{15}$$

which is the same as: $\Phi^* \star (\Phi^*)^{-1}\,\mathrm{d}\xi = \widehat{\star}\mathrm{d}\xi$, as in [10]. Hence, the commutation relation of Φ^* with $\star$ is:

$$\Phi^* \star \alpha = \Phi^* \star \left(\Phi^*\right)^{-1} \Phi^* \alpha = \widehat{\star}\Phi^* \alpha = \widehat{\star}\widehat{\alpha} \tag{16}$$

4.2.3 The Least-Squares Residual

Having defined the Hodge-⋆ and the inner product in curved domains one can easily define the LS residual in curved domains. In straight domains one has:

$$\mathscr{I} = \frac{1}{2}\int_{\Omega} (\star v - \mathrm{d}\phi, \star v - \mathrm{d}\phi)\,\mathrm{d}x\mathrm{d}y + \frac{1}{2}\int_{\Omega} (\mathrm{d}v - f, \mathrm{d}v - f)\,\mathrm{d}x\mathrm{d}y \tag{17}$$

and in curved domains one has:

$$\mathscr{I} = \frac{1}{2}\int_{\widehat{\Omega}} \left(\widehat{\star v} - \widehat{\mathrm{d}\phi}, \widehat{\star v} - \widehat{\mathrm{d}\phi}\right) \widehat{\mathrm{d}x\mathrm{d}y} + \frac{1}{2}\int_{\widehat{\Omega}} \left(\widehat{\mathrm{d}v} - \widehat{f}, \widehat{\mathrm{d}v} - \widehat{f}\right) \widehat{\mathrm{d}x\mathrm{d}y} \tag{18}$$

where we have used the property: $\int_{\Phi(\widehat{\Omega})=\Omega} \alpha \equiv \int_{\widehat{\Omega}} \widehat{\alpha}$.

But it was already seen that: $\widehat{\star v} = \widehat{\star}\widehat{v}$, $\widehat{\mathrm{d}\phi} = \mathrm{d}\widehat{\phi}$ and $\widehat{\mathrm{d}x\mathrm{d}y} = J\,\mathrm{d}\xi\mathrm{d}\eta$. Hence:

$$\mathscr{I} = \frac{1}{2}\int_{\widehat{\Omega}} \left(\widehat{\star}\widehat{v} - \mathrm{d}\widehat{\phi}, \widehat{\star}\widehat{v} - \mathrm{d}\widehat{\phi}\right) J\,\mathrm{d}\xi\mathrm{d}\eta + \int_{\widehat{\Omega}} \left(\mathrm{d}\widehat{v} - \widehat{f}, \mathrm{d}\widehat{v} - \widehat{f}\right) J\,\mathrm{d}\xi\mathrm{d}\eta \tag{19}$$

Now the discretization can be done in the local element and then transformed to the physical domain with the inverse of the pullback, as was back in the computation of the metric to pass k-forms in $\widehat{\Omega}$ to Ω. Figure 3 (left) shows the curved element mesh used to assess this scheme. Since the outer boundaries are the ones of a square, it is good to compare this case to the case of a straight element mesh. Figure 3 (right) shows the plot of the convergence of the error of ϕ for both the curved and straight elements. One can see an exponential convergence even for curved elements, although with a smaller rate when compared to the straight elements, as expected. This result, although requiring further investigation, seems promising, especially when compared to the results in Arnold et al. [4], where RT_0 elements fail to converge in non affine elements when implemented in a LS method.

5 Concluding Remarks

As can be seen in Fig. 1, the mimetic LS method results in smaller errors, especially for the **u** variable. Additionally, with regard to $\nabla\times\mathbf{u}$ one sees a great improvement in the mimetic LS method, where the errors are approximately 2 orders of magnitude smaller compared to conventional LS. The primal–dual differences, $\| \star \tilde{v}_h^1 + u_h^1 \|_0$,

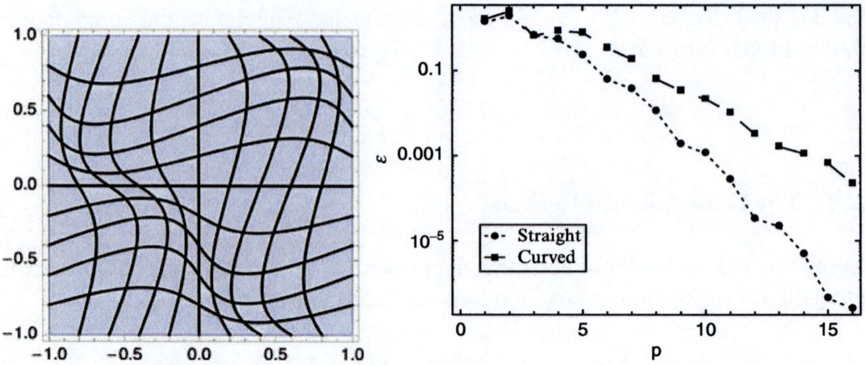

Fig. 3 On the *left*, the curved computational mesh. On the *right*, the plot of the L^2 error of ϕ as a function of the order p used for the approximation space, for the curved elements mesh (*dashed line*) and for the straight elements mesh (*dotted line*)

appears to be a good estimator for the error of the numerical solution, as can be seen in Fig. 1.

Acknowledgements The authors would like to acknowledge Foundation for Science and Technology (Portugal) for the funding given by the PhD grant SFRH/BD/36093/2007.

References

1. Bochev, P. and Hyman, J.: Principles of mimetic discretizations of differential operators. IMA **142**, 89–119 (2006)
2. Bochev, P. and Gunzburger, M.: On least-squares finite element methods for the Poisson equation and their connection to the Dirichlet and Kelvin principles. SIAM J. Num. Anal. **43**, 340–362 (2006)
3. Bossavit, A.: On the geometry of electromagnetism. J. Jpn. Soc. Appl. Electromagn. Mech. **6**, 318–326 (1998)
4. Arnold, D. N., Boffi, D., and Falk, R. S.: Quadrilateral **H**(div) finite elements, SIAM, J. Num. Anal. **42**, 2429–2451 (2005)
5. Burke, W. L.: Applied differential geometry, Cambridge University Press, Cambridge (1985)
6. Hou, B.: Differential geometry for physicists. World Scientific, Singapore, 1997
7. Flanders, H.: Differential forms with applications to the physical sciences. Academic Press, New York, 1963
8. Gerritsma, M., Bouman, M. and Palha, A.: Least-Squares spectral element method on a staggered grid, to appear in Large Scale Scientific Computing, LNCS **5910**, 659–666 (2010)
9. Gerritsma, M.: Edge functions for spectral element methods. Submitted to the proceedings of ICOSAHOM 2009 (this issue), 2010
10. Bouman, M., Palha, A., Kreeft, J., and Gerritsma, M.: A conservative spectral element method for arbitrary domains, Submitted to the proceedings of ICOSAHOM 2009 (this issue), 2010
11. Palha, A. and Gerritsma, M.: Mimetic least-squares spectral/*hp* finite element method for the Poisson equation, to appear in Large Scale Scientific Computing, LNCS **5910**, 667–675 (2010)
12. Kopriva, D. A. and Kolias, J. H.: A conservative staggered-grid Chebyshev multidomain method for compressible flows. J. Comput. Phys. **125**, 244–261 (1996)

13. Mattiussi, C.: An analysis of finite volume, finite element, and finite difference methods using some concepts from algebraic topology. J. Comp. Phys. **133**, 289–309 (1997)
14. Proot, M. M. J. and Gerritsma, M. I.: Mass and momentum conservation of the least-squares spectral element method for the Stokes problem. J. Sci. Comput. **27** (1–3), 389–401 (2007)
15. Tonti, E.: On the formal structure of physical theories. Consiglio Nazionale delle Ricerche, Milano (1975)

Uncertainty Propagation for Systems of Conservation Laws, High Order Stochastic Spectral Methods

G. Poëtte, B. Després, and D. Lucor

Abstract The application of the stochastic Galerkin-generalized Polynomial Chaos approach (sG-gPC) (Wiener, Am. J. Math. 60:897–936, 1938; Cameron and Martin, Ann. Math. 48:385–392, 1947; Xiu and Karniadakis, SIAM J. Sci. Comp. 24(2):619–644, 2002) for Uncertainty Propagation through NonLinear Systems of Conservation Laws (SLC) is known to encounter several difficulties: dimensionality (see, e.g., Nobile et al., SIAM J. Numer. Anal. 46(5):2309–2345, 2008; Blatman and Sudret, C. R. Méc. 336:518–523, 2008; Witteveen and Bijl, Comp. Struct. 86(23–24):2123–2140, 2008), non linearities (see, e.g., Debusshere et al., J. Sci. Comp. 26:698–719, 2004; Witteveen and Bijl, 47th AIAA Aerospace Sciences Meeting and Exhibit, 2006–2066, 2006), discontinuities (see Wan and Karniadakis, SIAM J. Sci. Comp. 27(1–3), 2006; Lin et al., J. Comp. Phys. 217:260–276, 2006; Le Maître and Knio, J. Comp. Phys. 197:28–57, 2004; Le Maitre et al., J. Comp. Phys. 197:502–531, 2004; Abgrall, Rapport de Recherche INRIA, 2007). In this paper, we first illustrate on a simple SLC (p-system) the difficulties occuring when dealing with non linearities and discontinuities. We will then present a new non adaptive *high order* uncertainty propagation method based on the entropy of the system of conservation laws, efficient on NonLinear systems and discontinuous solutions. Convergence tests are performed and *spectral convergence* is reached.

G. Poëtte (✉)
CEA/DAM/DIF, 91297 Arpajon cedex, France
and
IJLRA, 4 place Jussieu, 75292 Paris cedex 5, France
e-mail: gael.poette@gmail.com

B. Després
CEA/DAM/DIF, 91297 Arpajon cedex, France
e-mail: bruno.despres@cea.fr

D. Lucor
IJLRA, 4 place Jussieu, 75292 Paris cedex 5, France
e-mail: didier.lucor@lmm.jussieu.fr

J.S. Hesthaven and E.M. Rønquist (eds.), *Spectral and High Order Methods for Partial Differential Equations*, Lecture Notes in Computational Science and Engineering 76,
DOI 10.1007/978-3-642-15337-2_27, © Springer-Verlag Berlin Heidelberg 2011

1 Mathematical Framework

We are interested in the resolution of stochastic SLC thanks to Polynomial Chaos theory. We first briefly recall the main principles of theories of *1-D* SLC and of generalized Polynomial Chaos (gPC).

1.1 SLC in a Nutshell

We are interested in Hyperbolic SLC. The general form for 1-D SLC is

$$\partial_t u + \partial_x f(u) = 0, \text{ with } u : \begin{array}{c} \mathscr{D}\times]0,T[\longrightarrow \mathscr{U} \subset \mathbb{R}^n \\ (x,t) \longrightarrow u(x,t) \end{array}. \quad (1)$$

The field f is the *flux*. Hyperbolicity ensures the existence and the stability of the solution of the SLC [14, 15]. To prove hyperbolicity, two theorems are used in practice [14]:

Theorem 1 (Hyperbolicity of a 1-D SLC). *The SLC* (1) *is hyperbolic iff the Jacobian of the flux, $A = \nabla_u f(u)$, is diagonalizable in $\mathbb{R}^n$ in a complete basis of eigenvectors $\forall u \in \mathscr{U}$.*

Definition 1 (Mathematical entropy of a 1-D SLC). A real function $u \in \mathscr{U} \subset \mathbb{R}^n \longrightarrow s(u) \in \mathbb{R}$ is a *mathematical entropy* for (1) if it exists a function $u \in \mathscr{U} \subset \mathbb{R}^n \longrightarrow g(u) \in \mathbb{R}$, the *entropy flux*, such that

$$\begin{array}{l} \partial_t s(u) + \partial_x g(u) = 0, \text{ for smooth solutions of (1)}, \\ \partial_t s(u) + \partial_x g(u) \leq 0, \text{ for discontinuous solutions of (1)}. \end{array} \quad (2)$$

Theorem 2 (Hyperbolicity of a 1-D SLC (entropy formulation)). *If a SLC* (1) *owns a strictly convex mathematical entropy then the SLC is hyperbolic.*

Note that Theorem 2 implies the conditions of Theorem 1.

1.2 gPC in a Nutshell

The following convergence theorem is at the basis of gPC theory. It is a generalization by [17, 20] of Cameron Martin's theorem [3]:

Theorem 3 (Cameron–Martin [3]). *Let $(\Omega, \mathscr{F}, \mathscr{P})$ be a probability space. Let $\Xi = (\Xi_1, \ldots, \Xi_d)^t$ be a random vector (rv) of independent components of respective probability measures $(\mathrm{d}\mathscr{P}_{\Xi_i})_{i\in\{1,\ldots,d\}}$. We denote by $\mathrm{d}\mathscr{P}_\Xi$ the tensorized* *measure. Let* $(\phi_k^i)_{k\in\mathbb{N},i\in\{1,\dots,d\}}$ *be the gPC basis.*[1] *We denote by* $(\phi_k)_{k\in\mathbb{N}}$ *the tensorization of* $(\phi_k^i)_{k\in\mathbb{N},i\in\{1,\dots,d\}}$*. Let* $u(\Xi(\omega))$ *be a unknown rv. Then*

$$\int_{\omega\in\Omega} u^2(\Xi(\omega))\mathrm{d}\mathscr{P}_\Xi(\omega) < \infty \Longrightarrow \Pi^P u(\Xi(\omega))$$
$$= \sum_{k=0}^{P} u_k\phi_k(\Xi(\omega)) \underset{P\to\infty}{\overset{L^2(\Omega,\mathscr{F},\mathscr{P})}{\longrightarrow}} u(\Xi(\omega)) \tag{3}$$

where $u_k = \displaystyle\int_{\omega\in\Omega} u(\Xi(\omega))\phi_k(\Xi(\omega))\mathrm{d}\mathscr{P}_\Xi(\omega)$.

The application of the gPC approach will consist in developing our vector of unknown u on the polynomial basis truncated to order P potentially high (depending on the regularity of the solution). In the following, we apply sG-gPC to a simple SLC.

2 Application of sG-gPC to the p-System in Lagrangian Coordinates

We want to apply the sG-gPC method to the p-system in Lagrangian coordinates given by

$$\begin{cases} \partial_t\tau - \partial_x u = 0, \\ \partial_t u + \partial_x p = 0, \end{cases} \tag{4}$$

where τ denotes the specific volume, u is the velocity, here x is the mass coordinate ($\partial x = \frac{1}{\tau}\partial y$ where y denotes the position) and $p(\tau)$ is the closure.[2] Physically, this system describes the adiabatic evolution of a gas. The aim of this section is to emphasize certain important difficulties.

Theorem 4 (Hyperbolicity). *The SLC* (4) *is hyperbolic if the eos satisfies* $p'(\tau) < 0$.

Proof. It is simple verifying Theorem 1's hypothesis studying the Jacobian of the flux.

Suppose p is as in Theorem 4 and given initial and boundary conditions. Suppose, for example, the initial condition (IC) are uncertain,[3] modelled by a uniform law[4] on

[1] I.e. orthonormal polynomial basis with respect to $(\mathrm{d}\mathscr{P}_{\Xi_i})_{i\in\{1,\dots,d\}}$, see [17,20] for more details.

[2] Equation of state (eos).

[3] Note that the derived equations (5) and (6) also apply when considering uncertain boundary conditions or even uncertain model parameters: here, we decided to consider uncertain IC.

[4] This simplifies the study without loss of generalities.

$[-1, 1]$ and let's apply sG-gPC. The approach consists in considering the polynomial basis[5] orthonormal with respect to the uniform measure $d\mathscr{P}(\xi) = \frac{1}{2}\mathscr{I}_{[-1,1]}(\xi)d\xi$ and developing the vector of unknown $(\tau, u)^t$ on the P-truncated polynomial basis, see (5):

$$\begin{cases} \tau \approx \tau^P = \sum_{k=0}^{P} \tau_k \phi_k, \\ u \approx u^P = \sum_{k=0}^{P} u_k \phi_k, \\ \forall k \in \{0, \ldots, P\}, p_k = \int p \phi_k d\mathscr{P}_\Xi. \end{cases} \tag{5}$$

We inject the development (5) in (4) and perform a Galerkin projection to obtain the *high order* truncated system (6):

$$\begin{cases} \partial_t \begin{pmatrix} \tau_0 \\ \ldots \\ \tau_P \end{pmatrix} - \partial_x \begin{pmatrix} u_0 \\ \ldots \\ u_P \end{pmatrix} = 0, \\ \partial_t \begin{pmatrix} u_0 \\ \ldots \\ u_P \end{pmatrix} + \partial_x \begin{pmatrix} p_0 \\ \ldots \\ p_P \end{pmatrix} = 0. \end{cases} \tag{6}$$

The system (6) is a new SLC of high order which has to be closed and then studied.

Theorem 5 (Hyperbolicity of (6)). *The SLC (6) is hyperbolic iff the matrix A^P of general term $A^P_{i,j} = \frac{\partial p_i}{\partial \tau_j}$ is definite negative.*

Proof. The Jacobian matrix of the flux of (6) is given by (7)

$$J^P = \begin{pmatrix} 0_{P+1,P+1} & -I_{P+1,P+1} \\ A^P & 0_{P+1,P+1} \end{pmatrix}, \tag{7}$$

where $0_{P+1,P+1}$ and $I_{P+1,P+1}$ denote the null and identity matrices of size $(P+1) \times (P+1)$. We denote by A^P the matrix of general term $A^P_{i,j} = \frac{\partial p_i}{\partial \tau_j}$. The characteristic polynomial of J^P is given by (8)

$$Q_{J^P}(\lambda) = \left| -\lambda^2 I_{P+1,P+1} - A^P \right|, \tag{8}$$

where $|B|$ denotes the determinant of matrix B. From the expression of Q_{J^P}, we deduce the sufficient and necessary condition of hyperbolicity of Theorem 5.

[5] In this particular case, it corresponds to the orthonormalized Legendre polynomials $(\phi_k)_{k\in\mathbb{N}}$.

The first difficulty encountered when studying (6) is linked to the important size, $2\times(P+1)$, of the system:[6] we, here, were able to state the conditions of hyperbolicity of Theorem 5 mainly because of the simplicity of the SLC (4). For other systems,[7] the study is not conceivable due to the high-order truncation and complexity.

The second difficulty is linked to the definition of the closure of (6) satisfying the conditions of Theorem 5. This issue is dealt with in the next section.

2.1 Closure of (6) or Treatment of Non Linearities

The problem of closure of (6) is known in the literature as the "Treatment of non linearities" step, see for example [4, 19]. It consists in defining $\forall k \in \{0,\dots,P\}$, $p_k(\tau_0,\dots,\tau_P)$ according to a given $p(\tau)$. Let's consider two eos for (4):

$$p(\tau)=\frac{1}{\tau^2} \text{ and } p(\tau)=-\ln(\tau). \tag{9}$$

The first one is a perfect gas closure (with $\gamma=2$) and the second is used to model metals.[8]

Several methods have been suggested in the literature in order to close (6). We suggest to study three of them:[9]

1. First Method, suggested in [4]: let's consider eos $p(\tau)=\frac{1}{\tau^2}$. The method consists in a Galerkin projection of $\tau^2 p=1$. The quantities τ and p are developed on the gPC basis so that $(p_k)_{k\in\{0,\dots,P\}}$ satisfies:

$$\begin{pmatrix} & \dots & \\ \dots & \sum_{k=0}^{P}\sum_{l=0}^{P}\tau_k\tau_l c_{k,l,i,j} & \dots \\ & \dots & \end{pmatrix}\begin{pmatrix} p_0 \\ \dots \\ p_P \end{pmatrix}=\begin{pmatrix} 1 \\ \dots \\ 0 \end{pmatrix} \tag{10}$$

where $c_{i,j,k,l}=\int \phi_i\phi_j\phi_k\phi_l d\mathscr{P}_\Xi$, $\forall(i,j,k,l)\in\{0,\dots,P\}^4$. The main problem with such a method is that it can not be applied to eos $p(\tau)=-\ln(\tau)$. This leads to the second one.

[6] Note that this size $2\times(P+1)$ increases exponentially fast as the stochastic dimensions d and the truncation order in every direction grows, see [12, 13]. This problem is known as the Curse of dimensionality: it is an important problem but it will not be dealt with in this paper as the other presented difficulties (non linearities/discontinuities) also occur in 1-D stochastic problems.

[7] Euler, MHD system, for example.

[8] Note that both eos satisfy the condition of Theorem 4 giving birth to well posed systems iff $\tau>0$.

[9] Note that according to the respective authors, the three methods converges as $P\to\infty$.

2. Second Method, suggested in [4]: the method is based on a Taylor expansion of p around its mean, τ is then developed on the polynomial basis to obtain:

$$p(\tau) \approx p(\tau_0) + \sum_{k=0}^{P} \frac{\mathrm{d}^k p}{\mathrm{d}\tau^k}(\tau_0) \frac{\left(\sum_{k=1}^{P} \tau_k \phi_k\right)^k}{k!}. \tag{11}$$

The coefficients $(p_k)_{k\in\{0,\dots,P\}}$ are obtained by integration of (11) against $(\phi_k)_{k\in\{0,\dots,P\}}$.

3. Third Method, suggested in [19]: it consists in considering (12)

$$p \approx p^P = p\left(\sum_{k=0}^{P} \tau_k \phi_k\right), \tag{12}$$

so that the closure of (6) is given by (13)

$$\forall k \in \{0,\dots,P\}, p_k = \int p\left(\sum_{l=0}^{P} \tau_l \phi_l\right) \phi_k \mathrm{d}\mathscr{P}_{\Xi}. \tag{13}$$

In this case, the closure can be studied analytically $\forall P \in \mathbb{N}$, see [12]. We showed that A^P satisfies the conditions of Theorem 5 iff $\sum_{k=0}^{P} \tau_k \phi_k > 0$ (C_1). Under C_1, (6) is hyperbolic: Section 2.2 will show that C_1 is not always satisfied.

The question we tackle is whether the closure methods (1)–(3) enable to preserve the physical and mathematical properties of (6). The results of the study are given in Table 1. The calculations are not recalled but can be found in [12]. Table 1 shows several difficulties: (1) do not ensure the positiveness of p (physical problem), (2) do not ensure hyperbolicity (mathematical problem), (3) ensures hyperbolicity iff C_1 is respected: in the next section we show that C_1 can be violated.

Table 1 Conclusions of the study of the three methods for treatment of non linearities

$p(\tau) = \frac{1}{\tau^2}$	(1)	(2)	(3)
System $P = 1$	Hyperbolic under C_1	Weakly hyperbolic	Hyperbolic under C_1
System $P = 2$	Hyperbolic under C_1	Weakly hyperbolic	Hyperbolic under C_1
System (higher orders: $\forall P$)	?	?	Hyperbolic under C_1
Pressure $P = 1$	Negative	Negative	Positive
Pressure $P = 2$	Negative	Negative	Positive
Pressure (higher orders: $\forall P$)	?	?	Positive
$p(\tau) = -\ln(\tau)$	(1)	(2)	(3)
System $P = 1$	Not applicable	Weakly hyperbolic	Hyperbolic under C_1
System $P = 2$	Not applicable	Weakly hyperbolic	Hyperbolic under C_1
System (higher orders: $\forall P$)	Not applicable	?	Hyperbolic under C_1

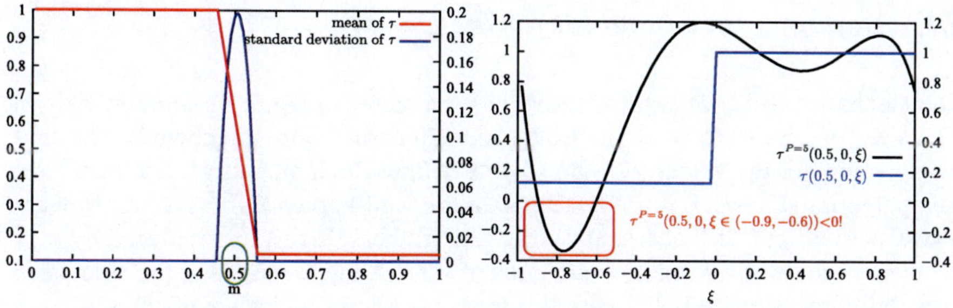

Fig. 1 Stochastic Riemann problem for the p-system, application of sG-gPC. *Left* of the interface $\tau = 1, u = 0$. *Right* of the interface $\tau = 0.125, u = 0$

2.2 Discontinuous Solutions and Gibbs Phenomenon

SLC are known for developing discontinuous solutions in finite times (see [9] for example): let's consider a discontinuous uncertain IC,[10] an uncertain Riemann problem: a light fluid (left) and a heavy fluid (right) at rest are separated by an interface whose initial position $x_{interface}$ is uncertain, modelled by an uniform law on $[-1, 1]$. Figure 1 (left) shows the mean and standard deviation (std) of τ at $t = 0$. Figure 1 (right) shows the IC in the random space at $m = 0.5$. The analytical solution presents a discontinuity with respect to (w.r.t.) the random parameter ξ. sG-gPC is such that for $P = 5$, $\exists I \subset [-1, 1]$ of strictly positive measure such that $\xi \in I \Longrightarrow \sum_{k=0}^{5} \tau_k \phi_k < 0$. Consequently, C_1 is not satisfied and the system is not hyperbolic (even with the method (3): numerically, the code crashes.

Several methods have been investigated in order to deal with Gibbs phenomenon [1, 2, 6–8, 18]. All these methods are adaptive and do not ensure the well-posedness in the case of a system of conservation laws. In the next sections, we present a non adaptive Polynomial Chaos method consistent with systems of conservation laws.

3 The Intrusive Polynomial Moment Method (IPMM)

The method has already been presented in [13]. We here briefly recall the main principles. The method is inspired on both Kinetic theory (Kt) [11] and Moment theory (Mt) [5, 10].

[10] Note that the same problem can occur dynamically for $t > 0$. Note that it can also happen when dealing with uncertain boundary conditions or uncertain model parameters. In this section, we chose to illustrate the difficulty through uncertain IC.

3.1 Analogy with Kt and Mt for the Closure

The stochastic SLC we want to solve has the same structure of some models in Kt. In Kt, one wants to solve the Boltzmann equation whose unknown is $f(x,t,v)$ where v denotes the velocity. To do so, one defines the moments of f with respect to a polynomial basis $1, v, v^2, \dots$ to obtain the well-known Euler system, Navier–Stokes system etc. In uncertainty propagation, the velocity v is replaced by a rv Ξ, the moments of the solution are defined w.r.t. the gPC basis and the associated probability measure. In both cases, this leads to systems which are not closed.

In Kt, the system is closed by application of Mt: it consists in solving the underdetermined inverse problem

$$\text{Find } f \in L^2(\Omega, \mathcal{F}, \mathcal{P}) \, / \quad \begin{cases} \displaystyle\int f l_0 \mathrm{d}\mathcal{P} = f_0, \\ \dots, \\ \displaystyle\int f l_k \mathrm{d}\mathcal{P} = f_k, \\ \dots, \\ \displaystyle\int f l_P \mathrm{d}\mathcal{P} = f_P, \end{cases} \tag{14}$$

where $(l_i)_{i\in\{0,\dots,P\}}$ are real functions defined on Ω, basis of $L^2(\Omega, \mathcal{F}, \mathcal{P})$ and where $f_0, \dots, f_P$ are the datas of the problem, called the moments of f, with f being the unknown.

We would like to do the same for Uncertainty Propagation: in this case, $(l_i)_{i\in\{0,\dots,P\}} = (\phi_i)_{i\in\{0,\dots,P\}}$ corresponding to the gPC basis. The distribution f^P solution of (14) is not unique: in Kt, one introduces the closure entropy (Shannon) $\eta(f) = \displaystyle\int f \ln(f) \mathrm{d}\mathcal{P}$. This results in a well posed moments problem: find f^P as the minimum of η under the constraints (14), solved, in practice, by standard methods of optimization under constraints.

Let's now go back to our Uncertainty Propagation/gPC formalism: let's consider an arbitrary closure entropy[11] η. We now want to solve the following system

$$\begin{cases} \partial_t u_0(\lambda_0, .., \lambda_P) + \partial_x f_0(\lambda_0, .., \lambda_P) = 0, \\ \dots \\ \partial_t u_k(\lambda_0, .., \lambda_P) + \partial_x f_k(\lambda_0, .., \lambda_P) = 0, \\ \dots \\ \partial_t u_P(\lambda_0, .., \lambda_P) + \partial_x f_P(\lambda_0, .., \lambda_P) = 0. \end{cases} \tag{15}$$

where $(\lambda_k)_{k\in\{0,\dots,P\}}$ minimizes (closure)

[11] A closure entropy is a strictly convex vector field ensuring the uniqueness of its minimum.

$$T(\lambda_0,..,\lambda_P) = -\int \eta(u^P(\lambda_0,..,\lambda_P))\mathrm{d}\mathscr{P} + \sum_{k=0}^{P}\int u^P(\lambda_0,..,\lambda_P)\lambda_k\phi_k\mathrm{d}\mathscr{P} \\ -\sum_{k=0}^{P} u_k\lambda_k. \tag{16}$$

We suggest to study more precisely the closure of system (15) given by (16). Performing Functional Variation w.r.t. u^P on (16) leads to:[12]

$$\nabla_u\eta(u^P(\lambda_0,..,\lambda_P)) = \sum_{k=0}^{P}\lambda_k\phi_k, \text{ i.e. } u^P(\lambda_0,..,\lambda_P) = (\nabla_u\eta)^{-1}\left(\sum_{k=0}^{P}\lambda_k\phi_k\right). \tag{17}$$

Consequently, in the gPC formalism, IPMM consists in developing the new variable $\nabla_u\eta(u^P(\lambda_0,\ldots,\lambda_P))$, called *associate variable*, on the gPC basis rather than the main variable u.

The closure entropy η represents the main degree of freedom of IPMM. Let's consider different entropies:

- If we choose $\eta(u) = \frac{u^2}{2}$, then $u^P = (\nabla_u\eta)^{-1}(\lambda) = u$: in this case, the associate variable is the main variable and IPMM degenerate into sG-gPC.
- Let's go back to the p-system problem: by choosing $\eta(\tau) = \tau\ln(\tau) - \tau$, then $(\nabla_\tau\eta)^{-1}(\lambda) = \tau^P(\lambda) = e^\lambda > 0$ and the positiveness of τ is ensured for $P \in \mathbb{N}$ (see [12]).
- Now consider the case $\eta = s$, i.e., the closure entropy is chosen as the mathematical entropy of the studied SLC. Then $\nabla_u s(u) = \lambda = v$, called the *entropic variable* [11].

Consider the last case, for which the entropic variable is developed on the gPC basis. This variable has the property of symmetrizing the non truncated SLC: this leads to Theorem 6.

Theorem 6 (Hyperbolicity of (15) closed by (16)). *Let's consider the P-truncated system (15) closed by (16) in the special case $\eta = s$. Then $(S,G) = \left(\int s\mathrm{d}\mathscr{P}, \int g\mathrm{d}\mathscr{P}\right)$ is a mathematical entropy for the SLC (15)–(16) and the system*[13] *is hyperbolic $\forall P \in \mathbb{N}$.*

Proof. See [12, 13].

[12] The inversion of $\nabla_u\eta$ is possible as by hypothesis, η is strictly convex.

[13] According to Theorem 2.

Thanks to an analogy with both Kt and Mt, we have shown that it is possible to preserve the mathematical properties of the P-truncated SLC. This ensures the existence and stability of the solution of the new SLC. In the next sections, we present other properties of IPMM through numerical examples.

4 Numerical Tests

We solve the stochastic Burgers and Euler equations with IPMM and sG-gPC. We show that the issues encountered in Sect. 2 are solved by construction of the truncated system.

4.1 Comparison Between sG-gPC and IPMM: Burgers

Burgers' equation is given by

$$\partial_t u + \partial_x \frac{u^2}{2} = 0, \tag{18}$$

for which every strictly convex function is a mathematical entropy. This equation is convenient to compare sG-gPC and IPMM as, being a scalar equation [14], every strictly convex function is a mathematical entropy. Consequently, even the truncated system obtained by sG-gPC is hyperbolic. Besides, analytical solutions can be computed.

We consider an IC with two states separated by a slope, blue curve of Fig. 2 (left). We suppose the position of the slope is uncertain modelled by an uniform rv Ξ on $[-1, 1]$. Figure 2 shows the time evolution of one realization of Ξ. At time $t = 0.09$, for every realizations of Ξ, a shock has formed in the physical space and propagates

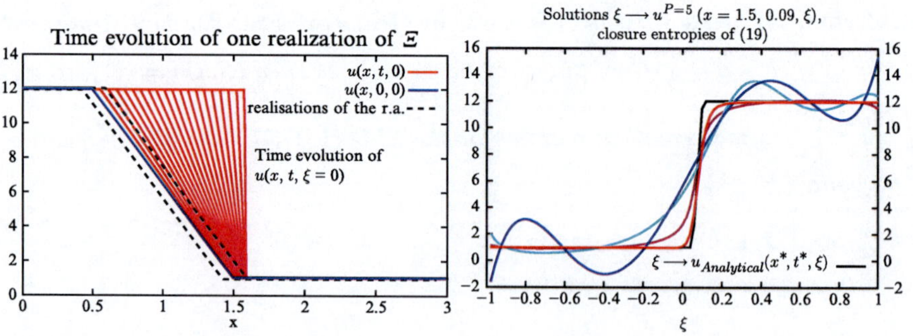

Fig. 2 *Right*: the different colors refer to the different entropies: $s_0(u)$, $s_1(u)$, $s_2(u)$, $s_2(u)$. The *black curve* corresponds to the analytical solution

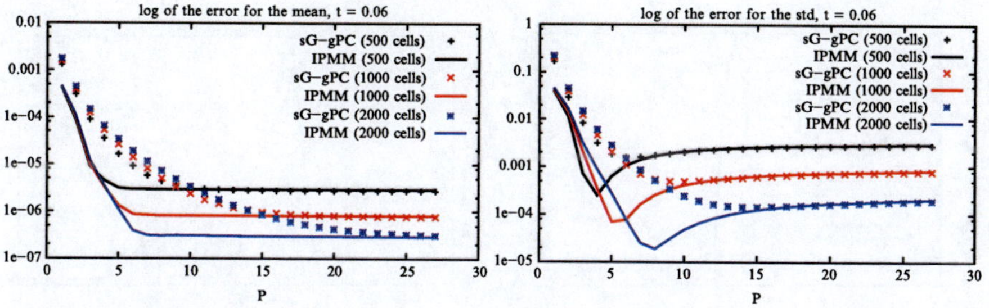

Fig. 3 Error (log-scale) on the mean and std on the whole physical space at time $t = 0.09$. Spectral convergence is reached for IPMM, for early P. For higher P, a threshold due to the spatial discretization is reached. Note that the non-monotonical behaviour of the IPMM curve for the std remains an open question

in the random space: see the analytical solution[14] $\xi \longrightarrow u(x = 1.5, t = 0.09, \xi)$ of Fig. 2 (right).

The truncated system is solved with a Roe scheme.[15] The following entropies are tested:

$$\begin{cases} s_0(u) = \frac{u^2}{2}, \\ s_1(u) = -ln(u - u_-), \\ s_2(u) = -ln(u - u_-) - ln(u_+ - u), \\ s_2(u) = (u - u_-)\ln(u - u_-) + (u_+ - u)\ln(u_+ - u) - 2u + u_- - u_+ . \end{cases} \tag{19}$$

The solutions are presented on Fig. 2: the polynomial order is the same for every entropies, $P = 5$. Convergence tests have been performed (with closure entropy s_2) for several spatial discretizations, see Fig. 3. Spectral convergence is reached with IPMM.

4.2 *Stochastic Riemann Problem: Euler System*

We consider the Euler system

$$\begin{cases} \partial_t \rho + \partial_x \rho u = 0, \\ \partial_t \rho u + \partial_x(\rho u^2 + p) = 0, \\ \partial_t \rho e + \partial_x(\rho u e + p u) = 0, \end{cases} \tag{20}$$

---

[14] Black curve of Fig. 2.

[15] See [12, 13]: here the stress is put on the stochastic resolution, not on the numerical scheme.

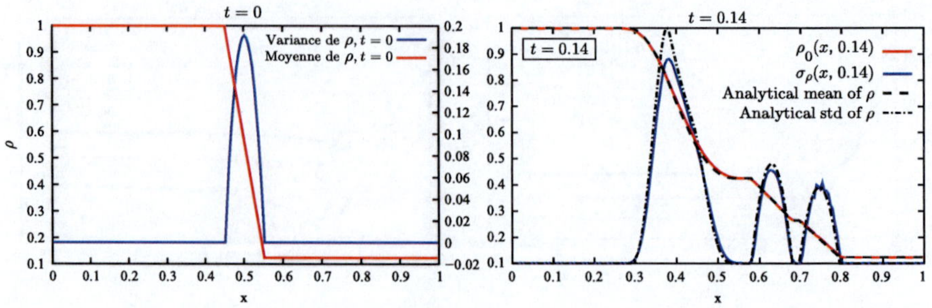

Fig. 4 Stochastic Riemann problem. The IC are the same as in the previous section with, in addition, $p = 1$ on the *right* and $p = 0.1$ on the *left* of the interface. The *left column* shows the IC in mean and std for ρ. The *left column* shows the mean and std of ρ at time $t = 0.14$. The computation has 200 cells, $P = 20$. Numerical integration is performed thanks to a 1-D Clenshaw–Curtis rule with level 7. The reference solutions, mean and std of ρ, are obtained by analytical integrations of analytical formulae available for this Riemann problem, see [10]

where ρ is the mass density, u is the velocity, e is the specific total energy and p is the pressure of the fluid. The SLC is closed by a perfect gas eos $p = (\gamma - 1)\rho\epsilon$ where $\epsilon = e - \frac{u^2}{2}$ is the specific internal energy and $\gamma = 1.4$. The system is hyperbolic iff $\epsilon > 0$. It exists a mathematical entropy (s, g) for (20) given by

$$\begin{cases} s(\rho, \rho u, \rho e) = -\rho \ln\left(\rho^{-\gamma}\left(\rho e - \frac{(\rho u)^2}{2\rho}\right)\right) \\ g(\rho, \rho u, \rho e) = \frac{\rho u}{\rho} s\left(\rho, \rho u, \rho e\right). \end{cases} \tag{21}$$

This mathematical entropy is chosen in the following as a closure entropy for IPMM so that hyperbolicity of the truncated system is ensured.

We consider the same Riemann problem as in Sect. 2 for which sG-gPC failed. The numerical results are displayed in Fig. 4. The numerical scheme used for Fig. 4 is described in [12, 13] and is not detailed here as the stress is on IPMM.

5 Conclusions

IPMM is a High-Order Spectral Method. It is numerically stable, conservative and non adaptive. The mathematical properties (hyperbolicity) are preserved by construction, the preservation of the physical properties ($\rho > 0, \epsilon > 0, p > 0$) is a corollary.

References

1. R. Abgrall. A Simple, Flexible and Generic Deterministic Approach to Uncertainty Quantifications in Non Linear Problems: Application to Fluid Flow Problems. *Rapport de Recherche INRIA*, 2007
2. R. Archibald, A. Gelb, R. Saxena, and D. Xiu. Discontinuity Detection in Multivariate Space for Stochastic Simulations. *J. Comp. Phys.*, 228(7):2676–2689, 2009
3. R.H. Cameron and W.T. Martin. The Orthogonal Development of Non-Linear Functionals in Series of Fourier–Hermite Functionals. *Ann. Math.*, 48:385–392, 1947
4. B.J. Debusshere, H.N. Najm, P.P. Pébay, O.M. Knio, R.G. Ghanem, and O.P. Le Maître. Numerical Challenges in the Use of Polynomial Chaos Representations for Stochastic Processes. *J. Sci. Comp.*, 26:698–719, 2004
5. M. Junk. Maximum Entropy for Reduced Moment Problems. *Math. Mod. Meth. Appl. Sci.*, 10:1001–1025, 2000
6. O. Le Maitre, M. Reagan, H. Najm, R. Ghanem, and O. Knio. Multi-Resolution Analysis of Wiener-Type Uncertainty Propagation Schemes. *J. Comp. Phys.*, 197:502–531, 2004
7. G. Lin, C.-H. Su, and G.E. Karniadakis. Predicting Shock Dynamics in the Presence of Uncertainties. *J. Comp. Phys.*, 217:260–276, 2006
8. O.P. Le Maître and O.M. Knio. Uncertainty Propagation using Wiener–Haar Expansions. *J. Comp. Phys.*, 197:28–57, 2004
9. A. Majda. *Compressible Fluid Flow and Systems of Conservation Laws*. Applied Mathematical sciences, 53, Springer, New York, 1984
10. L.R. Mead and N. Papanicolaou. Maximum Entropy in the Problem of Moments. *J. Math. Phys.*, 25(8):2404–2417, 1984
11. I. Müller and T. Ruggeri. *Rational Extended Thermodynamics, 2nd ed.* Tracts in Natural Philosophy, 37, Springer, New York, 1998
12. G. Poëtte. *Propagation d'Incertitudes pour les Systèmes de Lois de Conservation, Méthodes Spectrales Stochastiques*. Phd thesis, Université Pierre et Marie Curie, Institut Jean Le Rond D'Alembert, 2009
13. G. Poëtte, B. Després, and D. Lucor. Uncertainty Quantification for Systems of Conservation Laws. *J. Comp. Phys.*, 228(7):2443–2467, 2009
14. D. Serre. *Systèmes Hyperboliques de Lois de Conservation, partie I*. Diderot, Paris, 1996
15. D. Serre. *Systèmes Hyperboliques de Lois de Conservation, partie II*. Diderot, Paris, 1996
10. E.F. Toro. *Riemann solver and numerical methods for fluid dynamics*. Springer, Berlin, 1997
17. X. Wan and G.E. Karniadakis. Beyond Wiener–Askey Expansions: Handling Arbitrary PDFs. *SIAM J. Sci. Comp.*, 27(1–3):455–464, 2006
18. X. Wan and G.E. Karniadakis. Multi-Element generalized Polynomial Chaos for Arbitrary Probability Measures. *SIAM J. Sci. Comp.*, 28(3):901–928, 2006
19. J.A.S. Witteveen and H. Bijl. Using Polynomial Chaos for Uncertainty Quantification in Problems with Non Linearities. *47th AIAA Aerospace Sciences Meeting and Exhibit*, AIAA 2006–2066, 2006
20. D. Xiu and G.E. Karniadakis. The Wiener–Askey Polynomial Chaos for Stochastic Differential Equations. *SIAM J. Sci. Comp.*, 24:619–644, 2002

Reduced Basis Approximation for Shape Optimization in Thermal Flows with a Parametrized Polynomial Geometric Map

Gianluigi Rozza, Toni Lassila, and Andrea Manzoni

Abstract Reduced basis approximations for geometrically parametrized advection-diffusion equations are investigated. The parametric domains are assumed to be images of a reference domain through a piecewise polynomial map; this may lead to nonaffinely parametrized diffusion tensors that are treated with an empirical interpolation method. An a posteriori error bound including a correction term due to this approximation is given. Results concerning the applied methodology and the rigor of the corrected error estimator are shown for a shape optimization problem in a thermal flow.

1 Introduction

We consider the parametrized advection-diffusion equation in a bounded and piecewise smooth domain $\Omega_o(\boldsymbol{\mu}) \subset \mathbb{R}^2$, whose shape depends on a vector of geometrical parameters $\boldsymbol{\mu}$ residing in a low-dimensional parameter space $\mathcal{D} \subset \mathbb{R}^P$ (e.g., $P \leq 10$). The weak form of the equation reads as follows: for any given $\boldsymbol{\mu} \in \mathcal{D}$, find $u \in H^1(\Omega_o(\boldsymbol{\mu}))$ s.t. $u = u_D$ on Γ_D and

$$\int_{\Omega_o(\boldsymbol{\mu})} (\varepsilon \nabla u \cdot \nabla v + v\mathbf{b} \cdot \nabla u)\, d\Omega = \int_{\Omega_o(\boldsymbol{\mu})} f v \, d\Omega \qquad \forall v \in H^1(\Omega_o(\boldsymbol{\mu})) \quad (1)$$

G. Rozza (✉) and A. Manzoni
MATHICSE-CMCS Modelling and Scientific Computing, École Polytechnique Fédérale de Lausanne, Switzerland
e-mail: gianluigi.rozza@epfl.ch, andrea.manzoni@epfl.ch

T. Lassila
Institute of Mathematics, Helsinki University of Technology, P.O. Box 1100, 02015 TKK, Finland
e-mail: toni.lassila@tkk.fi

J.S. Hesthaven and E.M. Rønquist (eds.), *Spectral and High Order Methods for Partial Differential Equations*, Lecture Notes in Computational Science and Engineering 76, DOI 10.1007/978-3-642-15337-2_28,

where $\mathbf{b}$ is a given divergence-free constant vector field, Γ_D denotes the Dirichlet boundary, while homogeneous Neumann conditions are imposed on $\Gamma_N = \partial\Omega \backslash \Gamma_D$. Our interest is to solve (1) in a way that is:

- *Efficient* in the sense that for any $\boldsymbol{\mu}$ a numerical solution is obtained in real-time for arbitrarily fine discretizations.
- *Reliable* in the sense that for any $\boldsymbol{\mu}$ the obtained solution is verifiable within some prescribed tolerance from the finite element solution computed using a very fine mesh for discretization. Any error bound should be *rigorous*, that is to say, it should be a safe upper bound for the true error.

To that end, we employ the reduced basis (RB) method originally developed for nonlinear structural mechanics in the 1980s and more recently systematized for elliptic and parabolic, coercive and noncoercive, PDEs. The method is analyzed in detail in [9, 13] and in their references; previous works on reduced basis methods for the advection-diffusion equation include [3, 10, 15]. This method is a model reduction scheme for parametric PDEs based on the use of "snapshot" finite element solutions of the PDE (for certain values of the parameters) as global approximation basis functions. Our objective is to use the efficient evaluation of the RB solutions in a multi-query context, required for example in shape optimization of PDE modelled systems.

2 Reduced Basis Approximation of Parametric Advection-Diffusion Equations

We assume that the parametric domains $\Omega_o(\boldsymbol{\mu})$ are obtained by mapping from a reference domain Ω with $T(\boldsymbol{x}, \boldsymbol{\mu})$ a piecewise polynomial map w.r.t. both arguments, as $\Omega \mapsto \Omega_o(\boldsymbol{\mu}) := T(\Omega, \boldsymbol{\mu})$. Problem (1) is thus traced back to the reference domain as

$$\int_{\Omega} (\varepsilon \nu_T \nabla u \cdot \nabla v + v \chi_T \mathbf{b} \cdot \nabla u)\, d\Omega = \int_{\Omega} \eta_T f v \, d\Omega \quad \forall\, v \in \mathcal{X} \equiv H^1(\Omega), \tag{2}$$

where the parametric transformation tensors $\nu_T(\mathbf{x}, \boldsymbol{\mu})$, $\chi_T(\mathbf{x}, \boldsymbol{\mu})$ and $\eta_T(\mathbf{x}, \boldsymbol{\mu})$ are obtained from a change of coordinates (with the Jacobian of T denoted by J_T) as

$$\nu_T = J_T^{-T} J_T^{-1} |J_T|, \quad \chi_T = J_T^{-1} |J_T|, \quad \eta_T = |J_T|. \tag{3}$$

We may rewrite (2) as

$$\mathcal{A}(u, v; \boldsymbol{\mu}) = f(v; \boldsymbol{\mu}) \quad \forall\, v \in \mathcal{X}, \tag{4}$$

where the parametric bilinear form $\mathcal{A}(\cdot, \cdot; \boldsymbol{\mu})$ is coercive and the linear functional $f(\cdot; \boldsymbol{\mu})$ is continuous. The standard Galerkin finite element (FE) approximation of

(4) is to find $u^{\mathcal{N}} \in \mathcal{X}^{\mathcal{N}}$ s.t. $\mathcal{A}(u^{\mathcal{N}}, v; \boldsymbol{\mu}) = f(v; \boldsymbol{\mu})$ for all $v \in \mathcal{X}^{\mathcal{N}}$, where $\mathcal{X}^{\mathcal{N}}$ is a FE space constructed by using, e.g., piecewise linear shape functions on a discrete mesh [11]. Here we denote by $\mathcal{N}$ the dimension of the FE space, which is assumed to be large enough that the repeated assembly and solution of the FE system is too expensive for a multi-query context.

In order to find an approximation to $u^{\mathcal{N}}$ in an efficient and reliable way, we use Galerkin projection on a reduced subspace of basis functions. Let $\boldsymbol{\mu}^1, \ldots, \boldsymbol{\mu}^N$ be a collection of parameters and define the reduced basis approximation space as $\mathcal{X}_N^{\mathcal{N}} := \text{span}\{u^{\mathcal{N}}(\boldsymbol{\mu}^n) : n = 1, \ldots, N\}$, where each $u^{\mathcal{N}}(\boldsymbol{\mu}^n) \in \mathcal{X}^{\mathcal{N}}$ is a FE solution for a given parameter value $\boldsymbol{\mu}^n$. The reduced basis formulation reads as follows: find $u_N^{\mathcal{N}} \in \mathcal{X}_N^{\mathcal{N}}$ s.t. $\mathcal{A}(u_N^{\mathcal{N}}, v; \boldsymbol{\mu}) = f(v)$, for all $v \in \mathcal{X}_N^{\mathcal{N}}$. In practice, an orthonormalization procedure is required to build a basis $\{\Phi_n\}_{n=1}^N$ for the RB space $\mathcal{X}_N^{\mathcal{N}}$ that guarantees algebraic stability [9]. As long as the parametric bilinear form is affinely parametrized [13], that is to say of the form

$$\mathcal{A}(u, v; \boldsymbol{\mu}) = \sum_{q=1}^{Q_a} \Theta_a^q(\boldsymbol{\mu}) a^q(u, v) + \sum_{q=1}^{Q_b} \Theta_b^q(\boldsymbol{\mu}) b^q(u, v) \tag{5}$$

for some integers Q_a, Q_b, where $\Theta_a^q(\boldsymbol{\mu}) = \beta_k^{i,j}(\boldsymbol{\mu})$, $\Theta_b^q(\boldsymbol{\mu}) = \gamma_k^{i,j}(\boldsymbol{\mu})$, q is a condexed index for i, j, k and

$$a^{q(i,j,k)}(u, v) = \varepsilon \int_{\Omega} \xi_k^{i,j}(\mathbf{x}) \frac{\partial u}{\partial x_i} \frac{\partial v}{\partial x_j}\, d\Omega, \quad b^{q(i,j,k)}(u, v) = \int_{\Omega} \zeta_k^{i,j}(\mathbf{x}) b_i \frac{\partial u}{\partial x_j} v\, d\Omega,$$

$$\nu_T^{i,j}(\mathbf{x}, \boldsymbol{\mu}) = \sum_{k=1}^{K_{ij}^a} \beta_k^{i,j}(\boldsymbol{\mu}) \xi_k^{i,j}(\mathbf{x}), \qquad \chi_T^{i,j}(\mathbf{x}, \boldsymbol{\mu}) = \sum_{k=1}^{K_{ij}^b} \gamma_k^{i,j}(\boldsymbol{\mu}) \zeta_k^{i,j}(\mathbf{x}),$$

the solution of the reduced basis problem splits into two stages. In the so-called *offline stage* we assemble and store once and for all the parameter-independent system matrices $\underline{A}^q$ and $\underline{B}^q$ of components

$$[\underline{A}^q]_{m,n} = a^q(\Phi_n, \Phi_m), \qquad [\underline{B}^q]_{m,n} = b^q(\Phi_m, \Phi_n) \tag{6}$$

using the global reduced basis functions Φ_k, and similarly for the right-hand-sides. Then in the *online stage* for a given parameter $\boldsymbol{\mu}$ the parametric coefficients $\Theta_a^q(\boldsymbol{\mu})$, $\Theta_b^q(\boldsymbol{\mu})$ are evaluated and the reduced basis matrix $\underline{A}_N = \sum_{q=1}^{Q_a} \Theta_a^q(\boldsymbol{\mu}) \underline{A}^q + \sum_{q=1}^{Q_b} \Theta_b^q(\boldsymbol{\mu}) \underline{B}^q$ is assembled, and similarly for the right-hand-side. This linear system of dimension $N \times N$ is dense, but inexpensive to solve: the online complexity is independent of the FE solution dimension $\mathcal{N}$ and thus we fulfill the first requirement of efficiency.

The following greedy algorithm for choosing the parameters $\boldsymbol{\mu}^n$ has been used [9, 13, 16]. Let $\Xi_{\text{train}} \subset \mathcal{D}$ be a finite training sample of parameter points chosen according to a uniform or log-uniform distribution. Define the parameter-independent norm $||v||_X := \sqrt{\mathcal{A}(v, v; \bar{\boldsymbol{\mu}}) + \lambda ||v||_{L^2(\Omega)}}$ for some $\bar{\boldsymbol{\mu}} \in \mathcal{D}$ and $\lambda > 0$ large enough such that the resulting norm is well-defined. Given the first parameter value $\boldsymbol{\mu}^1$ and a sharp, inexpensive a posteriori error bound $\Delta_n(\boldsymbol{\mu})$ for the norm $|| \cdot ||_X$ such that $||u^{\mathcal{N}}(\boldsymbol{\mu}) - u_n^{\mathcal{N}}(\boldsymbol{\mu})||_X \leq \Delta_n(\boldsymbol{\mu})$ for all $\boldsymbol{\mu} \in \Xi_{\text{train}}$, we choose the remaining parameter values as the solutions $\boldsymbol{\mu}^n = \arg\max_{\boldsymbol{\mu} \in \Xi_{\text{train}}} \Delta_{n-1}(\boldsymbol{\mu})$, for $n = 2, \ldots, N$. The quality of the reduced basis approximation depends crucially on the quality of the a posteriori error estimator. The standard RB error estimator in literature [9] for problems that satisfy the affinity assumption (5) is

$$\Delta_N(\boldsymbol{\mu}) := \frac{||r(\cdot, \boldsymbol{\mu})||_{X'}}{\alpha_{\text{LB}}(\boldsymbol{\mu})} \geq ||u_N^{\mathcal{N}} - u^{\mathcal{N}}||_X = ||e(\boldsymbol{\mu})||_X, \tag{7}$$

where $||r(\cdot, \boldsymbol{\mu})||_{X'}$ is the dual norm of the residual $r(v, \boldsymbol{\mu}) = f(v) - \mathcal{A}(u_N^{\mathcal{N}}, v; \boldsymbol{\mu})$ and $\alpha_{\text{LB}}(\boldsymbol{\mu})$ is a computable lower bound for the discrete coercivity constant

$$0 < \alpha_{\text{LB}}(\boldsymbol{\mu}) \leq \alpha(\boldsymbol{\mu}) = \inf_{u \in \mathcal{X}^{\mathcal{N}}} \frac{\mathcal{A}(u, u; \boldsymbol{\mu})}{||u||_X^2}. \tag{8}$$

For efficient and reliable methods of computing both $||r(\cdot, \boldsymbol{\mu})||_{X'}$ and $\alpha_{\text{LB}}(\boldsymbol{\mu})$ we refer the reader to [2, 4, 6, 13]. In the greedy basis construction algorithm we usually fix a priori an error tolerance $\varepsilon_{\text{tol}}^{RB}$ and then we continue the process until the condition $\Delta_N(\boldsymbol{\mu}) \leq \varepsilon_{\text{tol}}^{RB}$ for all $\boldsymbol{\mu} \in \Xi_{\text{train}}$ is achieved.

If the affinity assumption does not hold, we rely on the empirical interpolation method (EIM) [1], which is an interpolation method for parametric functions based on adaptively chosen interpolation points and global shape functions. When the geometric transformation $T(\mathbf{x}, \boldsymbol{\mu})$ is polynomial the advection tensor $\chi_T(\mathbf{x}, \boldsymbol{\mu})$ is polynomial and therefore always affinely parametrized, while the diffusion one $\nu_T(\mathbf{x}, \boldsymbol{\mu})$ is a nonaffine tensor. To approximate each component $\nu_T^{i,j}(\mathbf{x}, \boldsymbol{\mu})$ of the tensor we use a different set of interpolation points and thus look for an affine approximation

$$\tilde{\nu}_T^{i,j}(\mathbf{x}, \boldsymbol{\mu}) := \sum_{m=1}^{M_{ij}} \vartheta_m^{i,j}(\boldsymbol{\mu}) \xi_m^{i,j}(x) = \nu_T^{i,j}(\mathbf{x}, \boldsymbol{\mu}) + \varepsilon^{i,j}(x; \boldsymbol{\mu}), \tag{9}$$

with the error terms under some tolerance, i.e., $||\varepsilon^{i,j}(\cdot; \boldsymbol{\mu})||_\infty < \varepsilon_{\text{tol}}^{\text{EIM}} \; \forall \boldsymbol{\mu} \in \mathcal{D}$.

For the reliability of the methodology we need to guarantee an a posteriori error bound between the "truth" finite element solution and the reduced basis approximation. The snapshot solutions $u^{\mathcal{N}}(\boldsymbol{\mu}^n)$ should be obtained by a FE stable method: for the advection-diffusion equation we can use a Galerkin formulation with either Galerkin least-squares (GLS) or streamline upwind (SUPG) stabilizers [11]. For more details on coupling the stabilizer with the reduced basis framework, see [3, 10]. To simplify things we choose the physical Peclet number $\text{Pe} = \varepsilon^{-1}$ small enough such that the finite element approximations are always guaranteed to be stable without adding any stabilizing terms. By applying the coercivity property it holds that

$$\alpha(\mu)||e(\mu)||_X^2 \le \mathcal{A}(e(\mu), e(\mu); \mu). \tag{10}$$

Using the ideas from [1, 8] we can prove an a posteriori error estimate of the form (7) also in the nonaffine case. Defining the trilinear forms $a_{i,j}(u, v, \varphi) := \varepsilon \int_\Omega \varphi \frac{\partial u}{\partial x_i} \frac{\partial v}{\partial x_j} d\Omega$ and the residual of the reduced basis solution $u_N^{\mathcal{N}}$ as

$$r_N(v; \mu) := f(v) - \sum_{q=1}^{Q_b} \Theta_b^q(\mu) b^q(u_N^{\mathcal{N}}, v) - \sum_{i,j=1}^{2} a_{i,j}(u_N^{\mathcal{N}}, v, \tilde{v}_T^{i,j}),$$

we have the following a posteriori error bound:

$$||e(\mu)||_X \le \frac{||r_N(\cdot; \mu)||_{X'}}{\alpha_{\mathrm{LB}}(\mu)} + \sup_{w \in \mathcal{X}} \frac{\sum_{i,j=1}^{2} a_{i,j}(u_N^{\mathcal{N}}, w, v_T^{i,j} - \tilde{v}_T^{i,j})}{\alpha_{\mathrm{LB}}(\mu)||w||_X}. \tag{11}$$

By using the definitions of the error $e(\mu)$ and the residual we get

$$\mathcal{A}(e(\mu), e(\mu); \mu) \le r_N(e(\mu); \mu) - \sum_{i,j=1}^{2} a_{i,j}(u_N^{\mathcal{N}}(\mu), e(\mu), v_T^{i,j} - \tilde{v}_T^{i,j});$$

by property (10) it follows that

$$||e(\mu)||_X \le \frac{1}{\alpha(\mu)} \left(\frac{r_N(e(\mu); \mu)}{||e(\mu)||_X} - \frac{\sum_{i,j=1}^{2} a_{i,j}(u_N^{\mathcal{N}}, e(\mu), v_T^{i,j} - \tilde{v}_T^{i,j})}{||e(\mu)||_X} \right)$$

and (11) is obtained by taking sups. The correction term originating from the empirical interpolation is of order $O(|u_N^{\mathcal{N}}|_1)$ and therefore does not vanish as $N \to \infty$ if the number of terms in the empirical interpolation approximation is kept fixed. We need to choose the tolerance $\varepsilon_{\mathrm{tol}}^{\mathrm{EIM}} \ll ||r_N(\cdot; \mu)||_{X'}/(\varepsilon\, |u_N^{\mathcal{N}}|_1)$, so that the correction term does not dominate the error estimate. To obtain an error estimate computable online without $\mathcal{N}$-dependence we use the estimator $\Delta_N^{corr}(\mu)$ proposed in [1]:

$$||e(\mu)||_X \lesssim \frac{||r_N(\cdot; \mu)||_{X'}}{\alpha_{\mathrm{LB}}(\mu)} + \sup_{w \in \mathcal{X}} \frac{\sum_{i,j=1}^{2} \tilde{\varepsilon}_{M_{ij}}^{i,j}\, a_{i,j}(u_N^{\mathcal{N}}, w, \xi_{M_{ij}+1}^{i,j})}{\alpha_{\mathrm{LB}}(\mu)||w||_X}, \tag{12}$$

where $\tilde{\varepsilon}_{M_{ij}}^{i,j} := |v_T^{i,j}(z^{M_{ij}+1}, \mu) - \tilde{v}_T^{i,j}(z^{M_{ij}+1}, \mu)|$ is a one-point estimate for the error $||\varepsilon^{i,j}(\cdot, \mu)||_{L^\infty(\Omega)}$ computed using the $(M_{ij}+1)$th interpolation point $z^{M_{ij}+1}$. By "$\lesssim$" we mean that the bound is no longer fully rigorous.

3 Numerical Example

We consider an optimal heat exchange problem. A NACA0012 airfoil is placed in a thermal flow; our control variables are the vertical position of the airfoil and its shape (for small perturbations). The reference geometry is shown in Fig. 1. The objective is to obtain the correct desired average temperature $\overline{u}_{\text{target}}$ at the outflow given a fixed angle of attack σ_0 for the airfoil:

$$
\begin{aligned}
&\min_{\boldsymbol{\mu}\in\mathcal{D}} \left[\overline{u}_{\text{target}} - \frac{1}{|\Gamma_{\text{out}}|}\int_{\Gamma_{\text{out}}} u(\mathbf{x})\, d\Gamma\right]^2 + \lambda\,[\sigma(\boldsymbol{\mu}) - \sigma_0]^2, \\
&\text{s.t.} \quad \int_{\Omega_o(\boldsymbol{\mu})} (\varepsilon\nabla u\cdot\nabla v + v\mathbf{b}\cdot\nabla u)\, d\Omega_o = \int_{\Omega_o(\boldsymbol{\mu})} f v\, d\Omega_o \quad \forall v\in H^1(\Omega_o(\boldsymbol{\mu})) \\
&\text{with } u = T_0 \text{ on } \Gamma_{\text{in}}\cup\Gamma_{\text{free}},\ u = T_1 \text{ on } \Gamma_{\text{surf}},\quad u = T_2 \text{ on the airfoil.}
\end{aligned}
\tag{13}
$$

We parametrize the geometry around the airfoil using free-form deformations (FFD) [14]: a 6×6 lattice of control points is placed around the airfoil and the closest four control points are allowed to move in the x_2-direction. This results in a polynomial geometric map $T(\mathbf{x},\boldsymbol{\mu})$ with $P = 4$ parameters built using Bernstein polynomials. In Fig. 1 we also display the control points and the deformation of the reference shape as the control points are moved. For more details on the FFD parametrization setup we refer the reader to [7]. For the finite element computations $\mathcal{N} = 15{,}718$ degrees of freedom are used.

To solve the optimization problem (13), the algorithm based on sequential quadratic programming (SQP) provided in Matlab has been used; convergence to the optimal solution has been reached after 25 functional evaluations. In order to evaluate effectively the state equation in the constraint of (13) we replace the FE solution $u^{\mathcal{N}}$ with the RB approximation $u_N^{\mathcal{N}}$. The nonaffinely parametrized diffusion tensor has been approximated with the empirical interpolation method using a toler-

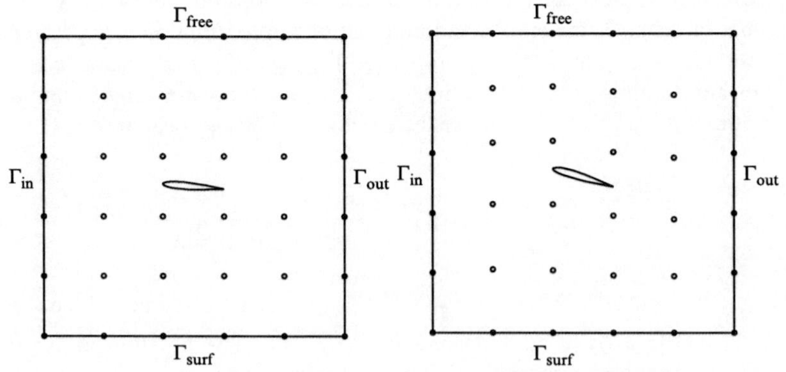

Fig. 1 Reference domain Ω and a deformed configuration $\Omega_o(\boldsymbol{\mu})$ using FFDs

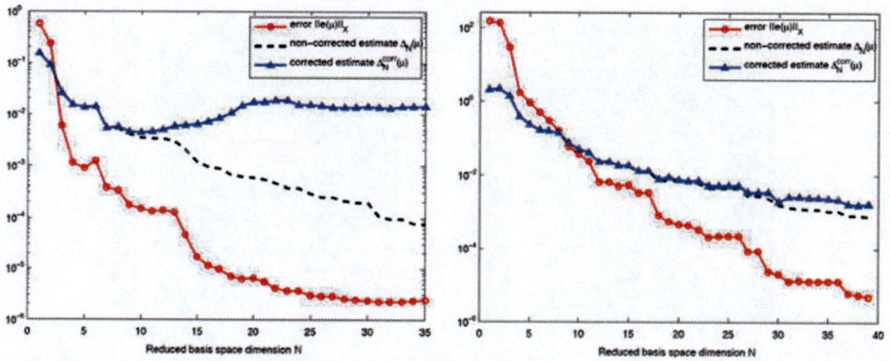

Fig. 2 Convergence of $\sup_{\mu\in\mathcal{D}} \|e(\boldsymbol{\mu})\|_X$ versus the corrected $\Delta_N^{corr}(\boldsymbol{\mu})$ and non-corrected $\Delta_N(\boldsymbol{\mu})$ error estimates for $M = 108$ (*left*) and $M = 208$ (*right*), respectively

ance of $\varepsilon_{\text{tol}}^{\text{EIM}} = 10^{-4}$ and, then, $\varepsilon_{\text{tol}}^{\text{EIM}} = 10^{-6}$, resulting in $M = \sum_{i,j=1}^{2} M_{ij} = 108$ and $M = 208$ terms, respectively, in the affine expansion. After this the reduced basis offline stage consists of assembly of the matrices (6), performing the successive constraint method [6] for estimation of the lower bound $\alpha_{LB}(\boldsymbol{\mu})$ of the coercivity constant, and finally a greedy procedure for choosing the reduced basis snapshots and the corresponding basis functions. The maximum number of basis functions used was $N = 38$. In Fig. 2 we show the error estimates (with and without the correction term from the empirical interpolation) $\Delta_N(\boldsymbol{\mu})$ and $\Delta_N^{corr}(\boldsymbol{\mu})$ as functions of N, compared to the true error $e(\boldsymbol{\mu})$ in the worst-case, for $M = 108$ and $M = 208$. In the first case the approximation performed by EIM is too poor and the correction term wider than in the second test, with a more accurate and rigorous error estimator. Moreover, in the first test we still have some *plateau* effect to be reduced [12]. For $M = 208$ we observe a reduction of 140:1 in the time to solve the RB system versus the assembling and solution of the FE system, while the reduction in the linear system size is 400:1.

Including the cost of the offline stage, we estimate that after 500 parametric PDE solutions we have passed the break-even point where RB computations are more efficient. The use of FFD also reduces the number of shape parameters: compared to a local boundary variations approach by moving individual mesh nodes we obtain a reduction of 238:1 in the number of geometric parameters. In Fig. 3 the optimal design for two particular configurations, together with the field solutions, are shown.

4 Conclusions

A reduced basis approximation for a shape optimization problem in a thermal flow has been presented. Recovering the assumption of parametric affinity is important to obtain a reduction in the online computational costs. When the geometric

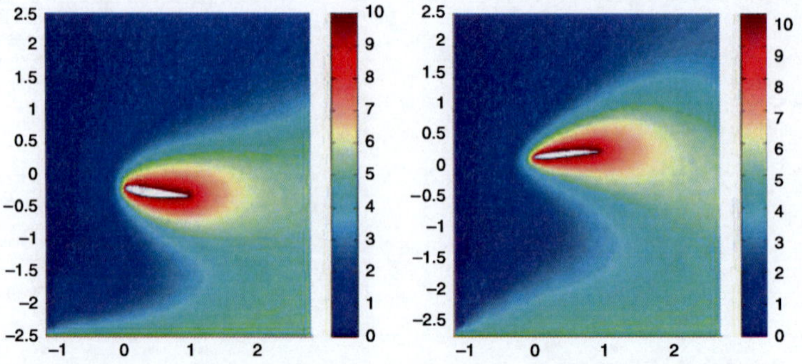

Fig. 3 Optimal design of the airfoil for the cases $\sigma_0 = 7°$, $\overline{u}_{\text{target}} = 4.1$ (*left*) and $\sigma_0 = -5°$, $\overline{u}_{\text{target}} = 4.5$ (*right*)

transformation map is polynomial, only the diffusive transformation tensor needs to be treated with the empirical interpolation method. This leads to a correction term in the a posteriori error bounds. We have demonstrated that the correction term is rigorous. In the proposed shape optimization problem of an airfoil in thermal flow with four shape parameters we observed a reduction of 140:1 in the computational time to solve the RB system versus the assembling procedure and the solution of the FE system.

Acknowledgements We are grateful to the family of Prof. Fausto Saleri for granting the use of the MLife library as a basis for the numerical FE simulations presented in this paper. We also acknowledge Prof. A. Quarteroni for insights and suggestions and the research group of Prof. Anthony Patera of MIT for the rbMIT [5] package, used for the RB computations.

References

1. M. Barrault, Y. Maday, N.C. Nguyen, and A.T. Patera. An 'empirical interpolation' method: application to efficient reduced-basis discretization of partial differential equations. *C. R. Math. Acad. Sci. Paris*, 339(9):667–672, 2004
2. Y. Chen, J. Hestaven, Y. Maday, and J. Rodriguez. A monotonic evaluation of lower bounds for inf-sup stability constants in the frame of reduced basis approximations. *C. R. Acad. Sci. Paris, Ser. I*, 346:1295–1300, 2008
3. L. Dedè. *Adaptive and reduced basis methods for optimal control problems in environmental applications*. PhD thesis, Politecnico di Milano, 2008
4. D.B.P. Huynh, D. Knezevic, Y. Chen, J. Hestaven, and A.T. Patera. A natural-norm successive constraint method for inf-sub lower bounds. *Scientific Computing Group*, Brown University, No. 2009–23
5. D.B.P. Huynh, N.C. Nguyen, G. Rozza, and A.T. Patera. Rapid reliable solution of the parametrized partial differential equations of continuum mechanics and transport, 2008. http://augustine.mit.edu
6. D.B.P Huynh, G. Rozza, S. Sen, and A.T. Patera. A successive constraint linear optimization method for lower bounds of parametric coercivity and inf-sup stability costants. *C. R. Acad. Sci. Paris. Sér. I Math.*, 345:473–478, 2007

7. T. Lassila and G. Rozza. Parametric free-form shape design with PDE models and reduced basis method. *Comput. Meth. Appl. Mech. Eng.*, 199(23–24):1583–1592, 2010
8. N.C. Nguyen. A posteriori error estimation and basis adaptivity for reduced-basis approximation of nonaffine-parametrized linear elliptic partial differential equations. *J. Comp. Phys.*, 227:983–1006, 2007
9. A.T. Patera and G. Rozza. *Reduced Basis Approximation and A Posteriori Error Estimation for Parametrized Partial Differential Equations*. Version 1.0, Copyright MIT 2006, to appear in (tentative rubric) MIT Pappalardo Graduate Monographs in Mechanical Engineering., 2009. Available at http://augustine.mit.edu
10. A. Quarteroni, G. Rozza, and A. Quaini. Reduced basis methods for optimal control of advection-diffusion problem. In *Advances in Numerical Mathematics, W. Fitzgibbon, R. Hoppe, J. Periaux, O. Pironneau, and Y. Vassilevski, Editors*, pages 193–216, 2007
11. A. Quarteroni and A. Valli. *Numerical Approximation of Partial Differential Equations*. Springer, Berlin, 2008
12. G. Rozza. Reduced basis methods for Stokes equations in domains with non-affine parameter dependence. *Comput. Vis. Sci.*, 12(1):23–35, 2009
13. G. Rozza, D.B.P. Huynh, and A.T. Patera. Reduced basis approximation and a posteriori error estimation for affinely parametrized elliptic coercive partial differential equations. *Arch. Comput. Methods Eng.*, 15:229–275, 2008
14. T.W. Sederberg and S.R. Parry. Free-form deformation of solid geometric models. *Comput. Graph.*, 20(4):151–160, 1986
15. T. Tonn and K. Urban. A reduced-basis method for solving parameter-dependent convection-diffusion problems around rigid bodies. In *Proc. ECCOMAS CFD*, 2006
16. K. Veroy, C. Prud'homme, D.V. Rovas, and A.T. Patera. A posteriori error bounds for reduced-basis approximation of parametrized noncoercive and nonlinear elliptic partial differential equations. In *Proc. 16th AIAA Comput. Fluid Dyn.*, 2003

Constrained Approximation in *hp*-FEM: Unsymmetric Subdivisions and Multi-Level Hanging Nodes

Andreas Schröder

Abstract In conform hp-finite element schemes on irregular meshes, one has to ensure the finite element functions to be continuous across edges and faces in the presence of hanging nodes. A key approach is to constrain the appropriate shape functions using so-called connectivity matrices. In this work the connectivity matrices for hierarchical tensor product shape functions are explicitly determined. In particular, the presented approach includes both unsymmetric subdivisions and multi-level hanging nodes *not* using hierarchical or multi-level information of subdivisions. Moreover, the problem of edge and face orientations is considered.

1 Introduction

In adaptive finite element schemes, local refinements are typically realized by subdivisions of mesh elements. Using conform finite element schemes, one has to ensure the finite element functions to be continuous across edges and faces. In the presence of hanging or irregular nodes, this is done through constraint of the local basis functions associated to them and to adjacent irregular edges and faces, which is known as constrained approximation. A natural approach is to use connectivity matrices in the assembly process. Let $\mathscr{T} := \{T_0, T_1, \ldots, \}$ be a mesh subordinate to $\Omega \subset \mathbb{R}^k$, $k \in \{2, 3\}$, where $\overline{T}_i \cap \overline{T}_j$ is empty or a vertex, an edge or a face of T_i or T_j, $i \neq j$. Furthermore, let $\Psi_T : \hat{T} \to T \in \mathscr{T}$ be a bijective and sufficiently smooth mapping for some reference element $\hat{T}$, e.g., $\hat{T} := [-1, 1]^k$ for quadrangles or hexahedrons, and let $\mathscr{P}_T$ be a finite polynomial space on $\hat{T}$. Thus, the space of piecewise continuous polynomials is defined as $\mathscr{S} := \{v \in C^0(\Omega) \mid \forall T \in \mathscr{T} : v_{|T} \circ \Psi_T \in \mathscr{P}_T\}$. We denote the global basis functions of $\mathscr{S}$ by $\{\varphi_i\}_{0 \leq i < n}$ and the local basis functions of $\mathscr{P}_T$ by $\{\eta_{T,i}\}_{0 \leq i < n_T}$. The matrices $\pi_T \in \mathbb{R}^{n \times n_T}$, $T \in \mathscr{T}$, connecting the local and global basis functions are called *connectivity matrices* and are given by

A. Schröder
Department of Mathematics, Humboldt-Universität zu Berlin, 10099 Berlin, Germany
e-mail: andreas.schroeder@mathematik.hu-berlin.de

J.S. Hesthaven and E.M. Rønquist (eds.), *Spectral and High Order Methods for Partial Differential Equations*, Lecture Notes in Computational Science and Engineering 76, DOI 10.1007/978-3-642-15337-2_29,

$\varphi_{i|T} = \sum_{j=0}^{n_T-1} \pi_{T,ij}\eta_{T,j} \circ \Psi_T^{-1}$. The assembly of the stiffness matrix K and the load vector F is thus given by $K := \sum_{T\in\mathcal{T}} \pi_T K_T \pi_T^\top$ and $F := \sum_{T\in\mathcal{T}} \pi_T F_T$ for the local stiffness matrices $K_T \in \mathbb{R}^{n_T \times n_T}$ and local load vectors $F_T \in \mathbb{R}^{n_T}$, $T \in \mathcal{T}$.

A fundamental problem in finite element implementations is to provide connectivity matrices through suitable data structures as their computation is highly dependent on the choice of shape functions and refinement patterns. Moreover, the edge and face orientations have to be taken into account. If mesh elements containing hanging nodes are subdivided, multi-level hanging nodes occur. This significantly complicates the computation of the connectivity matrices and, in particular, their implementation. Therefore, most finite element codes do not allow for more than one hanging node per edge or face.

In the literature, connectivity matrices, their calculation and several data structures are described. In [1], the constraints are stated for integrated Legendre shape functions on quadrangles. Also, the extension to multi-level *hp*-refinement is discussed. The constraints are inserted via data structures representing a sparse data format for connectivity matrices. In [3], some data structure arrays for quadrangles storing the constraint information are proposed which also describe connectivity matrices in sparse data format. Similar approaches are suggested in [2, 4, 7, 11, 12]. A broad overview on data structures and algorithms for constrained approximation in 2D and 3D is given in the comprehensive monographs by Demkowicz et al. [5, 6].

The aim of this work is to compute the connectivity matrices for hierarchical tensor product shape functions including both unsymmetric subdivisions and multi-level hanging nodes. The basic idea is to consider an irregular face as a subset of a regular face regardless of whether it results from a multi-level, symmetric or unsymmetric subdivision and to compute the entries of the connectivity matrices from this information only. Hence, no hierarchical or multi-level information of the subdivisions is needed. This simplifies the implementation greatly. A further emphasis of this work is on edge and face orientations and on implementation aspects based on some simple data structures for the storage of mesh elements.

2 Tensor Product Shape Functions of Legendre Type

Tensor product shape functions based on integrated Legendre or Gauss–Lobatto polynomials are a widely used family of shape functions for higher-order FEM. Using Gegenbauer polynomials $\{G_i^\varrho\}_{i\in\mathbb{N}_0}$ defined as $(i+1)G_{i+1}^\varrho(x) = 2(i+\varrho)xG_i^\varrho(x) - (i+2\varrho-1)G_{i-1}^\varrho(x)$ with $\varrho \in \mathbb{R}$, $G_0^\varrho(x) := 1$ and $G_1^\varrho(x) := 2\varrho x$, we obtain integrated Legendre ($\beta_i := 1$) or Gauss–Lobatto ($\beta_i := \sqrt{(2i-1)/2}$) polynomials $\xi_0(x) := \frac{1}{2}(1-x)$, $\xi_1(x) := \frac{1}{2}(1+x)$, and $\xi_i(x) := \beta_i G_i^{-1/2}(x)$ for $i = 2, \dots, p$. Tensor product shape functions are constructed on the unit cube $[-1,1]^k$ via

$$\eta_\alpha(x) := \prod_{r=0}^{k-1} \xi_{\alpha_r}(x_r), \; x \in \mathbb{R}^k$$

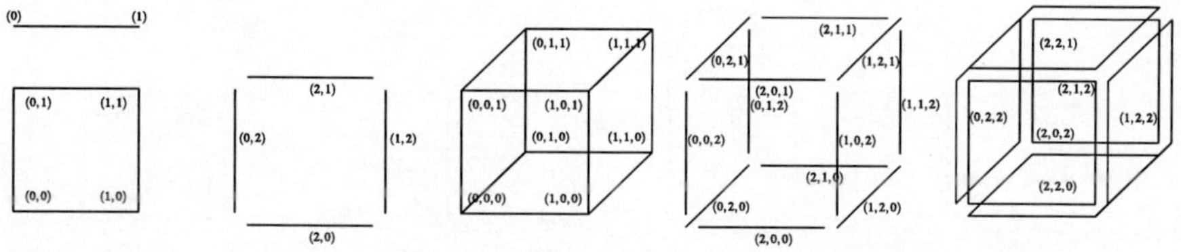

Fig. 1 Index tuple identifying nodes, edges and faces of the reference cube

for a k-tuple α with $\alpha_r \in \{0,\dots,p_r\}$, $0 \le r < k$ and local polynomial degrees $p_0,\dots,p_{k-1} \ge 1$, cf. [8, Chap. 3]. Usually, the shape functions are separated into nodal, edge, face and inner modes. For this purpose, we associate a node, an edge or a face to a k-tuple with values in $\{0,1,2\}$ as shown in Fig. 1 and the unit cube itself to the k-tuple $(2,\dots,2)$. In the following, let b be such a k-tuple. Typically, one also introduces additional local polynomial degrees for edges and faces, for instance, to ensure the minimum rule, cf. [11]. We denote these degrees by $p_r^b \in \{1,\dots,p_r\}$ for all $r = 0,\dots,k-1$ with $b_r = 2$. With these preparations at hand, the modes associated to b are $\{\eta_\alpha\}_{\alpha\in I^b}$ with

$$I^b := \{\alpha \mid \alpha_r := b_r \text{ if } b_r \in \{0,1\}, \text{ otherwise } \alpha_r \in \{2,\dots,p_r^b\}\}.$$

Also serendipity shape functions with reduced number of face and inner modes (cf. [8]) can be captured using this notation. Let q^b be a polynomial degree which is assigned to b and let ℓ be the dimension of the object associated to b. With $p_r^b := q^b - 2(\ell-1)$, the index set is given by $\tilde{I}^b := \{\alpha \in I^b \mid \sum_{r=0,\, b_r=2}^{k-1} \alpha_r \in \{2\ell,\dots,q^b\}\}$. In most finite element implementations, a mesh element $T \in \mathcal{T}$ is represented by a special data structure which enables the storage of information like coordinates, polynomial degrees, global numbering or to generate some information about the combinatorial structure of the mesh element. A simple data structure is given by the representation of T through $G_T = (G_T^0,\dots,G_T^{k-1}) \in (\mathcal{G}_0)^{\sigma_0} \times \dots \times (\mathcal{G}_{k-1})^{\sigma_{k-1}}$ where $\sigma_\ell := \sigma_\ell^k := 2^{k-\ell} k!/(\ell!(k-\ell)!)$ denotes the number of ℓ-dimensional adjacent objects in a k-dimensional cube. The set $\mathcal{G}_0 \subset \mathbb{R}^k$ represents the set of all nodes of $\mathcal{T}$, $\mathcal{G}_\ell \subset (\mathcal{G}_0)^{2^\ell}$ of all edges or faces of $\mathcal{T}$, $0 \le \ell < k$, respectively. For completeness, we define $\mathcal{G}_k := \{G_T^0 \mid T \in \mathcal{T}\}$. A natural orientation of edges and faces is shown in Fig. 2a, which is equivalently given by the matrices

$$\mathcal{I}^{1,2} := \begin{pmatrix} 0&1&3&0\\ 1&2&2&3 \end{pmatrix}, \quad \mathcal{I}^{1,3} := \begin{pmatrix} 0&1&3&0&0&1&2&3&4&5&7&4\\ 1&2&2&3&4&5&6&7&5&6&6&7 \end{pmatrix}, \quad \mathcal{I}^{2,3} := \begin{pmatrix} 0&1&3&4&0&0\\ 1&2&2&5&3&1\\ 2&6&6&6&7&5\\ 3&5&7&7&4&4 \end{pmatrix}.$$

Here, the entries of the j-th column denotes the node indices of the edge or face with index j. We assume that for all $1 \le \ell < k$ and $0 \le \nu < \sigma_\ell$, there exists a unique $0 \le i < 2^\ell$ such that

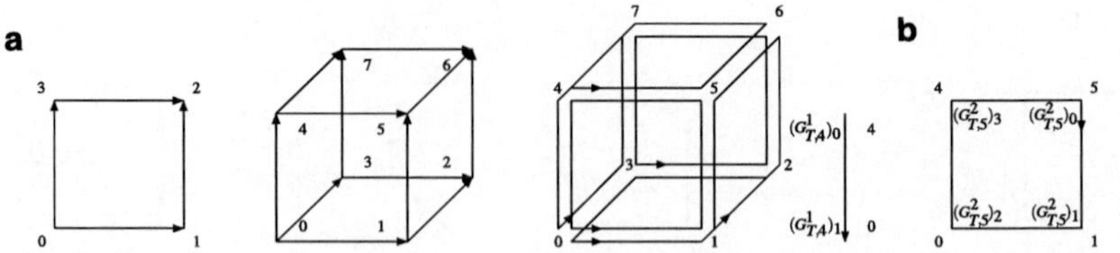

Fig. 2 (**a**) Edge and face orientations in the reference element, (**b**) non matching orientations

$$(G_T^0)_{\mathscr{I}_{0,\nu}^{\ell,k}} = ((G_T^\ell)_\nu)_i. \tag{1}$$

We denote this index by $g(G_T, \ell, \nu)$. Furthermore, we assume that for all $1 \leq \ell < k$ and $0 \leq \nu < \sigma_\ell$ there exists a unique $\delta \in \{-1, 1\}$ such that

$$(G_T^0)_{\mathscr{I}_{i,\nu}^{\ell,k}} = ((G_T^\ell)_\nu)_{(g(G_T,\ell,\nu)+\delta i) \bmod 2^\ell} \tag{2}$$

for all $0 \leq i < 2^\ell$. Given $h(G_T, \ell, \nu) := \delta$, we obtain $g(G_T, 1, 4) = 1$, $h(G_T, 1, 4) = -1$, $g(G_T, 2, 5) = 2$ and $h(G_T, 2, 5) = -1$ in Fig. 2b. Conditions (1) and (2) ensure, that the edges and faces consist of the nodes given by G_T^0 and that they can be transferred to the reference edge or face by rotation or reflections, respectively.

The approximation space $\mathscr{S}$ is defined through a degree distribution which is given by the global polynomial degrees $p(G)_0, \ldots, p(G)_{\ell-1}$. Here, $G \in \mathscr{G}_\ell$, $1 \leq \ell \leq k$ represents a non-hanging or regular edge, face or a mesh element in $\mathscr{T}$. In the case that G represents a face, we associate $p(G)_0$ to the direction given by the nodes G_0 and G_1, and $p(G)_1$ to the direction given by G_1 and G_2. In the following, let $M(G, \beta)$ be a suitable global numbering where β is an ℓ-tuple with $\beta_r \in \{2, \ldots, p(G)_r\}$, $0 \leq r < \ell$ which denotes the modes associated to G.

In the following, let $b(G)$ be the k-tuple associated to $G = (G_T^\ell)_\nu$ for some $0 \leq \ell < k$, $0 \leq \nu < \sigma_\ell$ or to $G = G_T^0$ with $\ell = k$. Furthermore, let $\alpha \in I^{b(G)}$. To construct continuous functions, we have to adjust the edge and face modes to the orientation of G given by the mappings g and h. This adjustment may be done switching the entries in α or using a sign number $\mu(\alpha)$. For this purpose, we specify the local polynomial degrees $p_r^{b(G)}$, the ℓ-tuple $\beta(\alpha)$ and the sign number $\mu(\alpha)$. In the case $\ell = 1$, we set $p_r^{b(G)} := p(G)_0$, $\beta(\alpha)_0 := \alpha_r$ and $\mu(\alpha) = h(G_T, 1, \nu)^{\beta(\alpha)_0}$ for the unique $r \in \{0, \ldots, k-1\}$ with $b(G)_r = 2$. In the case $\ell = 2$, we have unique $r_0, r_1 \in \{0, \ldots, k-1\}$ with $b(G)_{r_0} = b(G)_{r_1} = 2$ and $r_0 < r_1$. Here, we distinguish four cases depending on the values of $f(G_T, \nu) := (g(G_T, 2, \nu) + (h(G_T, 2, \nu) - 1)/2) \bmod 4 \in \{0, \ldots, 3\}$. For $j = 0, 1$, we define $p_{r_j}^{b(G)} := p(G)_j$, $\beta(\alpha)_j = \alpha_{r_j}$ if $f(G_T, \nu) \in \{0, 2\}$, and $p_{r_{(j+1) \bmod 2}}^{b(G)} := p(G)_j$ and $\beta(\alpha)_{(j+1) \bmod 2} = \alpha_{r_j}$ otherwise. Furthermore, we set

$$\mu(\alpha) := (\lambda_0 h(G_T, 2, \nu))^{\beta(\alpha)_0} \lambda_1^{\beta(\alpha)_1}$$

with $\lambda_i := 1$, $i = 0, 1$, if $f(G_T, \nu) \in \{i, i+1\}$, and $\lambda_i := -1$ otherwise. For completeness, we define $p_j^{b(G)} := p(G)_j$, $j = 0, 1, 2$, and $\beta(\alpha) = \alpha$, if $\ell = 3$, and $\mu(\alpha) := 1$ if $\ell \in \{0, k\}$. Using all these preparations, the connectivity matrices are given by

$$\pi_{T, M(G, \beta(\alpha)), m_T(\alpha)} := \mu(\alpha) \tag{3}$$

where $m_T(\alpha)$ is a suitable local numbering, cf. [8, Chap. 4.1.5.1]. All entries which are not captured by (3) are set to 0. Note that we implicitly assume that Ψ_T maps the vertices of the unit cube onto the nodes G_ν^0 in the same order as given in Fig. 1. This is, e.g., done by $\Psi_T := \sum_{0 \le \nu < 2^k} \eta_{b((G_T^0)_\nu)} (G_T^0)_\nu$.

3 Constraints Coefficients and Multi-Level Hanging Nodes

To calculate the connectivity matrices for elements with irregular nodes, edges or faces, we introduce a further data structure $G_F = (G_F^0, \dots, G_F^{k-2}) \in \mathscr{G}_{k-1} \times \mathscr{G}_1^{\tilde{\sigma}_1} \dots \times \mathscr{G}_{k-2}^{\tilde{\sigma}_{k-2}}$, $\tilde{\sigma}_\ell := \sigma_\ell^{k-1}$, which represents an edge $F \subset \mathbb{R}^2$ or a face $F \subset \mathbb{R}^3$ of $\mathscr{T}$ for $k = 2, 3$, respectively. Based on G_F, we define $\tilde{b}(G)$ as the $k-1$-tuple with values in $\{0, 1, 2\}$ which is associated to the node or edge $G = (G_F^\ell)_\nu$ as depicted in Fig. 1 or to $G = G_F^0$ with $\tilde{b}(G) := (2, \dots, 2)$. Furthermore, let $\hat{F} \subset \mathbb{R}^k$ be the unique regular edge or face of $\mathscr{T}$ with $F \subset \hat{F}$. We assume that there exists numbers $v_r, w_r \in \mathbb{R}$, $0 \le r < k-1$, such that

$$\Phi(\Psi_{\hat{F}}^{-1}((G_{\hat{F}}^0)_\nu)) = \Psi_{\hat{F}}^{-1}((G_F^0)_\nu), \tag{4}$$

for all $0 \le \nu < 2^\ell$ with $\Psi_F := \sum_{0 \le \nu < 2^{k-1}} \eta_{\tilde{b}((G_F^0)_\nu)} (G_F^0)_\nu$ and $\Phi(x)_r = v_r x_r + w_r$, $v_r \in (0, 1]$. Note that Ψ_F maps $[-1, 1]^{k-1}$ onto F and that Φ is a compression. Furthermore, we assume that

$$g(G_F, \ell, \nu) = g(G_{\hat{F}}, \ell, \nu), \quad h(G_F, \ell, \nu) = h(G_{\hat{F}}, \ell, \nu) \tag{5}$$

for all $1 \le \ell < k-1$ and $0 \le \nu < \tilde{\sigma}_\ell$. The conditions (4) and (5) ensure that G_F and $G_{\hat{F}}$ have the same orientation and $\Psi_F^{-1}(\hat{F})$ is paraxial, cf. Fig. 3a.

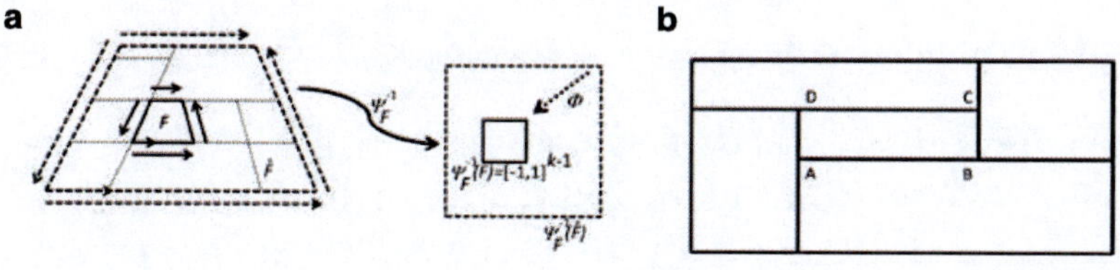

Fig. 3 (**a**) Orientation of F, $\hat{F}$ and their edges, (**b**) irregular 2D mesh for which the generation of $\mathscr{C}$ results in an infinite loop

Define $p(G_F^0)_r := p(G_{\hat{F}}^0)_r$, $0 \le r < k-1$, and $p((G_F^1)_\nu)_0 := \max\{p((G_{\hat{F}}^1)_\nu)_0, p(G_{\hat{F}}^0)_{\nu \bmod 2}\}$ if $(G_F^1)_\nu$ represents an irregular edge. Given assumption (4), the basic problem is to compute the so-called *constraints coefficients* $\kappa_{\hat{\gamma},\gamma}$, which are given by

$$\eta_{\hat{\gamma}} \circ \Phi = \sum_{G \in \mathrm{adj}(G_F),\, \gamma \in I^{\tilde{b}(G)}} \kappa_{\hat{\gamma},\gamma} \eta_\gamma$$

for $\hat{\gamma} \in I^{\tilde{b}(\hat{G})}$, $\mathrm{adj}(G_F) := \{G_F^0\} \cup \{(G_F^\ell)_\nu \mid 0 \le \ell < k-2,\ 0 \le \nu < \tilde{\sigma}_\ell\}$ and $\hat{G} \in \mathrm{adj}(G_{\hat{F}})$. Due to the tensor product structure and the properties of Φ, the coefficients are determined by $\kappa_{\hat{\gamma},\gamma} = \prod_{r=0}^{\ell-1} \bar{\kappa}_{\hat{\gamma}_r,\gamma_r}(v_r, w_r)$, where the coefficients $\bar{\kappa}_{ij}(v,w)$ solve the problem

$$\xi_i(vx + w) = \sum_{j=0}^{p} \bar{\kappa}_{ij}(v,w)\xi_j(x),\ x \in \mathbb{R} \tag{6}$$

for $v, w \in \mathbb{R}$, cf. [9]. A simple method to calculate the coefficients in (6) is to solve the linear equation $\xi_i(vx_s + w) = \sum_{j=0}^{p} \bar{\kappa}_{ij}(v,w)\xi_j(x_s)$ with suitable test points $x_s \in (-1,1)$, $s = 0, \ldots, p$, cf. [11,12]. In most finite element codes, the constraints coefficients are calculated for $v = 0.5$ and $w \in \{-0.5, 0.5\}$ describing symmetric subdivisions. In [9], an explicit and recursive formula for $\bar{\kappa}_{ij}(v,w)$ and arbitrary v and w is derived for the integrated Legendre and Gauss–Lobatto polynomials. This formula enables us to efficiently calculate the constraints coefficients for arbitrary subdivisions fulfilling condition (4).

To calculate the entries of the connectivity matrices, two preprocessing steps have to be accomplished. The first step is to iterate through all faces F of $\mathscr{T}$, all $G \in \mathrm{adj}(G_F)$ and all $\gamma \in I^{\tilde{b}(G)}$. If G is associated to a regular node, edge or face, we set $\mathscr{B}(G, \beta(\gamma)) := \{(G, \beta(\gamma), 1)\}$. Otherwise, we set

$$\mathscr{B}(G,\beta(\gamma)) := \left\{(\hat{G}, \beta(\hat{\gamma}), \kappa_{\hat{\gamma},\gamma}) \mid \hat{G} \in \mathrm{adj}(G_{\hat{F}}),\ G \ne \hat{G},\ \hat{\gamma} \in I^{\tilde{b}(\hat{G})},\ \kappa_{\hat{\gamma},\gamma} \ne 0\right\}.$$

The second step is to combine the constraints coefficients through

$$\mathscr{C}(G,\beta) := \left\{(\hat{G},\hat{\beta},\kappa) \mid (\hat{G},\hat{\beta},\kappa) \in \mathscr{B}(G,\beta),\ \hat{G} \text{ regular}\right\} \quad \biguplus_{\substack{(\hat{G},\hat{\beta},\kappa) \in \mathscr{B}(G,\beta),\\ \hat{G} \text{ irregular}}} \kappa \mathscr{C}(\hat{G},\beta)$$

with $\kappa\{(G_0,\beta_0,\kappa_0),(G_1,\beta_1,\kappa_1),\ldots\} := \{(G_0,\beta_0,\kappa\kappa_0),(G_1,\beta_1,\kappa\kappa_1),\ldots\}$ and

$$\begin{aligned}
\mathscr{C}_0 \uplus \mathscr{C}_1 :=& \{(G,\beta,\kappa) \mid (G,\beta,\kappa) \in \mathscr{C}_0,\ \nexists \kappa' :\ (G,\beta,\kappa') \in \mathscr{C}_1\} \\
&\cup \{(G,\beta,\kappa) \mid (G,\beta,\kappa) \in \mathscr{C}_1,\ \nexists \kappa' :\ (G,\beta,\kappa') \in \mathscr{C}_0\} \\
&\cup \{(G,\beta,\kappa+\kappa') \mid (G,\beta,\kappa) \in \mathscr{C}_0, \exists \kappa' :\ (G,\beta,\kappa') \in \mathscr{C}_1\}.
\end{aligned}$$

Using these sets, the entries of the connectivity matrix for a mesh element $T \in \mathscr{T}$ are computed using an extension of (3): For $G = (G_T^{\ell})_\nu \in \mathscr{G}^\ell$ and $\alpha \in I^{b(G)}$, we set

$$\pi_{T,M(\hat{G},\hat{\beta}),m_T(\alpha)} := \mu(\alpha)\kappa$$

for all $(\hat{G},\hat{\beta},\kappa) \in \mathscr{C}(G,\beta(\alpha))$.

Note that there are some (possibly artificial) cases for which the recursive definition of $\mathscr{C}$ results in an infinite loop over the hanging nodes. A 2D-example for such a situation is given in Fig. 3b for hanging nodes A, B, C and D. For implementation purposes, we need the data structures G_T and G_F to represent mesh elements and faces. Furthermore, we need a mapping which gives us the regular face $\hat{F}$ for an irregular face F with $F \subset \hat{F}$. Such a mapping is easily generated during the refinement process of a regular coarse mesh. The proposed approach may be extended to higher-dimensional mesh elements ($k \geq 4$), given an appropriate definition of $p_r^{b(G)}$, $\beta(\alpha)$ and $\mu(\alpha)$.

4 Numerical Results

In this section, we give some numerical results on the application of unsymmetric subdivisions and multi-level hanging nodes in 2D and 3D. The problem under consideration is Poisson's problem $-\Delta u = f$ on an L-shaped domain and on a cube. The right-hand side f and the boundary conditions are chosen so that u has a corner singularity in the re-entrant corner of the L-shaped domain and at one corner of the cube, respectively. We use serendipity shape functions and adapt the finite element mesh with symmetric (symm.) as well as unsymmetric (unsymm.) subdivisions at the corner and an increasing polynomial degree distribution. Figure 4d shows such an unsymmetric refinement for a cube with polynomial degrees marked by grey scales. Moreover, we use an automatic hp-adaptive scheme based on two a posteriori error estimators η_T and $\tilde{\eta}_T$ which estimate the local discretization error on $T \in \mathscr{T}$ for different degree distributions $p_T \leq \tilde{p}_T$. Using well-known a priori estimates, we estimate the local regularity ϱ_T of u with $\varrho_T \approx \frac{\log(\tilde{\eta}_T/\eta_T)}{\log(p_T/\tilde{p}_T)} + 1$. We increase the polynomial degree if $\varrho_T \geq \tilde{p}_T$, and refine T otherwise. For more details, see [10]. We use this strategy for symmetric (Fig. 4a,b – adaptive) as well as unsymmetric subdivisions (Fig. 4c – unsymm. 2). In Fig. 4c, only the polynomial degree is adapted whereas in Fig. 4a,e, both the polynomial degree and the mesh are adapted with multi-level hanging nodes. For all these hp-adaptive refinements, we obtain exponential convergence rates (Fig. 4f for the L-shaped domain and Fig. 4h for the cube). Additional refinements of all mesh elements with multi-level hanging nodes can be applied to ensure 1-irregularity of the mesh. However, in our numerical experiments with automatic hp-adaptive schemes on the L-shaped domain, the exponential convergence is lost, see Fig. 4g. This is due to the fact that only mesh elements at the corner are refined in the first steps of the adaptive refinement so that multi-level hanging nodes do not occur; but thereafter some mesh elements

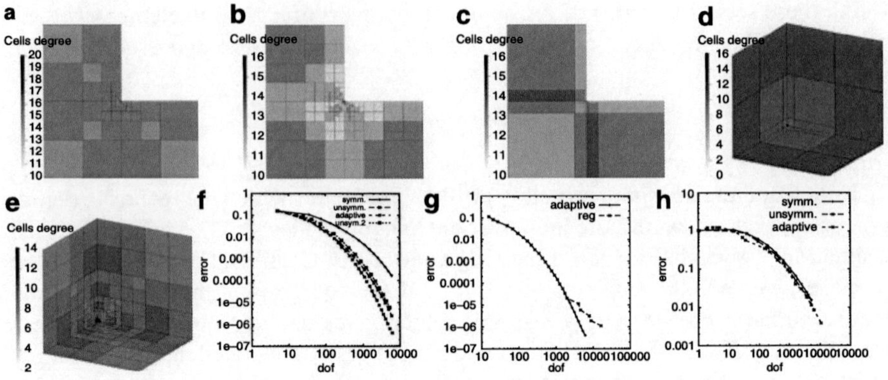

Fig. 4 hp-adaptive meshes in 2D and 3D and convergence rates

are refined which are not at the re-entrant corner so that multi-level hanging nodes are generated. The additional refinements for the elimination of multi-level hanging nodes leads to further multi-level hanging nodes on the next layer and so on. In the end, an almost global refinement is performed, which results in the decrease of the convergence rate, see Fig. 4b. This underlines the benefit of schemes which are able to handle multi-level hanging nodes.

References

1. Ainsworth, M., Senior, B.: Aspects of an adaptive hp-finite element method: Adaptive strategy, conforming approximation and efficient solvers. Comput. Methods Appl. Mech. Eng. **150**(1–4), 65–87 (1997)
2. Bangerth, W., Kayser-Herold, O.: Data structures and requirements for hp finite element software. ACM Transactions on Mathematical Software **36**(1), 4:1–4:31 (2009)
3. Demkowicz, L., Gerdes, K., Schwab, C., Bajer, A., Walsh, T.: HP90: A general and flexible Fortran 90 hp-FE code. Comput. Vis. Sci. **1**(3), 145–163 (1998)
4. Demkowicz, L., Oden, J.T., Rachowicz, W., Hardy, O.: Toward a universal h-p adaptive finite element strategy, Part 1: Constrained approximation and data structure. Comp. Methods Appl. Mech. Eng. **77**, 79–112 (1989)
5. Demkowicz, L.F.: Computing with hp-adaptive finite elements. Vol. 1: One- and two-dimensional elliptic and Maxwell problems. Chapman & Hall/CRC (2007)
6. Demkowicz, L.F., Kurtz, J., Pardo, D., Paszyński, M., Rachowicz, W., Zdunek, A.: Computing with hp-adaptive finite elements. Vol. II: Frontiers: Three-dimensional elliptic and Maxwell problems with applications. Chapman & Hall/CRC (2008)
7. Frauenfelder, P., Lage, C.: Concepts – an object-oriented software package for partial differential equations. M2AN, Math. Model Numer. Anal. **36**(5), 937–951 (2002)
8. Karniadakis, G.E., Sherwin, S.J.: Spectral/hp element methods for computational fluid dynamics. 2nd ed. Numerical Mathematics and Scientific Computation. Oxford University Press, Oxford (2005)
9. Schröder, A.: Constraints coefficients in hp-FEM. Kunisch, Karl (ed.) et al., Numerical mathematics and advanced applications. Proceedings of ENUMATH 2007, Springer, Berlin, 183–190 (2008)

10. Schröder, A.: Error control in h- and hp-adaptive FEM for Signorini's problem. J. Numer. Math. **17**(4), 299–318 (2009)
11. Šolín, Segeth, K., Delezel, I.: Higher-order finite element methods. Studies in Advanced Mathematics. CRC Press, Boca Raton (2004)
12. Šolín, P., Červený, J., Doležel, I.: Arbitrary-level hanging nodes and automatic adaptivity in the hp-FEM. Math. Comput. Simul. **77**(1), 117–132 (2008)

High Order Filter Methods for Wide Range of Compressible Flow Speeds

H.C. Yee and Björn Sjögreen

Abstract This paper extends the accuracy of the high order nonlinear filter finite difference method of Yee and Sjögreen [*Development of Low Dissipative High Order Filter Schemes for Multiscale Navier-Stokes/MHD Systems*, J. Comput. Phys., **225** (2007) 910–934] and Sjögreen and Yee [*Multiresolution Wavelet Based Adaptive Numerical Dissipation Control for Shock-Turbulence Computation*, RIACS Technical Report TR01.01, NASA Ames research center (Oct 2000); Also J. Scient. Comput., **20** (2004) 211–255] for compressible turbulence with strong shocks to a wider range of flow speeds without having to tune the key filter parameter. Such a filter method consists of two steps: a full time step using a spatially high-order non-dissipative base scheme, followed by a post-processing filter step. The post-processing filter step consists of the products of wavelet-based flow sensors and nonlinear numerical dissipations. For low speed turbulent flows and long time integration of smooth flows, the existing flow sensor relies on tuning the amount of shock-dissipation in order to obtain highly accurate turbulent numerical solutions. The improvement proposed here is to solve the conservative skew-symmetric form of the governing equations in conjunction with an added flow speed and shock strength indicator to minimize the tuning of the key filter parameter. Test cases illustrate the improved accuracy by the proposed ideas without tuning the key filter parameter of the nonlinear filter step.

1 Original High Order Filter Method

Consider the 3-D compressible Euler equations in Cartesian geometry,

$$U_t + \nabla \cdot \mathbf{F} = \mathbf{0};\; U = \begin{pmatrix} \rho \\ \mathbf{m} \\ e \end{pmatrix};\; \mathbf{F} = \begin{pmatrix} \rho \mathbf{u} \\ \rho \mathbf{u}\mathbf{u}^T + p \\ \mathbf{u}(e+p) \end{pmatrix}. \tag{1}$$

H.C. Yee (✉)
NASA Ames Research Center, Moffett Field, CA 94035, USA
e-mail: Helen.M.Yee@nasa.gov

J.S. Hesthaven and E.M. Rønquist (eds.), *Spectral and High Order Methods for Partial Differential Equations*, Lecture Notes in Computational Science and Engineering 76,
DOI 10.1007/978-3-642-15337-2_30,

Here the velocity vector $\mathbf{u} = (u, v, w)^T$, the momentum vector $\mathbf{m} = (\rho u, \rho v, \rho w)$, ρ is the density, and e is the total energy.

For turbulence with shocks, instead of solely relying on very high order high-resolution shock-capturing methods for accuracy, our filter schemes [8, 10–13] take advantage of the effectiveness of the nonlinear dissipation contained in good shock-capturing schemes as stabilizing mechanisms at locations where needed. The high order filter method consists of two steps, a full time step of spatially high order base scheme and a post-processing nonlinear filter step. The nonlinear filter consists of the *product* of an artificial compression method indicator or wavelet flow sensor and the nonlinear dissipative portion of a high-resolution shock-capturing scheme (e.g., any TVD, MUSCL, ENO, or WENO scheme). By design, the flow sensors, spatial base schemes and nonlinear dissipation models are standalone modules. Therefore, a whole class of low dissipative high order schemes can be derived with ease. Unlike standard shock-capturing and/or hybrid shock-capturing methods, the nonlinear filter method requires one Riemann solve per dimension, independent of time discretizations. The nonlinear filter method is more efficient than its shock-capturing method counterparts employing the same order of the respective methods. An advantage of the wavelet flow sensor of the filter method is that for problems with physical dissipation the more scales that are resolved, the less the filter is utilized, thereby gaining accuracy and computation time. In the limit when all scales are resolved, we are left with a "pure" centered high order spatial scheme without added numerical dissipation.

For viscous gas dynamics the same order of spatial centered base scheme for the convection terms and the viscous terms are employed. For all of the time-accurate test cases, the classical fourth-order Runge–Kutta time discretization is employed. In a Cartesian grid, denote the grid indices for the three spatial directions as (j, k, l). The spatial base scheme to approximate the x inviscid flux derivatives $F(U)_x$ (with the grid indices k and l for the y- and z-directions suppressed) is written as

$$\frac{\partial F}{\partial x} \approx D_{08} F_j, \tag{2}$$

where D_{08} is the standard eighth-order accurate centered difference operator.

After the completion of a full Runge–Kutta time step of the base scheme step, the second step is to adaptively apply a nonlinear filter. The nonlinear filter can be obtained e.g., in the x-direction by taking the full seventh-order WENO scheme (WENO7) for the inviscid flux derivative in the x-direction and subtracting $D_{08} F_j$. The final update of the solution is (with the numerical fluxes in the y- and z-directions suppressed)

$$U^{n+1}_{j,k,l} = U^{*}_{j,k,l} - \frac{\Delta t}{\Delta x}[H_{j+1/2} - H_{j-1/2}]. \tag{3}$$

The nonlinear filter numerical fluxes usually involve the use of field-by-field approximate Riemann solvers. If Roe type of approximate Riemann solver [7] is employed, for example, the x-filter numerical flux vector $H_{j+1/2}$ is

$$H_{j+1/2} = R_{j+1/2}\overline{H}_{j+1/2},$$

where $R_{j+1/2}$ is the matrix of right eigenvectors of the Jacobian of the inviscid flux vector in terms of the Roe's average states evaluated at the U^* solution from the base scheme step. Denote the elements of the vector $\overline{H}_{j+1/2}$ by $\overline{h}^l_{j+1/2}, l = 1, 2, \ldots, 5$. The nonlinear portion of the filter $\overline{h}^l_{j+1/2}$ has the form

$$\overline{h}^l_{j+1/2} = \frac{\kappa}{2}\omega^l_{j+1/2}\phi^l_{j+1/2}. \tag{4}$$

Here $\omega^l_{j+1/2}$ is the wavelet flow sensor to activate the nonlinear numerical dissipation $\phi^l_{j+1/2}$ and the original formulation for κ is a positive parameter that is less than or equal to one. Some tuning of the parameter κ is needed for different flow types. It is the purposes of this work to develop a new κ to be a local variable depending on the local Mach number for low speed flows and depending on local shock strength for high speed flows.

The dissipative portion of the nonlinear filter $\phi^l_{j+1/2} = g^l_{j+1/2} - b^l_{j+1/2}$ is the dissipative portion of WENO7 for the local lth-characteristic wave. Here $g^l_{j+1/2}$ and $b^l_{j+1/2}$ are numerical fluxes of WENO7 and the eighth-order central scheme for the lth characteristic, respectively. Hereafter, we denote this filter scheme as WENO7fi. For all of the computations, a three-level second-order Harten multiresolution wavelet decomposition of the computed density and pressure is used as the flow sensor [8].

A summary of the three basic steps to obtain $\omega^l_{j+1/2}$ can be found in Sjögreen and Yee [8] and Yee and Sjögreen [11]. For example, the flow sensor $\omega^l_{j+1/2}$ to turn on the shock-capturing dissipation using the cut off procedure is a vector (if applied dimension-by-dimension) consisting of "**1's**" and "**0's**."

2 Improved High Order Filter Method

The improvements proposed here for the original high order filter method are to solve the conservative skew-symmetric form of the governing equations [2] in conjunction with a new flow speed indicator to minimize the tuning of the key filter parameter κ in (4). It works well for Mach speeds below 1.5. Before presenting representative test cases, a relevant summary on the recent improvements is described.

Studies found that employing the entropy splitting [13] of the inviscid flux derivative can stabilize the central base scheme for smooth flows. Indirectly, less numerical dissipation is needed when the split form is used. Unfortunately, entropy splitting is not suitable for problems with moderate and strong shocks as the split form is not conservative. The conservative skew-symmetric splitting [2, 9] of the inviscid flux derivative can also stabilize the central base scheme. In addition, it is suitable for smooth flows and for problems containing strong shocks. In this study, in order to stabilize (minimize the use of added numerical dissipation for accuracy) the base scheme step for a wider range of flow conditions, the conservative skew-symmetric splitting is utilized. See [9] for a comparison of different skew-symmetric splittings of the inviscid flux derivative.

Previous numerical experiments on a wide range of flow conditions indicated that the filter scheme improves the overall accuracy of the computation compared with standard shock-capturing schemes of the same order. Studies found that the improved accuracy is more pronounced if the parameter κ in (4) is tuned according to the flow type. For hypersonic flows with strong shocks, κ is set to 1. For high subsonic and supersonic flows with strong shocks, κ is in the range of $(0.3, 0.9)$. For low speed turbulent flows without shocks or long time integration of smooth flows, κ can be one to two orders of magnitude smaller than 1. In other words, κ should be flow location and shock strength dependent. The proposed new flow sensor to be discussed later will take these two factors into consideration. Here a simple minded modification of κ is illustrated with representative numerical examples.

Inspired by Li and Gu's method to overcome the shortcomings of "low speed Roe scheme" [6], we modified their flow speed indicator formula to obtain a modified κ denoted by $\overline{\kappa}$ for (4) to minimize the tuning of the original κ for low Mach number flows. $\overline{\kappa}$ has the form:

$$\overline{\kappa} = f_1(M)\kappa, \tag{5}$$

with

$$f_1(M) = \min\left(\frac{M^2}{2}\frac{\sqrt{4 + (1 - M^2)^2}}{1 + M^2}, 1\right). \tag{6}$$

Here M is the maximum Mach number of the entire computational domain at the initial stage of the time evolution (i.e., the free stream Mach number M_∞). $f_1(M)$ has the same form as [6] except there is an extra factor "$\frac{M}{2}$" added to the first argument on the right-hand-side of the original form $f(M)$ in (18) of [6]. The added factor provides a similar value of the tuning κ observed from numerical experimentation. With the flow speed indicator $f_1(M)$ in front of κ, the same κ used for the supersonic shock problem can be used without any tuning for the very low speed turbulent flow cases. This sensor is evaluated only once before the first time step. Later,

$$\overline{f_1(M)} = \max(\min(\frac{M^2}{2}\frac{\sqrt{4 + (1 - M^2)^2}}{1 + M^2}, 1), \epsilon),$$

where ϵ is a small threshold value to avoid completely switching off the dissipation. A function which retains the majority of $f_1(M)$ but includes larger Mach number for not very strong shocks is

$$f_2(M) = (Q(M, 2) + Q(M, 3.5))/2$$

or

$$\overline{f_2(M)} = \max((Q(M, 2) + Q(M, 3.5))/2, \epsilon),$$

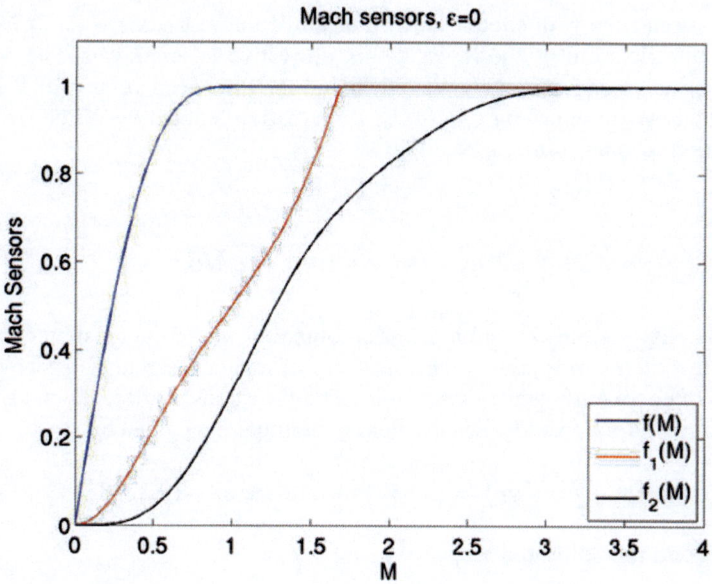

Fig. 1 Mach number sensors. $f(M)$ (*blue*) function by Li and Gu, $f_1(M)$ (*red*) modified $f(M)$, and $f_2(M)$ (*black*) (includes low supersonic Mach numbers)

where

$$Q(M,a) = \begin{cases} P(M/a) & M < a \\ 1 & \text{otherwise} \end{cases}.$$

The polynomial

$$P(x) = x^4(35 - 84x + 70x^2 - 20x^3)$$

is monotonically increasing from $P(0) = 0$ to $P(1) = 1$ and has the property that $P'(x)$ has three continuous derivatives at $x = 0$ and at $x = 1$. Numerical experiments indicate that setting $\kappa = 0.7$ works well for a wide range of flow speeds below hypersonic. The next section illustrates several representative test cases. It is noted that if the original $f(M)$ were used instead of $f_1(M)$ or $f_2(M)$ in (5), the amount of nonlinear filter dissipation can be too large for very low speed turbulent flows (for the same fixed κ). See Fig. 1 for details.

3 Numerical Results

Three different flow types are considered for the numerical experiments. A 1-D supersonic shock/turbulence interaction problem, a 3-D low speed turbulence problem without shocks (Taylor–Green vortex [1]), and a high speed compressible isotropic turbulence with shocklets [5]. For all of the test cases, $\kappa = 0.7$ and the skew-symmetric form of the inviscid flux derivative is employed. The accuracy comparison is among WENO7, WENO7fi and the improved version of WENO7fi discussed above by replacing κ in (4) by $\overline{\kappa}$ (hereafter denoted by WENO7fiM). All computations use uniform Cartesian grids.

3.1 1-D Shock/Turbulence Interaction Problem

This 1-D compressible inviscid ideal gas problem is one of the most computed test cases in the literature to assess the capability of a shock-capturing scheme in the presence of shock/turbulence interactions. The flow consists of a shock at Mach 3 propagating into a sinusoidal density field with initial data given by

$$(\rho_L,\ u_L,\ p_L) = (3.857143,\ 2.629369,\ 10.33333) \tag{7}$$

to the left of a shock located at $x = -4$, and

$$(\rho_R,\ u_R,\ p_R) = (1 + 0.2\sin(5x),\ 0,\ 1) \tag{8}$$

to the right of the shock, where ρ is the density, u is the velocity and p is the pressure. The computational domain is $[-5, 5]$ and the computation stops at time equal to 1.8.

Figure 2 shows the comparison among WENO7 and WENO7fiM using a very coarse uniform grid of 201 with the reference solution. The reference solution is obtained with WENO5 using a 16,000 grid. The two schemes give the similar accuracy near shock waves but with a large difference in accuracy in the fluctuation region where WENO7fiM is more accurate than WENO7. The result by WENO7fi is the same as WENO7fiM since $\overline{f(M)}$ is nearly 1. Note that in order for WENO5 to

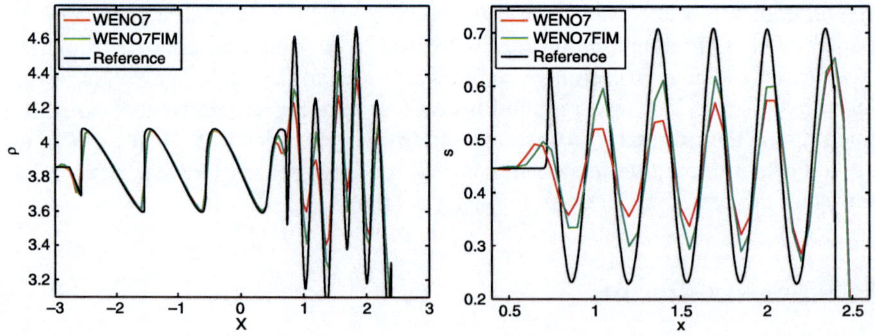

Fig. 2 1-D shock-turbulence interaction: Enlarge region of density profiles (*left*) and entropy profiles (*right*) by WENO7 (*red*) and WENO7fiM (*green*) using a 201 grid. The solid black line is the reference solution

obtain a similar accuracy as WENO7fi, nearly three times the number of grid point is needed.

3.2 *Taylor–Green Vortex*

The second test case solves the 3-D Euler equations of gas dynamics with $\gamma = 5/3$ and with initial data

$$\rho(0,x,y,z) = 1 \tag{9}$$
$$u(0,x,y,z) = \sin(x)\cos(y)\cos(z) \tag{10}$$
$$v(0,x,y,z) = -\cos(x)\sin(y)\cos(z) \tag{11}$$
$$w(0,x,y,z) = 0 \tag{12}$$
$$p(0,x,y,z) = 100 + \frac{1}{16}((\cos(2z)+2)(\cos(2x)+\cos(2y))-2) \tag{13}$$

on the computational domain $[0,2\pi]\times[0,2\pi]\times[0,2\pi]$. Here u, v, w are the three velocity components. The mean pressure is sufficiently high to make the problem essentially incompressible. This is known as a Taylor–Green vortex [1]. The computation stops at a total time equal to 10. The boundary conditions are periodic. The initial data are smooth, but the scales in the solution become smaller and smaller with time. The enstrophy (the square of the L^2 norm of the curl of the velocity) is often used as a measure of the content of small scales in the solution.

Figure 3 shows the temporal evolution of the mean kinetic energy, $<\rho u_i, u_i>/2$, and enstrophy, $<\omega_i, \omega_i>/2$, where $\omega = \nabla\cdot \times u$ is the vorticity, normalized by their initial values. The three schemes give very different accuracy using the same 64^3 grid. WENO7 is the least accurate and WENO7fiM is the most accurate. For the

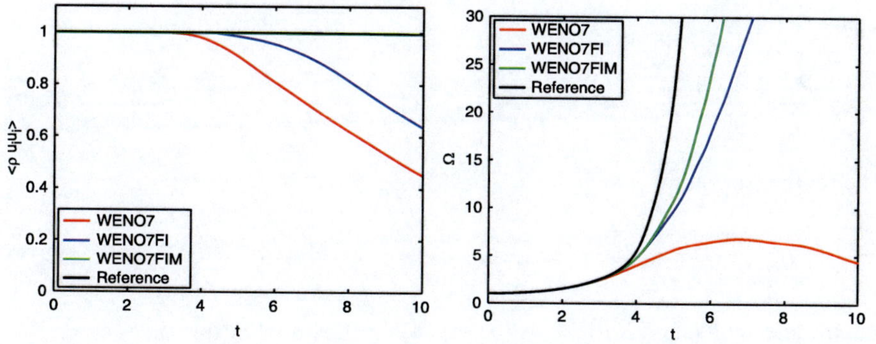

Fig. 3 Taylor–Green vortex: Kinetic energy (*left*) and enstropy (*right*) by WENO7 (*red*), WENO7fi (*blue*) and WENO7fiM (*green*) using a 64^3 grid. The *solid black line* is the reference solution by WENO7fi using a 256^3 grid

computed kinetic energy, the solution by WENO7fiM (green line) using a 64^3 grid is in-distinguishable from the reference solution (black line). The results indicate that with $\overline{\kappa}$, the same κ used for the first test case with the Mach 3 shock can be used for this nearly incompressible test case with high accuracy.

3.3 *Compressible Isotropic Turbulence with Shocklets*

The third test case is a 3-D viscous decaying isotropic turbulence with eddy shocklets Given a sufficiently high turbulent Mach number, $M_t = 0.6$, and a high Taylor scale Reynolds number, $Re_\lambda = 100$, eddy shocklets develop spontaneously from the turbulent motions. This problem tests the ability of the methods to handle randomly distributed shocklets, as well as the accuracy for broadband motions in the presence of shocks.

The gas constant is $\gamma = 1.4$, and the viscosity is assumed to follow a power-law

$$\frac{\mu}{\mu_{ref}} = \left(\frac{T}{T_{ref}}\right)^{3/4}. \tag{14}$$

Here $\mu_{ref} = 0.005$ and $T_{ref} = 1$. The heat conduction coefficient is

$$\kappa(T) = \frac{\gamma R}{Pr(\gamma - 1)} \mu_s(T) \tag{15}$$

where the Prandtl number, Pr, is 0.7. The important parameters are M_t and Re_λ, defined as

$$M_t = \frac{\sqrt{3} u_{rms}}{< c >}, \quad Re_\lambda = \frac{< \rho > u_{rms} \lambda}{< \mu >}, \quad u_{rms} = \sqrt{\frac{< u_i u_i >}{3}}, \tag{16}$$

where

$$\lambda = \frac{\lambda_x + \lambda_y + \lambda_z}{3}, \quad \lambda_x^2 = \frac{< u^2 >}{< u_x^2 >}, \quad \lambda_y^2 = \frac{< v^2 >}{< v_y^2 >}, \quad \lambda_z^2 = \frac{< w^2 >}{< w_z^2 >}. \tag{17}$$

The root mean square velocity is

$$u_{rms}^2 = \frac{1}{3}(< u^2 + v^2 + w^2 > -(< u >^2 + < v >^2 + < w >^2)), \tag{18}$$

and the speed of sound is $c^2 = \gamma p/\rho$. See [4] for the initial disturbance setup.

Figure 4 shows root mean square (RMS) of density, pressure and temperature by WENO7 and WENO7fiM using a 64^3 grid compared with the reference solution by WENO7fi using a 256^3 grid. Again WENO7fiM is more accurate than WENO7.

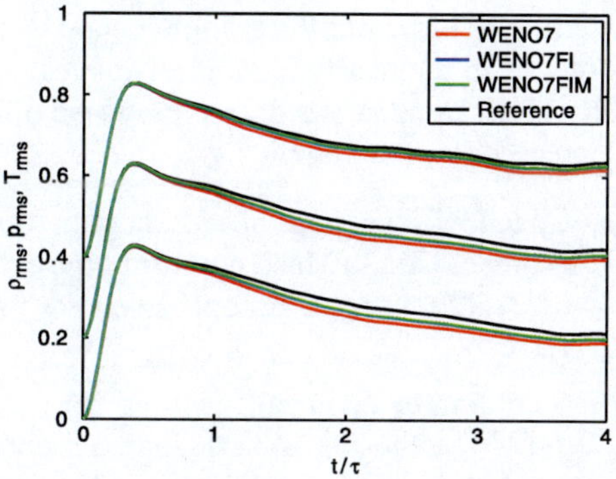

Fig. 4 Isotropic turbulence with shocklets: Comparison of RMS quantities by WENO7 (*red*) and WENO7fiM (*green*) using a 64^3 grid. The *solid black line* is the reference solution by WENO7fi using a 256^3 grid

However, the accuracy improvement by WENO7fiM is not as pronounced as in the first two test cases. The result by WENO7fi is the same as WENO7fiM since $f_1(M)$ is 1. The simple minded improvement proposed here to minimize the use of added numerical dissipation has been demonstrated for three less complicated flow types. Solving the conservative skew-symmetric form of the governing equation in conjunction with an added flow speed indicator has been shown to improve accuracy using the same key filter parameter κ.

4 New Flow Sensor for a Wide Spectrum of Flow Speed and Shock Strength

As evident from the numerical examples, a new κ in front of the wavelet flow sensor (4) is desirable for providing the location, and correct amount of numerical dissipation to be employed by high order numerical schemes for as wide a spectrum of flow speed as possible with the least number (and effort) of tuning parameters. Thus, the new κ has to be a local variable depending on the local Mach number for low speed flows and depending on local shock strength for high speed flows. The level of increasing complexity for the new κ can be investigated by the following stages. The modified $\overline{\kappa}$ proposed earlier is a good choice for smooth and/or nearly incompressible flows even though $\overline{\kappa}$ is based merely on the freestream Mach number of the flow. Thus, for up to low supersonic speed, for efficiency, the first level of improvement is to make a time-dependent global κ based on the maximum Mach number of the entire flow field at each time evolution. The second level of improvement is to make a time-dependent local κ based on $f_1(M)$ or $f_2(M)$. For each non-zero wavelet indicator $\omega^l_{j+1/2}$, a local κ is determined to provide an appropriate amount of numerical dissipation (between (0, 1)) to be filtered by the shock-capturing dissipation. For strong shocks, the shock strength should come into play. One measure of the shock strength can be based on the numerical Schlieren formula [3] for the chosen variables that exhibit the strongest shock strength. In the vicinity of turbulent fluctuation locations, the local kappa will be kept to the same order as in the nearly incompressible case except in the vicinity of high shear and shocklets. In other words, we proposed different new κ according to the following increased level of complexity:

- Up to low supersonic speeds, at each time step, a global κ is computed according to the maximum Mach number of the entire flow field and the value is determined by $f_1(M)$ or $f_2(M)$ proposed earlier for non-zero $\omega^l_{j+1/2}$.
- Up to low supersonic speeds, at each time step, a local $\kappa^l_{j+1/2}$ is computed according to the local Mach number and the value is determined by $f_1(M)$ or $f_2(M)$ (at the $j+1/2$ grid index) proposed earlier for non-zero $\omega^l_{j+1/2}$. In other word, the filter numerical flux indicated in (4) is replaced by:

$$\overline{h}^l_{j+1/2} = \frac{1}{2}[\kappa^l_{j+1/2}\omega^l_{j+1/2}\phi^l_{j+1/2}]. \tag{19}$$

- Same as above except now the final value of $\overline{\kappa^l_{j+1/2}}$ is determined by the previous local kappa if the local Mach number is below 0.4. Above local Mach number 0.4, at discontinuities detected by the wavelet sensor, the local kappa is determined by the shock strength (normalized between (0, 1)) based on the Schlieren formula near discontinuities. At turbulent fluctuation locations, determined by the Ducros et al. sensor, the local kappa is kept to the same order as in the nearly incompressible case except in the vicinity of high shear and shocklet locations where a slightly larger kappa would be used.

Due to a space limitation, the results by the local $\kappa^l_{j+1/2}$ that take the local flow speed and shock strength into consideration will be reported in an expanded version of this paper [14]. Preliminary study with more complex shock turbulence problems and the applicability of even wider flow types indicates the necessity of the local $\kappa^l_{j+1/2}$. The financial support from the NASA Fundamental Aeronautics (Hypersonic) program for the first author is gratefully acknowledged. The authors thank the DOE/SciDAC turbulence Science Team (S. Lele, PI) for providing insights into the last two test cases. Work by the second author was performed under the auspices of the U.S. Department of Energy by Lawrence Livermore National Laboratory under Contract DE-AC52-07NA27344.

References

1. M.E. Brachet, D.I. Meiron, S.A. Orszag, B.G. Nickel, R.H. Morf, *Small-Scale Structure of the Taylor-Green Vortex*, J. Fluid Mech., **130**, (1983), 411–452
2. F. Ducros, F. Laporte, T. Souleres, V. Guinot, P. Moinat, B. Caruelle, *High-order Fluxes for Conservative Skew-Symmetric-like Schemes in Structured Meshes: Application to Compressible Flows*, J. Comput. Phys., **16**(1) (2000) 114–139
3. A. Hadjadj, A. Kudryavtsev, *Computation and Flow Visualization in High Speed Aerodynamics*, J. Turbul., **6**(16) (2005) 33–81

4. E. Johnsen, J. Larson, A.V. Bhagatwala, W.H. Cabot, P. Moin, B.J. Olson, P.S. Rawat, S.K. Shankar, B. Sjögreen, H.C. Yee, X. Zhong, S.K. Lele, *Assessment of High-Resolution Methods for Numerical Simulations of Compressible Turbulence with Shock Waves*, J. Comput. Phys., **229** (2010) 1213–1237
5. S. Lee, S.K. Lele, P. Moin, *Eddy Shocklets in Decaying Compressible Turbulence*, Phys. Fluids **3** (1991) 657–664
6. X.-S Li, C.-W. Gu, *An All-Speed Roe-Type Scheme and its Asymptotic Analysis of Low Mach Number Behaviour*, J. Comput. Phys., **227** (2008) 5144–5159
7. P.L. Roe, *Approximate Riemann solvers, parameter vectors, and difference schemes*, J. Comput. Phys., **43** (1981) 357–372
8. B. Sjögreen, H.C. Yee, *Multiresolution Wavelet Based Adaptive Numerical Dissipation Control for Shock-Turbulence Computation*, RIACS Technical Report TR01.01, NASA Ames research center (Oct 2000); Also J. Scient. Comput., **20** (2004) 211–255
9. B. Sjögreen, H.C. Yee, *On Skew-Symmetric Splitting of the Euler Equations*, Proceedings of the EUNUMATH-09 Conference, June 29–July 2, 2009 (to appear)
10. H.C. Yee, B. Sjögreen, *Efficient Low Dissipative High Order Scheme for Multiscale MHD Flows, II: Minimization of Div(B) Numerical Error*, RIACS Technical Report TR03.10, July, 2003, NASA Ames Research Center; Also J. Scient. Comput., (2005), doi: 10.1007/s10915-005-9004-5
11. H.C. Yee, B. Sjögreen, *Development of Low Dissipative High Order Filter Schemes for Multiscale Navier-Stokes/MHD Systems*, J. Comput. Phys., **225** (2007) 910–934
12. H.C. Yee, N.D. Sandham, M.J. Djomehri, *Low Dissipative High Order Shock-Capturing Methods Using Characteristic-Based Filters*, J. Comput. Phys., **150** (1999) 199–238
13. H.C. Yee, M. Vinokur, M.J. Djomehri, *Entropy Splitting and Numerical Dissipation*, J. Comput. Phys., **162** (2000) 33–81
14. H.C. Yee, B. Sjögreen, A. Hadjadj, *Flow Sensors in Controlling the Amount of Numerical Dissipations for a Wide Spectrum of Flow Speed and Shock Strength* (in preparation)

hp-Adaptive CEM in Practical Applications

Adam Zdunek and Waldemar Rachowicz

Abstract A reduced order *hp*-FE sub-domain based scattering matrix methodology suitable for calculating the Radar Cross-Section (RCS) for electrically huge jet engine air intakes is presented. The efficiency gain in degrees of freedom obtained by using *hp*-version FEM instead of classical low order *h*-version FEM is shown to be roughly one order of magnitude. The model reduction achieved by changing from inter-facial FE-d.o.f:s to guided wave participation factors implies another gain in degrees of freedom which becomes very substantial for air intakes with electrically large homogeneous sections. It is shown that the modal reduction can be made without significant loss of accuracy in the cavity-RCS by comparing results obtained using the scattering matrix approach with coupled full wave *hp*-version finite element-infinite element (FE+IE) and finite element-boundary element (FE+BE) models.

1 Introduction

The scattering characterisation of aircraft configurations remain one of the most challenging problems in computational Electromagnetics (CEM). Reliable target identification is a very important task in both military and civilian applications. It is based on accurate Radar Cross-Section (RCS) calculations. Measurements have revealed that the jet engine inlets are one of the major contributors to the overall RCS signature of an airplane. Jet engine inlets are often electrically huge semi-open channels ($L/\lambda \sim 200$, $D/\lambda \sim 20$). In the military version they are often curved and

A. Zdunek (✉)
Swedish Defence Research Agency, FOI, SE-164 90 Stockholm, Sweden
e-mail: zka@foi.se

W. Rachowicz
Cracow University of Technology, PK, Pl-31 155 Cracow, Poland
e-mail: waldek@ices.utexas.edu

J.S. Hesthaven and E.M. Rønquist (eds.), *Spectral and High Order Methods for Partial Differential Equations*, Lecture Notes in Computational Science and Engineering 76,
DOI 10.1007/978-3-642-15337-2_31, © Springer-Verlag Berlin Heidelberg 2011

partially coated with radar absorbing materials (RAM) to hide jet engine fan disks at the far end.

Highly efficient parallelised computational sub-domain techniques have to be used to predict the RCS of real jet engine air inlets [1]. Different models, discrete and for simple cross-sections possibly analytical models, may be used and mixed along the channel to obtain the requested efficiency, accuracy and reliability of the RCS prediction. We use the *hp*-version of the Finite Element Method (FEM) for modelling sections with material inhomogeneities and for modelling the scattering on the terminating section consisting of disks with irregular jet engine fan blades. That is, we investigate if we can benefit from the fast (asymptotically exponential) convergence that characterises the adaptive *hp*-FEM. For that purpose we use the *hp3d*-code equipped with automatic adaptivity developed by Demkowicz and associates [4, 5]. The cavity RCS is predicted using the novel and efficient half-space modelling technique described in [6]. The modelling of the exterior problem may include either *H(curl)*-conforming *hp*-version infinite elements (IE) or *hp*-version boundary elements (BEM) (under development), respectively. We also provide cavity RCS predictions based on uncoupled interior problem analysis and the Kirchhoff aperture integration procedure. The diffraction at the aperture rim is neglected in this case. The efficiency gain and modelling error introduced using this simplified approach is discussed. Calderon operators of the boundary admittance type are used to represent and couple sub-domains. These may be built using different discretizations. In order to assemble sub-domain contributions from different discretizations we project the contributions of coupling neighbours onto an auxiliary complete basis at the common interface. A choice we investigate for a homogeneous channel cross-section is the waveguide modal basis. The use of truncated projections of discrete *hp*-version FEM type Calderon sub-domain operators is investigated. It will be shown that efficiency can be gained in this way without significant loss of accuracy and reliability.

We illustrate our methodology using a generic but non-trivial cylindrical channel test cases described in [2] and shown in Fig. 1. In this conference paper we focus on the simpler PEC channel denoted *TCA2*. The RCS for the RAM coated version *TCA4* are left aside here. The cone at the bottom represents a simplified jet engine hub. Its sharp apex as well as the RAM interface causes singularities in the electromagnetic solution field. The efficiency and accuracy of uniform p- and h-extensions are studied in terms of the quantity of interest, RCS.

2 The Scattering Matrix Based Approach [1–3]

Consider a homogeneous channel with termination, shown in Fig. 1. In reality the channel is most often so large so it has to be subdivided into sub-domains $\Omega = \cup\Omega_d$, $d = 1, 2, \ldots, N$ each of which can be regarded as a waveguide. The ports of these waveguides are denoted Σ_p, $p = 1, 2, \ldots, N + 1$. The electromagnetic fields on these ports can be expanded in terms of a complete set of orthogonal guided waves.

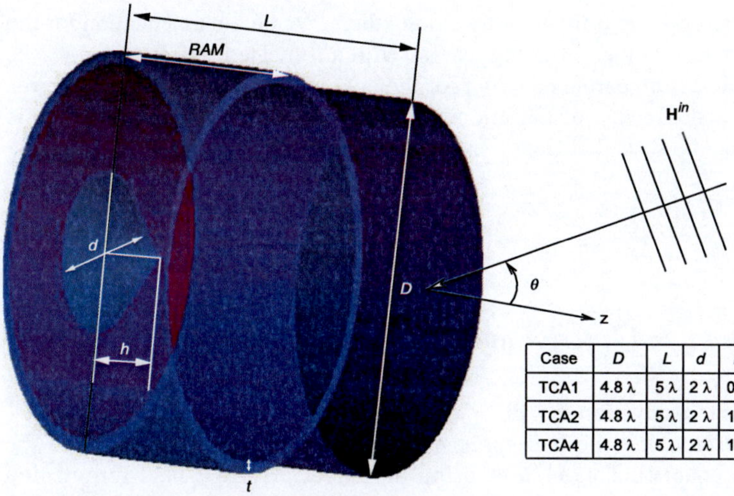

Case	D	L	d	h	RAM	t
TCA1	4.8 λ	5 λ	2 λ	0 λ	–	0 λ
TCA2	4.8 λ	5 λ	2 λ	1 λ	–	0 λ
TCA4	4.8 λ	5 λ	2 λ	1 λ	2.8 λ	0.15 λ

Fig. 1 The TCA1, 2 and 4 test cases. Straight circular PEC cylinder. TCA1 with flat termination. TCA2 and TCA4 have a conical termination

It is assumed that the Neumann-to-Dirichlet operator NtD for the apertures of sections of the channel is known. Further, an incident field $\mathbf{H}^{in}$ is given. It satisfies the Maxwell equations, but not necessarily the boundary conditions on the channel ports and bottom. We look for the resulting perturbation $\mathbf{H}^{sc}$ that satisfies Maxwell's equations and is such that the total field $\mathbf{H}^{in} + \mathbf{H}^{sc}$ also satisfies the boundary conditions. This is achieved by fulfilling the following requirement:

$$D(\mathbf{H}^{in} + \mathbf{H}^{sc}) = NtD(N(\mathbf{H}^{in} + \mathbf{H}^{sc})) \tag{1}$$

where $D(\mathbf{H})$ denotes the Dirichlet-trace of the total magnetic field $\mathbf{H}$, and $N(\mathbf{H})$ denotes the Neumann-trace of the same field on the aperture, respectively. In passing, it is noted that (1) is the definition of the NtD-operator for the aperture. On account of the linearity of the operators D, N and NtD, we have after putting known contributions on the right hand-side,

$$D(\mathbf{H}^{sc}) - NtD(N(\mathbf{H}^{sc})) = -\left(D(\mathbf{H}^{in}) - NtD(N(\mathbf{H}^{in}))\right). \tag{2}$$

In (2) $\mathbf{H}^{in}$ is given. It is expanded in guided waves $(\mathbf{e}_{t,n}, k_{z,n})_{n=1}^{\infty}$, where $\mathbf{e}_{t,n}$ are eigen-functions (TE- and TM-modes) and $k_{z,n}$ are the associated eigen-values, i.e. the propagation constants. Using the coordinate system depicted in Fig. 1 and suppressing sequence index n, in-coming travelling waves propagate in the negative z-direction as $\exp[-ik_z z]$. The participation factors are computed as usual by $L_2(\Sigma)$-projection. We assume a similar expansion of $\mathbf{H}^{sc}$ in terms of out-going travelling waves. These propagate in the positive z-direction as $\exp[+ik_z z]$. By $L_2(\Sigma)$-projecting the left and right hand-sides of (2) on suitable eigen-functions, and using the appropriate orthogonality among them, we obtain an equation for the participation factors of the out-going waves. With a truncated set of guided waves (2) provides a linear algebraic equation for the coefficients of the out-going waves in terms of the coefficients of the in-coming waves. This linear transformation is recognised to be the so-called Generalised Scattering Matrix (GSM) [3].

3 Results

3.1 Fully Coupled Interior and Exterior Model

We first present results for the PEC test channel TCA2 as described in Fig. 1 obtained by discretizing the half-space exterior to the channel aperture and the interior of the channel. This is done using the direct FE+IE **E**-field formulation described in [6]. That is, we use a half-space so-called PEC ground plane model with a Physical Optics (PO) correction for the aperture. The monostatic RCS for the TCA2 channel is computed at 300 MHz vs. the azimuth angle θ, for vertical and horizontal polarisations, respectively. The over-kill solution used as a reference for the full formulation is shown Fig. 2. It is obtained with a relatively coarse mesh of brick elements with a uniform polynomial ansatz of order $p = 7$. These brick elements are iso-parametric so the cylindrical geometry is described very accurately.

In Fig. 3 we present the error in RCS as a function of θ for uniform $p = 5$ and $p = 6$ extensions, relative to the over-kill uniform $p = 7$ solution shown in Fig. 2. Note that this fully coupled model includes aperture rim diffraction (as opposed to the scattering matrix model discussed in Sect. 2 which does not). We conclude that uniform $p = 6$ is acceptable for engineering purposes. That is for design or RCS characterisation purposes, where the accuracy of say 0.5 dBsm is regarded sufficient for the majority of incidence angles. Moreover, in Fig. 3 it can be seen that the corresponding uniform *p=6* mesh is sufficient for engineering purposes.

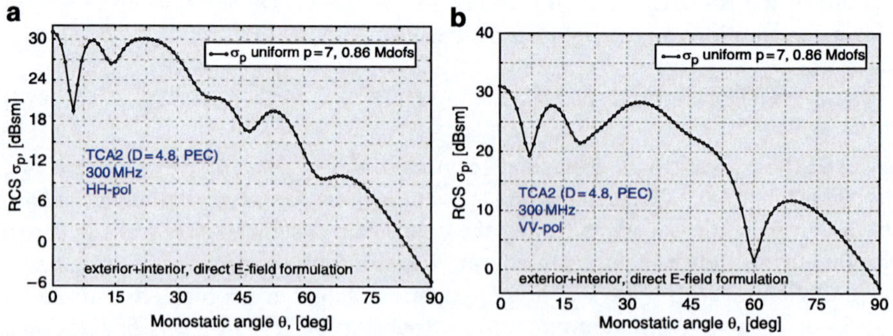

Fig. 2 Monostatic RCS vs. azimuth angle θ at 300 MHz. Reference solution, fully coupled interior and exterior model. **(a)** Horizontal polarisation. **(b)** Vertical polarisation

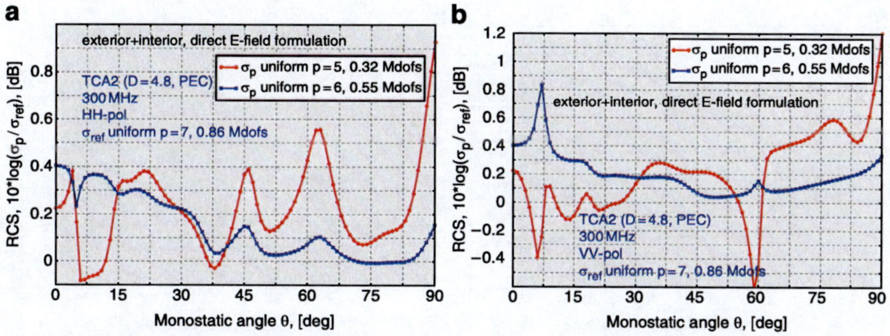

Fig. 3 Error in monostatic RCS vs. azimuth angle θ at 300 MHz. Uniform p-extensions. Fully coupled interior and exterior model. (**a**) Horizontal polarisation. (**b**) Vertical polarisation

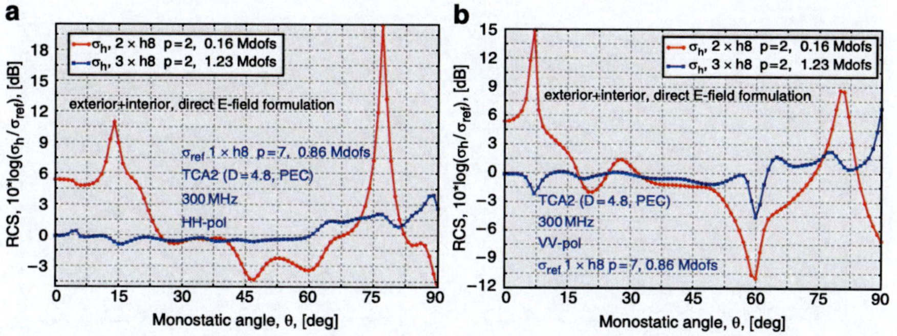

Fig. 4 Error in monostatic RCS vs. azimuth angle θ at 300 MHz. Uniform h-refinements. Fully coupled interior and exterior model. (**a**) Horizontal polarisation. (**b**) Vertical polarisation

In Fig. 4 we present the error in RCS as a function of θ using a mesh which is uniformly h-refined. We use the over-kill solution with uniform $p = 7$ as a reference. A refinement of type $3 \times h8$ means that each brick is split in 8 recursively three times. It is noted that RCS obtained with uniform $p = 7$ and uniform $3 \times h8$ $p = 2$ meshes, respectively, differ markedly at grazing incidence.

In Fig. 5 RCS results for channels with flat termination and conical termination, are compared, cf Fig. 1. It is seen that the presence of the cone changes the RCS over a wide range of incidence angles. At normal incidence it lowers the RCS significantly, as expected. The RCS-curves for $p = 5$ and $p = 6$ cannot be distinguished within the graphics.

In Fig. 6 we show RCS predictions that are made with the direct **E**-field based model [6], compared to preliminary results obtained with a newly developed *hp*-FE+BE (MFIE) formulation. The novel MFIE results are probably not fully converged (appropriate tuning of the required accuracy of the BE-formulation is still under investigation). The difference from the results obtained with the *hp*-FE+IE model increases as incidence angle increases. The investigation of the cause of the

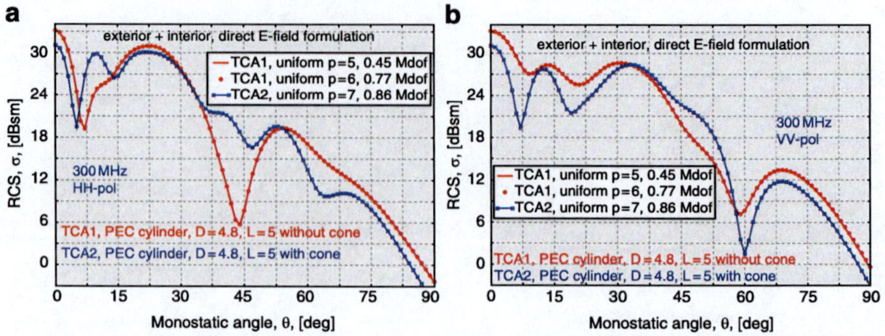

Fig. 5 Monostatic RCS vs. azimuth angle θ at 300 MHz. Fully coupled interior and exterior model. RCS results for TCA1 and TCA2 channels compared. (**a**) Horizontal polarisation. (**b**) Vertical polarisation

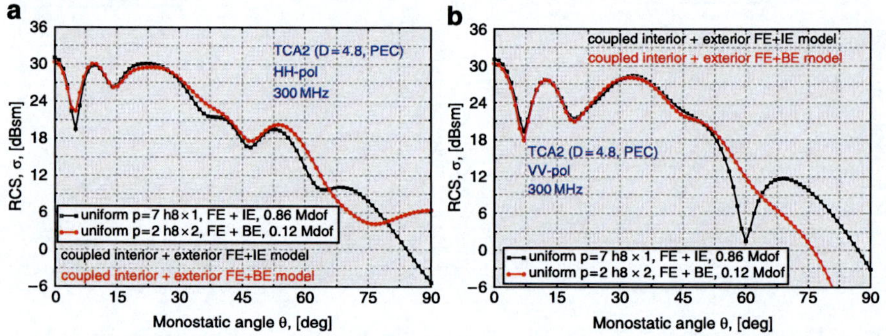

Fig. 6 Monostatic RCS vs. azimuth angle θ at 300 MHz. Results obtained with *hp*-FE+BE (MFIE) and *hp*-FE+IE [6] models, respectively. (**a**) Horizontal polarisation. (**b**) Vertical polarisation

discrepancy is out of the scope of this report. It will be investigated elsewhere using finer discretizations and a more efficient solver. The RCS values are the largest in magnitude for incidence angles close to normal. These results are therefore of greater practical value from the engineering point of view. In passing it should be noted that using a fully *hp*-adaptive approach with RCS as the quantity of interest will probably yield meshes that vary with the incidence angle.

3.2 Scattering Matrix Based Interior Only Model

In Fig. 7 we compare RCS results obtained with a low order *h*-type discretization with those obtained with our $p = 7$ reference solution for the scattering matrix model (with one sub-domain), respectively. It is concluded that there is a large

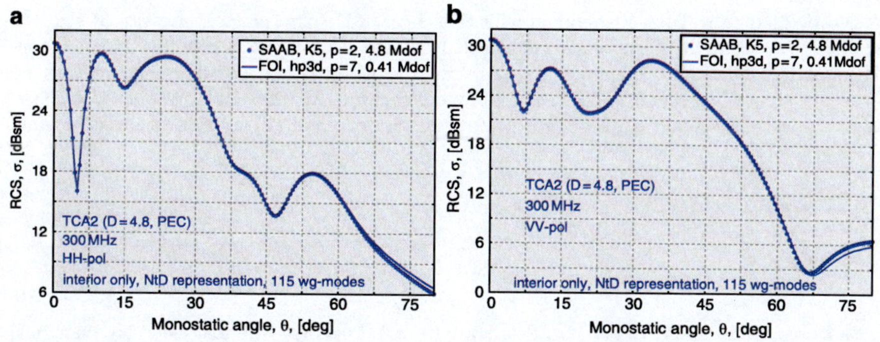

Fig. 7 Monostatic RCS vs. azimuth angle θ at 300 MHz. (**a**) Horizontal polarisation. (**b**) Vertical polarisation

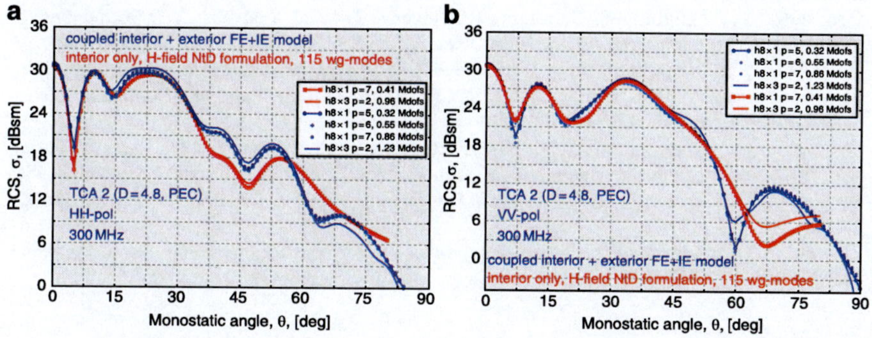

Fig. 8 Monostatic RCS vs. azimuth angle θ at 300 MHz. Comparison of RCS predicted with a fully coupled interior and exterior model [6], with RCS obtained with a scattering matrix based interior only model. (**a**) Horizontal polarisation. (**b**) Vertical polarisation

potential efficiency gain using *hp*-FEM. Note that the *h*-FEM reference solution has 4.8 Mdof, while the uniform *p*-FEM one at $p = 7$ uses only 0.41 Mdof. That is, the uniform *p*-FEM mesh gives an order of magnitude less physical degrees of freedom than the *h*-FEM reference solution giving virtually the same RCS accuracy. Also note that 115 wave-guide modes, i.e. modal degrees of freedom, are sufficient to capture the RCS of the TCA2 channel for close to normal incidence angles, cf. Figs. 7 and 8. The number of physical degrees of freedom of the corresponding aperture FE-mesh is readily a couple of orders larger.

Finally, in Fig. 8 we show RCS predictions that are made with the fully coupled interior plus exterior model [6], compared to those obtained with the scattering matrix based, uncoupled, model. The fully coupled model includes edge diffraction at the aperture. Here the aperture rim is sharp. In reality it is not. The RCS for the two models is close up to $\theta = 35°$for horizontal polarisation, and up to $\theta = 45°$ for vertical polarisation. This is contrary to common belief that the uncoupled model delivers good RCS predictions up to $\theta = 60°$ regardless of polarisation.

Acknowledgements The collaboration with Prof. Leszek Demkowicz at University of Texas at Austin, ICES, using the *hp3d* code, is acknowledged. The work reported herein is done in cooperation with MSc. Tomas Lundin at SAAB Aerospace Cooperation, and it is supported by the Swedish Defence Materiel Administration, FMV, contract FoT-25 2006–2008. Waldemar Rachowicz acknowledges support of the Polish Ministry of Science and Higher Education under grant N N519 405234.

References

1. Barka, A., Soudais, P., Volpert, D.: Scattering from 3-d cavities with plug and play numerical scheme combining ie, pde, and modal techniques. IEEE Trans. Antennas Propagat. **48**(5), 704–712 (2000)
2. Burkholder, R.J.: Comparison study of electromagnetic cavity scattering computations. Tech. Rep. 739353-2, The Ohio State University ElectroScience Lab (2001)
3. Conciauro, G., Guglielmi, M., Sorrentino R.: Advanced Modal Analysis, CAD Techniques for Waveguide Components and Filters. Wiley, New York (2000)
4. Demkowicz, L.: Computing with *hp*-Adaptive Finite Elements, One and Two Dimensional Elliptic and Maxwell Problems, *Chapman & Hall/CRC Applied Mathematics and Nonlinear Science Series*, vol. 1. Chapman & Hall, FL (2007)
5. Demkowicz, L., Kurz, J., Pardo, D., Paszyński, M., Rachowicz, W., Zdunek, A.: Computing with *hp*-Adaptive Finite Elements, Three-Dimensional Elliptic and Maxwell Problems with Applications, *Chapman & Hall/CRC Applied Mathematics and Nonlinear Science Series*, vol. 2. Chapman & Hall, FL (2007)
6. Zdunek, A., Rachowicz, W.: Cavity radar cross section prediction. IEEE Trans. Antennas Propagat. **56**(6), 1752–1763 (2008)

Anchor Points Matter in ANOVA Decomposition

Zhongqiang Zhang, Minseok Choi, and George Em Karniadakis

Abstract We focus on the analysis of variance (ANOVA) method for high dimensional approximations employing the Dirac measure. This anchored-ANOVA representation converges exponentially fast for certain classes of functions but the error depends strongly on the anchor points. We employ the concept of "weights per dimension" to construct a theory that leads to the optimal anchor points. We then present examples of a function approximation as well as numerical solutions of the stochastic advection equation up to 500 dimensions using a combination of anchored-ANOVA and polynomial chaos expansions.

1 Introduction

We consider an N-dimensional function f, which can be decomposed as

$$f(x_1, x_2, \cdots, x_N) = f_0 + \sum_{j_1=1}^{N} f_{j_1}(x_{j_1}) + \sum_{j_1<j_2}^{N} f_{j_1,j_2}(x_{j_1}, x_{j_2}) + \cdots + f_{j_1,j_2,\cdots,j_N}(x_{j_1,j_2,\cdots,j_N}), \quad (1)$$

where f_0 is a constant, and f_S are $|S|$-dimensional functions, called the $|S|$-order terms. (Here $|S|$ denotes the cardinality of the index set S). This is the so-called ANOVA model. Here we consider the domain $I^N = [0,1]^N$, in a tensor-product form. The terms in the ANOVA decomposition are computed as follows

$$f_0 = \int_{[0,1]^N} f(\mathbf{x}) d\mu(\mathbf{x}). \quad (2a)$$

Z. Zhang, M. Choi, and G.E. Karniadakis (✉)
Division of Applied Mathematics, Brown University, Providence, RI 02912, USA
e-mail: handyzang@gmail.com, minseok_choi@brown.edu, gk@dam.brown.edu

J.S. Hesthaven and E.M. Rønquist (eds.), *Spectral and High Order Methods for Partial Differential Equations*, Lecture Notes in Computational Science and Engineering 76,
DOI 10.1007/978-3-642-15337-2_32,

$$f_S = \int_{[0,1]^{N-|S|}} f(\mathbf{x})d\mu(\mathbf{x}_{-S}) - \sum_{T\subset S} f_T(\mathbf{x}_T). \tag{2b}$$

We note that there are several forms of ANOVA decomposition associated with different measures. Here we focus on the one using the Dirac measure, $d\mu(x) = \delta(x-c)\,dx$ ($c \in [0,1]$), which leads to the *anchored-ANOVA decomposition*. The point "c", which is often arbitrarily selected, is called the "anchor point". Another type is based on the Lebesgue measure, $d\mu(x) = \rho(x)\,dx$; this is the unanchored-ANOVA decomposition. See [1, 5] for details.

All the distinct ANOVA terms are mutually orthogonal with respect to the corresponding measure. Hence, for every term f_S with $S \subseteq \{1,2,\cdots,N\}$, we have

$$\int_{[0,1]} f_S(\mathbf{x}_S)d\mu(x_j) = 0, \qquad \text{if} \quad j \in S,$$

and

$$\int_{[0,1]^N} f_S(\mathbf{x}_S) f_T(\mathbf{x}_T) d\mu(\mathbf{x}) = 0, \qquad \text{if} \quad S \neq T.$$

The order at which we truncate the ANOVA model is called *effective dimension*, beyond which the difference between the ANOVA model and the truncated expansion in a certain measure is very small, see [2, 9, 10, 12]. It is not difficult to show that the variance of f can be a sum of variances of the ANOVA terms

$$\sigma^2(f) = \int_{[a,b]^N} f^2(\mathbf{x})d\mathbf{x} - \left(\int_{[a,b]^N} f(\mathbf{x})d\mathbf{x}\right)^2 = \sum_{\emptyset\neq S\subseteq\{1,2,\cdots,N\}} \int_{[a,b]^{|S|}} f_S^2(\mathbf{x}_S)d\mathbf{x}_S. \tag{3}$$

or in compact form

$$\sigma^2(f) = \sum_{\emptyset\neq S\subseteq\{1,2,\cdots,N\}} \sigma_S^2(f). \tag{4}$$

The effective dimension of f in the superposition sense is the smallest integer d_s satisfying

$$\sum_{0<|S|\le d_s} \sigma_S^2(f) \ge p\sigma^2(f), \tag{5}$$

where $S \subset \{1,2,\cdots,N\}$. This implies that we will ignore terms in the ANONA model corresponding to more than d_s interactions. The effective dimension is measured in the L^2-norm. Note that p is a proportionality constant with $0 < p < 1$ and close to 1, e.g., $p = 0.99$ in [2].

2 Weights and Effective Dimension

In order to obtain an estimate of the effective dimension, we adopt proper weights, which weight in some sense the contribution of each dimension. The concept of weights here is analogous to the concept employed in analyzing the Quasi Monte

Carlo (QMC) method [11]. In particular, the idea is to define appropriate weights so that their minimization also leads to minimization of errors in QMC, see [3, 8]. In general, the weights should be in the interval of [0,1]. In addition, most of the weights should be less than one in order to have a low effective dimension for a nominally high-dimensional function.

Assuming a function in tensor product form, the weights in [11] were determined by the mean and the variance of the corresponding one-dimensional functions. This can be easily seen from the definition of the mean effective dimension [9]. Specifically, given a tensor product function

$$f(\mathbf{x}) = \prod_{k=1}^{N} f_k(x_k),$$

the mean and the variance of the function are

$$\mu_k = \int_0^1 f_k(x_k)\,dx_k < \infty, \quad k = 1, 2, \cdots, N,$$
$$\lambda_k^2 = \int_0^1 \left(f_k(x_k) - \mu_k\right)^2 dx_k < \infty, \quad k = 1, 2, \cdots, N.$$

The ANOVA terms and the corresponding variances are [9]:

$$f_S = \prod_{k\in S} \left(f_k(x_k) - \mu_k\right) \cdot \prod_{k\notin S} \mu_k, \tag{6}$$
$$\sigma_S^2(f_S) = \prod_{k\in S} \lambda_k^2 \prod_{k\notin S} \mu_k^2.$$

Then, the weights γ_k's are defined as follows:

$$\gamma_k = \frac{\lambda_k^2}{\mu_k^2} \text{ if } \mu_k \neq 0 \text{ for } k = 1, 2, \cdots, N.$$

In the unanchored ANOVA (i.e., using the Lebesgue measure), the effective dimension has a more clear meaning. The truncation error, when the effective dimension is ν, by definition, is estimated as

$$\left\| f - \sum_{|S|\le\nu} f_S \right\|_{L^2}^2 \le (1-p)(\|f\|^2 - (\int_{I^N} f\,dx)^2),$$

where we use the equality $\|f\|^2 = (\int_{I^N} f\,dx)^2 + \sigma^2(f)$. Hence, we have

$$\left\| f - \sum_{|S| \le \nu} f_S \right\|^2 \le (1-p)(1 - (\frac{\int_{I^N} f\, dx}{\|f\|})^2) \|f\|^2$$
$$= (1-p)(\int_{I^N} f\, dx)^2 (\prod_{k=1}^{N} (1+\gamma_k) - 1). \quad (7)$$

Remark 2.1. From (7), we have that

$$\frac{\left\| f - \sum_{|S| \le \nu} f_S \right\|}{\|f\|} \le \sqrt{1-p}(1 - \prod_{k=1}^{N} (1+\gamma_k)^{-1})^{\frac{1}{2}} < 0.1,$$

by choosing $p = 0.99$. In fact, when p is chosen as 0.99 the effective dimension is not always an integer. The estimate above corresponds to the worst case and, in fact, the error can be far better; see [9] for specific examples.

Remark 2.2. From the definition of weights, we have that

$$\left\| f - \sum_{|S| \le \nu} f_S \right\|^2 = (\int_{I^N} f\, dx)^2 \sum_{m=\nu+1}^{N} \sum_{|S|=m} \prod_{k \in S} \gamma_k.$$

According to (7),

$$\sum_{m=\nu+1}^{N} \sum_{|S|=m} \prod_{k \in S} \gamma_k \le (1-p)(\prod_{k=1}^{N} (1+\gamma_k) - 1). \quad (8)$$

As already mentioned, when a function is of low effective dimension, the dominating weights are much smaller than one. In fact, if $\mu_k \neq 0$ and $\gamma_k < 1$ for all $k = 1, 2, \cdots, N$, the mean effective dimension is [9]

$$d_s = \frac{\sum_{k=1}^{N} \frac{\gamma_k}{\gamma_k + 1}}{1 - \prod_{k=1}^{N} \frac{1}{\gamma_k + 1}} = \frac{N - \sum_{k=1}^{N} \frac{1}{\gamma_k + 1}}{1 - \prod_{k=1}^{N} \frac{1}{\gamma_k + 1}}. \quad (9)$$

While the previous discussion concerns the ANOVA version with Lebesque measure, it is by analogy that we can extend the concept of weights to the anchored-ANOVA as well. To this end, we define the weights using the L^∞-norm, as follows:

$$\gamma_k = \frac{\|f_k - f_k(c_k)\|_\infty}{|f_k(c_k)|}, \quad \text{when} \quad f(c) \neq 0. \quad (10)$$

Lemma 2.3. *Assuming that the anchored-ANOVA is truncated at the $\tilde{\nu}$th order, and that $p_{\tilde{\nu}}$ satisfies*

$$\sum_{m=\tilde{\nu}+1}^{N} \sum_{|S|=m} \prod_{k\in S} \gamma_k = (1-p_{\tilde{\nu}})(\prod_{k=1}^{N}(1+\gamma_k)-1).$$

Then, the relative error in L^{∞}-norm can be estimated as

$$\frac{\left\| f - \sum_{|S|\le\tilde{\nu}} f_S \right\|_{L^\infty}}{\|f\|_{L^\infty}} \le (1-p_{\tilde{\nu}})(\prod_{k=1}^{N}(1+\gamma_k)-1)(\prod_{k=1}^{N}\frac{|f_k(c_k)|}{\|f_k\|_{L^\infty}}). \tag{11}$$

Also, for one-signed functions, if the anchored points $c = (c_1, c_2, \cdots, c_N)$ are selected such that

$$f_k(c_k) = \frac{1}{2}\max_{[0,1]} f_k(x_k) + \frac{1}{2}\min_{[0,1]} f_k(x_k).$$

Then, $\gamma_k = \left| \frac{\max_{[0,1]} f_k(x_k) - \min_{[0,1]} f_k(x_k)}{\max_{[0,1]} f_k(x_k) + \min_{[0,1]} f_k(x_k)} \right|$, and it minimizes the weights defined in (10).

The minimized weights, in turn, minimize the error estimate in the last lemma.

Proof. Recalling the results from the ANOVA using Lebesgue measure with the same weights, we have

$$\begin{aligned}
\frac{\left\| f - \sum_{|S|\le\nu} f_S \right\|_{L^\infty}}{\|f\|_{L^\infty}} &= \frac{\left\| f - \sum_{|S|\le\nu} f_S \right\|_{L^\infty}}{\prod_{k=1}^{N}|f_k(c_k)|} \frac{\prod_{k=1}^{N}|f_k(c_k)|}{\|f\|_{L^\infty}} \\
&\le \sum_{m=\tilde{\nu}+1}^{N} \sum_{|S|=m} \prod_{k\in S} \gamma_k (\prod_{k=1}^{N}\frac{|f_k(c_k)|}{\|f_k\|_{L^\infty}}) \\
&\le (1-p_{\tilde{\nu}})(\prod_{k=1}^{N}(1+\gamma_k)-1)(\prod_{k=1}^{N}\frac{|f_k(c_k)|}{\|f_k\|_{L^\infty}}).
\end{aligned}$$

This proves the error estimate. The following will complete the proof of how to minimize weights.

Suppose that f_k does not change sign over the interval [0, 1]. Without loss of generality, let $f_k > 0$. Denote the maximum and the minimum of f_k by M_k and m_k, respectively, and assume that $f_k(c_k) = \alpha_k M_k + (1-\alpha_k)m_k$ where $\alpha_k \in [0,1]$. Then

$$\|f_k - f_k(c_k)\|_\infty = \max(M_k - f_k(c_k), f_k(c_k) - m_k) = (M_k - m_k)\max(1-\alpha_k, \alpha_k),$$

and the weight γ_k is

$$\frac{\|f_k - f_k(c_k)\|_\infty}{|f_k(c_k)|} = \frac{(M_k - m_k)\max(1-\alpha_k, \alpha_k)}{\alpha_k M_k + (1-\alpha_k)m_k}.$$

Let us consider the function of $g(\alpha_k) = \frac{(1-y)\max(1-\alpha_k,\alpha_k)}{\alpha_k+(1-\alpha_k)y}$, where $\alpha_k \in [0,1]$, $y = \frac{m_k}{M_k} \in (0,1)$ and see how to choose α_k. Notice that

$$g'(\alpha_k) = \begin{cases} \frac{y-1}{(\alpha_k+(1-\alpha_k)y)^2} < 0 \text{ if } \alpha_k \in (0,\frac{1}{2}), \\ \frac{(1-y)y}{(\alpha_k+(1-\alpha_k)y)^2} > 0 \text{ if } \alpha_k \in (\frac{1}{2},1). \end{cases}$$

From this we know that $g(\frac{1}{2})$ reaches the minimum of $g(\alpha_k)$ with $\alpha_k \in (0,1)$. Then, $\alpha_k = \frac{1}{2}, \gamma_k = g(\frac{1}{2}) = \frac{1-\frac{m_k}{M_k}}{1+\frac{m_k}{M_k}} < 1$.

Actually, according to the definition of weights,

$$\Big(\prod_{k=1}^{N}(1+\gamma_k)-1\Big)\Big(\prod_{k=1}^{N}\frac{|f_k(c_k)|}{\|f_k\|_{L^\infty}}\Big) = \prod_{k=1}^{N}\frac{|f_k(c_k)| + \|f_k - f_k(c_k)\|_{L^\infty}}{\|f_k\|_{L^\infty}} - \prod_{k=1}^{N}\frac{|f_k(c_k)|}{\|f_k\|_{L^\infty}}.$$

If $\alpha_k > \frac{1}{2}$,

$$\begin{aligned}\prod_{k=1}^{N}(1+\gamma_k)-1)\Big(\prod_{k=1}^{N}\frac{|f_k(c_k)|}{\|f_k\|_{L^\infty}}\Big) &= \prod_{k=1}^{N}\frac{\alpha_k M_k + (1-\alpha_k)m_k + (M_k-m_k)\max(1-\alpha_k,\alpha_k)}{M_k} \\ &\quad -\prod_{k=1}^{N}\frac{\alpha_k M_k + (1-\alpha_k)m_k}{M_k} \\ &= \prod_{k=1}^{N}\Big(2\alpha_k\Big(1-\frac{m_k}{M_k}\Big)+\frac{m_k}{M_k}\Big) - \prod_{k=1}^{N}\Big(\alpha_k\Big(1-\frac{m_k}{M_k}\Big)+\frac{m_k}{M_k}\Big).\end{aligned}$$

Hence, the first term in the last inequality increases faster than the last term, since $2\alpha_k(1-\frac{m_k}{M_k}) + \frac{m_k}{M_k} > \alpha_k(1-\frac{m_k}{M_k}) + \frac{m_k}{M_k}$ for $\alpha_k > \frac{1}{2}$. If $\alpha_k < \frac{1}{2}$,

$$\prod_{k=1}^{N}(1+\gamma_k)-1)\Big(\prod_{k=1}^{N}\frac{|f_k(c_k)|}{\|f_k\|_{L^\infty}}\Big) = 1 - \prod_{k=1}^{N}\Big(\alpha_k\Big(1-\frac{m_k}{M_k}\Big)+\frac{m_k}{M_k}\Big).$$

Thus $\alpha_k = \frac{1}{2}$ is the best choice when it minimizes the error estimate. Notice here the choice of $\alpha_k = \frac{1}{2}$ also minimizes the weight. This ends the proof.

Remark 2.4. Weights and corresponding ancor points can also be defined in the L^1-norm using appropriate quadrature formulas, e.g., see [6].

3 Numerical Examples

Here we present two examples, first in approximating a high-dimensional function and subsequently in solving the stochastic advection equation.

Example 1. We consider the Genz function [4] $f_5 = \prod_{j=1}^{N} \exp(-c_j |x_j - w_j|)$ with the parameters $c_j = \exp(-0.2j)$ and w_j following a uniform distribution.

$$w = (0.695106, 0.851463, 0.413355, 0.410178, 0.226185, \\ 0.7078, 0.478756, 0.183078, 0.0724332, 0.483279)$$

The centered point refers to $(\frac{1}{2}, \frac{1}{2}, \cdots, \frac{1}{2})$, while the optimal point is the point chosen according to the Lemma 2.3. Both results in Table 1 demonstrate exponential accuracy in terms of the truncation dimension but using the optimal anchor points leads to accuracy close to three orders better than using the centered point.

Example 2. Next we consider the stochastic advection equation

$$\frac{\partial u}{\partial t} + V(t;\xi)\frac{\partial u}{\partial x} = 0$$

in the interval $[-1, 1]$ with periodic boundary conditions and initial condition $u(x, t = 0) = sin(\pi(x + 1))$. The advection velocity is a stochastic process with zero mean and is represented using a Karhunen-Loeve expansion, i.e., $V(t, \xi) = \sum_{k=0}^{M} \sqrt{\lambda_k}\phi_k(t)\xi_k$, with ξ_k being uncorrelated and also independent variables following a uniform distribution. The eigenpairs (λ_k, ϕ_k) are derived from the covariance kernel of the form $\exp[-|t_1 - t_2|/L]$, where L is the correlation length. Here we consider three values of L corresponding to different truncations, i.e., $(L, M) = (1, 4); (0.1, 10); (0.005; 500)$ selected so that 90% of the energy is captured by the coefficients of the truncated expansion. In the simulations we employ a Fourier-collocation in space and a probabilistic collocation method in random space using Legendre-chaos (8th-order).

In Fig. 1 we plot the mean solution at $t = 0.5$ in order to compare the effect of the anchor point on the convergence of the ANOVA expansion. We see that for the optimum point $c_1 = (0, 0, \ldots, 0)$ the solution converges to the exact solution when

Table 1 Error in the mean: $N = 10$

Truncation order	Centered point	Optimal point
1	6.6207×10^{-2}	3.7949×10^{-3}
2	5.2552×10^{-3}	8.8265×10^{-5}
3	2.3796×10^{-4}	1.2680×10^{-6}
4	6.2412×10^{-6}	1.1568×10^{-8}
5	9.0972×10^{-8}	6.6648×10^{-11}

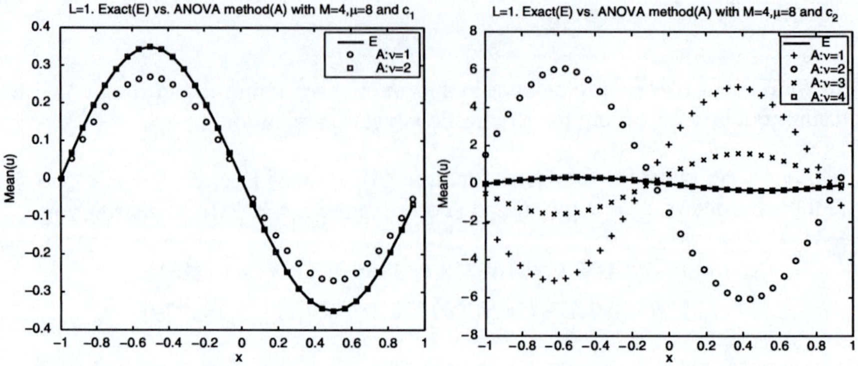

Fig. 1 Mean solution using the optimum anchor point c_1 (*left*) and a different point c_2 (*right*). Here $M = 4$; $L = 1$

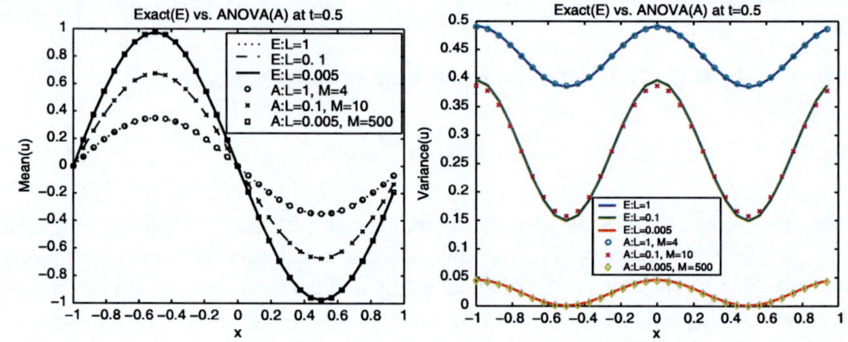

Fig. 2 Mean solution (*left*) and Variance (*right*) using the optimum anchor point c_1 for different values of the correlation length ($L = 1, 0.1, 0.005$) and corresponding truncation dimension ($\nu = 2, 2, 1$)

$\nu = 2$ but for another point $c_2 = (1, 1, \ldots, 1)$ the solution converges to the exact solution only if $\nu = M = 4$, i.e., for the full expansion. Here the exact solution is computed as in [7]. Using the optimum point we can now vary the correlation length L and produce accurate solutions in the high-dimensional space for small values of L and up to $M = 500$ dimensions as shown in Fig. 2.

Acknowledgements This work was supported by an OSD/AFOSR MURI grant and also by a DOE Computational Mathematics grant.

References

1. Bieri, M. and Schwab, C.: Sparse high order FEM for elliptic sPDEs, Tech. Report 22, ETH, Switzerland, May 2008
2. Caflisch, R.E., Morokoff, W. and Owen, A.: Valuation of mortgage-backed securities using Brownian bridges to reduce effective dimension, J. Comput. Finance, **1**, 27–46 (1997)
3. Dick, J., Sloan, I. H., Wang, X. and Wozniakowski, H.: Liberating the weights, Journal of Complexity, **20**, 593–623 (2004)
4. Genz, A.: A package for testing multiple integration subroutines, Numerical Integration: Recent developments, software and applications, Ed. Reidel, pp. 337–340 (1987)
5. Griebel, M.: Sparse grids and related approximation schemes for higher dimensional problems, In: Foundations of Computational Mathematics (FoCM05), Santander, L. Pardo, A. Pinkus, E. Suli, and M. Todd, eds., Cambridge University Press, Cambridge, 106–161 (2006)
6. Griebel, M. and Holtz, M.: Dimension-wise integration of high-dimensional functions with applications to finance, INS Preprint No. 0809, University of Bonn, Germany (2009)
7. Jardak, M., Su, C.H. and Karniadakis, G.E.: Spectral polynomial chaos solutions of the stochastic advection equation, Journal of Scientific Computing, **17**, 319–338 (2002)
8. Larcher, G.,Leobacher, G. and Scheicher, K.: On the tractability of the Brownian bridge algorithm, Journal of Complexity, **19**, 511–528 (2003)
9. Larcher, G.,Leobacher, G. and Scheicher, K.: The dimension distribution and quadrature test functions, Statistica Sinica, **13**, 1–17 (2003)
10. Paskov, S.H. and Traub, J.F.: Faster valuation of financial derivatives, Journal of Portfolio Management, **22**, 113–120 (1995)
11. Sloan, I.H. and Wozniakowski, H.: When are Quasi-Monte Carlo algorithms efficient for high dimensional integrals? Journal of Complexity, **14**, 1–33 (1998)
12. Wang, X. and Fang, K.-T.: The effective dimension and quasi-Monte Carlo integration, Journal of Complexity, **19**, 101–124 (2003)

An Explicit Discontinuous Galerkin Scheme with Divergence Cleaning for Magnetohydrodynamics

Christoph Altmann

Abstract The explicit space-time expansion discontinuous Galerkin scheme (Gassner et al., J. Sci. Comp. 34(3):260–286, 2008) is applied for solving ideal and viscous magnetohydrodynamic equations. Based on a Taylor expansion in space and time about the barycenter of each cell at the old time level, this predictor-corrector strategy enables each cell to have its own time step whereas the high order of accuracy in time is retained. Thus, it may significantly speed up computations. The discontinuous Galerkin method together with the local time-stepping algorithm allows for an efficient local sub-cycling for a divergence cleaning using a hyperbolic transport correction (Dedner et al., J. Comput. Phys. 175(2):645–673, 2002). Convergence tests and test problems are performed to challenge the capabilities of the space-time expansion scheme.

1 Introduction

Discontinuous Galerkin (DG) schemes gained significantly in popularity, since they combine flexibility in handling complex geometries, ability of performing h/p-adaptivity and efficiency in massively parallel calculations. These aspects turn them into an ideal candidate for modern numerical calculations in various fields of interest, including magnetohydrodynamics (MHD) and plasma physics. The recently developed [3] explicit space-time expansion discontinuous Galerkin scheme (STE-DG) provides a perfect basis. With an adapted hyperbolic divergence correction method that saves computational costs by using the high order explicit local time stepping functionality of the STE-DG scheme, we can efficiently handle MHD calculations.

C. Altmann
Universität Stuttgart, Institut für Aerodynamik und Gasdynamik, Pfaffenwaldring 21, 70569 Stuttgart, Germany
e-mail: altmann@iag.uni-stuttgart.de

J.S. Hesthaven and E.M. Rønquist (eds.), *Spectral and High Order Methods for Partial Differential Equations*, Lecture Notes in Computational Science and Engineering 76,
DOI 10.1007/978-3-642-15337-2_33,

2 STE-DG Discretization

We will first summarize the basics of the STE-DG scheme and its local time stepping functionality. For a detailed view on these topics, the reader is referred to the corresponding articles [2] or [3]. For the sake of simplicity, we will consider the advection-diffusion equation

$$u(\mathbf{x})_t + \nabla \cdot \mathbf{f}(u(\mathbf{x}), \nabla u(\mathbf{x})) = 0. \quad (1)$$

We initiate the DG discretization as usual by subdividing our domain Ω into non-overlapping spatial grid cells Q_i and introduce our numerical DG solution. To obtain the weak formulation of the STE-DG scheme, we multiply by a spatial test function $\phi = \phi(\mathbf{x})$ and integrate over an arbitrary space-time cell $Q_i^n := Q_i \times [t^n, t^{n+1}]$. Please note that performing a space-time integration is substantially different from the classical purely spatial dependent DG formulation. In addition to the classical integration by parts for deriving the weak formulation, Gassner et al. [2] introduced a new variational formulation for diffusion problems by performing a second integration by parts. This generates a new diffusion surface integral which depends on the solution itself. With the definition of suitable numerical fluxes like the HLLC flux of Li [7] and a suitable numerical state for the second diffusion surface integral as described in [2], we get an adjoint consistent formulation and thus a discretization with optimal order of convergence. This is accomplished by solving the so-called diffusive generalized Riemann problem (dGRP). See [2] for more details on this method of diffusion flux treatment.

Finally, the variational formulation of our advection-diffusion equation results in

$$\begin{aligned} &\int_{Q_i^n} (u_i)_t \phi \, d\mathbf{x}dt - \int_{Q_i^n} \mathbf{f}^a \cdot \nabla \phi \, d\mathbf{x}dt + \int_{Q_i^n} \mu \nabla u_i \cdot \nabla \phi \, d\mathbf{x}dt + \\ &\int_{\partial Q_i^n} \mathbf{g}^{\mathbf{a}} \cdot \mathbf{n} \phi \, ds dt - \int_{\partial Q_i^n} \mathbf{g}^{\mathbf{d}} \cdot \mathbf{n} \phi \, ds dt + \int_{\partial Q_i^n} g^s \, [\nabla \phi \cdot \mathbf{n}]^- \, ds dt = 0 , \end{aligned} \quad (2)$$

where the test functions ϕ run through all the basis functions. Here, the term $\mathbf{g}^{\mathbf{a}}$ denotes the numerical advection flux, the term $\mathbf{g}^{\mathbf{d}}$ denotes the numerical diffusion flux and $g^s := \mu u_i - [\mu u_i]^-$ denotes the additional scalar diffusion flux.

For nonlinear flux functions, the space-time integrals in (2) have to be computed in an approximate way. While this could be done using Gaussian quadrature formulae in space and time, we need to find a way to get approximate values at the space-time Gauss points. This has to be done in an explicit way, since we are interested in an explicit scheme. A possible solution is a Taylor series expansion, as described in the next section.

2.1 *Space-Time Expansion*

The concept of the STE-DG scheme is a Taylor series expansion at the barycenter to get a predictive approximate solution in the space-time cell [3]:

$$u(\mathbf{x},t) = u(\mathbf{x}_i, t_n) + \sum_{j=1}^{N} \frac{1}{j!} \left((t - t_n) \frac{\partial}{\partial t} + (\mathbf{x} - \mathbf{x}_i) \cdot \nabla \right)^j u \bigg|_{(\mathbf{x},t)=(\mathbf{x}_i,t_n)}, \quad (3)$$

about the barycenter $\mathbf{x}_i$ of the grid cell Q_i at time t_n. This expansion provides approximate values for u and ∇u at all space-time points $(\mathbf{x}, t) \in \Omega_i^n$. Since no neighbor information was put into this formulation, we can consider this an predictor of our solution that will be corrected later with neighboring information by the inter-cell flux exchange. The reader is referred to [4] for more information on this approach. While the pure space derivatives at $(\mathbf{x}_i, t_n)$ are readily available within the DG framework, the time and mixed space-time derivatives have to be computed using the Cauchy–Kowalevsky (CK) procedure. This procedure will replace them with pure spatial derivatives by the differential equation directly. More information about the CK procedure can be found in [3]. The resulting framework allows for a natural consistent explicit local time stepping.

2.2 *Local Time Stepping*

A major disadvantage of a conventional explicit DG scheme is its severe time step restriction to ensure stability. While for uniform grids, this time step is in the range of the "physical time step", needed to capture the right evolution of the unsteady phenomena, it becomes obstructive for unstructured grids with very small grid cells. The grid cell with the most restrictive local time step defines the time step for all grid cells. But, this drop in efficiency can be avoided: Due to the locality of the explicit STE-DG scheme, each grid cell may run with its own time step in a time-consistent manner, while the high order of accuracy of the numerical scheme is preserved. The local time step is determined exclusively by the in-cell time step restriction and is completely independent of the time steps of neighboring cells. This local time stepping algorithm minimizes the total number of time steps for a computation with fixed end time. However, when the difference of time levels of adjacent grid cells is very small compared to the local time steps, we locally synchronize the time levels of those cells and therefore reduce the number of flux evaluations to gain efficiency, as done in [3] or [4]. Common global time levels, needed, e.g., at the end of the computation, can easily be introduced. This procedure has absolutely no influence on the accuracy of the underlying numerical scheme, as convergence tests, e.g., in [3] verify.

3 Divergence Correction and Local Time Stepping

When dealing with MHD, the divergence free ($\nabla \cdot \mathbf{B} = 0$) constraint has to be maintained. Not doing so, numerical schemes may generate divergence errors that can have a negative influence on the solution. Since our scheme is running with a local time stepping mechanism, all divergence cleaning methods that are based on operations affecting all cells at the same time (e.g., projection methods) cannot be applied. Dedner et al. [1] presented a hyperbolic divergence cleaning that is easy to implement and still yields the desired effect on the divergence errors. It adds a divergence correction variable to the MHD equation system that can be solved in a very fast and straight-forward way. Since the effect of this variable on the whole system is similar to a Lagrangian multiplier, they called this method the Generalized Lagrange Multiplier (GLM) divergence correction method. With that addition, the viscous MHD equations as shown in [11] will look like

$$\begin{aligned}
\tfrac{\partial}{\partial t}\rho &= -\nabla\cdot(\rho\mathbf{v})\\
\tfrac{\partial}{\partial t}(\rho\mathbf{v}) &= -\nabla\cdot\left(\rho\mathbf{v}\mathbf{v}^t - \mathbf{B}\mathbf{B}^t + \left(p+\tfrac{1}{2}|\mathbf{B}|^2\right)I - \tau\right)\\
\tfrac{\partial}{\partial t}E &= -\nabla\cdot\Big((E+p)\,\mathbf{v} + \left(\tfrac{1}{2}|\mathbf{B}|^2 I - \mathbf{B}\mathbf{B}^t\right)\cdot\mathbf{v}\\
&\qquad -\mathbf{v}\tau \;+\; \eta\left(\mathbf{B}\cdot\nabla\mathbf{B} - \nabla\left(\tfrac{1}{2}|\mathbf{B}|^2\right)\right) - \mu\tfrac{1}{Pr}\nabla T\Big)\\
\tfrac{\partial}{\partial t}\mathbf{B} &= -\nabla\times(\mathbf{B}\times\mathbf{v} + \eta\nabla\times\mathbf{B} + \psi I)\\
\tfrac{\partial}{\partial t}\psi &= -\nabla\cdot\left(c_h^2\mathbf{B}\right) - \tfrac{c_h^2}{c_p^2},
\end{aligned} \tag{4}$$

with the divergence constraint $\nabla\cdot\mathbf{B} = 0$. Here, $\tau := \mu(\nabla\mathbf{v} + (\nabla\mathbf{v})^T - \frac{2}{3}(\nabla\cdot\mathbf{v})\,I)$ is the viscous stress tensor and $Pr = \frac{c_p\mu}{\gamma}$ the Prandtl number. The pressure (perfect gas) is derived to $p = \rho RT = (\gamma-1)\left(E - \frac{1}{2}\rho\mathbf{v}^2 - \frac{1}{2}\mathbf{B}^2\right)$. Setting viscosity μ and resistivity η to zero would result in the ideal MHD system. The additional variable ψ is introduced and propagates the divergence error out of the computational domain. With the addition of $-\frac{c_h^2}{c_p^2}$ for the ψ equation on the right hand side of (4), we will not only transport the errors out of the computational domain but also damp them. The damping effect can be scaled by setting the value of c_p. We are generally using a value of 0.18 for c_p, as proposed in [1], which is a good compromise between damping and hyperbolic transport. With that modification, the divergence cleaning method is then called mixed GLM method. For MHD equations, it was proposed in [1] to set the hyperbolic transport speed c_h to the fastest system wave, whereas errors are at least spread with the same velocity as they may be generated. Furthermore, it is ensured that the correction subsystem is not affecting the time step of the overlying MHD calculation.

Since the presented STE-DG scheme relies on local time steps, each cell allows for a divergence cleaning at a different speed. We therefore need to adapt the original setting of the GLM system based on global time steps. This is done by adjusting the calculation of the numerical flux of the correction system. By taking into account the corresponding local Riemann problem, the numerical flux then reads in one space dimension as:

$$\begin{pmatrix} B_{x,m} \\ \psi_m \end{pmatrix} = \begin{pmatrix} B_{x,l} \\ \psi_l \end{pmatrix} + \begin{pmatrix} \frac{1}{2}(B_{x,r} - B_{x,l}) - \frac{1}{2c_h}(\psi_r - \psi_l) \\ \frac{1}{2}(\psi_r - \psi_l) - \frac{c_h}{2}(B_{x,r} - B_{x,l}) \end{pmatrix}. \tag{5}$$

Here, B_x denotes the x component of the magnetic field vector **B**. The correction speed c_h in the above formula will be cell local. Consider two adjacent elements Q_i and Q_j with different correction speeds $c_{h,i}$ and $c_{h,j}$. The flux out of element Q_i into element Q_j will then be calculated with the speed $c_{h,i}$, the flux from element Q_j to Q_i with $c_{h,j}$. A similar approach within a different environment was already shown in [5]. Please keep in mind that for vanishing divergence errors, our proposed DG scheme is exactly conservative at all time. This flux can be added directly to the numerical flux of the MHD equations. Since we are solving 1D Riemann problems when calculating inter-cell fluxes from one element's side to its neighboring side, (5) is all we need even for a multi-dimensional scheme.

By increasing the divergence error transportation speed c_h, one will be able to remove errors even faster. Due to the increased number of time steps we then have to perform for the cleaning, this method has as a sub-cycling behavior: For each physical time step, we perform several divergence cleaning steps. When using this method, a global time step scheme would become computationally inefficient. However, by making use of local time steps, these limitations can be minimized. The local time stepping hereby ensures a computationally efficient operation.

An important question is how to choose the correction speed. It seems clear that strong divergence errors should be treated with high correction speeds while only small errors do not require going beyond the largest system speed of the MHD system. We are therefore directly making use of the divergence error itself by calculating the L_2 projection norm of the cell divergence and correlate it to the corresponding correction speed. By applying two different speed steps for the correction, the setting seems to be most efficient:

$$c_h = \begin{cases} c_{\min} & \| \nabla \cdot B \|_{L_2} \leq 5.0e-3 \\ 0.5\, c_{\max} \text{ for } 5.0e-3 \leq & \| \nabla \cdot B \|_{L_2} \leq 5.0e-2 \\ c_{\max} & \| \nabla \cdot B \|_{L_2} \geq 5.0e-2 \end{cases}, \tag{6}$$

where $c_{\min}$ denotes the maximum physical speed within the cell, determined by the CFL condition and $c_{\max}$ a user defined maximum correction speed, $c_{\max} \geq c_{\min}$. The introduction of speed steps is balancing areas of slightly varying divergence errors, building more consistent sub-cycling zones. This strategy showed to be superior to a direct mapping of the local divergence to the correction speed, since aliasing and resolution issues may badly affect the divergence calculation.

For analyzing efficiency, the divergence L_2-norm of a test problem with initially non-zero divergence is plotted over the CPU time for different correction settings in Fig. 1. Runs were performed by setting the divergence correction speed to be equal to the maximum CFL determined system speed, to be seven times this speed and to be in between these both speeds by using the setting (6) described above. It is obvious that the divergence correction with variable speed outperforms both other settings. This is due to the local time stepping functionality acting on the different correction speed steps.

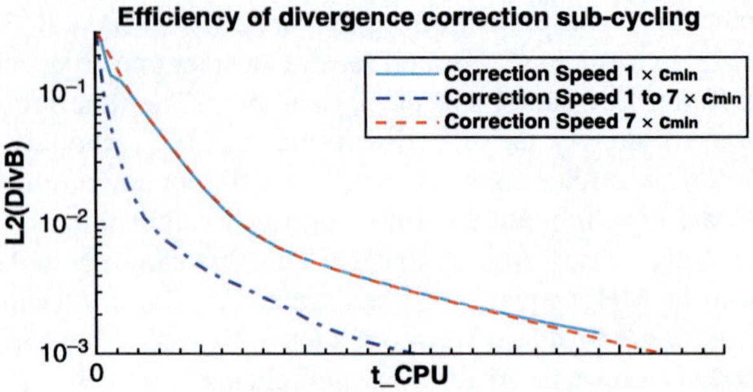

Fig. 1 Efficiency of the mixed GLM divergence correction with different correction speed settings

4 Numerical Results

In this section we focus on numerical results of the STE-DG scheme for MHD equations. For all test cases, $\gamma = \frac{5}{3}$.

4.1 Convergence Test

We were using the so-called manufactured solution technique for performing convergence tests of the STE-DG scheme for viscous MHD. To be able to find an exact solution, we choose an arbitrary smooth analytical function and insert it directly into the viscous MHD equations. For our test, we have chosen

$$\begin{aligned} \rho &= \sin(\beta)+2; & u &= 1; & v &= 1; \\ e &= \sin(\beta)+2; & B_x &= \sin(\beta)+2; & B_y &= -\sin(\beta)+2, \end{aligned} \tag{7}$$

with $\beta = 2\pi \sum_{j=1}^{dim} x_j - 4t$, where x_j are the spatial coordinates and t the time. Viscosity μ and resistivity η were set to 0.05. The resulting right hand side is then added as a source term into the code. Table 1 shows the convergence order of the STE-DG scheme for viscous MHD using third and fourth order DG polynomials. The L_2 error norm of the energy is used for the calculation. Table 1 shows that we do not obtain the desired order of accuracy when the divergence correction is neglected. In addition, also the absolute values of the error norms are significantly worse in that case. By using our proposed correction method, we were able to achieve the accuracy for both odd and even orders for the viscous MHD equations.

4.2 Orszag–Tang Vortex

The vortex system of Orszag and Tang [8] for ideal MHD is an ambitious test problem for almost any numerical scheme. In our case, the computational domain

Table 1 Order of accuracy of the STE-DG scheme with and without divergence correction

Nb cells	Nb DOF	$\|Eng\|_{L_2}$ w divB corr.	$\mathcal{O}_{L_2}$	$\|Eng\|_{L_2}$ w/o divB corr.	$\mathcal{O}_{L_2}$
		$\mathcal{P}3$ STE-DG			
4	160	1.82E-01		2.70E-01	
8	640	9.29E-03	4.3	1.98E-02	3.8
16	2,560	4.98E-04	4.2	9.60E-04	4.4
32	10,240	2.95E-05	4.1	6.58E-05	3.9
		$\mathcal{P}4$ STE-DG			
4	240	3.63E-02		4.34E-02	
8	960	9.82E-04	5.2	1.22E-03	5.2
16	3,840	3.07E-05	5.0	5.19E-05	4.6
32	15,360	9.36E-07	5.0	2.28E-06	4.5

is $[0;1] \times [0;1]$ with periodic boundaries. The initial condition of the problem is given by

$$\begin{aligned} \rho &= \gamma\, e; & u &= -\sin(2\pi y); & v &= \sin(2\pi x); \\ e &= \tfrac{10}{24}\pi; & B_x &= -\tfrac{1}{\sqrt{4\pi}}\sin(2\pi y); & B_y &= \tfrac{1}{\sqrt{4\pi}}\sin(4\pi x). \end{aligned} \tag{8}$$

The Mach number is set to 1.0. Ideal MHD calculations on a 50×50 and a 100×100 grid run up to $t = 0.5$. By then, several shocks have crossed the computational domain and a vortex system is formed near the center. Figure 2a, b show a very good agreement with reference results in the literature, e.g., [6] or [10]. For various numerical schemes without divergence cleaning, this test problem fails or will at least produce severe errors, see [6]. For the 100×100 calculation, we were able to keep the L_2 norm of the divergence errors below $1 \cdot 10^{-2}$. To capture the shock profiles, we are using artificial viscosity similar to [9]. One can see that the 100×100 calculation resolves the small-scale structures much better.

5 Conclusions

We have briefly presented the key ingredients of the STE-DG scheme to handle ideal and viscous MHD equations. Using the scheme's explicit local time stepping ability we were able to advance the GLM divergence correction method by using varying propagation speeds for performing the correction. The results of the presented test problems indicate that the scheme reaches the estimated order of accuracy and can handle MHD equations with divergence cleaning efficiently. An efficient adjustment of the divergence correction speed was also presented. Nevertheless, its settings may still be improved and are currently under investigation.

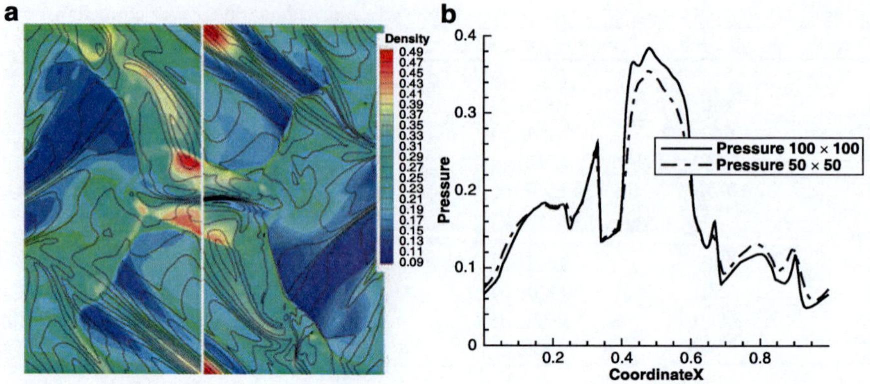

Fig. 2 (**a**) Density plot for the Orszag–Tang vortex at $t = 0.5$ on a 50×50 grid (*left*) and a 100×100 grid (*right*). Contour levels of the magnetic field magnitude are also shown. (**b**) 1D profile of the Orszag–Tang vortex at $y = 4.277$ for both calculations

References

1. Dedner, A., Kemm, F., Kröner, D., Munz, C.-D., Schnitzer, T. and Wesenberg, M. Hyperbolic divergence cleaning for the MHD equations. *J. Comput. Phys. 175*, 2 (2002), 645–673
2. Gassner, G., Lörcher, F. and Munz, C.-D. A contribution to the construction of diffusion fluxes for finite volume and discontinuous Galerkin schemes. *J. Comput. Phys. 224*, 2 (2007), 1049–1063
3. Gassner, G., Lörcher, F. and Munz, C.-D. A discontinuous Galerkin scheme based on a space-time expansion II. Viscous flow equations in multi dimensions. *J. Sci. Comp. 34*, 3 (2008), 260–286
4. Gassner, G. Discontinuous Galerkin Methods for the Unsteady Compressible Navier–Stokes Equations. *Dissertation*, Universität Stuttgart, 2009
5. Käser, M. and Dumbser, M. An arbitrary high-order discontinuous Galerkin method for elastic waves on unstructured meshes I. The two-dimensional isotropic case with external source terms. *Geo. J. Int. 166*, 2 (2006), 855–877
6. Li, F. and Shu, C.-W. Locally divergence-free discontinuous Galerkin methods for MHD equations. *J. Sci. Comp. 22–23*, 1 (2005), 413–442
7. Li, S. An HLLC Riemann solver for magneto-hydrodynamics. *J. Comput. Phys. 203*, 1 (2005), 344–357
8. Orszag, S. A. and Tang, C. M. Small-scale structure of two-dimensional magnetohydrodynamic turbulence. *J. Fluid Mech.*, (1979), 90–129
9. Persson, P.-O. and Peraire, J. Sub-cell shock capturing for discontinuous Galerkin methods. *Proc. of the 44th AIAA Aerospace Sciences Meeting and Exhibit*, (January 2006)
10. Ryu, D., Miniati, F., Jones T. W. and Frank, A. A divergence-free upwind code for multidimensional magnetohydrodynamic flows. *Astrophys. J.*, 509 (1998), 244–255
11. Warburton, T. C. and Karniadakis, G. E. A discontinuous Galerkin method for the viscous MHD equations. *J. Comput. Phys. 152*, 2 (1999), 608–641

High Order Polynomial Interpolation of Parameterized Curves

Tormod Bjøntegaard, Einar M. Rønquist, and Øystein Tråsdahl

Abstract Interpolation of parameterized curves differs from classical interpolation in that we interpolate each spatial variable separately. A difficult challenge arises from the option of *reparameterization*: a presumably good interpolation (e.g., at the Gauss points) of a given parameterization does not necessarily give the best approximation of the curve, as there may exist a reparameterization better suited for polynomial interpolation. The reparameterization can be done implicitly by choosing different sets of interpolation points along the exact curve. We present common interpolation methods, and propose a new method, based on choosing the interpolation points in such a way that the interpolant is tangential to the exact (reparameterized) curve at these points. The new method is compared to the traditional ones in a series of numerical examples, and results show that classical interpolation is sometimes far from optimal in the sense of the Kolmogorov n-width, i.e., the best approximation using n degrees-of-freedom.

1 Introduction

The topic of polynomial interpolation of parameterized curves appears in practical applications in high order methods for solving partial differential equations in deformed domains [3, 4]. The accuracy of the numerical solution is directly influenced by the accuracy of the geometry representation [7]. If the distortions are not too large, this representation can readily be achieved via a Gordon-Hall transfinite interpolation procedure [6]. For a deformed quadrilateral domain, this algorithm requires that we first construct an accurate representation

$$(x_N(\xi), y_N(\xi)), \qquad x_N, y_N \in \mathbb{P}_N(-1,1) \tag{1}$$

T. Bjøntegaard, E.M. Rønquist (✉), and Ø. Tråsdahl
Department of Mathematical Sciences, Norwegian University of Science and Technology, NO-7491 Trondheim, Norway
e-mail: ronquist@math.ntnu.no

J.S. Hesthaven and E.M. Rønquist (eds.), *Spectral and High Order Methods for Partial Differential Equations*, Lecture Notes in Computational Science and Engineering 76, DOI 10.1007/978-3-642-15337-2_34, © Springer-Verlag Berlin Heidelberg 2011

of each of the four boundary curves. This is merely an *approximation* of the exact curve, and an easy way to achieve a good approximation is through interpolation. Then the approximation problem simplifies to the problem of choosing a set of interpolation points.

In this paper we explore different ways of choosing these interpolation points. We compare previously proposed methods with a new method. The methods will be introduced in the context of plane curves, and then later extended to space curves. The accuracy of the different interpolation methods will be compared in numerical experiments.

2 Interpolation Methods for Plane Curves

The starting point is a given curve $y(x)$ in the plane, defined by the parameterization

$$(x(\eta), y(\eta)), \qquad \eta \in [-1, 1]. \tag{2}$$

We assume that $y(x)$ is C^1, so that there is a unique tangent vector at each point on the curve. Our numerical approximation is an interpolant based on a representation by high order polynomials (1). A nodal basis for the polynomial $x_N(\xi)$ is

$$x_N(\xi) = \sum_{j=0}^{N} x_j\, \ell_j(\xi),$$

and similarly for $y_N(\xi)$. Here, $\ell_j(\xi)$ is the jth Lagrangian interpolant through the Gauss–Lobatto–Legendre (GLL) points $\xi_i, i = 0, \ldots, N$, with the property that $\ell_j(\xi_i) = \delta_{ij}$. Hence, the expansion coefficients x_j and y_j are coordinates somewhere on the exact curve, i.e., $x_j = x(\eta_j)$ and $y_j = y(\eta_j)$ for some $\eta_j \in [-1, 1]$. We impose the restriction that the two end points of the numerical curve are interpolation points, i.e., $\eta_0 = -1$ and $\eta_N = 1$. However, we do not require the *internal* interpolation points η_j, $j = 1, \ldots, N-1$ to be the internal GLL points, as there always exists a *reparameterization* $(\tilde{x}(\xi), \tilde{y}(\xi))$, $\xi \in [-1, 1]$, such that the interpolation points are mapped from the GLL points in the reference domain, i.e., $x_j = \tilde{x}(\xi_j)$ and $y_j = \tilde{y}(\xi_j)$. The two parameterizations are connected by the relationship

$$\tilde{x}(\xi) = x(\eta(\xi)) = x\Big(\sum_{j=0}^{N} \eta_j \ell_j(\xi)\Big), \tag{3}$$

and correspondingly for y. The reparameterization is not unique [8], and we have here chosen η to be a polynomial of degree N in ξ. Equation (3) *implicitly* defines the reparameterization from the choice of interpolation points.

There are some widely known methods for choosing the values η_j. We will first describe them briefly, and then introduce two alternative methods.

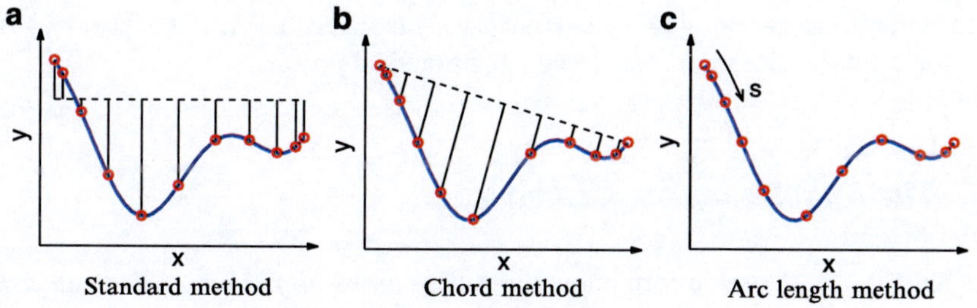

Fig. 1 Three common methods for choosing interpolation points

2.1 Common Interpolation Methods

The three most common interpolation methods all rely in some way on an affine mapping of the GLL points from the reference domain to the physical domain [4]. The first, which we will refer to as the *standard method*, uses an affine mapping $x_N(\xi)$ such that the interpolation points are distributed according to a GLL distribution along the x-axis. This implies that y_N will not only be a polynomial as a function of ξ, but also as a function of x_N.

We can also choose a GLL distribution along the chord between the two end points of the curve; see Fig. 1. This is the *chord method*, which coincides with the standard method when the chord is parallel to the x-axis.

The last method is based on a GLL distribution in the arc length variable s, and is called the *arc length method*.

2.2 The L^2-Method

The three previous methods each have special types of curves where they work well. However, we do not know how good the resulting interpolants are compared to the best possible interpolant.

In order to be able to define an optimal interpolant, we restrict our study to curves that can be described by a function $y(x)$ for $x \in [a, b]$. Then the L^2-norm can be used to measure the interpolation error, and we define the optimal set $\{\eta_j\}_{j=1}^{N-1}$ of internal interpolation points to be the one that minimizes the functional

$$\mathscr{J} = ||y - y_N||_{L^2}^2 = \int_a^b (y(x) - y_N(x))^2 \, dx. \tag{4}$$

We can differentiate $\mathscr{J}$ with respect to each independent variable η_j, $j = 1, \ldots, N-1$, and use Newton's method to search for the minimum. We will refer to this method as the L^2-method; see [2] for more details. The resulting minimizer can be viewed as the solution to the Kolmogorov n-width problem applied to the interpolation of curves. Note that we are searching for the *global* minimum of (4). Newton's method uses a local search, and is therefore dependent on a good initial guess.

In general, we are not guaranteed that $x_N(\xi)$ is invertible (i.e., that $y_N(x_N)$ is a function), but this does not seem to be a big practical problem.

2.3 The Equal-Tangent Method

The functional $\mathscr{J}$ uses information about the curves on the entire interval $[a, b]$, which makes the L^2-method slow and complicated as N increases. We therefore propose a method which uses information about the curves only at the interpolation points. The idea behind the equal-tangent method is to require that the exact and numerical curves are tangential at the $N-1$ internal interpolation points. This can be achieved if we are able to find the roots $\eta_1, \eta_2, \dots, \eta_{N-1}$ of the nonlinear system

$$\frac{\mathrm{d}x_N}{\mathrm{d}\xi}(\xi_j)\frac{\mathrm{d}y}{\mathrm{d}\eta}(\eta_j) - \frac{\mathrm{d}y_N}{\mathrm{d}\xi}(\xi_j)\frac{\mathrm{d}x}{\mathrm{d}\eta}(\eta_j) = 0, \qquad j = 1, \dots, N-1. \tag{5}$$

The left hand side represents an inner product between a tangent vector to the interpolant and a normal vector to the exact curve. In order to solve this system of equations, we will apply a Newton method. This requires that we differentiate the left hand side of (5) with respect to the $N-1$ independent variables η_j at the internal interpolation points.

We remark that the solution of (5) may not be unique; in such cases the particular solution obtained will depend on the initial guess. The existence of a solution in the general case has not been proven, however we have not yet encountered a counter-example. The method works well on a wide range of curves, and we will show a few examples here; see [2] for more details.

2.4 Numerical Results

The following examples are chosen to illustrate the behavior of the various methods in different situations. They are all given as functions $y(x)$; a parameterization (2) is readily achieved, using an affine mapping $x(\eta)$.

The interpolation error is measured in the discrete L^2-norm, where the integral in (4) is evaluated using GLL quadrature [2].

Case 1. The first example we consider is described by the function $y(x) = \frac{1}{1+16x^2}$, $x \in [-1, 1]$. Classical interpolation theory tells us that this function is particularly difficult to interpolate [5], and as the standard method yields a polynomial $y_N(x_N)$, we expect it to converge very slowly. Figure 2a confirms this, and it shows that the arc length method is even worse. Compared to this, the convergence rate of our proposed method is striking. By construction, the L^2-method is supposed to be the best, but it is only best for $N < 9$; as mentioned earlier, this is due to the complexity of computing the global minimizer of (4) as N increases.

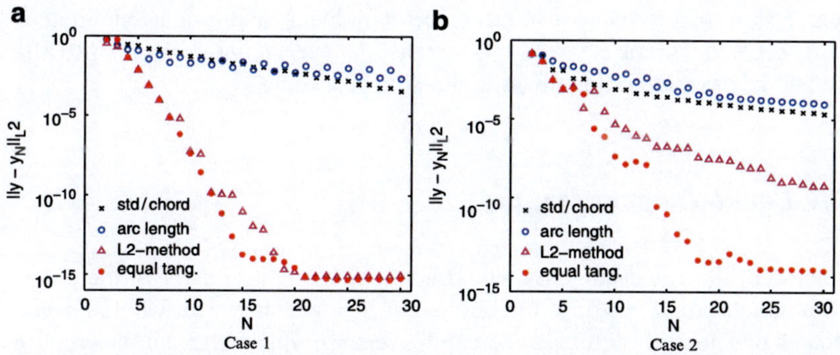

Fig. 2 Interpolation error, measured in the discrete L^2-norm. *Left*: for the Runge function, the standard method and the arc length method both converge slowly, but exponentially. The equal-tangent method and the L^2-method both converge much faster. *Right*: for this function of limited regularity, the standard method and the arc length method both yield algebraic convergence. The equal-tangent method, on the other hand, converges exponentially. The L^2-method performs reasonably well, but has difficulty finding the global minimum

Case 2. From classical interpolation theory we know that approximation of functions of limited regularity with polynomials results in algebraic convergence [1]. Consider the function $y(x) = 1 - |x|^3$, defined on $x \in [-1, 1]$. Both the standard method and the arc length method converge algebraically. The equal-tangent method, however, converges exponentially; see Fig. 2b. The L^2-method again converges fast only up to a certain value of N.

3 Interpolation of Space Curves

We now consider curves in space, defined by a given parameterization

$$(x(\eta), y(\eta), z(\eta)), \qquad \eta \in [-1, 1].$$

In order to be able to compare all methods, we restrict ourselves to curves where both y and z can be described by functions of x. Then, the standard, chord and arc length methods can all be extended in a natural way.

3.1 *The L^2-Method*

With our restriction on curves, we can define a functional similar to (4), extended to include the error in the z-variable:

$$\mathscr{J} = \int_a^b \left[(y(x) - y_N(x))^2 + (z(x) - z_N(x))^2 \right] \mathrm{d}x. \tag{6}$$

The integral is transformed to the reference variable ξ and evaluated numerically using GLL quadrature. With this extension, everything is similar to the two-dimensional case, including the minimization procedure.

3.2 The Equal-Tangent Method

The extension of the equal-tangent method is not as straightforward. In the plane, there is a unique normal vector to the curve, but in space there is a whole *normal plane*. Hence, one normal vector is not enough to ensure equal tangents. We propose a method where we use one normal vector from each coordinate plane,

$$\mathbf{n}_1 = \begin{bmatrix} 0 \\ -z'(\eta) \\ y'(\eta) \end{bmatrix}, \qquad \mathbf{n}_2 = \begin{bmatrix} z'(\eta) \\ 0 \\ -x'(\eta) \end{bmatrix}, \qquad \mathbf{n}_3 = \begin{bmatrix} -y'(\eta) \\ x'(\eta) \\ 0 \end{bmatrix}.$$

This is one more than we need to span the normal plane, but it gives symmetry in the space variables. Numerical experiments indicate that this may add to the robustness of the method. In order to realize the condition of orthogonality for all the three normal vectors, we *square* the inner products and take the sum

$$\sum_{i=1}^{3} (\mathbf{t}_N \cdot \mathbf{n}_i)^2 = 0, \tag{7}$$

where $\mathbf{t}_N = (x_N'(\xi), y_N'(\xi), z_N'(\xi))^T$. Newton's method applied to (7) do not result in the same set of equations as Newton's method applied to (5) for curves in the plane ($z(\eta) = 0$). However, both systems have the same sets of exact solutions.

3.3 Numerical Results

Case 3. The curve we are looking at is a distorted helix, spiraling along the x-axis with a varying radius. It is defined by the parameterization

$$\begin{aligned} x(\eta) &= -\frac{5}{2} + \frac{7}{4}(\eta + 1), \\ y(\eta) &= \frac{1}{2} e^{-(1+\eta)} \cos(2\pi\eta), \\ z(\eta) &= \frac{1}{8}(\eta + 2)\sin(2\pi\eta), \end{aligned}$$

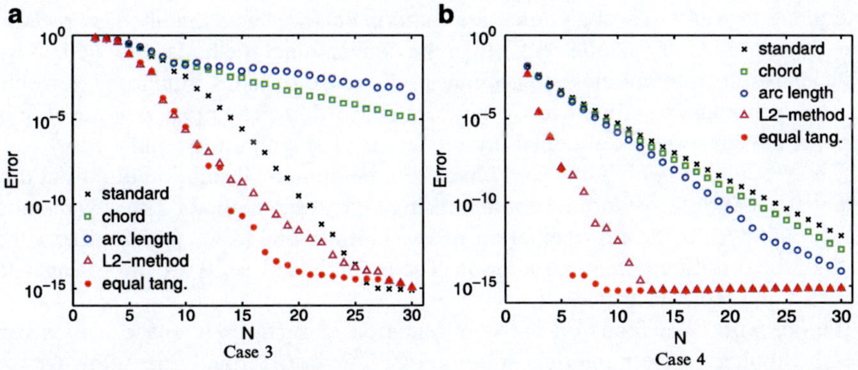

Fig. 3 Interpolation error, defined as the square root of (6). *Left*: the curve is well suited for the standard method, but we are still able to achieve faster convergence with the equal-tangent method and the L^2-method. *Right*: the parameterization consists of square and cubic roots, which makes the standard method a non-optimal choice. In particular, the curve can be reparameterized using polynomials of degree less than or equal to 6, which is detected by the equal-tangent method

for $\eta \in [-1, 1]$. Figure 3a shows that the situation is much the same as it was in two dimensions: the equal-tangent method is the best, and it almost coincides with the L^2-method up to $N = 11$. The standard method works well in this case due to the construction of the example, while the chord and arc length methods converge very slowly.

Case 4. Consider the curve parameterized by

$$x(\eta) = \eta + 1,$$
$$y(\eta) = \sqrt{\left(\eta + \frac{9}{4}\right)^{1/3} - 1},$$
$$z(\eta) = \left(\eta + \frac{9}{4}\right)^{2/3} - 1.$$

If we let $\eta(\xi) = ((\alpha\xi + \beta)^2 + 1)^3 - 9/4$ for suitable constants α and β, we get a reparameterization where $\tilde{y}(\xi)$ is affine and $\tilde{x}(\xi)$ and $\tilde{z}(\xi)$ are polynomials of degrees 6 and 4, respectively. Hence, the best distribution of interpolation points should give an *exact* representation of the curve from $N = 6$. Figure 3b shows that the equal-tangent method indeed finds this optimal solution, with no a priori knowledge of the optimal distribution of interpolation points.

4 Conclusions and Future Work

We have looked at interpolation of parameterized plane and space curves using high order polynomials. We have proposed a new method, iterative in nature, based on a requirement that the interpolant be tangential to the exact curve at the internal interpolation points. Through numerical experiments we show that the new method can give significantly smaller error than the conventional methods, and we believe it yields results that are close to optimal in the sense of the Kolmogorov n-width, i.e., the best approximation using n degrees-of-freedom. The most extreme case is exponential convergence obtained for a function $y(x)$ with finite regularity.

The motivation behind this study has been the numerical solution of partial differential equations in deformed domains using high order methods. The new method can be applied to the representation of the domain boundary, which affects the error of the resulting numerical solution. The preliminary results are promising, and reported in a separate article [2].

Future work will focus on the representation of surfaces in space, which can then be applied to the numerical solution of PDEs in deformed three-dimensional domains.

Acknowledgements The work has been supported by the Research Council of Norway under contract 185336/V30. The support is gratefully acknowledged.

References

1. C. Bernardi and Y. Maday. Spectral methods. In P.G. Ciarlet and J.L. Lions (eds), *Handbook of Numerical Analysis, Vol. V: Techniques of Scientific Computing (Part 2)*, pages 209–485. Elsevier, Amsterdam, 1997
2. T. Bjøntegaard, E.M. Rønquist, and Ø. Tråsdahl. High Order Interpolation of Curves in the Plane. Technical report, Norwegian University of Science and Technology, http://www.math.ntnu.no/preprint/numerics/2009/N11-2009.pdf, 2009
3. C. Canuto, M.Y. Hussaini, A. Quarteroni, and T.A. Zang. *Spectral Methods. Evolution to Complex Geometries and Applications to Fluid Dynamics*. Springer, New York, 2007
4. M.O. Deville, P.F. Fischer, and E.H. Mund. *High-Order Methods for Incompressible Fluid Flow*. Cambridge University Press, Cambridge, 2002
5. B. Fornberg. *A Practical Guide to Pseudospectral Methods*. Cambridge University Press, Cambridge, 1996
6. W.J. Gordon and C.A. Hall. Construction of curvilinear co-ordinate systems and applications to mesh generation. *International Journal for Numerical Methods in Engineering*, 7:461–477, 1973
7. Y. Maday and E.M. Rønquist. Optimal error analysis of spectral methods with emphasis on non-constant coefficients and deformed geometries. *Computer Methods in Applied Mechanics and Engineering*, 80(1–3):91–115, 1990
8. B. O'Neill. *Elementary Differential Geometry*. Academic Press, New York, 2006

A New Discontinuous Galerkin Method for the Navier–Stokes Equations

M. Borrel and J. Ryan

Abstract We introduce a new discontinuous Galerkin (EDG) method to solve the compressible Navier–Stokes equations where jumps across element boundaries are eliminated in the computation of the viscous fluxes using an L^2 projection of the discontinuous solution on the basis of overlapping elements (elastoplast). This method is related to the recovery method presented by Van Leer and Lo (AIAA paper, 2007-4003), and similarly it is compact and stable without introducing penalty terms. A comparison on a 1D convection-diffusion problem in terms of accuracy and stability with other viscous DG schemes is given. Finally, the first 2D results both on Cartesian and unstructured grids illustrate stability, precision and versatility of this method.

1 Introduction

Discontinuous Galerkin (DG) methods have become the subject of considerable research over the last decade due to their ability to give high order solutions in complex applications. Albeit well suited to the discretization of first order hyperbolic problems such as wave propagation phenomena, their extension to elliptic problems such as diffusion, is far less natural and still an up-to-date subject.

We can classify these extensions into two categories. In the first one, the scheme is devised through a mixed formulation by introducing an equation for the gradient that allows to take into account the jump of the solution at interfaces. The scheme needs to be stabilized by either interior penalty terms or numerical viscosity terms with parameters to be adjusted. Depending on the formulation, the resulting scheme is either compact or non compact.

Among the main contributors to this first category, we can cite Bassi and Rebay with their BR1 and BR2 methods for the compressible Navier–Stokes equations

M. Borrel and J. Ryan (✉)
Onera, BP 72 – 29, av. de la Division Leclerc, 92322 Chatillon, France
e-mail: borrel@onera.fr, ryan@onera.fr

J.S. Hesthaven and E.M. Rønquist (eds.), *Spectral and High Order Methods for Partial Differential Equations*, Lecture Notes in Computational Science and Engineering 76, DOI 10.1007/978-3-642-15337-2_35, © Springer-Verlag Berlin Heidelberg 2011

[1, 2], Cockburn and Shu with the LDG method [3], Peraire and Person with the CDG method [4], Brezzi et al. [5,6] with the symmetric interior penalty (IP) method. In [7], Munz et al. show the link between their diffusive generalized Riemann solver and the IP approach.

A second category is based on local reconstruction or recovery of the solution to smooth the discontinuities. Van Leer [8] was the first to propose a recovery method where the viscous fluxes at element boundaries are computed by merging the adjacent elements and defining on this new element a locally smooth P_{2k} recovered solution that is in the weak sense indistinguishable from the piecewise discontinuous P_k solution. This method eliminates the introduction of penalty terms and the tuning of parameters. An impediment is the construction of the local merging basis and the need to solve a linear problem at each interface which can be awkward if we use an adaptive strategy on unstructured grids.

In this paper, we develop a new DG method for the compressible Navier–Stokes equations where jumps across element boundaries are eliminated in the computation of the viscous fluxes using an L^2 projection of the piecewise P_k discontinuous solution on the P_k basis of overlapping rectangular elements: so, we propose to label this method the elastoplat DG method. This method is a sequel to the shift cell technique that uses the Green formula [9] that reconstructs the gradient by projection on the shift cell basis. The main motivation for developing the elastoplast method, which is closely related to Van Leer's recovery method, is to devise a simpler numerical procedure easily implemented on unstructured grids. This paper is devoted to a presentation of the method and an evaluation of its performances.

2 Numerical Discretization

2.1 DG Formulation and Time Stepping

The governing equations to be solved are the 2D time-dependent Navier–Stokes systems for a Newtonian compressible flow which express conservation of mass, momentum and energy,

$$\partial_t \mathcal{W} + \nabla \cdot \mathbf{F}_C(\mathcal{W}) - \nabla \cdot \mathbf{F}_D(\mathcal{W}, \nabla\mathcal{W}) = \mathbf{0} \qquad (1)$$

where $\mathcal{W} = (\rho, \rho\overrightarrow{\mathrm{U}}, \rho\mathrm{E})$ is the conservation variable vector with classical notation, $\mathbf{F}_C$ and $\mathbf{F}_D$ are the convective and diffusive fluxes.

Pressure is given with the perfect gas state law with a constant specific heat ratio $\gamma = 1.4$. Finally, we assume the gas to be calorifically perfect with the Prandtl number $Pr = 0.72$.

These equations are solved in a domain Ω discretized by a Cartesian or an unstructured partition $\mathcal{T}_h = \bigcup \Omega_i$ and the associated function space V_h,

$$V_h = \{\phi \in L^2(\Omega) \mid \phi/\Omega_i \in P_k\} \tag{2}$$

where P_k is the space of polynomials of degree k.

The DG formulation based on a weak formulation after a first integration by parts is of the form: find W in $(V_h)^4$ such that for all Ω_i in $\mathcal{T}_h$,

$$\forall \phi \in V_h, \quad \int_{\Omega_i} \partial_t W \, \phi \, dx = \int_{\Gamma_i} (F_C - F_D)\, \phi \, d\gamma - \int_{\Omega_i} (F_C - F_D) \nabla \phi \, dx \tag{3}$$

Here, the numerical fluxes F_C,F_D and W are approximations of $\mathbf{F}_C$, $\mathbf{F}_D$ and $\mathcal{W}$. The inviscid flux F_C is determined using the HLLC [10] or LLF techniques and we will detail in the next section the viscous flux computation.

If we neglect locally the dependancy of μ on temperature, the viscous term F_D can be split into a linear and a nonlinear part,

$$F_D = L(\nabla \vec{U}, \nabla T) + N(\vec{U} \cdot \tau) \tag{4}$$

where T is the temperature and $\vec{U}$ the velocity. A second integration by parts can be done on $L(\nabla \vec{U}, \nabla T)$ thus giving the following ultra weak formulation,

$$\begin{aligned} \forall \phi \in V_h, \quad \int_{\Omega_i} \partial_t W \, \phi \, dx = & \int_{\Gamma_i} (F_C - F_D + L(\vec{U}, T))\, \phi \, d\gamma \\ & - \int_{\Omega_i} (F_C + L(\vec{U}, T) - N(\vec{U} \cdot \tau)) \nabla \phi \, dx \end{aligned} \tag{5}$$

Finally, this formulation results in a system of coupled ordinary differential equations of the form,

$$M \, \partial_t W^h = R(W^h) \tag{6}$$

where W^h is the vector containing the degrees of freedom associated to W expressed in a basis of V_h. Here, M is the mass matrix, which is diagonal in our computations due to our choice of the basis functions, while R is the residual vector which is a nonlinear function of W^h. We have chosen the explicit time stepping RK3 of Shu–Osher [12] to solve (6).

2.2 The Elastoplast Method (EDG)

The simple idea of the elastoplast method is to reconstruct numerically the solution W^h over each edge of the cell in a rectangular cell R_e overlapping this edge (see Fig. 1). Reconstruction is done through an L^2 projection on a DG basis of R_e of the same order k as the original solution using on either side of the edge an equal number of Gauss quadrature points, the sum of which provides at least the order of the original solution.

More precisely, for any interface Γ between Ω_1, Ω_2, for all ϕ_r^p in the DG-P_k basis of R_e, $W_r = \sum_p W_r^p \phi_r^p$ where W_r^p is defined by:

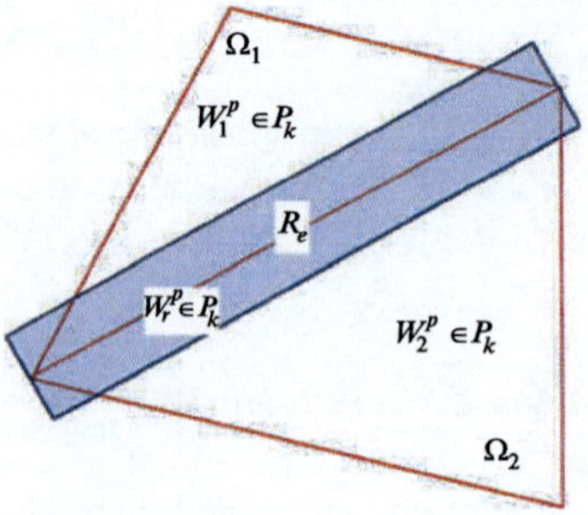

Fig. 1 General elastoplast technique

$$W_r^p = \frac{1}{(\phi_r^p, \phi_r^p)} \left(\int_{\Omega_1 \cap R_e} W_1^p \phi_r^p dx + \int_{\Omega_2 \cap R_e} W_2^p \phi_r^p dx \right) \tag{7}$$

where W_1^p,W_2^p are the local DG-P_k solutions in Ω_1, Ω_2, W_r the reconstructed solution in R_e and $(\cdot,\cdot)$ the L^2 product in R_e, (i.e.:$(f, g) = \int_{R_e} fg dx$). All integrals are numerically computed using a n-point Gaussian quadrature rule, such that $2 * n - 1 \geq k$, k order of the local DG formulation.

This method is general to any polygonal cells.

2.2.1 DG Basis and Implementation

If (x_0, y_0) is the Cartesian or unstructured cell (Ω_i) center, the local DG-P_k basis is built by orthogonalising in $L^2(\Omega_i)$ the function set $(x - x_0)^i \ (y - y_0)^j$, $(0 \leq i \leq k, 0 \leq j \leq k)$ where k is the DG discretization order.

Note on the unstructured grid implementation: All L^2 products on a triangle are done by mapping a triangle in (x, y) space to the standard 2D square: $\{(\xi, \eta) \mid -1 \leq \xi, \eta \leq 1\}$. Thus numerical orthogonalisation uses the same rectangular Gauss quadrature procedures (see [11]).

This allows for a common solver for both structured or unstructured grids and projection onto the overlapping cell basis is simplified as all data have the same structure. It is also to be noticed that the use of a rectangular overlapping cell R_e naturally gives rise to a diagonal mass matrix, thus saving storage costs.

3 Numerical Results

All computations are DG-P_2 and no limiters were used, except for the mixing layer test case. For both Cartesian and unstructured computations, we used the same functional space V_h.

Table 1 Comparison of experimental order of convergence between EDG and alternative methods

# cells	EDG	Order	BR1	Order	BR2 ($\eta = 2$)	Order	Recovery	Order
10	4.92 e–05		9.62 e–05		1.42 e–04		1.35 e–05	
20	3.13 e–06	3.97	5.70 e–06	4.07	8.81 e–06	4.01	8.15 e–07	4.05
40	1.96 e–07	3.99	3.51 e–07	4.02	5.50 e–07	4.00	5.02 e–08	4.01
80	1.22 e–08	3.99	2.18 e–08	4.00	3.43 e–08	4.00	3.12 e–09	4.00
D	0.06		0.03		0.04		0.07	

3.1 1D Diffusion

Comparison between some classical formulations are shown in Table 1 of the experimental order of convergence for the unsteady diffusion problem $u_t = \nu\ u_{xx}$ with sinusoidal initial data $u(x, 0) = sin(2\ \pi\ x), x \in [0, 1]$.

In all these computations, for each method, the time step was chosen as large as possible satisfying a diffusion stability requirement $\nu \Delta t / \Delta x^2 \leq D$. The L_∞ norm of the error is computed at time $T_{end} = 1$ and $\nu = 1$.

The behavior of the presented EDG method compares well in terms of stability and accuracy with the most popular methods such as Bassi and Rebay's methods (BR1, BR2) and Van Leer's recovery method.

3.2 Couette Thermal Flow

This is a simple case to verify the discretization of viscous effects in a laminar flow between two parallel walls at distance L = 1., a fixed lower wall (U1 = 0.) and a moving upper wall at velocity (U2 = 1.). Both walls are isothermal at different temperatures (lower wall T1 = 293., upper wall T2 = 294.). As the plates have infinite length, there is no physically relevant length scale in the streamwise direction. The analytical steady solution in this case is $U(x, y) = y$ and $\dfrac{\partial^2 T}{\partial y^2} = -\dfrac{\mu}{\kappa}$ where μ and κ are viscosity and thermal diffusivity coefficients. In our computation $\mu = 1/Re$, with the Reynolds number Re = 500. and $\kappa = \dfrac{\gamma \mu}{Pr}$. This case was computed with the full Navier–Stokes solver using an 9×3 mesh, imposing periodic boundary conditions in the streamwise direction and flow variables are imposed in both North and South fictitious cells. Convergence plotted in Fig. 2 is rather slow in terms of number of time iterations due to the use of an explicit time stepping, but is quite regular which shows that no numerical uncertainties remain in the numerical procedures. At convergence, the computed solution is exact. On Fig. 3, solution is also plotted for a 4×3 mesh, showing mesh independence.

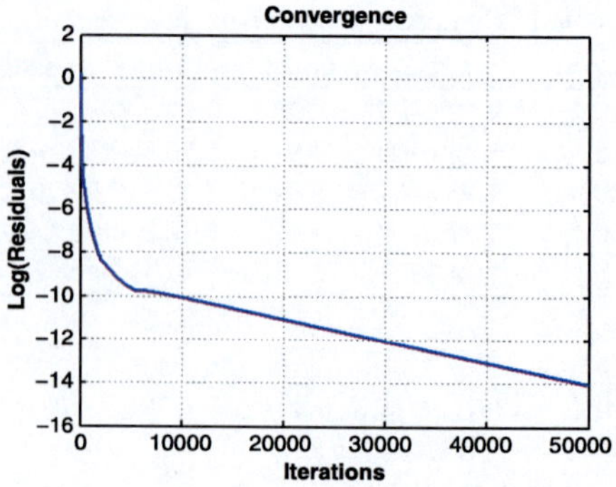

Fig. 2 Numerical convergence

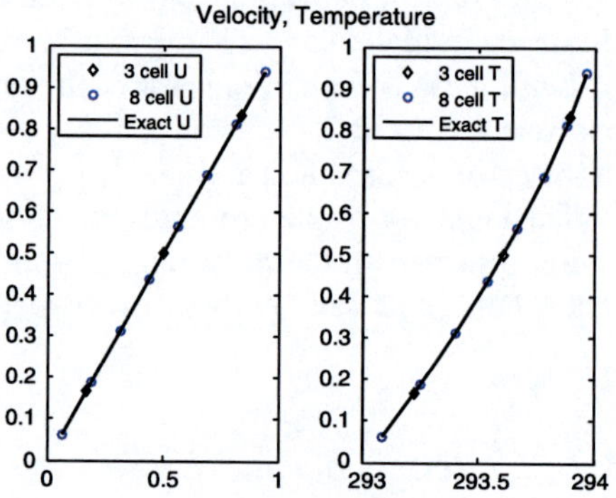

Fig. 3 Numerical vs. Couette profiles

3.3 Blasius Boundary Layer

This other classical test case consists in a boundary layer which develops over a plane plate immersed in an subsonic uniform flow. The numerical solution is compared with the incompressible Blasius solution obtained in the reduced wall variables (as given in [13]). The infinity Mach number is $M = 0.5$ and the Reynolds number with respect to the length 1 of the plane plate is $Re = 10{,}000$. Wall boundary conditions are imposed using a fictitious cell technique where values of the DG variable are such that antisymmetric conditions are imposed on momentum, and symmetric conditions on density and energy. Reconstruction of values on the interface are computed as for inner edges.

Our computational domain is such that the west boundary is situated upstream of the leading edge and the East boundary cuts the plate at L = 1. North boundary is situated at L/2 from the plate. This domain is discretized with a 29 × 19 Cartesian grid with a moderate refinement at the wall ($\delta y = 4.e - 03$) and at the leading edge ($\delta x = 4.e - 03$). At section x = 0.5, where the profiles are plotted, there are approximately 12 meshes inside the boundary layer. The convergence plotted in Fig. 4 is again rather slow in terms of number of time iterations but is still regular which shows that the no-slip boundary treatment does not modify the scheme

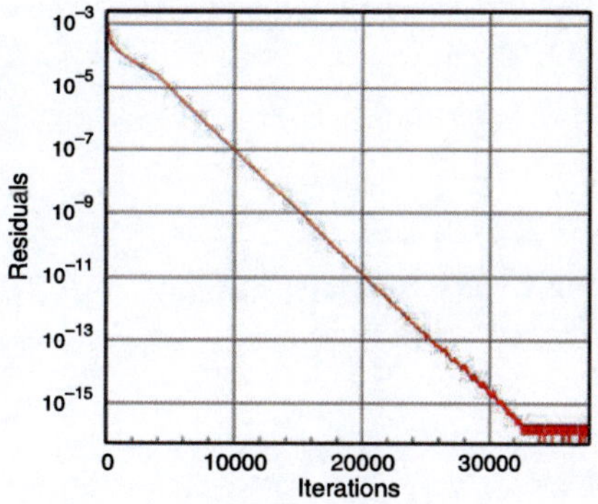

Fig. 4 Numerical convergence

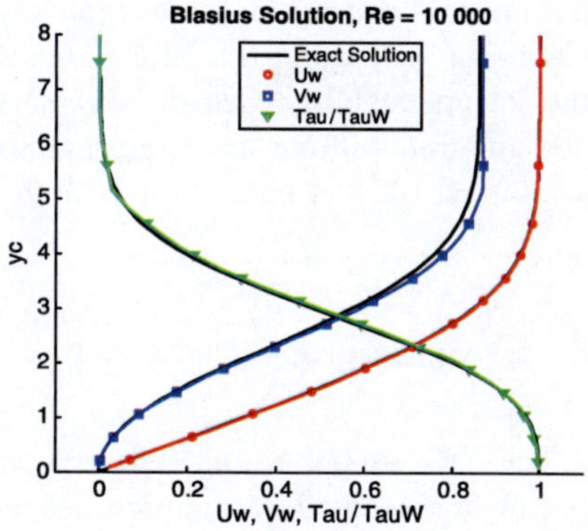

Fig. 5 Numerical vs. Blasius profiles

stability. This is confirmed with the close comparison of the profiles of the velocity components and the friction coefficients with the Blasius solution (Fig. 5).

Note: The exact solution is for an infinite plate. Our computed solutions are self similar around x = 0.5 but for stations close to the leading edge or close to the East boundary where an outlet boundary is imposed, velocity in the wall normal differs from the Blasius profile as mentioned in Hirsch's book [13].

3.4 *Supersonic Mixing Layer*

The fourth case has been developed to test the behavior of the scheme for shear layer vortices. We have chosen a supersonic configuration with Mach numbers of 2 and 4 for the bottom and upper flows in order that the outflow boundary remains supersonic everywhere. The shear layer is initialized just on one interface at $t = 0$ and a small Gaussian perturbation in time is imposed for the velocity components at the inflow boundary $x = 0$. Cockburn and Shu's generalized slope limiter (see [3]) was used for this computation. Figures 6 and 7 plotting entropy contours compare

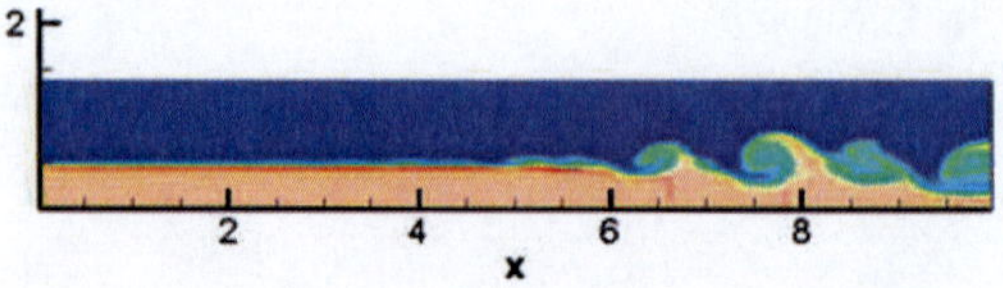

Fig. 6 Euler mixing layer, LLF flux

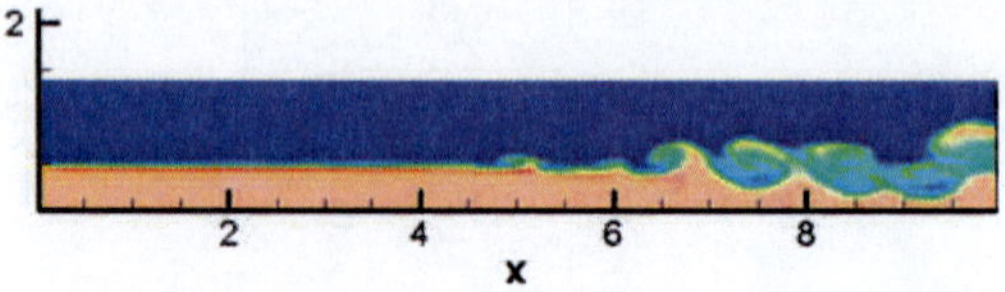

Fig. 7 NS mixing layer, LLF flux

the inviscid and viscous flows (at Reynold number Re = 200,000) on a 100×70 Cartesian grid using the LLF convective Flux after 50,000 time steps enhancing the scheme stability. (It can be remarked that the initial shear layer in an inviscid computation without any perturbations remains unchanged: this is the case if we use the HLLC flux but not the LLF flux.)

4 Conclusions

A new DG scheme to discretize the viscous fluxes in the Navier–Stokes equations has been presented. Discontinuities are removed near each interface by projection on an overlapping rectangular element (elastoplast). The main advantage of the proposed scheme is its simplicity to implement either on a Cartesian or an unstructured grid. Close to Van Leer's recovery method, it is compact, stable and accurate, without penalty terms, First numerical results indicate that the accuracy and stability compare well with that of alternative schemes such as the BR2 or the recovery method. Finally, the proposed scheme seems to be a well adapted method for DNS or LES and especially if we use a dynamic AMR technique with hybrid meshes.

References

1. F. Bassi, S. Rebay, *A high-order accurate discontinuous finite element method for the numerical solution of the compressible Navier-Stokes equations*, J. Comput. Phys., 131 (1997), 267–279
2. F. Bassi, S. Rebay, G. Mariotti, S. Pedinotti, M. Savini, *A high-order accurate discontinuous finite element method for viscous and turbomachinery flows*, Proceedings of the 2nd European Conference on turbomachinery Fluid Dynamics and Thermodynamics, Belgium (1997), pp. 99–108
3. B. Cockburn, C.-W. Shu, *Runge-Kutta discontinuous Galerkin methods for time-dependent convection-diffusion systems*, SIAM J. Numer. Anal., 35 (1998), 2440–2463
4. J. Peraire, P.-O. Persson, *The compact discontinuous Galerkin (CDG) method for elliptic problems*, SIAM J. Sci. Comput., 30 (2008), 1806–1824

5. J. Douglas, Jr., T. Dupont, *Interior penalty proceedures for elliptic and parabolic Galerkin methods*, Lecture Notes in Phys., 58 (1976), 207–216
6. F. Brezzi, G. Manzini, D. Marini, P. Pietra, A. Russo, *Discontinuous Galerkin approximations for elliptic problems*, Numer. Meth. Partial Didd. Eqns., 16 (2000), 365–378
7. G. Gassner, F. Lörcher, C.-D. Munz, *A contribution to the construction of diffusion fluxes for finite volume and Discontinuous Galerkin schemes* J. Comput. Phys., 224 (2007), 1049–1064
8. B. van Leer, M. Lo, *A Discontinuous Galerkin Method for Diffusion Based on Recovery*, AIAA paper 2007-4003, 18th CFD Conf., Miami
9. C. Drozo, M. Borrel, A. Lerat, *Discontinuous Galerkin schemes for the compressible Navier-Stokes equations*, Proceedings of the 16th ICNMFD, Arcachon, *Springer Ed.*, 266–271 (1998)
10. E. F. Toro, M. Spruce, W. Speares, *Restoration of the contact surface in the HLL-Riemann solver*, Shock Waves, 4 (1994), 25–34
11. H. T. Rathod, K. V. Nagaraja, B. Venkatesudu, N. L. Ramesh, *Gauss Legendre quadrature over a triangle* J. Indian Inst. Sci., 84 (2004), 183–188
12. C.-W. Shu, S. Osher, *Efficient implementation of essentially non-oscillatory shock-capturing schemes* J. Comput. Phys., 77 (1988), 439–471
13. C. Hirsch, *Numerical Computation of Internal and External Flows* Volume 1 (2nd ed.) 2007, Elsevier, Amsterdam

A $P_n^{\alpha,\beta}$-Based Method for Linear Nonconstant Coefficients High Order Eigenvalue Problems

F.I. Dragomirescu

Abstract A weighted residual method based on generalized Jacobi polynomials is proposed to solve a class of eigenvalue problems governing the linear stability of the mechanical equilibria of certain fluids occurring in complex circumstances. One concrete natural convection problem of great interest from the applications point of view is numerically investigated. Fairly accurate approximations of the lower part of the spectrum are given in comparison with other numerical evaluations existing in the literature.

1 Introduction

The approximation of a function by a finite sum of basis functions is the basic idea in spectral methods. It is also well known the fact that the choice of the expansion basis functions influences the superior approximation properties of spectral methods when compared to finite difference and finite element methods. The expansion functions must have a basic property: they must be easy to evaluate. That is why, most of the times, trigonometric and polynomials are used in the discretization process. Another property concerns the completeness of these families of functions such that each function of the given space can be represented as a limit of a linear combination of such functions. Chebyshev, Legendre, Jacobi polynomials, sine and cosine functions satisfy this property. Orthogonality of the expansion functions is however one of their most important property. In fact, orthogonality of classical polynomials is the key for the study of many properties of these polynomials and their intensive applications. Over a finite interval much is known about expansion properties and periodic Fourier expansions or polynomial expansions are intensively studied [4,5]. Usually the most used in polynomials based spectral methods are the Chebyshev

F.I. Dragomirescu
"Politehnica" University of Timisoara, 300006 Timisoara, Romania
e-mail: ioana.dragomirescu@mat.upt.ro

J.S. Hesthaven and E.M. Rønquist (eds.), *Spectral and High Order Methods for Partial Differential Equations*, Lecture Notes in Computational Science and Engineering 76,
DOI 10.1007/978-3-642-15337-2_36,

and Legendre polynomials. In many scientific papers [1,7,9] it was emphasized that for spectral methods the use of Chebyshev polynomials has been proving advantageous, offering a great accuracy in the discretization process. However, there has been a renewed interest in using Jacobi polynomials in spectral methods lately due to the complicated eigenvalue problems for fluid flows with just simply nonconstant, degenerated or singular coefficients.

In this paper generalized Jacobi polynomials (GJP) basis of functions are used to investigate a certain class of eigenvalue problems, in particular, a benchmark model is used, i.e. an eigenvalue problem governing the linear stability of the mechanical equilibria of a fluid layer heated from below in the presence of a linear variable gravity field decreasing across the layer for both the rigid and the free boundary case. The aim of the paper is to emphasize the influence of the parameters α, β, indexes of the GJP, on the numerical evaluations of the critical eigenvalue for this type of high order eigenvalue problem. The paper is organized as follows. The present section stands like a motivation of such a study. In the second section of the paper some details on the physical problem and its importance are given. The mathematical problem, i.e. the eigenvalue problem, is also defined. The third section provides detailed exposition of the numerical results. Some general conclusions concerning the application of Jacobi polynomials to solve eigenvalue problems with nonconstant coefficients will be drawn in the last section of the paper.

In an article from 2003, Shen [10] pointed out that basis functions which are in fact compact combinations of Legendre polynomials can be viewed as GJP with negative indexes. The use of such polynomials not only simplifies the numerical analysis, but also leads to very good numerical algorithms for high odd-order differential equations. There is no need for construction of special quadratures involving derivatives at the end-points as in the collocation approach [3]. For systems of ordinary differential equations with constant coefficients one property can be emphasized: the linear systems obtained with this algorithm are well conditioned and sparse. Guo, Shen and Wang [8] also introduced a family of GJP with real indexes which are in fact orthogonal with respect to the corresponding Jacobi weights and some of their basic properties were investigated. In [3] some efficient basis choices using GJP are presented. These approximations functions are used in a Jacobi–Galerkin method in order to solve a general one-dimensional fourth-order equation subject to various boundary conditions. The presented elliptic equation defined on $\Omega = I^d$ [3] is usually subject to the first type boundary conditions $u|_{\partial\Omega} = \frac{\partial u}{\partial n}|_{\partial\Omega} = 0$ or either the type second boundary conditions $u|_{\partial\Omega} = \frac{\partial^2 u}{\partial n^2}|_{\partial\Omega} = 0$ where $I = (-1, 1)^d$, $d = 1$ or 2, and n is the outward normal vector on $\partial\Omega$ [3]. These two types of boundary conditions correspond to the rigid and the free boundary case, respectively. The advantages of using such GJP is that each of the expansion functions automatically satisfy all given boundary conditions of the underlying problem.

2 Physical and Mathematical Preliminaries

In [2, 3, 8, 10] the GJP based methods were always applied for eigenvalue problems defined by systems of ordinary differential equations with constant coefficients. It is our purpose to study the applicability of such methods to eigenvalue problems defined by systems of ordinary differential equations with non-constant coefficients and when more than one physical parameter is involved. Although their study started decades ago, due to the lack of spectral theory for nonselfadjoint operators with non-constant coefficients and also to the mathematical problems involved (the completeness of normal modes, the principle of exchange of stabilities) usually this type of problems are investigated as particular ones. Let us consider a class of such two-point boundary value problems governing the linear stability of the mechanical equilibria of the fluid for several type of fluid motions [6]

$$\begin{cases} (D^2 - a^2)^2 U = F(z)V, \\ (D^2 - a^2)V = -a^2 R G(z) U, \end{cases} \tag{1}$$

$$U = DU = V = 0 \text{ at } z = 0, 1 \tag{2}$$

where D denotes the differentiation with respect to the variable z, i.e. $D := \frac{d}{dz}$, $F(z)$, $G(z)$ are known indefinitely derivable functions, a is the wavenumber and R is the eigenparameter (the most important physical parameter of the problem). The unknown functions U and V stand for the amplitudes of the perturbations fields encountered in various convection problems. In (1) and (2) the vector (U, V) represents the eigenvector and R is the corresponding eigenvalue. In the general case, the basic mathematical problem is: given $F(z)$ and $G(z)$ determine the smallest value of R for $a > 0$ such that a solution of (1) and (2) exists. In the following we will restrict our attention on a particular physical case of natural convection in the presence of a variable gravity field.

The fluid and atmosphere dynamics introduces variations in the gravity field. Many gravity fields can be encountered in applications, some of them extremely important in crystal growth or other convective flows from biomechanics, chemistry, so on [11]. The onset of convection in a horizontal layer of fluid heated from below in a presence of a gravity field linearly decreasing across the layer is numerically investigated here. Consider the horizontal layer of fluid situated between $z = 0$ and $z = h$ with the gravity field acting in the vertical direction and assumed orthogonal on the fluid layer. The linear stability of the corresponding conduction stationary solution against normal mode perturbations governed in this case by a two-point problem of the type (1) and (2) with $F(z) = 1 - \epsilon z$, $G(z) = 1$, $U = W$ the amplitude of the vertical component of the velocity and $V = \Theta$ the amplitude of the temperature perturbation, R representing the Rayleigh number and ϵ the scale parameter defining the variation in the gravity field. The boundary conditions on a free surface are that the perturbations of the stress components are zero. This implies that a free surface behaves as a rigid one with tangential slip but without any tangential stress [6]. In our case, for normal modes perturbations, in the free boundaries case, the boundary conditions reduce to

$$W = D^2 W = \Theta = 0 \text{ at } z = 0, 1. \tag{3}$$

Straughan provided in [11] numerical evaluation of the Rayleigh number in the case of rigid boundaries by using the energy method. Stability bounds were found by us in the case of rigid boundaries [4] using spectral methods based on trigonometric series for more than a linearly decreasing variable gravity field and, moreover, the eigenvalue problem (1) and (2) was studied using expansion series based on shifted polynomials (Legendre and Chebyshev polynomials) [5].

The necessary spaces to characterize the discretization process here are introduced in the following. Let us denote $S_N = \text{span}\{P_0^{\alpha,\beta}, P_1^{\alpha,\beta}, \dots, P_N^{\alpha,\beta}\}$ the space of the Jacobi polynomials $P_k^{\alpha,\beta}$, $k > 0$ defined by the Rodrigues formula [3]

$$P_n^{\alpha,\beta} = \frac{(-1)^n}{2^n n!}(1-x)^{-\alpha}(1+x)^{-\beta} D^n[(1-x)^{\alpha+n}(1+x)^{\beta+n}], \tag{4}$$

with α, β two complex parameters.

The classical Jacobi polynomials associated with the real parameters $\alpha, \beta > -1$ are a sequence of orthogonal polynomials, i.e. $\int_{-1}^{1} P_m^{(\alpha,\beta)}(x) P_n^{(\alpha,\beta)}(x) w^{\alpha,\beta}(x) dx = \gamma_n^{\alpha,\beta} \delta_{n,m}$ with $w^{\alpha,\beta}(x) = (1-x)^{\alpha}(1+x)^{\beta}$ the Jacobi weight function, $\delta_{n,m}$ the Kronecker symbol and $\gamma_n^{\alpha,\beta} = \frac{2^{\lambda}\Gamma(n+\alpha+1)\Gamma(n+\beta+1)}{(2n+\lambda)\Gamma(n+1)\Gamma(n+\lambda)}$, $\lambda = \alpha + \beta + 1$. Jacobi polynomials can also be defined using (4) for $\alpha, \beta < -1$. However, the main property used in numerical applications, the orthogonality in $L^2_{w^{\alpha,\beta}}$ for all α, β of these polynomials it is no longer valid.

In order for spectral or pseudo-spectral methods to provide numerical solutions having a high rate of convergence, boundary conditions must be well taken into consideration. The boundary conditions can be either imposed separately or a restriction of the numerical operator to exactly enforce the boundary conditions has to be imposed. Although the first approach is more general and widely used, enforcing the boundary conditions ensures that for any given order the boundary conditions are exact to machine precision. That is why, a careful restriction of the polynomial space in which the solution is sought so that it automatically satisfies the boundary conditions must be constructed. A suitable transformation imply a mapping of the physical domain onto the standard interval of definition of the Jacobi polynomials, i.e. $x = 2z - 1$, such that the boundary conditions are written at -1 and 1. Let us consider the spaces

$$W_{N_i} = \{\phi_k \in S_N : \phi_k = D^i \phi_k = 0 \text{ at } x = -1, 1\}, \quad i = 1, 2$$

of ϕ_k functions which we will define later on. We will expand the eigenfunctions in GJP series $W(x) = \sum_{k=1}^{N} W_k \phi_k(x)$, $\Theta(x) = \sum_{k=1}^{N} \Theta_k \phi_k(x)$. The derivatives of the eigenfunctions from (1) are obtained by differentiating these expansions. Replacing the above expansions expressions in the system (1) and imposing the condition that the obtained equations be orthogonal on the trial set of function $\{\psi_j\}_{j=1,\dots,N} \in S_N$, not necessarily from W_{N_i} we obtain an algebraic system in the expansion functions only which can now be solved, i.e. the eigenvalues are obtained by imposing the condition that non-vanishing coefficients W_k, Θ_k exist.

3 Numerical Results

In the following, several classes of GJP defined in [8] for various values of α and β are used. In [8] index sets are classified taking into account if α or β are less or greater than -1. Starting from the classical Jacobi polynomials $P_n^{\alpha,\beta}$ the GJP are defined [8]

$$j_n^{\alpha,\beta}(x) = \begin{cases} (1-x)^{-\alpha}(1+x)^{-\beta} P_{n_1}^{-\alpha,-\beta}, \ \alpha \le -1, \beta \le -1, n_1 = n - [-\alpha] - [-\beta], \\ (1-x)^{-\alpha} P_{n1}^{-\alpha,\beta}, \ \alpha \le -1, \beta > -1, n_1 = n - [-\alpha], \\ (1+x)^{-\beta} P_{n_1}^{\alpha,-\beta}, \ \alpha > -1, \beta \le -1, n_1 = n - [-\beta], \\ P_n^{\alpha,\beta}(x), \alpha > -1, \beta > -1. \end{cases} \tag{5}$$

It is worth pointing out that the major advantage of these GJP is that they are mutually $L^2_{w^{\alpha,\beta}}(-1,1,)$- orthogonal. Other important properties are also deduced in [8].

3.1 *The Rigid Boundaries Case*

Let us introduce the functions $\phi_k \in W_{N_1}, k = 1, 2, \dots,$ [3]

$$\phi(x) = (1-x^2)^2 \cdot P_k^{\alpha,\beta}(x), k = 1, \dots, N$$

which fulfills the boundary conditions (2). Using the properties of the Jacobi polynomials it is easy to verify that the functions $\phi_k(x)$, $0 < k < N-4$, are linearly independent and the dimension of the corresponding generated space W_{N_1} is equal to $N-3$. In fact, these function can be viewed as GJP of the form (5) since we can write

$$\phi(x) = (1-x)^2(1+x)^2 P_n^{\alpha,\beta}(x) = P_n^{\alpha',\beta'}(x) \text{ with } \alpha' \text{ and } \beta' \text{ real indexes.}$$

Numerical evaluations of the Rayleigh number for various values of the wavenumber and various linearly decreasing gravity fields are presented in Tables 1 and 2 for both equal or different indexes α, β in comparison with previous results obtained also by us for either trigonometric expansion functions or shifted Legendre

Table 1 Numerical evaluations of the Rayleigh number for various values of the parameters ϵ and a and various parameters $\alpha = \beta$

ϵ	a^2	$Ra_{\alpha,\beta=-1/2}$	$Ra_{\alpha,\beta=0}$	$Ra_{\alpha,\beta=1/2}$	Ra_{trig}[4]	Ra_{SLP}[5]
0	9.711	1730.0	1748.5	1743.9	1715.079356	1749.975727
0.01	9.711	1738.8	1757.2	1752.8	1723.697848	1758.769253
0.2	9.711	1922.2	1942.3	1937.5	1905.643719	1944.243122
0.2	12.0	1951.3	1969.6	1965.1	1937.927940	1977.079049
0.2	14.5	2037.1	2053.9	2049.9	2026.289430	3475.507241
1	10.0	3434.5	3470.8	3461.8	3431.318766	3475.507241

Table 2 Numerical evaluations of the Rayleigh number for various values of the parameters ϵ and a and various parameters $\alpha \neq \beta$

ϵ	a^2	$R_{\alpha=3/2,\beta=0}$	$R_{\alpha=0,\beta=3/2}$	$R_{\alpha=0,\beta=1/2}$	$R_{\alpha=1/2,\beta=0}$
0	9.711	1754.1	1754.1	1788.7	1788.7
0.01	9.711	1762.8	1763.0	1797.7	1797.7
0.2	9.711	1947.4	1950.1	1987.9	1986.7
0.2	12.0	1973.5	1976.0	2010.6	2009.2
0.2	14.5	2056.7	2059.7	2092.2	2091.0
1	10.0	3459.1	3503.2	3559.2	3539.0

polynomials. It is clear that for the same small value of the spectral parameter, $N = 3$, the numerical results obtained here are fairly accurate, but not the best. In the case of GJP with $\alpha = \beta$ an increasing value of N leads to very good numerical results with a computational time similar to the one from the trigonometric Fourier series case for smaller N. In fact it is common that a weight is introduced such that the ultraspherical polynomials are proportional to $P_n^{\alpha-1/2,\alpha-1/2}(x)$ [3] in order to obtain better results. In Fig. 1 neutral curves for the classical case of Rayleigh–Bénard convection are represented pointing out a numerical convergence of the algorithm for an increasing spectral parameter N.

3.2 The Free Boundary Case

Following [8], suitable expansion functions in the free boundaries case can be

$$\phi_k(x) = j_n^{-2,-2}(x) = \frac{4(k-2)(k-3)}{(2k-3)(2k-5)}\left(L_{k-4}(x) - \frac{2(2k-3)}{2k-1}L_{k-2}(x) + \frac{2k-5}{2k-1}L_k(x)\right) \tag{6}$$

where $L_k(z)$ is the Legendre polynomial of the kth used in general to approximate the solutions of fourth-order equations with Dirichlet boundary conditions [10]. Although there are no boundary conditions on the second order derivatives of Θ at $x = \pm 1$ we considered the same trial functions set for this unknown function as for W. Various approximation basis, taking into account the differentiation order

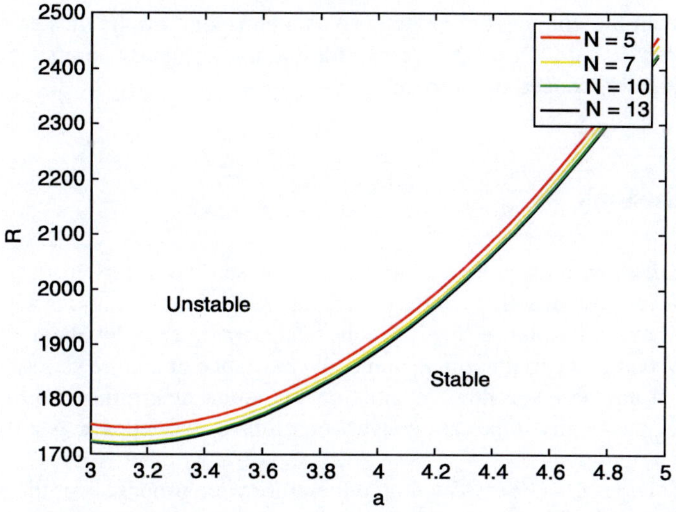

Fig. 1 Neutral curves for various values of the spectral parameter N in the classical case of Rayleigh–Bénard convection ($\epsilon = 0$) for $\alpha = \beta = 0$

for each unknown function in the system (1) and the boundary conditions can also be considered.

A much more flexible and general strategy is considered in [7]. Mainly it transforms the original problem into one containing lower order derivatives in the differential equation as well as in the boundary conditions. We suggested a method which takes into account the particularities of the problem (parity of order of differentiation, symmetries, etc.) and the problem was reduced from a fourth-order eigenvalue problem (an even order problem) with Dirichlet and hinged boundary conditions to a system of second order equations supplied exclusively with Dirichlet boundary conditions. These boundary conditions are much more simple to handle than the hinged ones. This procedure simplified considerably the construction of test and trial functions in tau and Galerkin methods as well as the formulation of differentiation matrices in the collocation (pseudospectral) method [7].

Taking into account the parity of the differentiation order in (1) and (2) a new variable was introduced $\Psi := (D^2 - a^2)W$ and thus, the two-point boundary value problem was rewritten as the second order system in the unknown functions W, Ψ, Θ completed with Dirichlet type boundary conditions $W = \Psi = \Theta = 0$ at $x = -1, 1$. Thus GJP of the form [3]

$$\phi_k(x) = j_n^{-1,-1}(x) = \frac{2(n-1)}{2n-1}(L_{n-2}(x) - L_n(x)) \qquad (7)$$

can be used as basis functions to approximate the solution. Our example demonstrates rather strikingly that the D^2 strategy leads to a big improvement of the numerical results. The implemented GJP based method led to larger algebraic problems than the collocation methods so as a consequence the qz step was more expensive in this case. However, we remark that the necessary computational time was significantly reduced in this case.

4 Conclusions

This paper dealt with the analytical and numerical study of a certain class of eigenvalue problems governed by systems of ordinary differential equations with variable coefficients and when more than one physical parameter is involved. When this parameter exceeds a certain critical value, the existence of a large variety of pattern bifurcated from the basic flow depends on the strata determined in the parameter space. A physically important convection problem was chosen as a benchmark model.

Several classes of GJP existing in the literature were proposed for the numerical investigation of the eigenvalue problem governing the stability of the considered

Table 3 The numerical evaluations of the Rayleigh number and the relative error for various values of the parameters a and ϵ

ϵ	a^2	Ra	Relative error
0	4.92	673.143	2.3%
0.01	4.92	676.52	2.3%
0.03	4.92	683.351	2.3%
0.33	4.92	805.99	2.3%
0.2	5.0	747.803	2.3%
0.2	9.0	844.65	1.8%
0.5	7.5	949.74	1.8%
0.5	9.0	1012.89	1.8%
0.75	10.0	1273.41	1.7%

(a) using $j_n^{-2,-2}(x)$ polynomials

ϵ	a^2	Relative error
0	4.92	0.0076%
0.01	4.92	0.453%
0.03	4.92	0.003%
0.33	4.92	0.4%
0.2	5.0	0.1%
0.2	9.0	0.1%
0.5	7.5	1%
0.5	9.0	1%
0.75	10.0	4%

(b) using $j_n^{-1,-1}(x)$ polynomials

Table 4 The numerical evaluations of the Rayleigh number for the $j_n^{-1,-1}$ polynomials and for various values of the parameters a and ϵ

ϵ	a^2	$R(N=5)$	$R(N=6)$	$R(N=8)$	$R(CC)$	$R(TS)$
0	4.92	658.486	658.692	657.56	657.5133	657.51
0.01	4.92	661.929	662.03	660.84	660.8173	660.81
0.03	4.92	668.69	668.64	667.50	667.5254	667.52
0.33	4.92	793.21	792.54	790.90	787.2880	787.28
0.2	5.0	733.22	733.06	731.54	730.5647	730.56
0.2	9.0	832.15	831.98	830.82	829.3918	829.39
0.5	7.5	948.27	945.87	944.64	930.9239	930.92
0.5	9.0	1013.5	1010.3	1009.4	994.4721	994.47
0.75	10.0	1327.4	1314.1	1312.0	1251.0924	1251.09

flow. The accuracy and effectiveness of these GJP method for these type of problems were tested for various values of the parameters corresponding to the GJP indexes. It is proved that the D^2 strategy applied to the problem is by far superior with respect to accuracy when compared with the same GJP based method for the direct formulation. The following physical conclusion was also emphasized: the stability domain increases as the gravity field is linearly decreasing across the layer.

References

1. Canuto, C., Hussaini, M.Y., Quarteroni, A., Zang, T.A., Spectral methods. Evolution to complex geometries and applications to fluid dynamics, Springer, Berlin, 2007
2. Doha, E.H., Bhrawy, A.H., *Efficient spectral-Galerkin algorithms for direct solution of second-order differential equations using Jacobi polynomials*, Numer. Algor., **42** (2006), 137–164
3. Doha, E.H., Bhrawy, A.H., *Efficient spectral-Galerkin algorithms for direct solution of fourth-order differential equations using Jacobi polynomials*, Appl. Num. Math., **58** (2008), 1224–1244
4. Dragomirescu I., *Approximate neutral surface of a convection problem for variable gravity field*, Rend. Sem. Mat. Univ. Pol. Torino, **64** (3) (2006), 331–342
5. Dragomirescu, F.I., *Shifted polynomials in a convection problem*, Appl. Math. Inf. Sci. J., **2** (2) (2008), 163–172
6. Drazin, P.G., Reid, W. H., Hydrodynamic stability, Cambridge University Press, London, 1981
7. Gheorghiu, C. I., Dragomirescu, I. F., *Spectral methods in linear stability. Applications to thermal convection with variable gravity field*, Appl. Numer. Math., **59** (2009), 1290–1302
8. Guo, B.Y., Shen, J., Wang, L.L., *Generalized Jacobi polynomials/functions and their applications*, Appl. Numer. Math., doi:10.1016/j.apnum.2008.04.003
9. Shen, J., Wang, L. L., *Legendre and Chebyshev dual-Petrov-Galerkin methods for hyperbolic equations*, Comput. Methods Appl. Mech. Eng., **196** (37–40) (2007), 3785–3797
10. Shen, J., *Efficient spectral-Galerkin mthod I. direct solvers for second- and fourth-order equations by using Legendre polynomial*, SIAM J. Sci. Comput. **15** (1994), 1489–1505
11. Straughan, B., *The energy method, stability, and nonlinear convection*, Springer, Berlin, 2003

Spectral Element Discretization of Optimal Control Problems

Loredana Gaudio and Alfio Quarteroni

Abstract In this work we consider the numerical solution of a distributed optimal control problem associated with an elliptic partial differential equation. We approximate the optimality system by the spectral element method and derive a posteriori error estimates with respect to the cost functional. Then we use an hN adaptive refinement technique to reduce this error: the error indicator is used to mark what elements must be refined. The choice between an h or N refinement is based on the use of a predicted error reduction algorithm. Numerical results show the way this algorithm works.

Keywords A posteriori error estimates · Mesh refinement · Optimal control · Spectral element method

Introduction

We present an hN adaptive algorithm for a linear optimal control problem discretized by spectral element method. The use of adaptive algorithms to reduce the error on the cost functional is generally accepted in the context of finite element methods, see e.g., [1, 2]. Very few results exist on the use of spectral elements discretization of optimal control problems. In [3, 6] error estimates are obtained for the control, state and adjoint variables in the natural norms of the corresponding spaces. However, these results do not guarantee an error bound on the cost functional,

L. Gaudio (✉)
MOX-Dipartimento di Matematica "F. Brioschi," Politecnico di Milano, Piazza L. da Vinci 32, 20133 Milano, Italy
e-mail: loredana.gaudio@polimi.it

A. Quarteroni
École Polytechnique Fédérale de Lausanne (EPFL), FSB, Chaire de Modelisation et Calcul Scientifique (CMCS), Station 8, 1015, Lausanne, Switzerland, and MOX-Dipartimento di Matematica "F. Brioschi," Politecnico di Milano, Piazza L. da Vinci 32, 20133 Milano, Italy
e-mail: alfio.quarteroni@epfl.ch

J.S. Hesthaven and E.M. Rønquist (eds.), *Spectral and High Order Methods for Partial Differential Equations*, Lecture Notes in Computational Science and Engineering 76,
DOI 10.1007/978-3-642-15337-2_37, © Springer-Verlag Berlin Heidelberg 2011

a quantity of interest in many applications. The purpose of this paper is to derive a posteriori error estimates for the error on the objective functional besides those on energy norm error estimates in the context of spectral approximation, then to use them to guide an hN adaptive design mesh. Starting by an initial conforming spectral element mesh we solve the optimal control problem and estimate the error on the cost functional. When necessary we adapt the mesh to improve the discretization error and we solve again the optimal control problem on the new mesh until convergence within error tolerance.

The paper is organized as follows. In Sect. 1 we introduce the model linear optimal control problem. In Sect. 2 we introduce the spectral element approximation space and the discrete problem formulation. In Sect. 3, the error on the cost functional is estimated by the sum of two contributions: the iteration and discretization errors. In Sect. 4, some numerical results are presented to show how the algorithm works.

1 Linear Optimal Control Problem

Let $\Omega \subset \mathbb{R}^2$ be a bounded open set with a Lipschitz boundary $\partial\Omega$ and V, U be the Hilbert spaces of state and control functions, respectively. On the product space $V \times U$ we introduce a functional J that represents the quantity of physical interest, the objective of the control problem. The state problem describes, for each given control variable $u \in U$, the way the system evolves. The model problem considered features a distributed observation and a distributed control problem, in which:

- The functional J is quadratic:

$$J(y,u) = \frac{1}{2}\|Cy - z_d\|^2_{L^2(\Omega)} + \frac{1}{2}\alpha\|u - u_d\|^2_{L^2(\Omega)},$$

  where for a given Hilbert space of observations Z, $z_d \in Z$ is an assigned desired function, $C : V \to Z$ a bounded operator, $\alpha > 0$ is a penalization factor, $u_d \in U$ a given desired control (possibly zero);
- The state problem is an elliptic partial differential equation:

$$A(y(u), u; f) = 0,$$

  where A is the linear differential operator defined on the domain Ω and f is a given source term. If we introduce the bilinear form $a : V \times V \to \mathbb{R}$, $a(u,v) = < Au, v >_{V',V}$, with $< \cdot, \cdot >_{V',V}$ the duality pairing between V and V', then the variational formulation of the problem is

$$\text{find } y \in V : \quad a(y,v) =< f, v >_{V',V} + < Bu, v >_{V',V} \quad \forall v \in V,$$

  where $B : U \to V'$ is a bounded linear functional. We assume a to be a bilinear continuous coercive form to ensure the well posedness of the state problem for each control.

Our optimal control problem reads as follows: look for $(y,u) \in V \times U$ s.t.

$$\min_{(y,u)} J(y(u),u),$$
$$\text{sbj to } A(y(u),u;f) = 0.$$

Under the assumptions on the bilinear form a and on the functional, it is well-know that this problem is well-posed, see e.g., [7].

Our approach to solve the problem is to introduce a Lagrangian functional $\mathcal{L}$ and to transform the optimal control problem as the search for the saddle-point of $\mathcal{L}$. We define $\mathcal{L} : V \times V \times U \to \mathbb{R}$ as

$$\mathcal{L}(y,p,u) := J(y,u) + < p, A(y,u) >_{V',V},$$

where p is the Lagrange multiplier, also called the adjoint variable.

If $x = (y,p,u)$ is the optimal solution then $\nabla\mathcal{L}(x)[\phi,\mu,\psi] = 0$ where the derivative is of Fréchet type. Upon taking the derivatives with respect to each variable, this yields the KKT (Karush–Kuhn–Tucker) optimality system:

$$\begin{cases} \nabla_p\mathcal{L}(x)[\phi] = 0 \;\forall\phi \in V \longmapsto \text{state problem,} \\ \nabla_y\mathcal{L}(x)[\mu] = 0 \;\forall\mu \in V \longmapsto \text{adjoint problem,} \\ \nabla_u\mathcal{L}(x)[\psi] = 0 \;\forall\psi \in U \longmapsto \text{optimality conditions.} \end{cases}$$

For the model problem at hand the KKT system is find $(y,p,u) \in V \times V \times U$:

$$\begin{cases} a(y,v) & = < f + Bu, v >_{V',V} & \forall v \in V, \\ a^*(p,v) & = < C'\Lambda_Z(Cy - z_d), v >_{V',V} & \forall v \in V, \\ < B'p + \alpha\Lambda_U u, \tilde{v} >_{U',U} = 0 & & \forall \tilde{v} \in U, \end{cases}$$

where $a^*(\cdot,\cdot)$ is the adjoint bilinear form of a, whereas $\Lambda_Z : Z \to Z'$ and $\Lambda_U : U \to U'$ are the Riesz inclusion operators, see [10]. To solve this problem we use an iterative method: given u^0, we solve the state and the adjoint problem according to the optimality conditions, then we update the derivative functional $\nabla_u\mathcal{L}(x^j)$. If $\|\nabla_u\mathcal{L}(x^j)\| \le tol$ (for an assigned tolerance) we stop else we update the control variable u by a steepest-descent method $u^{j+1} = u^j - \tau\nabla_u\mathcal{L}(x^j)$, whit τ being a relaxation parameter.

2 SEM Discretization

At each step of the iterative method used for the solution of the KKT system we solve the state and the dual problem by a spectral element method. Let us decompose Ω into K spectral elements: $\overline{\Omega} = \cup_{k=1}^{K}\overline{\Omega}_k$, such that $\forall\Omega_k$ there exists a bijective

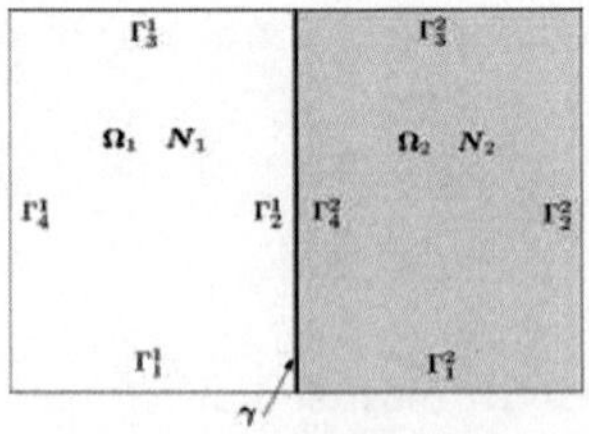

Fig. 1 Decomposition of Ω, $K = 2$

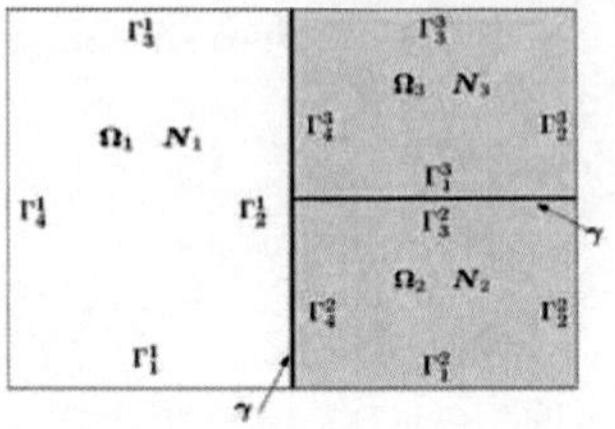

Fig. 2 Decomposition of Ω, $K = 3$

transformation $\varphi_k : \widehat{\Omega} \longrightarrow \Omega_k$, $\widehat{\Omega} = (-1,1)^2$. We denote with $\overrightarrow{\delta} = \{\delta_k\}_{k=1}^K$ the vector of discretization parameters, $\delta_k = (h_k, N_k)$, h_k being the diameter of Ω_k and N_k the degree of the polynomial in Ω_k. For each couple of neighboring elements, say Ω_k, Ω_m, three different situations may occur:

1. Either geometric and polynomial conformity, that is $\gamma = \overline{\Omega}_k \cap \overline{\Omega}_m$ is a common (full) side of Ω_k and Ω_m, and $\overline{N} = N_k = N_m$. In this case the $\overline{N}+1$ LGL nodes on γ are called active nodes and I_γ^a is the corresponding set of such nodes;
2. Geometrical conformity but polynomial non-conformity, that is $N_k \neq N_m$. Then, I_γ^a is the set of the $\overline{N}+1$ LGL nodes, called active nodes, on γ, where $\overline{N} = \min(N_k, N_m)$;
3. Full non-conformity, both geometrical and polynomial. In this case one chooses the longest edge and we call it γ. Then on γ we choose the smallest value $\overline{N}$ of the polynomial degree among those of all the spectral elements sharing γ. Finally, I_γ^a is the set of the $\overline{N}+1$ LGL active nodes on γ.

For the sake of illustration, two examples are shown in Figs. 1 and 2, where we denote with Γ_q^k the qth side (according to the local side numbering) of the element Ω_k. With reference to Fig. 1 we have $K = 2$, $N_1 \neq N_2$, $\gamma = \Gamma_2^1 = \Gamma_4^2$, $\overline{N} = \min(N_1, N_2)$. With reference to Fig. 2 we have $K = 3$, $N_1 \neq N_2 \neq N_3$. In Fig. 2 we have two different interfaces on which we enforce pointwise continuity. The former is $\gamma = \Gamma_3^2 = \Gamma_1^3$ for which we set $\overline{N} = \min(N_2, N_3)$, the latter is $\gamma = \Gamma_2^1$ on which we set $\overline{N} = \min(N_1, N_2, N_3)$.

On the non-conforming interface γ, we enforce the C^0 continuity by matching the active and passive (non-active) unknowns on γ. The passive degrees of freedom (d.o.f), corresponding to the passive nodes, will be defined as a linear combination of the active d.o.f, corresponding to the active nodes. Namely as before, we define the set of active d.o.f. I_γ^a then, for each passive d.o.f v_i on γ the following equality is enforced:

$$v_i = \sum_{j \in I_\gamma^a} c_{ij} v_j, \tag{1}$$

the constraints coefficients c_{ij} are chosen in such a way to have continuity on the interface and the v_j, with $j \in I_\gamma^a$, are the active d.o.f on γ. If we introduce the unconstrained spectral element space X_δ:

$$X_\delta := \{v_\delta : v_\delta \circ \varphi_k \in \mathbb{Q}_{N_k}(\widehat{\Omega}) \text{ and } v_{\delta_k} = 0 \quad \text{on } \Gamma_D \cap \partial\Omega_k, \forall k = 1, \ldots, K\}$$

where $\mathbb{Q}_N(\Omega)$ is the set of polynomials of two variables with degree $\leq N_k$ with respect to each variable and Γ_D is the Dirichlet boundary. Then, the constrained SEM space is:

$$V_\delta := \{v_\delta \in X_\delta : \text{ for every passive d.o.f. } v_i, \text{ (1) is satisfied}\}.$$

With this formulation we could have nonconformity for both mesh and functional space, this is a natural situation that may arise after every step of an adaptive algorithm, when only some elements are refined. To ensure comparable mesh diameters between neighboring elements only one hanging node for side is allowed. So in addition to the elements marked by the a posteriori indicator, some further refinements could be made. The described formulation is not optimal to solve the PDE on a nonconforming interface, other formulations like mortar methods as well as DG could improve the convergence of the method, see e.g., [4]. Now given V_δ, U_δ the finite discretization of the state and control space V, U, respectively, we search $(y_\delta, p_\delta, u_\delta) \in V_\delta \times V_\delta \times U_\delta$:

$$\begin{cases} a(y_\delta, v_\delta) & = < f + Bu_\delta, v_\delta >_{V',V} & \forall v_\delta \in V_\delta, \\ a^*(p_\delta, v_\delta) & = < C'\Lambda_Z(Cy_\delta - z_d), v_\delta >_{V',V} & \forall v_\delta \in V_\delta, \\ < B'p_\delta + \alpha\Lambda_U u_\delta, v_\delta >_{U',U} = 0 & & \forall v_\delta \in U_\delta. \end{cases} \tag{2}$$

We highlight that for the problem at hand, the control is discretized on the same mesh of the state. By the optimality condition, u has the same regularity as the solution of the adjoint problem, the latter depends on the data regularity of the state problem.

3 Iteration and Discretization Error Estimates

After the discrete KKT system (2), we analyze the accuracy on the functional that we have achieved. By proceeding as done in [5], we split the functional error into two parts:

$$\left|J(y,u) - J(y_\delta^j, u_\delta^j)\right| \le \underbrace{\left|J(y,u) - J(y^j, u^j)\right|}_{\epsilon_{\text{iter}}^{(j)}} + \underbrace{\left|J(y^j,u^j) - J(y_\delta^j, u_\delta^j)\right|}_{\epsilon_{\text{dis}}^{(j)}},$$

where (y,u) is the exact optimal control solution, (y^j, u^j) are the hypothetical continuous solutions at the iterative step j and (y_δ^j, u_δ^j) is the discrete optimal control solution. The first part represents the iteration error and the second the discretization error. We will estimate each term as follows.

Theorem 1. *For linear control problems, the iteration error at the j-th iteration has the following expression:*

$$\epsilon_{iter}^{(j)} = |J(y,u) - J(y^j, u^j)| = \frac{1}{2}(\nabla_u \mathcal{L}(x^j), u - u^j).$$

Corollary 1. *If a steepest-descent iterative method with constant relaxation parameter τ is used, $\epsilon_{iter}^{(j)}$ can be estimated as: $\left|\epsilon_{iter}^{(j)}\right| \simeq \frac{1}{2}\tau \left\|\nabla_u \mathcal{L}(x^j)\right\|^2$.*

See [5] for the proofs. Then the first part of the error is minimized during the iterative solution of the KKT system, accordingly with the stopping criterium, $\|\nabla_u \mathcal{L}(x^j)\| \le tol_{iter}$. For the $\epsilon_{\text{dis}}^{(j)}$ we use a dual weighted estimation.

Theorem 2. *Assume the mesh to be γ shape regular, that is $\exists \gamma > 0 : \gamma^{-1} h_k \le h_{k'} \le \gamma h_k$ if k and k' are such that $\overline{\Omega}_k \cap \overline{\Omega}_{k'} \ne \emptyset$, with polynomial degrees of neighboring elements comparable $\gamma^{-1}(N_k + 1) \le N_{k'} + 1 \le \gamma(N_k + 1)$. Then for the spectral element discretization we have:*

$$\epsilon_{dis}^{(j)} \le C \sum_{k=1}^{K} \rho_k^y \frac{h_k}{N_k} \|\nabla p_\delta^j\|_{L^2(\omega_k^1)} + \rho_k^p \frac{h_k}{N_k} \|\nabla y_\delta^j\|_{L^2(\omega_k^1)} + \rho_k^u \frac{h_k}{N_k} \|\nabla u_{\delta}^j\|_{L^2(\omega_k^1)},$$

where
$\rho_k^y := \|R(y_\delta^j, u_\delta^j)\|_{\Omega_k} + (\frac{h_k}{N_k})^{-\frac{1}{2}} \|r(y_\delta^j)\|_{\partial\Omega_k}$, $\rho_k^u := \|\alpha u_\delta + p_\delta\|_{\Omega_k}$,
$\rho_k^p := \|R(p_\delta^j, y_\delta^j)\|_{\Omega_k} + (\frac{h_k}{N_k})^{-\frac{1}{2}} \|r(p_\delta^j)\|_{\partial\Omega_k}$.

Here $R(\cdot,\cdot)$ ($r(\cdot,\cdot)$, respectively) are the interior (edge, respectively) residuals associated with either the state or adjoint elliptic operators, and $\omega_k^1 = \cup_{m \in I_k} \Omega_m$, where I_k is the set of index of the elements sharing at least one vertex with Ω_k.

Proof. According to [2] for the Galerkin element discretization we have

$$\epsilon_{\text{dis}}^{(j)} \le \sum_{k=1}^{K} \{\rho_k^y \omega_k^p + \rho_k^p \omega_k^y + \rho_k^u \omega_k^u\},$$

where $\omega_k^y := \|y^j - I_\delta y^j\|_{\Omega_k} + (\frac{h_k}{N_k})^{\frac{1}{2}} \|y^j - I_\delta y^j\|_{\partial\Omega_k}$, $\omega_k^p := \|p^j - I_\delta p^j\|_{\Omega_k} + (\frac{h_k}{N_k})^{\frac{1}{2}} \|p^j - I_\delta p^j\|_{\partial\Omega_k}$, $\omega_k^u := \|u^j - I_\delta u^j\|_{\Omega_k}$. $I_\delta y^j$, $I_\delta p^j$, $I_\delta u^j$ are hN-Clément interpolant of y^j, p^j, u^j, respectively, see [8]. Now using the estimates for the Clément interpolant, see e.g., [8], each term in the weights ω_k^y, ω_k^p, ω_k^u can be estimated by the norms of the gradients in the ω_k^1 domain associated to Ω_k. Collecting the previous estimates concludes the proof.

The choice between h or N refinement strategy is made according to a predictable error estimates. For both the state and the adjoint equations we construct a posteriori residual estimates and a predictable estimates. Then we define the total residual and total predictable estimates as the sum of the two contributions by the state and adjoint problems. Comparing this total estimates following the algorithm proposed in [9] we choose between a spatial h (each element is subdivided into four sub-elements by joining the midpoints) or a functional N (increasing N by one) refinement.

4 Numerical Results

We present some numerical results to show how the algorithm works. Let $\Omega = (0, 1)^2$ and consider an initial conform mesh. More particularly Ω is subdivided in $K = 4$ spectral elements and on each element we use a uniform degree $N = 2$. The state equation is:

$$\begin{cases} -\Delta y = u & \text{in } \Omega, \\ \quad y \;\; = 0 & \text{on } \partial\Omega. \end{cases}$$

For the quadratic functional J we fix $\alpha = 0.1$ and $z_d = \exp(-(x^2 + y^2)/0.04)$. We solve both the optimization and the adaptive process in an iterative way, the two tolerances are $tol_{iter} = tol_{dis} = 1e - 7$, an we start with an initial control $u_0 = 1$. In the adaptive process we admit at maximum $it_{dis}^{max} = 2$ iterations because changing the approximation of the functional J the optimal control calculated on the old mesh could be very different from the one on the new mesh. In Table 1 we report the results obtained during the process.

In the figures below we report an intermediate mesh in Fig. 3 and the final mesh in Fig. 4. For each element we plot the degrees of freedom and the local polynomial degree. In Fig. 5 the final control function and in Fig. 6 the associated final state function.

5 Conclusions

In this note we have presented a spectral element method for the discretization of an elliptic optimal control problem and the use of hN adaptivity to reduce the error on the cost functional. The proposed estimate for the discretization part has driven to an automatic design of either the mesh and the polynomial degrees in a configuration

Table 1 The error estimates at each optimization and adaptive step and the number of elements refined in h, N at each adaptive step

it_{iter}	it_{dis}	ϵ_{iter}	ϵ_{dis}	#{ref h}	#{ref N}
1	0	1.4725e-008	1.2677e-005	0	0
1	1	1.3425e-005	1.9999e-006	1	0
1	2	5.5197e-006	3.5798e-007	1	0
2	0	1.8497e-008	8.0828e-007	0	0
2	1	1.0809e-007	7.6918e-007	1	0
2	2	6.3807e-007	3.0858e-007	2	6
3	0	3.0720e-008	3.4665e-007	0	0
3	1	4.3163e-007	1.3016e-006	3	1
3	2	4.1335e-007	5.3046e-008	5	0
4	0	1.7597e-008	7.0768e-008	0	0

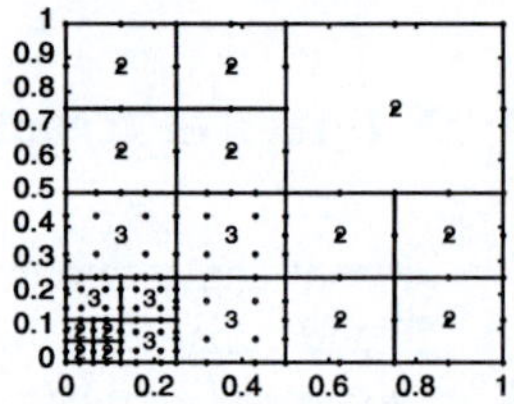

Fig. 3 After 3 steps of the algorithm

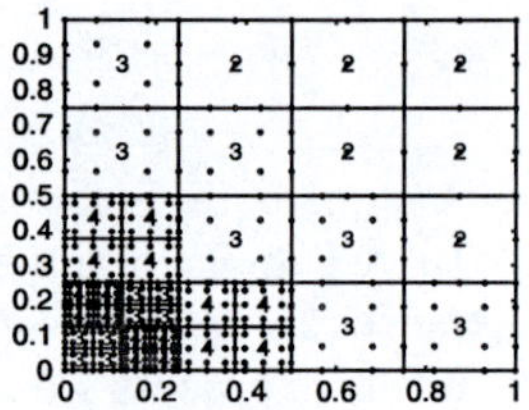

Fig. 4 Final mesh and degrees for every element

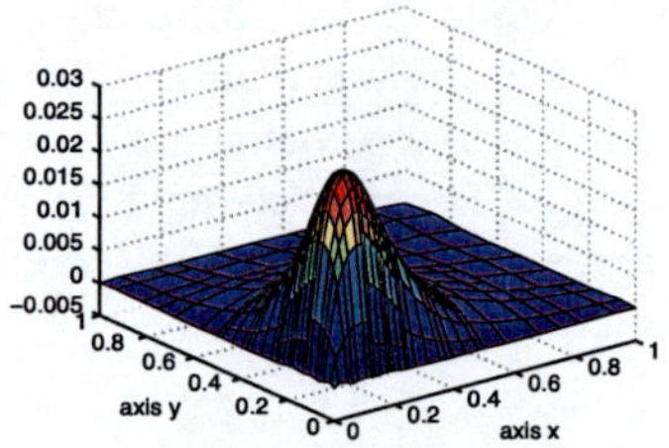

Fig. 5 Final control function

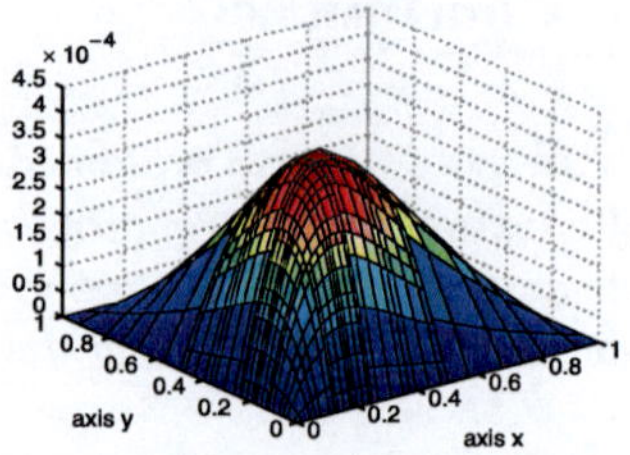

Fig. 6 Final state function

strictly dependent on the problem considered. More information are used near the corner where the desired functions and the control variables change more rapidly.

Acknowledgements We thank Dr. Luca Dedè for helpful discussions and comments. We acknowledge the support of the Italian MIUR through the project COFIN07 "Mathematical and numerical modelling for cardiovascular and fluid dynamics applications."

References

1. R. Becker, H. Kapp and R. Rannacher. Adaptive finite element methods for optimal control of partial differential equations: basic concept. *SIAM J. Control Optim.*, **39**, 113–132 (2000)
2. R. Becker and R. Rannacher. An optimal control approach to a posteriori error estimation in finite element methods. *Acta Numer.*, **10**, 1–102 (2001)
3. Y. Chen, N. Yi and W. Liu. A Legendre-Galerkin spectral method for optimal control problems governed by elliptic equations. *SIAM J. Numer. Anal.*, **46**, 2254–2275 (2008)
4. C. Bernardi, N. Debit and Y. Maday. Coupling finite element and spectral methods: first results. *Math. Comp.*, **54**, 21–39 (1990)
5. L. Dedè and A. Quarteroni. Optimal control and numerical adaptivity for advection-diffusion equations. *Math. Model. Numer. Anal.*, **39**, 1019–1040 (2005)
6. R. Ghanem and H. Sissaoui. A posteriori error estimate by a spectral method of an elliptic optimal control problem. *J. Comput. Math. Optim.*, **2**, 111–125 (2006)
7. J.-L. Lions, Optimal control of systems governed by partial differential equations. Springer, New York (1971)
8. J.M. Melenk. *hp*-interpolation of nonsmooth functions and an application to *hp*-a posteriori error estimation. *SIAM J. Numer. Anal.*, **43**, 127–155(2005)
9. J.M. Melenk and B.I. Wohlmuth. On residual-based a posteriori error estimation in hp-FEM. *Adv. Comput. Math.*, **15**, 311–331 (2001)
10. A. Quarteroni. Numerical models for differential problems. Springer, Milan (2009)

Applications of High Order Methods to Vortex Instability Calculations

Leo M. González, Vassilis Theofilis, and Fernando Meseguer-Garrido

Abstract Highly resolved solutions of the two-dimensional incompressible Navier-Stokes and continuity equations, describing the evolution of vortex systems, have been obtained accurately and efficiently by spectral collocation methods. Such solutions have formed the basic state for subsequent three-dimensional BiGlobal eigenvalue problem (EVP) linear instability analyses, which monitor the modal response of these vortical systems to small-amplitude perturbations, periodic along the homogeneous axial spatial direction, without the need to invoke an assumption of azimuthal spatial homogeneity. A spectral/hp methodology has been adapted to study instability of vortical flows and has been validated on the isolated Batchelor vortex. Subsequently, a stability analysis of an aircraft wake model, composed of two counter-rotating vortices, has been performed by the present spectral/hp element methodology.

1 Introduction

Work spanning several decades exists, which focuses on the problem of inviscid or viscous instability of vortical flows. Short of resorting to a direct numerical simulation methodology analysis [15], an approach hardly appropriate for parametric studies, practically all instability work has dealt with basic flows that correspond to vortices either in isolation or in the presence of a shear flow that models the presence

L.M. González (✉)
Naval Architecture Department (ETSIN), Technical University of Madrid (UPM), Arco de la Victoria s/n, 28040, Madrid, Spain
e-mail: leo.gonzalez@upm.es

V. Theofilis
School of Aeronautics, Universidad Politécnica de Madrid Pza. Cardenal Cisneros 3, 28040 Madrid, Spain
e-mail: vassilis@torroja.dmt.upm.es

F. Meseguer-Garrido

J.S. Hesthaven and E.M. Rønquist (eds.), *Spectral and High Order Methods for Partial Differential Equations*, Lecture Notes in Computational Science and Engineering 76, DOI 10.1007/978-3-642-15337-2_38,

of a second co- or counter-rotating vortex. By contrast, Hein and Theofilis [6] and Jacquin et al. [7] have first employed the BiGlobal instability analysis concept [14] in order to analyze three-dimensional instability of arbitrary vorticity distributions on the plane normal to the axial direction, treating the latter spatial direction as homogeneous, but without resorting to the assumption of spatial homogeneity in the azimuthal; the basic states analyzed in those works were constructed analytically with the aid of the Batchelor vortex model. Interestingly, validations studies on the Batchelor vortex [6] have demonstrated the stringent resolution requirements placed on the stability analysis by the tight structure of the amplitude functions of the small-amplitude perturbations developing in the core of the basic flow vortex. The use of a regular Cartesian tensor-product spectral collocation computational mesh [6] has adversely influenced the convergence of the results presented (though convergence has been achieved), since a large portion of the available (mapped) Chebyshev collocation points utilized have been wasted in resolving the innocuous far-field. It thus becomes natural to depart from the structured-mesh technologies used in the earlier analyses and focus on numerical methodologies for BiGlobal instability analysis which rely on unstructured meshes. The work of Broadhurst, Sherwin and Theofilis [1] was the first step in this direction, employing a spectral/hp element methodology [8]. Building upon earlier low order version work by González et al. [5], here attention is turned again to a spectral/hp element methodology approach for the solution of the incompressible BiGlobal eigenvalue problem (EVP). In order to eliminate potential influences of the basic state on the quality of the instability results, two-dimensional direct numerical simulation, based on spectral collocation and an eigenvalue decomposition algorithm, has been employed to obtain the basic state [13]. Results delivered by this methodology, when the DNS is initialized on different models of aircraft wing loading, have been found to be in excellent agreement with those produced by a different, appropriate and efficient, DNS methodology for this class of problems [2, 15]. Here, the initial conditions are provided by a counter-rotating pair of vortices; after an initial transient period a dipole is formed, which descends and diffuses according to the imposed Reynolds number. Snapshots in time of this flowfield are extracted at predefined characteristic times [9, 11] and are analyzed with respect to their three-dimensional instability. Turning to the BiGlobal EVP, the present spectral/hp-based methodology has been validated against instability results of a single-Batchelor vortex [10]. This exercise has delivered information, firstly regarding efficient meshing strategies, and secondly on resolution requirements for in-core serial solution of the EVP.

Section 2 discusses the theoretical background of both the basic flow and the eigenvalue problem, as well as the boundary conditions of the stability problem and aspects of the numerical solution of both basic and perturbed flow. Subsequently, results are presented in Sect. 3, on the instability of both the isolated Batchelor vortex and that of the numerically-obtained dipoles. Conclusions and projected future directions of the present research are discussed in Sect. 4.

2 Theory

The Navier–Stokes equations governing incompressible flows are written in primitive-variables formulation. A particular non-parallel solution of the Navier–Stokes equations known as basic flow $(\bar{u}_i, \bar{p})$ is perturbed by small-amplitude velocity $\tilde{u}_i$ and kinematic pressure $\tilde{p}$ perturbations, as follows

$$u_i = \bar{u}_i + \varepsilon \tilde{u}_i + c.c. \qquad p = \bar{p} + \varepsilon \tilde{p} + c.c., \tag{1}$$

where $\varepsilon \ll 1$ and $c.c.$ denotes conjugate of the complex quantities $(\tilde{u}_i, \tilde{p})$. Linearizing around the basic flow, the equations for the perturbation quantities are obtained

$$\frac{\partial \tilde{u}_i}{\partial t} + \bar{u}_j \frac{\partial \tilde{u}_i}{\partial x_j} + \tilde{u}_j \frac{\partial \bar{u}_i}{\partial x_j} = -\frac{\partial \tilde{p}}{\partial x_i} + \frac{1}{Re} \frac{\partial^2 \tilde{u}_i}{\partial x_j^2}, \tag{2}$$

$$\frac{\partial \tilde{u}_i}{\partial x_i} = 0, \tag{3}$$

The boundary condition used for this system is $\tilde{u}_i = 0$ on the domain boundary.

2.1 *The Basic Flows*

A detailed discussion of the numerical approach used to recover the basic states may be found elsewhere [13]; here a brief summary is exposed. A Cartesian coordinate system is considered, taking $(x_1, x_2, x_3) \equiv (x, y, z)$ and $(u_1, u_2, u_3) \equiv (u, v, w)$. The basic flow is calculated by time-marching the vorticity transport equation,

$$\zeta_t + \bar{v}\zeta_y + \bar{w}\zeta_z - \nu\nabla^2\zeta = 0, \tag{4}$$

where $\zeta = -\bar{w}_y + \bar{v}_z$ is the basic flow vorticity, $\nabla^2 = \frac{\partial^2}{\partial y^2} + \frac{\partial^2}{\partial z^2}$, and the streamfunction, ψ, is related with the vorticity through

$$\nabla^2\psi + \zeta = 0. \tag{5}$$

An equation analogous to (4) is solved for the axial component of velocity $\bar{u}$, which is decoupled from the system (4) and (5). Dimensional time t^* is non-dimensionalized as

$$t = \frac{t^* \nu}{2\pi a_0^2} \tag{6}$$

The initial conditions for the flows analyzed are composed of an isolated or system of Lamb–Oseen vortices to which an axial flow has been superimposed. The initial vorticity ζ_0 of a single such vortex and the circulation of the flow Γ are defined in [5]. Additionally, an initial axial velocity, $\bar{u}_0$

$$\bar{u}_0 = U_0\, e^{-\frac{r^2}{a_0^2}}, \tag{7}$$

is defined, U_0 being the peak value of the axial velocity; the latter is a free parameter measuring the jet strength [3, 9]. The vortex radius a is based on the vorticity polar moments a_y, a_z on the half-plane defined in [12].

In the present case of a system of two vortices, the initial distance between the centroids is denoted by b. The flow Reynolds number is defined by $Re = \Gamma/\nu$, ν being the kinematic viscosity. The numerical solution of the basic flow problem is detailed in [5].

2.2 *The BiGlobal Eigenvalue Problem (EVP)*

The Ansatz used to describe the small-amplitude perturbations is

$$\tilde{u}_i = \hat{u}_i(y,z)e^{i(\alpha x - \omega t)}, \tag{8}$$
$$\tilde{p} = \hat{p}(y,z)e^{i(\alpha x - \omega t)}, \tag{9}$$

where a temporal formulation has been adopted, considering α is a real wavenumber parameter associated with the axial periodicity length through $L_x = \frac{2\pi}{\alpha}$ and ω is the complex eigenvalue sought. Substitution into (2) and (3) results in

$$\mathrm{i}\,\alpha\hat{u} + \hat{v}_y + \hat{w}_z = 0, \tag{10}$$
$$\mathscr{L}\hat{u} - \bar{u}_y\hat{v} - \bar{u}_z\hat{w} - \mathrm{i}\,\alpha\hat{p} = -\mathrm{i}\,\omega\hat{u}, \tag{11}$$
$$\left(\mathscr{L} - \bar{v}_y\right)\hat{v} - \bar{v}_z\hat{w} - \hat{p}_y = -\mathrm{i}\,\omega\hat{v}, \tag{12}$$
$$\left(\mathscr{L} - \bar{w}_z\right)\hat{w} - \bar{w}_y\hat{v} - \hat{p}_z = -\mathrm{i}\,\omega\hat{w}, \tag{13}$$

where $\mathscr{L} = 1/Re\left(-\alpha^2 + \partial^2/\partial y^2 + \partial^2/\partial z^2\right) - \mathrm{i}\,\alpha\bar{u} - \bar{v}\partial/\partial y - \bar{w}\partial/\partial z$. The complex generalized eigenvalue problem for the determination of ω may thus be obtained,

$$A\begin{pmatrix}\hat{u}\\ \hat{v}\\ \hat{w}\\ \hat{p}\end{pmatrix} = -\mathrm{i}\,\omega B\begin{pmatrix}\hat{u}\\ \hat{v}\\ \hat{w}\\ \hat{p}\end{pmatrix}. \tag{14}$$

Here an iterative Arnoldi method has been used for the solution of the EVP. In both the isolated Batchelor vortex validation case and the dipole analyses that follow, the spatial discretization of (14) is performed using the spectral/hp element method. The polynomial order considered for the pressure is $P - 2$, while the polynomial order for the velocity is P.

3 Results

3.1 Basic Flow

A system of two counter-rotating vortices has been defined by the parameters $a_0 = a(t = 0) = 0.25, b_0 = b(t = 0) = 1/0.134$. A wide square integration domain has been taken, the extent of which, L is taken to fulfill $L \gg b_0$, such that the periodic boundary conditions do not affect the results of the simulations. The equations of motion (4) and (5) have been integrated in time, until certain predetermined criteria, indicated below, are met. Resolutions upward of $N_y = Nz = 400$ Fourier collocation points per spatial direction have been used. In the multiparametric problem at hand a constant initial circulation of unity has always been considered, while basic flow was run for Reynolds number value, $Re = 3{,}180$, the initial axial velocity values, $U_0 = \frac{0.5}{\pi}$. The vortex aspect ratio, $E = a_z/a_y$, settles to a linear growth after a short initial transient, the latter indicating the short period during which the initially-imposed analytical vorticity distribution adjusts itself to the equations of motion. Stopping the simulations when this ratio reaches the value $a/b = 1/4$ defines one time, $t_0 \approx 0.09369$, at which one of the subsequent instability analyses was performed. By the end of the simulation, it has been verified that the circulation is constant to within 2×10^{-3}. The self-advection speed of the vortex pair is subtracted such that the vortices remain in the computational domain.

3.2 Instability Analyses

Prior to analyzing the system of vortices obtained in the previous Section, it is instructive to present results obtained in the well-studied Batchelor vortex instability problem [10]. In this case the basic flow is analytically-constructed, $(\overline{u}, \overline{v}, \overline{w}) = (\exp(-r^2), -qz(1 - \exp(-r^2))/r^2, qy(1 - \exp(-r^2))/r^2)$, where r is the radial cylindrical coordinate. This non-dimensional baseflow only depends on the swirl parameter q, which is the quotient between the circulation velocity and the axial velocity.

The resulting system is solved in a large domain of 20 times the vortex radius, such that homogeneous Dirichlet boundary conditions may be imposed on all amplitude functions at the boundary of the domain. While in this limiting case the classic instability analysis which exploits periodicity in the azimuthal direction [10] could have been used, here the BiGlobal EVP (10)–(13) has been solved without the need to resort to this assumption. Stability results at $Re = 1{,}200, q = 0.8$ are shown for $P = 7$ in Fig. 1. The lack of linearity for the largest values of m and α in the frequency representation could be improved by mesh refinement.

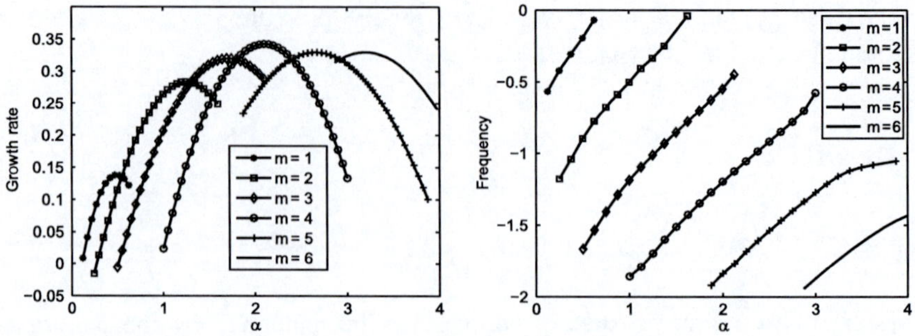

Fig. 1 Dependence of amplification rate (*left*) and frequency (*right*) vs. wavelength α for the different unstable modes of an isolated Batchelor vortex at $Re = 1{,}200$, $q = 0.8$. The value of m indicates the number of equal signed lobes in the structure

The (high-order) spectral/hp is capable of delivering accurate results in flows where small fluid structures compared to the fluid domain are searched. Spectral convergence has been obtained in the EVP calculation and a polynomial order $P = 7$ has been used for the present calculations. The mesh required in order for such results to be obtained has $h = 842$ elements (542 triangles and 300 quadrilaterals). In the vicinity of the basic flow vortex this mesh comprises a finely-resolved structured core of dimension one (in radius units) embedded into an unstructured mesh, the density of which decreases monotonically to the end of the calculation domain.

The extent of the domain is calculated on the basis of the following considerations: although the vortex core radius is of order 2π, homogeneous Dirichlet boundary conditions are imposed in the far-field. As identified by Delbende, Chomaz and Huerre [3], the amplitude functions of the perturbation components are expected to decay exponentially as $r \to \infty$ according to $e^{-\alpha r}$. In this respect, the far field boundaries must be situated sufficiently far away from the vortex core, especially for small α values, in order for the error that the Dirichlet boundary condition induces to be negligible. The axial disturbance velocity component, $\hat{u}$, of a Batchelor vortex at $Re = 1{,}200, q = 0.8$ for $\alpha = 2.125$ (most unstable case for $\alpha \in [0, 4]$) and $\alpha = 2.875$ are shown in Fig. 2.

Based on this experience, analogous grids have been calculated for the BiGlobal instability analysis of an aircraft wake model based on a vortex dipole. Again, the extent of the domain is the square $y, z \in [-40, 40]$.

A linear superposition for the axial vorticity of the leading eigenmode upon the basic state of the most unstable mode at $Re = 3{,}800, \alpha = 3$ is shown in Fig. 3. A key observation in these and all other results obtained but not shown here is the spatial inhomogeneity of the amplitude functions along the azimuthal direction may be appreciated. Features of the eigenmodes, known from classic instability analyses which invoke azimuthal homogeneity as assumption, may be seen in these results. Specifically, remnants of elliptic instability in the vortex core are visible. Even more significantly, the braids surrounding the vortex core are essential parts

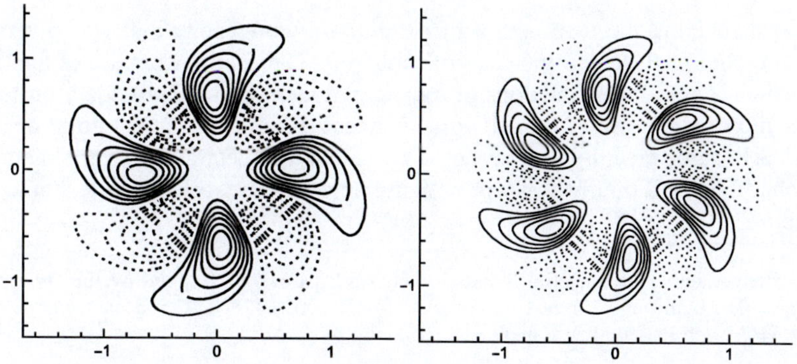

Fig. 2 The amplitude function of the normalized axial perturbation, $\hat{u}$, in the Batchelor vortex at $Re = 1,200, q = 0.8, \alpha = 2.125, m = 4$ (*left*) and $\alpha = 2.875, m = 6$ (*right*), obtained through numerical solution of (10)–(13). *Dashed lines* denote negative values

Fig. 3 Axial vorticity superposition upon the steady laminar basic state at $Re = 3{,}180$, $U_0 = 0.5/\pi$, of its most amplified eigenmode at amplitude 1%. Axial spatial direction reconstructed using $L_x = 2\pi/\alpha$, with $\alpha = 3.0$; eigenvalue $\omega_i = 0.04603$ (growth rate),$\omega_r = 0.06072$ (frequency)

of the amplitude functions, of the same (or larger) magnitude as the structures in the vortex cores. Such structures are clearly out of reach of instability approaches invoking periodicity along the azimuthal direction; use of the BiGlobal instability concept is mandatory for their recovery.

4 Conclusions and Outlook

A spectrally-accurate two-dimensional DNS methodology has been utilized in order to obtain the time-evolution of a pair of counter-rotating vortices, the initialization of which used the Batchelor model. A snapshot of this flow field was considered as quasi-steady basic state and subsequently analyzed with respect to their instability against three-dimensional disturbances. The BiGlobal analysis context employed permitted relaxing the assumption of azimuthal homogeneity that was invariably used in earlier analyses of this class of problems. The spatial structure of the (two-dimensional) amplitude functions of the eigenmodes obtained provide a-posteriori justification for the use of the BiGlobal concept. From a numerical point of view, what has become clear by the present work is that the spectral/hp, as applied to the BiGlobal EVP solution [4], can provide reliable results working on technological problems.

Acknowledgements The material is based upon work partially sponsored by the Universidad Politcnica de Madrid and contract numbers AST4-CT-2005-012338 (European Union Specific Targeted Research Project: Far Wake).

References

1. Broadhurst, M., Theofilis, V., Sherwin, S.J.: Spectral element stability analysis of vortical flows. 6th IUTAM Laminar-Turbulent Transition Symposium, pp. 153–158. Bangalore, India, Dec 13–17, 2004 (2006)
2. Capart, R., Winckelmans, G.: On characterization of the rollup produced by different span loadings. Tech. Rep. AW-UCL-111-002D, Universite Catholique de Louvain (2002)
3. Delbende, I., Chomaz, J.M., Huerre, P.: Absolute/convective instabilities in a batchelor vortex: a numerical study of the linear response. J. Fluid Mech. **355**, 229–254 (1998)
4. Gonzalez, L.M., Theofilis, V., Gomez-Blanco, R.: Finite-element numerical methods for viscous incompressible biglobal linear instability analysis on unstructured meshes. AIAA J. **45**(4), 840–855 (2007)
5. Gonzalez, L.M., Theofilis, V., Gomez-Blanco, R.: Eigenmodes of a counter-rotating vortex dipole. AIAA J. **46**(11), 2796–2805 (2008)
6. Hein, S., Theofilis, V.: On instability characteristics of isolated vortices and models of trailing-vortex systems. Comput. Fluid. **33**, 741–753 (2004)
7. Jacquin, L., Fabre, D., Sipp, D., Theofilis, V., Vollmers, H.: Instability and unsteadiness of aircraft wake vortices. Aero. Sci. Techn. **7**, 577–593 (2003)
8. Karniadakis, G., Sherwin, S.: Spectral/hp element methods for computational fluid dynamics, Oxford, Oxford University Press, (2005)
9. Lacaze, L., Ryan, K., Le Dizes, S.: Elliptic instability in a strained batchelor vortex. J. Fluid Mech. **577**, 341–361 (2007)
10. Mayer, E.W., Powell, K.G.: Viscous and inviscid instabilities of a trailing vortex. J. Fluid Mech. **245**, 91–114 (1992)
11. Roy, C., Schaeffer, N., Dizès, S.L., Thompson, M.: Stability of a pair of co-rotating vortices with axial flow. Phys. Fluid. **20**, 1–8 (2008)
12. Sipp, D., Jacquin, L., Cosssu, C.: Self-adaptation and viscous selection in concentrated two-dimensional vortex dipoles. Phys. Fluid. **12**, 245–248 (2000)
13. Theofilis, V.: Direct numerical simulation of the roll-up process of a trailing-vortex system, using experimentally obtained data at realistic reynolds numbers. Tech. Rep. IB-224-2002-C-12, DLR (2002). AWIATOR Final Report (confidential)
14. Theofilis, V.: Advances in global linear instability analysis of nonparallel and three-dimensional flows. Prog. Aero. Sci. **39**, 249–315 (2003)
15. Winckelmans, G., Leonard, A.: Contributions to vortex particle methods for the computation of three-dimensional incompressible unsteady flows. J. Comput. Phys. **109**(2), 247–273 (1993)

Entropy Viscosity Method for High-Order Approximations of Conservation Laws

J.L. Guermond and R. Pasquetti

Abstract A stabilization technique for conservation laws is presented. It introduces in the governing equations a nonlinear dissipation function of the residual of the associated entropy equation and bounded from above by a first order viscous term. Different two-dimensional test cases are simulated – a 2D Burgers problem, the "KPP rotating wave" and the Euler system – using high order methods: spectral elements or Fourier expansions. Details on the tuning of the parameters controlling the entropy viscosity are given.

1 Introduction

High-order methods, especially spectral methods, are very efficient for solving Partial Differential Equations (PDEs) with smooth solutions since the approximation error goes exponentially fast to zero as the polynomial degree of the approximation goes to infinity, i.e. spectral accuracy is observed. Unfortunately this property breaks down for non-smooth solutions such as those that arise from solving nonlinear conservation laws. This type of equations generates shocks which in turn induce the so-called Gibbs phenomenon. The problem is not new and many sophisticated algorithms have been developed to address this issue. Particularly popular among these methods are the so-called monotone and Total Variation Diminishing (TVD) schemes that aim at enhancing the accuracy far from the shocks and promoting non-oscillatory behavior at the shocks. These techniques are mainly based on Essentially Non Oscillatory polynomial reconstructions (ENO) and the use of flux/slope limiters

J.L. Guermond (✉)
Department of Mathematics, Texas A & M University, College Station, TX, USA
(on leave from LIMSI, CNRS)
e-mail: guermond@math.tamu.edu

R. Pasquetti
Lab. J. A. Dieudonné (CFD group), UMR CNRS 6621, Nice-Sophia Antipolis University, Nice
e-mail: richard.pasquetti@unice.fr

J.S. Hesthaven and E.M. Rønquist (eds.), *Spectral and High Order Methods for Partial Differential Equations*, Lecture Notes in Computational Science and Engineering 76,
DOI 10.1007/978-3-642-15337-2_39,

whose goal is to bound the fluxes. One may consult [3] for an overview on this class of methods, which were mainly developed for Finite Volume approximations.

It is remarkable that few methods have been proposed for solving nonlinear conservation laws with high order methods. Among them, in the frame of spectral methods the well known "spectral vanishing viscosity" technique [6] introduces a dissipation term only active in the high frequency range of the spectral approximation. In the same spirit, but on the basis of a hp-finite element approximation and a Discontinuous Galerkin method, it was also recently proposed to introduce a dissipation term, based on a viscosity controlled by a smoothness indicator [5]. The goal of the present paper is to present a somewhat different viscosity method, which was recently introduced in [1] by the authors. Here again the key idea consists of augmenting the PDE with a dissipation term, but the viscosity is based on the residual of the associated entropy equation. Here we propose a simplified formulation of the method and extend it to two-dimensional problems. The technique is implemented with Fourier polynomials and the Spectral Element Method (SEM).

The paper is organized as follows. We describe the entropy viscosity method in Sect. 2. An application to the two-dimensional inviscid Burgers equation with Fourier polynomials is described in Sect. 3 and convergence tests are reported. The method is adapted to the SEM setting in Sect. 4 and is illustrated on a nonlinear conservation law exhibiting a rotating composite wave. In Sect. 5 we adapt the entropy viscosity method to the two-dimensional Euler system and solve a classical benchmark problem using the Fourier approximation.

2 The Entropy Viscosity Method

It is well known that the relevant weak solution of the *scalar* conservation law

$$\partial_t u(\mathbf{x},t) + \nabla \cdot \mathbf{f}(u(\mathbf{x},t)) = 0\,, \quad \mathbf{x} \in \Omega\,, \quad t \in \mathbb{R}^+ \tag{1}$$

with appropriate initial and boundary conditions, is the so-called entropy solution, which is also characterized by $u = \lim_{\nu \to 0} u_\nu$ where

$$\partial_t u_\nu + \nabla \cdot \mathbf{f}(u_\nu) = \nu \Delta u_\nu . \tag{2}$$

Let us recall the following points, see e.g. [3] and references herein: (a) Solving (2) rather than (1), with a "small" value of ν, yields the Von-Neumann-Richtmyer method, developed for the Euler equations in 1950! Such an approach is however well known to be too diffusive. (b) Linear techniques such as the Lax-Wendroff scheme are more accurate than the first-order viscosity regularization but they are not fully satisfactory since the solution is often polluted by spurious oscillations. To overcome this difficulty one usually resorts to TVD schemes. (c) High-order (>1) TVD (and so monotonicity preserving) schemes must be nonlinear, as stated by the Godunov theorem. (d) Nonlinear schemes with flux/slope limiters essentially add some *nonlinear viscosity dissipation*.

Starting from this last point, the entropy viscosity method introduces a nonlinear dissipation term $\nabla.(\nu_h \nabla u)$ in the right hand side of (1). Let $E(u)$ be a convex function and assume that there exists an entropy pair $(E(u), \mathbf{F}(u))$ such that

$$\partial_t E(u) + \nabla \cdot \mathbf{F}(u) \leq 0$$

characterizes the unique viscous limit to (1) (i.e. the entropy solution). Let $r_E(u) := \partial_t E(u) + \nabla \cdot \mathbf{F}(u)$ be the entropy residual. This quantity is a negative measure supported on the shocks, i.e. $r_E < 0$ at the shocks and $r_E = 0$ elsewhere.

Assume that the computational domain Ω is discretized, let h be the grid size and u_h the numerical solution. We propose to construct a local artificial nonlinear viscosity based on the entropy residual $r_E(u_h)$. To this end we first set

$$\nu_E(\mathbf{x}, t) := \alpha h^2(\mathbf{x}) \mathscr{R}(r_E(u_h)) / \| E(u_h) - \bar{E} \|_{\infty,\Omega} \tag{3}$$

where α is a proportionality coefficient, $\bar{E}$ is the space average of $E(u_h)$ (recall that E is defined up to a constant), $\|.\|_{\infty,\Omega}$ is the usual $L^\infty(\Omega)$ norm and $\mathscr{R}(r_E)$ is a positive function (or functional) of the residual r_E. The terms $h^2(\mathbf{x})$ and $\| E(u_h) - \bar{E} \|_{\infty,\Omega}$ are scaling factors. The aim of $\mathscr{R}(r_E)$ is to extract a useful information from the residual; Hereafter we use $\mathscr{R}(r_E) = |r_E|$. Note that in smooth parts of u, one may expect that $r_E(u_h)$ scales like the approximation error of the resolution method.

We now provide an upper bound for the entropy viscosity. For the one-dimensional scalar conservation equation $\partial_t u + f'(u)\partial_x u = 0$, the first-order Finite Difference upwind scheme (linear monotone scheme) is equivalent to the second-order centered Finite Difference augmented with a viscous dissipation with viscosity $\nu_{max} = \frac{1}{2} f'(u)h$. By analogy we set

$$\nu_{max}(\mathbf{x}, t) = \alpha_{max}\, h \max_{\mathbf{y} \in V_\mathbf{x}} |\mathbf{f}'(u_h(\mathbf{y}, t))|, \tag{4}$$

where α_{max} is a constant coefficient, and $V_\mathbf{x}$ is a neighborhood of $\mathbf{x}$ still to be defined and dependent on the approximation method. In practice the size of $V_\mathbf{x}$ is a few multiples of h in each direction. Finally the entropy viscosity is defined to be

$$\nu_h(\mathbf{x}, t) := \mathscr{S}(\min(\nu_{max}, \nu_E)) \tag{5}$$

where $\mathscr{S}$ is a smoothing operator. Smoothing may indeed be required because $r_E(u_h)$ is generally highly oscillatory, since when a shock occurs we actually try to approximate a Dirac distribution. Practical implementation details on the operator $\mathscr{S}$ and on the neighborhood $V_\mathbf{x}$, as well as details on how to tune the coefficients α and α_{max} are provided in the examples studied in next sections.

3 2D Burgers (Fourier)

Let $\Omega = (0,1)^2$ and consider the following inviscid Burgers problem, where $\mathbf{v} = (1,1)$ is a constant vector field:

$$\partial_t u + \nabla.(\frac{1}{2}u^2\mathbf{v}) = 0, \qquad u|_{t=0} = u_0(x,y) \tag{6}$$

where $u_0 = -0.2$ if $x < 0.5,\ y > 0.5$, $u_0 = -1$ if $x > 0.5,\ y > 0.5$, $u_0 = 0.5$ if $x < 0.5,\ y < 0.5$ and $u_0 = 0.8$ if $x > 0.5,\ y < 0.5$. The local velocity $\mathbf{f}'(u) = u\mathbf{v}$ is parallel to $\mathbf{v}$ and of amplitude u.

To be able to solve this problem with a pseudo-spectral Fourier method we transform it into a periodic problem by extending the computational domain to $(0,2)^2$ and by extending the initial condition by symmetry about the axes $\{x = 1\}$, $\{y = 1\}$.

We choose the entropy pair $E(u) = \frac{1}{2}u^2$, $\mathbf{F}(u) = \frac{1}{3}u^3\mathbf{v}$, and then follow the procedure described in Sect. 2. The entropy viscosity, ν_h, and the non-linear flux, $\frac{1}{2}u_h^2\mathbf{v} - \nu_h\nabla u_h$, are computed in the physical space (pseudo-spectral approach). For each Fourier node $\mathbf{x}$, the neighborhood $V_\mathbf{x}$ is composed of the $7{\times}7$ Fourier nodes surrounding $\mathbf{x}$. The smoothing operation is performed by doing two smoothing sweeps, each one based on a two-dimensional averaging rule involving 5 grid-point values, with weight 4 for the central point and 1 for the 4 closest points.

The time marching is done with the standard Runge-Kutta scheme (RK4). The entropy viscosity is taken constant in time during the time-step, say from time t_n to t_{n+1}, and so computed at time t_n. Using the second order backward finite difference approximation, the time derivative of the entropy is computed from the values of E at time t_n, t_{n-1} and t_{n-2}. Space derivatives result from the Fourier approximation.

We show in Fig. 1 computations done at time $t = 0.5$ with 192 Fourier modes in each direction, i.e. with 192^2 grid points in $(0,1)^2$. The non-linearity was de-aliased using the $\frac{3}{2}$ padding rule. The entropy viscosity control parameters are $\alpha = 0.2$ and

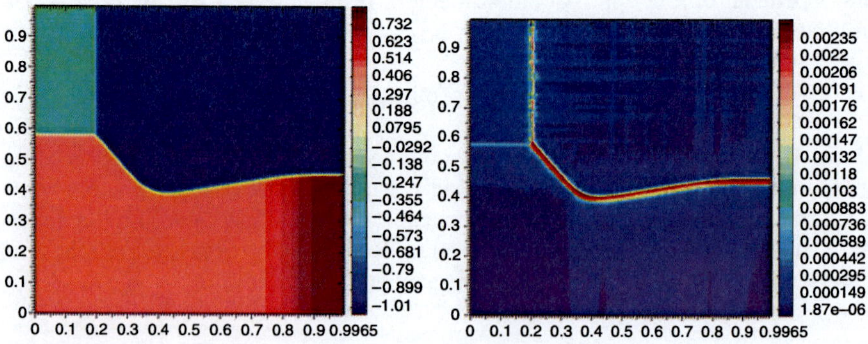

Fig. 1 Fourier-RK4 solution (*left*) and entropy viscosity (*right*)

Table 1 Errors and convergence rates for the 2D Burgers problem

h	L^1	Rate	L^2	Rate	L^∞
2.78E–2	1.92E–2	–	1.02E–1	–	1.47
1.39E–2	9.99E–3	0.94	7.28E–2	0.49	1.50
6.94E–3	5.34E–3	0.89	5.41E–2	0.43	1.50
3.47E–3	2.79E–3	0.95	3.80E–2	0.51	1.51

$\alpha_{max} = 1.5$. The approximate solution is shown in the left panel of Fig. 1, and the entropy viscosity is shown in the right panel. The shocks are well described and the entropy viscosity focuses in the shocks as expected.

The exact solution to (6) can be evaluated at time $t = 0.5$. Table 1 gives the relative error in the L^1- and L^2-norm for different grid sizes. One observes convergence rates close to optimality, i.e. order one in the L^1-norm and half order in the L^2-norm. Of course, no convergence is obtained in the L^∞ norm.

4 KPP Rotating Wave (SEM)

We now use the SEM method to solve the following two-dimensional nonlinear scalar conservation law:

$$\partial_t u + \nabla \cdot \mathbf{f}(u) = 0, \quad \mathbf{f}(u) = (\sin u, \cos u), \quad u|_{t=0} = \begin{cases} 3.5\pi & \text{if } |\mathbf{x}| < 1 \\ \frac{1}{4}\pi & \text{otherwise} \end{cases}$$

in the domain $\Omega = (-2,2)\times(-2.5,1.5)$ for $t \in (0,1)$. This problem has been proposed by Kurganov, Petrova and Popov [2] to test the convergence properties of some WENO schemes.

The local velocity is $\mathbf{v} = \mathbf{f}'(u) = (\cos u, -\sin u)$. We choose the entropy pair $E(u) = \frac{1}{2}u^2$, $\mathbf{F}(u) = (u \sin u + \cos u, u \cos u - \sin u)$. Then we follow the procedure defined in Sect. 2.

The domain is uniformly discretized using squares of side h and the approximation space is composed of the functions that are continuous and piecewise polynomial of partial degree at most N. The local shape functions are the Lagrange polynomials associated with the $(N+1)^2$ Gauss-Lobatto-Legendre (GLL) points. To define the entropy viscosity we follow the procedure described in Sect. 2, except that in (4) we have used the local grid size of the GLL mesh, say h_{GLL}, rather than h. The neighborhood $V_\mathbf{x}$ is defined as the corresponding spectral element of $\mathbf{x}$, during the assembling procedure. The smoothing is achieved inside each element on the GLL mesh, by one smoothing sweep based on a two-dimensional averaging rule involving 5 GLL grid-points. The entropy viscosity control parameters are $\alpha = 40$ and $\alpha_{max} = 0.8/N$. The time marching is done by using the standard Runge-Kutta scheme (RK4). The entropy viscosity is made explicit and computed by using the

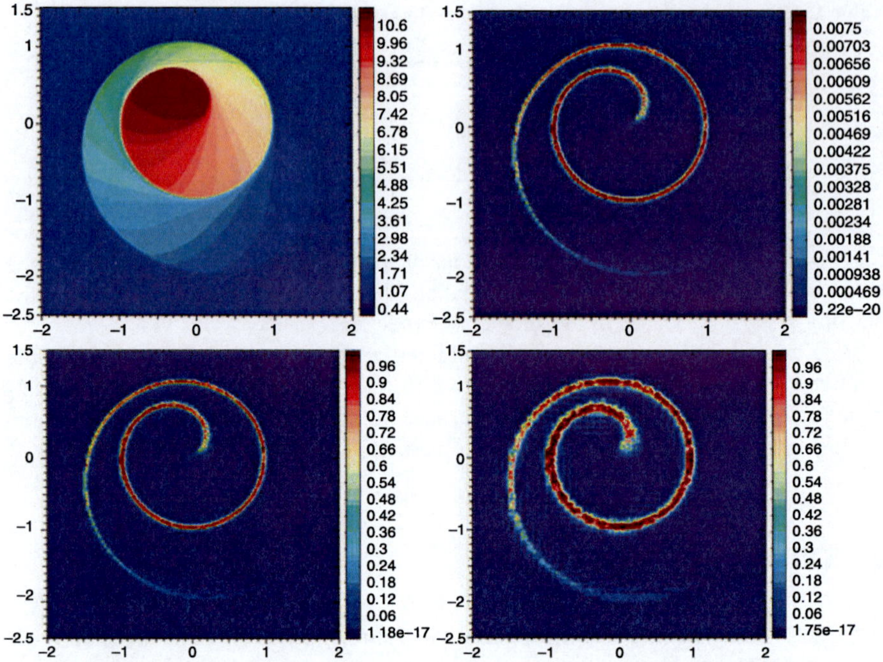

Fig. 2 SEM-RK4 solution for the KPP rotating wave. Solution (*top left*) and corresponding entropy viscosity (*top right*) for $N = 4$ and 96^2 cells. Ratio ν/ν_{max} for 96^2 cells (*bottom left*) and 48^2 (*bottom right*)

second order backward finite difference approximation for the time derivative of the entropy.

Results reported in the two top panels of Fig. 2 have be obtained with a grid composed of 96^2 square elements and with polynomials of degree $N = 4$ in each variable. The numerical solution is shown in the left panel; It exhibits the correct composite wave structure. The corresponding entropy viscosity is shown in the right panel; As expected, dissipation is added only where the shock develops.

We finish this section by providing more details on how to adjust the entropy viscosity parameters. The idea is that to be efficient, the viscosity must reach its maximum value in the shocks. Consequently, we propose the following two-step adjustment procedure:

1. Set $\alpha = \infty$ and increase α_{max} until obtaining a smooth solution (a good guideline is that $\alpha_{max} = \frac{1}{2}$ is the correct answer in one space dimension on uniform grids).
2. Once α_{max} is fixed, set α so that the entropy viscosity saturates in the shocks, i.e. $\max(\nu) = \nu_{max}$ in shocks.

The two bottom panels in Fig. 2 show the ratio ν/ν_{max} for two different discretizations. Observe that this ratio equals 1 in the shock.

5 2D Euler System (Fourier)

We finish this paper by explaining how the entropy viscosity method can be adapted to the compressible Euler equations:

$$\partial_t \mathbf{u} + \nabla \cdot \mathbf{f}(\mathbf{u}) = 0, \quad \mathbf{u} = \begin{pmatrix} \rho \\ \rho \mathbf{v} \\ E \end{pmatrix}, \qquad \mathbf{f} = \begin{pmatrix} \rho \mathbf{v} \\ \rho \mathbf{v} \otimes \mathbf{v} + p\mathbb{I} \\ \mathbf{v}(E+p) \end{pmatrix} \tag{7}$$

where $p = \rho T, T = (\gamma - 1)(E/\rho - \mathbf{v}^2/2)$. Usual notations are used: $\rho, \mathbf{v}, p, T, \gamma, E$ stand for density, velocity, pressure, temperature, ratio of specific heat, and total energy, respectively. The physical entropy functional $S(p,\rho) = \frac{\rho}{\gamma-1} \log(p/\rho^\gamma)$ is such that $r_S := \partial_t S + \nabla \cdot (\mathbf{v} S) \geq 0$.

To understand where and how the entropy dissipation must be set, it is helpful to follow the physics by considering the viscous fluxes appearing in the Navier-Stokes equations:

$$\mathbf{f}_{visc}(\mathbf{u}) = \begin{pmatrix} 0 \\ -\mu \nabla \mathbf{v} \\ -\mu \mathbf{v} : \nabla \mathbf{v} - \kappa \nabla T \end{pmatrix}.$$

The quantity μ is the dynamic viscosity and κ is the thermal conductivity.

First, we compute μ_S, except that there is no need to normalize by $\|S - \bar{S}\|_{\infty,\Omega}$ in (5): $\mu_S = \alpha\, h^2\, \rho(\mathbf{x},t)|r_S(\mathbf{x},t)|$. Then, estimating the maximum local wave speed to be $|\mathbf{v}| + \sqrt{\gamma T}$, we set $\mu_{max} = \alpha_{max}\, h\, \rho(\mathbf{x},t) \max_{\mathbf{y}\in V_\mathbf{x}}(|\mathbf{v}(\mathbf{y},t)| + \sqrt{\gamma T(\mathbf{y},t)})$. Finally, $\mu = \mathscr{S}(\min(\mu_{max}, \mu_S))$ and, taking κ to be proportional to μ, $\kappa = \beta\mu$.

We now validate this approach by solving the benchmark problem number 12 from [4]. It is a two-dimensional Riemann problem set in $\mathbb{R}^2$. In the restricted computational domain $(0,1)^2$ the initial set of data is defined as follows:

$$\begin{array}{lllll} p = 1., & \rho = 0.8, & \mathbf{v} = (0., 0.), & 0. < x < 0.5 & 0. < y < 0.5, \\ p = 1., & \rho = 1., & \mathbf{v} = (0.7276, 0.), & 0. < x < 0.5, & 0.5 < y < 1., \\ p = 1., & \rho = 1., & \mathbf{v} = (0., 0.7276), & 0.5 < x < 1., & 0. < y < 0.5, \\ p = 0.4, & \rho = 0.5313, & \mathbf{v} = (0., 0.) & 0.5 < x < 1., & 0.5 < y < 1.. \end{array}$$

The solution is computed at time $t = 0.2$. Proceeding as in Sect. 3, the problem is first made periodic by extending the computational domain to $(0,2)^2$, and the initial data are extended by symmetry about the axes $\{x = 1\}$ and $\{y = 1\}$.

The time marching algorithm, the definition of the smoothing operator, and the neighborhood $V_\mathbf{x}$ are the same as in Sect. 3. The nonlinear terms are de-aliased. The control parameters for the entropy viscosity are $\alpha = 20$, $\alpha_{max} = 0.5$ and $\beta = 2$. We show in Fig. 3 results obtained with 400 Fourier modes in each direction, i.e. with 400 grid-points in $(0,1)^2$. They compare well with those obtained with other more sophisticated shock capturing methods, see [4].

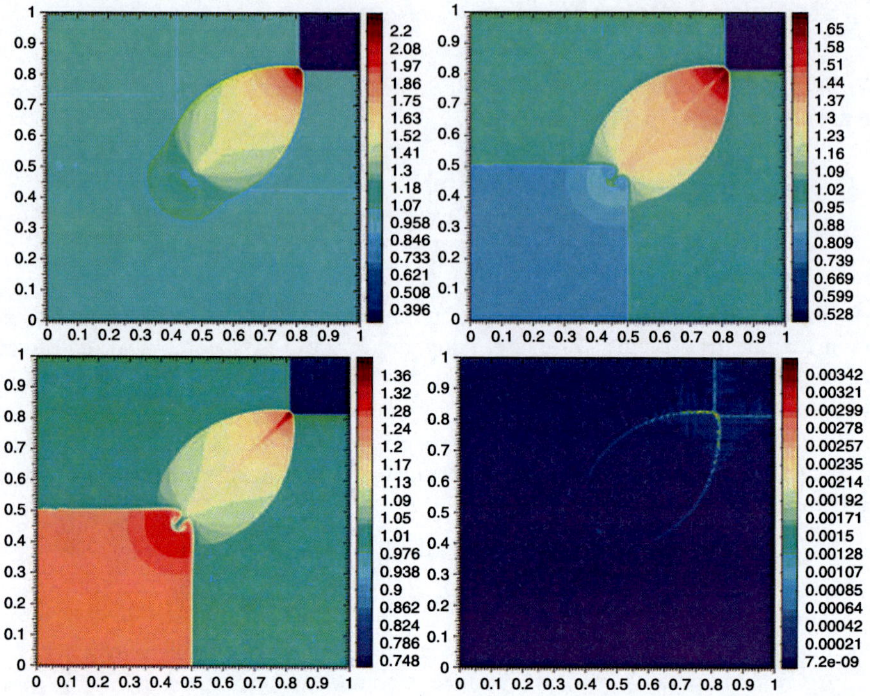

Fig. 3 Pressure (*top left*); Density (*top right*); Temperature (*bottom left*); Entropy viscosity μ (*bottom right*)

Acknowledgements This material is based upon work supported by the National Science Foundation grant DMS-0510650 and DMS-0811041 and partially supported by Award No. KUS-C1-016-04, made by King Abdullah University of Science and Technology (KAUST).

References

1. Guermond, J.L. and Pasquetti, R.: Entropy-based nonlinear viscosity for Fourier approximations of conservation laws. C.R. Acad. Sci. Paris, Ser. I, **346**, 801–806 (2008)
2. Kurganov, A., Petrova, G. and Popov, B.: Adaptive semidiscrete central-upwind schemes for nonconvex hyperbolic conservation laws. SIAM J. Sci. Comput., **29** (6), 1064–8275 (2007)
3. Leveque, R.J.: Numerical methods for conservation laws. Lectures in Mathematics, ETH Zürich. Birkhäuser, Berlin (1992)
4. Liska, R. and Wendroff, B.: Comparison of several difference schemes on 1D and 2D test problems for the Euler equations. SIAM J. Sci. Comput., **25** (3), 995–1017 (2004)
5. Persson, P.-O. and Peraire, J.: Sub-cell shock capturing for discontinuous Galerkin methods. AIAA-2006-0112, Reno, (January 2006)
6. Tadmor, E.: Convergence of spectral methods for nonlinear conservation laws. SIAM J. Numer. Anal., **26** (1), 30–44 (1989)

High-Order Accurate Numerical Solution of Incompressible Slip Flow and Heat Transfer in Microchannels

Kazem Hejranfar, Mir Hamed Mohafez, and Ali Khajeh-Saeed

Abstract A high-order accurate implicit operator scheme is used to solve steady incompressible slip flow and heat transfer in 2D microchannels. The present methodology considers the solution of the Navier–Stokes equations using the artificial compressibility method with employing the Maxwell and Smoluchowski boundary conditions to model the slip flow and temperature jump on the walls in microchannels. Since the slip and temperature jump boundary conditions contain the derivatives of the velocity and temperature profiles, using the compact method the boundary conditions can be easily and accurately implemented. The computations are performed for a 2D microchannel and a 2D backward facing step in the slip regime. The results for these cases for different conditions are compared with the available results which show good agreement. The effects of the Knudsen and Reynolds numbers on the flow field and heat transfer characteristics are also investigated.

1 Introduction

Recently, progress in micro-fabrication techniques has led to development of a large number of Micro-Electro-Mechanical Systems (MEMS) and microfluidic Technologies. In MEMS devices, the fluid mechanics and heat transfer of gas-phase microflows due to non-equilibrium effects such as rarefaction and gas-surface interactions can differ significantly from the macroscopic world and the no-slip boundary conditions of the Navier–Stokes equations are no longer valid. The applicability of the Navier–Stokes equations is determined by the Knudsen number Kn which is defined as the ratio of mean free path of gas molecules λ to the characteristic length of a microdevice L. Schaaf and Chambre [14] classified different flow regimes based on the Knudsen number. The continuum fluid flow occurs for $Kn < 0.01$, the slip flow regime is considered for $0.01 \le Kn \le 0.1$, the transition flow regime is

K. Hejranfar (✉), M.H. Mohafez, and A. Khajeh-Saeed
Aerospace Engineering Department, Sharif University of Technology, Tehran, Iran
e-mail: Khejran@sharif.edu, mirhamed.moha@gmail.com, khajehsaeed@gmail.com

J.S. Hesthaven and E.M. Rønquist (eds.), *Spectral and High Order Methods for Partial Differential Equations*, Lecture Notes in Computational Science and Engineering 76,
DOI 10.1007/978-3-642-15337-2_40,

accomplished for $0.1 \le Kn \le 10$ and the free molecular regime exists for $Kn > 10$. To model the flow field in the slip flow regime, $Kn \le 0.1$, the Navier–Stokes equations can be used together with appropriate slip and temperature jump boundary conditions between gas and substrate. Up to now, many efforts have been made for solving the flow field in microchannels by using the Navier–Stokes equations with employing slip and temperature jump boundary conditions (see [1, 3, 5, 8, 11] and others). These flow solvers for modeling the slip flow regime in microchannels are based on usual finite-difference methods or finite-volume, finite-volume-element and spectral-element methods.

In this paper, a fourth-order implicit compact operator method is used for solving 2D incompressible microchannel flows with heat transfer by using the artificial compressibility method. Compact methods, compared with the traditional finite-difference schemes of the same order of accuracy, are shown to be significantly more accurate with advantage of having good resolution properties without increasing excessively the computational stencil size. The computations are performed for a 2D microchannel and also a 2D backward facing step in slip flow regime. The results for different conditions are compared with the available results and the effects of the Knudsen and Reynolds numbers on the solutions are studied.

2 Problem Formulation

The problem considered to be solved is laminar incompressible slip flow and heat transfer in 2D microchannels. Herein, the thermophysical properties such as the viscosity and thermal conductivity of the fluid are assumed to be constant. Body forces and viscous heating are also assumed to be negligible. The 2D incompressible Navier–Stokes equations using the artificial compressibility method can be written in dimensionless and conservative form in Cartesian coordinates as:

$$\frac{\partial Q}{\partial t} + \frac{\partial E}{\partial x} + \frac{\partial F}{\partial y} = \frac{1}{Re}([N]\,\nabla^2 Q) \tag{1}$$

where Q is the solution vector and E and F are the inviscid flux vectors as follows:

$$Q = \begin{bmatrix} p \\ u \\ v \\ T \end{bmatrix}, \quad E = \begin{bmatrix} \beta u \\ u^2 + p \\ uv \\ uT \end{bmatrix}, \quad F = \begin{bmatrix} \beta v \\ uv \\ v^2 + p \\ vT \end{bmatrix}, \quad N = \begin{bmatrix} 0 & 0 & 0 & 0 \\ 0 & 1 & 0 & 0 \\ 0 & 0 & 1 & 0 \\ 0 & 0 & 0 & \frac{1}{\mathrm{Pr}} \end{bmatrix} \tag{2}$$

Here, β is the artificial compressibility parameter, (u, v) are the Cartesian velocity components, p is the pressure, T is the temperature, and Pr is the Prandtl number ($Pr = 0.72$ for air). $Re = \bar{u}\,H/\nu$ denotes the Reynolds number based on the mean inflow velocity $\bar{u}$, the channel height H and the kinematic viscosity ν.

3 Wall Boundary Conditions

The wall boundary conditions used in this study are the Maxwell [12] and Smoluchowski boundary conditions [15] that are given respectively as follows:

$$u_s - u_w = \frac{2-\sigma_v}{\sigma_v} Kn \left(\frac{\partial u}{\partial n}\right)_s + \frac{3}{2\pi}\frac{\gamma-1}{\gamma}\frac{Kn^2 Re}{Ec}\frac{\partial T}{\partial s} \tag{3}$$

$$T_s - T_w = \frac{2-\sigma_T}{\sigma_T}\frac{2\gamma}{\gamma+1}\frac{Kn}{\Pr}\left(\frac{\partial T}{\partial n}\right)_s \tag{4}$$

where s and n denote the local tangential and normal directions, respectively, u_s and T_s are the slip velocity and the gas temperature on the wall, respectively, u_w and T_w are the wall velocity and the wall temperature, respectively, σ_v and σ_T are the tangential momentum accommodation coefficient and the thermal accommodation coefficient, respectively. γ is the ratio of specific heats ($\gamma = 1.40$ for air) and Ec is the Eckert number. The second term in the right hand side of (3) denotes the thermal creep effect which accounts for fluid flow induced by the temperature gradient near the wall along the surface. This term is second order in Knudsen number [2] and becomes negligible for the moderate temperature gradients in slip flow regime.

4 Numerical Procedure

The numerical method applied is an alternating direction implicit compact operator scheme which has been used for computing 2D compressible and incompressible flows [6, 7, 9]. In [11], the algorithm has been extended to model the incompressible flows in 2D microchannels in slip flow regime. Herein, the algorithm is used and extended for solving incompressible microchannel flows with heat transfer. By implementing this scheme to the incompressible Navier–Stokes equations,

$$\begin{aligned}&\left(I - 3\tfrac{\Delta t}{h}A_{i-1,j}\right)\Delta Q^*_{i-1,j} + 4\Delta Q^*_{i,j} + \left(I + 3\tfrac{\Delta t}{h}A_{i+1,j}\right)\Delta Q^*_{i+1,j}\\ &= RHS_{i-1,j} + 4RHS_{i,j} + RHS_{i+1,j}\end{aligned} \tag{5}$$

$$\begin{aligned}&\left(I - 3\tfrac{\Delta t}{h}B_{i,j-1}\right)\Delta Q_{i,j-1} + 4\Delta Q_{i,j} + \left(I + 3\tfrac{\Delta t}{h}B_{i,j+1}\right)\Delta Q_{i,j+1}\\ &= \Delta Q^*_{i,j-1} + 4\Delta Q^*_{i,j} + \Delta Q^*_{i,j+1}\end{aligned} \tag{6}$$

where $RHS = \Delta t\left[-\frac{\partial E}{\partial x} - \frac{\partial F}{\partial y} + \frac{1}{Re}([N]\,\nabla^2 Q)\right]$ and A and B are known as the flux Jacobian matrices. Note that the spatial derivatives in RHS are computed with the fourth-order compact scheme. Note also that computing RHS in (5) with satisfying slip and temperature jump conditions can be easily and accurately performed. For example, for calculating the first derivative of the temperature in the y-direction, one can use the following fourth-order compact relations [10]

$$T'_{j-1} + 4T'_j + T'_{j+1} = \tfrac{3}{h}\left(T_{j+1} - T_{j-1}\right) + O(h^4). \tag{7}$$

For boundary treatment, the following third-order compact formula can be used

$$\begin{aligned} T'_1 + 2T'_2 &= \tfrac{1}{6h}\left(-15T_1 + 12T_2 + 3T_3\right) + O(h^3) \\ T'_{jmax} + 2T'_{jmax-1} &= \tfrac{1}{6h}\left(15T_{jmax} - 12T_{jmax-1} - 3T_{jmax-2}\right) + O(h^3). \end{aligned} \tag{8}$$

Since the thermal boundary condition (4) contains the first derivative of the temperature, ($T_s = \alpha_T T'_s$ where $\alpha_T = \frac{2-\sigma_T}{\sigma_T}\frac{2\gamma}{\gamma+1}\frac{Kn}{\mathrm{Pr}}$), then one can combine (4) and (8) to perform the appropriate boundary formula:

$$\begin{aligned} \left(1 + \tfrac{15}{6h}\alpha_T\right) T'_1 + 2T'_2 &= \tfrac{1}{6h}\left(12T_2 + 3T_3\right) + O(h^3) \\ \left(1 + \tfrac{15}{6h}\alpha_T\right) T'_{jmax} + 2T'_{jmax-1} &= \tfrac{1}{6h}\left(12T_{jmax-1} - 3T_{jmax-2}\right) + O(h^3) \end{aligned} \tag{9}$$

Now, the compact relation (7) together with the boundary formula (8) can be used to obtain T' through the wall-normal direction with the fourth-order accuracy with no especial treatment at the wall boundary. The same trend can be used for evaluating the first derivative of the velocity, u'.

Equations (5) and (6) along with suitable boundary conditions form a block-tridiagonal system of equations with a block size of 4×4 in the I- or J-sweeps to obtain $\Delta Q_{i,j}$ and then calculate the solution vector $Q^{n+1}_{i,j} = Q^n_{i,j} + \Delta Q_{i,j}$. Details of the numerical method implemented can be found in [9, 11].

5 Numerical Results and Discussion

At first, the results based on the compact scheme are presented for the 2D simple microchannel (see Fig. 1, $h = 0.0$) in the slip flow regime with the constant wall temperature condition. The velocity and temperature are uniform at the inlet and the fully developed condition for the velocity is assumed at the outlet. Figure 2 gives the distribution of the local Nusselt number on the wall along the microchannel for $Kn = 0.0, 0.05, 0.10$. The computed results are in agreement with those presented in

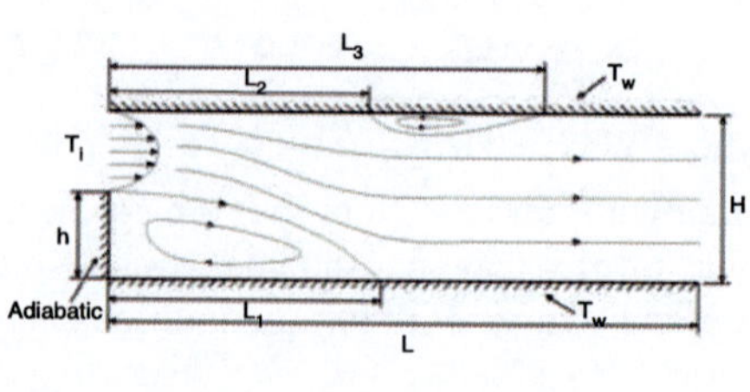

Fig. 1 Backward facing step geometry and boundary conditions used

Fig. 2 Rarefaction effect on local Nusselt number for 2D microchannel

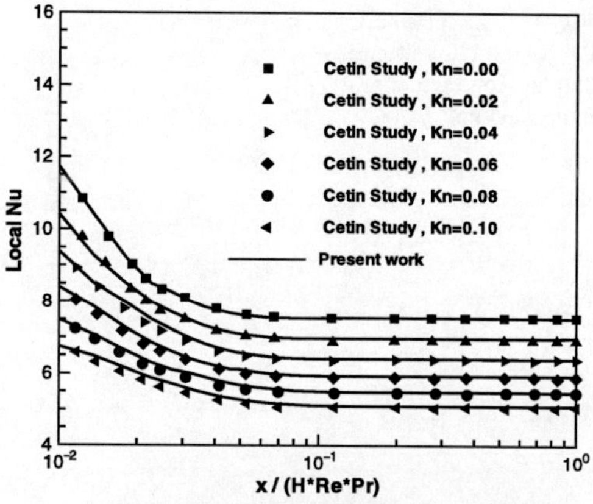

Fig. 3 Rarefaction effect on temperature contours for 2D microchannel for different *Kn*

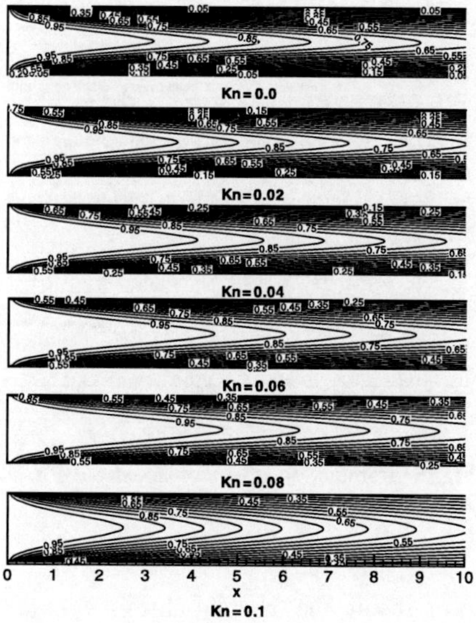

[4]. In the entrance region, large gradients occur that can produce very large velocity slip and temperature jump and their magnitudes are significantly reduced as the flow develops along the channel owing to weaker gradients. It is found that in the slip flow regime, the thermal developing length is less than 10% of the microchannel length. It is clear that by increasing *Kn* (the degree of rarefaction), the local and fully developed Nusselt numbers due to increasing gas temperature jump close to the wall are decreased. The figure also indicates that the distribution of *Nu* in the entrance region of the microchannel is smoothed. The local and fully developed friction factor, C_f, due to increasing the slip velocity and therefore decreasing the shear stress on the wall are also decreased (not shown here). In Fig. 3, the temperature contours ($\theta = \frac{T-T_w}{T_i-T_w}$, $\theta_w = 0.0$) along the microchannel show a decrease in the growth of the thermal boundary layer along the microchannel due to the rarefaction effect.

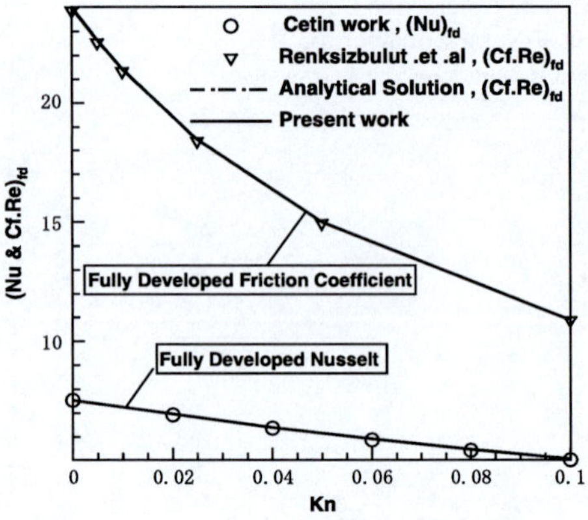

Fig. 4 Comparison of fully developed Nusselt number and friction factor for 2D microchannel

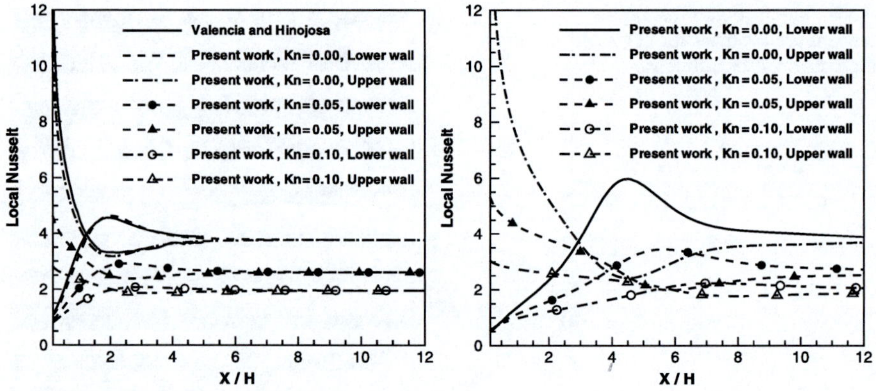

Fig. 5 Rarefaction effect on local *Nu* for backward facing step with $Re = 100$ (*upper*) and $Re = 400$ (*lower*) in slip flow regime

As a result, the thermal entrance length increases as *Kn* increases (see also Fig. 2). A similar trend has also be found for the hydrodynamic entrance length. Figure 4 gives the rarefaction effect on the fully developed *Nu* and C_f. It is obvious that by increasing *Kn*, both the values of these variables are decreased. A good agreement between the present results and those of [4, 13] is exhibited.

Now, the results for the 2D backward facing step in slip flow regime (see Fig. 1, $h = H/2$) are presented. This geometry contains complex flow features associated with separation and reattachment. In Fig. 5, the computed local *Nu* on the walls for $Kn = 0.0, 0.05, 0.10$ with $Re = 100,\ 400$ are shown. The results for the noslip case for $Re = 100$ are in good agreement with those of [16]. Very large reduction in heat transfer in the entrance region in slip flow regime are due to rarefaction. It is clear that for constant wall temperature condition by increasing *Kn*, the variation of *Nu* along both the walls is smoothed and the maximum and minimum values of *Nu* are reduced and their positions are shifted to the downstream. These effects are more pronounced for higher *Re*. Figure 6 depicts the associated flow field and heat

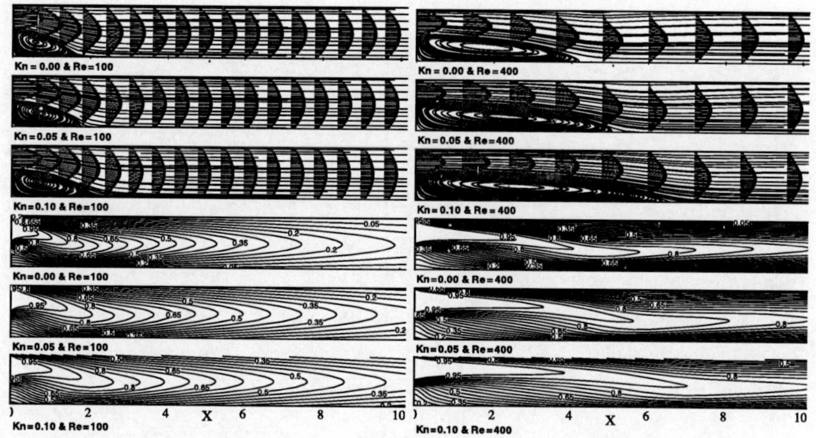

Fig. 6 Computed flow field and temperature contours for backward facing step with $Re = 100$ (*upper*) and 400 (*lower*) in slip flow regime

Fig. 7 Rarefaction effect on wall shear distribution for backward facing step with $Re = 400$

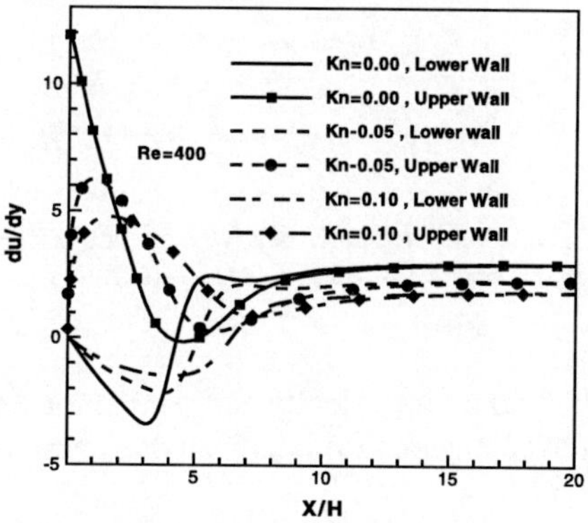

transfer characteristics shown by the velocity profiles and temperature contours. The study demonstrates that by increasing *Kn* or *Re*, the reattachment zone near the lower wall is stretched and the center of recirculating region is moved to the right accordingly. It is found that the small recirculating zone on the upper wall for $Re = 400$ is disappeared due to rarefaction which can clearly be concluded from the wall shear distribution in Fig. 7. Figure 8 illustrates the u-velocity and temperature profiles in the slip flow regime at $x/H = 1,\ 10$ for $Re = 100,\ 400$. The figure shows that the height of reattachment zone is slightly increased by increasing *Kn* and the position of the maximum velocity and temperature is moved toward the upper wall. In the recirculating region, the absolute velocity slip and the temperature jump on the lower wall are less than those of the upper wall. By increasing *Kn*, the absolute slip velocity and temperature jump on both the walls are increased. For higher *Re*,

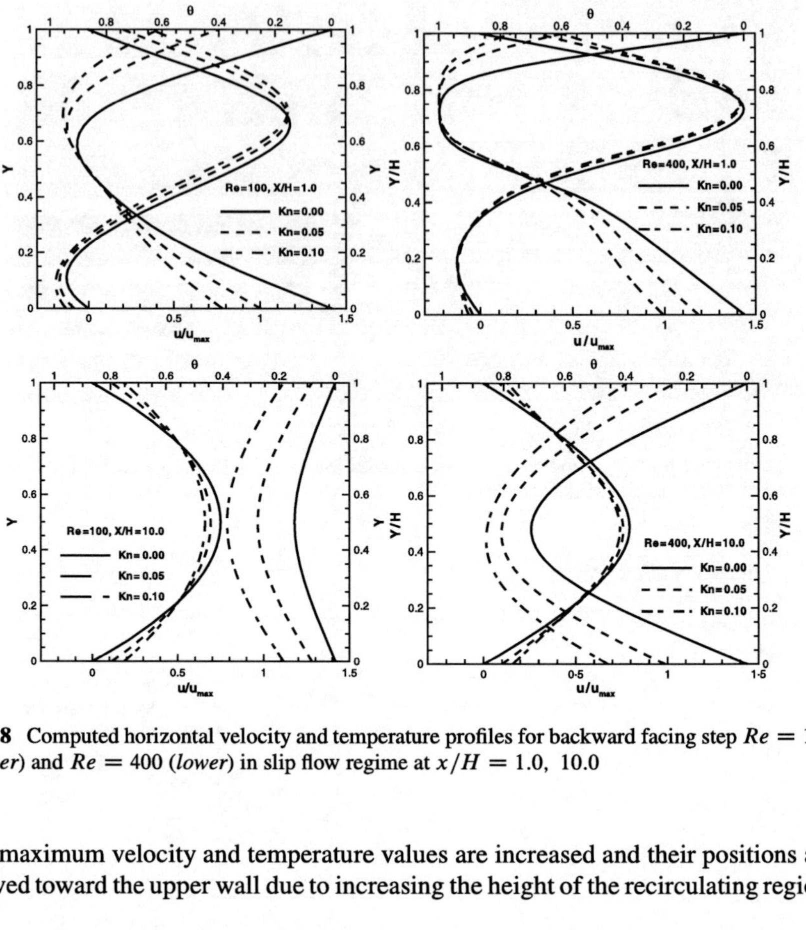

Fig. 8 Computed horizontal velocity and temperature profiles for backward facing step $Re = 100$ (*upper*) and $Re = 400$ (*lower*) in slip flow regime at $x/H = 1.0,\ 10.0$

the maximum velocity and temperature values are increased and their positions are moved toward the upper wall due to increasing the height of the recirculating region.

6 Concluding Remarks

A fourth-order compact implicit operator scheme is employed for solving 2D incompressible microchannel flows with heat transfer. Beside the high-order accuracy of the numerical method used, the implementation of slip and temperature jump conditions can also be done with more ease and accuracy. The present study shows that rarefaction has significant effects on the flow field and heat transfer characteristics. Note that gas flows in microchannels are usually experience density variations as a function of the pressure drop. The algorithm presented can be extended for solving compressible flows in microchannels.

Acknowledgements The authors would like to thank Sharif University of Technology for the financial support of this study.

References

1. Ahmed, I. and Beskok, A.: Rarefaction, Compressibility, and Viscous Heating in Gas Microfilters, J. Thermophys. Heat Transf. **16**, 161–170 (2002)
2. Beskok, A., Em Karniadakis, G. and Trimmer, W.: Rarefaction and Compressibility Effects in Gas Microflows, J. Fluid Eng. **118**, 448-456 (1996)
3. Barber, R. W. and Emerson, D. R.: Tech. Rep, Comput. Sci. Eng. Dept., CLRC Darebury Laboratory, (2000)
4. Cetin, B.: Analysis of Single Phase Convective Heat Transfer in Microtubes and Microchannels, MSc Thesis, Mechanical engineering of Middle East Technical University, Ankara, Turkiye (2005)
5. Darbandi, M., Rikhtegar, F. and Schbeider, G. E.: AIAA Paper 2007–3991, (2007)
6. Ekaterinaris, J. A.: Implicit High-Order Accurate in Space Algorithms for the Navier–Stokes Equations, AIAA J. **38**, 1594–1602 (2000)
7. Ekaterinaris, J. A.: Implicit, High-Resolution, Compact Schemes for Gas Dynamics and Aeroacoustics, J. Comp. Phys. **156**, 272–299 (1991)
8. Hadjiconstantinou, N. G. and Simek O.: Constant-wall-temperature Nusselt Number in Micro and Nano-channels, J. Heat Transf. **44**, 4225–4234 (2002)
9. Hejranfar, K. and Khajeh-Saeed, A.: Implementing a High-Order Accurate Implicit Operator Scheme for Solving Steady Incompressible Viscous Flows using Artificial Compressibility Method, Int. J. Num. Meth. Fluid, 2010. DOI: 10.1002/fld.2288
10. Lele, S. K.: Compact Finite Different Schemes with Spectral-Like Resolution. J. Comp. Phys., **103**, 16–42 (1992)
11. Hejranfar, K., Mohafez, M. H. and Khajeh-Saeed, A.: A High-Order Accurate Implicit Operator Scheme for Solving Incompressible Microchannel Slip Flows Using Artificial Compressibility Method, The 17th Annual (International) Conference on Mechanical Engineering (ISME2009), Iran (2009)
12. Maxwell, J. C.: On Stress in Rarefied Gases Arising From Inequalities of Temperature. Philos. Trans. R. Soc. Part 1, 231–256 (1897)
13. Renksizbulut, M., Niazmand, H. and Tercan, G.: Flow and Heat Transfer in Rectangular Microchannels with Constant Wall Temperature. Int. J. Ther. Sci. **45**, 870–881 (2006)
14. Schaaf, S. A. and Chambre, P. L.: Flow of Rarefied Gaseous. Prinston University Press, Prinston, NJ (1961)
15. Smoluchowski, von M.: Ueber Warmeleitung in Verdunnten Gasen, Annalen der Physik und Chemie. **64**, 101–130 (1898)
16. Valencia, A. and Hinojosa, L.: Numerical Solutions of Pulsating Flow and Heat Transfer Characteristics in a Channel with Backward-Facing Step. Int. J. Heat Mass Transf. **32**, 143–148 (1997)

Spectral Methods for Time-Dependent Variable-Coefficient PDE Based on Block Gaussian Quadrature

James V. Lambers

Abstract Krylov subspace spectral (KSS) methods have previously been applied to the variable-coefficient heat equation and wave equation, as well as systems of coupled equations such as Maxwell's equations, and have demonstrated high-order accuracy, as well as stability characteristic of implicit time-stepping schemes, even though KSS methods are explicit. KSS methods compute each Fourier coefficient of the solution using techniques developed by Gene Golub and Gérard Meurant for approximating elements of functions of matrices by Gaussian quadrature in the spectral, rather than physical, domain. In this paper, we review the most effective type of KSS method, that relies on block Gaussian quadrature, and compare its performance to that of Krylov subspace methods from the literature.

1 Introduction

In [13] a class of methods, called block Krylov subspace spectral (KSS) methods, was introduced for the purpose of solving parabolic variable-coefficient PDE. These methods are based on techniques developed by Golub and Meurant in [4] for approximating elements of a function of a matrix by Gaussian quadrature in the *spectral* domain. In [12], these methods were generalized to the second-order wave equation, for which these methods have exhibited even higher-order accuracy.

It has been shown in these references that KSS methods, by employing different approximations of the solution operator for each Fourier coefficient of the solution, achieve higher-order accuracy in time than other Krylov subspace methods (see, for example, [8]) for stiff systems of ODE, and they are also quite stable, considering that they are explicit methods. They are also effective for solving systems of coupled equations, such as Maxwell's equations [14].

J.V. Lambers
Department of Mathematics, University of Southern Mississippi, 118 College Dr #5045, Hattiesburg, MS 39406-0001, USA
e-mail: James.Lambers@usm.edu

J.S. Hesthaven and E.M. Rønquist (eds.), *Spectral and High Order Methods for Partial Differential Equations*, Lecture Notes in Computational Science and Engineering 76,
DOI 10.1007/978-3-642-15337-2_41,

In this paper, we review block KSS methods, as applied to various types of PDE, and compare their performance to other Krylov subspace methods from the literature. Section 2 reviews the main properties of block KSS methods, as applied to the parabolic problems for which they were designed. Section 3 discusses implementation details, and demonstrates why KSS methods need to explicitly generate only one Krylov subspace, although information from several is used. In Sect. 4, we discuss modifications that must be made to block KSS methods in order to apply them to systems of coupled wave equations, such as Maxwell's equations. Numerical results are presented in Sect. 5, and conclusions are stated in Sect. 6.

2 Krylov Subspace Spectral Methods

We first review block KSS methods, which are easier to describe for parabolic problems. Let $S(t) = \exp[-Lt]$ represent the exact solution operator of the problem

$$u_t + Lu = 0, \quad t > 0, \tag{1}$$

with appropriate initial conditions and periodic boundary conditions. The operator L is a second-order, self-adjoint, positive definite differential operator.

Let $\langle \cdot, \cdot \rangle$ denote the standard inner product of functions defined on $[0, 2\pi]$. Block Krylov subspace spectral methods, introduced in [13], use Gaussian quadrature on the spectral domain to compute the Fourier coefficients of the solution. These methods are time-stepping algorithms that compute the solution at time $t_1, t_2, \ldots$, where $t_n = n\Delta t$ for some choice of Δt. Given the computed solution $\tilde{u}(x, t_n)$ at time t_n, the solution at time t_{n+1} is computed by approximating the Fourier coefficients that would be obtained by applying the exact solution operator to $\tilde{u}(x, t_n)$,

$$\hat{u}(\omega, t_{n+1}) = \left\langle \frac{1}{\sqrt{2\pi}} e^{i\omega x}, S(\Delta t)\tilde{u}(x, t_n) \right\rangle. \tag{2}$$

In [4] Golub and Meurant describe a method for computing quantities of the form

$$\mathbf{u}^T f(A)\mathbf{v}, \tag{3}$$

where $\mathbf{u}$ and $\mathbf{v}$ are N-vectors, A is an $N \times N$ symmetric positive definite matrix, and f is a smooth function. Our goal is to apply this method with $A = L_N$ where L_N is a spectral discretization of L, $f(\lambda) = \exp(-\lambda t)$ for some t, and the vectors $\mathbf{u}$ and $\mathbf{v}$ are obtained from $\hat{\mathbf{e}}_\omega$ and $\mathbf{u}^n$, where $\hat{\mathbf{e}}_\omega$ is a discretization of $\frac{1}{\sqrt{2\pi}} e^{i\omega x}$ and $\mathbf{u}^n$ is the approximate solution at time t_n, evaluated on an N-point uniform grid.

The basic idea is as follows: since the matrix A is symmetric positive definite, it has real eigenvalues

$$b = \lambda_1 \geq \lambda_2 \geq \cdots \geq \lambda_N = a > 0, \tag{4}$$

and corresponding orthogonal eigenvectors $\mathbf{q}_j$, $j = 1, \ldots, N$. Therefore, the quantity (3) can be rewritten as

$$\mathbf{u}^T f(A)\mathbf{v} = \sum_{j=1}^{N} f(\lambda_j)\mathbf{u}^T \mathbf{q}_j \mathbf{q}_j^T \mathbf{v}. \tag{5}$$

which can also be viewed as a Riemann–Stieltjes integral

$$\mathbf{u}^T f(A)\mathbf{v} = I[f] = \int_a^b f(\lambda)\, d\alpha(\lambda). \tag{6}$$

As discussed in [4], the integral $I[f]$ can be approximated using Gaussian quadrature rules, which yields an approximation of the form

$$I[f] = \sum_{j=1}^{K} w_j f(\lambda_j) + R[f], \tag{7}$$

where the nodes λ_j, $j = 1, \ldots, K$, as well as the weights w_j, $j = 1, \ldots, K$, can be obtained using the symmetric Lanczos algorithm if $\mathbf{u} = \mathbf{v}$, and the unsymmetric Lanczos algorithm if $\mathbf{u} \neq \mathbf{v}$ (see [6]).

In the case $\mathbf{u} \neq \mathbf{v}$, there is a possibility that the weights may not be positive, which destabilizes the quadrature rule (see [1] for details). Instead, we consider

$$\begin{bmatrix}\mathbf{u} & \mathbf{v}\end{bmatrix}^T f(A) \begin{bmatrix}\mathbf{u} & \mathbf{v}\end{bmatrix}, \tag{8}$$

which results in the 2×2 matrix

$$\int_a^b f(\lambda)\, d\mu(\lambda) = \begin{bmatrix} \mathbf{u}^T f(A)\mathbf{u} & \mathbf{u}^T f(A)\mathbf{v} \\ \mathbf{v}^T f(A)\mathbf{u} & \mathbf{v}^T f(A)\mathbf{v} \end{bmatrix}, \tag{9}$$

where $\mu(\lambda)$ is a 2×2 matrix function of λ, each entry of which is a measure of the form $\alpha(\lambda)$ from (6).

In [4] Golub and Meurant showed how a block method can be used to generate quadrature formulas. We will describe this process here in more detail. The integral $\int_a^b f(\lambda)\, d\mu(\lambda)$ is now a 2×2 symmetric matrix and the most general K-node quadrature formula is of the form

$$\int_a^b f(\lambda)\, d\mu(\lambda) = \sum_{j=1}^{K} W_j f(T_j) W_j + error, \tag{10}$$

with T_j and W_j being symmetric 2×2 matrices. By diagonalizing each T_j, we obtain the simpler formula

$$\int_a^b f(\lambda)\, d\mu(\lambda) = \sum_{j=1}^{2K} f(\lambda_j)\mathbf{v}_j\mathbf{v}_j^T + error, \tag{11}$$

where, for each j, λ_j is a scalar and $\mathbf{v}_j$ is a 2-vector.

Each node λ_j is an eigenvalue of the matrix

$$\mathcal{T}_K = \begin{bmatrix} M_1 & B_1^T & & & \\ B_1 & M_2 & B_2^T & & \\ & \ddots & \ddots & \ddots & \\ & & B_{K-2} & M_{K-1} & B_{K-1}^T \\ & & & B_{K-1} & M_K \end{bmatrix}, \tag{12}$$

which is a block-triangular matrix of order $2K$. The vector $\mathbf{v}_j$ consists of the first two elements of the corresponding normalized eigenvector. To compute the matrices M_j and B_j, we use the block Lanczos algorithm, which was proposed by Golub and Underwood in [5].

We are now ready to describe block KSS methods. For each wave number $\omega = -N/2+1,\ldots,N/2$, we define $R_0(\omega) = \begin{bmatrix} \hat{\mathbf{e}}_\omega & \mathbf{u}^n \end{bmatrix}$ and compute the QR factorization $R_0(\omega) = X_1(\omega)B_0(\omega)$. We then carry out block Lanczos iteration, applied to the discretized operator L_N, to obtain a block tridiagonal matrix $\mathcal{T}_K(\omega)$ of the form (12), where each entry is a function of ω.

Then, we can express each Fourier coefficient of the approximate solution at the next time step as

$$[\hat{\mathbf{u}}^{n+1}]_\omega = \left[B_0^H E_{12}^H \exp[-\mathcal{T}_K(\omega)\Delta t] E_{12} B_0\right]_{12} \tag{13}$$

where $E_{12} = \begin{bmatrix} \mathbf{e}_1 & \mathbf{e}_2 \end{bmatrix}$. The computation of (13) consists of computing the eigenvalues and eigenvectors of $\mathcal{T}_K(\omega)$ in order to obtain the nodes and weights for Gaussian quadrature, as described earlier.

This algorithm has local temporal accuracy $O(\Delta t^{2K-1})$ [13]. Furthermore, block KSS methods are more accurate than the original KSS methods described in [10], even though they have the same order of accuracy, because the solution $\mathbf{u}^n$ plays a greater role in the determination of the quadrature nodes. They are also more effective for problems with oscillatory or discontinuous coefficients [13].

Block KSS methods are even more accurate for the second-order wave equation, for which block Lanczos iteration is used to compute both the solution and its time derivative. In [12, Theorem 6], it is shown that when the leading coefficient is constant and the coefficient $q(x)$ is bandlimited, the 1-node KSS method, which has second-order accuracy in time, is also unconditionally stable. In general, as shown in [12], the local temporal error is $O(\Delta t^{4K-2})$ when K block Gaussian nodes are used.

3 Implementation

KSS methods compute a Jacobi matrix corresponding to *each* Fourier coefficient, in contrast to traditional Krylov subspace methods that normally use only a single Krylov subspace generated by the initial data or the solution from the previous time step. While it would appear that KSS methods incur a substantial amount of additional computational expense, that is not actually the case, because nearly all of the Krylov subspaces that they compute are closely related by the wave number ω, in the 1-D case, or $\omega = (\omega_1, \omega_2, \ldots, \omega_n)$ in the n-D case.

In fact, the only Krylov subspace that is explicitly computed is the one generated by the solution from the previous time step, of dimension $(K + 1)$, where K is the number of block Gaussian quadrature nodes. In addition, the averages of the coefficients of L^j, for $j = 0, 1, 2, \ldots, 2K - 1$, are required, where L is the spatial differential operator. When the coefficients of L are independent of time, these can be computed once, during a preprocessing step. This computation can be carried out in $O(N \log N)$ operations using symbolic calculus [11, 15].

With these considerations, the algorithm for a single time step of a 1-node block KSS method for solving (1), where $Lu = -pu_{xx} + q(x)u$, with appropriate initial conditions and periodic boundary conditions, is as follows. We denote the average of a function $f(x)$ on $[0, 2\pi]$ by $\overline{f}$, and the computed solution at time t_n by u^n.

$\hat{u}^n = \textbf{fft}(u^n)$, $v = Lu^n$, $\hat{v} = \textbf{fft}(v)$
for each ω **do**
  $\alpha_1 = -p\omega^2 + \overline{q}$ (in preprocessing step)
  $\beta_1 = \hat{v}(\omega) - \alpha_1 \hat{u}^n(\omega)$
  $\alpha_2 = \langle u^n, v\rangle - 2\,\mathrm{Re}\,[\hat{u}^n(\omega)\overline{v(\omega)}] + \alpha_1 |u^n(\omega)|^2$
  $e_\omega = [\langle u^n, u^n\rangle - |\hat{u}^n(\omega)|^2]^{1/2}$
  $T_\omega = \left[\begin{array}{cc} \alpha_1 & \beta_1/e_\omega \\ \beta_1/e_\omega & \alpha_2/e_\omega^2 \end{array}\right]$
  $\hat{u}^{n+1}(\omega) = [e^{-T_\omega \Delta t}]_{11}\hat{u}^n(\omega) + [e^{-T_\omega \Delta t}]_{12} e_\omega$
end
$u^{n+1} = \textbf{ifft}(\hat{u}^{n+1})$

It should be noted that for a parabolic problem such as (1), the loop over ω only needs to account for non-negligible Fourier coefficients of the solution, which are relatively few due to the smoothness of solutions to such problems.

4 Application to Maxwell's Equations

We consider Maxwell's equation on the cube $[0, 2\pi]^3$, with periodic boundary conditions. Assuming nonconductive material with no losses, we have

$$\mathrm{div}\,\hat{\mathbf{E}} = 0, \quad \mathrm{div}\,\hat{\mathbf{H}} = 0, \tag{14}$$

$$\mathrm{curl}\,\hat{\mathbf{E}} = -\mu \frac{\partial \hat{\mathbf{H}}}{\partial t}, \quad \mathrm{curl}\,\hat{\mathbf{H}} = \varepsilon \frac{\partial \hat{\mathbf{E}}}{\partial t}, \tag{15}$$

where $\hat{\mathbf{E}}$, $\hat{\mathbf{H}}$ are the vectors of the electric and magnetic fields, and ε, μ are the electric permittivity and magnetic permeability, respectively.

Taking the curl of both sides of (15) yields

$$\mu\varepsilon\frac{\partial^2\hat{\mathbf{E}}}{\partial t^2} = \Delta\hat{\mathbf{E}} + \mu^{-1}\operatorname{curl}\hat{\mathbf{E}} \times \nabla\mu, \tag{16}$$

$$\mu\varepsilon\frac{\partial^2\hat{\mathbf{H}}}{\partial t^2} = \Delta\hat{\mathbf{H}} + \varepsilon^{-1}\operatorname{curl}\hat{\mathbf{H}} \times \nabla\varepsilon. \tag{17}$$

In this section, we discuss generalizations that must be made to block KSS methods in order to apply them to a non-self-adjoint system of coupled equations such as (16). Additional details are given in [14].

First, we consider the following 1-D problem,

$$\frac{\partial^2\mathbf{u}}{\partial t^2} + L\mathbf{u} = 0, \quad t > 0, \tag{18}$$

with appropriate initial conditions, and periodic boundary conditions, where $\mathbf{u}$: $[0, 2\pi] \times [0, \infty) \to \mathbb{R}^n$ for $n > 1$, and $L(x, D)$ is an $n \times n$ matrix where the (i, j) entry is an a differential operator $L_{ij}(x, D)$ of the form

$$L_{ij}(x, D)u(x) = \sum_{\mu=0}^{m_{ij}} a_\mu^{ij}(x)D^\mu u, \quad D = \frac{d}{dx}, \tag{19}$$

with spatially varying coefficients a_μ^{ij}, $\mu = 0, 1, \ldots, m_{ij}$.

Generalization of KSS methods to a system of the form (18) can proceed as follows. For $i, j = 1, \ldots, n$, let $\overline{L}_{ij}(D)$ be the constant-coefficient operator obtained by averaging the coefficients of $L_{ij}(x, D)$ over $[0, 2\pi]$. Then, for each wave number ω, we define $L(\omega)$ be the matrix with entries $\overline{L}_{ij}(\omega)$, i.e., the symbols of $\overline{L}_{ij}(D)$ evaluated at ω. Next, we compute the spectral decomposition of $L(\omega)$ for each ω. For $j = 1, \ldots, n$, let $\mathbf{q}_j(\omega)$ be the Schur vectors of $L(\omega)$. Then, we define our test and trial functions by $\phi_{j,\omega}(x) = \mathbf{q}_j(\omega) \otimes e^{i\omega x}$.

For Maxwell's equations, the matrix A_N that discretizes the operator

$$A\hat{\mathbf{E}} = \frac{1}{\mu\varepsilon}\left(\Delta\hat{\mathbf{E}} + \mu^{-1}\operatorname{curl}\hat{\mathbf{E}} \times \nabla\mu\right)$$

is not symmetric, and for each coefficient of the solution, the resulting quadrature nodes λ_j, $j = 1, \ldots, 2K$, from (11) are now complex and must be obtained by a straightforward modification of block Lanczos iteration for unsymmetric matrices.

5 Numerical Results

In this section, we compare the performance of block KSS methods with various methods based on exponential integrators [7, 9, 18].

5.1 Parabolic Problems

We first consider a 1-D parabolic problem of the form (1), where the differential operator L is defined by $Lu(x) = -pu''(x) + q(x)u(x)$, where $p \approx 0.4$ and

$$q(x) \approx -0.44+0.03\cos x-0.02\sin x+0.005\cos 2x-0.004\sin 2x+0.0005\cos 3x$$

is constructed so as to have the smoothness of a function with three continuous derivatives, as is the initial data $u(x, 0)$. Periodic boundary conditions are imposed.

We solve this problem using the following methods:

- A 2-node block KSS method. Each time step requires construction of a Krylov subspace of dimension 3 generated by the solution, and the coefficients of L^2 and L^3 are computed during a preprocessing step.
- A preconditioned Lanczos iteration for approximating $e^{-\tau A}\mathbf{v}$, introduced in [16] for approximating the matrix exponential of sectorial operators, and adapted in [18] for efficient application to the solution of parabolic PDE. In this approach, Lanczos iteration is applied to $(I + hA)^{-1}$, where h is a parameter, in order to obtain a restricted rational approximation of the matrix exponential. We use $m = 4$ and $m = 8$ Lanczos iterations, and choose $h = \Delta t/10$, as in [18].
- A method based on exponential integrators, from [7], that is of order 3 when the Jacobian is approximated to within $O(\Delta t)$. We use $m = 8$ Lanczos iterations.

Since the exact solution is not available, the error is estimated by taking the ℓ_2-norm of the relative difference between each solution, and that of a solution computed using a smaller time step $\Delta t = 1/64$ and the maximum number of grid points.

The results are shown in Fig. 1. As the number of grid points is doubled, only the block KSS method shows an improvement in accuracy; the preconditioned Lanczos method exhibits a slight degradation in performance, while the explicit fourth-order exponential integrator-based method requires that the time step be reduced by a factor of 4 before it can deliver the expected order of convergence; similar behavior was demonstrated for an explicit 3rd-order method from [8] in [10].

The preconditioned Lanczos method requires 8 Lanczos iterations to match the accuracy of a block KSS method that uses only 2. On the other hand, the block KSS method incurs additional expense due to (1) the computation of the moments of L, for each Fourier coefficient, and (2) the exponentiation of separate Jacobi matrices for each Fourier coefficient. These expenses are mitigated by the fact that the first takes place once, during a preprocessing stage, and both tasks require an amount

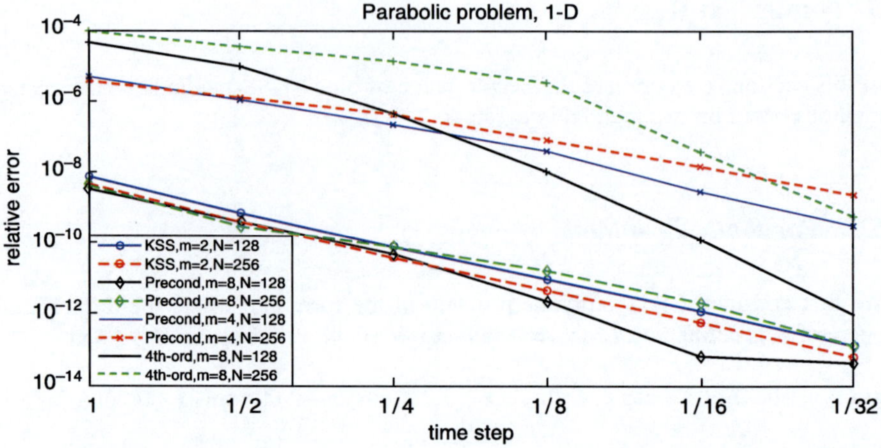

Fig. 1 Estimates of relative error at $t = 0.1$ in solutions of (1) computed using preconditioned exponential integrator [18] with 4 and 8 Lanczos iterations, a 4th-order method based on an exponential integrator [9], and a 2-node block KSS method. All methods compute solutions on an N-point grid, with time step Δt, for various values of N and Δt

of work that is proportional not to the number of grid points, but to the number of non-negligible Fourier coefficients of the solution.

5.2 *Maxwell's Equations*

We now apply a 2-node block KSS method to (16), with initial conditions

$$\hat{\mathbf{E}}(x, y, z, 0) = \mathbf{F}(x, y, z), \quad \frac{\partial \hat{\mathbf{E}}}{\partial t}(x, y, z, 0) = \mathbf{G}(x, y, z), \tag{20}$$

with periodic boundary conditions. The coefficients μ and ε are given by

$$\begin{aligned} \mu(x, y, z) = {} & 0.4077 + 0.0039 \cos z + 0.0043 \cos y - 0.0012 \sin y \\ & +0.0018 \cos(y + z) + 0.0027 \cos(y - z) + 0.003 \cos x \\ & +0.0013 \cos(x - z) + 0.0012 \sin(x - z) + 0.0017 \cos(x + y) \\ & +0.0014 \cos(x - y), \end{aligned} \tag{21}$$

$$\begin{aligned} \varepsilon(x, y, z) = {} & 0.4065 + 0.0025 \cos z + 0.0042 \cos y + 0.001 \cos(y + z) \\ & +0.0017 \cos x + 0.0011 \cos(x - z) + 0.0018 \cos(x + y) \\ & +0.002 \cos(x - y). \end{aligned} \tag{22}$$

The components of **F** and **G** are generated in a similar fashion, except that the x- and z-components are zero.

We use a block KSS method that uses $K = 2$ block quadrature nodes per coefficient in the basis described in Sect. 4, that is 6th-order accurate in time, and a cosine method based on a Gautschi-type exponential integrator [7, 9]. This method is second-order in time, and in these experiments, we use $m = 2$ Lanczos iterations to approximate the Jacobian. It should be noted that when m is increased, even to a substantial degree, the results are negligibly affected.

Figure 2 demonstrates the convergence behavior for both methods. At both spatial resolutions, the block KSS method exhibits approximately 6th-order accuracy in time as Δt decreases, except that for $N = 16$, the spatial error arising from truncation of Fourier series is significant enough that the overall error fails to decrease below the level achieved at $\Delta t = 1/8$. For $N = 32$, the solution is sufficiently resolved in space, and the order of overgence as $\Delta t \to 0$ is approximately 6.1.

We also note that increasing the resolution does not pose any difficulty from a stability point of view. Unlike explicit finite-difference schemes that are constrained by a CFL condition, KSS methods do not require a reduction in the time step to offset a reduction in the spatial step in order to maintain boundedness of the solution, because their domain of dependence includes the entire spatial domain for any Δt.

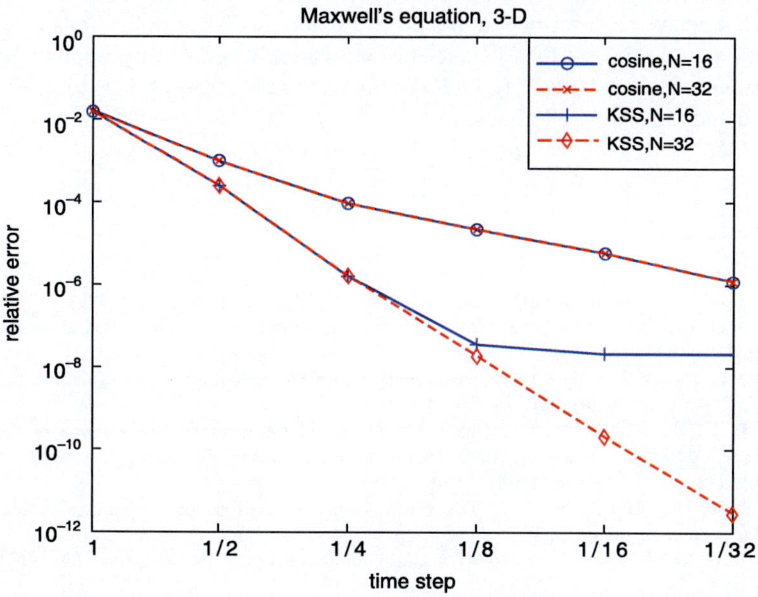

Fig. 2 Estimates of relative error at $t = 1$ in solutions of (16), (20) computed using a cosine method based on a Gautschi-type exponential integrator [7, 9] with 2 Lanczos iterations, and a 2-node block KSS method. Both methods compute solutions on an N^3-point grid, with time step Δt, for various values of N and Δt

The Gautschi-type exponential integrator method is second-order accurate, as expected, and delivers nearly identical results for both spatial resolutions, but even with a Krylov subspace of much higher dimension than that used in the block KSS method, it is only able to achieve at most second-order accuracy, whereas a block KSS method, using a Krylov subspace of dimension 3, achieves sixth-order accuracy. This is due to the incorporation of the moments of the spatial differential operator into the computation, and the use of Gaussian quadrature rules specifically tailored to each Fourier coefficient.

6 Summary and Future Work

We have demonstrated that block KSS methods can be applied to Maxwell's equations with smoothly varying coefficients, by appropriate generalization of their application to the scalar second-order wave equation, in a way that preserves the order of accuracy achieved for the wave equation. Furthermore, it has been demonstrated that while traditional Krylov subspace methods based on exponential integrators are most effective for parabolic problems, especially when aided by preconditioning as in [18], KSS methods perform best when applied to hyperbolic problems, in view of their much higher order of accuracy. Future work will extend the approach described in this paper to more realistic applications involving Maxwell's equations by using symbol modification to efficiently implement perfectly matched layers (see [2]), and various techniques (see [3, 17]) to effectively handle discontinuous coefficients.

References

1. Atkinson, K.: *An Introduction to Numerical Analysis, 2nd Ed.* Wiley, NY (1989)
2. Berenger, J.: A perfectly matched layer for the absorption of electromagnetic waves. *J. Comp. Phys.* **114** (1994) 185–200
3. Gelb, A., Tanner, J.: Robust reprojection methods for the resolution of the gibbs phenomenon. *Appl. Comput. Harmon. Anal.* **20** (2006) 3–25
4. Golub, G. H., Meurant, G.: Matrices, Moments and Quadrature. *Proceedings of the 15th Dundee Conference*, June–July 1993, Griffiths, D. F., Watson, G. A. (eds.), Longman Scientific & Technical, England (1994)
5. Golub, G. H., Underwood, R.: The block Lanczos method for computing eigenvalues. *Mathematical Software III*, Rice, J. (ed.), pp. 361–377 (1977)
6. Golub, G. H, Welsch, J.: Calculation of Gauss quadrature rules. *Math. Comp.* **23** (1969) 221–230
7. Hochbruck, M., Lubich, C.: A Gautschi-type method for oscillatory second-order differential equations, *Numerische Mathematik* **83** (1999) 403–426
8. Hochbruck, M., Lubich, C.: On Krylov subspace approximations to the matrix exponential operator. *SIAM J. Numer. Anal.* **34** (1997) 1911–1925
9. Hochbruck, M., Lubich, C., Selhofer, H.: Exponential integrators for large systems of differential equations. *SIAM J. Sci. Comput.* **19** (1998) 1552–1574

10. Lambers, J. V.: Krylov subspace spectral methods for variable-coefficient initial-boundary value problems. *Electron. Trans. Numer. Anal.* **20** (2005) 212–234
11. Lambers, J. V.: Practical implementation of Krylov subspace spectral methods. *J. Sci. Comput.* **32** (2007) 449–476
12. Lambers, J. V.: An explicit, stable, high-order spectral method for the wave equation based on block Gaussian quadrature. *IAENG J. Appl. Math.* **38** (2008) 333–348
13. Lambers, J. V.: Enhancement of Krylov subspace spectral methods by block Lanczos iteration. *Electron. Trans. Numer. Anal.* **31** (2008) 86–109
14. Lambers, J. V.: A spectral time-domain method for computational electrodynamics. *Adv. Appl. Math. Mech.* **1** (2009) 781–798
15. Lambers, J. V.: Krylov subspace spectral methods for the time-dependent Schrödinger equation with non-smooth potentials. *Numer. Algorithm* **51** (2009) 239–280
16. Moret, I., Novati, P.: RD-rational approximation of the matrix exponential operator. *BIT* **44** (2004) 595–615
17. Vallius, T., Honkanen, M.: Reformulation of the Fourier nodal method with adaptive spatial resolution: application to multilevel profiles. *Opt. Expr.* **10**(1) (2002) 24–34
18. van den Eshof, J., Hochbruck, M.: Preconditioning Lanczos approximations to the matrix exponential. *SIAM J. Sci. Comput.* **27** (2006) 1438–1457

The Spectral Element Method Used to Assess the Quality of a Global C^1 Map

A.E. Løvgren, Y. Maday, and E.M. Rønquist

Abstract In this work we focus on C^1 maps from a reference domain to a family of deformed domains. The regularity of a map affects the approximation properties of the mapped mesh, and we use the regularity as a measure of the quality of the mesh. To compare the regularities of different maps we consider the convergence of the spectral element method when a Laplace problem is solved on the resulting meshes.

1 Introduction

We are interested in the numerical solution of partial differential equations defined on a family of deformed domains. Examples of such situations include dependent problems where the geometry changes with time, and the reduced basis element method where solutions on selected sample domains are used to generate global basis functions on topologically similar domains [8]. In such cases it is of interest to have access to a global mapping between a reference domain and a family of deformed domains.

A.E. Løvgren (✉)
Center for Biomedical Computing, Simula Research Laboratory, P.O. Box 134, 1325 Lysaker, Norway
e-mail: emill@simula.no

Y. Maday
Laboratoire J.-L. Lions, Université Pierre et Marie Curie-Paris6, UMR 7598, Paris, F-75005 France and Division of Applied Mathematics, Brown University, 182 George Street, Providence, RI 02912, USA
e-mail: maday@ann.jussieu.fr

E.M. Rønquist
Department of Mathematical Sciences, Norwegian University of Science and Technology, 7491 Trondheim, Norway
e-mail: ronquist@math.ntnu.no

J.S. Hesthaven and E.M. Rønquist (eds.), *Spectral and High Order Methods for Partial Differential Equations*, Lecture Notes in Computational Science and Engineering 76,
DOI 10.1007/978-3-642-15337-2_42,

We let (ξ, η) denote the coordinates of the reference domain, and $\mathbf{x} = (x, y)$ the corresponding coordinates on the deformed domain. The Jacobian of the map from the reference domain to the deformed domain is then defined as

$$\mathcal{J} = \begin{bmatrix} \frac{\partial x}{\partial \xi} & \frac{\partial x}{\partial \eta} \\ \frac{\partial y}{\partial \xi} & \frac{\partial y}{\partial \eta} \end{bmatrix}. \quad (1)$$

The Jacobian is used in the Piola transformation to map vector fields between the domains. In order for the vector fields to be continuous after they are mapped, the Jacobian must also be continuous. Thus the map from the reference domain to the deformed domain must be C^1-continuous.

To compare the regularities of different mapping strategies we consider the convergence of the spectral element method when a Laplace problem is solved on the deformed domain. Since the convergence of the spectral element method relies on the regularity of both the solution and the geometry [10], the corresponding convergence rate gives a good indication of the regularity of the underlying global map when the solution itself is analytic.

2 Methods

In this section we present different strategies to construct global C^1 maps from a generic reference domain to a domain found as a deformation of the reference domain. Depending of the topology of the reference domain, and the severity of the deformation, some of the methods are better suited than others, while some might not apply at all.

Transfinite Extension

For rectangles and triangles, the method of transfinite extension was developed by Gordon and coworkers [1, 5], and we present in some detail the method applied to a rectangular reference domain. Although the method is restricted to simple domains, complex domains can be decomposed into a union of simpler subdomains and transfinite interpolation can be applied to map a reference grid to each subdomain. Globally the map will not be C^1, but on each subdomain the grid will be excellent.

The basic idea of transfinite extension is to interpolate between opposing sides in the reference rectangle, using proper weight functions, linear or non-linear. Given the reference domain $\widehat{\Omega} = [0, 1]^2$, with coordinates (ξ, η), we let $\{\widehat{\Gamma}_i\}_{i=1}^4$ denote the different parts of the boundary (numbered counterclockwise), such that $\widehat{\Gamma}_1 = (0, \eta)$ is the left boundary, and $\widehat{\Gamma}_{4+i} = \widehat{\Gamma}_i$.

We assume that the boundaries of a deformed rectangle $\Omega = \Phi(\widehat{\Omega})$, are well defined as one-to-one maps of the reference boundaries, e.g., $\Gamma_1 = \mathbf{x}(0, \eta)$, and

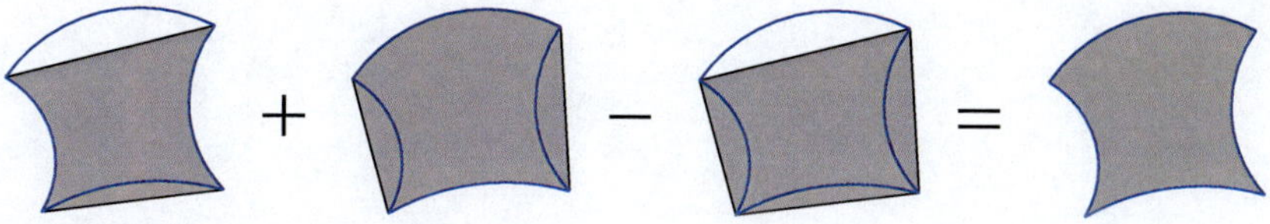

Fig. 1 Transfinite interpolation

compute the interior coordinates $\mathbf{x}(\xi,\eta)$ of the map Φ. Given regular weight functions $\{\phi_i(\xi,\eta)\}_{i=1}^4$ such that $\phi_i = 1$ on $\widehat{\Gamma}_i$ and $\phi_i = 0$ on $\widehat{\Gamma}_{i+2}$, the transfinite interpolation from $\widehat{\Omega}$ to Ω is defined by

$$\mathbf{x}(\xi,\eta) = \phi_1(\xi,\eta)\mathbf{x}(0,\eta) + \phi_3(\xi,\eta)\mathbf{x}(1,\eta) \tag{2}$$

$$+\phi_2(\xi,\eta)\mathbf{x}(\xi,0) + \phi_4(\xi,\eta)\mathbf{x}(\xi,1) \tag{3}$$

$$-\sum_{i=1}^{4}\phi_i(\xi,\eta)\phi_{i+1}(\xi,\eta)\mathbf{x}_i(1), \tag{4}$$

where $\mathbf{x}_i(1)$ is the value of $\mathbf{x}$ in the corner between Γ_i and Γ_{i+1}. The first line (2) preserves the left and right boundaries exactly, and is illustrated by the shaded area to the left in Fig. 1. Similarly, the second line preserves the top and bottom boundaries, and is seen as the second shaded area in Fig. 1. The sum of these interpolations covers parts of the interior of the domain twice, and in addition regions outside the domain are included. The final step of the procedure is thus to subtract the interpolation of the corners, given by (4) and shown by the third shaded area in Fig. 1.

By introducing $\pi_1(\xi,\eta) = \eta$ as the projection of any interior point (ξ,η) on the reference domain to the left boundary $\widehat{\Gamma}_1$, and similarly for the other boundaries, we get the more compact form of the transfinite interpolation

$$\mathbf{x}(\xi,\eta) = \sum_{i=1}^{4}[\phi_i(\xi,\eta)\mathbf{x}_i(\pi_i(\xi,\eta)) - \phi_i(\xi,\eta)\phi_{i+1}(\xi,\eta)\mathbf{x}_i(1)], \tag{5}$$

where $\mathbf{x}_i(t)$ is the coordinates of Γ_i for $t \in [0,1]$. In general the weight functions ϕ_i are one-dimensional, e.g., $\phi_1(\xi,\eta) = 1-\xi$, and only functions of the distance between two opposing sides, but alternative definitions are also possible. See [6] for examples and applications to mesh generation. We also mention that the extension to $3D$ when the reference domain is the unit cube is straight forward.

Generalized Transfinite Extension

Based on the transfinite extension described above, we generalize the method in order to use any domain with more than four corners as a reference domain. The most crucial difference is that we can no longer use one-dimensional weight

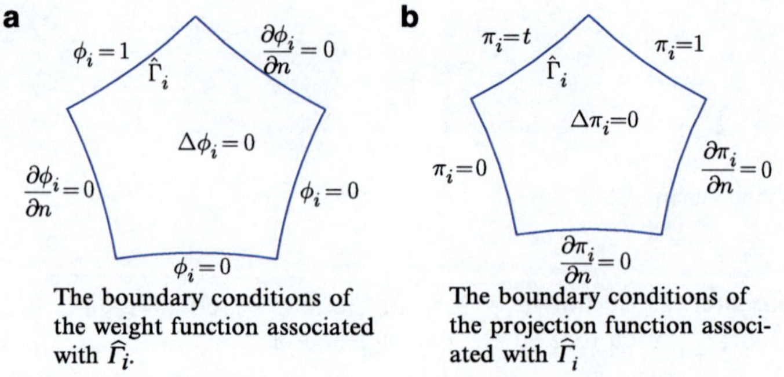

Fig. 2 Illustration of the boundary conditions for the harmonic weight and projection functions used in the generalized transfinite extension scheme

functions, and we need the projection of the reference coordinates onto each part of the boundary. The generalized transfinite extension method was first presented in [9], and we give the details here as well.

On an n-sided reference domain, where $n \geq 4$, we denote each side $\widehat{\Gamma}_i$, $i = 1, \ldots, n$, and number the sides in a counterclockwise manner. Associated with each side is a weight function ϕ_i, and a projection function π_i, both defined over $\widehat{\Omega}$. To define the weight functions, we let $\phi_i = 1$ on $\widehat{\Gamma}_i$, and solve the Laplace problem

$$\Delta \phi_i = 0 \quad \text{in } \widehat{\Omega}, \tag{6}$$

with homogeneous Neumann boundary conditions on the two sides of $\widehat{\Omega}$ adjacent to $\widehat{\Gamma}_i$, and homogeneous Dirichlet boundary conditions on the remaining sides; see Fig. 2a. On the reference square these harmonic weight functions will coincide with one-dimensional, linear weight functions, as seen in Fig. 3a, but on a general non-convex reference domain, the weight functions will be non-affine C^1 functions; see Fig. 3b, c.

In the generalized transfinite extension scheme, we also need the projection from the interior onto each side $\widehat{\Gamma}_i$. On the unit square these projections are given by the reference coordinates as $\pi_1(\xi, \eta) = \pi_3(\xi, \eta) = \eta$ and $\pi_2(\xi, \eta) = \pi_4(\xi, \eta) = \xi$. On a general domain we compute the projection function π_i onto the side $\widehat{\Gamma}_i$ by solving the Laplace problem

$$\Delta \pi_i = 0 \quad \text{in } \widehat{\Omega}, \tag{7}$$

with linear Dirichlet boundary condition along $\widehat{\Gamma}_i$, distributed from 0 to 1 with respect to arc-length. On the sides adjacent to $\widehat{\Gamma}_i$ we set π_i equal to either 0 or 1, and on the remaining sides we use homogeneous Neumann boundary conditions; see Fig. 2b. On the unit square this procedure will reproduce the linear distribution of the reference coordinate corresponding to $\widehat{\Gamma}_i$, as seen in Fig. 3d, while on general reference domains we again get non-affine C^1 functions; see Fig. 3e, f.

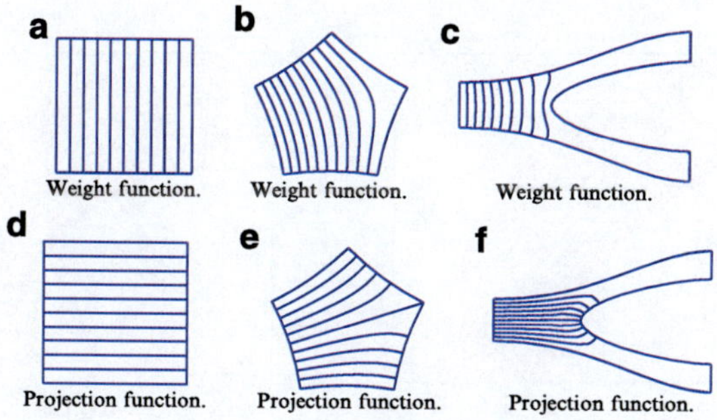

Fig. 3 Contour lines of the weight and projection functions, $\phi_1(\xi,\eta)$ and $\pi_1(\xi,\eta)$, associated with the left side of a rectangle (*left*), a curved pentagon (*middle*), and a bifurcation (*right*). The weight functions decrease from 1 to 0 going from *left* to *right*, the projection functions decrease linearly with respect to arc-length from 1 to 0 going from *top* to *bottom* along the left boundary

In general we have to solve a total of $2n$ elliptic problems on the reference domain, but if the reference domain has rotational symmetries, like the curved pentagon in Fig. 2, it is sufficient to solve two elliptic problems related to one of the sides, and then rotate the solutions to fit the other sides.

Since the boundaries of a general reference domain is not given by varying one single reference coordinate, as was the case for the square reference domain, we assume that each boundary Γ_i of the deformed domain can be expressed as a function of the arc-length, t, of the reference boundary $\widehat{\Gamma}_i$, and let $\mathbf{x}_i(t)$ represent map from $\widehat{\Gamma}_i$ to Γ_i. Furthermore we let $\widehat{\Gamma}_{n+1} = \widehat{\Gamma}_1$, and define the generalized transfinite extension as

$$\mathbf{x}(\xi,\eta) = \sum_{i=1}^{n} [\phi_i(\xi,\eta)\mathbf{x}_i(\pi_i(\xi,\eta)) - \phi_i(\xi,\eta)\phi_{i+1}(\xi,\eta)\mathbf{x}_i(1)]. \tag{8}$$

Again, $\mathbf{x}_i(1)$ denotes the corner between Γ_i and Γ_{i+1}.

We note that, as for transfinite interpolation on the unit square, the value of the extension $\mathbf{x}$ in any point $(\xi,\eta) \in \widehat{\Omega}$ only depends on the values of the boundary functions, $\mathbf{x}_i(t)$, in isolated points on the boundary of the reference domain.

Harmonic Extension

Where the transfinite extension only applies to reference domains with corners, the harmonic extension can be defined on any closed domain. Again we let $\widehat{\Omega}$ be the reference domain with coordinates (ξ,η), and Ω the deformed domain with coordinates $\mathbf{x} = (x, y)$. In order to find $\mathbf{x}$, we solve

$$\begin{aligned} \Delta\mathbf{x}(\xi,\eta) &= 0 \quad \text{in } \widehat{\Omega} \\ \mathbf{x} \qquad &= \mathbf{x}_b \quad \text{on } \partial\widehat{\Omega}, \end{aligned} \tag{9}$$

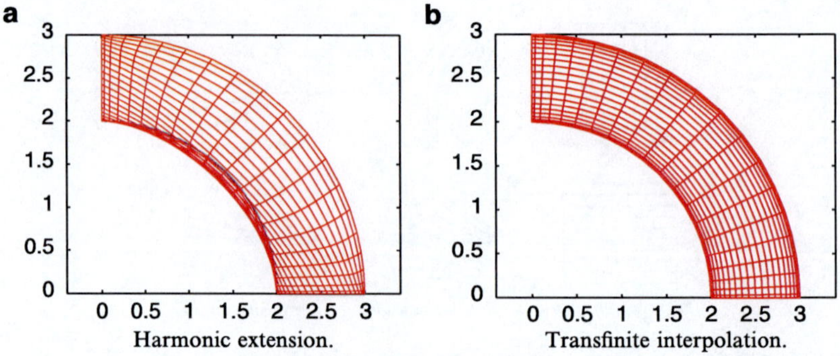

Fig. 4 The result of using different methods for mapping a square to an axisymmetric bend

where $\mathbf{x}_b = (x_b, y_b)$ are the boundary coordinates of Ω defined as a one-to-one map from the boundary of the reference domain, $\partial\widehat{\Omega}$. We note that this is not a coupled system for x and y, and that each spatial dimension is solved separately.

This is by far the easiest method to set up, since it only requires a standard Laplace solver and is independent of the reference domain. It has, however, some limitations with respect to large deformations, as can be seen in Fig. 4a. The square reference domain $(-1, 1)^2$ is here mapped to an axisymmetric bend using the harmonic extension method, and we see that some of the points belonging to the interior are actually mapped outside the boundary of the deformed domain. In comparison, the transfinite interpolation method described above, yields an optimal distribution of points the (r, θ)-plane on the axisymmetric bend; see Fig. 4b.

Transfinite Barycentric Interpolation

For convex reference domains with piecewise differentiable boundaries, Gordon and Wixom [7] introduced pseudo-harmonic extension. On a bounded and convex domain $\Omega \subset \mathbb{R}^2$ the extension u is defined as

$$u(\xi,\eta) = \frac{1}{2\pi}\int_0^{2\pi}\left[\frac{d_2(\theta)}{d_1(\theta)+d_2(\theta)} f(Q_1(\theta)) + \frac{d_1(\theta)}{d_1(\theta)+d_2(\theta)} f(Q_2(\theta))\right] d\theta, \tag{10}$$

where Q_1 and Q_2 are the intersections between $\partial\Omega$ and the line through the point (ξ, η) at inclination θ, and d_1 and d_2 are the distances from (ξ, η) to these intersection points. Note that on the unit disk it is shown that the extension defined in (10) is the solution of the Laplace problem (9). At each point (ξ, η), the extension u defined in (10) depends on the value of f along the *entire* boundary of Ω. For comparison, the extension defined through the generalized transfinite extension scheme (8) only depends on $2n$ boundary points.

The pseudo-harmonic extension behaves similarly to the harmonic extension, and for a given reference domain, the necessary weight and distance functions can be computed once, allowing for rapid computation of large series of deformed domains. The pseudo-harmonic extension method was also generalized to non-convex domains by Belyaev [2].

The mean-value coordinates was introduced by Floater [4], and can be seen as part of the same general barycentric construction that includes the pseudo-harmonic extension. The mean-value interpolation only has linear precision, but with a simple formula for computing weights and distance functions it is easier to implement than the pseudo-harmonic extension. As for the pseudo-harmonic extension, the largest benefit is when computing multiple deformed geometries from one reference geometry.

3 Regularity

We consider the mapping from one of the two domains to the left in Fig. 5 to the deformed domain to the right in the same figure. To evaluate the different extension methods we apply the spectral element method and decompose the reference domain into several subdomains. The discrete space is defined by

$$X_N = \{v \in H^1, v_{|\Omega_k} \circ \Phi_k \in \mathbb{P}_N(\widehat{\Lambda})\}, \tag{11}$$

where $\mathbb{P}_N$ is the space of all polynomials of degree less than or equal to N in each spatial direction on $\widehat{\Lambda} = [-1, 1]^2$. We solve a Laplace problem on the deformed domain, where the interior points are mapped through the different extension methods described in the previous section. It is known that the convergence of the spectral element approximation depends on the regularity of both the solution and the geometry. In order to reveal the regularity of the maps, we choose a known analytic solution to the Laplace problem, $u = e^x \sin(y)$. On a regular mesh the spectral element approximation, u_N, of this solution converges exponentially, but due to the lower regularity of the mapped meshes we only get algebraic convergence rate with respect to the polynomial degree N, i.e., $|u_N - u|_{H^1} \leq N^{-m}$.

When the straight pentagon to the left in Fig. 5 is used as a reference domain, we see from Table 1 that the harmonic extension gives the best convergence rate. For the pentagon in the middle of Fig. 5 all corners have the angle $\frac{\pi}{2}$, and we see that the convergence rates of the harmonic extension and the generalized transfinite

Fig. 5 Two different reference domains (*left* and *middle*) and a deformed domain (*right*)

Table 1 The convergence rate of the error in the spectral element solution. The difference in the convergence rate is solely due to the different meshes used

Method	Reference domain	
	Uniform pentagon	Curved pentagon
Transfinite extension	3.7	5.0
Harmonic extension	4.3	5.0
Pseudo-harmonic extension	2.5	
Mean value extension	2.5	

extension have improved. In addition they are now equally good. For more details and results we refer to Løvgren et al. [9]. The large benefit of using the generalized transfinite extension compared to the harmonic extension defined in (9) is that all the weight and projection functions are computed only once. For each new given boundary $\partial\Omega$ we only need to find the corresponding boundary functions $\mathbf{x}_i(t)$, and perform the linear combination of the functions in (8).

Again we stress the need for global C^1 maps when vector fields are mapped from one domain to another while preserving the (in)compressibility of the field. This is crucial when the reduced basis element method is applied to fluid flow problems [8].

Acknowledgements This work has been supported by the Research Council of Norway through a Centre of Excellence grant to the Center for Biomedical Computing.

References

1. R. E. Barnhill, G. Birkhoff, W. J. Gordon. Smooth interpolation in triangles. *J. Approx. Theor.*, **8**, 114–128 (1973)
2. A. Belyaev. On transfinite barycentric coordinates. In *Eurographics Symposium on Geometric Processing*, 89–99 (2006)
3. C. Bernardi and Y. Maday. Polynomial approximation of some singular functions. *Appl. Anal.*, **42**, 1–32 (1991)
4. M. S. Floater. Mean value coordinates. *Comp. Aided Geom. Design*, **20**, 19–27 (2003)
5. W. J. Gordon and C. A. Hall. Transfinite element methods: Blending-function interpolation over arbitrary curved element domains. *Numer. Math.*, **21**, 109–129 (1973)
6. W. J. Gordon and C. A. Hall. Construction of curvilinear co-ordinate systems and applications to mesh generation. *Int. J. Numer. Meth. Eng.*, **7**, 461–477 (1973)
7. W. J. Gordon and J. A. Wixom. Pseudo-harmonic interpolation on convex domains. *SIAM J. Numer. Anal.*, **11**, 909–933 (1974)
8. A. E. Løvgren, Y. Maday, and E. M. Rønquist. The reduced basis element method for fluid flows. *Analysis and Simulation of Fluid Dynamics, Advances in Mathematical Fluid Mechanics*, 129–154 (2007)
9. A. E. Løvgren, Y. Maday, and E. M. Rønquist. Global C^1 maps on general domains. *Math. Models Meth. Appl. Sci.*, **19**, 803–832 (2009)
10. Y. Maday and E. M. Rønquist. Optimal error analysis of spectral methods with emphasis on non-constant coefficients and deformed geometries. *Comput. Meth. Appl. Mech. Eng.*, **80**, 91–115 (1990)

Stabilization of the Spectral-Element Method in Turbulent Flow Simulations

J. Ohlsson, P. Schlatter, P.F. Fischer, and D.S. Henningson

Abstract The effect of over-integration and filter-based stabilization in the spectral-element method is investigated. There is a need to stabilize the SEM for flow problems involving non-smooth solutions, e.g., turbulent flow simulations. In model problems such as the Burgers' equation (similar to Kirby and Karniadakis, J. Comput. Phys. 191:249–264, 2003) and the scalar transport equation together with full Navier–Stokes simulations it is noticed that over-integration with the full 3/2-rule is not required for stability. The first additional over-integration nodes are the most efficient to remove aliasing errors. Alternatively, filter-based stabilization can in many cases alone help to stabilize the computation.

1 Introduction

The spectral-element method (SEM) has mainly been applied to relatively low Reynolds numbers, with a focus on laminar and, to some extent, transitional flows (see, e.g., [19–21]). However, for fully turbulent flows at moderate Reynolds numbers ($Re \sim 10^3$–10^4), there has been less attention [1, 4, 5, 22], which can probably be ascribed to the anxiety about the stability of the SEM at these Reynolds numbers. The cause of this instability is thought to be the accumulation of aliasing errors, which are strongly enhanced in a turbulent flow simulation. Our belief is that as soon as these errors are reduced or eliminated in an appropriate way, the stability of the method can be fully assured for all Re. The reduction or elimination of aliasing errors can be accomplished either by so-called over-integration

J. Ohlsson (✉), P. Schlatter, and D.S. Henningson
Linné Flow Centre, KTH Mechanics, Stockholm, Sweden
e-mail: johan@mech.kth.se

P.F. Fischer
MCS, Argonne National Laboratory, Argonne, USA
e-mail: fischer@mcs.anl.gov

J.S. Hesthaven and E.M. Rønquist (eds.), *Spectral and High Order Methods for Partial Differential Equations*, Lecture Notes in Computational Science and Engineering 76,
DOI 10.1007/978-3-642-15337-2_43,

(see, e.g., [3, 11, 13]), spectral vanishing viscosity (SVV) techniques [10, 15, 17, 23], or filter-based stabilization as proposed in [7]. In the framework of the weak form, the nonlinearity of the governing Navier–Stokes equations gives rise to the integration of three polynomials of order N. Using Gaussian quadrature, this requires approximately $M = 3/2N$ points in each direction in order to get an exact integration, which is similar as the well-known 3/2-rule in pseudo-spectral methods. In this work, we specifically consider the number of Gauss–Lobatto–Legendre (GLL) points, M, needed for stability, which may be considerably less. This is examined first by an eigenvalue analysis of the (linearized) viscous Burgers' equation and the linear scalar transport equation; then these ideas are applied to the full Navier–Stokes equations and evaluated a posteriori.

2 Equations and Discretization

Our interest lies in understanding the cause of the instability of SEM at high Reynolds numbers. In order to achieve this, simpler model problems in $\mathbb{R}^1$ and $\mathbb{R}^2$ will be analyzed, eventually leading to the full Navier–Stokes in $\mathbb{R}^3$. Following [11] we proceed in $\mathbb{R}^1$ by analyzing the viscous Burgers' equation on the interval $\Omega = [-1, 1]$, written here in non-conservative form,

$$\frac{\partial u}{\partial t} + u\frac{\partial u}{\partial x} = \nu \frac{\partial^2 u}{\partial x^2} \tag{1}$$

with initial condition $u(0, x) = u^0(x) = -\sin(\pi x)$ and periodic boundary conditions. To account for a nontrivial velocity field we need to consider a problem in $\mathbb{R}^2$, here being the scalar transport equation,

$$\frac{\partial q}{\partial t} + \mathbf{c} \cdot \nabla q = 0 \tag{2}$$

where q may be a scalar concentration of any kind convected by the velocity field $\mathbf{c}$. For simplicity, we assume that $\Omega = [-1, 1]^2$. Finally, the incompressible Navier–Stokes equations in $\mathbb{R}^3$,

$$\frac{\partial \mathbf{u}}{\partial t} + \mathbf{u} \cdot \nabla \mathbf{u} = -\nabla p + \frac{1}{Re}\nabla^2 \mathbf{u} \quad \text{in } \Omega, \qquad \nabla \cdot \mathbf{u} = 0 \quad \text{in } \Omega \tag{3}$$

are considered, where $\mathbf{u}$ is the velocity, p is the pressure and $Re = UL/\nu$ the Reynolds number based on characteristic velocity and length scales, U and L respectively. Discretization in space proceeds by the high-order weighted residual spectral-element technique, extensively described in [8], whereas temporal discretization is based on high-order splitting techniques [12].

3 Stabilization of Turbulent Flow Simulations

The 3/2-rule in pseudo-spectral methods gives the criteria for the evaluation of the non-linear terms in the Navier–Stokes equations to be free from aliasing errors. The corresponding over-integration in SEM follows the same idea, since the polynomial expansion in Legendre space is indeed truncated at N. But since the SEM operates in physical space, it might be more straightforward to view the over-integration as the action taken in order for the evaluation (Gaussian quadrature) of the integrals arising from the weak formulation to be exact, as pointed out in [11]. Either view yields the same conclusion: 3/2 times more points are needed for the non-linear terms in order to avoid aliasing errors. If additionally curvature is taken into account, even more points are required depending on the polynomial order of the curvature.

The first sign of aliasing errors is the occurrence of "spectral blocking", i.e., the accumulation of energy in the highest modes. The filter-based stabilization technique proposed in [7] has the property of suppressing the highest mode, thereby preventing aliasing errors to occur. In a well-resolved calculation, the solution will be smooth, and the amount of energy in the high wavenumber coefficients will be exponentially small. The filter, which operates only on the highest wavenumbers, has the desirable property of not influencing the well-resolved parts of the flow – it only impacts the under-resolved regions, which is precisely what is needed for turbulence. The success of the filter-based stabilization technique was demonstrated in [7]. Considering the 1D case in a domain $\Omega = [-1, 1]$ and $\mathbb{P}_N(\Omega)$ is the space of polynomials of maximum degree N defined on Ω, the filter operator, Π_{N-1}, was originally proposed as the interpolation operator in physical space (but can alternatively and formally equivalent be defined as a filter operator in modal space [2]), $\Pi_{N-1} : \mathbb{P}_N(\Omega) \to \mathbb{P}_{N-1}(\Omega) \to \mathbb{P}_N(\Omega)$. With the use of a relaxation parameter α such that $0 < \alpha < 1$ the filter operator F_α is defined as

$$F_\alpha = \alpha \Pi_{N-1} + (1-\alpha)I \qquad 0 \leq \alpha \leq 1 \tag{4}$$

with I being the identity matrix. Acting with F_α on the velocity vector at each time-step, such that $\underline{u}^{n+1} = F_\alpha \underline{\tilde{u}}^{n+1}$ where $\underline{\tilde{u}}^{n+1}$ is the unfiltered field at the current time-step, allows for a smooth damping of the highest mode with effectively no changes to the existing solver. As pointed out in [16], due to the opposite parity of the Legendre polynomials L_{N-1} and L_N and the fact that Π_{N-1} preserves parity, the amplitude of the highest mode is not dissipated but rather transferred to the third highest mode.

4 Analysis of Model Problems

4.1 1D: Stabilization of the Burgers' Equation

In order to perform a quantitative analysis of the Burgers' equation, the nonlinear problem was transformed into a linear problem by defining the convective operator

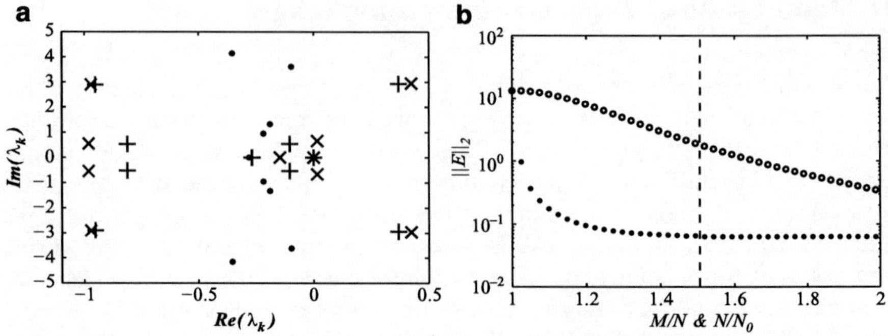

Fig. 1 (**a**) Eigenvalues of the right hand side operator of the linear model problem for (*crossed*) unstabilized, (*plus*) filtered and *filled circle* over-integrated case with $M = N + 4$ points for the convective operator. (**b**) L_2-error with respect to the "exact" numerical solution, where the overall resolution is increased (*circle*) and the nonlinear term is computed with increased number of points (*filled circle*). *dashed lined* indicates where $M/N = 3/2$

based on a constant solution (in time) when the gradients are large, mimicking the conditions in a highly fluctuating turbulent velocity field. The distribution of the eigenvalues of the resulting problem, $\frac{du}{dt} = \mathrm{A}\underline{u}$, are shown in Fig. 1a for the unstabilized, filtered and over-integrated cases. Here, $M = N + 4$ GLL points are used to compute the convective operator (compared to $M = N + 1$ for the other terms), which apparently has a strong influence on the eigenvalues. In particular, the unstable eigenvalues (compare to the unstabilized case) have been completely moved over to the real negative half-plane. Hence, for a marginally resolved simulation subject to large velocity fluctuations, adding only three extra points for the convective term can help to stabilize the numerical method. The filtered case improves the situation by moving the unstable eigenvalues slightly in the negative real direction. In addition to rendering a simulation stable, one would also like to make sure that the solution is not polluted by aliasing errors. Here, the error is investigated by means of the L_2-error for a various number of extra points, M, for the nonlinear term and reported in Fig. 1b. $M/N_0 = 1$ corresponds to equal number of points for the viscous term and for the nonlinear term. As predicted by theory and shown in [11], beyond $M/N_0 = 1.5$ (indicated by the dashed vertical line) the error stays constant. An increased resolution for all terms yields an exponentially decrease of the error as expected. However, the resolution has to be more than doubled in order to get the same error as if 1.5 times more points is added only for the nonlinear term. Notice also that by performing over-integration with only one extra point decreases the error by one order of magnitude.

4.2 2D: Recovery of Skew-Symmetry for the SEM Convection Operator in the Scalar Transport Equation

In high Reynolds number flows, structures are not readily dissipated but rather convected over long distances and times, thus accurate integration of the convective term is essential to obtain reliable results. Here, we investigate how this can be achieved in the scalar transport equation in $\mathbb{R}^2$, given by (2). In the case $\mathbf{c}$ is solenoidal and the domain is closed or periodic, the weak form predicts the convective term in (2) to be skew symmetric, i.e., $c(v,q) = -c(q,v)$. This is easily seen by casting the convective term in (2) in the weak form by multiplying by a test function, v, integrating over the domain, Ω, and using integration by parts, so that

$$\begin{aligned} c(v,q) = \int_{\Omega} v\mathbf{c}\cdot\nabla q d\mathbf{x} &= \int_{\partial\Omega} vq\mathbf{c}\cdot\hat{\mathbf{n}}dA - \int_{\Omega} \nabla\cdot(v\mathbf{c})q d\mathbf{x} \\ &= \int_{\partial\Omega} vq\mathbf{c}\cdot\hat{\mathbf{n}}dA - \int_{\Omega} q\mathbf{c}\cdot\nabla v d\mathbf{x} - \int_{\Omega} \nabla\cdot\mathbf{c}vq d\mathbf{x} = -c(q,v). \end{aligned} \tag{5}$$

The last equality holds as long as the first and the last term on the left hand side are identically zero. The first term vanishes due to the boundary conditions on v and q (homogeneous Dirichlet, periodicity or symmetry) and the last because of the incompressibility constraint, $\nabla\cdot\mathbf{c} \equiv 0$. The remaining equality states the skew-symmetric property of the convective operator. In a discretized form this can only be true if skew-symmetry of the involved matrices is preserved. As we shall see, over-integration may play a crucial role to assure this property. Since the eigenvalues of a skew-symmetric operator are purely imaginary, quadrature errors are easily detected by eigenvalues of the discretized operator with real part $\neq 0$. These errors are reduced by over-integration of the convective term as described earlier. In the case $M = 3(N+1)/2$ the numerical quadrature is exact for all polynomials $\mathbf{c} \in \mathbb{P}_N$. If, however, $\mathbf{c}$ has a polynomial order less than this, recovery of this skew-symmetry – and hence the elimination of the quadrature errors – can be obtained by performing over-integration with $M \ll 3(N+1)/2$, shown by the following examples. We consider the case $\mathbf{c} \in \mathbb{P}_1$, shown in Fig. 2a as a vortical convective field given by $\mathbf{c}_1 = (-y,x)$ and in Fig. 2b as a stagnation point given by $\mathbf{c}_2 = (-x,y)$. Both cases identically fulfil $\nabla\cdot\mathbf{c} \equiv 0$. Although the convective field appears as a first order polynomial in both cases, the particular tensor product structure of the spectral-element method distinguishes between the vortical and the stagnation point velocity fields. For both these cases, each component of the velocity field is separable, i.e., $\mathbf{c}_1 = (1\cdot a(y), b(x)\cdot 1)$ and $\mathbf{c}_2 = (a(x)\cdot 1, 1\cdot b(y))$ and it follows that the double integral in (5) can be separated (for both components, x and y) in one symmetric part and one skew-symmetric part. The symmetric part will be symmetric regardless of exact integration, and does not contribute to the skew-symmetric properties of the convective operator. In this respect, it suffices to examine whether the skew-symmetric part is integrated correctly or not. For "rotational" velocity

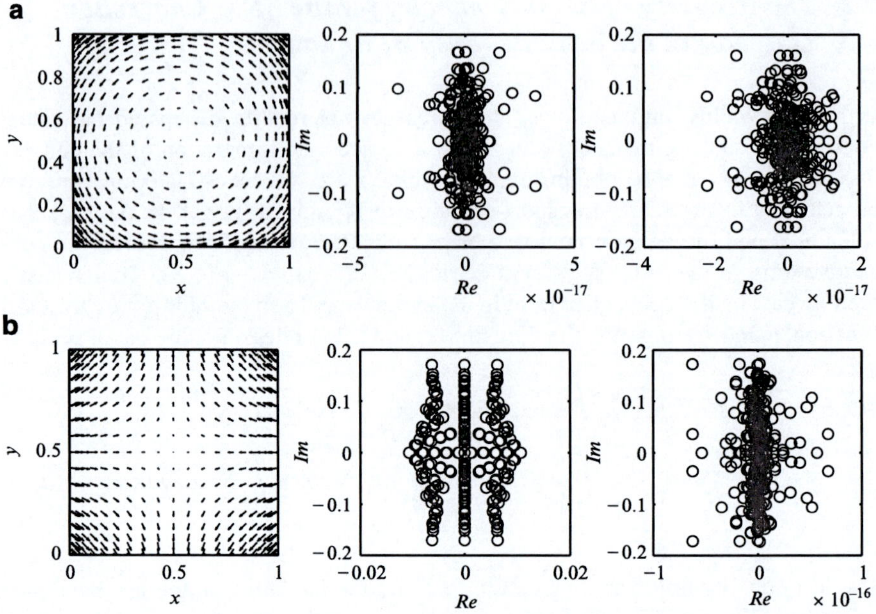

Fig. 2 (**Row a**) Vortical convective field with associated eigenvalue distribution of the operator when $M = N + 1$ and when $M = N + 2$. (**Row b**) same as (**row a**) for a stagnation convective field

fields such as the vortex, the skew-symmetric part will indeed be integrated exactly, since the integrand, p, is a polynomial $p = v^N q^{N-1} \in \mathbb{P}_{2N-1}$. The conclusion is thus that skew-symmetry (i.e., purely imaginary eigenvalues) is obtained using the original $M = N + 1$ grid, which is shown in Fig. 2a. In the latter case the skew-symmetric part for both components, x and y, cannot be integrated exactly since the integrand, p, will be a polynomial $p = v^N c^1 q^{N-1} \in \mathbb{P}_{2N}$. However, by adding one extra point for the integration so that $M = N + 2$, skew-symmetry can again be recovered, as can be seen in Fig. 2b.

5 Application to the Navier–Stokes Equations

In the following, numerical simulations of the incompressible Navier–Stokes equations (3) are performed and evaluated a posteriori. Equation (3) are solved using the Legendre polynomial based SEM code `nek5000` [6].

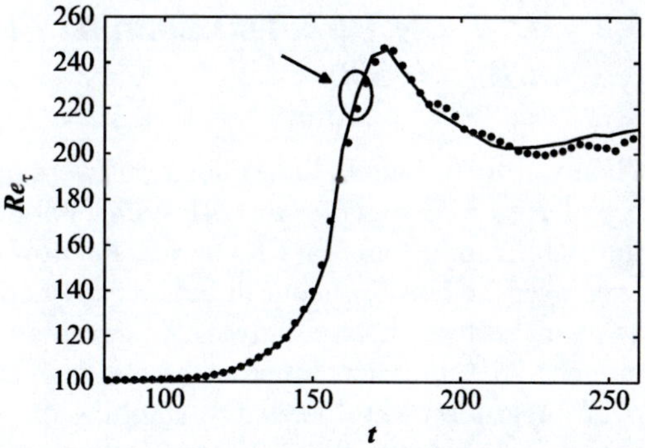

Fig. 3 Evolution of Re_τ for the (stable) transitional channel flow simulation. The *arrow* shows where the numerical instability occurred. *dotted line* DNS by Schlatter et al.

5.1 3D: Subcritical K-type Transition Simulations

Direct numerical simulations (DNS) of subcritical K-type transition at $Re_b = 3{,}333$ (similar to [18]) and a resolution of 91^3 grid points were performed to further highlight the fact that it is indeed the appearance of intermittent turbulence which might render a SEM simulation unstable. The initial disturbances of this classical transition scenario consists of a 2D TS wave (streamwise wave number of $\alpha = 1.12$ and amplitude 3%) together with two 3D oblique waves (wave numbers $\alpha = 1.12$ and $\beta = 2.1$ and amplitude 0.05%) taken from the solution of the Orr–Sommerfeld equation and superimposed on a plane Poiseuille flow profile (see [9] and [18]). The disturbances grow in time, t, and eventually lead to turbulent breakdown. The laminar stage up to $t \approx 160$ is followed by the highly fluctuating transitional stage with an overshoot in the skin friction and finally fully turbulent phase, seen in Fig. 3 showing the skin friction Reynolds number, Re_τ, as a function of time. Unlike [11], who was able to simulate transition in a triangular duct without any stabilization, we found that performing the simulation without any filtering or over-integration of the nonlinear term would yield a numerical instability exactly at the time just before the skin friction peaks ($t = 165$). Adding one extra point to compute the nonlinear term helped to continue the simulation exactly to the skin-friction peak ($t = 169$). However, adding four more points could stabilize the simulation through transition and continue stably in the following fully turbulent stage. This is exactly half the number of points predicted by the 3/2-rule. It should be pointed out that an increase of the spatial resolution ($91 \rightarrow 127$ points in each direction) could not help to stabilize the simulation, which would experience the instability at approximately the same time, just before the peak of the skin-friction. The filtering alone was also able to stabilize the simulation through the skin-friction peak and during the fully turbulent phase.

5.2 3D: Fully Turbulent Channel Flow Simulations at $Re_\tau = 590$

Finally, fully turbulent flow simulations were performed at a friction Reynolds number of $Re_\tau = 590$ similar to [14] in channel geometry in order to see the effect of the stabilization tools in a moderate Re flow. All statistical quantities were averaged over the homogeneous directions x, z and t (for sufficiently long time) as well as over the two channel halves. An acceptable resolution was chosen of approximately 75% in each direction of the fine DNS resolution in [14]. Filtering or over-integration were needed to stabilize the calculation. In one of the two cases shown in Fig. 4, the full 3/2-rule (*dashed*) was used to stabilize the computation, whereas (*solid*) could be rendered stable with only four extra points. No filtering was used for either of these cases. The obtained mean flow results as well as fluctuations show very good agreement with results obtained in [14], and no particular difference can be noticed between the two cases. As an alternative, only filtering could be used to stabilize the computation, shown in Fig. 4 (*thin solid*). Here, as little as 5% filtering of the last mode could ensure a stable computation and good results compared to the reference data. The obtained shape factors, defined as $H_{12} = \frac{\delta^*}{\theta} = \int_{-1}^{1}\left(1 - \frac{U(y)}{U_{CL|\mathrm{lam}}}\right)\mathrm{d}y / \int_{-1}^{1}\frac{U(y)}{U_{CL|\mathrm{lam}}}\left(1 - \frac{U(y)}{U_{CL|\mathrm{lam}}}\right)\mathrm{d}y$ (see, e.g., [18]), where δ^* is the displacement thickness, θ is the momentum thickness, $U_{CL|\mathrm{lam}}$ is the laminar centerline velocity, $U(y)$ is the mean velocity profile and the integration is made between the two walls located at $y = \pm 1$ were $H_{12}^{3(N+1)/2} = 1.583$, $H_{12}^{N+5} = 1.589$ and $H_{12}^{only\ filt} = 1.589$ compared to the reference data $H_{12} = 1.574$ [18]. The obtained skin friction Reynolds number, Re_τ, based on friction velocity, u_τ, and channel half height, h, were $Re_\tau^{3(N+1)/2} = 586.1$, $Re_\tau^{N+5} = 585.7$ and $Re_\tau^{only\ filt} = 588.6$ compared to the reference data $Re_\tau = 587.2$ [14, 18]. Thus,

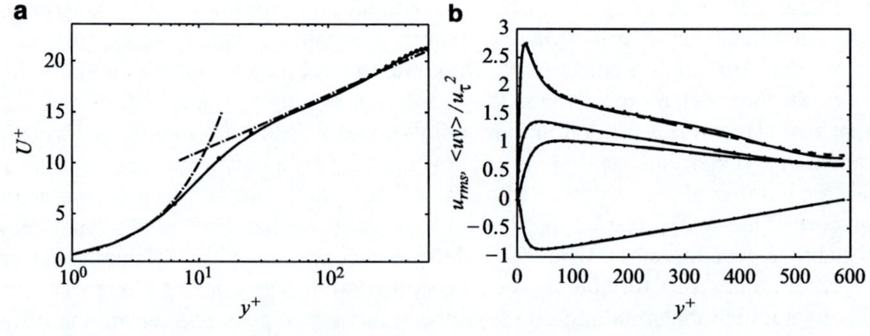

Fig. 4 Turbulent channel flow simulations at $Re_\tau = 590$ with polynomial order 15 (a resolution of 288 in the homogeneous directions and 192 in the wall-normal direction) showing (**a**) mean velocity profile, (**b**) Reynolds stresses *dotted line* DNS data from [14], *dashed-dot-dashed line* log law, *solid line* $M = N + 5$, *dashed line* $M = 3(N + 1)/2$, and *thin line* only filtering (5%)

both these turbulent quantities show a difference on the order of a few per mille, compared to the reference data.

6 Conclusions

Stabilization techniques for the spectral-element method was investigated through two model problems: Burgers' equation in 1D similar to [11] and the scalar transport equation in 2D together with transitional and turbulent Navier–Stokes channel flow simulations in 3D.

The general results from the 1D problem show consistently with [11] that applying over-integration with the full 3/2-rule to an equation with a quadratic non-linearity indeed enhances both the accuracy and stability of the solution. In addition, it could be seen in both model problems and in the full Navier–Stokes simulations that for such equations over-integration with the full 3/2-rule is not needed for stability. Stability was achieved already with $\leq 25\%$ more GLL points, with the first over-integration point being the most efficient to remove aliasing errors. Filter-based stabilization can in most cases alone help to stabilize the computation and is normally not needed together with over-integration, although this combination can be essential for significantly under-resolved cases. The present study suggest that by the use of these techniques stability can be achieved at any Re.

References

1. H. M. Blackburn and S. Schmidt. Spectral element filtering techniques for large eddy simulation with dynamic estimation. *J. Comput. Phys.*, 186(2):610–629, 2003
2. J. P. Boyd. Two comments on filtering (artificial viscosity) for chebyshev and legendre spectral and spectral element methods: preserving boundary conditions and interpretation of the filter as a diffusion. *J. Comput. Phys.*, 143(1):283–288, 1998
3. C. Canuto, M. Y. Hussaini, A. Quarteroni, and T. A. Zang. *Spectral Methods in Fluid Dynamics*. Springer, Berlin, 1988
4. D. C. Chu and G. E. Karniadakis. A direct numerical simulation of laminar and turbulent flow over riblet-mounted surfaces. *J. Fluid Mech.*, 250:1–42, 1993
5. S. Dong, G. E. Karniadakis, A. Ekmekci, and D. Rockwell. A combined direct numerical simulation-particle image velocimetry study of the turbulent near wake. *J. Fluid Mech.*, 569:185–207, 2006
6. P. Fischer, J. Kruse, J. Mullen, H. Tufo, J. Lottes, and S. Kerkemeier. NEK5000 – Open Source Spectral Element CFD solver. https://nek5000.mcs.anl.gov/index.php/MainPage, 2008
7. P. Fischer and J. Mullen. Filter-based stabilization of spectral element methods. *C.R. Acad. Sci. Paris*, t. 332, Serie I:p. 265–270, 2001
8. P. F. Fischer. An overlapping schwarz method for spectral element solution of the incompressible Navier–Stokes equations. *J. Comput. Phys.*, 133(1):84–101, 1997
9. N. Gilbert and L. Kleiser. Near-wall phenomena in transition to turbulence. In S. J. Kline and N. H. Afgan, editors, *Near-Wall Turbulence*, pages 7–27, New York, USA, 1990. 1988 Zoran Zarić Memorial Conference

10. G.-S. Karamanos and G. E. Karniadakis. A spectral vanishing viscosity method for large-eddy simulations. *J. Comput. Phys.*, 163(1):22–50, 2000
11. R. M. Kirby and G. E. Karniadakis. De-alising on non-uniform grids: algorithms and applications. *J. Comput. Phys.*, 191:249–264, 2003
12. Y. Maday, A. T. Patera, and E. M. Rønquist. An operator-integration-factor splitting method for time-dependent problems: application to incompressible fluid flow. *J. Sci. Comput.*, 5(4): 263–292, 1990
13. Y. Maday and E. M. Rønquist. Optimal error analysis of spectral methods with emphasis on non-constant coefficients and deformed geometries. *Comput. Methods Appl. Mech. Eng.*, 80 (1–3):91–115, 1990
14. R. D. Moser, J. Kim, and N. Mansour. Direct numerical simulation of turbulent channel flow up to $Re_\tau = 590$. *Phys. Fluids*, 11(4):943–945, 1999
15. R. Pasquetti. Spectral vanishing viscosity method for large-eddy simulation of turbulent flows. *J. Sci. Comput.*, 27(1–3):365–375, 2006
16. R. Pasquetti and C. J. Xu. Comments on "Filter-based stabilization of spectral element methods". *Note in J. Comput. Phys.*, 182:646–650, 2002
17. P. Schlatter, S. Stolz, and L. Kleiser. Relaxation-term models for LES of turbulent and transitional wall-bounded flows. In *DLES-5*, 2003
18. P. Schlatter, S. Stolz, and L. Kleiser. LES of transitional flows using the approximate deconvolution model. *Int. J. Heat Fluid Flow*, 25(3):549–558, 2004
19. S. Sherwin and G. Karniadakis. A triangular spectral element method; applications to the incompressible Navier–Stokes equations. *Comput. Methods Appl. Mech. Eng.*, 123:189, 1995
20. A. Tomboulides and S. Orszag. Numerical investigation of transitional and weak turbulent flow past a sphere. *J. Fluid Mech.*, 416:45–73, 2000
21. H. M. Tufo and P. F. Fischer. Fast parallel direct solvers for coarse grid problems. *J. Parallel Distrib. Comput.*, 61(2):151–177, 2001
22. C. E. Wasberg, T. Gjesdal, B. A. Pettersson Reif, and Ø. Andreassen. Variational multiscale turbulence modelling in a high order spectral element method. *J. Comput. Phys.*, 228(19):7333–7356, 2009
23. C. Xu and R. Pasquetti. Stabilized spectral element computations of high Reynolds number incompressible flows. *J. Comput. Phys.*, 196(2):680–704, 2004

The Spectral-Element and Pseudo-Spectral Methods: A Comparative Study

J. Ohlsson, P. Schlatter, C. Mavriplis, and D.S. Henningson

Abstract Turbulent and transitional channel flow simulations have been performed in order to assess the differences concerning speed and accuracy in the pseudo-spectral code `simson` and the spectral-element code `nek5000`. The results indicate that the pseudo-spectral code is 4–6 times faster than the spectral-element code in fully turbulent channel flow simulations, and up to 10–20 times faster when taking into account the more severe CFL restriction in the spectral-element code. No particular difference concerning accuracy could be noticed either in the turbulent nor the transitional cases, except for the pressure fluctuations at the wall which converge slower for the spectral-element code.

1 Introduction

The simulation of fluid flows – sensitive and often complicated – puts large requirements on the numerical method. Due to the nonlinear nature of the flow, accuracy may be one of the most important ingredients. In particular for direct simulation of complex multiscale flows, such as transitional and turbulent flows, high order methods are preferred. However, the choice of methods, e.g., fully spectral, multidomain spectral such as spectral element, or compact differences, is not clear as trade-offs exist between computational efficiency, geometrical flexibility and accuracy. Proper comparisons in terms of speed and accuracy are sorely needed. In order to quantify differences and similarities between high-order methods in a more systematic way, we have chosen to compare two well established codes based on the Chebyshev-Fourier pseudo-spectral method (`simson` [1]) and the spectral-element method (`nek5000` [2]). While the grid is essentially prescribed by the order for the

J. Ohlsson (✉), P. Schlatter, and D.S. Henningson
Linné Flow Centre, KTH Mechanics, Stockholm, Sweden
e-mail: johan@mech.kth.se

C. Mavriplis
Department of Mechanical Engineering, University of Ottawa, Ottawa, Canada
e-mail: catherine.mavriplis@uottawa.ca

J.S. Hesthaven and E.M. Rønquist (eds.), *Spectral and High Order Methods for Partial Differential Equations*, Lecture Notes in Computational Science and Engineering 76,
DOI 10.1007/978-3-642-15337-2_44,

pseudo-spectral method, a more flexible point distribution is possible in the spectral element method. In order to concentrate the comparison on relative efficiency, we have chosen canonical test cases like turbulent and transitional channel flow, in which the effect of the point distribution might be considerably less crucial for achieving high accuracy.

2 Study Setup

The study is divided into two parts: Part A is concerned with the computational efficiency in terms of the wall-clock time per time step and part B deals with accuracy, aiming at establishing a way to compare the number of grid-points needed to compute a given turbulent or transitional quantity with comparable accuracy. In the first part of the study, turbulent channel flow simulations at a Reynolds number $Re_\tau = 180$, based on friction velocity, u_τ, and channel half height, h, were considered in a domain of size comparable to that by Moser et al. [6]. Two different resolutions called r1 and r2 were simulated ($\sim$43 and 95 grid points in each direction respectively). For the spectral-element code this was achieved by fixing the polynomial order (seventh) and varying the number of elements (6 and 12 in each direction). It was noted that by using polynomial order 7 instead of 11 for the spectral-element simulations increases the speed by $\sim$15% per time step. In order to make the comparison as fair as possible, the order of the temporal scheme was synchronized so that a third order time discretization was used in both codes. Also, the scaling was adapted so that the Re in both codes were based on $Re_b = 2{,}800$, based on bulk velocity, u_b, and channel half height, h. Timings were made in serial mode (one core AMD 3.0 GHz) on the same computer. Dealiasing was used in both codes.

In the second part of the study K-type transition similar to Schlatter et al. [7] and turbulent channel flow similar to Moser et al. [6] at $Re_\tau = 180$, based on friction velocity, u_τ, and channel half height, h, were simulated for a number of different resolutions, given in Table 1. It should be pointed out that the two lowest resolution spectral-element cases had to be stabilized by a filtering procedure described in [4].

Snapshots from each of these two cases are shown in Fig. 1. Important measures such as the time and amplitude for the skin-friction peak were computed for the transitional cases, whereas mean velocity profile, Reynolds stresses, pressure and pressure fluctuations together with integral quantities such as Re_τ, shape factor and "point measures", i.e., $\max(u_{rms})$, were computed and compared for the turbulent cases.

Table 1 Overview of the different resolutions in terms of degrees of freedom (dof) used in the present study. Two different polynomial orders were used for the spectral-element simulations. The number of degrees of freedom was matched as closely as possible for all cases

Fully spectral	24^3	40^3	80^3	128^3	160^3
Spectral-element (7th/11th)	29^3	$43^3/45^3$	$85^3/78^3$	$127^3/122^3$	$155^3/155^3$

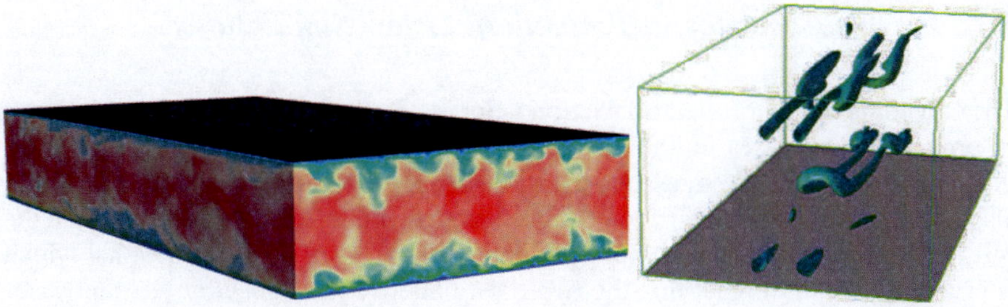

Fig. 1 The canonical flow cases investigated: (**a**) snapshot of turbulent channel flow at $Re_\tau = 180$ showing pseudocolor of streamwise velocity and (**b**) temporal K-type transition showing the hairpin vortex (isosurfaces of $\lambda_2 = -0.1$) emerging at $t = 135$

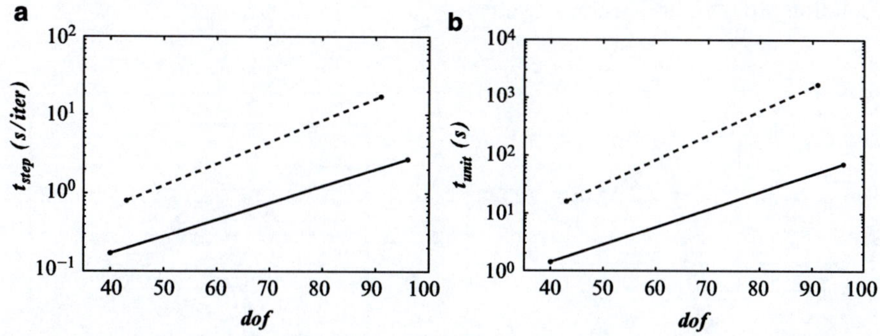

Fig. 2 Comparison with respect to time advancement in a turbulent channel flow simulation between the spectral-element code (polynomial order 7) (*dashed line*) and the pseudo-spectral code (*solid line*) for the two different resolutions r1 and r2 (degrees of freedom in each direction). (**a**) Wall-clock time per iteration, (**b**) wall-clock time per unit time

3 Results

3.1 Part A: Efficiency

The wall-clock time per iteration, i.e., one full time step using the largest possible time step for the spectral-element code and one full Runge Kutta time step (containing four sub-steps) for the pseudo-spectral code, was measured and is reported in Fig. 2a below. It can be seen that the lines diverge, i.e., the spectral code gets relatively faster for larger problem sizes, due to the increasingly efficient fast Fourier transforms (FFT). In particular, the spectral code is 4–6 times faster for these two problem sizes. In addition, we show wall-clock time per unit time in Fig 2b, where it can be seen that the spectral code is 10–20 times faster due to the more severe CFL restriction in the spectral-element code, arising from the clustering of the Gauss-Lobatto-Legendre points close to each element boundary.

3.2 Part B: Accuracy in Transitional Flow Simulations

The Reynolds number based on friction velocity, Re_τ, was computed as a function of time, t, for all cases in Table 1 during K-type transition [7] and are shown in Fig. 3. We note that the most underresolved cases lead to a premature transition, also noted by other authors, followed by an overprediction of the skin-friction in the fully turbulent phase. This is more pronounced in the fully spectral results, which is probably due to the fact that the two most underresolved spectral-element cases had to be stabilized by the filter, which in some sense acts like a simple subgrid scale (SGS) model. For higher resolutions, the two codes converge (from below) to the correct Re_τ for essentially the same number of degrees of freedom, as also seen in Fig. 4. This behavior indicates that the initial stages of transition are essentially a low-order phenomenon, not requiring full resolution. Thus, the third highest resolution (80^3) yields accurate results.

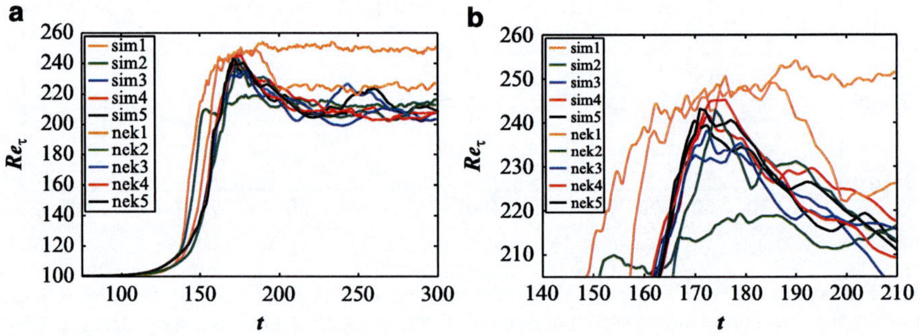

Fig. 3 (**a**) Skin-friction Reynolds number Re_τ as a function of time, t, computed for all cases shown in Table 1, where sim1-sim5 and nek1-nek5 corresponds to increasing resolutions of the pseudo-spectral and spectral-element codes respectively, (**b**) close-up view of the peak in (**a**)

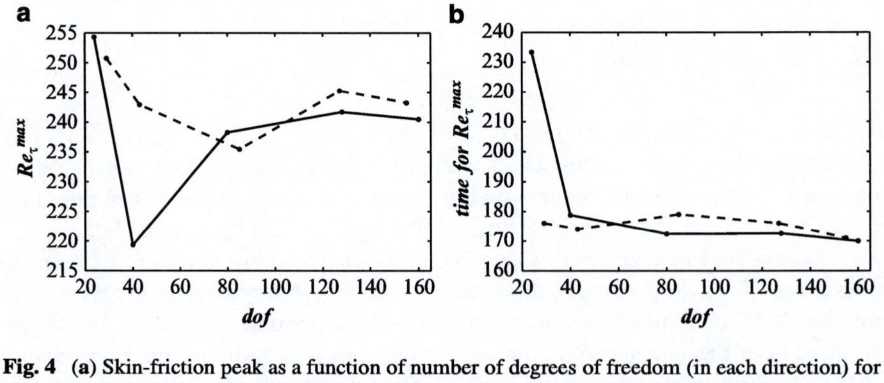

Fig. 4 (**a**) Skin-friction peak as a function of number of degrees of freedom (in each direction) for the spectral-element code (*dashed line*) and the pseudo-spectral code (*solid line*). (**b**) Time when the peak in (**a**) occurs

3.3 Part B: Accuracy in Turbulent Flow Simulations

The turbulent mean velocity profile at $Re_\tau = 180$ is shown for all cases in Fig. 5. The most underresolved cases in both the fully spectral Fig. 5a and the spectral-element Fig. 5b code show the same tendency to underpredict the velocity in the log region, which is related mainly to the scaling given by an overpredicted friction velocity, u_τ, since indeed $u^+ = \langle u \rangle / u_\tau$. In a close-up view of the log region (Fig. 5c), where only the three highest resolution cases (80^3, 128^3, 160^3 and seventh order for the spectral-element code) are shown, convergence is seen for the two codes for the same number of degrees of freedom (shown by an arrow). The 128^3 cases are converged and the 160^3 cases do not improve the results further. The spatial distribution of the Reynolds stresses is examined in Fig. 6. While the fully spectral results capture the peaks correctly when compared to the direct numerical results (DNS) of Moser et al. [6], even for the most underresolved cases, the skin-friction Reynolds number is heavily over-predicted as noted in the transitional simulations. This is in contrast to the spectral-element results, where the peaks are overpredicted for all normal stress components but the skin-friction Reynolds

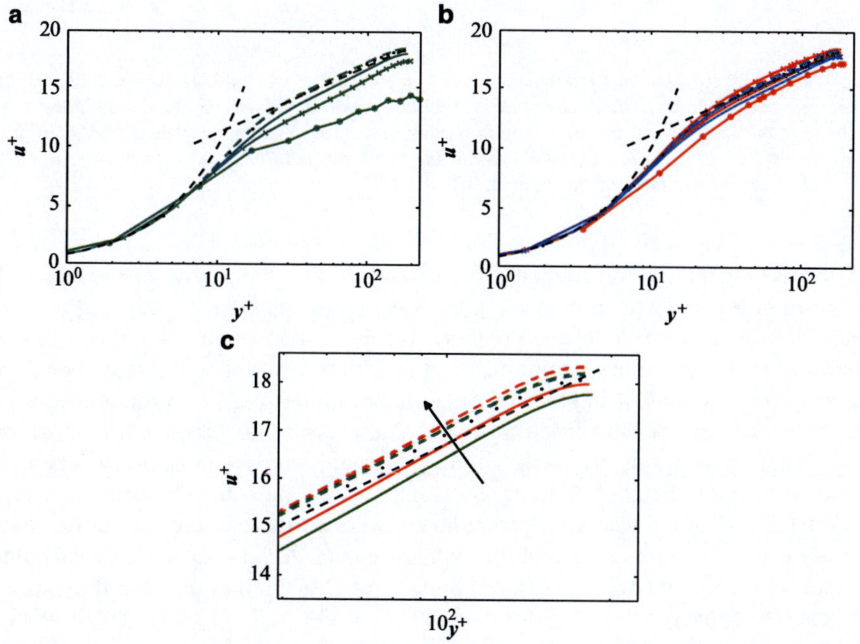

Fig. 5 Turbulent mean velocity profiles for (**a**) the spectral code (*green*), (**b**) the spectral-element code (*blue*: 11th order, *red*: seventh order) and (**c**) the three highest resolutions of the spectral and spectral-element (seventh order) simulations. *dashed-star-dashed line* (24^3), *dashed-crossed-dashed line* (40^3), *solid line* (80^3), *dashed line* (128^3), *dashed-dot-dashed line* (160^3), *dotted line* direct numerical simulation (DNS) of Moser et al. [6]

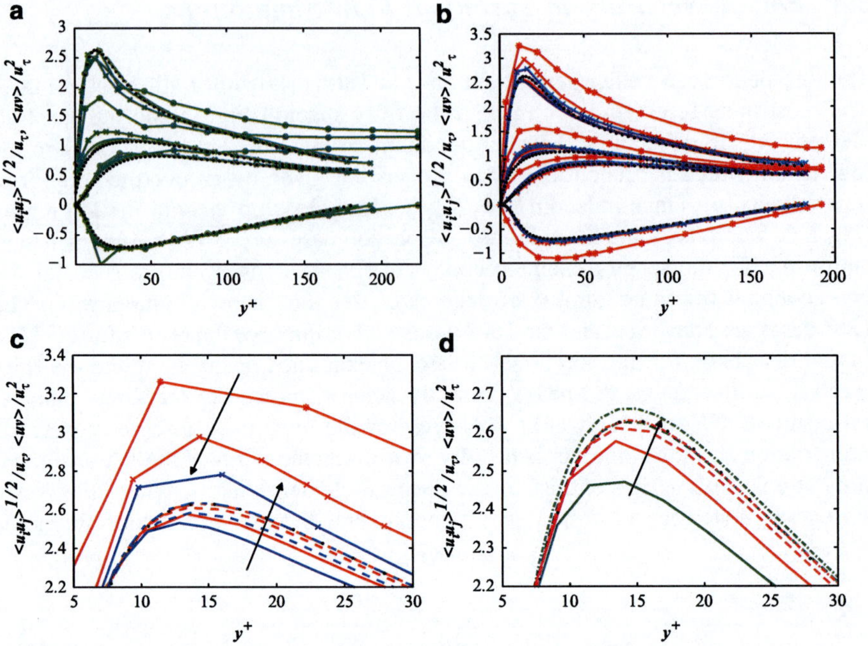

Fig. 6 Reynolds stresses for (**a**) the spectral code (*green*), (**b**) the spectral-element code (*blue*: 11th order, *red*: seventh order), (**c**) close-up view of the u_{rms} peak in (**b**) and (**d**) the three highest resolutions of the spectral and spectral-element (seventh order) simulations. *dashed-star-dashed line* (24^3), *dashed-crossed-dashed line* (40^3), *solid line* (80^3), *dashed line* (128^3), *dashed-dot-dashed line* (160^3), *dotted line* DNS of Moser et al. [6]

number is only mildly overpredicted. A close-up view of the spectral-element results is shown in Fig. 6c, where the peak u_{rms} is shown to converge in a zig-zag pattern (indicated by arrows): first, overpredicted for the lowest resolutions, then, underpredicted for intermediate resolutions, and finally, converging to the reference data for the same number of degrees of freedom. A similar but less pronounced zig-zag pattern is seen for the spectral results. The pressure fluctuations (Fig. 7) from the spectral simulations are fairly good at the wall, whereas those in the channel center are overpredicted. The spectral-element results show the opposite behavior: the fluctuations at the wall are overpredicted, whereas those in the core of the flow are in fairly good agreement with the reference data. A close-up view reveals that the spectral-element code needs more points (roughly double) than the fully spectral code to converge the pressure fluctuations at the wall (Fig. 7d), which would make the spectral code around 40 times faster. The reasons for this may be that in a $\mathbb{P}_N - \mathbb{P}_{N-2}$ spectral-element method [5] the number of degrees of freedom for the pressure is less than the velocities and thus less than for the corresponding pressure resolution in the spectral simulation. Another reason may be the absence of a pressure node at the wall in the spectral-element $\mathbb{P}_N - \mathbb{P}_{N-2}$ formulation, leading

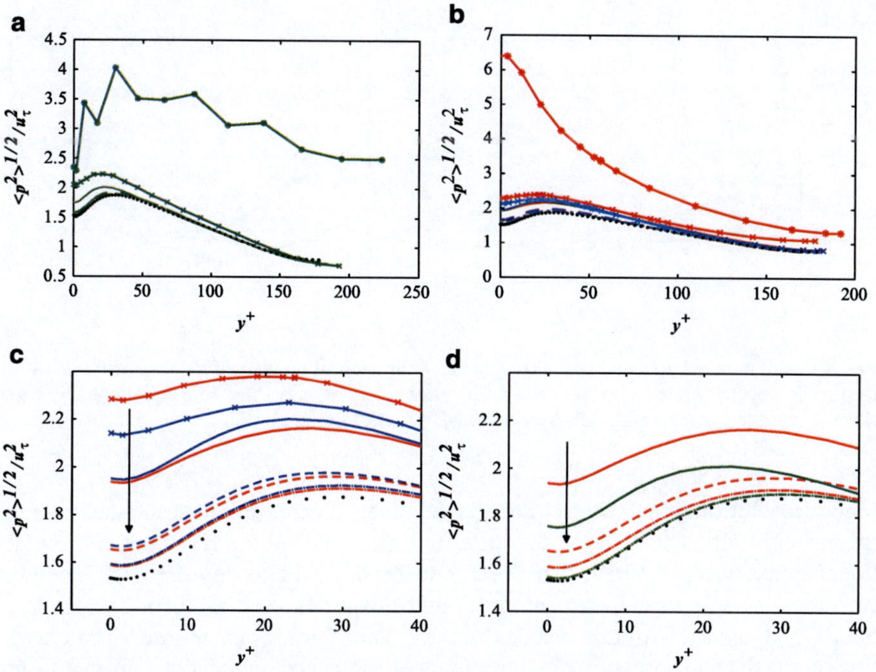

Fig. 7 Pressure fluctuations for (**a**) the spectral code (*green*), (**b**) the spectral-element code (*blue*: 11th order, *red*: seventh order), (**c**) close-up view of (**b**) close to the wall and (**d**) the three highest resolutions of the spectral and spectral-element (seventh order) simulations close to the wall. *dashed-star-dashed line* (24^3), *dashed-crossed-dashed line* (40^3), *solid line* (80^3), *dashed line* (128^3), *dashed-dot-dashed line* (160^3), *dotted line* DNS of Moser et al. [6]

to reduced control of the pressure at the wall. Finally, we compare turbulent integral quantities such as actual Re_τ and "point measures" such as $\max(u_{rms})$. The actual Re_τ (given as a simulation result when constant mass-flux is prescribed) is shown in Fig. 8a, where an overestimation of the Re_τ for the lowest resolution cases in the fully spectral simulations already mentioned can be seen. Similarly, the zig-zag pattern described in Sect. 3.3 for the peak u_{rms} is seen in Fig. 8b. For both quantities, convergence seems to follow the same "slope" for the two codes, as well as for the two different orders in the spectral-element simulations.

4 Conclusions

The present results indicate that the pseudo-spectral code is 4–6 times faster than the spectral-element code in fully turbulent channel flow simulations. Taking into account the more severe CFL restriction in the spectral-element method due to the clustering of the points near the element boundaries, this number rises to 10–20. For

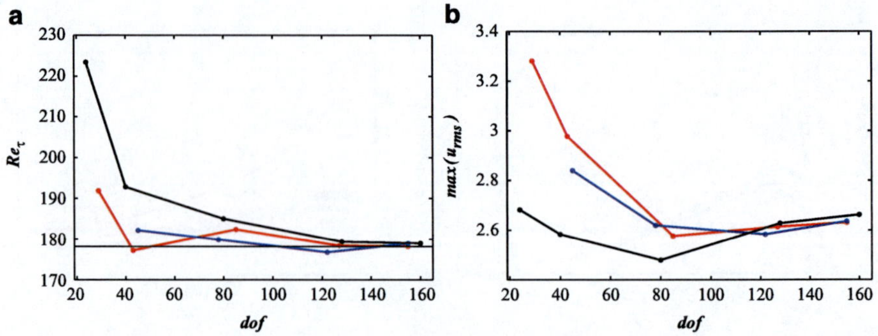

Fig. 8 (**a**) Computed Re_τ and (**b**) $\max(u_{rms})$ as a function of number of degrees of freedom for the pseudo-spectral code (*black*) and and different polynomial orders for the spectral-element code (*blue*: 11th order, *red*: seventh order). *solid line* DNS of Moser et al. [6]

higher resolutions, the spectral code is relatively faster, due to implementionally increasingly efficient FFTs as the order increases. Of all the turbulent and transitional quantities computed, there seems to be no favor to any particular method (or order in the spectral-element code) and quantities such as shape factor, Re_τ, $\max(u_{rms})$ and skin-friction peak exhibit the same "convergence-rate". The exception seems to be the pressure fluctuations close to the wall which did converge faster in the spectral code. Taking this into account the spectral code may be around 40 times faster. Moreover, by using polynomial order 7 instead of 11 for the spectral-element simulations increases the speed by $\sim$15% per time step which, in addition to the larger time step that can be achieved with a lower order, seems to be a better choice in general.

Naturally, for canonical flows such as channel flows fully spectral methods are superior due to their near optimal point distribution. But it should be noted that being faster on one CPU does not necessarily mean a faster code. For example, taking into account that a spectral-element code performs spectrally in local elements, this method has an enormous parallel scalability and might be faster than spectral codes for very large cases. Moreover, as we progress away from canonical flows towards more complex geometry flows (see, e.g., Fischer et al. [3]) such as real aircraft wing geometries, the geometrical flexibility of the spectral element approach will be favored over the pseudo-spectral approach.

References

1. M. Chevalier, P. Schlatter, A. Lundbladh, and D. S. Henningson. A Pseudo-Spectral Solver for Incompressible Boundary Layer Flows. Technical Report TRITA-MEK 2007:07, Royal Institute of Technology (KTH), Department of Mechanics, Stockholm, 2007
2. P. Fischer, J. Kruse, J. Mullen, H. Tufo, J. Lottes, and S. Kerkemeier. NEK5000 – Open Source Spectral Element CFD solver. https://nek5000.mcs.anl.gov/index.php/MainPage, 2008
3. P. Fischer, J. Lottes, D. Pointer, and A. Siegel. Petascale algorithms for reactor hydrodynamics. *J. Phys. Conf. Series*, 125:1–5, 2008

4. P. Fischer and J. Mullen. Filter-based stabilization of spectral element methods. *C.R. Acad. Sci. Paris*, t. 332, Serie I: pp. 265–270, 2001
5. Y. Maday and A. Patera. Spectral element methods for the Navier–Stokes equations. In A.K. Noor, editor, *State of the Art Surveys in Computational Mechanics*, pp. 71–143, ASME, New York, 1989
6. R. D. Moser, J. Kim, and N. Mansour. Direct numerical simulation of turbulent channel flow up to $Re_\tau = 590$. *Phys. Fluids*, 11(4):943–945, 1999
7. P. Schlatter, S. Stolz, and L. Kleiser. LES of transitional flows using the approximate deconvolution model. *Int. J. Heat Fluid Flow*, 25(3):549–558, 2004

Adaptive Spectral Filtering and Digital Total Variation Postprocessing for the DG Method on Triangular Grids: Application to the Euler Equations

S. Ortleb, A. Meister, and Th. Sonar

Abstract With respect to the possible presence of discontinuities in the solutions of nonlinear wave propagation problems high order methods have to be provided with a dose of supplementary numerical dissipation, otherwise the approximate solution may severely suffer from the presence of Gibbs oscillations. To prevent these oscillations from rendering the scheme unstable we apply the spectral filtering framework to the DG method on triangular grids. The corresponding spectral filter has been derived in [18] from a spectral viscosity formulation and is applied adaptively in order to restrict artificial viscosity to shock locations. Furthermore, the image processing technique of DTV filtering is shown to be a useful postprocessor. Numerical experiments are carried out for the two-dimensional Euler equations where we show results for the Shu-Osher shock–density wave interaction problem as well as the interaction of a moving vortex with a stationary shock.

1 Introduction

If the discontinuous Galerkin method [5, 13] is applied to hyperbolic problems where the entropy solution may develop discontinuities, minmod-type limiters are often employed which reduce the polynomial degree in so-called troubled cells and thus disregard the information contained in higher oder coefficients. Limiters starting from the higher order coefficients and modifying them only when it is needed have been suggested in [2,15,26] and WENO or HWENO reconstructions in regions marked by a shock sensor have been applied in [20–22,27], but these techniques are computationally expensive. Spectral methods on the other hand may apply spectral

S. Ortleb (✉) and A. Meister
Fachbereich Mathematik, Universtität Kassel, Heinrich Plett Str. 40, 34132 Kassel, Germany
e-mail: ortleb@mathematik.uni-kassel.de, meister@mathematik.uni-kassel.de

Th. Sonar
Institut Computational Mathematics, Technische Universität Braunschweig, Pockelsstr. 14, 38106 Braunschweig, Germany
e-mail: t.sonar@tu-bs.de

J.S. Hesthaven and E.M. Rønquist (eds.), *Spectral and High Order Methods for Partial Differential Equations*, Lecture Notes in Computational Science and Engineering 76,
DOI 10.1007/978-3-642-15337-2_45,

viscosity [17, 25] in order to stabilize the calculation. As suggested in [9], the stabilization can be carried out within the spectral filtering framework resulting in a computationally very efficient implementation with successful applications to wave propagation problems as in [6,7]. In [18] we showed that these promising techniques usually applied to Fourier and Chebychev spectral methods can also be transferred to the discontinuous Galerkin discretization on triangular grids. The crucial advantage of this novel scheme is a reduced computational cost compared to reconstructions over large stencils. While the results in [18] were limited to scalar equations the present work deals with their extension to systems of conservation laws with specific interest in solving the Euler equations. As modal filtering may degrade the accuracy of the approximation when applied to a large number of time steps, see [12], we employ spatially adaptive filters in order to restrict the introduced artificial damping to the vicinity of shock locations. The resulting approximations of these spectral viscosity solutions at a final time or at certain intermediate times where a truthful pointwise solution is desired will still suffer from Gibbs oscillations. Popular remedies in the 1D case are postprocessing techniques such as Gegenbauer reprojection [10] requiring the detection of discontinuities. As edge detection as well as the necessary parameter specification for the Gegenbauer technique will become difficult for conservation laws in higher dimensions, we use the digital total variation filter [4] which was developed in the context of image processing and has been applied to a Chebychev pseudospectral method in [23].

2 The Discontinuous Galerkin Scheme with Spectral Filtering

We consider two-dimensional hyperbolic conservation laws of the form

$$\frac{\partial}{\partial t}\mathbf{u}(\mathbf{x},t) + \frac{\partial}{\partial x_1}\mathbf{f}_1(\mathbf{u}(\mathbf{x},t)) + \frac{\partial}{\partial x_2}\mathbf{f}_2(\mathbf{u}(\mathbf{x},t)) = 0, \qquad (\mathbf{x},t) \in \Omega \times \mathbb{R}_+, \tag{1}$$

where $\Omega \subset \mathbb{R}^2$ is an open polygonal domain and $\mathbf{u}(\mathbf{x},t) \in \mathbb{R}^n$. Furthermore, initial conditions $\mathbf{u}(\mathbf{x},0) = \mathbf{u}_0(\mathbf{x})$ and appropriate boundary conditions are assumed to be given. Let $\mathscr{T}^h$ be a conforming triangulation of the closure $\overline{\Omega}$ of the computational domain and let V^h be the piecewise polynomial space defined by $V^h = \{v_h \in L^\infty(\Omega) \mid v_h|_{\tau_i} \in \mathscr{P}^N(\tau_i) \quad \forall\, \tau_i \in \mathscr{T}^h\}$, where $\mathscr{P}^N(\tau_i)$ denotes the space of all polynomials on τ_i of degree $\leq N$. For the discontinuous Galerkin space discretization an approximation $\mathbf{u}_h : \Omega \times \mathbb{R}_+ \to \mathbb{R}^n$, $\mathbf{u}_h(\cdot,t) \in (V^h)^n$ is constructed satisfying the semidiscrete equation

$$\frac{\mathrm{d}}{\mathrm{d}t}\int_{\tau_i} \mathbf{u}_h \mathbf{v}\mathrm{d}\mathbf{x} = \int_{\tau_i}\left(\mathbf{f}_1(\mathbf{u}_h)\cdot\frac{\partial \mathbf{v}}{\partial x_1} + \mathbf{f}_2(\mathbf{u}_h)\cdot\frac{\partial \mathbf{v}}{\partial x_2}\right)\mathrm{d}\mathbf{x} - \int_{\partial\tau_i}\mathbf{H}(\mathbf{u}_-^i,\mathbf{u}_+^i,\mathbf{n})\cdot\mathbf{v}\,\mathrm{d}\sigma, \tag{2}$$

for any $\tau_i \in \mathcal{T}^h$, $\mathbf{v} \in (V^h)^n$, where $\mathbf{H}$ represents a suitable numerical flux function and $\mathbf{u}_-^i$, $\mathbf{u}_+^i$ denote the solution within τ_i and within an adjacent element, respectively. The integrals in (2) are approximated by quadrature formulae of sufficiently high order. In order to represent the approximate solution on each triangle, we use the well-known Koornwinder-Dubiner basis [8,13,14] which is given by the polynomials $\Phi_{lm}(r,s) = P_l^{0,0}\left(2\frac{1+r}{1-s}-1\right)\left(\frac{1-s}{2}\right)^l P_m^{2l+1,0}(s)$ for $l,m \in \mathbb{N}_0$, $0 \le l+m \le N$, on the reference element $T = \{(r,s) \in \mathbb{R}^2 \mid -1 \le r,s;\ r+s \le 0\}$, where $P_n^{\alpha,\beta}$ denotes the one-dimensional Jacobi polynomial of degree n associated to the weight function $w(x) = (1-x)^\alpha(1+x)^\beta$. Denoting by $\Lambda_i : \tau_i \to T$ an orientation-preserving affine transformation which maps the specific triangle τ_i to the reference element T, the approximation $\mathbf{u}_h|_{\tau_i}$ can be expanded as $\mathbf{u}_h(\Lambda_i^{-1}(r,s),t) = \sum_{l+m\le N} \hat{\mathbf{u}}_{lm}^i(t)\Phi_{lm}(r,s)$. The resulting system of ordinary differential equations for the coefficients $\hat{\mathbf{u}}_{lm}^i$ may then be solved by appropriate time integration schemes depending on the specific application under consideration. In our calculations we used a 4th order low storage RK scheme, see [3]. This basic DG scheme is now supplemented by a modal filter which is applied to the Koornwinder-Dubiner coefficients after each time step. The filter is obtained by an extension of the spectral viscosity method to multidomain Koornwinder-Dubiner expansions, see [18], and results in a modification of the vector of Koornwinder-Dubiner coefficients by an exponential filter of the form $\hat{\mathbf{u}}_{lm}^{i,\text{mod}} = \exp\left(-\alpha_i s_i \eta^{2p}\right)\hat{\mathbf{u}}_{lm}^i$, $\eta = \frac{l+m}{N+1}$, with shock indicator s_i, filter order $2p$ and filter strength $\alpha_i \sim \frac{N\Delta t}{h_i}$. As usual, Δt and h_i denote the time step size and the shortest distance of the barycenter of τ_i to the element boundary $\partial\tau_i$, respectively. Our experiments indicated to use rather low order filters because of their adaptive application, hence a forth order filter is chosen in this work. We focussed on two different shock indicators that were also investigated in [1,26], i.e. the resolution-based indicator $s_i = \min\left\{1,\ 5000(5N^4+1)\sum_{l+m=N}\left(\hat{u}_{lm}^i\right)^2 / \sum_{l+m<N}\left(\hat{u}_{lm}^i\right)^2\right\}$ first suggested in [19] as well as the jump indicator [16] $s_i = \min\left\{1,\ 1000\int_{\partial\tau_i} \left|[[u^i]]/\{u^i\}\right| \cdot \mathbf{n} / |\partial\tau_i|\, \mathrm{d}\sigma\right\}$, where u^i and $\hat{u}_{lm}^i$ denote the specific components of $\mathbf{u}_h|_{\tau_i}$ and $\hat{\mathbf{u}}_{lm}^i$ that are chosen for shock indication, the common notation $[[u^i]] = u_-^i\mathbf{n}_- + u_+^i\mathbf{n}_+$, $\{u^i\} = \frac{1}{2}(u_-^i + u_+^i)$ is employed and the integral is solved by high-order Gaussian quadrature. However, as there were no visible differences between the numerical solutions obtained with the two indicators we dropped their further investigation and used the resolution indicator requiring less computatational work.

3 The Digital Total Variation Filter

Whereas the adaptive spectral filter is used in every time step, the DTV filter serves as a pure postprocessing step with the advantage of an already incorporated edge detection. DTV filtering applies to general graphs $[V, E]$ with a finite set of nodes

V and edges E. If two nodes $\alpha, \beta \in V$ are linked by an edge, we write $\alpha \sim \beta$ and the set of nodes linked to $\alpha \in V$ is denoted by $N_\alpha = \{\beta \in V \mid \alpha \sim \beta\}$. Let $\mathbf{u}^0$ be composed of the oscillatory nodal values u_α^0 of one component of the solution $\mathbf{u}_h(\cdot, t_{\text{out}})$ at time t_{out}. Following [4] the DTV filter is implemented as an iterative procedure $u_\alpha^{n+1} = \sum_{\beta \in N_\alpha} h_{\alpha\beta}(\mathbf{u}^n)u_\beta^n + h_{\alpha\alpha}(\mathbf{u}^n)u_\alpha^0, \quad n = 0, 1, \ldots,$ where the filter coefficients are given by $h_{\alpha\beta}(\mathbf{u}) = \omega_{\alpha\beta}(\mathbf{u}) / \left(\lambda + \sum_{\gamma \in N_\alpha} \omega_{\alpha\gamma}(\mathbf{u})\right)$, $h_{\alpha\alpha}(\mathbf{u}) = \lambda / \left(\lambda + \sum_{\gamma \in N_\alpha} \omega_{\alpha\gamma}(\mathbf{u})\right)$ for appropriate non-negative weights $\omega_{\alpha\beta}$ measuring the local variation of the given data and a non-negative, user-dependent parameter λ that balances the competing tasks of removing spurious oscillations and retaining relevant information of the noisy initial data. The weights are chosen by $\omega_{\alpha\beta}(\mathbf{u}) = \frac{1}{|\nabla_\alpha \mathbf{u}|_a} + \frac{1}{|\nabla_\beta \mathbf{u}|_a}$, where $|\nabla_\alpha \mathbf{u}|_a = \left[\sum_{\beta \in N_\alpha} (u_\beta - u_\alpha)^2 + a\right]^{1/2}$ is the regularized local variation at node α equipped with a small regularization parameter $a > 0$ to avoid a zero denominator.

To our knowledge, convergence of the DTV filter has not been proven yet. However, the results in [4, 23] indicate that its iterative application leads to a steady image. Based on these previous investigations, we evaluated the DG solution at cartesian grid points and carried out DTV iterations until a steady state was reached.

4 Numerical Experiments

We consider the 2D Euler equations for polytropic ideal gases, i.e. (1) with

$$\mathbf{u} = \begin{pmatrix} \varrho \\ \varrho v_1 \\ \varrho v_2 \\ \varrho E \end{pmatrix}, \quad \mathbf{f}_1 = \begin{pmatrix} \varrho v_1 \\ \varrho v_1^2 + p \\ \varrho v_1 v_2 \\ v_1(\varrho E + p) \end{pmatrix}, \quad \mathbf{f}_2 = \begin{pmatrix} \varrho v_2 \\ \varrho v_1 v_2 \\ \varrho v_2^2 + p \\ v_2(\varrho E + p) \end{pmatrix},$$

where ϱ, v_1, v_2, p denote the density, the two components of velocity and the pressure, respectively. The total energy E ist related to these quantities by the equation of state $p = (\gamma - 1)\varrho\left(E - \frac{1}{2}(v_1^2 + v_2^2)\right)$, where γ denotes the ratio of specific heats and is set to $\gamma = 1.4$ in all our tests. The adaptivity indicators for spectral filtering were always based on the density component.

Shock–Density Wave Interaction

First, we consider the shock–density wave interaction problem by Shu and Osher [24] which is extended to two space dimensions. The initial conditions are given by

$$(\varrho, v_1, v_2, p) = \begin{cases} (3.857143, 2.629369, 0, 10.333333) & \text{if } x < -4, \\ (1 + 0.2 \cdot \sin(5x), 0, 0, 1) & \text{if } x \geq -4. \end{cases}$$

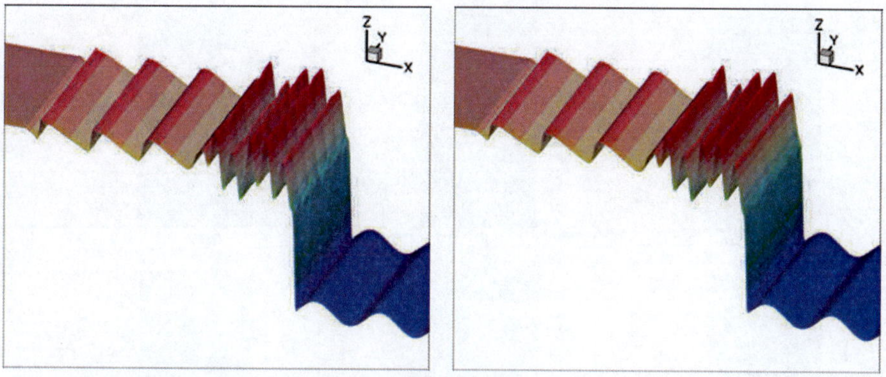

Fig. 1 SV filtered and DTV postprocessed density solution, $N = 5$

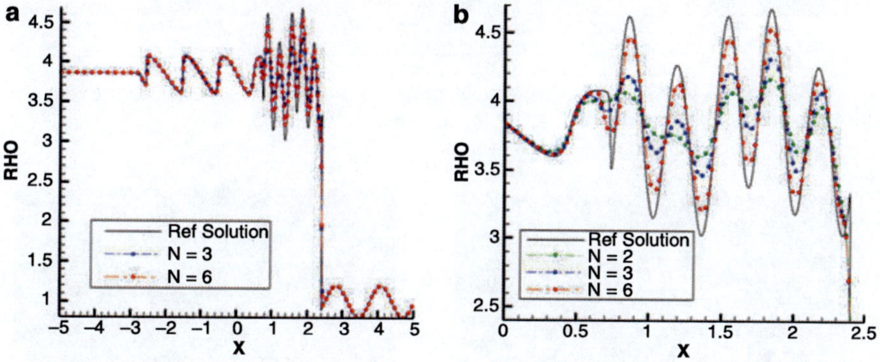

Fig. 2 DTV postprocessed solutions (**a**) for $N = 3, 6$; (**b**) for $N = 2, 3, 6$, close-up view

Figure 1 shows the density distribution computed by the DG scheme with spectral filtering for a polynomial degree of $N = 5$ on the computational domain $\Omega = [-5, 5] \times [0, 0.5]$ at a final time of $t_{\text{out}} = 1.8$ as well as the DTV postprocessed solution (100 iterations with $\lambda = 5$ on 500×25 cartesian grid points). For the DG scheme, a grid consisting of 1,250 triangles has been used where two edge points have an average distance of 0.1. An additional iterative application of the spectral filter has been implemented in this test case in order to enforce positive physical quantities ϱ and p at cell interfaces. Thus, the DG scheme with modal filtering produces only small overshoots which are removed by the DTV postprocessor.

Figure 2 depicts 1D cuts of the DTV postprocessed solutions for $N = 2, 3, 6$. The reference solution is obtained by solving the corresponding 1D problem by a second order FV scheme with TVD reconstruction on a grid of 30,000 cells. Here we clearly see that a higher order scheme is necessary to resolve the high frequencies and that the DTV postprocessed solutions are completely free of spurious oscillations.

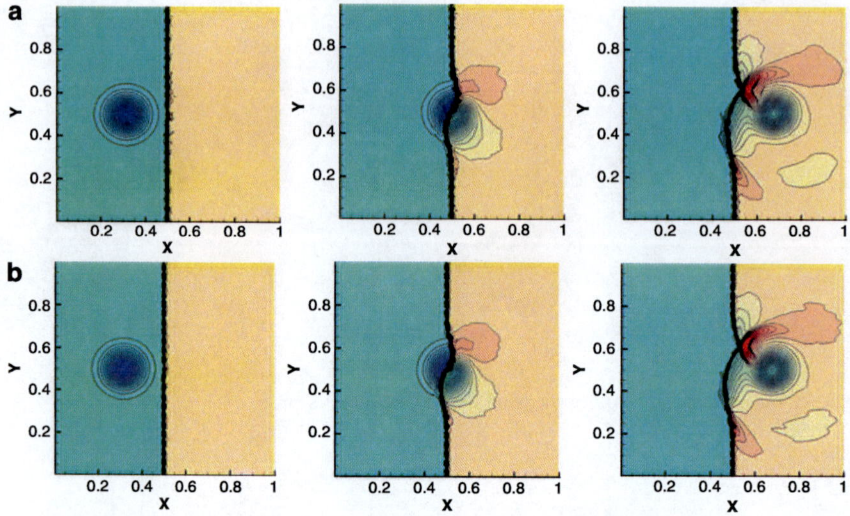

Fig. 3 DG solutions with spectral filtering at output times $t_{out} = 0.05, 0.2, 0.35$. Pressure contours with 46 contour levels from 0.85 to 1.35 for **(a)** $N = 3$ and **(b)** $N = 5$

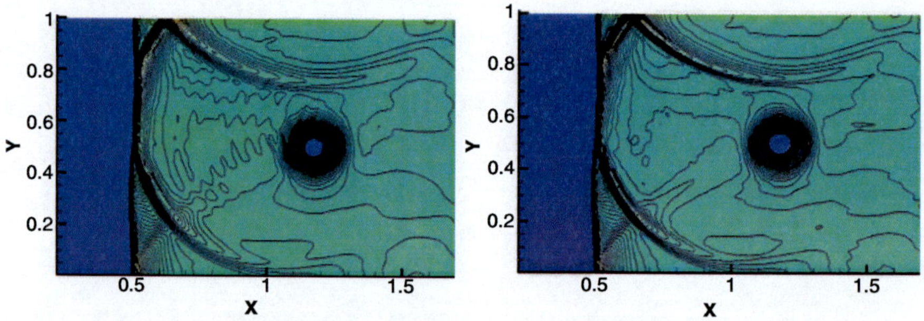

Fig. 4 DG solutions with spectral filtering for $N = 3$ and $N = 5$, pressure contours for $t_{out} = 0.8$. Ninety contour levels from 1.09 to 1.37

2D Shock–Vortex Interaction

This test case is taken from [11] and describes the interaction between a stationary shock and a vortex. The computational domain is $\Omega = [0, 2] \times [0, 1]$. A stationary Mach 1.1 shock normal to the x-axis is positioned at $x = 0.5$. Its left state is $(\varrho, v_1, v_2, p) = (1, 1.1\sqrt{\gamma}, 0, 1)$ and its right state is defined by the Rankine-Hugoniot conditions. An isentropic vortex, i.e. $p/\varrho^\gamma = const$, is superposed to the flow and centers at $(x_c, y_c) = (0.25, 0.5)$. The vortex is described as a perturbation of the velocity and the temperature by $v_1' = \varepsilon\tau e^{\alpha(1-\tau^2)} \sin\theta$, $v_2' = -\varepsilon\tau e^{\alpha(1-\tau^2)} \cos\theta$ and $T' = -(\gamma - 1)\varepsilon^2 e^{2\alpha(1-\tau^2)}/(4\alpha\gamma)$, where $\tau = r/r_c$,

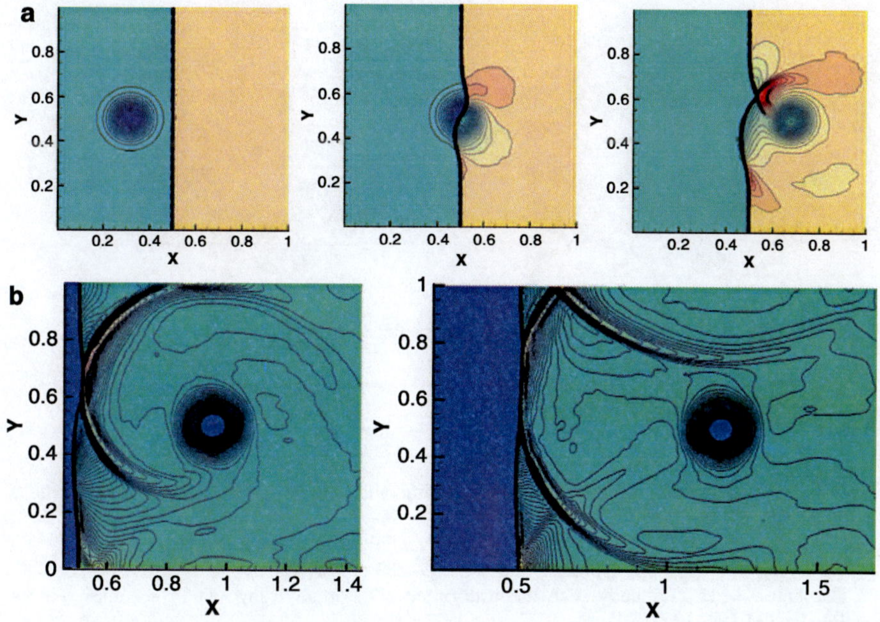

Fig. 5 DTV solutions (λ = 10,400×200 grid points). (**a**) Pressure contours with 46 contour levels from 0.85 to 1.35, $t_{\text{out}} = 0.05, 0.2, 0.35$; (**b**) Ninety contour levels from 1.09 to 1.37, $t_{\text{out}} = 0.6, 0.8$

$r = \sqrt{(x-x_c)^2 + (y-y_c)^2}$ and the remaining parameters are set to $r_c = 0.05$, $\varepsilon = 0.3$ and $\alpha = 0.204$. Figure 3 shows the pressure contours obtained by the DG scheme with spectral filtering for $N = 3$ and $N = 5$ at output times of $t_{\text{out}} = 0.05, 0.2, 0.35$. The computational grid consists of 2122 triangles with more resolution at the shock location (average point distances of 0.05 away from and 0.025 close to the shock). The difference – both in capturing the shock and in resolving the vortex – is more pronounced in Fig. 4 showing pressure contours for a later output time of $t_{\text{out}} = 0.8$. For $N = 5$, a slight improvement in capturing the shock is visible. Hence, for long time integration a numerical scheme with low dissipation such as the DG scheme with spectral filtering is necessary. In Fig. 5 the DTV postprocessed solutions for $N = 5$ are shown where we observe sharp shock profiles. For $t_{\text{out}} = 0.35$, Fig. 6 depicts the interacting shock and vortex before and after postprocessing.

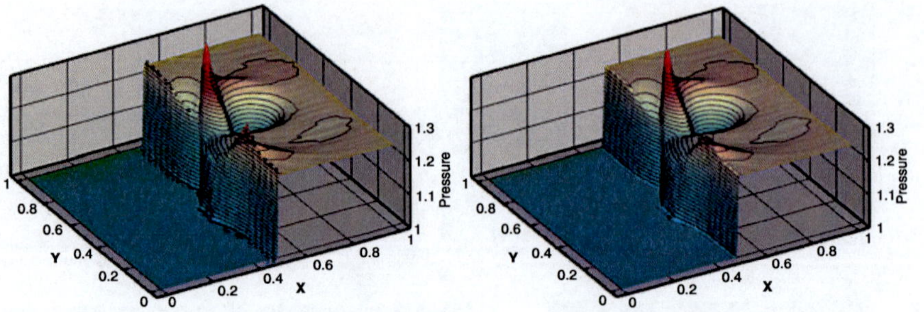

Fig. 6 SV filtered and DTV postprocessed solutions for $N = 5$, $t_{\mathrm{out}} = 0.35$

References

1. Barter, G.E., Darmofal, D.L.: Shock Capturing with Higher-Order, PDE-Based Artificial Viscosity. AIAA 2007-3823 (2007)
2. Biswas, R., Devine, K.D., Flaherty, J.E.: Parallel, adaptive finite element methods for conservation laws. Appl. Numer. Math. **14**, 255–283 (1994)
3. Carpenter, M.H., Kennedy, C.A.: Fourth-order 2N-storage Runge-Kutta schemes. NASA Report TM 109112 (1994)
4. Chan, T.F., Osher, S., Shen, J.: The digital TV filter and nonlinear denoising. IEEE Trans. Image Process. **10**, 231–241 (2001)
5. Cockburn, B., Shu, C.-W.: Runge-Kutta discontinuous Galerkin methods for convection-dominated problems. J. Sci. Comp. **16**, 173–261 (2001)
6. Don, W.S.: Numerical study of pseudospectral methods in shock wave applications. J. Comput. Phys. **110**, 103–111 (1994)
7. Don, W.S., Gottlieb, D., Jung, J.H.: A multidomain spectral method for supersonic reactive flows. J. Comput. Phys. **192**, 325–354 (2003)
8. Dubiner, M.: Spectral methods on triangles and other domains. J. Sci. Comput. **6**, 345–390 (1991)
9. Gottlieb, D., Hesthaven, J.S.: Spectral methods for hyperbolic problems. J. Comput. Appl. Math. **128**, 83–131 (2001)
10. Gottlieb, D., Shu, C.-W.: On the Gibbs phenomenon and its resolution. SIAM Rev. **39**, 644–668 (1998)
11. Jiang, G.-S., Shu, C.-W.: Efficient implementation of weighted ENO schemes. J. Comput. Phys. **126**, 202–228 (1996)
12. Kanevsky, A., Carpenter, M.H., Hesthaven, J. S.: Idempotent filtering in spectral and spectral element methods. J. Comput. Phys. **220**, 41–58 (2006)
13. Karniadakis, G.E., Sherwin, S.: Spectral/hp element methods for computational fluid dynamics, 2nd edn. Oxford University Press, London (2005)
14. Koornwinder, T.: Two-variable analogues of the classical orthogonal polynomials. In: Askey, R. (ed.) Theory and Applications of Special Functions. Academic Press, San Diego (1975)
15. Krivodonova, L.: Limiters for high-order discontinuous Galerkin methods. J. Comput. Phys. **226**, 879–896 (2007)
16. Krivodonova, L., Xin, J., Remacle, J.-F., Chevaugeon, N.,Flaherty, J.E.: Shock detection and limiting with discontinuous Galerkin methods for hyperbolic conservation laws. Appl. Numer. Math. **48**, 323–338 (2004)
17. Maday, Y., Ould Kaber, S.M., Tadmor, E.: Legendre pseudospectral viscosity method for nonlinear conservation laws. SIAM J. Numer. Anal. **30**, 321–342 (1993).

18. Meister, A., Ortleb, S., Sonar, Th.: On spectral filtering for discontinuous Galerkin methods on unstructured triangular grids. Preprint: Mathematische Schriften Kassel (2009) http://cms.uni-kassel.de/unicms/fileadmin/groups/w_180000/prep/prep0904.pdf
19. Persson, P.-O., Peraire, J.: Sub-cell shock capturing for discontinuous Galerkin methods. AIAA-2006-0112 (2006)
20. Qiu, J., Shu, C.-W.: Hermite WENO schemes and their application as limiters for Runge Kutta discontinuous Galerkin method: one dimensional case. J. Comput. Phys. **193**, 115–135 (2004)
21. Qiu, J., Shu, C.-W.: Hermite WENO schemes and their application as limiters for Runge Kutta discontinuous Galerkin method: two dimensional case. Comp. Fluid **34**, 642–663 (2005)
22. Qiu, J., Shu, C.-W.: Runge-Kutta discontinuous Galerkin method using WENO limiters. SIAM J. Sci. Comput. **26**, 907–929 (2005)
23. Sarra, S.A.: Digital total variation filtering as postprocessing for Chebyshev pseudospectral methods for conservation laws. Numerical Algorithms **41**, 17–33 (2006)
24. Shu, C.-W., Osher, S.: Efficient implementation of essentially non-oscillatory shock capturing schemes II. J. Comput. Phys. **83**, 32–78 (1989)
25. Tadmor, E.: Convergence of spectral methods for nonlinear conservation laws. SIAM J. Numer. Anal. **26**, 30–44 (1989)
26. Yang, M., Wang, Z.J.: A parameter-free generalized moment limiter for high-order methods on unstructured grids. AIAA-2009-605 (2009)
27. Zhu, J., Qiu, J., Shu, C.-W., Dumbser, M.: Runge-Kutta discontinuous Galerkin method using WENO limiters II: Unstructured meshes. J. Comput. Phys. **227**, 4330–4353 (2008)

BDDC and FETI-DP Preconditioners for Spectral Element Discretizations of Almost Incompressible Elasticity

Luca F. Pavarino and Olof B. Widlund

1 Introduction

We construct and study a BDDC (Balancing Domain Decomposition by Constraints) algorithm, see [1, 2], for the system of almost incompressible elasticity discretized with Gauss–Lobatto–Legendre (GLL) spectral elements. Related FETI-DP algorithms, see, e.g., [3–5], could be considered as well. We show that sets of primal constraints can be found so that these methods have a condition number that depends only weakly on the polynomial degree, while being independent of the number of subdomains (scalability) and of the Poisson ratio and Young's modulus of the material considered (robustness).

Earlier work on domain decomposition algorithms for mixed elasticity and Stokes systems can be found in [6–14]. Previous works on BDDC algorithms for GLL spectral elements have focused on the scalar elliptic case only, see [15, 16].

2 Almost Incompressible Elasticity and Spectral Elements

The Continuous Problem Given a domain $\Omega \subset R^d, d = 2, 3$, and a nonempty subset $\partial\Omega_D$ of its boundary, we consider, for the case of constant material properties, a mixed formulation of linear elasticity for almost incompressible materials as, e.g., in [17]: find $(\mathbf{u}, p) \in \mathbf{V} \times U$ such that

$$\begin{cases} \mu a(\mathbf{u},\mathbf{v}) + \ b(\mathbf{v},p) \ = \mathbf{F}(\mathbf{v}) & \forall \mathbf{v} \in \mathbf{V} \\ b(\mathbf{u},q) \ - \frac{1}{\lambda} c(p,q) = \ 0 & \forall q \in U. \end{cases} \quad (1)$$

L.F. Pavarino
Department of Mathematics, Università di Milano, Via Saldini 50, 20133 Milano, Italy
e-mail: luca.pavarino@unimi.it

O.B. Widlund (✉)
Courant Institute of Mathematical Sciences, 251 Mercer Street, NY 10012, USA
e-mail: widlund@cims.nyu.edu

J.S. Hesthaven and E.M. Rønquist (eds.), *Spectral and High Order Methods for Partial Differential Equations*, Lecture Notes in Computational Science and Engineering 76,
DOI 10.1007/978-3-642-15337-2_46,

Here,

$$a(\mathbf{u},\mathbf{v}) := 2\int_{\Omega} \varepsilon(\mathbf{u}) : \varepsilon(\mathbf{v})\, dx,\ b(\mathbf{v},q) := -\int_{\Omega} \operatorname{div}\mathbf{v}\, q\, dx,\ c(p,q) := \int_{\Omega} pq\, dx,$$

$\mathbf{V} := \{\mathbf{v} \in H^1(\Omega)^3 : \mathbf{v}|_{\partial\Omega_D} = 0\}$, $U := L^2(\Omega)$ (or $L^2_0(\Omega)$ if $\partial\Omega_D = \partial\Omega$). $\mathbf{F}$ represents the applied forces and $\mu := \frac{E}{2(1+\nu)}$ and $\lambda := \frac{E\nu}{(1+\nu)(1-2\nu)}$, where ν and E are the Poisson ratio and Young's modulus, respectively. We approach the incompressible limit when $\nu \to 1/2$.

We will assume that the domain Ω can be decomposed into N nonoverlapping elements Ω_i, each of which is an image, $\Omega_i = \phi_i(\Omega_{\text{ref}})$, of a reference square or cube $\Omega_{\text{ref}} = (-1,1)^d$, $d = 2,3$, under an affine mapping ϕ_i. In between these elements, we have the interface Γ. We will in fact consider the case when the Lamé parameters μ_i and λ_i are constant, but potentially different, in the different elements; our analysis is reduced to developing bounds for individual elements Ω_i. The global problem can be obtained by assembling contributions to the bilinear forms from the different elements. Our estimates will be independent of the values of the Lamé parameters.

GLL Spectral Elements The space of displacements $\mathbf{V}$ is discretized, component by component, by continuous, piecewise tensor product polynomials of degree n:

$$\mathbf{V}_n := \{\mathbf{v} \in \mathbf{V} : v_k|_{\Omega_i} \circ \phi_i \in Q_n(\Omega_{\text{ref}}),\ i = 1,2,\ldots,N,\ k = 1,2,\ldots,d\}.$$

The pressure space U is discretized by discontinuous, piecewise tensor product polynomials of degree $n-2$:

$$U_n := \{q \in U : q|_{\Omega_i} \circ \phi_i \in Q_{n-2}(\Omega_{\text{ref}}),\ i = 1,2,\ldots,N\}.$$

We use Gauss–Lobatto–Legendre (GLL(n)) quadrature, which also allows for the construction of a very convenient nodal tensor-product basis for $\mathbf{V}_n$. We denote by $\{\xi_i\}_{i=0}^n$ the set of GLL(n) points of $[-1,1]$, by σ_i the quadrature weight associated with ξ_i, and by $l_i(x)$ the Lagrange interpolating polynomial of degree n that vanishes at all the GLL(n) nodes except at ξ_i, where it equals 1. Each element of $Q_n(\Omega_{\text{ref}})$ is expanded in this GLL(n) basis, and any L^2–inner product of two scalar components u and v is replaced, in the three-dimensional case, by

$$(u,v)_{n,\Omega} = \sum_{s=1}^{N} \sum_{i,j,k=0}^{n} (u \circ \phi_s)(\xi_i,\xi_j,\xi_k)(v \circ \phi_s)(\xi_i,\xi_j,\xi_k)|J_s|\sigma_i\sigma_j\sigma_k\,,$$

where $|J_s|$ is the determinant of the Jacobian of ϕ_s. The mass matrix based on these basis elements and GLL(n) quadrature is then diagonal. Similarly, a very convenient basis for U_n consists of the tensor-product Lagrangian nodal basis functions associated with the internal GLL(n) nodes; i.e., the endpoints -1 and $+1$ are excluded.

The $Q_n - Q_{n-2}$ method satisfies a nonuniform inf-sup condition

$$\sup_{\mathbf{v} \in \mathbf{V}_n} \frac{(\mathrm{div}\mathbf{v}, q)}{\|\mathbf{v}\|_{H^1}} \geq \beta_n \|q\|_{L^2} \quad \forall q \in U_n\,, \tag{2}$$

where $\beta_n = cn^{-(d-1)/2}$, $d = 2, 3$, and the constant $c > 0$ is independent of n and q; see [18]. It is also known that β_n decays quite slowly for practical values of n, e.g., $n \leq 16$.

Discrete System and Positive Definite Reformulation The discrete system, obtained from the GLL spectral elements, is assembled from the saddle point matrices of individual elements Ω_i:

$$\begin{bmatrix} \mu_i A^{(i)} & B^{(i)T} \\ B^{(i)} & -1/\lambda_i\, C^{(i)} \end{bmatrix}.$$

Since we are using discontinuous pressures, all pressure degrees of freedom can be eliminated element by element to obtain reduced positive definite stiffness matrices

$$K^{(i)} = \mu_i A^{(i)} + \lambda_i B^{(i)T} C^{(i)-1} B^{(i)},$$

that can be subassembled into a global positive definite stiffness matrix K.

The load vector of the full system can similarly be assembled from contributions from the elements.

3 The BDDC Algorithm

We will associate each spectral element with a subdomain; using several elements per subdomain would of course also be possible. We split the set of basis functions into interior functions, with the subscript I, and the remaining interface basis functions, with the subscript Γ.

Subspaces We will use the framework of [5] Let $V^{(i)}$ be the local space of spectral element displacements defined on Ω_i and that vanish on $\partial\Omega_i \cap \partial\Omega_D$. We split this space as the direct sum of its interior and interface subspaces $V^{(i)} = V_I^{(i)} \bigoplus V_\Gamma^{(i)}$ and we define the associated product spaces by $V_I := \prod_{i=1}^N V_I^{(i)}$, $V_\Gamma := \prod_{i=1}^N V_\Gamma^{(i)}$. The functions in V_Γ are generally discontinuous across Γ, while our spectral element functions are continuous across Γ. Therefore, we also define the subspace

$$\widehat{V}_\Gamma := \{\text{functions of } V_\Gamma \text{ that are continuous across } \Gamma\}.$$

We will also need an intermediate subspace $\widetilde{V}_\Gamma := V_\Delta \bigoplus \widehat{V}_\Pi$ defined by further splitting the interface degrees of freedom into primal (with the subscript Π) and dual (with the subscript Δ) degrees of freedom. Here:

(a) $\widehat{V}_\Pi$ is a global subspace consisting of selected continuous functions, the *primal* variables; these can be the subdomain vertex basis functions of $\widehat{V}$ and/or edge/face basis functions with constant value at the nodes of the associated edge/face. We will assume that, after a change of basis, each primal variable correspond to an explicit degree of freedom; cf. [5, Sect. 3.3]. This simplifies the presentation and also adds to the robustness of the algorithms; see [19].

(b) $V_\Delta = \prod_{i=1}^N V_\Delta^{(i)}$ is the product space of the subspaces $V_\Delta^{(i)}$ of *dual* interface functions that vanish at the primal degrees of freedom.

Restriction and Scaling Operators In order to define our preconditioners, we need certain restriction and interpolation operators represented by matrices with elements in the set $\{0, 1\}$:

$$R_{\Gamma\Delta} : \widetilde{V}_\Gamma \longrightarrow V_\Delta, \quad R_{\Gamma\Pi} : \widetilde{V}_\Gamma \longrightarrow \widehat{V}_\Pi,$$
$$R_\Gamma^{(i)} : \widehat{V}_\Gamma \longrightarrow V_\Gamma^{(i)}, \quad R_\Delta^{(i)} : V_\Delta \longrightarrow V_\Delta^{(i)}, \quad R_\Pi^{(i)} : \widehat{V}_\Pi \longrightarrow V_\Pi^{(i)}.$$

With these operators, we build the following operators:

$$R_\Gamma : \widehat{V}_\Gamma \longrightarrow V_\Gamma, \quad \text{the direct sum of the } R_\Gamma^{(i)};$$
$$\widetilde{R}_\Gamma : \widehat{V}_\Gamma \longrightarrow \widetilde{V}_\Gamma, \quad \text{the direct sum } R_{\Gamma\Pi} \oplus R_\Delta^{(i)} R_{\Gamma\Delta}.$$

We will also need the standard counting functions of Neumann–Neumann methods and in particular their pseudoinverses $\delta_i^\dagger(x)$, defined at each node x on the interface $\Gamma_i := \partial\Gamma_i \cap \Gamma$ of subdomain Ω_i by

$$\delta_i^\dagger(x) := \mu_i(x) / (\sum_{j \in \mathcal{N}_x} \mu_j(x)), \tag{3}$$

where $\mathcal{N}_x$ is the set of indices of the subdomains having the node x on their boundary; see also [20, Sect. 6.2.1] for alternatives. We define scaled local restriction operators $R_{D,\Gamma}^{(i)}$ and $R_{D,\Delta}^{(i)}$ by multiplying the sole nonzero element of each row of $R_\Gamma^{(i)}$ and $R_\Delta^{(i)}$ by $\delta_i^\dagger(x)$. Then, let

$$R_{D,\Gamma} := \text{ the direct sum of } R_{D,\Gamma}^{(i)}, \quad \widetilde{R}_{D,\Gamma} := \text{ the direct sum } R_{\Gamma\Pi} \oplus R_{D,\Delta}^{(i)} R_{\Gamma\Delta}.$$

Schur Complement After reordering the interior displacements first and then those of the interface, resulting in $(\mathbf{u}_I, \mathbf{u}_\Gamma)$, the local spectral element stiffness matrix for subdomain Ω_i can be rewritten:

$$K^{(i)} = \begin{bmatrix} K_{II}^{(i)} & K_{\Gamma I}^{(i)T} \\ K_{\Gamma I}^{(i)} & K_{\Gamma\Gamma}^{(i)} \end{bmatrix}.$$

By eliminating the interior displacement variables, we obtain the local Schur complement $S^{(i)}$, of the subdomain Ω_i, as

$$S^{(i)} = K_{\Gamma\Gamma}^{(i)} - K_{\Gamma I}^{(i)} K_{II}^{(i)-1} K_{\Gamma I}^{(i)T}$$

and then, by subassembly, the Schur complement

$$\widehat{S} = \sum_{i=1}^{N} R_{\Gamma}^{(i)T} \widehat{S}^{(i)} R_{\Gamma}^{(i)}. \tag{4}$$

The BDDC Preconditioner The splitting of the interface displacements into dual (with subscript Δ) and primal (with subscript Π) interface displacements, induces the following partition of the local stiffness matrices:

$$K^{(i)} = \begin{bmatrix} K_{II}^{(i)} & K_{\Delta I}^{(i)T} & K_{\Pi I}^{(i)} \\ K_{\Delta I}^{(i)} & K_{\Delta\Delta}^{(i)} & K_{\Pi\Delta}^{(i)T} \\ K_{\Pi I}^{(i)} & K_{\Pi\Delta}^{(i)} & K_{\Pi\Pi}^{(i)} \end{bmatrix}.$$

The BDDC preconditioner for the Schur complement $\widehat{S}$ is defined by

$$M_{BDDC}^{-1} = \widetilde{R}_{D,\Gamma}^{T} \widetilde{S}^{-1} \widetilde{R}_{D,\Gamma}, \tag{5}$$

where

$$\widetilde{S}^{-1} = R_{\Gamma\Delta}^{T} \left(\sum_{i=1}^{N} \begin{bmatrix} 0 & R_{\Delta}^{(i)T} \end{bmatrix} \begin{bmatrix} K_{II}^{(i)} & K_{\Delta I}^{(i)T} \\ K_{\Delta I}^{(i)} & K_{\Delta\Delta}^{(i)} \end{bmatrix}^{-1} \begin{bmatrix} 0 \\ R_{\Delta}^{(i)} \end{bmatrix} \right) R_{\Gamma\Delta} + \Phi S_{\Pi\Pi}^{-1} \Phi^{T}. \tag{6}$$

The first term in (6) represents local Neumann solves on individual subdomain Ω_i with the primal variables constrained to vanish. The second term involves a coarse solve for the primal variables, with the coarse matrix

$$S_{\Pi\Pi} = \sum_{i=1}^{N} R_{\Pi}^{(i)T} \left(K_{\Pi\Pi}^{(i)} - \begin{bmatrix} K_{\Pi I}^{(i)} & K_{\Pi\Delta}^{(i)} \end{bmatrix} \begin{bmatrix} K_{II}^{(i)} & K_{\Delta I}^{(i)T} \\ K_{\Delta I}^{(i)} & K_{\Delta\Delta}^{(i)} \end{bmatrix}^{-1} \begin{bmatrix} K_{\Pi I}^{(i)T} \\ K_{\Pi\Delta}^{(i)T} \end{bmatrix} \right) R_{\Pi}^{(i)}$$

and a matrix Φ representing a change of variable given by

$$\Phi = R_{\Gamma\Pi}^{T} - R_{\Gamma\Delta}^{T} \sum_{i=1}^{N} \begin{bmatrix} 0 & R_{\Delta}^{(i)T} \end{bmatrix} \begin{bmatrix} K_{II}^{(i)} & K_{\Delta I}^{(i)T} \\ K_{\Delta I}^{(i)} & K_{\Delta\Delta}^{(i)} \end{bmatrix}^{-1} \begin{bmatrix} K_{\Pi I}^{(i)T} \\ K_{\Pi\Delta}^{(i)T} \end{bmatrix} R_{\Pi}^{(i)}.$$

Choice of Primal Constraints In our BDDC algorithm, we choose as primal variables the displacements at the subdomain (spectral element) vertices and the average of the normal displacements on each subdomain edge for the two-dimensional case. In our experiments, we also explore a richer choice where the vertex constraints are augmented by the edge averages of both displacement components.

As is clear from [4] and [5], the three dimensional case is much more complicated; we need to satisfy the two assumptions of [5]. One of these assumptions guarantees that the dual displacement component has a divergence-free extension and the other essentially guarantees that the algorithm performs well for compressible elasticity problems. One successful recipe involves using primal vertex constraints for all subdomain vertices, augmenting them with a primal constraint on the average of the normal displacement component over each face, and the averages, over the edges, of the two displacement components orthogonal to each edge. Following the discussion in [5, Sect. 7], one additional primal constraint per face is required and it can be chosen as the tangential average of the displacement over one of the edges of each face. Should the distribution of the Lamé parameters be particularly difficult around an edge, we in addition have to make such an edge *fully primal*, with five primal constraints associated with it; see further [4, Sect. 5]. Given the sufficiently rich sets of primal constraints just outlined, the following bound can be proven, see [21].

Theorem 1. *The BDDC preconditioned operator, and the FETI-DP operator using the same set of primal variables, have all eigenvalues* ≥ 1 *and a maximum eigenvalue bounded above by*

$$C\beta_n^{-2}(1+\log n)^2.$$

Here C *is independent of* n, N, *and the values of the Lamé parameters. The parameter* β_n *is the inf-sup parameter of the mixed* $Q_n - Q_{n-2}$ *spectral element method.*

4 Numerical Results in the Plane

We report on results of numerical experiments in MATLAB for the positive definite reformulation of the mixed elasticity system with homogeneous Dirichlet boundary conditions, discretized with GLL spectral elements. The domain is the reference square $\Omega = \Omega_{ref}$, subdivided into $N = N_x \times N_y$ square spectral elements (subdomains). The reduced interface system with the Schur complement matrix (4) is solved by the preconditioned conjugate gradient algorithm (PCG) with the BDDC preconditioner (5), zero initial guess and stopping criterion $\|r_k\|_2/\|r_0\|_2 \leq 10^{-6}$, where r_k is the residual at the $k-$th iterate. The right-hand side is random and uniformly distributed.

Table 1 reports the iteration counts (it) and maximum eigenvalue (λ_{max}) of the BDDC preconditioned operator with only vertex primal constraints (columns labeled V), vertex and normal edge average constraints (columns labeled V+1E), vertex and all (two) edge average constraints per edge (columns labeled V+2E)

Table 1 2D elasticity, $Q_n - Q_{n-2}$ GLL SEM: iteration counts (it) and maximum eigenvalue (λ_{max}) of BDDC preconditioned operator with only vertex primal constraints (V), vertex and normal edge average constraints (V+1E), vertex and all (two) edge average constraints (V+2E), as a function of the polynomial degree n (*top*) and the number of spectral elements N (*bottom*). Compressible material with $\nu = 0.4$ (*left*) and almost incompressible material with $\nu = 0.499999$ (*right*). $(\kappa\widehat{S})$ = condition number of the unpreconditioned Schur complement

$N = 3\times 3$	$\nu = 0.4$							$\nu = 0.499999$						
		V		V+1E		V+2E			V		V+1E		V+2E	
n	$(\kappa\widehat{S})$	it	λ_{max}	it	λ_{max}	it	λ_{max}	$(\kappa\widehat{S})$	it	λ_{max}	it	λ_{max}	it	λ_{max}
3	9.87	8	3.52	6	1.68	4	1.17	4.7e+5	27	1.6e+5	6	1.75	5	1.24
4	14.34	8	3.55	7	1.90	6	1.40	5.3e+5	31	1.4e+5	8	2.88	6	1.54
5	19.88	9	4.53	8	2.28	6	1.53	5.7e+5	41	1.7e+5	8	2.50	7	1.77
6	25.63	9	4.62	8	2.38	7	1.75	6.1e+5	40	1.6e+5	9	3.86	7	2.04
7	31.51	10	5.37	8	2.66	7	1.85	6.3e+5	44	1.8e+5	10	3.87	8	2.07
8	37.51	10	5.47	9	2.75	8	2.02	6.5e+5	45	1.8e+5	10	4.60	8	2.31
9	43.60	10	6.10	9	2.97	8	2.10	6.7e+5	47	1.9e+5	10	4.53	8	2.35
10	49.79	10	6.23	9	3.08	8	2.35	6.8e+5	49	1.9e+5	10	5.16	8	2.53
$n = 3$														
N														
6×6	47.53	14	4.99	9	2.17	6	1.58	2.7e+6	107	3.1e+5	8	2.39	6	1.53
9×9	115.1	17	5.73	10	2.52	6	1.62	6.1e+6	149	3.8e+5	10	2.84	6	1.77
12×12	206.3	18	6.26	10	2.56	6	1.88	1.1e+7	206	4.1e+5	10	2.88	7	1.89
15×15	322.6	18	6.53	11	2.63	6	1.81	1.7e+7	261	4.3e+5	10	2.97	7	1.92

for two values of Poisson ratio, $\nu = 0.4$ (compressible material, left) and $\nu = 0.499999$ (almost incompressible material, right). We also report the condition number $\kappa(\widehat{S})$ of the unpreconditioned Schur complement $\widehat{S}$, that becomes increasingly ill-conditioned as ν tends to 1/2. The minimum BDDC eigenvalue λ_{min} is not reported because it is always very close to 1. The results appear to indicate that the rate of convergence of our BDDC algorithm depends only weakly on the polynomial degree n and is independent of the number of subdomains N. The convergence rate is also independent of the Poisson ratio ν for both BDDC algorithms with vertex and edge constraints, with a better performance for the richer choice with all edge average constraints for each edge. On the other hand, the convergence rate of the algorithm with only vertex primal constraints degenerates badly when the material becomes almost incompressible.

References

1. Clark R. Dohrmann. A preconditioner for substructuring based on constrained energy minimization. *SIAM J. Sci. Comput.*, 25(1):246–258, 2003
2. Jan Mandel, Clark R. Dohrmann, and Radek Tezaur. An algebraic theory for primal and dual substructuring methods by constraints. *Appl. Numer. Math.*, 54:167–193, 2005
3. Charbel Farhat, Michel Lesoinne, Patrick LeTallec, Kendall Pierson, and Daniel Rixen. FETI-DP: a dual-primal unified FETI method – part I. A faster alternative to the two-level FETI method. *Int. J. Numer. Meth. Eng.*, 50(7):1523–1544, 2001

4. Axel Klawonn and Olof B. Widlund. Dual-Primal FETI methods for linear elasticity. *Comm. Pure Appl. Math.*, 59:1523–1572, 2006
5. Jing Li and Olof B. Widlund. FETI–DP, BDDC, and Block Cholesky methods. *Int. J. Numer. Meth. Eng.*, 66(2):250–271, 2006
6. Luca F. Pavarino and Olof B. Widlund. Iterative substructuring methods for spectral element discretizations of elliptic systems. II. Mixed methods for linear elasticity and Stokes flow. *SIAM J. Numer. Anal.*, 37(2):375–402, 2000
7. Luca F. Pavarino and Olof B. Widlund. Balancing Neumann–Neumann methods for incompressible Stokes equations. *Comm. Pure Appl. Math.*, 55(3):302–335, 2002
8. Paulo Goldfeld, Luca F. Pavarino, and Olof B. Widlund. Balancing Neumann-Neumann preconditioners for mixed approximations of heterogeneous problems in linear elasticity. *Numer. Math.*, 95(2):283–324, 2003
9. Clark R. Dohrmann. A substructuring preconditioner for nearly incompressible elasticity problems. Technical Report SAND2004-5393, Sandia National Laboratories, Albuquerque, NM, 2004
10. Jing Li. A dual-primal FETI method for incompressible Stokes equations. *Numer. Math.*, 102:257–275, 2005
11. Jing Li and Olof B. Widlund. BDDC algorithms for incompressible Stokes equations. *SIAM J. Numer. Anal.*, 44(6):2432–2455, 2006
12. Lourenço Beirão da Veiga, Carlo Lovadina, and Luca F. Pavarino. Positive definite balancing Neumann–Neumann preconditioners for nearly incompressible elasticity. *Numer. Math.*, 104 (3):271–296, 2006
13. Clark R. Dohrmann and Olof B. Widlund. Hybrid domain decomposition algorithms for compressible and almost incompressible elasticity. *Int. J. Numer. Meth. Eng.*, 82:157–183, 2010
14. Clark R. Dohrmann and Olof B. Widlund. An overlapping Schwarz algorithm for almost incompressible elasticity. *SIAM J. Numer. Anal.*, 47(4):2897–2923, 2009
15. Luca F. Pavarino. BDDC and FETI-DP preconditioners for spectral element discretizations. *Comput. Meth. Appl. Mech. Eng.*, 196(8):1380–1388, 2007
16. Axel Klawonn, Luca F. Pavarino, and Oliver Rheinbach. Spectral element FETI-DP and BDDC preconditioners with multi-element subdomains. *Comput. Meth. Appl. Mech. Eng.*, 198:511–523, 2008
17. Franco Brezzi and Michel Fortin. *Mixed and Hybrid Finite Element Methods*. Springer, Berlin, 1991
18. Christine Bernardi and Yvon Maday. *Spectral Methods*. In: P. G. Ciarlet and J.-L. Lions, editors, Handbook of Numerical Analysis, Volume V: Techniques of Scientific Computing (Part 2). North-Holland, Amsterdam 1997
19. Axel Klawonn and Oliver Rheinbach. A parallel implementation of Dual-Primal FETI methods for three dimensional linear elasticity using a transformation of basis. *SIAM J. Sci. Comput.*, 28(5):1886–1906, 2006
20. Andrea Toselli and Olof B. Widlund. *Domain Decomposition Methods – Algorithms and Theory*. Springer, Berlin, 2005
21. Luca F. Pavarino, Olof B. Widlund, and S. Zampini. BDDC preconditioners for spectral element discretizations of almost incompressible elasticity in three dimensions. To appear in *SIAM J. Sci. Comput.*

A Two-Dimensional DG-SEM Approach to Investigate Resonance Frequencies and Sound Radiation of Woodwind Instruments

Andreas Richter and Jörg Stiller

Abstract Investigations of musical instruments typically carried out in the frequency domain. In contrast, numerical investigations in the time domain offer several advantages. Physical effects and single travelling waves can both be observed directly. This allows studying how connections, tone holes and bore perturbations influence the wave propagation and act as sound sources. In contrast to methods, which are formulated in the frequency domain, time domain investigations also enable the examination of transient effects. However, these are accompanied by the disadvantage that acoustic waves have to be tracked over a long period of time. When using low-order methods, numerical dissipation and dispersion errors can can have highly distortive effects on the results. In order to overcome these issues we use a high-order discontinuous Galerkin formulation. Extrapolation boundary conditions in conjunction with a slope limiting procedure provide stable, compact and non-reflecting boundary conditions. This report presents the results of numerical investigations of woodwind instruments using the examples of the recorder and the bassoon.

1 Introduction

The interaction between the geometry and the timbre of the played woodwind instrument can easily be studied by decoupling the resonator from the excitation mechanism. Such investigations are typical research methods in the frequency domain. Our objective is to overcome certain limitations of frequency based approaches by working in the time domain. This offers several advantages: Acoustical waves

A. Richter (✉)
Technische Universität Dresden, Institute for Aerospace Engineering, 01062 Dresden
e-mail: andreas.richter4@tu-dresden.de

J. Stiller
Technische Universität Dresden, Institute of Fluid Mechanics, 01062 Dresden
e-mail: joerg.stiller@tu-dresden.de

J.S. Hesthaven and E.M. Rønquist (eds.), *Spectral and High Order Methods for Partial Differential Equations*, Lecture Notes in Computational Science and Engineering 76, DOI 10.1007/978-3-642-15337-2_47, © Springer-Verlag Berlin Heidelberg 2011

can be tracked directly, which helps gaining a better insight into the instrument. Furthermore, the approach allows studying effects basing on the superimposed and non-homogeneous mean flow as well as transient effects.

Impulse reflectometry is a common tool in the experimental investigation of acoustical systems [1, 3]. For that purpose, the acoustic system is excited by a short impulse, e.g., the Dirac impulse. All frequencies of the acoustic system are being activated evenly, but frequencies that do not correlate with the system are being damped rapidly. The remaining frequencies are resonance frequencies of the acoustic system and can be measured by simple microphone techniques. The perfomance of numerical investigations requires the tracking of acoustical waves over a long period of time. Due to numerical dissipation and dispersion errors traditional low-order schemes fail here. This effect is illustrated by solving the inviscid propagation of an acoustical wave in a periodic domain with an amplitude of 711 Pa. The configuration is given in Fig. 1a. Figure 1b shows the amplitude error after a wave run of nearly 10 m. The results base on a second and third order Finite Volume solver (Fluent^{TM}) as well as on a discontinuous Galerkin high-order Spectral Element solver. The third order Finite Volume scheme was achieved by blending a central differencing scheme and a second order upwind scheme, which gives not a proper third order convergence. Even though the order of the DG-SEM can be arbitrary, the third order scheme ($p = 2$) demonstrates the superiority of the method. Table 1 shows the amplitude error as a function of the non-dimensional wave number kh.

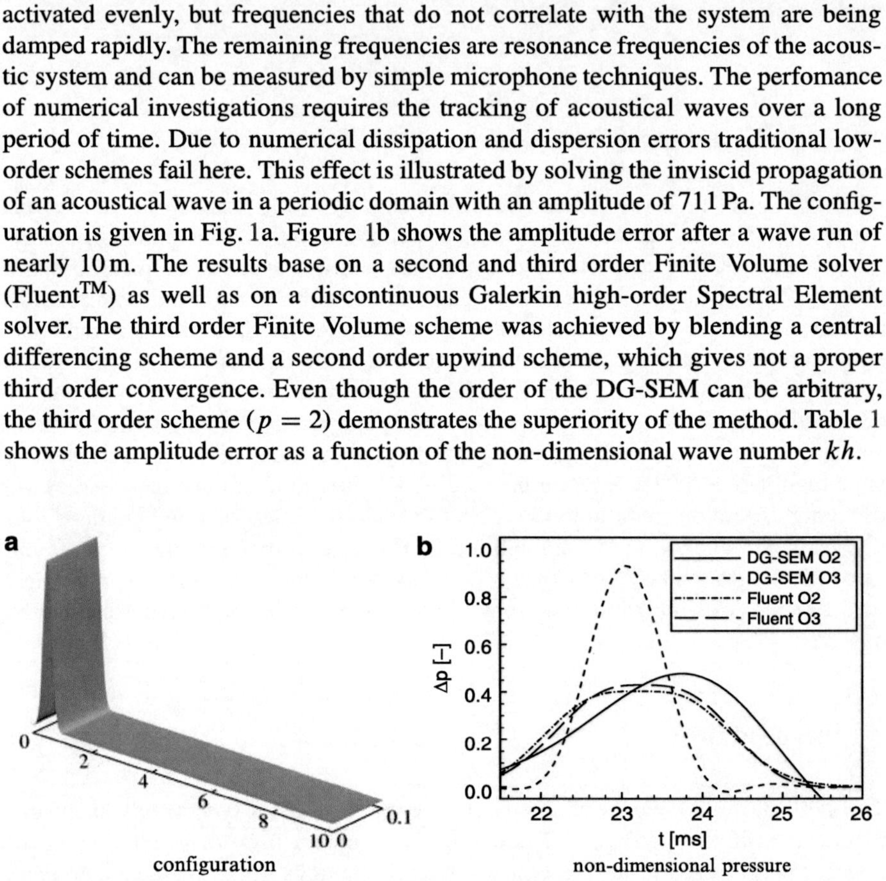

Fig. 1 Dissipation error estimation for a linear wave propagation; O2: 2nd order and O3: 3rd order scheme

Table 1 Error in the pressure amplitude as a function of the non-dimensional wave number kh; DG2/3: DG-SEM $O(2)$ and $O(3)$, Fl2/3: Fluent $O(2)$ and $O(3)$, respectively

kh	ε^A_{DG2}	ε^A_{DG3}	ε^A_{Fl2}	ε^A_{Fl3}
0.10	−52.66%	−6.85%	−59.62%	−57.01%
0.04	−23.96%	0.36%	−28.17%	−25.73%
0.02	−6.80%	0.02%	−11.37%	−9.67%
0.01	0.12%	3.10^{-4}%	−3.97%	−3.18%

2 Discontinuous Galerkin Method for the Euler Equations

2.1 Conservation Equations

The two-dimensional compressible, unsteady Euler equations are considered in the conservative form

$$\partial_t \mathbf{U} + \nabla \cdot \vec{\mathbf{F}} = 0 \tag{1}$$

where $\mathbf{U} = (\rho, \rho v_x, \rho v_y, \rho e)^{\mathrm{T}}$ are the conservative variables, $\vec{\mathbf{F}} = (\mathbf{F}_x, \mathbf{F}_y)^{\mathrm{T}}$ with $\mathbf{F}_x = (\rho v_x, \rho v_x^2 + p, \rho v_x v_y, \rho e_t v_x + p v_x)^{\mathrm{T}}$, $\mathbf{F}_y = (\rho v_y, \rho v_x v_y, \rho v_y^2 + p, \rho e_t v_y + p v_y)^{\mathrm{T}}$ the convective flux, respectively, the density ρ, the velocity $\vec{v} = (v_x, v_y)^{\mathrm{T}}$, the pressure p and the total energy per unit mass $e_t = \frac{1}{\gamma-1}\frac{p}{\rho} + \frac{1}{2}(v_x^2 + v_y^2)$.

2.2 Numerical Scheme

For the spatial discretization the computational domain Ω is partitioned into a set of non-overlapping elements $\{\Omega_e\}$. Conforming or non-conforming elements of arbitrary shape can be used. In the following, we will specifically research quadrilateral elements with possibly curved sides. Given a test function ϕ_e the weak form of 1 is

$$\int_{\Omega_e} \phi_e \, \partial_t \mathbf{U}_e \mathrm{d}\Omega_e = \int_{\Omega_e} \nabla\phi_e \cdot \vec{\mathbf{F}}(\mathbf{U}_e) \mathrm{d}\,\Omega_e - \int_{\Gamma_e} \phi_e \mathbf{H}(\mathbf{U}_e^{\pm}, \vec{n}) \mathrm{d}\Gamma_e \,. \tag{2}$$

This equation needs to be satisfied for any test function from the space of trial solutions spanned by the basis in Ω_e.

Special attention has to be paid to the numerical flux $\mathbf{H}$ since it has to be consistent and keep the transport properties of the system. We use the Roe flux [4,6]

$$\mathbf{H}(\mathbf{U}^{\pm}, \vec{n}) = \frac{1}{2}\Big[\mathbf{F}_n(\mathbf{U}^-) + \mathbf{F}_n(\mathbf{U}^+) - |\mathbf{A}_n(\bar{\mathbf{U}})|(\mathbf{U}^+ - \mathbf{U}^-)\Big]$$

where $\mathbf{A}_n(\mathbf{U}) = \mathbf{F}_n'(\mathbf{U})$ is the Jacobian of the normal convective flux $\mathbf{F}_n = \vec{\mathbf{F}} \cdot \vec{n}$ and $\bar{\mathbf{U}}$ the Roe average of $\mathbf{U}^-$ and $\mathbf{U}^+$.

In the spectral element framework functions $\varphi(\vec{x}, t)$ on element Ω_e are transformed to a standard element Ω^s and approximated using a set of polynomial base functions. Various choices are possible for the basis [2, 5]. Hereby, we use a tensor product basis of one-dimensional Lagrange polynomials $\{\pi_i\}$ based on the Gauss–Lobatto–Legendre (GLL) points to approximate φ

$$\varphi(\vec{\xi}, t) = \sum_{p,q=0}^{P} \varphi_{pq}(t) \pi_i(\xi) \pi_j(\eta),$$

with the Lagrange interpolation property $\pi_j(\xi_i) = \delta_{ij}$. For reasons of simplification, the polynomial degree and the point distributions were assumed to be constant although they can vary over the elements and may also be different with respect to the ξ and η directions. The weak formulation (2) is transformed into semi-discrete equations, with the differentiation matrix $d_{ij} = \pi'_j(\xi_i)$, the weights $\{w_k\}$ and a quadrature scheme that bases on the GLL points (ξ_k, η_l). These equations can be assumed as

$$\begin{aligned}\dot{\mathbf{U}}_{ij,e} &= \frac{1}{J_{ij,e}w_i}\Big(\sum_{k=0}^{P} w_k d_{ki}\mathbf{F}_{\xi,kj,e} - \pi_i(-1)J^W_{j,e}\mathbf{H}^W_{j,e} - \pi_i(+1)J^E_{j,e}\mathbf{H}^E_{j,e}\Big)\\ &+ \frac{1}{J_{ij,e}w_j}\Big(\sum_{l=0}^{P} w_l d_{lj}\mathbf{F}_{\eta,il,e} - \pi_j(-1)J^S_{i,e}\mathbf{H}^S_{i,e} - \pi_j(+1)J^N_{i,e}\mathbf{H}^N_{i,e}\Big)\end{aligned} \tag{3}$$

with $(\mathbf{F}_\xi, \mathbf{F}_\eta)^{\mathrm{T}} = J\,\mathbf{J}^{-\mathrm{T}}(\mathbf{F}_x, \mathbf{F}_y)^{\mathrm{T}}$, the Jacobian matrix $\mathbf{J}^{-1} = \nabla_{\vec{\xi}}\,\vec{x}$ and the Jacobian determinant J. The indices W, E, N, and S indicate the edges of Ω^s, e.g., $W \mathrel{\hat{=}} (\xi = -1, \eta)$ and J^W is the corresponding boundary Jacobian determinant.

The semi-discrete equations (3) are integrated in time using a 3-stage TVD Runge-Kutta method following Shu and Osher [9] with embedded slope limiting and boundary correction [8]. For the implementation of the full Navier-Stokes equations see [7].

3 The Influence of the Vocal Tract on the Recorder

3.1 Problem Description

While playing the recorder, a player driven volume flow leaves the wind channel and forms a jet. This jet strikes against the sharp edge, the so-called labium. Depending on the resonator's characteristic this jet is disturbed, forms periodical vortexes and excites the air column to oscillate. The vortex shedding is illustrated in Fig. 2. The playing condition for the note D6 (1,175 Hz) is computed by solving the two-dimensional, compressible Navier–Stokes equations

$$\partial_t \mathbf{U} + \nabla \cdot \vec{\mathbf{F}} = \nabla \cdot \vec{\mathbf{D}} \tag{4}$$

where $\vec{\mathbf{D}}$ is the diffusive flux, respectively. While playing the instrument, the player can modify the frequency and to some extent the instrument's timbre by modifying the blowing pressure and also his vocal tract. This mechanism is not yet fully understood. Numerical investigations have also showed the shedding of smaller vortexes at the exit of the wind channel. These vortexes strongly depend on the defined properties at the inflow of the channel.

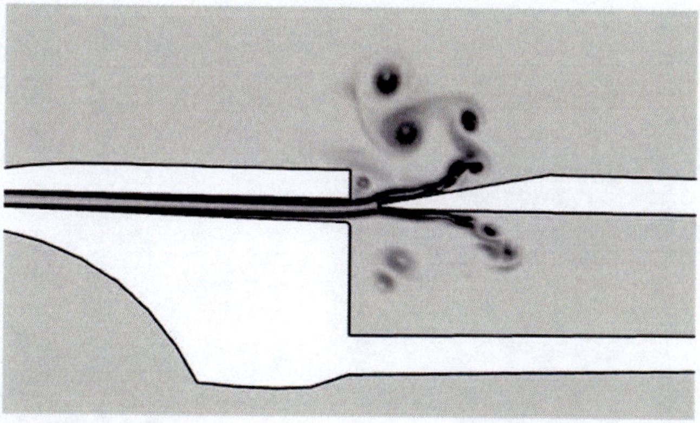

Fig. 2 Vortex shedding at the recorder labium, note D6 (1,175 Hz). 20,546 elements, $p = 3$. Vorticity magnitude from $0 \ldots [75, 000]1s$

3.2 *Influence of the Vocal Tract*

To investigate the acoustical resonator of the instrument we performed the impulse reflectometry numerically and excited the system with a half sine impulse with the wave length of $\lambda = 0.05\,\text{m}$. The pressure amplitude is 7.1 Pa, the reference temperature is 300 K. This impulse excites frequencies up to 10 kHz in a unique manner. The cutoff frequency is approximately 14 kHz. The impulse is released inside the recorder at the beginning of the main bore. Viscous effects don't influence the resonators resonance frequencies significantly. Hence, we have solved the Euler equations for this purpose.

Two vocal tract geometries have been defined. The diameter of these geometries is constant (1 cm) and corresponds to the human vocal tract, which varies in length. The first length is approximately the same as the resonator length, the second one is half as long. Fig. 3 shows the temporal development of the pressure field of vocal tract 1. After a short time acoustic waves are reflected at the instrument's tone holes and openings, they superpose and form a complex pressure field. Particularly interesting is the occurrence of many oscillations with similar amplitudes inside the vocal tract.

Figure 4 compares the estimated resonance frequencies and the reference model, which neglects the influence of the vocal tract. This is achieved by defining a non-reflecting boundary condition [8] at the inflow of the wind channel, which corresponds to a sound radiation into an infinite tube. The measuring position is situated inside the instrument and near the labium. The volumes of the vocal tract and the instrument are only weakly coupled by the wind channel. Although both figures show a significant influence of the vocal tract on the resonance frequencies, both the measured amplitudes of the resonator's fundamental frequency and the vocal tract are nearly equal. The fundamental frequency of vocal tract 2 is twice the instrument's fundamental frequency. Despite this, the amplitude of the instrument's

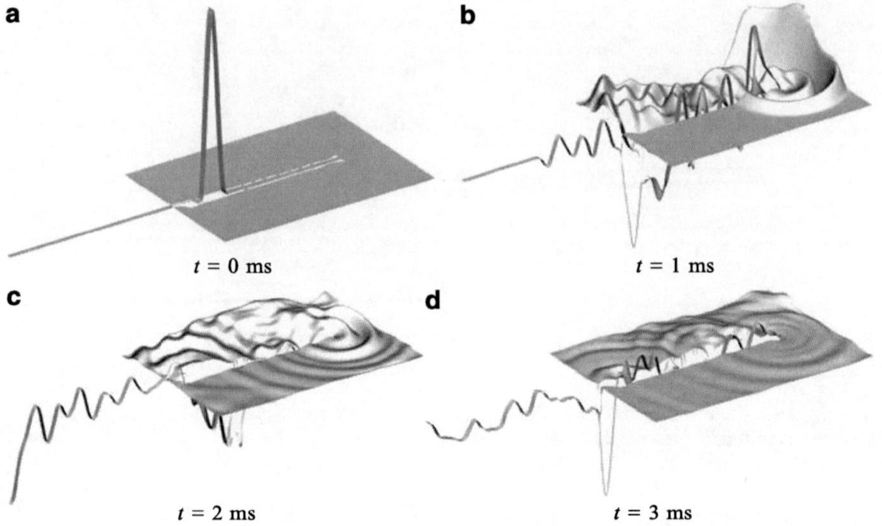

Fig. 3 Time history of the static pressure field of the recorder, note D6 (1,175 Hz). Pressure amplitude from $-7.1 \ldots 7.1$ Pa. 5,718 elements, $p = 3$

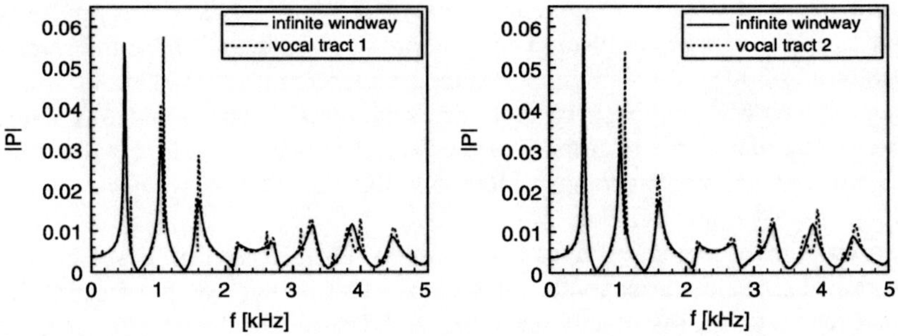

Fig. 4 Influence of the vocal tract on the resonance frequencies of the recorder, note C5 (523 Hz)

fundamental frequency is also increased. Both figures show additional resonance frequencies at higher modes. This demonstrates how strong the player can modify the instrument's timbre by producing additional resonance frequencies. Also, non-linear effects between the two resonators can amplify this effect.

4 Sound Radiation of the Bassoon

Unlike recorders, bassoons have a conical bore, a wider range of playable notes, a more distinct sound radiation characteristic and a timbre, which features a very rich overtone spectrum. In order to investigate the sound radiation pattern of the

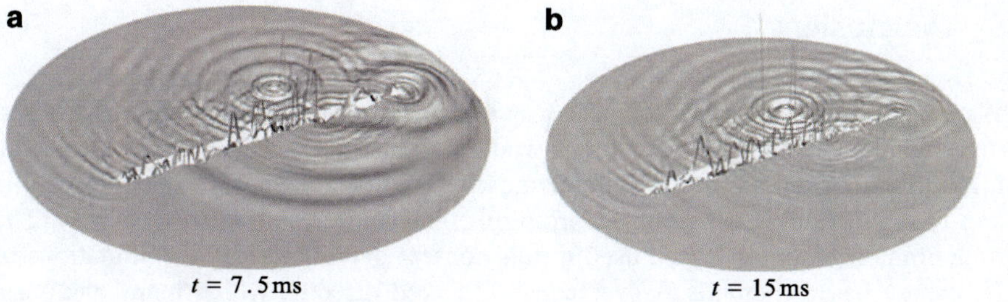

Fig. 5 Time history of the bassoon's static pressure field, note C2 (65 Hz). Pressure amplitude from −7.1 . . . 7.1 Pa. 104,418 elements, $p = 3$

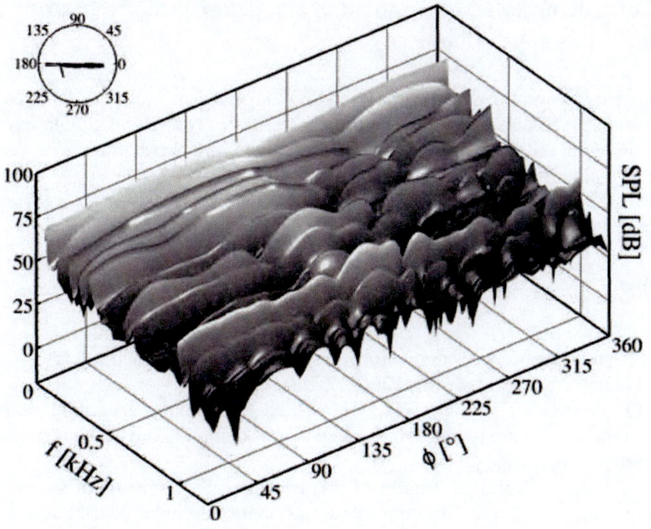

Fig. 6 Radiation characteristic of the bassoon for the note C2 (65 Hz)

instrument we have also performed the impulse reflectometry. Figure 5 illustrates an example of the temporal development of the basson's pressure field.

To estimate the radiation characteristics we measured the pressure signal at equidistant spaced points at the outer boundary of the circular computational domain. In Fig. 6 the radiation behaviour for the note C2 (65 Hz) is given. While the fundamental note nearly constantly radiates around the instrument, the following overtones show a distinct radiation characteristic. Interestingly, the radiation characteristic varies between the overtones. Additional calculations showed that the radiation pattern of one overtone also varies from one fingering to another. Consequently, it is necessary to investigate all fingerings and overtones in order to consider and possibly modify the instrument's radiation characteristic.

5 Conclusions

We have performed the widely-used impulse reflectometry numerically in order to investigate the acoustical behavior of woodwind instruments. Therefore, we used a high-order discontinuous Galerkin formulation to solve the non-linear, compressible and unsteady Euler equations according to the required numerical accuracy. The time domain based approach used in this context overcomes certain limitations of frequency based methods. Waves can be tracked directly, which helps gaining a better understanding of the resonator's behavior. It also allows transient phenomena and a non-homogeneous mean flow to be investigated.

The investigation of both the influence of the player's vocal tract on the resonance frequencies of a recorder and the radiation characteristics of a bassoon proved the practicability of this approach.

Acknowledgements This work was supported by the German National Science Foundation (Deutsche Forschungsgemeinschaft, DFG) within the project "Numerical simulation of the sound spectrum and the sound radiation in and around a recorder (Numerische Simulation des Klangspektrums und der Schallausbreitung in und um eine Blockflöte)".

References

1. Backus, J.: Input impedance curves for the reed woodwind instruments. Journal of the Acoustical Society of America **56**(4), 1266–1279 (1974)
2. Deville, M.O., Fischer, P.F., Mund, E.H.: High-Order Methods for Incompressible Fluid Flow. No. 9 in Cambridge Monographs on Applied and Computational Mathematics. Cambridge University Press, Cambridge (2002)
3. Dickens, P., Smith, J., Wolfe, J.: Improved precision in measurements of acoustic impedance spectra using resonance-free calibration loads and controlled error distribution. Journal of the Acoustical Society of America **121**(3), 1471–1481 (2007). DOI 10.1121/1.2434764. URL http://link.aip.org/link/?JAS/121/1471/1
4. Hirsch, C.: Numerical Computation of Internal and External Flows Vol. 1 & 2. Wiley, NY (1990)
5. Karniadakis, G., Sherwin, S.: Spectral/hp Element Methods for Computational Fluid Dynamics, Second Edition. Oxford University Press, New York (2005)
6. Leveque, R.: Finite Volume Methods for Hyperbolic Problems. Cambridge University Press, Cambridge (2002)
7. Richter, A., Brußies, E., Stiller, J., Grundmann, R.: Stabilized high-order DGM for aeroacoustic investigations. CFD Review 2008, Springer, Berlin (2009) (to appear)
8. Richter, A., Stiller, J., Grundmann, R.: Stabilized discontinuous Galerkin methods for flow-sound interaction. Journal of Computational Acoustics **15**(1), 123–143 (2007). DOI 10.1142/S0218396X0700324X
9. Shu, C.W., Osher, S.: Efficient implementation of essentially non-oscillatory shock-capturing schemes. Journal of Computational Physics **77**(2), 439–471 (1988)

Spectral Properties of Discontinuous Galerkin Space Operators on Curved Meshes

Thomas Toulorge and Wim Desmet

Abstract Grids with curved elements are necessary to fully benefit from the high order of accuracy provided by the Discontinuous Galerkin (DG) method, when dealing with complex geometries. We study the relation between the quadratic shape of simplex elements and the spectral properties of the semi-discrete space operators, with emphasis on consequences for the maximum allowable timestep for stability in Runge–Kutta DG methods. A strong influence of element curvature on the eigenvalue spectrum is put in evidence, but no explicit relation could be found to describe the evolution of the spectral radius in function of geometric properties of elements. Furthermore, we show that a correct estimation of stability bounds cannot be obtained by considerations on the norm of integration matrices involved in the DG Method.

1 Introduction

Among the numerous numerical methods used to solve hyperbolic partial differential equations on unstructured grids, the Discontinuous Galerkin (DG) Method is receiving increasing attention in different fields like Computational ElectroMagnetics, Computational Fluid Dynamic or Computational Aeroacoustics. Its ability to obtain solutions with arbitrarily high order of accuracy is a particularly interesting feature. Other advantages over concurrent high-order methods are the straightforward formulation of boundary conditions, as well as the compactness of the scheme, that allows efficient parallel computation.

Nevertheless, the benefits of using higher-order methods with a lower grid density are limited if coarse grids made up of straight elements (i.e., elements with straight edges in 2D or flat faces in 3D) fail to correctly discretize curved boundaries.

T. Toulorge (✉) and W. Desmet
Department of Mechanical Engineering, K.U. Leuven, Celestijnenlaan 300b, 3001 Heverlee, Belgium
e-mail: thomas.toulorge@mech.kuleuven.be, wim.desmet@mech.kuleuven.be

J.S. Hesthaven and E.M. Rønquist (eds.), *Spectral and High Order Methods for Partial Differential Equations*, Lecture Notes in Computational Science and Engineering 76, DOI 10.1007/978-3-642-15337-2_48, © Springer-Verlag Berlin Heidelberg 2011

In the framework of the non-linear Euler equations, the necessity of a higher-order treatment of curved wall boundaries was put in evidence by Bassy and Rebay [1] and is now generally accepted [2]. In the context of linear aeroacoustic propagation, the use of higher-order geometry description and its positive impact on accuracy was reported [3,4].

However, curved elements used for higher-order geometry representation influence the conditioning of DG space operators. When explicit methods such as Runge–Kutta (RK) schemes are used for time integration, this may lead to a more restrictive CFL condition: the maximum timestep allowed to maintain the stability of the simulation is then lower, and the global computation cost is increased. Dissipation and dispersion properties of the scheme are also locally affected. The aim of the work presented in this paper is to study the relation between geometric properties of curved simplex elements and the spectral properties of DG operators, in order to draw conclusions on stability and time-stepping in Runge–Kutta Discontinuous Galerkin (RKDG) methods.

2 Method

2.1 Discontinuous Galerkin Method

As a model for hyperbolic conservation laws, that are the natural target of DG methods, we consider the scalar advection equation over a domain with periodic boundary conditions:

$$\frac{\partial q}{\partial t} + \frac{\partial a_r q}{\partial x_r} = 0 \tag{1}$$

where q is the unknown, t is the time, x_r is the r-th space coordinate, and a_r is the r-th component of the constant advection vector $\mathbf{a}$. Einstein's summation convention is used over the r index.

For each element Ω resulting from the partition of the computational domain, a basis $\mathscr{B} = \{\varphi_j,\ j = 1 \ldots N_p\}$ is defined, in which the components φ_j are polynomials of order p supported on Ω, with $N_p = \frac{(p+1)(p+2)}{2}$ for triangular elements. An approximation q^{Ω} of q on Ω is obtained by a projection on this basis:

$$q^{\Omega} = \sum_{j=1}^{N_p} q_j^{\Omega} \varphi_j$$

Applying the Discontinuous Galerkin procedure to (1) results in:

$$\mathbf{M}^{\Omega} \frac{\partial q^{\Omega}}{\partial t} - \mathbf{K}_{\mathbf{r}}^{\Omega} a_r q^{\Omega} + \sum_i \mathbf{M}^{\partial \Omega_i} F^{\partial \Omega} = 0 \tag{2}$$

with:

$$\begin{aligned}
\mathbf{M}_{kj}^{\Omega} &= \int_{\Delta} \varphi_k \varphi_j \left| J^{\Omega} \right| d\Delta \\
\left(\mathbf{K}_{\mathbf{r}}^{\Omega}\right)_{kj} &= \int_{\Delta} \left(J^{\Omega}\right)_{sr}^{-1} \frac{\partial \varphi_k}{\partial \xi_s} \varphi_j \left| J^{\Omega} \right| d\Delta \\
\mathbf{M}_{kj}^{\partial\Omega_i} &= \int_{\partial\Delta_i} \varphi_k \varphi_j \left| J^{\partial\Omega_i} \right| d\partial\Delta_i
\end{aligned} \tag{3}$$

where $F^{\partial\Omega}$ is the Lax–Friedrichs approximation of the Riemann flux computed on the element boundary $\partial\Omega$. In (3), each element Ω is mapped onto a unique reference element Δ by a function $\mathscr{M}^{\Omega}$ with Jacobian matrix J^{Ω}. Likewise, each element boundary $\partial\Omega_i$ is mapped onto a unique edge of Δ by a function $\mathscr{M}^{\partial\Omega_i}$ with Jacobian matrix $J^{\partial\Omega_i}$. The basis $\mathscr{B}$ is then expressed in Δ with reference coordinates $\boldsymbol{\xi}$.

2.2 Stability Analysis

To deal with element curvature, quadratic functions of $\boldsymbol{\xi}$ are considered for the mappings $\mathscr{M}^{\Omega}$. In addition to the vertices of the element, control points on the edges of the element are used to define $\mathscr{M}^{\Omega}$, as in the classical Finite Element Methods. Elements can then be arranged to form periodic patterns, as shown in Fig. 1.

In 1D, a non-dimensional parameter γ is used to define the location of the middle control point on the segment (see Fig. 1a). Although the element cannot be geometrically curved, the quadratic mapping allows us to mimic the effect of curvature. In 2D, two non-dimensional parameters are needed to define the curvature of each edge (see Fig. 1b). Given the periodicity constrains, six parameters characterize the curvature of the whole pattern.

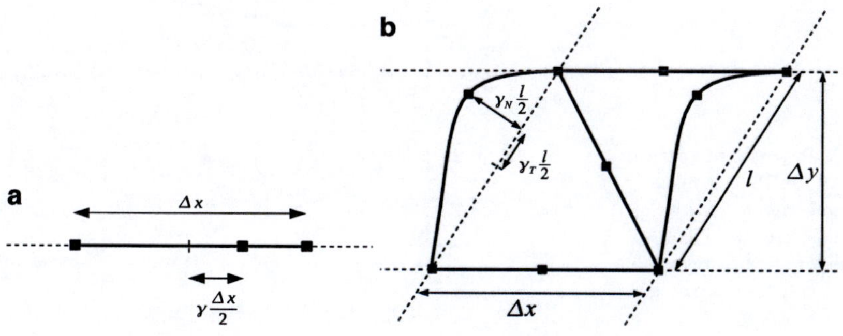

Fig. 1 Periodic patterns of elements used for the stability analysis in 1D (**a**) and 2D (**b**)

The global DG space operator $\mathbf{L}$ can then be assembled directly by considering a grid made up of repeated patterns, explicitly imposing periodic boundary conditions at the boundaries of the computational domain, and formulating the semi-discrete scheme:

$$\frac{\partial \tilde{q}}{\partial t} = \mathbf{L}\,\tilde{q}$$

where $\tilde{q}$ represents the solution on the whole computational domain. Alternatively, a Von Neumann-like procedure can be followed by looking for harmonic solutions, for which the periodicity of patterns can be exploited to diagonalize $\mathbf{L}$ with blocks $\mathbf{L_p}$, yielding:

$$\frac{\partial \hat{q}}{\partial t} = \mathbf{L_p}\,\hat{q}$$

where $\hat{q}$ represents the complex amplitude of the solution on a pattern, and the operator $\mathbf{L_p}$ depends on the position of that pattern in the global grid.

To evaluate the stability of the RKDG method with a given timestep Δt, we compute the eigenvalues $\lambda(\mathbf{L})$ and compare the spectrum $\lambda \cdot \Delta t$ to the stability region of the RK scheme, as shown in Fig. 2. Although the presence of the whole spectrum inside the absolute stability region is not a sufficient condition for the stability of the fully discrete scheme, it provides an excellent guideline for the choice of Δt [5].

3 Results

In this section, we study the dependence of the eigenvalue spectrum $\lambda(\mathbf{L})$ on the curvature of the elements. For this purpose, the spectral radius $\rho(\mathbf{L}) = \max|\lambda(\mathbf{L})|$ is the main quantity of interest.

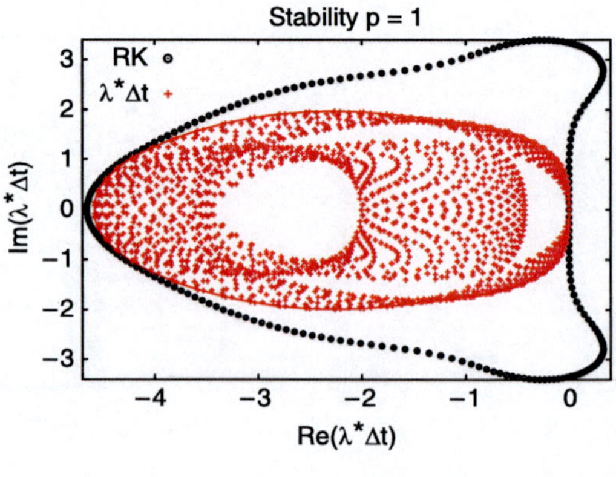

Fig. 2 Example of a stability plot for the 2D DG space operator at $p = 1$ with straight elements. The RK stability region corresponds to Carpenter's low-storage (4,5)-RK scheme

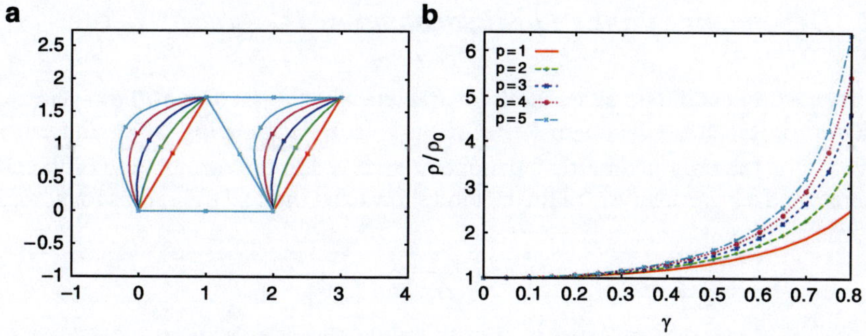

Fig. 3 Pattern shape (**a**) and normalized spectral radius (**b**) for $p = 1$ to $p = 5$ for a normal deformation of one edge in a 2D equilateral pattern

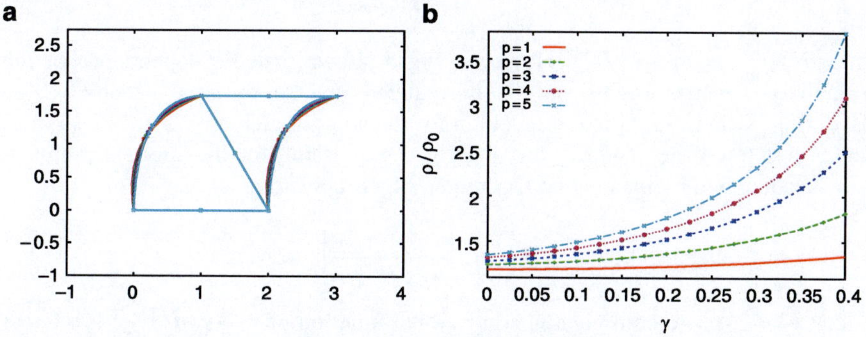

Fig. 4 Pattern shape (**a**) and normalized spectral radius (**b**) for $p = 1$ to $p = 5$ for a tangential deformation of one (already curved) edge in a 2D equilateral pattern

3.1 Qualitative Results in 2D

Figure 3 illustrates the effect of element curvature in 2D. As seen in Fig. 3a, an edge of a 2D pattern made up of equilateral triangles is deformed in the direction normal to the straight edge. The normalized spectral radius (i.e., $\frac{\rho}{\rho_0}$ where ρ_0 is the spectral radius for the straight element), plotted in Fig. 3b up to order $p = 5$, shows that the conditioning of **L** is severely influenced by element curvature. This influence become stronger when the order is increased.

For a deformation in the tangential direction, the impact seems to be milder, but significant, although the shape of the pattern undergoes little change (see Fig. 4).

These qualitative results show that element curvature can lead to a strong reduction of the timestep, above all at high order. Even little modifications in the global pattern shape can affect the conditioning of the operator in a significant manner, which suggests that local geometric properties of the element govern the effect.

3.2 Dependence on the Local Jacobian in 1D

An interesting candidate as governing parameter for the scaling of the eigenvalue spectrum is the local Jacobian $\left|J^{\Omega}\right|$ of the quadratic mapping $\mathcal{M}^{\Omega}$. Indeed, in 1D, $\left|J^{\Omega}\right|$ is the only geometrical parameter involved in the formulation of the DG scheme, and a basic analysis of the method shows that the spectral radius for straight elements scales as:

$$\rho_0(\mathbf{L}) \sim \frac{1}{\Delta x} \sim \frac{1}{\left|J^{\Omega}\right|}$$

One can thus conjecture that the spectral radius for curved elements scales as:

$$\rho(\mathbf{L}) \sim \frac{1}{\left|J^{\Omega}\right|_{\min}}$$

with $\left|J^{\Omega}\right|_{\min} = \min_{\Omega}\left|J^{\Omega}\right|$. However, Fig. 5 shows that the dependence of the normalized spectral radius on the normalized inverse minimal Jacobian (i.e., $|J_0^{\Omega}|/|J^{\Omega}|_{\min}$ where $\left|J_0^{\Omega}\right|$ is the Jacobian for the straight element) is not linear, particularly at low order. Indeed, the analytical expression for the spectral radius at $p = 1$, calculated by means of a Computer Algebra System, is:

$$\rho(\mathbf{L}) = \left|\frac{9}{4 \cdot \gamma^2 - 3}\right|$$

For $p = 2$, ρ is a complicated non-rational function of γ. As $\left|J^{\Omega}\right|_{\min}$ is a linear function of γ, there cannot be an simple dependence of ρ on $\left|J^{\Omega}\right|_{\min}$.

An interesting aspect of the problem is related to the value γ_{inf} of γ for which the spectral radius becomes infinitely large. $\gamma = 0.5$ is the value beyond which $\left|J^{\Omega}\right|_{\min} \leq 0$, the element then degenerates and the DG method cannot be used anymore. However, we measured $\gamma_{inf} > 0.5$, at least up to order $p = 10$. Moreover, γ_{inf} seems to decrease and come closer to 0.5 when the order is increased.

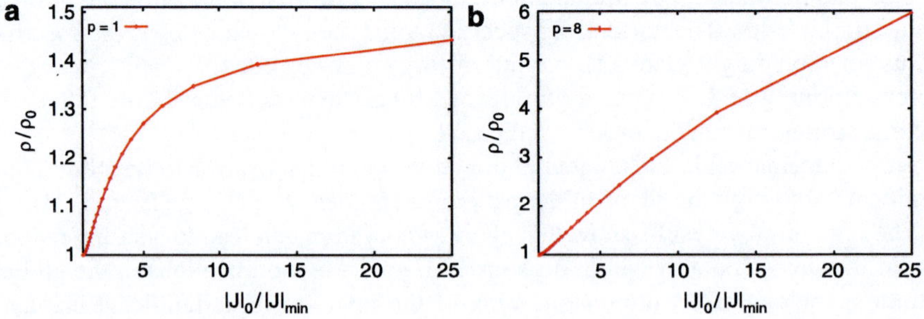

Fig. 5 Dependence of the normalized spectral radius on the normalized inverse minimal Jacobian for $p = 1$ **(a)** and $p = 8$ **(b)** in 1D

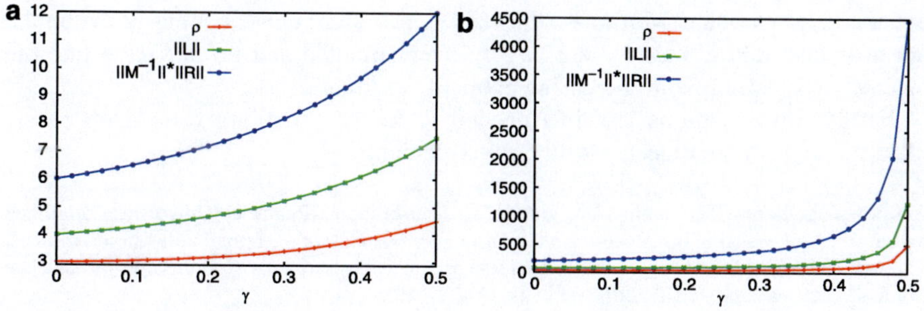

Fig. 6 Evolution of the spectral radius and quantities based on matrix 1-norms for $p = 1$ (**a**) and $p = 8$ (**b**) in 1D

3.3 Estimation Based on Integration Matrices in 1D

As the eigenvalue spectrum does not seem to scale with geometric properties in a simple manner, one can wonder whether the spectral radius can be estimated by simple considerations on the integration matrices $\mathbf{M}^{\Omega}$, $\mathbf{K}^{\Omega}$ and $\mathbf{M}^{\partial\Omega_i}$. In 1D, the geometric properties of the element are only involved through the mass matrix $\mathbf{M}^{\Omega}$. Let the operator $\mathbf{L}$ be decomposed as:

$$\mathbf{L}(\gamma) = \mathbf{M}^{-1}(\gamma)\,\mathbf{R}$$

where $\mathbf{M}$ is the block-diagonal matrix with blocks $\mathbf{M}^{\Omega}$, $\mathbf{M}^{\Omega}$ being the same for all elements. The following inequalities hold for any induced norm:

$$\rho(\mathbf{L}) \leq \|\mathbf{L}\| \leq \left\|\mathbf{M}^{-1}\right\| \|\mathbf{R}\|$$

Figure 6 shows that $\left\|\mathbf{M}^{-1}\right\|_1 \|\mathbf{R}\|_1$ is too large and grows faster than ρ with increasing γ, so that no estimation based on $\left\|\mathbf{M}^{-1}\right\|_1$ would be useful. Figure 6 indicates that even $\|\mathbf{L}\|_1$ is a bad approximation for ρ. This is due to the non-normality of $\mathbf{L}$, which gets stronger when the deformation and the order p are increased. Similar results are obtained with the infinity-norm and the 2-norm.

4 Conclusions

In this paper, the conditioning of DG space operators for quadratic elements has been studied, with the aim of relating the geometrical properties of curved elements and the maximum allowable timestep in RKDG methods.

The strong influence of element curvature on the scaling of the eigenvalue spectrum, specially at high order, has been put in evidence. However, 1D investigations have not succeeded in explaining the relation between the scaling of the eigenvalues and the local Jacobian. Moreover, they have shown that upper bounds based on the norm of integration matrices are largely overestimated and do not scale like the eigenvalue spectrum with increasing element deformation.

Further investigations are thus needed to find a satisfying estimation of the maximum allowable timestep with curved elements.

Acknowledgements The authors are grateful to Jean-François Remacle (Université Catholique de Louvain) for valuable discussions on the Discontinuous Galerkin Method with curved meshes. Thomas Toulorge acknowledges the financial support of the European Commission through the Marie–Curie Research and Training network "AETHER".

References

1. Bassi F and Rebay S (1997) High-order Accurate Discontinuous Finite Element Solution of the 2D Euler Equations. J Comput Phys 138:251–285
2. Krivodonova L and Berger M (2006) High-order Accurate Implementation of Solid Wall Boundary Conditions in Curved Geometries. J Comput Phys 211:492–512
3. Atkins HL (1997) Continued Development of the Discontinuous Galerkin Method for Computational Aeroacoustic Applications. AIAA Paper 97–1581
4. Toulorge T and Desmet W (2009) High-order Boundary Treatments for the Discontinuous Galerkin Method Applied to Aeroacoustic Propagation. 15th AIAA/CEAS Aeroacoustics Conference, Miami (FL), USA, May 11–13, 2009
5. Hesthaven JS and Warburton T (2008) Nodal Discontinuous Galerkin Methods: Algorithms, Analysis, and Applications. Volume 54, Springer Texts in Applied Mathematics, Springer, New York, p. 95

Post-Processing of Marginally Resolved Spectral Element Data

Carl Erik Wasberg

Abstract When derivatives of a spectral element solution are used in a different context, such as visualization or in calculations with a different numerical method, the discontinuity of the derivatives at the element interfaces is a potential problem. Asymptotically, the jumps in the derivatives decay spectrally fast, but it is not always possible or efficient use of computational resources to repeat the spectral element calculations with increased resolution. The usual way of treating the discontinuities is discussed here, however it is not always satisfactory. New methods based on polynomial interpolation across element interfaces and polynomial filtering are suggested, and illustrated by examples.

1 Introduction

Most spectral element formulations (see e.g. [3, 6]) are based on C^0-continuity across element interfaces, while the discontinuities in the derivatives decrease with increasing order of the polynomials representing the solution at each element. In practical applications, the discontinuity of the derivatives can cause problems, and examples and possible remedies are presented in this paper. In this context, the term "marginal resolution" is taken to mean that the solution itself is resolved, but the derivatives of a given order may be under-resolved.

Even though the solution produced by the spectral element method is the best solution of the weak form of the partial differential equation, there may be other considerations involved that necessitates smoothing of the derivatives across element interfaces. Some examples are the use of derived quantities involving derivatives in visualization, post-processing, and multi-physics systems where different solvers are combined. It is typical for these applications that quantities involving derivatives from the spectral element solution are transferred to another program, usually

C.E. Wasberg
Norwegian Defence Research Establishment (FFI), P.O. Box 25, 2027 Kjeller, Norway
e-mail: Carl-Erik.Wasberg@ffi.no

J.S. Hesthaven and E.M. Rønquist (eds.), *Spectral and High Order Methods for Partial Differential Equations*, Lecture Notes in Computational Science and Engineering 76,
DOI 10.1007/978-3-642-15337-2_49,

on a different grid. It is not always possible, or indeed the best way of spending the computational resources, to increase the resolution of the spectral element solution.

Spectral element methods with C^1-regularity across the interfaces have been proposed [8, 10], but we choose to keep the well-established C^0 spectral element method unchanged and only post-process the data. This approach also allows us to treat higher order derivatives.

2 Numerical Test Problems

The methods described in this work will be tested on two one-dimensional examples, described in the following. The focus is mainly on the calculation of second derivatives, because the potential problems are exposed more clearly by differentiating twice, and it illustrates the issue of sequential treatment. In addition, some of the motivation for this study comes from the work on flow noise described in [4], and this is used as the second example.

2.1 An Analytical Example

In the first example, a known function is approximated and differentiated. The exact formulas for the derivatives are used to calculate the errors in the constructed continuous derivatives. The function is

$$f(x) = \frac{1}{2 + cos(2\pi x)}, \qquad x \in [0, 2\rangle, \tag{1}$$

and is shown, together with its second derivative, in Fig. 1.

2.2 Turbulent Channel Flow

The second example illustrates how the discontinuous derivatives at element interfaces can lead to problems in applications. We consider direct numerical simulation

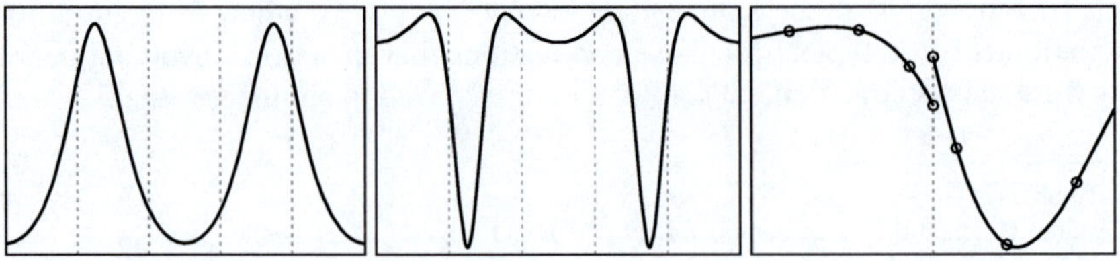

Fig. 1 The analytical example function (1). *Left:* $f(x)$. *Middle:* $f''(x)$. *Right:* Close-up of the spectral element representation of $f''(x)$ close to an element interface

Fig. 2 The dominant source term in Lighthill's approximation, with discontinuous derivatives at the 16 element interfaces normal to the x-direction. The horizontal line shows the location of the one-dimensional results reported in this paper

data of turbulent channel flow at $\mathrm{Re}_\tau = 180$, described in detail in [11]. The results compare well with the reference data from Moser et al. [9].

Lighthill [7] formulated source terms for a wave equation describing the sound propagation from turbulent flow. These source terms involve second derivatives of the velocity components:

$$\left(c_0^2\nabla^2 - \frac{\partial^2}{\partial t^2}\right)\rho = -\frac{\partial^2 T_{ij}}{\partial x_i \partial x_j}, \qquad T_{ij} = \rho u_i u_j + (p - c^2\rho)\delta_{ij} - \tau_{ij}. \tag{2}$$

Even though the flow is statistically well resolved, the discontinuities in the second derivatives at the element interfaces normal to the x-direction are clearly seen in the volume visualization of the term $(\partial^2(u^2)/\partial x^2)$ in Fig. 2. For the channel flow case, this is the dominant term of T_{ij}, and the spikes along the element interfaces act as artificial sound sources. When we return to this example, we study the calculation of this term along the horizontal line shown in Fig. 2.

3 Interface Averaging

The standard way of post-processing interface discontinuities is by simple or weighted averages. The calculation of a derivative $f = \frac{du}{dx}$ in a one-dimensional domain Ω can be formulated weakly as

$$\int_\Omega f v\, dx = \int_\Omega \frac{du}{dx} v\, dx, \qquad \forall v \in V, \tag{3}$$

for a suitable test space V. Following a standard spectral element discretization, the derivative at an interface point becomes

$$f^I = \frac{1}{\rho_N^L + \rho_0^R}\left(\rho_N^L f_N^L + \rho_0^R f_0^R\right), \qquad \rho_i^k = \rho_i l_k/2, \tag{4}$$

where the superscripts L and R denote the element to the left and right of the interface, respectively, ρ_i is the Gauss-Lobatto-Legendre weight at the ith point of an element, and l_k is the element size. The "Smoothing interface method" proposed by Meng et al. [8] also ends up with the expression (4) for the interface values.

The weighting in (4) depends on the element sizes and polynomial orders and is perhaps not the best choice, but these quantities do not vary in the examples shown here. Therefore, we simplify by not distinguishing between weighted and unweighted averages, and just call the method "Interface averaging (IA)".

We note that the interface averaging viewed as an operator is idempotent, but does not commute with differentiation, so for higher derivatives the order of the operations is important:

$$\left(\frac{d^2u}{dx^2}\right)^{\mathrm{IA}} \neq \left(\frac{d}{dx}\left(\frac{du}{dx}\right)^{\mathrm{IA}}\right)^{\mathrm{IA}} \equiv \left(\frac{d^2u}{dx^2}\right)^{\mathrm{SIA}} \tag{5}$$

If every differentiation is followed by interface averaging, we denote it "Sequential Interface Averaging (SIA)". For mixed derivatives, sequential treatment ensures that the result of each differentiation is continuous at the interfaces, and is thus essential to avoid special cases at edges and corners.

As expected from the spectral element theory, both the discontinuous derivatives and the interface averaging methods converge with spectral accuracy as the number of grid points is increased, as shown in Fig. 3. However, these methods are not always satisfactory at moderate polynomial orders, as illustrated at the right-hand part of Fig. 3, showing the second derivative. Note that all functions shown by lines are plotted on a very fine grid in order to shown the piecewise polynomial representation, and not only the values at the Gauss-Lobatto-Legendre points. This is relevant because the calculated derivatives are typically interpolated to a different grid, as mentioned in the Sect. 1.

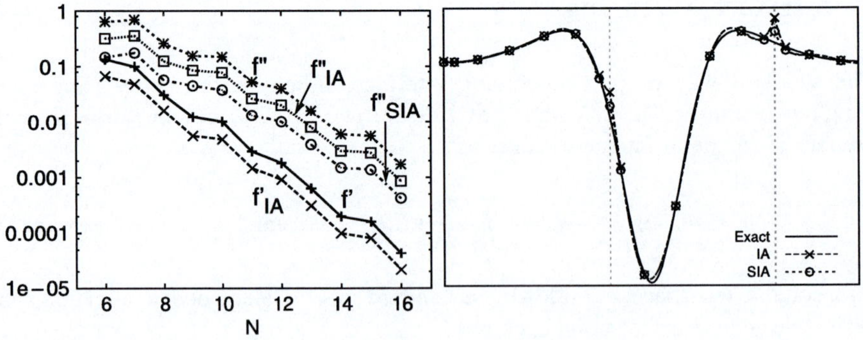

Fig. 3 *Left:* Max. error in the derivatives of (1), five elements, varying polynomial degree. *Right:* IA and SIA applied in the calculation of the second derivative of (1)

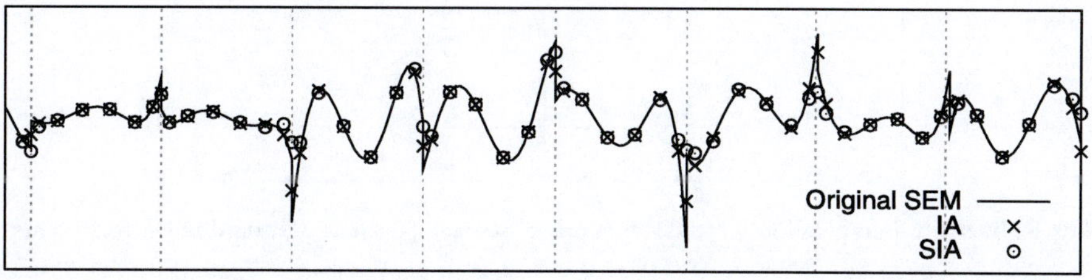

Fig. 4 Original solution, IA, and SIA applied to the dominant source term in Lighthill's approximation

From our other example, the calculation of sound generated by turbulent flow, we pick the horizontal line shown in Fig. 2. A variety of situations occur at the element interfaces along this line, and Fig. 4 shows the dominant source term here, in original and post-processed forms.

4 Improved Interface Treatment

The derivatives plotted in Figs. 3 and 4 indicate that there is more information in the data than the interface averaging methods can extract. From the plots it is possible to come up with better guesses for the interface values just by visual inspection, but the challenge is to formulate this in an algorithm.

Two general points can be made:

- Information from the interior of the elements adjacent to the interface should be used to obtain better results than just using the two one-sided values at the interface.
- When the one-sided values differ, it stems from marginal resolution in one of the elements, or in both. If possible, the "best" of the two values should be given most weight.

4.1 Polynomial Interpolation

The point about two-sided information is used here, as the value at the element interface is constructed from evaluation of an interpolating polynomial through points at both sides of the interface. We shall call this method "Interface Interpolation (II)" in the following.

The simplest interpolation method is just a straight line (first order polynomial) between the first grid values at both sides of the interface. This is called "II1", and is illustrated in Fig. 5 together with third order interpolation ("II3") using two neighbour points from each side of the interface.

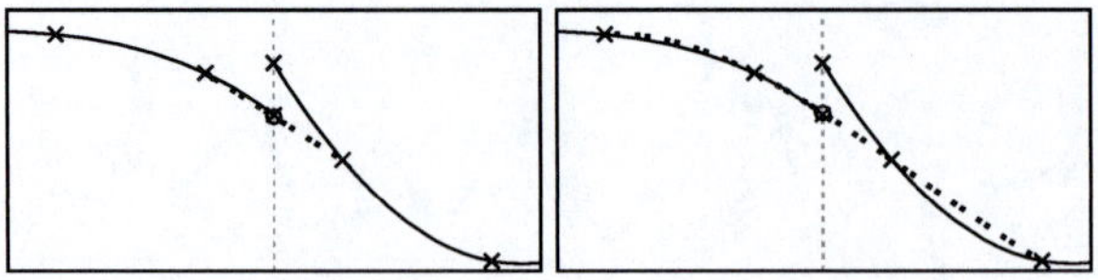

Fig. 5 Interface Interpolation. *Left:* II1, first order, two points. *Right:* II3, third order, four points

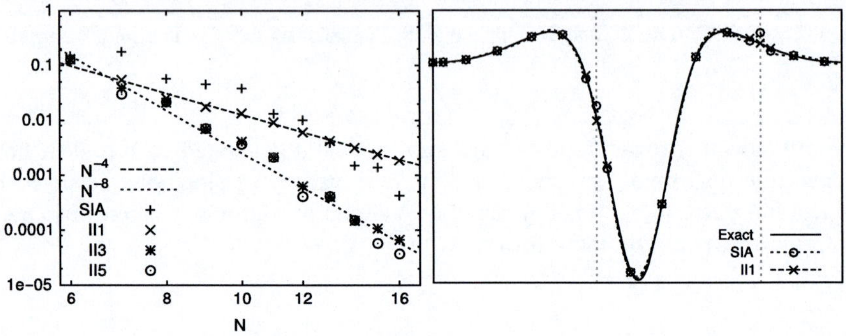

Fig. 6 *Left:* Max. error in the second derivative of (1), 5 elements, varying polynomial degree. *Right:* SIA and II1 applied in the calculation of the second derivative of (1)

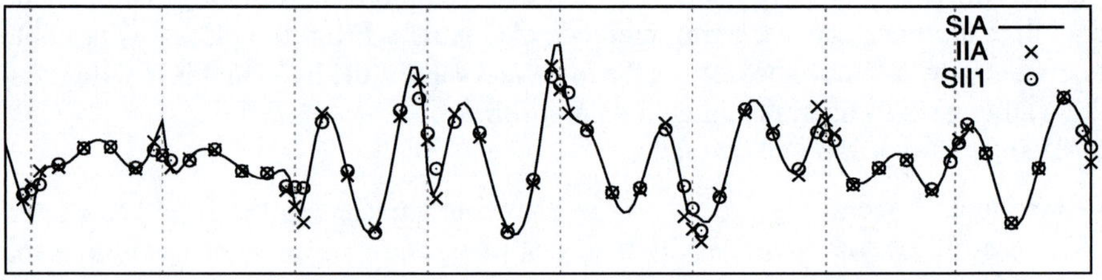

Fig. 7 SIA, II1, and SII1 applied to the dominant source term in Lighthill's approximation

Figure 5 also illustrates the second point made above, as the interpolants follow the solution from the "smoothest" side of the interface much more closely than from the other side. (Figure 1 shows that the solution has less structure, and is thus better resolved, in the first element than in the second.)

Sequential versions ("SII1", "SII3", etc.) of the Interface Interpolation for higher derivatives can be constructed just as for the averaging, by applying the interpolation after each differentiation. Higher order interpolants, based on more values or derivatives, can also be constructed, but in the experiments done here, there seem to be little gain in using higher than third order interpolation. Approximation results for the second derivative of (1) are shown in Fig. 6, while results from the turbulent channel case are shown in Fig. 7. These figures show the reduction in spikes at the element interfaces.

In discontinuous Galerkin methods, post-processing techniques involving convolutions have been shown to improve the accuracy for hyperbolic problems, see e.g. [2]. The present technique is related, but only applied at the interface points and after the spectral element solution is differentiated.

4.2 Filtering

Giving most weight to the one-sided derivative at the element interface from the "best" resolved element could be done by analyzing the Legendre coefficients of the one-dimensional representations used in the calculation of the derivatives, and try to identify the best resolved side from the rate of decay in the spectrum.

In this paper, however, we try a simpler alternative, namely to filter the fields before differentiation to try to reduce the influence of oscillations from marginally resolved elements. A simple filtering procedure that preserves the interface values, as described in [1], is the polynomial filtering of Fischer and Mullen [5].

For the two examples used above, filtering does not change the results much, but in another example from the turbulent channel flow, 25% polynomial filtering of the velocity field combined with the first order interface interpolation really improves the calculated derivatives. There is a strong front in one element, which gives rise to fine structures in the second derivative. A comparison with the Sequential Interface Averaging method is shown for a part of a horizontal plane in Fig. 8.

5 Conclusions

This work is based on the conjecture that there is more information in the spectral element solution about the derivatives than what is extracted by interface averaging, and it has been demonstrated that using some information from the interior of the elements can yield a better approximation of the derivative at the interface. There

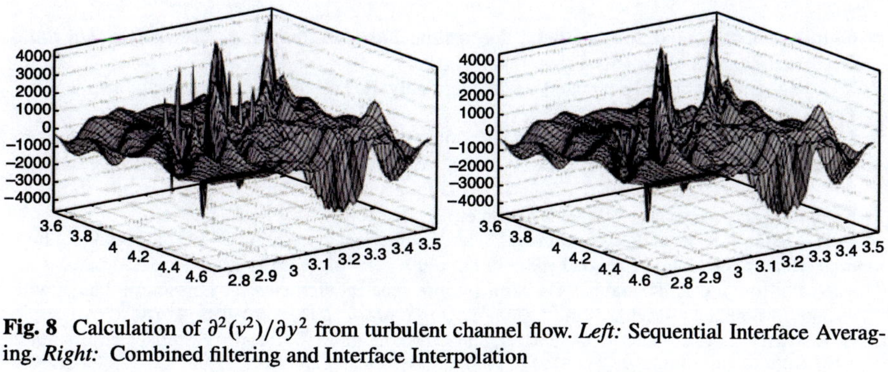

Fig. 8 Calculation of $\partial^2(v^2)/\partial y^2$ from turbulent channel flow. *Left:* Sequential Interface Averaging. *Right:* Combined filtering and Interface Interpolation

is a fine balance between the smoothing effect of using non-local information and retaining the accuracy. Interpolation of higher than third order does not improve the results in the examples shown here. However, for functions with large gradients, it may be advantageous to apply a polynomial filter to the spectral element solution before the interface corrections are calculated.

For some of the methods described here, spectral accuracy is demonstrated for a simple test example. However, as the title indicates, the focus is more on improving the results from a given calculation than studying the asymptotic convergence properties.

All the techniques discussed in this study only use information from the two elements that shares an interface. This greatly simplifies the parallel implementation. They are also inherently one-dimensional, as they are applied along lines in the differentiation direction, normal to the interfaces. The methods are therefore trivial to apply in higher dimensions by tensor products, and there is no need for special treatment of edges or corners.

Acknowledgements The author is grateful to Thomas Elboth, Thor Gjesdal, and Anders Helgeland for valuable input and discussions on this subject. The work is supported in part by Fugro and The Norwegian Research Council through grant PETROMAKS 175921/S30.

References

1. J. P. Boyd. Two comments on filtering (artificial viscosity) for Chebyshev and Legendre spectral and spectral element methods: Preserving the boundary conditions and interpretation of the filter as a diffusion. *J. Comput. Phys.*, 143:283–288, 1998
2. B. Cockburn, M. Luskin, C.-W. Shu, and E. Süli. Enhanced accuracy by post-processing for finite element methods for hyperbolic equations. *Math. Comput.*, 72(242):577–606, 2003
3. M. O. Deville, P. F. Fischer, and E. H. Mund. *High-Order Methods for Incompressible Fluid Flow*. Cambridge University Press, Cambridge, 2002
4. T. Elboth, C. E. Wasberg, A. Helgeland, Ø. Andreassen, and B. A. P. Reif. Flow noise simulations around a cylinder. In B. Skallerud and H. Andersson, editors, *MekIT'09 Fifth national conference on Computational Mechanics*. Tapir Academic Press, Trondheim, 2009
5. P. F. Fischer and J. S. Mullen. Filter-based stabilization of spectral element methods. *Comptes Rendus de l'Académie des sciences Paris, t.332, Série I - Analyse numérique*, pages 265–270, 2001
6. G. E. Karniadakis and S. J. Sherwin. *Spectral/*hp *Element Methods for Computational Fluid Dynamics*. Oxford University Press, USA, 2005
7. M. J. Lighthill. On sound generated aerodynamically. II. Turbulence as a source of sound. *Proc. Roy. Soc. A*, 222(1148):1–32, 1954
8. S. Meng, X. K. Li, and G. Evans. Smooth interfaces for spectral element approximations of Navier-Stokes equations. In P. Sloot, C. J. K. Tan, J. J. Dongarra, and A. G. Hoekstra, editors, *Computational science-ICCS 2002, Pt I, Proceedings*, volume 2329 of *Lecture Notes in Computer Science*, pages 910–919. Springer, Berlin, 2002
9. R. D. Moser, J. Kim, and N. N. Mansour. Direct numerical simulation of turbulent channel flow up to $\mathrm{Re}_\tau = 590$. *Phys. Fluids*, 11:943–945, 1999
10. T. N. Phillips and A. R. Davies. On semi-infinite spectral elements for Poisson problems with re-entrant boundary singularities. *J. Comput. Appl. Math.*, 21(2):173–188, 1988
11. C. E. Wasberg, T. Gjesdal, B. A. P. Reif, and Ø. Andreassen. Variational multiscale turbulence modelling in a high order spectral element method. *J. Comput. Phys.*, 228:7333–7356, 2009